Interrelationships of the
Platyhelminthes

The Systematics Association Special Volume Series

Series Editor

Alan Warren
Department of Zoology, The Natural History Museum,
Cromwell Road, London, SW7 5BD, UK.

The Systematics Association provides a forum for discussing systematic problems and integrating new information from genetics, ecology and other specific fields into taxonomic concepts and activities. It has achieved great success since the Association was founded in 1937 by promoting major meetings covering all areas of biology and palaeontology, supporting systematic research and training courses through the award of grants, production of a membership newsletter and publication of review volumes by its publishers Taylor & Francis. Its membership is open to both amateurs and professional scientists in all branches of biology who are entitled to purchase its volumes at a discounted price.

The first of the Systematics Association's publications, *The New Systematics*, edited by its then president Sir Julian Huxley, was a classic work. Over 50 volumes have now been published in the Association's 'Special Volume' series often in rapidly expanding areas of science where a modern synthesis is required. Its *modus operandi* is to encourage leading exponents to organise a symposium with a view to publishing a multi-authored volume in its series based upon the meeting. The Association also publishes volumes that are not linked to meetings in its 'Volume' series.

Anyone wishing to know more about the Systematics Association and its volume series are invited to contact the series editor.

Forthcoming titles in the series:

Major Events in Early Vertebrate Evolution
Edited by P. E. Ahlberg

The Changing Wildlife of Great Britain and Ireland
Edited by David L. Hawksworth

Other Systematics Association publications are listed after the index for this volume.

The Systematics Association Special Volume Series 60

Interrelationships of the Platyhelminthes

Edited by D. T. J. Littlewood
and R. A. Bray

Department of Zoology
The Natural History Museum
London
UK

CRC Press
Taylor & Francis Group
Boca Raton London New York

CRC Press is an imprint of the
Taylor & Francis Group, an **informa** business
A TAYLOR & FRANCIS BOOK

CRC Press
Taylor & Francis Group
6000 Broken Sound Parkway NW, Suite 300
Boca Raton, FL 33487-2742

First issued in paperback 2019

© 2001 D. T. J. Littlewood & R. A. Bray
CRC Press is an imprint of Taylor & Francis Group, an Informa business

Typeset in Sabon by Keyword Publishing Services

No claim to original U.S. Government works

ISBN-13: 978-0-7484-0903-7 (hbk)
ISBN-13: 978-0-367-39785-2 (pbk)

British Library Cataloguing in Publication Data
A catalogue record for this book is available from the British Library

Library of Congress Cataloging in Publication Data

Interrelationships of the Platyhelminthes / edited by D. T. J. Littlewood, R. A. Bray.
      p. cm.
      Includes bibliographical references.
      1. Platyhelminthes–Phylogeny. I. Littlewood, D. T. J. (D. Timothy J.), 1961–   II. Bray,
Rodney Alan.

QL391.P7 I62 2000
592'.4–dc21   00-056367

Cover design by Tim Littlewood using drawings kindly supplied by Janine Caira. Original drawings appear in:
Caira, J. N., and D. T. J. Littlewood. 2000. Worms, Platyhelminthes.
In S. A. Levin (ed.), *Encyclopedia of Biodiversity*, Vol. 5, pp. 863–899. Academic Press, London.

Visit the Taylor & Francis Web site at
http://www.taylorandfrancis.com

and the CRC Press Web site at
http://www.crcpress.com

Contents

Contributors

Jaume Baguñà, Departament de Genètica, Facultat de Biologia, Universitat de Barcelona, Diagonal 645, 08028 Barcelona, Spain.

Ian Beveridge, Department of Veterinary Science, The University of Melbourne, Parkville, Victoria 3052, Australia.

Walter A. Boeger, Departamento de Zoologia, Universidade Federal do Paraná, Caixa Postal 19073, Curitiba, PR 81531-990, Brazil.

Rodney A. Bray, Department of Zoology, The Natural History Museum, Cromwell Road, London SW7 5BD, UK.

Daniel R. Brooks, Department of Zoology, University of Toronto, Toronto, Ontario, Canada M5S 1A1.

Janine N. Caira, Department of Ecology and Evolutionary Biology, University of Connecticut, U-43, Rm. 312, 75 N. Eagleville Road, Storrs, Connecticut, USA.

Lester R. G. Cannon, Queensland Museum, Cultural Centre, South Brisbane, Queensland 4101. Australia.

Salvador Carranza, Department of Zoology, The Natural History Museum, Cromwell Road, London SW7 5BD, UK.

Thomas H. Cribb, Department of Microbiology and Parasitology, The University of Queensland, Brisbane, Queensland 4072, Australia.

Marco Curini-Galletti, Dipartimento di Zoologia e Antropologia Biologica, University of Sassari, via Muroni 25-07100, Sassari, Italy.

David W. Halton, Parasitology Research Group, School of Biology and Biochemistry, The Queen's University of Belfast, Belfast BT9 7BL, UK.

Claire J. Healy, Department of Ecology and Evolutionary Biology, University of Connecticut, U-43, Rm. 312, 75 N. Eagleville Road, Storrs, Connecticut, USA.

Jan Hendelberg, Department of Zoology, University of Göteborg, PO Box 463, SE 405 30 Göteborg, Sweden.

Elisabeth A. Herniou, Department of Zoology, The Natural History Museum, Cromwell Road, London SW7 5BD, UK.

Eric P. Hoberg, United States Department of Agriculture, Agricultural Research Service, Biosystematics and National Parasite Collection Unit, BARC East No. 1180, 10300 Baltimore Avenue, Beltsville, Maryland 20705, USA.

Kirsten Jensen, Department of Ecology and Evolutionary Biology, University of Connecticut, U-43, Rm. 312, 75 N. Eagleville Road, Storrs, Connecticut, USA.

Boris I. Joffe, Zoological Institute of the Russian Academy of Sciences, St Petersburg, Russia.

David A. Johnston, Department of Zoology, The Natural History Museum, Cromwell Road, London SW7 5BD, UK.

Ulf Jondelius, Department of Systematic Zoology, Evolutionary Biology Centre, Uppsala University, Norbyvägen 18D, SE-752 36 Uppsala, Sweden.

Jean-Lou Justine, Laboratoire de Biologie parasitaire, Protistologie, Helminthologie, EP 1790 CNRS 'Biologie et Évolution de Parasites', Muséum National d'Histoire Naturelle, 61 rue Buffon, F-75231 Paris cedex 05, France.

Elena E. Kornakova, Zoological Institute of the Russian Academy of Sciences, St Petersburg, Russia.

Delane C. Kritsky, Department of Health and Nutrition Sciences, College of Health Professions, Idaho State University, Campus Box 8090, Pocatello, Idaho 83209, USA.

D. Timothy J. Littlewood, Department of Zoology, The Natural History Museum, Cromwell Road, London SW7 5BD, UK and Division of Life Sciences, Franklin Wilkins Building, King's College London, 150 Stamford Street, London SE1 8WA, UK.

Eric S. Loker, Department of Biology, University of New Mexico, Albuquerque, New Mexico 87131, USA.

Kennet Lundin, Department of Zoology, Zoomorphology, University of Göteborg, Box 463, SE 405 30 Göteborg, Sweden.

Jean Mariaux, Département d'Herpétologie et Ichtyologie, Muséum d'Histoire Naturelle, 1 Route de Malagnou, CP 6434, CH 1211-Genève 6, Switzerland.

Michael Norén, Swedish Museum of Natural History and Stockholm University, PO Box 50007, SE-104 05 Stockholm, Sweden.

Peter D. Olson, Department of Zoology, The Natural History Museum, Cromwell Road, London SW7 5BD, UK.

Jordi Paps, Departament de Genètica, Facultat de Biologia, Universitat de Barcelona, Diagonal 645, 08028 Barcelona, Spain.

Jan Pawlowski, Department of Zoology and Animal Biology, University of Geneva, 154 route de Malagnou, CH-1224, Chêne-Bougeries and Muséum d'Histoire Naturelle, 1 Route de Malagnou, CP 6434, CH 1211-Genève 6, Switzerland.

Sylvie P. Pichelin, Department of Microbiology and Parasitology, The University of Queensland, Brisbane, Queensland 4072, Australia.

Olga I. Raikova, Zoological Institute of the Russian Academy of Sciences, 199034 St Petersburg, Russia.

Maria Reuter, Department of Biology, Åbo Akademi University, Artillerigatan, 6, SF-20520 Åbo, Finland.

Reinhard Rieger, Institut für Zoologie und Limnologie, Universität Innsbruck, Technikerstrasse 25, A-6020 Innsbruck, Austria.

Marta Riutort, Departament de Genètica, Facultat de Biologia, Universitat de Barcelona, Diagonal 645, 08028 Barcelona, Spain.

Klaus Rohde, School of Biological Sciences, University of New England, Armidale, New South Wales 2351, Australia.

David Rollinson, Department of Zoology, The Natural History Museum, Cromwell Road, London SW7 5BD, UK.

Iñaki Ruiz-Trillo, Departament de Genètica, Facultat de Biologia, Universitat de Barcelona, Diagonal 645, 08028 Barcelona, Spain.

Ronald Sluys, Institute for Biodiversity and Ecosystem Dynamics, Zoological Museum, University of Amsterdam, PO Box 94766, 1090 Amsterdam, The Netherlands.

Scott D. Snyder, Department of Biology and Microbiology, University of Wisconsin at Oshkosh, Oshkosh, Wisconsin 54901-8640, USA.

Wolfgang Sterrer, Bermuda Natural History Museum, Flatts FLBX, Bermuda.

Zdzislaw Swiderski, Institute of Parasitology, Polish Academy of Sciences, Twarda Street 51/55, 00-818, Warsaw, Poland.

Maximilian J. Telford, Department of Zoology, The Natural History Museum, Cromwell Road, London SW7 5BD, UK.

Joseph L. Thorley, School of Biological Sciences, University of Bristol, Bristol BS8 1UG, UK.

Vasyl V. Tkach, Department of Parasitology, Institute of Zoology, Ukrainian Academy of Sciences, 15 Bogdan Khmelnitsky Street, Kiev-30, MSP, 252601, Ukraine.

Seth Tyler, Department of Biological Sciences, University of Maine, 5751 Murray Hall, Orono, Maine 04469-5751, USA.

Nikki A. Watson, School of Biological Sciences, University of New England, Armidale, New South Wales 2351, Australia.

Mark Wilkinson, Department of Zoology, The Natural History Museum, Cromwell Road, London SW7 5BD, UK.

Willi E. R. Xylander, Staatliches Museum für Naturkunde Görlitz, Postfach 300154, D-02806 Görlitz, Germany.

Preface

Flatworms are ubiquitous on all continents and in all aquatic environments. They are dorso-ventrally flattened, apparently simple bilaterally symmetrical multicellular animals. Although they have no convincing fossil record, their simplicity and estimated position relative to other phyla suggests they are an ancient group of organisms that may have been among the first bilaterians to have crept on the planet. The majority of them are parasites of vertebrates, and few (if any) vertebrate species escapes parasitism from one or more species of flatworm. Some are serious pests and others cause devastating diseases. Few disciplines in biology are not flavoured by examples from the Platyhelminthes, and some disciplines dedicate themselves to their study. This volume focuses on the inter-relationships of the group, which in turn allows the wealth of comparative data on this group to be put into an evolutionary context.

The inspirations for this book are many and varied. First, the debt that platyhelminth workers owe to the volume published by Ehlers in 1985, Das Phylogenetische System der Plathelminthes, is widely acknowledged and we would like to reiterate this tribute. It introduced modern phylogenetic methods to the study of the phylum as a whole.

More immediately, the production of works summarizing recent knowledge of major taxa – in particular the volumes on *Interrelationships of fishes* (eds. Stiassny *et al.* 1996), *Origins and evolutionary radiation of the Mollusca* (ed. Taylor 1996) and *Arthropod relationships* (eds. Fortey and Thomas 1998) – suggested to us that a similar volume up-dating Ehlers' volume, but multi-authored, would be welcomed by the community of platyhelminth workers. It is a truism to observe that platyhelminthologists belong in two disparate camps: those who study free-living organisms, and the parasitologists. These camps have developed somewhat independently, although efforts to bring them together have been made, notably by the organizers of recent 'Turbellaria' congresses. Our goal was to bring together workers from these divergent areas (and the few workers who have feet in both camps) to produce a work unified by modern approaches to phylogenetic analysis. The impact of molecular phylogeny is being felt strongly on platyhelminth studies, and this is reflected in this volume, though we have been careful to balance this with contributions based on other types of data, such as morphology and ultrastructure. It becomes clear in reading the chapters of this book that consideration of all data, possibly with the use of combined evidence approaches, is a *sine qua non* of reliable phylogenetic inference.

To the uninitiated, systematics and phylogenetics may appear to be dry, rather uneventful areas of investigation. However, the issues surrounding the recognition of species, the search for homologous characters that allow a systematic framework to be attempted, the coding of characters and the classification of organisms, are all fraught with deeply divided philosophical approaches and a tendency towards a less than objective approach at a multitude of levels. Even the meekest, most upright and objective of practitioners cannot escape the pitfalls that the disciplines present. Such problems as the recognition of homology (even its definition; e.g. see Hall 1994), its recognition as an *a priori* or an *a posteriori* approach, its use in character coding (see papers in Scotland and Pennington 2000), the coding of inapplicable characters (e.g., Strong and Lipscomb 1999), character weighting (e.g., Neff 1996) or adding weight to particular characters (e.g., Soltis and Soltis 1998), choice of individual species as exemplars for broader taxonomic groups, and the combination of multiple independent data sets (e.g., de Queiroz *et al.* 1995; Huelsenbeck *et al.* 1996; Rodrigo 1996) are just a few of the topics that will remain hotly debated for some time yet. Molecular phylogeneticists suffer additional problems from assumptions about nucleotide evolution, black-box models that fall foul of heterogeneous data (e.g., through long-branch attraction), and the fact that the genes under study may be evolving along quite different paths from the organisms whence they came. Analytical methods available for molecular data can add more hurdles in the search for robust phylogenies. Distance and model-driven methods (e.g., maximum likelihood) can provide additional sources of incongruence, even for the same data set. In the light of such daunting debates it might be regarded that phylogeneticists are at least brave, if not foolhardy, to present their ideas and analyses in print. Whilst the methods of analysis may change and new interpretations may be possible with the information contained in this volume, we believe that the chapters here provide substantial foundation material upon which subsequent workers can build. Of course, the individual chapters themselves stand substantially on the shoulders of earlier workers, most of whom recognized that systematics and phylogenetics are the cornerstones of comparative biology. The extensive bibliography is testament to this.

Those readers wishing to learn more on the general biology of the flatworms should consult general zoology texts such as Hyman (1951), Brusca and Brusca (1990) and Westheide and Rieger (1996), and parasitology texts such as Mehlhorn (1988), Williams and Jones (1994), Kearn (1998), and Roberts and Janovy (1999). There are few single sources that deal with the biology of the free-living ('turbellarian') flatworms, but *Proceedings of the International Symposia on the Biology of Turbellaria* are variously published as special volumes in journals (e.g., 8th ISBT, *Hydrobiologia* 383, 1998). Rieger (1998) reviews 100 years' research on the group in this same volume, and it is an excellent starting point. Purely systematic treatments of groups can be found in Cannon (1986), Khalil *et al.* (1994), and Yamaguti (1963, 1971).

We have asked contributors to summarize their recent work and not, in most cases, to initiate new studies. A consequence of this is a lack of complete coverage. Some conspicuous gaps will be noticed, both in taxa (e.g., Polycladida) and in techniques (e.g., karyology). Our aim was, therefore, not completeness, but rather a reflection of current thinking. The chapters have been written with as much detail as is necessary for subsequent workers to be able to use and expand on. By this we mean that matrices are published in full, along with detailed character argumentation, and that any molecular data (including nucleotide alignments) are readily recoverable from the standard internet sources.

Organization of the book

The book has been split into four sections, rather dissimilar in length, that highlight the underlying goals. The first section takes a broader perspective on the status of the Platyhelminthes, its monophyly, placement in relation to other Metazoa and the nature of the basal taxa. The second section deals with the interrelationships of major free-living taxa, and the third on symbiotic and parasitic taxa. The final section encompasses contributions that view phylogeny and phylogenetic inference from the point of view of particular characters or techniques. Some of these chapters, we hope, will appeal to a broader audience of phylogeneticists and may guide similar studies on other groups.

Tyler (Chapter 1) places the phylum in context by discussing its wider relationships and, inevitably, touches on the question of its monophyly. He discusses enteropneusts as models for the origin of platyhelminths, pointing out many similarities, while not claiming they are in a direct line to the flatworms. Rather, he suggests that platyhelminths arose from a stem group of vermiform deuterostomes by progenesis, going as far as saying that progenesis was 'undoubtedly' involved in flatworm origins. The non-monophyly of the acoelomorphs and the rest of the phylum has, Tyler reckons, a 'low probability of being correct'. Following from this view, he reckons that *Xenoturbella* is probably a flatworm (contrast this with the recent papers by Ehlers and Sopott-Ehlers 1997c; Norén and Jondelius 1997; Israelsson 1997, 1999). Tyler's approach is perhaps the least cladistic in outlook, but few authors have the history, ability or nerve to take on such a difficult topic.

Raikova *et al.* (Chapter 2) tackle the question of the monophyly of the Platyhelminthes by a close study of the Acoela and Acoelomorpha, and in particular, what nervous system and sperm structure have to say. Using immunocytochemistry and confocal scanning laser microscopy, they report that although the acoelomorph brain differs from that of other flatworms, they feel that brain structure does not settle the question of the monophyly of the Acoelomorpha, as it is probable that the acoel and nemertodermatidan brains arose independently from a more primitive organ. Acoel spermatozoa have some similarities with that of the Trepaxonemata, but these must be homoplasious. Within the Acoela, the spermatozoa is variable and indicates that *Paratomella* is indeed sister to the rest of the taxon, i.e., the Euacoela (cf. Ax 1996).

Lundin and Sterrer (Chapter 3) report on the small, putatively basal, group Nemertodermatida, which only has 10–11 species. The phylogeny they present, based on 73 characters, shows a monophyletic group, consisting of two monophyletic families. They maintain that the Nemertodermatida and Acoela – which some evidence suggests are sister taxa – are separate from the Catenulida and Rhabditophora, and are probably more basal. They imply that they consider the Nemertodermatida as the earliest offshoot of the Platyhelminthes.

Rieger (Chapter 4) provides a summary of characters, hypotheses and ideas concerning the relationships of the Macrostomorpha. Few would argue as to their pivotal, basal position among the Rhabditophora, and understanding their radiation will become that much more important if catenulids and acoelomorphs do ultimately fall outside of the Platyhelminthes.

Proseriates may be locally so abundant that they characterize entire marine communities (e.g., the 'Otoplana' zone). Curini-Galletti (Chapter 5), however, using both molecular and morphological data, is not able to come to satisfactory conclusions on the monophyly of the group. He considers the present doubts as to relationships to reflect the 'intermediate state' of knowledge, where the 'certainty' of the past has been obscured by molecular data, but sampling is not yet dense enough for a new consensus to arise. Perhaps Curini-Galletti's reticence is a more realistic reflection on the general flux in this field.

Triclads are treated at two levels. Baguñà and colleagues (Chapter 6) provide an overview of the group using molecular evidence. The monophyly of the Tricladida is supported, but, concurring with other molecular data, from small subunit (SSU) ribosomal RNA genes (Jondelius *et al.*, Chapter 6; Littlewood and Olson, Chapter 25; Joffe and Kornakova, Chapter 26), they are not related to the Proseriata, invalidating the taxon Seriata. In fact, the sister-group of the triclads is, according to these data, the Prolecithophora. These two taxa, along with the small group of parasitic or symbiotic 'turbellarians' (*Urastoma*, Fecampiida and *Ichthyophaga*), form a clade. The Terricola and the family Dugesiidae share a strong molecular synapomorphy in the duplication event involving the SSU rDNA gene, a relationship supported by the partial cytochrome oxidase I (COI) sequences. Neither gene, however, validates the monophyly of the Terricola or the Dugesiidae, probably due to the level of sampling. A different approach is taken by Sluys (Chapter 7) who gives a 'morphological perspective' of the phylogeny of the family Dugesiidae. The problems associated with morphological cladistics are amply illustrated by this very careful and thorough study, where 'minimal changes in the data matrix may result in large topological changes' in trees. Characters which may appear reliable autapomorphies, such as the oviducal opening in *Romankenkius*, may not be found strong enough to force monophyly on the apparent clade. Although Sluys did not feel able to use differential weighting ('a complex and contentious issue'), he is able to argue from the characters for a 'guestimated' tree, which is somewhat longer than the most parsimonious tree.

The close relationship of the triclads and the Prolecithophora, discussed by Baguñà and colleagues is also found by Jondelius and colleagues (Chapter 8), in their study of the latter taxon, although they considered the relationship 'highly tentative'. They used SSU rDNA sequences of 16 prolecithophorans. The monophyly of the group was strongly supported. The symbiont *Urastoma* is not a prolecithophoran, but groups with the triclads, not with the Neodermata.

The Temnocephalida, with some 100 species, is the largest group of symbiotic 'turbellarians'. Its phylogeny is discussed by Cannon and Joffe (Chapter 9), and a morphological matrix is presented by Cannon (Chapter 9, appendix). Temnocephalids are considered rhabdocoels, related to the Dalyellidae. A temporal phylogeny is presented, with hypotheses of divergence dates based on current geographic distributions and plate tectonics. Readers interested in the diversity of other commensal and symbiotic turbellarians should consult Jennings (1971, 1974, 1997).

Boeger and Kritsky (Chapter 10) provide their latest synopsis and revision of their morphologically based matrix for the Monogenea (Boeger and Kritsky 1994, 1997). It represents a third generation of considered opinion, character coding and analysis, and not only does this evolution recognize their willingness to modify early ideas in the light of new data and new concepts, but once again indicates that matrices are dynamic entities in themselves. They consider the group (Monogenoidea) strongly monophyletic and contrast their findings with some of the recent early attempts at resolving monogenean relationships from molecular data (see also Mollaret *et al.* 2000).

Several chapters address aspects of perhaps the most morphologically distinct group of platyhelminths, the tapeworms. The autapomorphies of the Cestoda are enumerated by Xylander (Chapter 11), who also lists the autapomorphies of the Gyrocotylidea, Nephroposticophora, Amphilinidea and 'Cestoidea'. There is little doubt now that these gutless worms form a monophyletic group. Using morphological data, the sister-group relationship of the Gyrocotylidea and Nephroposticophora, and the Amphilinidea and 'Cestoidea' is strongly supported. Hoberg and colleagues (Chapter 12) present a combined morphological and molecular phylogeny of the Eucestoda (the same group as is known as 'Cestoidea' in Xylander). They found general concordance between molecular and morphological evidence. The Caryophyllidea is basal, suggesting that its monozooy is plesiomorphic. Monophyly of most recognized orders is supported, but the Tetraphyllidea is paraphyletic. The deep age of the group, initially radiating in the Palaeozoic, is recognized. Mariaux and Olson (Chapter 13) review the advance in cestode systematics as a result of molecular studies. They point out the preponderance of rDNA data used so far, but suggest that alternatives are, at present, limited. This probably reflects the case for many flatworm groups. Caira and colleagues (Chapter 14) present a very detailed morphological phylogeny of the orders Tetraphyllidea and Lecanicephalidea, and focus on individual species rather than adopting the potentially dangerous exemplar approach. Although the Lecanicephalidea has 'limited' support for its monophyly, the Tetraphyllidea is apparently paraphyletic. The chapter is also a useful example of ways in which inapplicable data can be handled, and how troublesome they are.

Rohde (Chapter 15) reviews the interrelationships of the Aspidogastrea, a small but fascinating group of trematodes that are the sister-group to the Digenea. For the first time, a matrix is developed from a number of morphological characters and the interrelationships of the group are estimated cladistically.

The largest group of platyhelminths may well be the Digenea, with about 18 000 described species. Three chapters approach this subclass in decreasing levels of inclusiveness. Cribb and colleagues (Chapter 16) present a combined evidence phylogeny of the whole subclass, using adult morphological and life-cycle characters and complete nuclear SSU rDNA sequences. The trees presented show the predominant effect of the gene data on the tree. On the whole, however, the confused state of higher digenean systematics is reflected in poor resolution at the broader levels. In contrast, several taxa at about the superfamily level appear well resolved by this study. The phylogeny is discussed in relation to life-cycle characters. Tkach and colleagues (Chapter 17) narrow the focus, treating the molecular phylogeny of the suborder Plagiorchiata. The molecular sequence used was from the 5' end of the large subunit (LSU) rDNA and was found a good target for resolving genus to superfamily level relationships. They found that when their results contradicted current systematic schemes, in some cases strong morphological and/or life-cycle evidence was found to support the molecular phylogeny. The suborder Plagiorchiata, as considered in previous phylogenies, is found to be polyphyletic. Finally, Snyder and colleagues (Chapter 18) focus on one family, probably the most intensively studied of all platyhelminths, the Schistosomatidae, which include the blood-flukes that cause human disease and misery in a frighteningly high proportion of the population of tropical regions. Considering there are several thousand schistosome sequences in the public domain, it is surprising that so little molecular phylogeny has been attempted on this unusual group. The biomedical community has been slow to realize in general the value of applied phylogeny. Furthermore, few have appreciated the potential value of blood-fluke phylogeny in the understanding of the development of heterogametic dioecy. In fact, the platyhelminths, with their high preponderance of hermaphroditic species, but with dioecy spread thinly among various groups, are likely to be a useful model of the evolution of gonochorism. This chapter also discusses the development of the genomics of schistosomes and how knowledge so gained might be used in studies of other platyhelminths.

The final section of the book focuses on characters and techniques. Rohde (Chapter 19) reviews the extensive literature on the protonephridia of platyhelminths, and provides statements of homology that are readily available for analysis for future wider data sets, although he wisely steers clear of estimating a phylogeny based on these characters alone. Watson (Chapter 20) and Justine (Chapter 21) review another suite of ultrastructural characters that has received as much, if not more, attention. Sperm morphology and spermiogenesis provide some strong synapomorphies for particular groupings. Watson has studied Rhabdocoela extensively, and here reviews the sperm data that she and others have accumulated on this complex group. Justine reviews sperm characters for the whole phylum and provides a matrix and analysis specific to these characters, highlighting their contribution to flatworm phylogenetics.

A scan of flatworm literature reveals a comprehensive body of knowledge on comparative neurobiology. Reuter and Halton (Chapter 22) review the extensive, often complex, and (most recently with the advance of confocal microscopy) visually appealing, work on the neurobiology of Platyhelminthes, only to conclude that such work is probably restricted in value for phylogenetics as a result of high levels of homoplasy in the system. Nevertheless, such work has yet to be extended to many groups and there may still be phylogenetic signals to be detected.

Beveridge (Chapter 23) describes the 'long history of the use of life-cycle characters' in cestode phylogeny. He argues that the knowledge of cestode life cycles, although abundant, is not spread evenly through the group, with a great preponderance of studies involving cyclophyllideans, and particularly hymenolepidids. The paucity of data from major groups, such as the Tetraphyllidea and Trypanorhyncha, as well as some (numerically) minor groups is also noted. However, he is able to discuss the value of probable major synapomorphies pointing out that, for example, four life-cycle characters support the monophyly of the Eucestoda and that life-cycle characters support several other clades in the group. Better organization of available data along with the search for new characters or more detailed study of others, particularly those utilizing ultrastructural features, are considered likely areas for progress.

Telford (Chapter 24) reviews the utility of embryology and developmental genetics as sources of synapomorphies, and concludes that current limited data demonstrated that the Rhabditophora, at least, are not basal Bilateria, whilst further evidence is required to satisfactorily place the acoelomorphs.

Littlewood and Olson (Chapter 25) take the well-used SSU rRNA gene database and add to it many new taxa, testing it for limits of resolution and signal whilst ever mindful that alone, it can only provide a gene tree. A solution based solely on these data seems surprisingly well resolved in spite of relatively poor sampling among many turbellarian groups.

Joffe and Kornakova (Chapter 26) take a closer look at a subset of SSU data through the eyes of morphologists, and argue a strong case for caution in accepting apparently robust molecular solutions at the expense of forfeiting well-argued morphological synapomorphies. Further, they promote the emphasis towards the utility of molecular data when gross molecular changes are themselves phylogenetic markers.

Finally, Wilkinson and colleagues (Chapter 27) utilize Platyhelminthes, the trees which have been propounded in the literature and a singular philosophical approach to investigate the possible pros and cons and even the likelihood of producing a supertree of the flatworms. Further pitfalls for the phylogeneticist are elucidated with examples from the flatworms, but the possibility of the production of a reliable phylogenetic supertree for the group remains fairly remote, awaiting better-understood analytical techniques. They make the point that use of 'black-box' parsimony techniques such as MRP (matrix representation with parsimony) is no substitute for increased sampling and better understanding of the groups.

Future work will clearly focus on denser sampling of taxa and the search for additional characters. Morphological approaches, in particular the examination of ultrastructure, and molecular methods will be equally important. New techniques, such as mapping mitochondrial gene order, have yet to be tried and tested throughout the phylum, although even preliminary studies (e.g., Le *et al.* 2000) suggest that no single source of phylogenetic signal exists.

Whether the readers of this book accept or deny the various statements of homology, character coding, analytical methodology or conclusions drawn, we hope at least that the volume is a useful resource in pursuing the interrelationships of a fascinating phylum.

Nomenclature

Throughout the volume, nomenclature is retained as used by the individual authors, but readers should notice the disagreements over the usage of certain terms, e.g., Monogenea/Monogenoidea or Eucestoda/Cestoidea. As editors we have not taken a position on these disputes, other than feeling that it would be better for the ending -oidea to be retained for superfamilies (International Commission on Zoological Nomenclature, 1999, Art. 29.2). A useful table juxtaposing the various usages in monogenean nomenclature was given by Mollaret *et al.* (2000).

Acknowledgements

One of the incentives to contribute to this volume was an opportunity to present papers at a meeting entitled *Interrelationships of the Platyhelminthes* held from the 14–16 July 1999 at The Linnean Society of London (see Olson 2000, for a report). The travel, registration and subsistence costs of the majority of speakers were largely underwritten by The Linnean Society of London with generous support also from The Systematics Association, who have published this volume, and The British Society for Parasitology. Additionally, The Wellcome Trust allowed one of us (D.T.J.L.; Wellcome 043965/Z/95/Z) to spend a sizeable sum of grant money in offsetting the costs of travel for some of the many participants/contributors from Australia. We are extremely grateful to all the institutions and individuals who made the meeting possible, and subsequently the volume a reality. In particular we thank John Marsden, Vaughan Southgate, Marquita Baird, Gina Douglas, Kathie Way and David Pescod for their unstinting support. Elisabeth Herniou graciously assisted throughout the meeting. Kathryn Hall and Katie Clouston assisted in the compilation of abstracts. Ulrich Ehlers provided a personal view on the progress of platyhelminth systematics since the publication of his seminal work, and Graham Kearn provided a delightful personal account of his heroes in the field. We are only too sorry that we were not able to include these contributions in the volume. It should be noted, however, that this book is not a conference proceedings volume, but a series of original, fully researched, chapters. Production costs for the colour plates in the volume were generously provided by a grant from The Natural Environment Research Council administered by The Linnean Society of London, and we thank both organizations.

All chapters have been peer reviewed by at least two, and mostly three, referees. We are indebted to the following individuals who kindly assisted in this task: Mike Akam, Jaume Baguñà, Ian Beveridge, David Blair, Barbara Boyer, Lester Cannon, Salvi Carranza, Tom Cribb, Tim Day, Boyko Georgiev, David Gibson, Bill Font, Gerhard Haszprunar, Jan Hendelberg, Robert Hirt, Eric Hoberg, Barrie Jamieson, Ulf Jondelius, Arlene Jones, Hugh Jones, Mal Jones, Jean-Lou Justine, Kennet Lundin, Barbara MacKinnon, Aaron Maule, Pete Olson, Robin Overstreet, Rod Page, Olga Raikova, Marta Riutort, Klaus Rohde, Mike Sanderson, Ernest Schockaert, Ronald Sluys, Scott Snyder, Wolfgang Sterrer, Max Telford, David Thompson, Vasyl Tkach, Seth Tyler, Claude Vaucher, Nikki Watson, Ian Whittington, Mark Wilkinson, Dave Williams and Leigh Winsor.

We also wish to thank our colleagues at The Natural History Museum for their support in this endeavour. The names are too few not to mention. In particular, we thank Eileen Harris for endless tea and biscuits, the staff canteen for driving us to 'The Anglesea Arms' where the plan for this volume was hatched, and our bosses David Rollinson and David Gibson for letting us get on with it. Pete Olson hovered attentively in the background and Steve Donovan and Phil Rainbow offered much needed words of encouragement.

Lastly, but by no means least, we are extremely grateful to the authors for providing excellent contributions and revisions on time with little or no threats from us.

D. Timothy J. Littlewood and Rodney A. Bray
The Natural History Museum, London

Section I
Early origins and basal taxa

Chapter 1

The early worm:
Origins and relationships of the lower flatworms

Seth Tyler

The evolutionary origin of the flatworms remains a major unsolved puzzle in metazoan phylogenetics. While three clearly monophyletic clades are recognized among the flatworms, namely Acoelomorpha, Catenulida, and Rhabditophora (Figure 1.1), how or even whether these three are related to each other and what their relationships are to other invertebrate groups is controversial (Smith *et al.* 1986). Perhaps, as has been argued recently from morphological characters (Haszprunar 1996) as well as from base-sequence data of the 18S ribosomal gene (Ruiz-Trillo *et al.* 1999), the Acoelomorpha, or at least the Acoela within it, could be the most primitive group of all bilaterally symmetrical animals, maybe even quite far removed from the other flatworms (Ruiz-Trillo *et al.* 1999). Similarly, the Catenulida may stand entirely separate from other flatworms by virtue of the very peculiar nature of their reproduction, among other things – the female organs are suppressed in many species and conceivably these animals develop by parthenogenesis of their quite aberrant, large spermatozoa (Sterrer and Rieger 1974; Schuchert and Rieger 1990a).

Deciphering the relationships among these clades depends substantially on knowing where flatworms came from, even if it is a matter of assigning separate origins for each of the clades. If they arose from a cnidarian-like diploblastic ancestor, for example, and are representative of an ancestor to such coelomate phyla as the Annelida, as is widely represented in textbooks, then we might assume that planula-like features of the acoels indicate their plesiomorphic standing in the phylum and that other platyhelminths would have descended from such a solid-bodied form by development of an epithelial gut (Hyman 1959; Salvini-Plawen 1978). If they arose from an ancestor like the Gnathostomulida (Ax 1985, 1996; Ehlers 1985a), then the Catenulida with sac-like gut, sparse epidermal ciliation, and lack of frontal glands could represent the plesiomorphic condition and the acoels would be derived. Alternatively, if flatworms arose by progenesis from a larva in the life cycle of a coelomate ancestor (Rieger 1980, 1986a), then we should expect the most plesiomorphic of the clades to be more like whatever that larva was, perhaps a trochophore-like larva of a spiralian (cf. Jägersten 1972; Ruppert 1978; Rieger 1986a). Many other possibilities remain.

A cladistic analysis of the Platyhelminthes and its relatives should eventually produce a stable phylogeny for these clades. For now, we are very far from a stable system; the cladistic analyses that have been published are tremendously varied and all in conflict largely because of differences in choice of characters, differences in the way characters are identified and evaluated (the most important part of any phylogenetic reconstruction: Neff 1986; Tyler 1988; cf. Jenner and Schram 1999), and the inherent difficulty of scoring morphological characters in meaningful ways across broad samplings of taxa. Rather than add yet another cladistic analysis, I argue here for consideration of specific characters that should be part of the ultimate phylogenetic system, particularly those that relate to assigning outgroups for the phylum and deciphering what the ancestor (or ancestors, if any of the clades has a separate origin) of the flatworms looked like. That is, what was the early worm?

Hypotheses of platyhelminth relationships

As a phylum, the Platyhelminthes has appeared in a wide range of positions in phylogenetic trees of the Metazoa, among them: 1) as the earliest bilaterians, derived from a planula-like ancestor and sister group to the rest of the Bilateria (Salvini-Plawen 1978); 2) as the next oldest bilaterians, sister group to the Gnathostomulida, with which they stand as sister group to other spiralians (Ax 1996); 3) as a sort of intermediate between pseudocoelomate phyla and the coelomate phyla, while also sister group to the Gnathostomulida (Schram 1991); 4) as the earliest protostomes, sister group to the coelomate protostomes (Kozloff 1990; Brusca and Brusca 1990; Nielsen 1995, 1998); 5) as descendants of other protostomes, sharing ancestry with non-coelomate protostomes and part of a sister group to panarthropods+nematodes (Eernisse *et al.* 1992); 6) as secondarily reduced descendants of an enterocoelic coelomate (the Archicoelomate theory; Remane 1959; Siewing 1980b); and 7) as descendants of an unspecified coelomate ancestor from which it arose by progenesis of that ancestor's larva (Rieger 1980, 1986a, 1994a,b); members of the Spiralia have served as models for this ancestor (Rieger 1988; Schuchert and Rieger 1990b).

At the heart of the differences among these hypotheses is the nature of the body cavity. The first three hypotheses consider the acoelomate body structure of platyhelminths to be plesiomorphic, and they appeal to the widely held notion that bilaterian phyla originated by stages of increasing complexity and size. The fourth rests at least partly on this same assumption. The fifth places the platyhelminths in the most derived position, relating their small size and apparent simplicity to reduction and regression. The last neatly dispenses with this issue by invoking abbreviated development of the body cavity. These positions are critical for relating the platyhelminth clades to each other because they specify which of their features are plesiomorphic. If lack of coelom and the sac-shaped gut are plesiomorphic, for instance, then these characters cannot be used to relate the clades.

Testing of these morphology-based hypotheses with an independent set of molecular characters should, we have long hoped, settle the issue of which is correct or whether another hypothesis applies. Unfortunately, the nucleic-acid sequences tested so far, those that code for ribosomal RNA, have proven unsuitable. Several studies with the 18S rDNA gene (Katayama *et al.* 1996; Carranza *et al.* 1997; Zrzavý *et al.* 1998; Littlewood *et al.* 1999a; Ruiz-Trillo *et al.* 1999), some of them including complete sequences for it, have produced only ambiguous results because of highly unequal substitution rates among flatworm species (Balavoine 1998). All of these studies would indicate polyphyly of the Platyhelminthes, and particularly the Acoela seems to fall out as a very early stem of

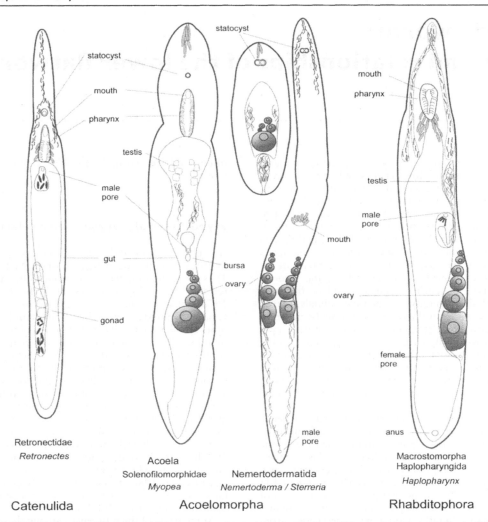

Figure 1.1 The three major clades of the Platyhelminthes represented by schematic drawings of basal genera in each. Both of the orders within the Acoelomorpha, the Acoela and the Nemertodermatida, are represented (by *Nemertoderma westbladi* and *Sterreria psammicola*, in the latter case). Depending on how these clades are related, one of these representations could be closest in form to the ancestor of the Platyhelminthes, or each could represent an independently derived group of convergently similar taxa.

the Bilateria (Ruiz-Trillo *et al.* 1999; supposedly supporting hypothesis 1 above) but only because of long-branch attraction in this very rapidly evolving sequence in acoels (Joffe and Kornakova 2001, this volume). As Balavoine (1998) opines, 'It is questionable whether any phylogenetic information concerning the position of the acoels relative to the Bilateria or the other flatworms can be recovered from such an abnormally evolving 18S rDNA sequence'. While the 18S gene has been useful for corroborating monophyly in some higher groups of the Platyhelminthes (Littlewood *et al.* 1999a), it cannot be applied to understanding the position of the Platyhelminthes as a whole. Similar results from 18S rDNA have emerged in studies of other phyla, showing, according to Cavalier-Smith (1998), 'how grossly misleading the rRNA tree can sometimes be'.

More promising for phylogenetic hypotheses in terms of molecular analysis are *Hox* genes (Balavoine 1997, 1998; Bayascas *et al.* 1998). Although data on these genes are yet fragmentary, the indication is that their arrangement in the HOX cluster shows a relationship of the Platyhelminthes (actually of triclad turbellarians, the only group for which such data are yet available) with other protostomes. In fact, the sequences indicate a sufficiently close affinity to annelids

that Balavoine (1997, 1998) interprets them as showing that platyhelminths are derived from a coelomate ancestor among the protostomes, and so he favors Rieger's hypothesis (number 6 above) of progenetic origin of flatworms.

Progenesis, that is, the programmed sexual maturation of an organism otherwise still in a juvenile or larval stage (Rieger 1980; Westheide 1987), is supported as well by ultrastructural features as the mode of origin of flatworms. In fact, it is scrutiny of these ultrastructural characters that inspired Rieger (1980, 1986) to propose it. Rieger's (1985, 1986a) critical analysis of the acoelomate condition showed that the plesiomorphic condition of musculature in the Bilateria would be epitheliomuscular cells in a coelomic lining; this is the plesiomorphic condition in both deuterostomes and protostomes and must stem directly from the epitheliomuscular cells of a cnidarian-like ancestor. (The fibre-form muscles of the Ctenophora are probably derived (Rieger 1994a), and it is unlikely that the ctenophores represent an intermediate between a cnidarian ancestor and the Bilateria.) If the flatworms (with their exclusively fibre-form muscles) were to be the basal bilaterians, standing intermediate between an epitheliomuscular-cell-possessing cnidarian-like ancestor and the coelomate phyla, then the coelomates would have had to develop their epitheliomuscular cells independently. Fibre-form muscles and acoelomate condition are seen in many small members of the coelomate phyla, and the larvae and juveniles of coelomate species also show acoelomate and pseudocoelomate conditions; so it is likely that all three conditions, acoelomate, pseudocoelomate, and coelomate, occurred in the life cycle of the bilaterian stem species. The plesiomorphic nature of epitheliomuscular cells in the Bilateria is even supported by molecular (nucleic acid-based) phylogenies (Collins 1998).

Rieger has used as models for the progenesis hypothesis several protostome coelomates, namely *Lobatocerebrum* spp., a turbellariomorph annelid (Rieger 1980, 1988) and the dwarf male of the echiuran *Bonellia viridis* (Schuchert and Rieger 1990b), as well as various interstitial polychaete annelids (Fransen 1980, 1988). Balavoine's (1998) finding comparable HOX clusters in annelids and planarians underscores the relevance of such models; but Rieger does not actually specify any one of them as pertaining directly to the Platyhelminthes.

It would be appropriate, nevertheless, to consider potential models among the deuterostomes. Many of the phylogenetic trees mentioned above place deuterostomes at the earliest branch of the bilaterian stem, a position supported by correlation of the burgeoning pool of rDNA-sequence data and patterns of early development (Valentine 1997; Aguinaldo and Lake 1998; Knoll and Carroll 1999), and if the flatworms (or only the Acoela) occupy a basal position as molecular data indicate to some, then there may be some point of convergence in these hypotheses. The stem species for the origin of the Bilateria in Rieger's hypothesis is a clonal organism, 'similar in organization to Pterobranchia or Bryozoa' (Rieger *et al.* 1991); and among deuterostomes, enteropneust hemichordates and holothurian echinoderms have larval stages entering the interstitial environment where the stage for progenesis is set. Moreover, enteropneust hemichordates are reported to develop an acoelomate condition as adults, ostensibly through obliteration of the coelom by muscles (Hyman 1959; Benito and Pardos 1997) and otherwise lacking parenchyma as do acoelomorphs (Rieger 1985). Others have drawn comparisons between enteropneusts and flatworms, particularly in structure of the epidermis and its ciliation (Pardos 1988; Rieger *et al.* 1991; Benito and Pardos 1997).

Enteropneust model

Tiny interstitial juveniles of an unidentified species of enteropneust from coral sands may serve as a model (Figure 1.2). These juveniles appeared in samples of coarse sand from shallow subtidal sites just inside the northern reef edge at Bermuda, June 1999. As they were immature, identification of the species remains unknown, but it is likely that they are the enteropneust in this sandy habitat at Bermuda, namely *Glossobalanus crozieri* van der Horst. They were strikingly turbellarian-like, entirely ciliated and moving by ciliary gliding, including occasional backward swimming reminiscent of retronectid catenulid and solenofilomorphid acoel turbellarians. The strongly ciliated and muscular pharynx was also reminiscent of catenulids and of solenofilomorphids (Figure 1.2A–D).

By electron microscopy (three specimens of a total of 24 embedments), a number of other features reminiscent of turbellarian flatworms and others specific to enteropneusts emerged. The epidermis was a simple ciliated epidermis, composed of a single layer of multiciliated cells without a cuticle (Figure 1.2B). Over the trunk, the epidermis was squamous, its cells broader than wide; on the proboscis and collar, they were tall and columnar. The entire epidermis was underlain by a distinct basal lamina, as was the simple ciliated, cellular gut, and between the two laminae were only fibre-form muscle cells and occasionally what may be another mesenchymal cell type, but coelomic space was restricted to the proboscis and posterior reaches of the trunk; the bulk of the trunk and the collar region were acoelomate. The muscles were of the usual invertebrate-smooth type and in a simple orthogonal arrangement, dominated by longitudinal muscles and with circular muscles best developed but still weak in the proboscis. Blood vessels, which appeared as spaces homogeneously filled with a finely granular material (blood) between basal laminae, were also readily apparent at certain sites.

The nervous system was a basiepithelial nerve plexus (basiepidermal nerve net), typical of enteropneusts; it was especially thick in the proboscis and collar and in local concentrations in the trunk that probably corresponded to longitudinal tracts. Circular tracts also were evident.

The pharynx was densely ciliated and muscle-lined and bore a single pair of gill slits dorsally. The rest of the gut, composed of simple ciliated epithelium, was fairly well differentiated into stomach, intestine, and hindgut (Figure 1.2A) and terminated at a supraterminal anus.

Similarities with the Platyhelminthes

A detailed analysis of the ultrastructure of this enteropneust has yet to be made (as Rieger (1980, 1981) did so elegantly for *Lobatocerebrum*), but preliminary observations do show a number of similarities with platyhelminths, the most critical of which may be the acoelomate structure of the trunk and collar. The body wall could be viewed as turbellarian-like, with a simple orthogonal grid of muscles and with a completely ciliated epidermis lacking cuticle, and it appears most reminiscent of that in the Catenulida by its sparse ciliation and the shape of the ciliary rootlets (Figures 1.2B,F,G and 1.3). While Benito and Pardos (1997) have compared the shape of the tips of cilia in the pharynx of the enteropneust *Balanoglossus minimus* with those of acoelomorph turbellarians, those in this intersitial juvenile enteropneust were unlike the shelfed tips of acoels; epidermal cilia were simply blunt-tipped (Figure 1.2E) and those of the pharynx had long, tapered tips.

The gut is comparable to that of most flatworms in having a simple cellular ciliated epithelium, but not in its differentiation into regions, including hindgut and anus. A few turbellarians do have anuses that amount to simple fusions between gut and body wall without hindgut (see summary by Knauss 1979). Among them, *Haplopharynx*, at the base of the rhabditophoran clade (Rieger 2001, this volume), has an anus that in its supraterminal position and connection to a strongly ciliated posterior region in the gut (Karling 1965), appears most comparable to that of the enteropneust; in the context of the hypothesis of progenetic origin of flatworms, further scrutiny is warranted. Other enteropneusts apparently do not have a hindgut differentiated from the rest of the intestine, and the anus appears as a poorly differentiated fusion with the epidermis (Benito and Pardos 1997); in this respect, they could be considered more comparable to the flatworms than this juvenile. The anus in *Haplopharynx*, then, in the scheme of flatworm phylogeny, would be a plesiomorphic similarity with that of coelomates, not a convergent one.

The pharynx of the enteropneust bears a striking resemblance in its general shape, muscularity, and dense ciliation to the pharynx simplex of lower turbellarians, including catenulid, acoel, and macrostomid turbellarians. Whether these

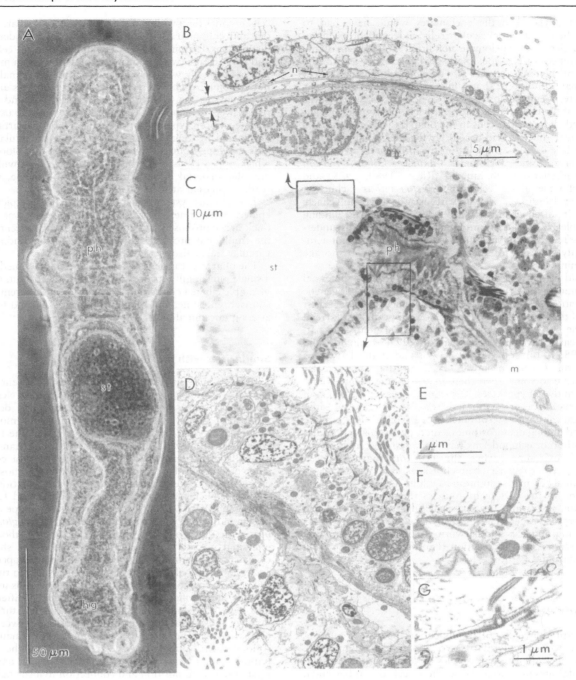

Figure 1.2 Juvenile enteropneust, probably *Glossobalanus crozieri* van der Horst, recently settled in interstices of coral sand. **A**) Whole mount live, dorsal aspect, phase-contrast microscopy; length = 540 μm. **B**) Dorsal trunk body wall and gut in sagittal section by electron microscopy (EM); arrows mark basal laminae of epidermis and gut. **C**) Sagittal, toluidine-blue-stained, 1.5-μm section of epoxy-embedded specimen by light microscopy showing acoelomate nature of collar and trunk. Upper and lower boxes enclose areas similar to those shown by EM in **B**) and **D**), respectively. Scale as in **B**). **D–G**) Transmission electron micrographs of sagittal sections. **D**) Ventral body wall and pharynx. **E**) Tip of epidermal cilium. **F**) and **G**) Bases of epidermal cilia with rootlets; to same scale. hg, hindgut; m, mouth; n, nerves of basiepithelial plexus; ph, pharynx; st, stomach.

pharynges are even homologous among the turbellarians, however, is problematic. Sterrer and Rieger (1974) believed that the pharynges simplex in catenulids and macrostomids were 'a heterogeneous assemblage of superficially similar structures of different origin', and Doe (1981), judging from comparative study of the ultrastructure of these pharynges, says those of the Macrostomida are sharply different from those of the Catenulida and Acoela in such characters as innervation and component glands. Both Doe (1981) and Crezée (1975) believed that the pharynges of acoel turbellarians were not only distinct from those of other turbellarians but that they developed independently in the various acoels that posses them (Solenofilomorphidae, Proporidae, Hallangiidae, Diopisthoporidae, etc.); most acoels lack a pharynx and have

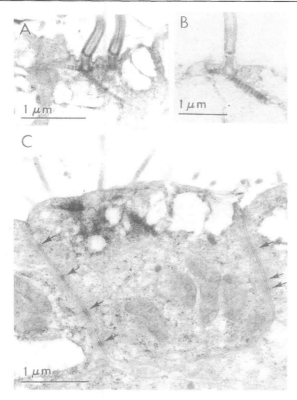

Figure 1.3 *Retronectes* sp. (Caterulida) from Seal Harbor, Maine, USA. Electron micrographs of sagittal sections. **A**), **B**) epidermal cilia and their rootlets. **C**) Epidermal cells joined by peculiar cell junctions (arrows) reminiscent of those in *Xenoturbella*.

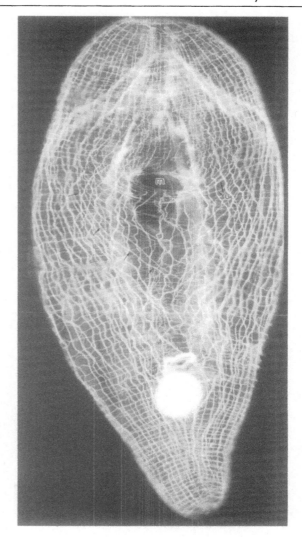

Figure 1.4 Body-wall musculature of *Praeaphanostoma* 'wadsworthi' (Acoela) from Castine, Maine, USA, a species lacking a pharynx but having complex involvement of ventral musculature, including U-shaped muscles (u) arcing behind mouth (m) and muscles that pivot at mouth between longitudinal orientations anteriorly and diagonal orientations posteriorly (arrows). (Fluorescence microscopy of whole mount stained with Alexa-488-phalloidin, which labels filamentous actin; courtesy of Matthew D. Hooge.)

only a simple pore joining epidermis and digestive tissue, and this was assumed to be the primitive condition.

We have found, however, that those acoels lacking a pharynx have a more derived body-wall musculature, deviating from the simple orthogonal grid of longitudinal and circular muscles found in the more primitive families (M. D. Hooge and S. Tyler, unpublished data), and that this musculature is arranged in a way to substitute for the food-engulfing functions of a pharynx (Figure 1.4; see also Tyler and Rieger 1999). The so-called 'pharynx-0', therefore, is probably not plesiomorphic; instead, possession of a pharynx is likely the plesiomorphic condition in acoels, as Ax (1963) proposed. It is not surprising, then, to find a pharynx in a model for the origin of the platyhelminths such as the enteropneust juvenile. Admittedly, however, probability of homology does remain low; further scrutiny for such features as innervation and position of glands is needed before strong weight can be placed on the general similarity. Still, it is easy to imagine how such a pharynx, with its strong musculature and dense ciliation, preadapted the progenetic ancestor of flatworms to feeding on diatoms and other meiobenthic organisms.

The nervous system of the juvenile enteropneust is, as is classically recognized for that of other enteropneusts as well, a basiepithelial plexus. Turbellarians also have a basiepithelial plexus as one component of their nervous system, which also comprises a subepithelial plexus and a submuscular plexus, the latter encompassing or connecting to more substantial longitudinal nerve cords (Rieger *et al.* 1991). Certain members of the Acoelomorpha have only the basiepithelial plexus, however, a condition often regarded as plesiomorphic; and

especially with their lack of a concentration of nervous tissue that could qualify as brain, they resemble the enteropneusts. Enteropneusts, then, serve as a good outgroup to the flatworms with respect to evolution of the nervous system.

Rieger appears careful to apply his examples of *Lobatocerebrum* and *Bonellia* only as *models* for the origin of acoelomate and pseudocoelomate organisms, not claiming that either is in a direct line with the ancestor of the Platyhel-minthes. In the same vein, I cite the enteropneusts as a model for this process.

The enteropneusts offer a few advantages not seen in *Lobatocerebrum* (and other spiralians) as a model for the transition between a coelomate ancestor and platyhelminth body plan: enteropneusts are clearly coelomate, and still show more or less limited acoelomate structure in some species, including this juvenile enteropneust; their epidermis has a

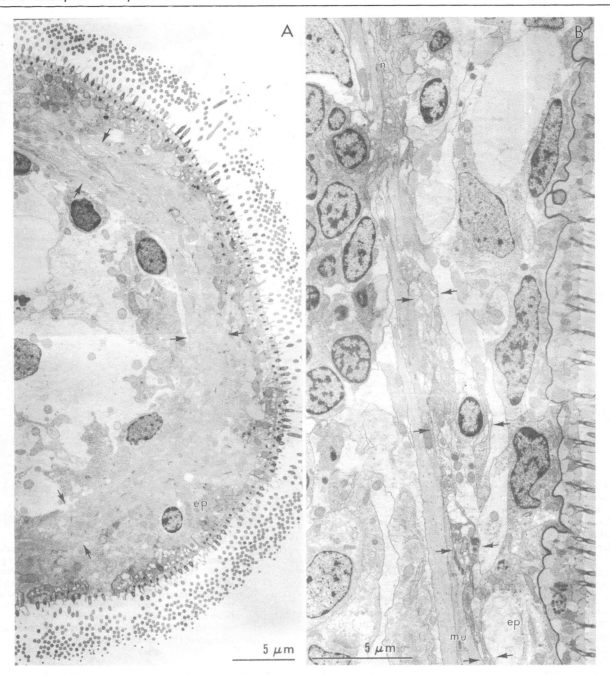

Figure 1.5 Electron micrographs of nervous systems in Acoelomorpha. **A**) Submuscular nerve ring (marked between pairs of arrows) in cross-section of *Diopisthoporus gymnopharyngeus* (Acoela) anterior to statocyst. **B**) Dorsal longitudinal nerve (between arrow pairs) extending from deeplying neuropile of brain (n) into basiepithelial nerve plexus in sagittal section of *Flagellophora apelti* (Nemertodermatida). ep, epidermis; mu, muscle.

similar structure, especially in lack of cuticle and monolayering of epidermal cells, but also in the multiciliated nature of the epidermal cells and the structure of the cilia and their rootlets; enteropneusts also lack the gliointerstitial system characteristic of annelids and molluscs and not found in platyhelminths (Rieger 1981a).

In other ways the enteropneust might be viewed as less appropriate a model, particularly in that as a deuterostome its development, by way of radial cleavage instead of spiral, is

different from that of protostomes; it also has a blood-vascular system which would be expected to be associated with a metanephridial excretory system (Ruppert and Smith 1988), both unknown in flatworms.

Embryonic cleavage in the Platyhelminthes is believed to be fundamentally spiral, as evidenced in the Polycladida and Acoela, even though most flatworms have what appear to be greatly derived patterns of cleavage (see review by Boyer and Henry 1998). Polyclads have a classic quartet spiral, homologous with that of annelids and molluscs (Boyer *et al.* 1998), and acoel turbellarians have a duet spiral pattern that has long been believed to be true spiral but progressing by duets by virtue, perhaps, of suppression of every other cleavage. Boyer, Martindale and Henry (Henry *et al.* 2000), inspired by claims of the acoels' basal-bilaterian status (Ruiz-Trillo *et al.* 1999), have recently reanalysed this pattern and concluded also that it is more bilateral than spiral. Such an interpretation brings the acoels into the realm of deuterostome patterns; at the same time, the production of micromeres by laeotropic cleavages ties them to the spiral pattern. Effectively, then, they straddle the border between deuterostomes and protostomes, an appropriate position for a protostome group derived by progenesis from a deuterostome ancestor.

While a blood-vascular system was developed in the juvenile enteropneust, I did not find evidence of podocytes in the proboscis, and thus excretion in this juvenile is likely to be handled by the pore canal which Ruppert and Balser (1986) equate with a protonephridium. Protonephridia are the excretory organs of the Catenulida and Rhabditophora, and in the Catenulida, at least, the protonephridial complex and its pore is unpaired in a mid-dorsal position as is the hydropore/pore-canal complex of the tornaria larva of enteropneusts, though details of its structure are different. In any event, by the progenesis hypothesis for the origin of flatworms, their protonephridia would be expected to be derived from protonephridia of the larva-like ancestor.

Yet another indicator of the relationship between platyhelminths and deuterostomes is their sharing of certain codons in the mitochondrial genome: platyhelminths (triclads) share with hemichordates the coding of isoleucine by the codon AUA and with echinoderms the coding of asparagine by AAA

(Castresana *et al.* 1998). Neither of these codons is seen in the other protostomes.

Relationships of the lower turbellarians

While the three monophyletic groups of flatworms easily fit the model as descendants of such a larva-like ancestor, there is still no synapomorphy that ties the three together. Even the nature of the acoelomate structure of the body could have three separate origins (Rieger 1985) albeit, probably, from such a deuterostome line as produced the enteropneusts. Most like the enteropneusts in their lack of any tissue other than muscle between epidermis and gut is the Acoelomorpha (Rieger 1985).

Also the nervous system, as a basiepithelial plexus, finds greatest similarity in the Acoelomorpha in that certain acoels are reported to have only such a plexus, with local concentrations at the anterior end serving as brain (see Rieger *et al.* 1991 for literature). Most acoels and the nemertodermatids have anterior ring-like thickenings of the nervous system (Figures 1.4 and 1.5) as do the enteropneusts, and so, too, the brain in Catenulida is ring-like (personal communication Boris Joffe 1999). Like the rhabditophorans, at the same time, the central nervous system of most acoelomorphs is sunk below the body wall and forms well-delimited brains consisting of a central neuropile and a rind of nerve cell bodies (Figure 1.6). In all three clades, paired longitudinal nerves at a variety of dorsal, lateral, and ventral positions extend back from the brain (Rieger *et al.* 1991; Reuter *et al.* 1998a) so characterization of the flatworms as 'gastroneuralians', as has been done in some phylogenetic hypotheses, is not valid. Differences between acoels and the rest of the flatworms in the distribution of

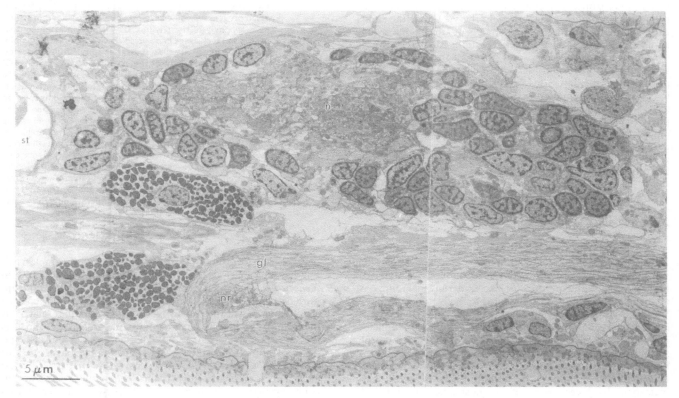

Figure 1.6 Electron micrograph of brain, with its central neuropile (n) surrounded by rind of peripheral cell bodies, in sagittal section of *Flagellophora apelti* (Nemertodermatida); gland necks (gl) of broom organ loop through ventral part of nerve ring (nr) from brain; st, statocyst.

neuropeptides have been cited as indicative of a unique position for the Acoela (Reuter *et al.* 1998b; Raikova *et al.* 1998a), but these features taken together with form of the brain in most acoels (with central neuropile and rind of nerve cells), could still be interpreted as falling in a continuum linking a basiepithelial plexus like that in enteropneusts with the more concentrated orthogonal arrangements in the higher flatworms (Reisinger 1972; Ehlers 1985a).

The gut in all three platyhelminth clades generally is sac-like, without hindgut or anus, and the probability that it originated, not by descent from the blind-ended gastrovascular cavity of coelenterates, but from specialization of the flow-through intestine of a coelomate ancestor gives it status as a shared derived character among flatworms. While the Gnathostomulida is also characterized by a sac-like gut, distinctive features of it (including a non-permanent functional anus; Knauss 1979) as well as other characters tying this phylum to the Rotifera show that it is only convergently similar. Lack of an anus is not simply a convergent feature of organisms of small size, since other small and smaller animals (nematodes, gastrotrichs, loriciferans) do have an anus and since many giant flatworms (e.g., triclads of Lake Baikal) do not. While most acoels have a central syncytium as their gut, this is undoubtedly derived from a cellular condition as found in the nemertodermatids (see Smith and Tyler 1985) as well as, perhaps, in such acoels as *Paratomella* (see Smith and Tyler 1985) and the solenofilomorphids, which also have a lumen to the intestine like other non-acoel flatworms (unpublished observations).

The hermaphroditic reproductive system, with its complicated copulatory and bursal organs and biflagellate spermatozoa, is also potentially a shared derived feature of Acoelomorpha and Rhabditophora, but it is difficult to define it in such a way as to qualify as a synapomorphy. Hermaphroditic systems in other phyla, notably the Gnathostomulida and Gastrotricha appear qualitatively different, unrelated to the platyhelminths, strengthening the view that the similarities among flatworms are not convergent (although they may be homoiologous [from parallel evolution; Riedl 1978]).

Mapping these characters and others related to structure of spermatozoa, protonephridia, and cilia in a strictly defined phylogeny of the flatworms is difficult, the arguments for various possibilities remaining essentially as outlined by Smith *et al.* (1986). For the reasons cited above, probability of monophyly remains good, at least (Figure 1.7A), but even if homology of quartet spiral cleavage speaks against this and shows paraphyly of the phylum (Figure 1.7B), it is clear that all three clades of the flatworms are more closely related to each other than to other metazoans (Figure 1.7D).

Xenoturbella

Another animal that has been proposed to be a 'neotenic hemichordate' is *Xenoturbella*, a strangely simple, 1- to 3-cm-long vermiform organism with a mouth at midbody leading to a blind gut and with enteropneust-like epidermis including basiepithelial nerve plexus. First assigned to the Platyhelminthes by Westblad (1949) and ranked in the Acoela by Hyman (1959), it has since been proposed to be a relative of the Enteropneusta and Holothuroidea (by Reisinger 1960), of the Acoelomorpha (by Franzén and Afzelius 1987; Rohde *et al.* 1988e), of protobranch bivalve molluscs (by Norén and Jondelius 1997; Israelsson 1997), and to be independently the most basal metazoan (by Jägersten 1959) or most basal bilaterian (by Ehlers and Sopott-Ehlers 1997c). Reisinger's (1960) evidence for ties to the deuterostomes remains plausible, especially in

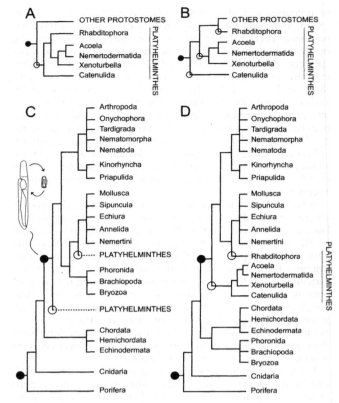

Figure 1.7 Cladograms of platyhelminth groups relative to other metazoans. Filled circles mark ancestors with biphasic life cycle (alternating between a macroscopic coelomate adult and a microscopic acoelomate larva); open circles mark points at which progenesis would have occurred. **A**) Cladogram assuming monophyly of the phylum. **B**) Cladogram assuming that quartet spiral cleavage of polyclad rhabditophorans is homologous with that of annelids and other spiralian protostomes, thus showing paraphyly of the phylum Platyhelminthes. **C**) Cladogram of Valentine (1997) modified to show two potential sites for a monophyletic Platyhelminthes (as in **A**), either as sister group to all protostomes or as sister group to only spiralian protostomes. **D**) The same cladogram modified to accommodate a paraphyletic Platyhelminthes (as in **B**), emphasizing homology of quartet spiral cleavage in rhabditophorans and other members of Spiralia and assuming that the bilateral/duet-spiral cleavage of Acoela is plesiomorphic. (Note that such emphasis on developmental characters places the lophophorate phyla as sister group to the deuterostomes, distinct from spiralian phyla. On both trees, C and D, controversial groups such as the Gastrotricha, Rotifera, and other small worm groups, are omitted.)

parallel with evidence for the progenetic origin of platyhelminths. Reisinger's (1960) comparison of epidermis in *Xenoturbella* with that of enteropneusts is apparently not supported at the ultrastructural level (Pedersen and Pedersen 1988), but the distinctive form of its extracellular matrix is most similar to that in deuterostomes, including enteropneusts (Pedersen and Pedersen 1986). The cytology of muscles and other features of the acoelomate body structure of *Xenoturbella*, as with the flatworms, show that it is derived from a coelomate ancestor. *Xenoturbella* could easily rank in a phylogenetic position intermediate between deuterostomes and flatworms. The nemertodermatid-like characters of its epidermis (Westblad 1949), the acoelomorph-like characters in cilia structure ascribed to it by Franzén and Afzelius (1987), Rohde *et al.* (1988e), Rieger *et al.* (1991), and Lundin (1998), the hemichordate-like characters in body wall and nervous system recognized by Reisinger (1960) and Pedersen and Pedersen (1986), and the holothurian-like features in myocytes and their junctions recognized by Ehlers and Sopott-Ehlers (1997c) all make sense in the scenario of progenesis. Yet another feature of *Xenoturbella* fitting this scenario is the structure of the gut; it is cellular but not ciliated, in a fashion reminiscent of lower acoelomorphs, namely nemertodermatids and solenofilomorphids (see also Lundin 1998). The logical position for *Xenoturbella*, as Lundin (1998) has shown, is sister group to the Acoelomorpha. A striking similarity with the Catenulida is the peculiar nature of intercellular junctions in the epidermis, comprising small densities, fairly regularly spaced along what appear to be rigidly parallel apposing cell membranes (Figure 1.3C; compare with Pedersen and Pedersen 1986: 233); as in *Xenoturbella*, these are especially numerous along basal interdigitations of epidermal cells. Because of such a tie and the distinctive nature of the Catenulida among flatworms with respect to reproductive organs, excretory organs, and cytology of the epidermis, as well as similarities with the hypothesized deuterostome-derived ancestor, the Catenulida is likely to occupy a more basal position, possibly as sister group to *Xenoturbella*+Acoelomorpha; therefore *Xenoturbella* must be considered a member of the Platyhelminthes (Figure 1.7A).

Phylogenetic position of the flatworms

Similarities between the enteropneust juvenile and flatworms warrant rethinking the relationships of the Platyhelminthes as a whole. Rieger's (1985, 1986a) demonstration of the link between fibre-form muscles in flatworms and those of coelomates and of that between the acoelomate nature of the body in flatworms and larvae and juveniles of coelomates remains incontrovertible. But it is not just spiralian coelomates that fit the bill. The similarities in epidermal structure, including ciliary fine structure, and the nature of the nervous system, including possession of a basiepithelial system, indicate probability of closer ties with the deuterostomes than with the higher spiralian protostomes. These similarities also warrant reconsidering what the original features of the stem platyhelminth(s) were; if a deuterostome is the outgroup of platyhelminth clades, then features like the pharynx and anus would be plesiomorphic in them, not convergent similarities as has been assumed. The probabilities of homology of these features remain to be thoroughly evaluated using proper criteria (Rieger and Tyler 1985; Tyler 1988), but there is sufficient reason already to regard the similarities as phylogenetically significant.

Certain qualitative molecular characters (*sensu* Joffe and Kornakova 2001, this volume) may show a closer tie with spiralian coelomates than deuterostomes – specifically those of gene order in the HOX cluster (Balavoine 1997, 1998), as already mentioned. Gene order in mitochondrial genomes may also be brought to bear on this question; so far, only one platyhelminth species has had some of its mtDNA sequenced, *Fasciola hepatica*, which is a relatively derived neodermatan, and while the sparse information available on it shows some similarity with mtDNA gene order in the annelid *Lumbricus*, the phylogenetic significance of this similarity is unknown (Boore 1999). Until a broader sampling of mtDNA genomes as well as of *Hox* genes is made, it is premature to rule out close ties to deuterostomes with these characters. More significant for the time being is that platyhelminths in all characters appear closer to coelomates of some kind, whether protostome or deuterostome.

I am not suggesting, in any case, that platyhelminths are enteropneusts, rather that flatworms could have originated from a stem group of vermiform deuterostomes by way of progenesis (Figure 1.7). Even the group to which this deuterostome ancestor belonged is probably extinct; otherwise clues to its existence would not have to rest on such details as ultrastructure of cells and organelles. It is possible that this group included the common ancestor to the Deuterostomia and Protostomia together, particularly if one places greater weight on similarities in body wall and form of the body cavity (Figure 1.7C, lower dotted line). Valentine (1997) considered this ancestor to have pseudocoelomate body structure, even though his phylogenetic tree has pseudocoelomate phyla in more derived positions; far more likely is that this ancestor is coelomate (hence the homology of the coelom in both deuterostomes and protostomes (Balavoine 1998; Dewel 2000)) and that its larva provided the progenetic origin for the flatworms. The pseudocoelomate phyla would also have arisen by progenesis (Rieger 1986a), but probably from a descendant in the protostome line.

On the other hand, if the cleavage patterns of ecdysozoan phyla and lophophorates are not derived from the spiral pattern (Valentine 1997) and the *Ubd-A*-like *Hox* genes of planarians are homologous with those of spiralians and different from those of other protostomes (Balavoine 1997), then it would be reasonable to view the platyhelminths as standing at the base of the spiralian line, as an early off-shoot of it. This position is supported especially by the homology of spiral cleavage in polyclads, annelids, and molluscs and the intermediate nature of the bilateral/duet-spiral cleavage of acoels (Haszprunar 1996). Retention of the coelom in the main spiralian line shows that they are not descendants of flatworms, but more likely descendants of the macroscopic adult phase of the same biphasic life cycle whose microscopic larva provided the ancestor to the flatworms.

Conclusions

While proof of the monophyly of the Platyhelminthes remains elusive, the three monophyletic clades composing it are clearly related, and hypotheses in which the Acoela is separated from the other flatworms by branches leading to two or more other protostome phyla (Zrzavý *et al.* 1998; Littlewood *et al.* 1999a; Ruiz-Trillo *et al.* 1999) have, in my estimation, low probability of being correct. A separate origin of the Acoelomorpha (and *Xenoturbella*) from that of the Rhabditophora remains possible (Figure 1.7B) even though the monophyly of the phylum (Figure 1.7A) seems more likely.

Rieger's (1980, 1986a) hypothesis of progenetic origin of acoelomate and pseudocoelomate phyla makes eminent sense. Progenesis is undoubtedly the evolutionary process by which the flatworms arose; there can be no other explanation for their histological structure. Progenesis has played a major role in much of the evolution of the interstitial fauna (Westheide 1987), and while some cases of progenetic origin provide sufficient intermediates that the process can be recognized (the interstitial annelids, for instance), that for the flatworms and the pseudocoelomate phyla has produced such a drastic revamping of morphology that the origin of a line becomes difficult to perceive, hence our longstanding uncertainty about the phylogenetic position of the Platyhelminthes. Like a bottleneck, progenesis leaves precious few morphological clues to the phylogenetic origin of the flatworms. Further testing for the sister group (or groups) of the flatworms should, however, find fruit in quantitative molecular characters such as gene order in the HOX cluster or mitochondrial genomes, and such testing especially needs to focus on the lower orders, the acoels, catenulids, and macrostomids. The origin of the flatworms could still lie entirely within the protostomes and within the spiralian clade (Balavoine 1998), but consideration of other possibilities, closer to the stem bilaterian ancestor, cannot yet be ruled out. In any case, that origin or origins must follow a course of progenesis.

ACKNOWLEDGEMENTS

I thank Reinhard Rieger and Julian P.S. Smith III for the insights they have shared with me since we all began our search for clues to platyhelminth phylogeny years ago, and I thank Matt Hooge for sharing his data and offering his own insights on invertebrate relationships. Supported by US National Science Foundation grant DEB-9419723. Contribution from the Bermuda Biological Station for Research Vol. 44, #1561.

Chapter 2

Contributions to the phylogeny and systematics of the Acoelomorpha

Olga I. Raikova, Maria Reuter and Jean-Lou Justine

The Acoelomorpha Ehlers, 1984, comprising Acoela and Nemertodermatida, were traditionally considered monophyletic and were included within the monophyletic Platyhelminthes (Ehlers 1985a,b; Rieger et al. 1991). However, a conflicting view existed regarding Acoelomorpha as one of the three groups (Acoelomorpha, Catenulida and Rhabditophora) constituting the non-monophyletic Platyhelminthes (Smith *et al.* 1986). Recent morphological and molecular data suggest that Acoela could be the sister group to the Bilateria (Haszprunar 1996; Ruiz-Trillo *et al.* 1999). The position of Nemertodermatida was even more difficult to determine in molecular studies (Littlewood *et al.* 1999a; Ruiz-Trillo *et al.* 1999). Therefore the search for additional characters is pertinent to our understanding of the phylogeny of the Acoelomorpha. This study summarizes data obtained so far from our recent studies on brain and sperm structure, and attempts to clarify the interrelationships of the Acoela and the Nemertodermatida on the one hand and the phylogeny within the Acoela on the other hand.

Patterns of 5-HT and neuropeptide immunoreactivity in the nervous system of the Acoelomorpha

The brains of Acoela and other flatworms were traditionally considered homologous, the acoelan brain being the earliest extant form of flatworm brain (Hyman 1951; Beklemischev 1963; Reisinger 1972; Ivanov and Mamkaev 1973; Ehlers 1985a). Recently, an alternative concept appeared which argued that the brain of Acoela has evolved independently of that of the Platyhelminthes and therefore cannot be called a 'true brain' (Haszprunar 1996). The current knowledge about the neuroanatomy of the Acoelomorpha has been based on studies using conventional staining techniques (Westblad 1937, 1949, 1950; Dörjes 1968; Reisinger 1976), the acetylcholinesterase method (Mamkaev and Kotikova 1972; Boguta 1976; Raikova *et al.* 1998a) and ultrastructural studies (Ferrero 1973; Crezée and Tyler 1976; Ehlers 1985a; Bedini and Lanfranchi 1991; Raikova 1991). The acoelan brain appeared to be either a ring of nerve fibres close to the statocyst or a terminal thickening of the subepidermal nerve plexus. However, it was evident that the brain structure displayed considerable variations among species (Ivanov and Mamkaev 1973). Two parts of the acoelan brain structure were distinguished: an outer part named 'the orthogonal brain', composed of thickened anterior parts of the longitudinal brain cords and an inner part named 'the endonal brain' (Ivanov and Mamkaev 1973). The endonal brain was considered homologous in all Platyhelminthes.

Using immunocytochemistry (ICC) and confocal scanning laser microscopy (CSLM), two parallel studies of the immunoreactivity patterns of the neurotransmitter serotonin (5-hydroxytryptamine; 5-HT) and the regulatory peptide FMRF-amide have recently been performed (Raikova *et al.* 1998a; Reuter *et al.* 1998b). The aim of these studies was to provide new data to solve the problem whether species of the taxon Acoela have a brain and an orthogon of the common flatworm type, i.e., a bilobed ganglionic brain and a regular pattern of longitudinal nerve cords connected by transverse commissures. The results revealed noticeable differences between the patterns of immunoreactivity in Acoela in comparison to those stained by the same reactions in other flatworm species. Using antibodies to 5-HT we found no ganglionic cell mass typical for other platyhelminths, only a symmetrical brain-like anterior structure composed of commissural fibres associated with a few cell bodies, i.e., a commissural brain (Raikova *et al.* 1998a), and three to five pairs of radially arranged longitudinal nerve cords, connected by an irregular network of transverse fibres. The FMRF-amide immunocytochemistry revealed a bilobed brain-like structure lacking neuropile and composed of two clusters of large nerve cells with short processes, resembling endocrine cells of higher animals, i.e., an endocrine brain (Reuter *et al.* 1998b). However, complementary immunocytochemical observations on two species *Paraphanostoma crassum* and *Avagina incola* (Figure 2.1a,b; opposite p. 20), using antibodies to the native flatworm neuropeptide GYIRF-amide (Johnston *et al.* 1995) instead of FMRF-amide, widens the view of the peptidergic part of the acoelan brain (Reuter *et al.* in press). As in the previously described acoels, clusters of cells, lying peripherally to the commissural 5-HT immunoreactive 'brain' characterize the peptidergic part of the acoelan brain in both species (Figure 2.1 a,b). The native flatworm neuropeptide GYIRF-amide reveals, however, more nerve processes than the FMRF-amide. Compared to 5-HT immunoreactive fibres, the peptidergic fibres are sparse and concentrated around the cell clusters. A few fibres run frontally and a pair of lateral longitudinal fibres run backwards parallel to the most prominent 5-HT immunoreactive fibres. No true neuropile has been observed with the antisera mentioned above.

Concerning the immunoreactivity patterns in the nervous system structure in the Acoela the following points can be stressed:

1. The neuroanatomical structure, 'the commissural brain', showing positive reactions for 5-HT, catecholamines and acetylcholinesterase, is highly specialized.

2. The symmetrical aggregation of FMRF-amide and GYIRF-amide immunoreactive elements in the anterior end merits the name brain, but the commissural and the peptidergic parts of the brain are not integrated into a brain of the common flatworm type.

3. No regular orthogon *sensu* Reisinger (1972) was visualized in Acoela, only the 'cordal nervous system' (Kotikova 1991) consists of a varying number of longitudinal nerve cords connected by an irregular network of fibres.

4. The longitudinal cords cannot be categorized into distinct main and minor nerve cords, as they can in other flatworms.

In order to solve the question whether these characteristics are apomorphic or plesiomorphic features and to check the suggested monophyly of Acoelomorpha, brain structure in two species of Nemertodermatida (*Nemertoderma westbladi* and *Meara stichopi*) has been recently studied by means of 5-HT, and FMRF-amide immunocytochemistry. Our immunocytochemical observations (Raikova *et al.* 2000) show that these two species of Nemertodermatida display considerable differences in the structure of their nervous system.

Common to the nervous system of *Meara stichopi* and most flatworms are the 5-HT immunoreactive surface nerve net and the deeper-lying cord-like longitudinal nerve bundles connected by transverse commissures (Figure 2.1c). Seemingly more primitive features are the presence of only two 5-HT immunoreactive cells at the frontal end and the loose structure of nerve bundles, made of separate fibres. No brain-like structure, either peptidergic or aminergic, has been detected in *Meara*. In *Nemertoderma westbladi* (Figure 2.1d), a ring-shaped 5-HT immunoreactive brain-like structure is reminiscent of the commissural brain of acoels. Longitudinal nerve fibres running in the posterior direction are, however, much more numerous in *Nemertoderma* than in acoels, and 5-HT immunoreactive nerve cells are even more scarce. No peptidergic cell clusters common for the Acoela were revealed, only some thick FMRF-amide reactive fibres.

Clearly, the organization of the nervous system in both taxa of the Acoelomorpha differs from that in most flatworms. The 'brain' structure of the Acoela, composed of a commissural 5-HT immunoreactive part and clusters of peptidergic 'brain' cells seems to be apomorphic for the group. The questions of the plesiomorphic or apomorphic nature of the nervous system structure in Nemertodermatida and the monophyly of the Acoelomorpha are not yet settled by our studies. The above stressed differences observed in Acoela and Nemertodermatida seem to reflect a great variability typical of archaic groups of organisms (Mamkaev 1986), thus indirectly indicating a basal phylogenetic position for the Acoelomorpha. It is likely that brain structures in *Meara*, *Nemertoderma* and Acoela have evolved independently, possibly from a common ancestor displaying an even more primitive nervous system pattern. As studies on the distribution of other neuroactive substances in the Acoelomorpha are lacking, the final conclusions concerning the phylogenetic implications of acoelomorph neuroanatomy need further research.

Spermiogenesis and sperm structure

Acoelomorpha

With regard to the sperm structure, no apomorphies are currently known for the Acoelomorpha. In fact, the Nemertodermatida (Tyler and Rieger 1975, 1977; Hendelberg 1977; Lundin and Hendelberg 1998) have a uniflagellar sperm with a proximo-distally-oriented axoneme, while all the Acoela have filiform spermatozoa with two incorporated axonemes. In acoelan spermatozoa the axonemes have a reverse, disto-proximal orientation (i.e., centriolar derivatives are situated at the extremity of the cell opposite the nucleus). Axonemal structure in Nemertodermatida is $9 + 2$, clearly a plesiomorphic character. In Acoela the axonemal structure is either $9 + 2$ or $9 + 0$ (without central microtubules) or $9 + 1$ (with one central electron-dense element). On the other hand, the Acoela and the Trepaxonemata (ref. in: Watson 1999b) share some sperm characters, which include filiform biflagellate sperm, inverted position of axonemes (centrioles distal), disto-proximal fusion of the axonemes with the sperm cell during spermiogenesis, modified centrioles, presence of longitudinal cytoplasmic microtubules and cytoplasmic granules.

All these characters could be homoplasies, but their number and unique occurrence among the Metazoa suggest possible homology. In fact, we think that it is only the inclusion of the Nemertodermatida within the Acoelomorpha that has prevented Ehlers (1985b) from considering all these characters synapomorphies of Acoela and Trepaxonemata. Nemertodermatids display a set of common characters with the Acoela in epidermal ciliation structure (Tyler and Rieger 1977; Ehlers 1985a; Lundin 1997, 1998). Therefore, acceptance of the monophyly of the Acoelomorpha and the monophyly of the Acoela leads us to consider all the similarities in sperm structure listed above as homoplasies.

Acoela

Acoela is a group displaying the greatest morphological variety. A rather small percentage of the described species have so far been studied by means of electron microscopy. Classification of the group is based mainly on the structure of the male copulatory organs (Dörjes 1968). However, light microscopy alone is often unable to test the homology of the characters used for systematics. For example, the family Childiidae Dörjes has the following diagnosis (Cannon, 1986): 'Mouth ventral. Male organ well developed, conical penis stylet always present, not inserted into seminal vesicle, (if present), male pore variable in position'. Our ultrastructural and immunocytochemical preliminary studies on the stylet structure in three childiids have shown that stylets are non-homologous structures (Raikova and Justine, in preparation). In *Actinoposthia beklemischevi* the stylet is formed of microtubules and contains tubulin, in *Paraphanostoma cycloposthium* the stylet is formed by fused secretory granules of spicule-forming cells, while in *Philactinoposthia* sp. the 'stylet spicules' were found to be needles, which do not contain tubulin, formed inside the spermatozoa. Clearly, the systematics of the Acoela needs revision. It has become very important to determine the characters that could provide a basis for this revision, and sperm structure seems a good candidate. In the case of spermatozoa we deal with structures of lesser complexity than the male copulatory organs. Sperm structure shows enough variability in the group, while closely related species display similar spermatozoa: for example, family Sagittiferidae (Raikova *et al.* 1998b).

The combination of ultrastructural studies, tubulin immunocytochemistry and fluorescent labelling of nuclei provides valuable information about the general morphology of the filiform spermatozoon and of the arrangement of the nucleus, axonemes and cytoplasmic microtubules along its length. This approach permits us to define several phylogenetically significant characters of spermiogenesis and spermatozoa within the Acoela.

Spermiogenesis in Acoela

Spermiogenesis has been studied in detail only in a few species of the Acoela. However, a number of common characters can be stressed:

1. During spermiogenesis, the nucleus remains in the proximal part of the spermatid. The shape of the nucleus changes from crescent-shaped to filiform (Figure 2.1e,f).
2. The originally free flagella fuse with the cytoplasm of the spermatid in the disto-proximal direction, each incorporated flagellum being originally surrounded by two membranes of one cytoplasmic canal (Figure 2.2A–C). The

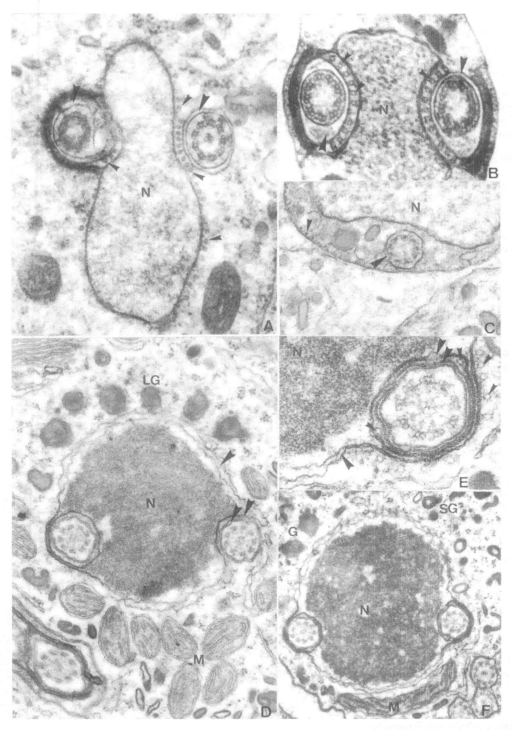

Figure 2.2 Spermiogenesis in Acoela: cytoplasmic canals in the course of spermatid elongation. **A)** *Symsagittifera bifoveolata* (×78 000). **B)** *Symsagittifera schultzei* (×89 000). In A and B, the single cytoplasmic canal around each flagellum is lined by microtubules on the side of the nucleus and strengthened by electron-dense material on the other side. **C)** *Actinoposthia beklemischevi*. The single cytoplasmic canal surrounds each flagellum. Note the absence of microtubules along the nucleus and the presence of cortical microtubules lining the cell membrane (from Raikova and Justine 1994; reproduced with permission of Éditions Elsevier, Paris) (×40 000). **D–F)** *Paratomella rubra* (from Raikova *et al.* 1997; reproduced with permission of John Wiley & Sons, Inc.). **D)** Early spermatid. Note lateral position of the axonemes, each one surrounded by two cytoplasmic canals. The third canal surrounds both the axonemes and the nucleus. Numerous small mitochondria are present on the ventral side of the nucleus. Large granules gather on the dorsal side (×48 000). **E)** Detail of the following (F), showing four membranes around the axoneme and two membranes, encircling also the nucleus. Note the presence of two microtubules between the membrane profiles and of the microtubules lining the outermost membrane (×91 000). **F)** Transverse section of a late spermatid. Note the presence of a single section of the mitochondrion on the ventral side (×33 000). M, mitochondria; N, nucleus; LG, large cytoplasmic granules; SG, small cytoplasmic granules. Large arrowheads point to cytoplasmic canals; small arrowheads point to microtubules.

term 'cytoplasmic canal' means 'an empty space (extra-cellular space) between two parts of the cytoplasm'.

3. During spermatid elongation in most species studied the membranes of the cytoplasmic canal are lined by transitory microtubules (Figure 2.2A,B).

This type of flagellar incorporation and spermatid elongation is characteristic for all of the Acoela studied so far, with an exception of *Paratomella rubra* (Raikova *et al.* 1997). The latter displays unique features both in spermiogenesis and in sperm structure.

In *Paratomella rubra* the flagellar incorporation takes place in the usual way: originally, free flagella fuse with the spermatid shaft by means of a lateral groove which forms a single canal around each flagellum. Further spermatid elongation is characterized by the presence of four membranes surrounding each axoneme and two membranes, surrounding both the axonemes and the nucleus (Figure 2.2D–F). These structures were interpreted as being three cytoplasmic canals situated one inside the other (Figure 2.3). In the course of spermiogenesis, numerous small mitochondria fuse into a single elongate mitochondrion, running alongside the nucleus (Figures 2.2D,F, 2.4 and 2.5A–C). Mature spermatozoa also shows a set of peculiar characters, different from the other acoels.

Apomorphies of Paratomella rubra, based on spermiogenesis and sperm structure

The following characters are considered apomorphic (Raikova *et al.* 1997):

1. The presence of three cytoplasmic canals around the incorporated flagella in the course of spermatid maturation (Figure 2.3).
2. The spermatozoon displays a highly ordered, bilaterally symmetrical pattern of organelles (Figures 2.4 and 2.5B).
3. Membrane-bound granules form regular rows in the sperm cytoplasm and are embedded into the nucleus in its distal part (Figures 2.4 and 2.5C; see also Figure 2.10b, opposite p. 21).
4. The spermatozoon has a single long 'ventral' mitochondrion instead of numerous small ones (Figures 2.4 and 2.5A,B).
5. The spermatozoon has a very long nucleus (see Figure 2.10a,b). The nucleus-to-sperm-length ratio (N/S ratio) in *Paratomella* is about 90%, while in the other acoels described so far it is less than 40%. However, this character state could represent a plesiomorphy.

Our data (Raikova *et al.* 1997) support the view of Ehlers (1992a), that *Paratomella* constitutes a sister-group to all the other acoels, the Euacoela Ehlers, 1992.

Sperm characters within the Euacoela

In the Euacoela Ehlers, 1992 the following characters of mature spermatozoa seem to be promising for phylogenetic and systematic purposes (the supposed primitive character state is listed first; the supposedly derived state is listed second; the polarity of some of these character states is not easy to determine):

- axoneme structure: $9 + 2$ versus modified ($9 + 0$ or $9 + $ '1');
- cytoplasmic microtubules: cortical versus axial;
- nucleus-to-sperm-length ratio: high versus low;

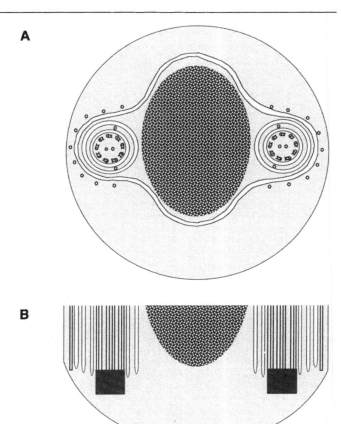

A

B

Figure 2.3 *Paratomella rubra*. Schematic drawing, illustrating the system of three sets of cytoplasmic canals during spermatid elongation (from Raikova *et al.* 1997; reproduced with permission of the Royal Swedish Academy of Sciences). **A)** transverse section; **B)** longitudinal section. Note two canals encircling the incorporated flagella, and the third canal encircling both the flagella and the nucleus.

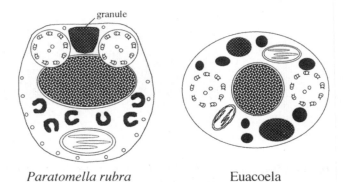

granule

Paratomella rubra Euacoela

Figure 2.4 Diagram illustrating the difference in mature sperm structure between *Paratomella rubra* and Euacoela. Note the ordered, bilaterally symmetrical pattern of organelles in *Paratomella*, the dorsal position of large granules, incorporated between the axonemes, the ventral position of the single mitochondrion and the 'gastrula-shaped' small granules. In Euacoela, numerous small mitochondria and round or oval granules of two size classes are randomly distributed in the sperm cytoplasm.

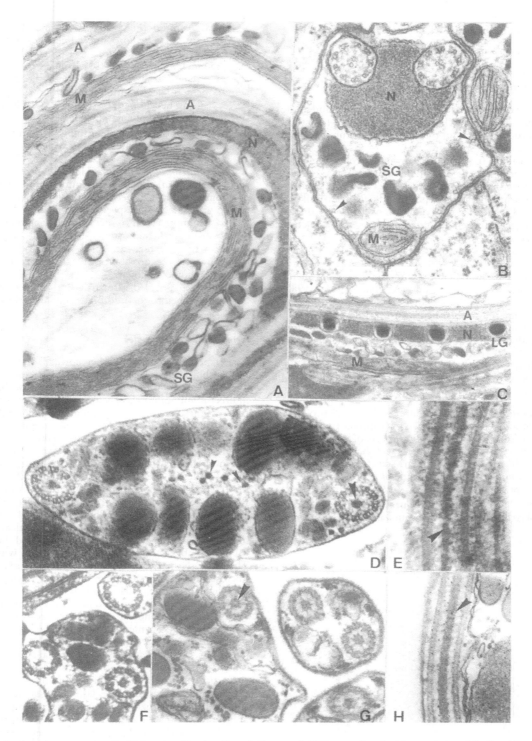

Figure 2.5 Details of acoel spermatozoa. **A–C)** *Paratomella rubra* (from Raikova *et al.* 1997; reproduced with permission of the Royal Swedish Academy of Sciences). **A)** Coiled spermatozoa. Note a single long mitochondrion (M) running alongside the nucleus (N) and axonemes (A) (×53 000). **B)** Transverse section of the bilaterally symmetrical spermatozoon, showing the nucleus (N), two membrane-bound axonemes on the dorsal side, a section of the mitochondrion (M) on the ventral side, small granules (SG) and cortical microtubules (arrowheads) (×64 000). **C)** Longitudinal section of the distal region of the spermatozoon. Note a row of large granules (LG) embedded in the nucleus (N) and the long mitochondrion on the ventral side (×24 000). **D–F)** *Symsagittifera schultzei*, 9 + '1' axonemal structure. **D)** Transverse section of the spermatozoon showing two 9 + '1' axonemes with electron-dense cores (large arrowheads) and axial cytoplasmic microtubules (small arrowheads) (From Raikova *et al.* 1998b; reproduced with permission of Kluwer Academic Publishers) (×81 000). **E)** Longitudinal section of the axoneme. Note the irregular contours of the dense axonemal core (large arrowhead) (×113 000). **F)** Distal region of a late spermatid. Note the 9 + 2 structure of one of the axonemes and 9 + 0 structure of the other (×73 000). **G), H)** *Paraphanostoma cycloposthium*, spermatozoa. **G)** Axonemes of 9 + '1' structure with hollow tubular core (large arrowhead) on a transverse section (×66 000). **H)** Longitudinal section of the axoneme, showing a rather regular structure of the core (large arrowhead), that is, however, different from the axonemal core in the Trepaxonemata (×69 000).

- nucleus position: proximal terminal versus more distal;
- overlap of flagella and nucleus: long versus short.

Sperm axoneme structure

Three variations of sperm axoneme structure were found in the Acoela (Figure 2.6; Table 2.1): a plesiomorphic 9 + 2 structure, a 9 + 0 structure (without two central microtubules of the axoneme) and a 9 + '1' structure.

In the case of 9 + 0 flagella, according to our numerous observations, sections of the proximal regions of the axonemes always show a normal 9 + 2 structure, the two central microtubules disappearing in the distal parts of the flagella (Raikova and Justine 1994). Therefore the notation of 9 + 2/9 + 0 axonemes was suggested for this type of axonemal structure (Raikova and Justine 1994).

The 9 + '1' axoneme structure (see Figure 2.6) was reported in a few species of Acoela (Table 2.1) and in the earliest works was presumed to be homologous with the 9 + '1' axoneme structure of the Trepaxonemata (Afzelius 1966, 1982; Baccetti and Afzelius 1976; Hendelberg 1977). It has been demonstrated that the trepaxonematan central cylinder does not contain tubulin (Iomini et al. 1995) and consists of two elements helically wound one around the other (Henley et al. 1969; Silveira 1969). Tubulin immunocytochemical studies of the central element of the 9 + '1' in Acoela are still lacking. However, it seems that the cases of 9 + '1' in Acoela described hitherto fall into several groups:

1. In *Symsagittifera schultzei* (Figure 2.5D–F) the proximal parts of the axonemes contain two microtubules masked by electron-dense material, while distal parts of the axonemes display a 9 + 0 pattern. Late spermatids have 9 + 2/9 + 0 axonemes (Figure 2.5F). Similar structure occasionally occurs in *Amphiscolops* sp. (Rohde et al. 1988d). Clearly this case represents a modified 9 + 2/9 + 0 pattern.
2. Another 9 + '1' type, where the central electron-dense structure displays a tubular core occurs in *Paraphanostoma*

cycloposthium (Figures 2.5G,H and 2.6), *Paramecynostomum diversicolor*, *Pseudomecynostomum westbladi*, *Mecynostomum auritum* (see Hendelberg 1977) and *Pseudactinoposthia* sp. (see Rohde et al. 1988d). In this case the central structure reaches the distal end of the axoneme. The longitudinal sections of the hollow core reveals no helical pattern (Figure 2.5H). It is thus different from the trepaxonematan type.

3. There are also reports describing acoelan 9 + '1' structure, very similar to that of Trepaxonemata, at least at the transversal sections (Afzelius 1982, Figure 3C – in *Mecynostomum auritum*; Hendelberg 1977 – in *Childia groenlandica*). These descriptions sometimes have been considered as erroneous (Raikova and Justine 1994), due to probable wrong species identification, but now we tend to give more credit to them considering that if it is difficult to distinguish one acoel species from another, it is quite easy to recognize an acoel among other turbellarians. However, the possibility of existence among the Acoela of the homoplasious 9 + '1' Trepaxonemata-like structure will need detailed ultrastructural studies of longitudinal sections before it could be definitively rejected.

Cytoplasmic microtubules

Acoel spermatozoa have either cortical or axial cytoplasmic microtubules (Figure 2.7; Table 2.1). The spermatozoa of the Trepaxonemata often have cortical microtubules. Tubulin indirect immunofluorescence reveals differences in tubulin epitopes in these microtubule populations (Figure 2.7). The anti-alpha-tubulin antibody DM 1A (Sigma) does not label the axial microtubules in the Acoela, but labels the cortical microtubules in the Acoela and the Trepaxonemata. The cortical microtubules in the Acoela are acetylated, while the axial microtubules in the Acoela and the cortical microtubules in the Trepaxonemata are not acetylated (Justine et al. 1998; Raikova and Justine 1999; Raikova et al. 1998b; Justine 1999). These data suggest that the cortical microtubules in the Acoela are non-homologous to the cortical microtubules in the Trepaxonemata, while the cortical and the axial microtubules in the Acoela are not homologous with each other. Therefore the presence of either cortical or axial cytoplasmic microtubules appears to be an important phylogenetic character (see also Justine 2001, this volume).

It is interesting to note that while the cortical microtubules always have a 'standard' appearance, resembling the microtubules of the axonemes, the ultrastructure of the axial microtubules is less conventional. Often they seem not hollow, but filled with electron-dense material (Figures 2.5D and 2.8A,C,D).

Cortical microtubules form a continuous layer underlying the sperm plasma membrane. Usually, they run from the distal end of the spermatozoon almost to its proximal end, finishing approximately at the same level as the axonemes (see Figure 2.10c). As for the axial microtubules, their arrangement is far more variable. In *Mecynostomum* sp. (Figure 2.8A), in the distal part of the spermatozoon the longitudinal microtubules form two groups underlying the plasma membrane on the 'ventral' and the 'dorsal' sides of the cell, while proximally they shift towards the central axis of the cell, partly encircling the nucleus and other cell organelles, and continue to the proximal end of the spermatozoon (Figure 2.10e). The axial microtubules often form a ring, resembling a third axoneme (Figure 2.8D), or two parallel lateral rows facing each other (Figure 2.8B,C). The microtubules in these rows are linked either by two lateral electron-dense structures or by a single central

Sperm axoneme structure

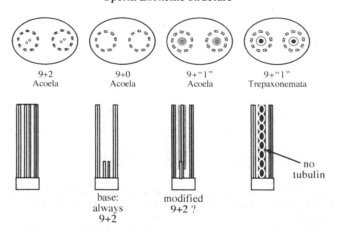

| 9+2 Acoela | 9+0 Acoela | 9+"1" Acoela | 9+"1" Trepaxonemata |

base: always 9+2   modified 9+2 ?   no tubulin

Figure 2.6 Diagram of sperm axoneme structure in Acoela and Trepaxonemata. Upper row, transverse sections; lower row, longitudinal sections of the axonemes. The acoelan 9 + 0 sperm structure is based on the 9 + 2 pattern, discernible at the proximal end of the axoneme. Note differences in the structure of the central core of the 9 + '1' axonemes in Acoela and Trepaxonemata. The latter has a helical structure and contains no tubulin.

Table 2.1 Acoela grouped according to sperm characters. The classification used is that of Dörjes (1968), except for the family Sagittiferidae Kostenko and Mamkaev 1990. The species name used in the corresponding paper, if different from the accepted one, is given in square brackets. All the cases when the cytoplasmic microtubules were not described (marked by * after the species name), have been put into the group of axial microtubules (for justification, see 'cytoplasmic microtubules', p. 18). The cases of 9 + '1' axonemes with the short proximal electron-dense central structure, of the type found in Symsagittifera schultzei (see 'sperm axoneme structure', p. 18), were classified with 9 + 0 axonemes and marked by: 9 + 0 (9 + '1') after the species name, while the other 9 + '1' cases were put in a separate group.

Sperm characters and species	Family	Reference
Cortical microtubules, 9 + 2 axonemes		
Paratomella rubra Rieger and Ott, 1971	Paratomellidae	Raikova *et al.* 1997
Hesiolicium inops Crezée and Tyler, 1976	Paratomellidae	Crezée and Tyler 1976
Kuma viridis (An der Lan, 1936)	Haploposthiidae	Hendelberg 1969, 1975
[=*Haploposthia viridis*]		
Haploposthia opisthorchis Mamkaev, 1967	Haploposthiidae	Raikova and Justine 1994
Actinoposthia beklemischevi Mamkaev, 1965	Childiidae	Raikova and Justine 1994
Pelophila lutheri (Westblad, 1946)	Childiidae	Hendelberg 1969, 1974
[=*Mecynostomum lutheri*]		
Diopisthoporus longitubus Westblad, 1940	Diopisthoporidae	Hendelberg 1977
Proporus venunosus (Schmidt, 1852)	Proporidae	Hendelberg 1977
Aphanostoma virescens Oersted, 1845	Convolutidae	Raikova and Justine 1994
Archaphanostoma agile (Jensen, 1878)	Convolutidae	Raikova and Justine 1994
[=*Baltalimania agile*]		
Axial microtubules, 9 + 2 axonemes		
Philocelis karlingi (Westblad, 1946)	Otocelididae	Hendelberg 1977
Mecynostomum sp.	Mecynostomidae	This chapter
[from Stradbroke Island, Australia]		
Philactinoposthia saliens (Graff, 1882)	Childiidae	Hendelberg 1969, 1974
[=*Convoluta saliens*]		Raikova and Justine 1999
Axial microtubules, 9 + '1' axonemes		
Childia groenlandica * (Levinsen, 1879)	Childiidae	Hendelberg 1977
Paraphanostoma cycloposthium Westblad, 1942	Childiidae	This chaper
*Pseudactinoposthia sp.**	Childiidae	Rohde *et al.* 1988d
*Mecynostomum auritum** (Schultze, 1851)	Mecynostomidae	Afzelius 1966, 1982; Hendelberg 1977
*Eumecynostomum westbladi** (Dörjes, 1968)	Mecynostomidae	Hendelberg 1977
[=*Pseudmecynostomum westbladi*]		
*Paramecynostomum diversicolor** (Oersted, 1845)	Mecynostomidae	Hendelberg 1969, 1977
Axial microtubules, 9 + 0 axonemes		
*Childia groenlandica** (Levinsen, 1879)	Childiidae	Henley 1968, 1974; Henley *et al.* 1968; Costello *et al.* 1969
Philactinoposthia sp.	Childiidae	This chapter
[from Stradbroke Island, Australia]		
Anaperus biaculeatus Boguta, 1970	Anaperidae	This chapter
Anaperus gardineri Graff, 1911	Anaperidae	Silveira 1967, 1969
Anaperus tvaerminnensis (Luther, 1912)	Anaperidae	Henley and Costello 1969; Hendelberg 1969, 1977
Convoluta boyeri Bush, 1984	Convolutidae	Boyer and Smith 1982
Convoluta philippinensis Bush, 1984	Convolutidae	Boyer and Smith 1982
Convoluta convoluta (Abildgaard, 1806)	Convolutidae	Mamkaev and Ivanov 1970
*Polychoerus carmelensis** Costello and Costello, 1938	Convolutidae	Henley 1974
*Polychoerus caudatus** Mark, 1892	Convolutidae	Henley 1974
*Oligochoerus limnophilus** Ax & Dörjes, 1966	Convolutidae	Klima 1967
Amphiscolops sp. 9 + 0 (9 + '1')	Convolutidae	Rohde *et al.* 1988
Symsagittifera schultzei (Schmidt, 1852) 9 + 0 (9 + '1')	Sagittiferidae	Raikova *et al.* 1998b
Symsagittifera psammophila (Beklemischev, 1957)	Sagittiferidae	Bedini and Papi 1970
[=*Convoluta psammophila*]		
Symsagittifera bifoveolata (Mamkaev, 1971)	Sagittiferidae	This chapter
Symsagittifera roscoffensis (Graff, 1891)	Sagittiferidae	This chapter

Tubulins in cytoplasmic microtubules

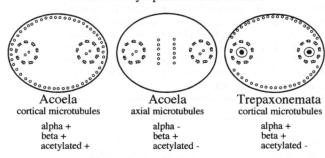

Acoela cortical microtubules	Acoela axial microtubules	Trepaxonemata cortical microtubules
alpha + beta + acetylated +	alpha - beta + acetylated -	alpha + beta + acetylated -

Figure 2.7 Diagram illustrating differences that the tubulin indirect immunofluorescence reveals in tubulin epitopes in the microtubule populations of flatworm spermatozoa. The antibodies used are anti-alpha-tubulin, clone DM 1A (Sigma); anti-beta-tubulin, clone TUB 2.1 (Sigma) and anti-acetylated tubulin, clone 6-11B-1 (Sigma).

structure (Figure 2.8C,E). In the former case, tubulin immunofluorescence shows two parallel lines between the axonemes (Figure 2.10m,q) and in the latter case a single line (Figure 2.10r). The microtubule arrangement along the sperm length (Figure 2.9) was studied in detail in *Convoluta saliens* (see Raikova and Justine 1999). Usually, axial microtubules hardly reach the distal end of the nucleus, and in sections through the nuclear region no cytoplasmic microtubules are present. Therefore in cases where no cytoplasmic microtubules were described (see Table 2.1) it is probable that the corresponding descriptions were based on nuclear region sections of the spermatozoon with axial microtubules.

Nucleus-to-sperm-length ratio

Using tubulin immunofluorescence and propidium iodide nuclear fluorescent staining, it is possible to visualize the exact position of nucleus, flagella and cytoplasmic microtubules in a filiform spermatozoon. This task is practically impossible by transmission electron microscopy (TEM) alone, and rather difficult even by *in vivo* observations of the spermatozoon (Hendelberg 1969). Illustrations of spermatozoa of different acoels obtained using this method are shown in Figure 2.10, and summarized in Figure 2.11.

The ratio of nuclear length to total sperm length (N/S ratio) varies considerably among species, ranging from 90% in *Paratomella rubra* to 20% in representatives of the family Sagittiferidae Kostenko and Mamkaev, 1990. It is noticeable that in species with cortical microtubules the N/S ratio is higher than in the animals with axial microtubules. An interesting exception is *Anaperus biaculeatus*, displaying the N/S ratio of 40%. 'Higher Acoels' of the family Convolutidae and Sagittiferidae display the lowest N/S values. In spermatids, the N/S ratio is always higher than in mature spermatozoa, due to the fact that at the end of spermiogenesis the distal (flagellar) part of the spermatid elongates more than the proximal (nuclear) part. We consider the high N/S ratio as a plesiomorphic character versus low N/S ratio as a derived character. However, the case of *Paratomella rubra* might represent an exception. This species has a great number of apomorphic characters, separating it from other acoels (see p. 16). It is possible that the plesiomorphic N/S ratio is about 40–50%, and that a very high ratio (90%) and a low ratio (20–25%) represent different apomorphic states.

Unfortunately, so far only very few species have been studied in this respect; therefore it is difficult to estimate the phylogenetic value of this character.

Position of the nucleus

In a majority of acoels the nucleus occupies a proximal terminal position. However, it seems that some exceptions could be encountered. In *Mecynostomum* sp. (Figures 2.10e,f and 2.11) the nucleus is shifted in a distal direction, while the flagella run up to the proximal end. This character is considered apomorphic.

Overlap of flagella and nucleus

The respective position of flagella and nucleus is easily visualized on immunofluorescent preparations (Figures 2.10 and 2.11). In *Paratomella rubra*, *Actinoposthia beklemischevi* and *Mecynostomum sp.*, the flagella almost reach the proximal end of the sperm, the region of overlap of nucleus and flagella being very long. In *Convoluta saliens* and in species of the genus *Symsagittifera* the overlap of flagella and nucleus is much shorter. In *Anaperus biaculeatus* the flagella hardly reach the distal end of the nucleus (Figure 2.10i,j). The long region of overlap of nucleus and flagella is considered primitive versus short overlap as a derived character state.

Other characters of acoel spermatozoa

Hendelberg, in his excellent studies of live spermatozoa (1969, 1977), described two major shapes of acoel spermatozoa: filiform spermatozoa and spermatozoa with undulating membranes, as well as different aberrant cases; and spermatozoa with terminal filaments and 'ciliated spermatozoa' where the sperm membrane appeared to be covered with cilia. Our observations demonstrate that in closely related species the general sperm shape varies greatly. *Symsagittifera schultzei* and *S. roscoffensis* share the same habitat and are so hardly distinguishable that Dörjes (1968) even questioned the validity of one of these species (*S. roscoffensis*). However, *S. schultzei* has extremely long filiform spermatozoa (Figure 2.10n) while *S. roscoffensis* has shorter spermatozoa with two undulating membranes and a distal 'fin', a widening of the cell, that sometimes is quite prominent (Figure 2.10o–q). In *Mecynostomum* sp. we have encountered the case of 'ciliated' spermatozoa, reminiscent of those of *Convoluta fulvomaculata*, described by Hendelberg (1969). In fact, light microscopic examination shows the spermatozoa of *Mecynostomum* to have a 'fir-tree' appearance. However, our TEM observations show that the presumed 'cilia' are in reality membrane-bound electron-dense projections of the sperm membrane, that appear thickened by apposition of electron-dense material both on the inner and on the outer surfaces (Figure 2.8A). The general morphology of the spermatozoa (axonemes and axial cytoplasmic microtubules) is not altered.

All the acoel spermatozoa described by now seem to have two types of membrane-bound cytoplasmic granules: smaller and larger ones (Silveira 1967; Henley 1968, 1974; Hendelberg 1977; Boyer and Smith 1982; Rohde *et al.* 1988d; Raikova and Justine 1994, 1999; Raikova *et al.* 1998b). The small dense granules are round or ovoid in shape in the representatives of Euacoela (Figures 2.2C, 2.4, 2.5D,F,G and 2.8A,C,D) and are 'gastrula-shaped' (Figures 2.2D,F, 2.4 and 2.5B) in *Paratomella rubra* (Raikova *et al.* 1997). Perhaps this peculiar shape of small

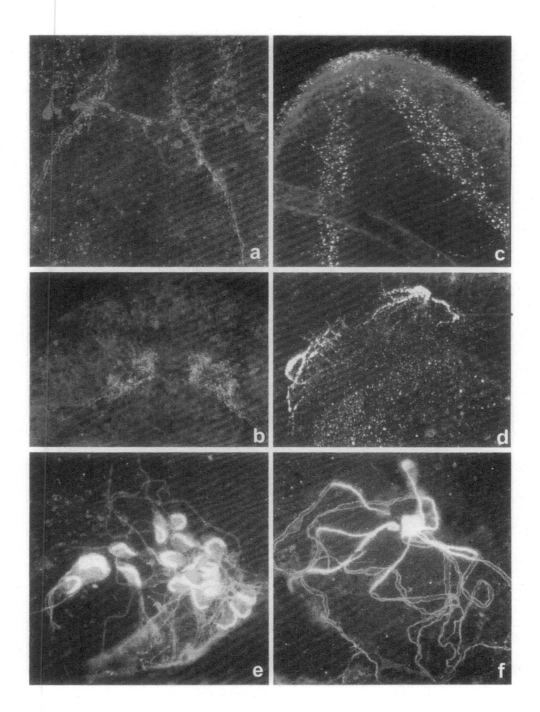

Figure 2.1

Brain structure and spermiogenesis in Acoelomorpha. *a)* Brain-like structure of *Paraphanostoma crassum* (Acoela). Double staining: 5-HT immunoreactive (IR) elements shown in red (TRITC), GYIRF-amide IR elements shown in green (FITC). Note two loose lateral clusters of GYIRF-amide IR cells at the periphery of 5-HT-IR commissural brain (x 200). *b)* Brain-like structure of *Avagina incola* (Acoela). Staining as in (a). Few GYIRF-amide IR cells form two compact lateral clusters (x 200). *c)* *Meara stichopi* (Nemertodermatida), anterior part of the nervous system. Note two loose longitudinal bundles of 5-HT-IR fibres, interconnected by thin commissures (x 200). *d)* *Nemertoderma westbladi* (Nemertodermatida). Ring-shaped 5-HT-IR brain-like structure (x 240). e, f) Spermiogenesis in *Convoluta saliens* (Acoela). Propidium iodide fluorescent staining of nuclei (yellow) and anti-alpha tubulin clone DM 1A (Sigma) immunocytochemistry (green).*e)* Cluster of early spermatids with free flagella and crescent shaped nuclei (x 410). f) Cluster of late spermatids with filiform nuclei in proximal position and two axonemes incorporated in spermatid shafts (x 430). (*e,f*, see also Raikova and Justine 1999; reproduced with permission of John Wiley & Sons Inc.)

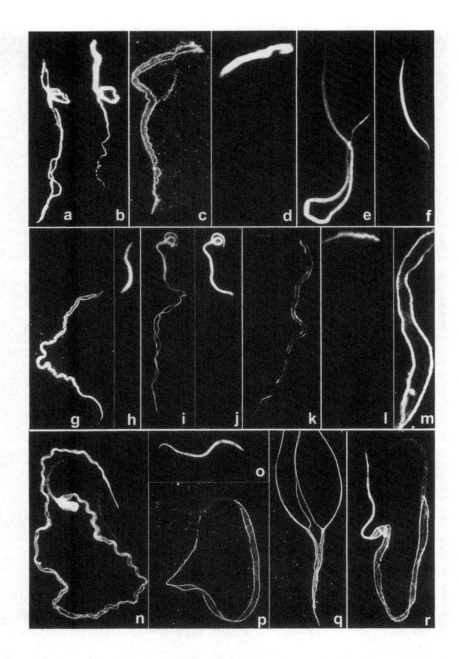

Figure 2.10

Light microscope immunocytochemistry of acoel spermatozoa. Anti-tubulin immunofluorescence (a, c, e, g, i, k, m, n, p, q, r) and propidium iodide (PI) fluorescent staining of nuclei (b, d, f, h, j, l, o). a, b) *Paratomella rubra*. Anti-alpha tubulin - PI immunostaining shows two widely spaced axonemes and a very long nucleus. The axonemes do not actually reach the proximal end of the nucleus. Note the region close to the thin distal end of the nucleus, where the nucleus appears as a row of beads (x 640). c, d) *Actinoposthia beklemischevi*. Anti-acetylated-alpha tubulin - PI immunostaining showing two axonemes almost reaching to the proximal end of the cell, cortical cytoplasmic microtubules and a rather long nucleus (x 800). e, f) *Mecynostomum* sp. Anti-beta tubulin immunostaining labels the whole length of the spermatozoon. Note the shifting of the nucleus from the proximal end of the sperm distally (x 290). g, h) *Philactinoposthia* sp. Anti-alpha tubulin - PI staining showing two axonemes reaching the distal end of the rather short nucleus (x 290). i, j) *Anaperus biaculeatus*. Anti-acetylated tubulin - PI staining. Only one the axonemes reaches the distal end of the long nucleus (x 230). k, l, m) *Convoluta saliens*. The spermatozoon has a short proximal nucleus (l, x 230). Anti-acetylated tubulin antibody shows only two axonemes, reaching the distal end of the nucleus (k, x 240), while anti-beta tubulin antibody reveals two parallel lines of axial microtubules between the axonemes (m, x 1030). n) *Symsagittifera schultzei*, long filiform spermatozoon . Anti-alpha tubulin staining show two axonemes reaching to the distal end of the nucleus (x 300). o, p, q) *Symsagittifera roscoffensis*. Spermatozoon with undulating membranes and distal 'fin'. Anti-acetylated tubulin - PI staining shows a short nucleus and two axonemes (o, p, x 310), while the anti-beta tubulin reaction of the 'fin' region reveals two lines of axial microtubules, converging proximally to run close to each other (q, x 420). Note the resemblance with figure 2.9, based on the reconstuction of ultra-thin sections of the spermatozoon of *Convoluta saliens*. r) *Symsagittifera psammophila*. Anti-beta tubulin antibody reveals a line of axial microtubules between the two axonemes (x 330).

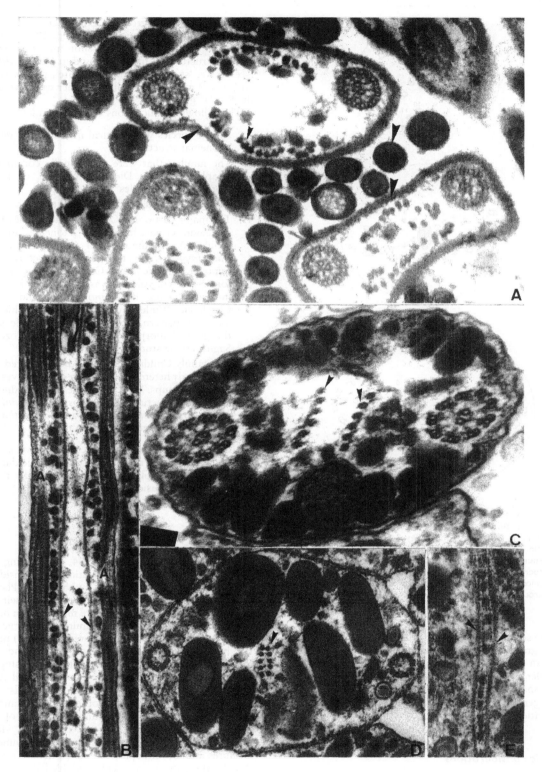

Figure 2.8 Arrangement of axial cytoplasmic microtubules in acoelan spermatozoa. **A)** *Mecynostomum* sp. Electron-dense axial microtubules form two groups underlying the plasma membrane on the 'ventral' and the 'dorsal' sides of the cell. Note the thickening of the sperm membrane (large arrows) by apposition of electron-dense material on the inner and the outer surfaces and the presence of membrane-bound electron-dense profiles outside the spermatozoa (×82 000). **B), C)** *Convoluta saliens* (from Raikova and Justine 1999; reproduced with permission of John Wiley & Sons, Inc.). Axial microtubules arranged in two lateral rows. **B)** Longitudinal section; note similarity with immunocytochemistry observation (Figure 2.10m) (×35 000). **C)** Transverse section (×108 000). **D), E)** *Symsagittifera psammophila*. **D)** Transverse section. Axial microtubules are arranged in a central ring (×43 000). **E)** Longitudinal section (from Raikova *et al.* 1998b; reproduced with permission from Kluwer Academic Publishers). Note the electron-dense granules inside the ring of microtubules (×50 000). Small arrowheads point to axial microtubules.

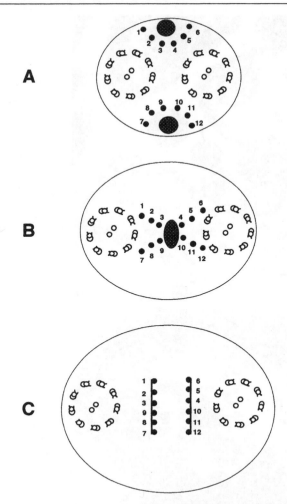

Figure 2.9 Diagram of axial microtubule pattern in the spermatozoon of *Convoluta saliens* (from Raikova and Justine 1999; reproduced with permission from John Wiley & Sons, Inc.) of transverse sections drawn from TEM observations; singlet microtubules are numbered for showing spatial pattern. **A)** Distal extremity of sperm cell, close to the centrioles. **B)** X pattern of microtubules. **C)** Microtubules grouped as two rows, with intermicrotubule links.

granules represents yet another apomorphy for *Paratomella*. As for the large granules, their size and appearance are quite variable and seem species-specific.

We consider both the general shape of spermatozoon and the size and shape of its cytoplasmic granules as characters of minor phylogenetic importance, and also quite useful for species identification.

Phylogenetic considerations

The characters discussed above could constitute a good basis for phylogenetic reconstructions within the Acoela. However, at present only two of the characters (sperm axoneme structure and cytoplasmic microtubules) have been studied in a sufficient number of species (Table 2.1). We have made an attempt to group acoels according to sperm characters (Table 2.1), and during this operation made two concessions. First, all cases when the cytoplasmic microtubules were not described have

been included in the group of axial microtubules (for justification, see 'cytoplasmic microtubules', p. 18). Second, the cases of 9 + '1' axonemes with the short proximal electron-dense central structure, of the type found in *Symsagittifera schultzei* (see above, 'sperm axoneme structure', p. 18), were classified with 9 + 0 axonemes, while the other cases were classified in the 9 + '1' group.

The following conclusions could be made:

1. All the species with cortical microtubules display a 9 + 2 axonemal structure.
2. The families that were traditionally considered primitive (Haploposthiidae, Diposthoporidae, etc.) fall into the (cortical microtubules 9 + 2) group.
3. All the Anaperidae fall into one group.
4. The major part of the Convolutidae (Dörjes 1968), including Sagittiferidae, have 9 + 0 axonemes and axial microtubules. As for the cases of *Aphanostoma virescens* and *Baltalimania agile*, their sperm structure suggests their removal from the family Convolutidae. *Convoluta saliens*, formerly a convolutid, was transferred by Dörjes (1968) into the genus *Philactinoposthia*, and to the family Childiidae. While having the plesiomorphic 9 + 2 sperm axoneme pattern, it displays considerable similarities in the general morphology of the spermatozoon with the Sagittiferidae studied.
5. The family Childiidae, as was demonstrated above, seems particularly heterogeneous. Its representatives fell into three different groups. Noticeably, the three childiids with different stylet structure (*Actinoposthia beklemischevi*, *Paraphanostoma cycloposthium* and *Philactinoposthia* sp.) also have different sperm structure.

Using the sperm characters discussed above, some aspects of the phylogeny within the Acoela could be elucidated. However, more extensive sampling is necessary in order to reach wider phylogenetic conclusions.

Summary

The Acoelomorpha Ehlers, 1984, comprising Acoela and Nemertodermatida, were traditionally considered monophyletic, and were included within the Platyhelminthes. Recent molecular investigations questioned their phyletic affinities. 'Brain' and sperm structure may provide new characters useful for phylogeny. Our immunocytochemical studies of 5-HT and FMRF-amide immunoreactivity (IR) indicate a deep rift between Acoela + Nemertodermatida and Catenulida + Rhabditophora. However, the interrelationships of Acoela and Nemertodermatida remain unclear. In all Acoela and in *Nemertoderma westbladi* the 5-HT-IR reveals no ganglionic cell mass, which is typical for other platyhelminths, but only a symmetrical brain-like structure composed of commissural fibres associated with a few cell bodies. In *Meara stichopi* (Nemertodermatida), two loose longitudinal bundles of 5-HT-IR fibres and an infraepidermal nerve net are observed. In the Acoela, the patterns of 5-HT- and FMRF-amide (or GYIRF-amide) IR differ from each other. In the Nemertodermatida, in contrast, FMRF-IR nerves follow the 5-HT-IR nerves.

Comparative ultrastructural and immunocytochemical studies of spermatozoa and spermiogenesis in several species of acoels permit the definition of several character states (presumed polarity of these characters is derived versus primitive): 1. Presence of several (versus one) cytoplasmic canals around the incorporated flagella in the course of spermatid maturation;

Sperm types in Acoela

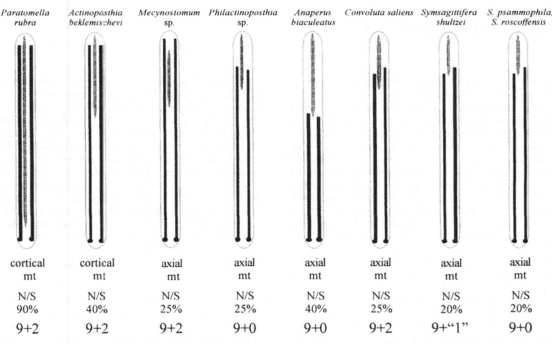

Figure 2.11 Diagram of sperm structure in different species of Acoela, studied by the tubulin indirect immunofluorescence method. Note the position of the nucleus (proximal terminal in all species studied except *Mecynostomum* sp.), and the varying length of overlap of flagella and nucleus. The presence of cortical/axial microtubules, the value of nucleus-to-sperm-length (N/S) ratio and the type of sperm axoneme structure are listed below each illustration.

2. Presence (versus absence) of bilateral symmetry in the pattern of sperm cell organelles; 3. Presence of a single mitochondrion (versus numerous small ones); 4. Presence of axial (versus cortical) microtubules in mature spermatozoon; 5. Presence of 9 + 0 or 9 + '1' axonemal pattern (versus typical 9 + 2); 6. Low nucleus-to-sperm-length ratio versus high; 7. Nucleus position: middle versus proximal terminal; 8. Short overlap of flagella and nucleus versus long. Data obtained so far support the position of *Paratomella* as a sister group for the other Acoels, the Euacoela Ehlers, 1992. These data also may provide a basis for a further phylogenetic analysis of the Euacoela.

ACKNOWLEDGEMENTS

The authors wish to thank Dr Jondelius for sampling of material, Dr Margaretha Gustafsson for many helpful suggestions to this text, and Dr Halton for providing anti-GYIRF-amide antibodies. We are deeply grateful to Tim Littlewood and Rod Bray for financial support to attend the Conference on 'Interrelationships of Platyhelminthes'. This work was partly funded by Russian Basic Research Foundation grant No. 99-04-49807 to O. Raikova.

Chapter 3

The Nemertodermatida

Kennet Lundin and Wolfgang Sterrer

The order Nemertodermatida Karling 1940, a small group of, currently, ten to eleven described marine species, has long been the subject of debate regarding its basal phylogenetic position in the Platyhelminthes. The first record of a nemertodermatid was published by Einar Westblad (1926), who reported collecting specimens of *Meara stichopi* in Norway in 1923. The species had by then already been discovered and informally named by Sixten Bock. However, the formal description of *M. stichopi* was published much later by Westblad (1949), after Bock's death. In the summer of 1927, Otto Steinböck and Erich Reisinger extracted a single small worm from mud dredged off the east coast of Greenland. On this specimen Steinböck (1931) based his description of *Nemertoderma bathycola*. A more thorough description of another species, subsequently named *N. westbladi* by Steinböck (1938), was given by Westblad (1937). Initially placed within the Acoela, the group was given suborder status by Karling (1940) and full order status by Ax (1961). New descriptions of species and genera were presented by Sterrer (1966, 1970; *N. psammicola*), Faubel and Dörjes (1978; *Flagellophora apelti*) and Riser (1987; *Nemertinoides elongatus*). In his revision of the Nemertodermatida, Sterrer (1998) also introduced the new genus *Ascoparia* (*A. neglecta, A. secunda*) and the new family Ascopariidae. Lundin (2000) made a cladistic analysis of the order and referred the species *Nemertoderma psammicola* to the new genus *Sterreria*.

Diagnosis

This diagnosis combines morphological characters at both light and electron microscope levels. All species are free-living with the exception of the endosymbiotic *Meara* spp. The mature size ranges from 0.2 mm to 10 mm.

The body has a complete and dense covering of locomotory cilia. The epidermal cilia have a distal cap consisting of two or three rounded plates. There is a shelf-like narrowing at about 0.5 μm from the ciliary tip and dense material in the ciliary shaft. A 'cup' is present at the junction between the cilium and the basal body. A bilayered dense plate constitutes the bottom of this cup. There is no aggregation of granules closely below the dense plate, as is usual among metazoans, but granules are scattered within the basal body. The single main rootlet arises on the anterior face of the basal body. The main rootlet is hollow and tube-shaped along the basal body. There is a short basal foot ('posterior rootlet') with two continuous microtubule bundles which lead to the main rootlets of the next ciliary row behind. Epitheliosomes are filled with a flocculent material. The lateral borders of ciliated epidermal cells distinctly interdigitate. Well-developed *zonula adhaerentes* are present between epidermal cells, but septate junctions are absent. The terminal web is prominent and in some species very thick, i.e., in *Nemertoderma* spp. and *Meara* spp. In these species, intracellular tonofilaments extend from the terminal web to the base of the epidermal cells. Nemertodermatids lack a basal lamina, but there is a narrow string of extracellular matrix between epidermal cells and the underlying muscle cells.

In a mode similar to the Acoela, worn ciliated epidermal cells are not cast off but withdrawn internally and digested (reported from *Meara stichopi* and a species of *Nemertoderma*). Epidermal mucus cells with flocculent contents occur in abundance. In *Nemertoderma* spp., *Meara* spp. and *Nemertinoides elongatus* these cells are large and swollen, displacing the ciliated epidermal cells to such a degree that the epidermis appears pseudostratified. The gland necks are supported by many microtubules, but there are no cilia at the epithelial surface of the gland neck. Some of the mucus cells are insunk below the muscle layer, at least in *M. stichopi*. A frontal gland complex is situated at the apical end of the body, with glands opening between unspecialized ciliated epidermal cells (Ehlers 1992b). The frontal glands and their openings are scattered in species of *Nemertoderma* and *Meara*. In *Nemertinoides elongatus*, *Flagellophora apelti* and *Sterreria psammicola*, the frontal gland necks are bundled apically (Smith and Tyler 1988; Ehlers 1992b; Sterrer 1998), a condition similar to some primitive acoels. The elongated epidermal rhabdoids of the *Meara* spp. are weakly positive for polyphenols (Smith J.P.S. *et al.* 1994). There is a complete lack of protonephridia in all nemertodermatids.

Meara spp., *N. westbladi* and *N. bathycola* have a simple mouth pore where the ciliated epidermis and the gastrodermis contact each other without any structural specialization. *N. elongatus* and *S. psammicola* have a pharynx simplex, where the ciliated epidermis invaginates to form a tube-shaped channel leading to the gut cavity. A mouth is lacking in Ascopariidae (*Ascoparia* spp. and *F. apelti*), but the presence of digestive glands opening near the expected location of the mouth suggests a temporary opening. A mouth is also lacking in mature, ovigerous specimens of *N. westbladi* and *N. bathycola* (possibly also in *S. psammicola*, personal communication Seth Tyler 1999).

The gut lumen is almost occluded by narrow processes of the gastrodermal cells, which lack ciliation (the single cross-section of a cilium in the gut lumen of *Flagellophora* sp. in Figure 8, Tyler and Rieger 1977, may be of a sensory cell, personal communication Reinhard M. Rieger 1999). Distinct granular club glands ('körnerkolben') are present among the gastrodermal phagocytes. In some species, e.g., *S. psammicola*, the gland cells are only found near the mouth.

There is a bi-lithophorous statocyst with few parietal cells. The surface of the lithocyte has a blistered appearance in Nemertodermatidae (*Nemertoderma* spp., *Meara* spp., *S. psammicola* and *N. elongatus*), but is smooth in Ascopariidae. The nervous system is entirely basiepithelial, except neural masses in association with the statocyst. This subepidermal neural mass is very small and surrounds the statocyst in most species, with the exception of *F. apelti* where it is larger, bilobed and placed caudally to 'the statocyst' (condition unknown for *Ascoparia* spp.). A narrow, peripheral nerve ring has been reported from *N. westbladi* (Raikova *et al.* 2001, this volume) and *F. apelti* (Tyler 2001, this volume). Epidermal monociliary receptor cells with a straight, simple rootlet have been observed in *S. psammicola* (with collar of microvilli, see Ehlers 1985a) and *M. stichopi* (uncollared; Lundin, unpublished results). Receptors with up to four cilia have been observed in a *Nemertoderma* sp. (Smith and Tyler 1988).

Nemertodermatida reproduce only sexually. The male antrum is a simple ciliated invagination of the epidermis, without eversible penis or any hard parts. A few gland cells open at the bottom of the antrum. There is no true seminal vesicle, but a cavity without epithelial lining in association with the antrum contains spermatozoa to be ejected. Autosperm pass from testes in well-defined passages between the gastrodermal cells. A dorsally located female pore with a bursa is present in Ascopariidae. The oocytes are entolecithal, with peripheral egg-shell granules devoid of polyphenols (Smith *et al.* 1988). Mature eggs are released through the mouth. Duet-cleavage similar to that of acoels has been reported from *M. stichopi* (personal communication Olle Israelsson 1999). The spermatozoon has a single flagellum with $9 + 2$ microtubule configuration. Mitochondria or mitochondrial derivatives are located along part of the flagellum. The spermatozoon of some species has an apically oriented acrosome (*F. apelti*) or acrosomal derivative (*M. stichopi*). The mature spermatozoon is radially symmetrical in the Nemertodermatidae, and bilaterally symmetrical in the Ascopariidae. Allosperm in the Nemertodermatidae are often strongly spiralized. The testis follicles are surrounded by lining cells which most often form a thin layer, but a thick layer is present in *N. westbladi*. These lining cells apparently are derived from gut cells, at least in *F. apelti* (cf. Rieger *et al.* 1991: 93–95).

Nemertodermatids have been found in the North Atlantic, Mediterranean, Red Sea and South-West Pacific (cf. Sterrer 1998). There are no reports from the Indian Ocean, the South Atlantic or from remaining areas of the Pacific Ocean, but this is probably because no specific investigation has been carried out in these areas.

Phylogeny

A phylogenetic analysis of the Nemertodermatida based mostly on morphology was presented by Lundin (2000). Conclusions made from the previous analysis and the availability of additional data justified a revision of the matrix (Tables 3.1 and 3.2). Such a refined matrix should be useful in future analyses with combinations of different data sets.

From the original matrix (see Lundin 2000, with notes on coding and references therein), seven characters were deleted, the character states for 23 of the remaining 65 characters were recoded, and eight new characters were added. Hence the total

Table 3.1 Character matrix. Alternative possible states 0/1 shown as 'P'. State not applicable coded as '–'. Missing data coded as '?'.

Taxa	Characters							
	1 1234567890	*2* 1234567890	*3* 1234567890	*4* 1234567890	*5* 1234567890	*6* 1234567890	*7* 1234567890	*123*
Meara stichopi	0011111010	0011111111	1010111200	0021000??0	0010010101	020010310–	0000001?00	303
Nemertoderma westbladi	00?1121000	0011111111	1010111200	1021000??1	0010011101	0200112?0–	0000001 1?0	000
Nemertoderma bathycola	00?1121000	00111?1111	1010111200	PP2100???1	0010011101	0200112?0–	00??001??0	000
Sterreria psammicola	0010231001	1011111111	1010111100	PP2100?01P	0110110101	020010 1?0–	00??0011?0	101
Nemertinoides elongatus	00?0131P01	?0111?1111	1010111101	PP2100???0	0110010101	0300104?0–	0???0011?0	101
Flagellophora apelti	00?0240100	0011111111	1010111102	??2000???2	0010P10011	0100011?20	0011011000	202
Ascoparia neglecta	???0240100	??????????	?0??????02	??2000???2	001011001?	0100011?21	0???0110?0	212
Ascoparia secunda	???0220100	??????????	?0??????02	??2000???2	0010P1001?	0100011?21	0???0110?0	212
Childia groenlandica	0110050010	?021110111	111011001–	1110111110	0021–10200	1111–0110–	11111–1010	00–
Convoluta convoluta	0110050010	1021110111	111011001–	PP10111110	0021–10200	1001–01110	11111–1010	50–
Paratomella rubra	0110020010	30211?0?11	111011001–	PP101010?0	0021–10200	0011–1110–	10112–1010	40–
Stenostomum sp.	1–00000000	0100000000	000100002–	220–––?000	1100001010	1300110?0–	2–––1-0–21	40–
Macrostomum sp	1–00020022	2100000000	000100012–	220–––?200	1100001202	1011–01010	2–––1–011	00–

Table 3.2 List of characters and character states. For explanations and references see the Diagnosis section (p. 24) and Lundin (2000). Additional references are noted in the character list.

1. Epidermal cells: interdigitate laterally (0), cuboidal, not interdigitating (1)
2. Interdigitating epidermal cells: not insunk (0), insunk (1)
3. Worn epidermal cells: cast off (0), withdrawn and digested (1)
4. Intracellular tonofilament bundles: absent (0), present (1)
5. 'Vacuolated' epidermal mucus glands: absent (0), large flocculent glands (1), small flocculent glands (2)
6. Frontal gland complex: absent (0), intraepidermal gland cells (1), dispersed sunken glands (2), paired long sunken glands (3), median long sunken glands (4), common apical pore, sunken glands (5)
7. Bottle-shaped epidermal glands: absent (0), present (1)
8. Hooklet-shaped epidermal glands: absent (0), present (1)
9. Epidermal rhabdoids or rhabdites: absent (0), rhabdoids (1), rhabdites (2)
10. Rhammoids or rhammites: absent (0), rhammoids (1), rhammites (2)
11. Adhesive glands (Tyler 1976; Ehlers 1992a): absent (0), simple, collar of cilia (1), duo-gland system (2), haptocilia (3)

Continued overleaf

Table 3.2 Continued.

12. Epitheliosomes: flocculent (0), electron-dense ultrarhabdites (1)
13. Epidermal extracellular matrix (ECM): distinct (0), reduced (1), absent (2)
14. Distal ciliary cap: absent (0), present (1)
15. Distal ciliary shelf: absent (0), present (1)
16. Dense material in distal shaft: absent (0), present (1)
17. Proximal ciliary cup: absent (0), present (1)
18. Dense (basal) plate: homogeneous (0), bilayered (1)
19. Aggregation of granules in basal body: prominent (0), reduced (1)
20. Scattered lower granules: absent (0), present (1)
21. Main ciliary rootlets: two (0), one (1)
22. Lateral rootlets: absent (0), present (1)
23. Knee on main rootlet(s): absent (0), present (1)
24. Basal foot (posterior rootlet): present (0), absent (1)
25. Main rootlet frontal of basal body: absent (0), present (1)
26. Main rootlet proximally: rounded, homogeneous (0), tube- or trough-shaped (1)
27. Glycocalyx layer: weak (0), distinct (1)
28. Terminal web (cell web): weak (0), distinct but narrow (1), thick (2)
29. Peripheral nervous system: basiepidermal plexus only (0), insunk nerve cords (1); orthogon (2)
30. Anterior part of plexus: diffuse thickening (0), distinct thicker cords (1), insunk behind statocyst (2)
31. Aminergic brain (personal communication Olga Raikova 1999): concentration of longitudinal fibres (0), commissural brain (1), bilobed ganglionic brain with central neuropile (true brain) (2)
32. Peptidergic brain (personal communication Olga Raikova 1999): fibres (0), Paired lateral groups of cells (1), bilobed ganglionic brain with central neuropile (true brain) (2)
33. Statocyst: absent (0), one statolith (1), two statoliths (2)
34. Lithocyte surface: smooth (0), blistered appearance (1)
35. Statocyst parietal cells: few (0), two (1)
36. Tubular structure on lithocyte: absent (0), present (1)
37. Monociliary epidermal receptor cell: simple rootlet only (0), trough-shaped rootlet (1)
38. Collared epidermal receptor cell: simple rootlet (0), 'swallow-nest' rootlet (1), 'tubular body' rootlet (2)
39. Microvilli in receptor collar: low 7–10 (0), high 15–25 (1)
40. Mouth opening: present (0), sometimes closed (mature specimens) (1), no opening found (2)
41. Mouth position: in first half to midpart of gut (0), anterior to gut (1)
42. Pharynx simplex: absent (0), present (1)
43. Gut cavity: present and open (0), partly occluded (1), digestive parenchyma (2)
44. Granular club cells in gastrodermis: present (0), absent (1)
45. Location of club cells: scattered (0), near mouth (1)
46. Cilia on gut epithelium: present (0), absent (1)
47. Shape of testes: longitudinal follicles (0), compact, globular (1)
48. Position of testes: at base of gut (0), at body-wall muscle cells (1), in parenchyma (2)
49. Testes, paired or unpaired : paired (0), single median (1)
50. Testes lining cells: absent (0), present (1), thick tunica (2)
51. Male antrum: ciliated epidermal invagination (0), tube without cilia (1)
52. Male pore position on body: ventral (0), subterminal (1), supraterminal (2), dorsal (3)
53. Penis stylet: absent (0), present (1)
54. Position of ovaries in body (Rieger *et al.* 1991): at base of gut (0), in parenchyma (1)
55. Position of ovaries at gut: ventral side of gut (0), dorsal side of gut (1)
56. Ovary, paired or unpaired: paired (0), single, median (1)
57. Oocyte maturation: diffuse (0), posteriorly (1), anteriorly (2), ventrally, in circle (3), ventrally, pairwise (4)
58. Cleavage pattern: quartet spiral (0), duet spiral (1)
59. Female pore and bursa: absent (0), ventral side of body (1), dorsal side of body (2)
60. Female pore ciliation: absent (0), present (1)
61. Sperm flagellum: single (0), double (1), absent (2)
62. Microtubule configuration: 9 + 2 (0), derived 9 + 1, 9 + 0 (1)
63. Flagellum incorporated in cell: in channel with microtubule lining (0), in cytoplasm (1)
64. Anchoring fibre apparatus: present (0), absent (1)
65. Position of sperm mitochondria: along flagellum (0), several in cytoplasm (1), single long in cytoplasm (2)
66. Shape of basal mitochondria: tubular middle piece, (0), lateral derivative (1)
67. Allosperm in various body tissues: absent (0), present (1)
68. Allosperm nucleus strongly spiralized: absent (0), present (1)
69. Acrosome (or presumed derivative): apical acrosome (0), dense bodies in cytoplasm (1), other (2)
70. Protonephridia: absent (0), present (1)
71. Body shape (e.g., Sterrer 1998): ovoid to fusiform (0), filiform (1), bottle-shaped (2), shoe-shaped (3), slender (4), convolute (5)
72. Posterior end of body (Sterrer 1998): rounded (0), pointed as a short tail (1)
73. Habitat (e.g., Sterrer 1998): mud, deepwater (0), sand, shallow (1), mixed substrate and depth (2), endosymbiotic (3)

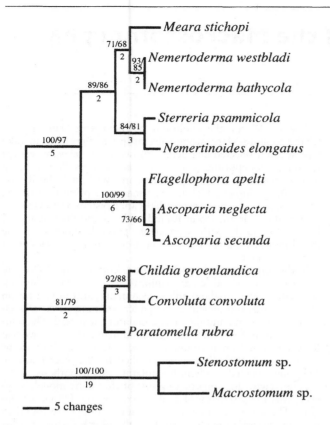

Figure 3.1 The single most parsimonious tree in the analysis shown as a phylogram. Bootstrap and jack-knife values (bs/jk) are shown above branches and Bremer support values below.

number of parsimony-informative characters in the revised matrix became 73. The outgroup taxon *Xenoturbella bocki* was replaced by the acoel *Paratomella rubra*. The analysis was conducted using PAUP* 4.0 beta 2 (Swofford 1998) and the character matrix was set up using MacClade version 3.0.1 (Maddison and Maddison 1992). All character states in the analysis were treated as unordered, unweighted and without restriction in state changes. Clade support was assessed by bootstrap and jack-knife (1000 replicates). Bremer-support was determined by stepwise filtering of trees.

The analysis resulted in a single most parsimonious tree (Figure 3.1) with 130 steps and a consistency index (CI) of 0.84. The tree topology is similar to Lundin (2000), but with better support. The Nemertodermatida, Nemertodermatidae and Ascopariidae are shown as monophyletic taxa. (See Lundin (2000) for apomorphic character state changes for each clade.)

Conclusion

Present structural, immunocytochemical and embryological evidence favours the idea that the Nemertodermatida and Acoela are sister taxa (e.g., Tyler and Rieger 1975, 1977; Lundin and Hendelberg 1996; Lundin 1997). By contrast, the two orders have been assigned widely separate positions in phylogeny estimates of the Platyhelminthes based on sequence data from 18S rDNA (Carranza *et al.* 1997; Jondelius 1998; Ruiz-Trillo *et al.* 1999). However, the 28S rDNA study by Litvaitis and Rohde (1999) shows a close relationship between Nemertodermatida and Acoela. Unpublished molecular data (personal communication Ulf Jondelius 1999) indicate that the Nemertodermatida and Acoela group together basally in the Platyhelminthes, albeit not as sister taxa. 'Total-evidence' analyses of platyhelminth phylogeny combining molecular and morphological data (Zrzavý *et al.* 1998; Littlewood *et al.* 1999a) show no clear consensus of the phylogenetic relationship between the Nemertodermatida and Acoela, possibly because the molecular (18S rDNA) and morphological data sets used lead to contradictory results. Zrzavý *et al.* (1998) conclude that 'nemertodermatids are one of the most basal triploblastic clades, but . . . their position is highly unstable'. Another recent total evidence tree (Giribet *et al.* 2000) identifies Nemertodermatida as the only acoelomates branching outside the infrakingdom Platyzoa, as defined by Cavalier-Smith (1998), i.e., Rotifera, Acanthocephala, Gastrotricha, Gnathostomulida and Platyhelminthes.

In balancing such discrepancies we maintain that the Nemertodermatida and Acoela are separate from the Catenulida and Rhabditophora (cf. Smith *et al.* 1986; Carranza *et al.* 1997; Lundin and Hendelberg 1998; Reuter *et al.* 1998b; Raikova *et al.* 1998a; Rieger 1998; Zrzavý *et al.* 1998; results from analysis presented herein) and probably have more basal phylogenetic positions. There is still reason to consider the Nemertodermatida as the earliest offshoot of the Platyhelminthes.

Chapter 4

Phylogenetic systematics of the Macrostomorpha

Reinhard M. Rieger

In most molecular and morphological studies of flatworms, the Macrostomorpha are interpreted as the plesiomorphic sister group to all higher rhabditophoran platyhelminth taxa (Tyler 2001, this volume; Littlewood and Olson 2001, this volume). The taxon was established by Doe (1986a) to encompass the Haplopharyngida and the Macrostomida *sensu* Karling (1974). Doe used three synapomorphies to characterize this group: the duo-gland adhesive organs (Tyler 1976, 1977); the pharynx simplex coronatus (Doe 1981); and the aciliary spermatozoa (Ehlers 1984). The Macrostomorpha are therefore a very well-defined monophyletic group in the lower platyhelminths.

Paratomy has been thought likely to be a plesiomorphic feature of these animals for some time (e.g., Rieger 1971b), and the discovery of the macrostomid genus *Myomacrostomum* (Rieger 1986b) generated further evidence for this notion. Paratomy is now held to be plesiomorphic even for all the Platyhelminthes (Ehlers 1985a).

While so far only two, perhaps three, species of the marine genus *Haplopharynx* Meixner 1938, (see Ax 1971; Rieger 1977) have been described, the genus *Macrostomum* O. Schmidt 1882 contains over 100 species, and these occur widely in marine, brackish, and freshwater environments (see Ferguson 1939–1940, 1954; Papi 1950; Luther 1960; Schmidt and Sopott-Ehlers 1976; Faubel *et al.* 1994; Ax and Armonies 1987, 1990; Ax 1994b; Ladurner *et al.* 1997; and taxonomic summary of the Macrostomida by Tyler http://www.umesci.maine.edu/biology/turb/). The Macrostomida are divided presently into three subtaxa: the primarily marine Microstomidae, with most species in the genus *Microstomum* O. Schmidt 1848, which also have well-known freshwater species (e.g., *M. lineare* (Müller 1773)); the very heterogeneous Macrostomidae, which will need taxonomic revision; and the marine, rarely brackish-water-living species of the Dolichomacrostomidae (about one-quarter of all described species in the taxon Macrostomorpha) which are now further subdivided, mainly because of differences in the very complicated male and female genital structures, into the taxa Karlingiinae, Dolichomacrostominae, and monotypic Bathymacrostominae (see Rieger 1971b and Faubel 1977 for further discussion).

Two hypotheses of character evolution have been proposed for the Macrostomorpha: one by Tyler (1976, 1977) concerning adhesive organs, the other by Doe (1981) concerning pharyngeal ultrastructure. In most older systems of the Macrostomida – that is, Macrostomorpha without Haplopharyngida – the genus *Macrostomum* was thought to represent the plesiomorphic condition of the taxon (e.g., Graff 1882, 1904–08; Luther 1905, 1947, 1960; Bresslau 1928–33; Reisinger 1933; Papi 1953; Ax 1961, 1963, 1995, 1996). A different evolutionary interpretation was first sketched by Rieger (1971a,b,c). It was based on the assumption that asexual reproduction, in the form of paratomy, was a plesiomorphic trait in the taxon and not, as had been argued by Ax and Schulz (1959), a derived feature. This claim could be substantiated with the descriptions of *Myomacrostomum unichaeta* Rieger 1986 and *Myomacrostomum bichaeta* Rieger 1986. The discussion of phylogeny of the free-living platyhelminths by Smith *et al.* (1986) was partly based on this assumption of the plesiomorphic nature of asexual reproduction in Platyhelminthes in general.

The model for the evolution of the taxon 'Turbellaria'-Rhabditophora-Macrostomorpha proposed here tries to take into account selected characters known to be relevant, except for the protonephridial system. The latter is peculiar in being highly variable in the group (Rohde and Watson 1998; Rohde 2001, this volume) which in itself may be a primitive feature. The origin of the protonephridial canal system and cyrtocyte (terminal cell) in the lower Rhabditophora and the Catenulida has to be dealt with separately; pulsatile bodies of the Acoelomorpha may be the key for understanding the evolution of the cyrtocyte.

In the phylogenetic hypotheses I present here, the following characters are of particular significance: the pharynx simplex coronatus, the duo-gland adhesive system, paratomy, location of shell glands and cement glands, microanatomy of the aflagellar spermatozoon, and the male copulatory structure. The reinvestigation of Westblad's macrostomid material at the Swedish Riksmuseet and my own unpublished observations on various macrostomids including some supplied by W. Sterrer and S. Tyler have played a role in formulating this hypothesis.

A key factor for constructing this phylogenetic model, however, was the discovery of a new taxon in the Haplopharyngida. The seven specimens of this new group were found in medium-coarse and coarse sand off the North Carolina coast collected in April and October 1970 and extracted in June of 1970 and in January of 1971. Serial sections of two specimens were prepared, the others were studied alive. Here, I provide a brief overview of the interesting body plan of this new group. The description of the new taxon will be published elsewhere (in honour of the late Professor Dr Tor Karling and his wife Annalisa).

A new taxon

The animals are about 1–1.5 mm long and rather sensitive to $MgCl_2$ anaesthetic, because of large vacuolized parenchymal cells, which are also common in the Karlingiinae, another interstitial group (Rieger 1969, 1971a). The rostrum is long and features adenal rhabdite glands throughout the epidermis and large rhammite glands which open in two distinct groups at the anterior tip of the body (Figures 4.1 and 4.2). Highly complex rod-like structures are the main kind of secretion granule in the rhammite strands (Figures 4.3, 4.8 and 4.9). The epidermis consists of large epithelial cells. Structures at the caudal end of living animals suggest the presence of a facultative anus. A large entolecithal egg is present at the end of the unpaired ovary; shell glands (Figure 4.5) occur in the epidermis in the region of the mature egg, in front, to the sides, and in back of it. Just behind the mouth, a male copulatory apparatus is situated on the ventral side, visible both in squeeze preparations and in the sectioned specimens (Figure 4.4). In sectioned specimens, so far only paired seminal vesicles were found to be linked to the copulatory apparatus.

It is mainly the location of the male pore (just behind the mouth), the penial structure (intracellular needles surrounding the male canal), and the lack of a sclerotic, tubular stylet that set this new taxon apart from all other macrostomorphan

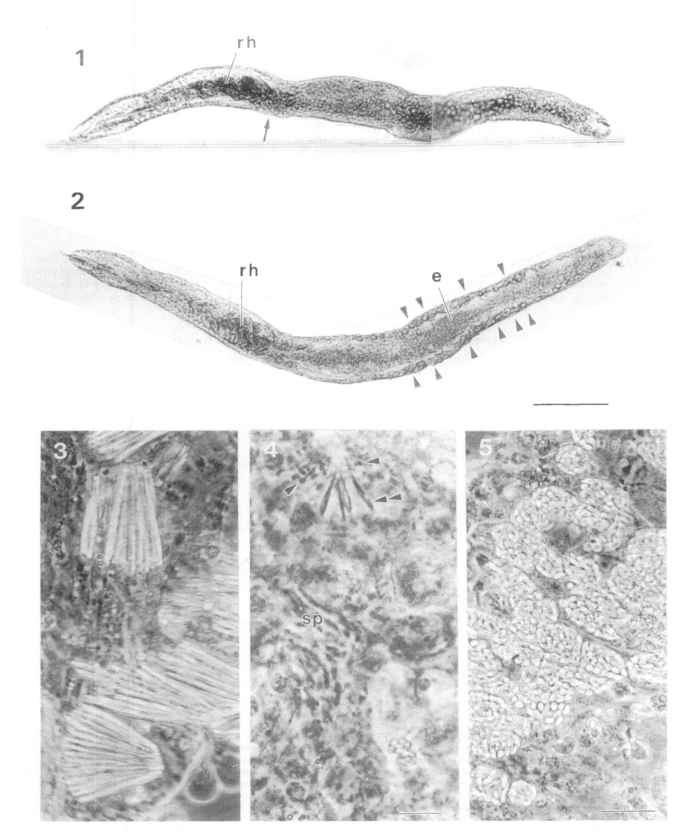

Figure 4.1–4.5 New taxon. **4.1** Lateral view, head at left, rh = rhammites, arrow = mouth region. **4.2** Dorsal view of same specimen, embedded in paraffin–celloidin, rh = rhammites, e = egg, arrowheads mark shell/cement glands. **4.3** Rhammite glands. **4.4** Male copulatory organ with stylet needles in squeeze preparation, arrowheads small penial spines, double arrow heads large penial needles; sp = spermatozoa. **4.5** Shell/cement glands in epidermis, lateral to mature egg. Scale bars: **4.1** and **4.2** 200 μm; **4.3** and **4.5** 30 μm; **4.4** 10 μm.

species. Two types of needles were obvious in the live preparations (Figure 4.4). Such a copulatory organ is most similar to that of the genus *Haplopharynx*. In its basic construction it resembles that of the Paratomellidae and Hofsteniidae, both most likely representing early side branches in the phylogenetic tree of the Acoela (see literature in Steinböck 1966, 1967; Ehlers 1992a; Sopott-Ehlers and Ehlers 1999). Also species of the Polycladida and Proseriata have such a penial organ, and such a distribution leads, therefore, to the interpretation that it is a plesiomorphous trait for the Rhabditophora in general (Smith *et al.* 1986). By the ultrastructure work of Doe (1982, 1986a,b) and Brüggemann (1985) we are now able to explain the evolution of the penial structures from this new genus to *Haplopharynx* and further to the higher macrostomids.

Phylogeny based on the adhesive system, according to Tyler (1976, 1977) (see Figure 4.6)

The detailed analysis of the adhesive structures by Tyler (1976) and its phylogenetic consequence set the stage for the now

generally accepted proposal of three monophyletic groups within the 'Turbellaria' (see also Doe 1981: 181 and especially Rieger 1981b) finally leading to the new system of the Platyhelminthes by Ehlers (1984, 1985a). In Tyler's hypothesis, an adhesive organ in the plesiomorphic condition in the Macrostomorpha had several glands, as is characteristic of the Haplopharyngida. In the subtaxon Macrostomida, the genus *Myozona* was apparently primitive in having more than one of the two gland types, the viscid gland. This interpretation allows the most parsimonious explanation of the evolution of the duo-gland adhesive system for all rhabditophorans, starting with multiple viscid and releasing gland cells in the Haplopharyngida and Polycladida and leading to a reduction to one cell of each type in the Macrostomorpha and to a branching of the gland necks of both gland cell types in the Neoophora (namely, those taxa investigated, the Proseriata, Tricladida, and Rhabdocoela).

It might be debatable whether the additional association of a rhabdite gland with some of the adhesive papillae, as seen in all dolichomacrostomids, *Myomacrostomum*, and *Bradynectes*, warrants the basal position of the genera *Microstomum* and

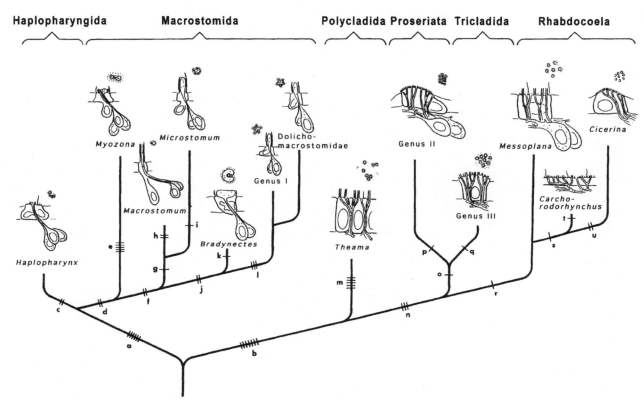

Figure 4.6 Scheme of the evolution of adhesive organs in the Turbellaria (exclusive of Acoela, Nemertodermatida, Lecithoepitheliata, Prolecithophora and Catenulida). Duo-gland adhesive organs in representative genera are diagrammed. Cross bars on a branch in the scheme designate characters that are believed to have been present in the ancestor to that branch, or are characters that distinguish a branch from its sister branch. Each bar represents one character, and each set of characters is marked by a letter. *a* Viscid and releasing gland necks emerge in a common collar of microvilli; 2–5 gland cells per anchor cell; one papilla on each anchor cell; each gland has necks to only one anchor cell; tonofilament core is central in microvillus; no sensory processes in anchor cell. *b* Viscid and releasing gland necks emerge separately; gland necks branch; more than one papilla on each anchor cell; a single gland may have necks to more than one anchor cell; microvillous collar around viscid gland neck only; sensory processes and rhabdite gland necks in anchor cell; adjacent viscid gland necks share microvilli of collar; gland necks enter sides of anchor cell. *c* Two viscid and one releasing gland per anchor cell, or three viscid and two releasing glands; diffuse cell web. *d* Specialization of anchor cell surface (all microvilli and/or cilia participate in collar); elongate microvilli. *e* Insunk anchor cell; split anchor cell; two viscid and one releasing gland per anchor cell; all microvilli with core; modified cilia. *f* Only one viscid and one releasing gland; no cilia on anchor cell. *g* Insunk anchor cell. *h* Narrow papilla; single cycle of elongate microvilli in collar. *i* Releasing gland neck branches. *j* Rhabdite gland associated with some adhesive organs; all microvilli with core. *k* Special form of cell web. *l* Viscid gland neck star-shaped in papilla; releasing gland neck branches in papilla; single cycle of microvilli in collar.

Macrostomum (and *Psammomacrostomum*) next to *Myozona*. Equally, the apomorphy 'insunk anchor cell' could be interpreted also as a convergent, i.e. parallel, adaptation, in the genera *Macrostomum* and *Microstomum*. More significantly, a cluster of three linked apomorphies seems to connect the branch to *Myomacrostomum* (= Genus I) with that to the dolichomacrostomids, forming a monophyletic unit: 1) terminal viscid gland neck in papilla star-shaped; 2) releasing cell in papilla branched; and 3) single row of cored microvilli forming the papilla. The position of the genus *Bradynectes* basal to this monophylum seems the best explanation, also, for the irregularly expanded viscid neck in this genus. The apomorphy

'special form of cell web' that characterizes the lineage to *Bradynectes* and the dolichomacrostomids is sound and, therefore, marks it as a distinct phyletic branch.

Phylogeny based on construction of the pharynx, according to Doe (1981) (see Figure 4.7)

The most significant difference in the phylogenetic tree of the Macrostomorpha based on pharyngeal features (Doe 1981) is the basal position of the genus *Microstomum* in the Macrostomida. Besides a number of cytological features of

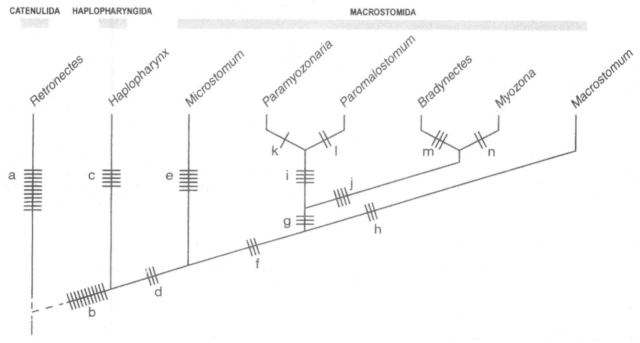

Figure 4.7 Scheme of evolution of pharynx in the Catenulida and Macrostomida. Cross bars on branch designate characters that are believed to have been present in the ancestor to that branch or are characters that separate a branch from its sister branch. Each bar represents one character, and each letter represents a set of characters. *a* No prominent commissure behind mouth; pharynx nervous system with no prominent nerve ring; recessed monociliated sensory cells; transition zone of two rings of insunk epithelial cells with epidermal-type rootlets; no or one gland cell type in pharynx; gland ring of one gland cell type at border of transition zone and pharynx proper; cytological features including microvilli with dense cores, nuclei with heterochromatin in isolated patches and mitochondria with parallel arranged cristae; muscle layers not clearly established with most of the circular muscles inside longitudinal muscles, pharynx longitudinal muscles derived from body wall longitudinal and circular muscles; mouth position between anterior third and half of body; pharynx proper with elongated caudal rootlets and shortened rostral rootlets; pharynx proper with intraepithelial to partially insunk nuclei. *b* Post-oral nerve commissure with more than 12 axons; pharynx nerve ring with more than 30 axons; unrecessed pharynx sensory cells with variable number of cilia; transition zone of 1–5 circles of insunk ciliated epithelial cells with modified rootlets and cell web or terminal web; two or more gland cell types in pharynx; gland ring of one or more gland types proximal to nerve ring in region of border of transition zone and pharynx proper; cytological features including microvilli without cores, nuclei with outer rim of heterochromatin and mitochondria with irregularly arranged cristae; with at least two layers of muscles present, circular muscles inside to longitudinal muscles and longitudinal muscles derived from regular and special body wall circular muscles; mouth occurring in first third of body and pharynx directed dorsally; pharynx proper with elongated caudal rootlets and shortened rostral rootlets; pharynx proper with intraepithelial nuclei. *c* Transition zone is single ring of unciliated cells; type I gland cell granules of protein and polysaccharide; type I gland cells restricted to gland ring; no cell web in transition zone; one gland cell type in gland ring and one type in pharynx proper. *d* Type I gland cell granules solely protein; two or more gland cell types in gland ring, pharynx proper with partially or completely insunk nuclei. *e* Pharynx oriented dorsally; transition zone and pharynx proper rostral rootlet retained and caudal rootlet absent; two gland cell types present in pharynx; no gland type restricted to gland ring; cell web absent from transition zone; longitudinal muscles inside to circular muscles. *f* Transition zone with vertical or caudal rootlet retained and rostral rootlets reduced or absent; four or more gland cell types present in pharynx, always including types I–IV; one or more gland cell types restricted to gland ring always including type I. *g* Transition zone is multiple layers of cells; no rostral rootlets in transition zone cells; gland cells in ring not in repetitive groups. *h* Transition zone is single circle of cells; rostral rootlets present in transition zone cells; gland cells in ring in repetitive groups. *i* Four to five gland cell types in pharynx; terminal web in transition zone; 50–100 axons in nerve rings; gland ring distal to proximal pharynx proper cells. *j* Six gland cell types; cell web in transition zone; less than 50 axons in nerve ring; gland ring proximal to proximal pharynx proper cells. *k* Epidermis at mouth with normal rootlets. *l* Epidermis at mouth with shortened rootlets; longitudinal muscles inside to circular muscles. *m* Longitudinal muscles inside to circular muscles; proximal ring of pharynx proper cells with cell web and modified rootlet orientation; transition zone is single circle of cells. *n* Proximal ring of pharynx proper cells typical; modified sensory cell cytology.

the pharyngeal epidermis, the diversity of pharyngeal glands was used by Doe (1981) to argue for this position, hypothesizing that the simpler condition of only two types of pharyngeal gland cells in the genus *Microstomum* is more plesiomorphic than the several pharyngeal gland types in others. Additional information about the function of the different types of pharyngeal glands might clarify this position; for example, predators may need fewer types of pharyngeal cells than diatom-eating forms. In the genus *Myozona*, for example, where diatom feeding seems to be common (e.g., Marcus 1949), a relatively high number of pharyngeal gland cell types (five) were found in both investigated species, while the Microstomidae and Dolichomacrostomidae, in which predators are particularly common (e.g., Westblad 1923, 1953; Pawlak 1969; Rieger 1971a,b; own unpublished observations on more than 30 undescribed species) have fewer. On the other hand, the genus *Macrostomum* comprises both predatory and diatom-feeding species, and of the three representatives for which we know of pharyngeal-gland fine structure, the predatory species (feeding on nematodes) has as many (seven) gland cell types as the maximum number of gland cell types found in the two diatom-feeding species (Doe 1981). Clearly, more data on feeding strategies and foregut structure are needed to evaluate the phylogenetic significance of this character.

As far as the genus *Microstomum* is concerned, it may be safe to conclude that some of the features that Doe (1981: 183) described as derived characters (e.g., orientation of pharynx, preoral gastrodermal region) are related to fission by paratomy and can, therefore, be interpreted as plesiomorphic (see below).

While the genus *Myozona* appears to be the most primitive form of the Macrostomida in terms of characters of the adhesive system, it is in a derived position in Doe's phylogenetic tree, together with *Bradynectes* and the dolichomacrostomids. In arguing for the derived position of *Myozona* not only the number of pharyngeal gland cell types (five) is significant, but also the basic similarity in structure to those of *Bradynectes* and the genus *Macrostomum*.

A new phylogenetic model for the Macrostomorpha

Using characters of the duo-gland adhesive system, pharynx simplex coronatus, and rhammites (Figures 4.8 and 4.9), as well as of the characters pertaining to parts of the female and of the male system and of reduction of fission, I propose an alternative phylogenetic tree for the Macrostomorpha, with two versions of the evolution of the female canal system

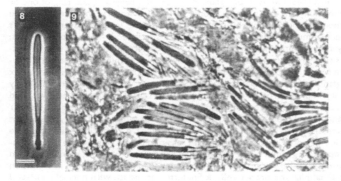

Figure 4.8, 4.9 Rhammites in the new taxon. **4.8** Single rhammite. **4.9** Rhammites, in squeeze preparation, phase contrast. Scale bars: **4.8** 10 μm; **4.9** 30 μm.

(Figures 4.10 and 4.11). The most important characters are elaborated separately below.

Rhammites and the frontal glands

Following Karling (1965), Doe (1986b) specified the glandular/muscular proboscis as an apomorphy for the Haplopharyngida. New data on the truly gigantic rhammites of the new taxon (Figures 4.8 and 4.9) corroborate the speculation that the ancestor of the Haplopharyngida (and possibly all Macrostomorpha) had such a proboscis organ. This structure may be the plesiomorphic condition from which the frontal gland complex (Klauser *et al.* 1986) of higher macrostomids was derived. In *Haplopharynx* and in the new taxon, the proboscis gland necks pass by the brain dorsally and ventrally. Microstomids, the plesiomorphic sister group to the other Macrostomida, do not feature larger rhammites (own unpublished observations), but it is notable that the glands opening at the anterior tip of the animal do not penetrate the brain either. In all other macrostomids the dorsal strands of the rhammite glands clearly penetrate the neuropile of the brain, making this one of the best characters to identify all higher macrostomorphans (see Rieger 1971b).

Considering the high frequency of apical glandular-muscular organs among the Acoelomorpha (the frontal organ with frontal pore and, in *Flagellophora* of the Nemertodermatida, the proboscis; see Rieger *et al.* 1991; Sterrer 1998), it is tempting to speculate, as did Tor Karling (1965), that this character and the proboscis of nemerteans are parallel relics of an early evolutionary trend in the Bilateria.

Reduction of fission and the reversal of the genital pores

With the description of *Myomacrostomum unichaeta* and *M. bichaeta*, I have shown that the muscle ring on the gut of the genera *Myozona*, *Myozonaria*, and *Paramyozonaria* could be a derivative of the musculature in the fission plane in a paratomizing ancestor (Rieger 1986b); in particular, some specimens of *Myomacrostomum bichaeta* featured a 'normal' paratomizing division plane in the same place that other specimens had a muscle ring on the gut. Muscle rings were probably preserved for different reasons in the genus *Myozona* (for breaking of diatom frustules, see Marcus 1949) and the genera *Myozonaria* and *Paramyozonaria* (possibly for discharge of undigested food in these elongated, vermiform animals; see Karling 1965, 1966c; Rieger 1971a). Evidently in the dolichomacrostomids, as well as in the genus *Myozona* (see last summary of literature in Sopott-Ehlers and Schmidt 1974) and in probable relatives of *Myozona* constituting a species group with muscular penial organ (*Psammomacrostomum*, *Anthomacrostomum*, *Siccomacrostomum*, and *Dunwichia*; see below), paratomy was lost, leaving a gastrodermal muscle ring in *Myozona* as a rudiment of the division plane.

This view of the evolution in the Macrostomorpha can also explain why the male genital organs are located in front of the female organs in the haplopharyngids (see Rieger 1971a for detail) but behind them in the macrostomid taxa. Both conditions could have arisen from an ancestor with two zooids that had differential development of the gonads; as paratomy was reduced in the line to the Haplopharyngida, female organs would be the only remnants of the posterior zooid; the male organs would develop from the anterior zooid (Figure 4.12). In the case of the Macrostomida both male and female organs

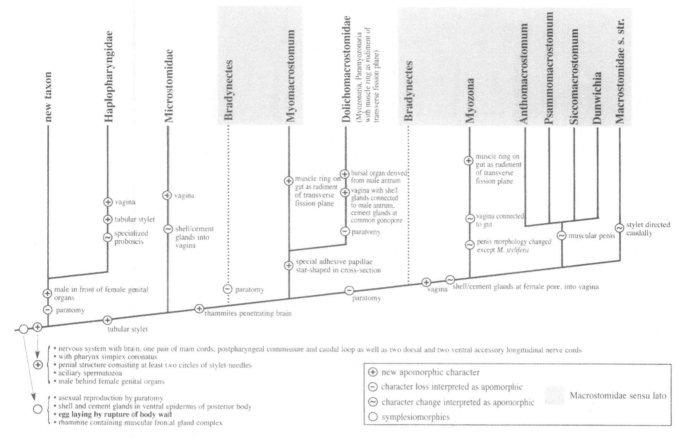

Figure 4.10 Phylogenetic tree of the Macrostomorpha, alternative I, assuming lack of female pore and vagina in the stem species.

may be derived from sexual organs in the caudal zooid only (Figure 4.13).

With respect to paratomy, the most plesiomorphic taxon in the Macrostomida is the Microstomidae, comprising the genera *Microstomum* and *Alaurina*. This position is supported also by characters of its pharynx simplex coronatus (see above). Evidently, then, the sunken nature of the anchor cell in the adhesive organs in *Microstomum* is convergent with that of *Macrostomum*. In the proposed phylogenetic trees, paratomy is seen to have been lost more than once.

The interpretation of paratomy as a plesiomorphic trait allows actually two different placements in the phylogenetic tree for *Bradynectes* (Figures 4.10 and 4.11) and possibly *Myozona*. In the case of *Myozona* (and conceivably of related genera with muscular penis), one position is more in correspondence with the character phylogeny of the pharynx simplex (Figures 4.10 and 4.11). Another position, in correspondence with the adhesive organs, would place this clade between the Microstomidae and all other Macrostomida. Decisions between the possibilities will have to await further character analysis.

The female canal system, shell and cement glands

According to my observations, neither the new taxon nor other members of the Haplopharyngida have a fully developed vagina. In *Haplopharynx rostratus* and the new taxon, large gland cells containing small, often naviculated to ovoid granules, are located slightly insunken in the epidermis in the posterior half of the animal (Figures 4.2 and 4.5). In the new taxon these gland cells occur at the level of the largest ovum (see above). In *Haplopharynx rostratus* from Rovinj, Istria they apparently form a circular 'clitellum' around the caudal end of the body just behind the largest oocyte (Figure 4.14). Meixner (1938) and Karling (1965) report a complete female canal system in *Haplopharynx rostratus*. Karling, however, mentions that the histological material he had on hand for the descriptions did not allow a 'histologische Analyse des weiblichen Apparates', and he thought that Meixner's figure (1938, his Figure 35) showing female gonad, canal and pore was most likely adapted from live observations. Thus, a female pore in *H. rostratus* is probably known only from two live observations.

The only female part depicted by Ax (1971) for *H. quadristimulus* is an unpaired ovary with a large egg behind the male structures and possibly shell glands behind the egg. My own unpublished, sectioned material from two North Carolina forms of *H. quadristimulus* suggest that it also lacks a female pore. It does have epidermal glands with secretory granules similar to the shell granules in the dolichomacrostomids and *Bradynectes*.

Accordingly, it is conceivable that originally *Haplopharynx* and the new taxon lacked a female canal system, but had accumulations of epidermal glands serving as shell/cement glands located near the caudal end of an unpaired entolecithal ovary. The function of shell glands and cement glands may not yet have been distinct at this stage. Egg laying would have occurred through rupture of the body wall. A similar location of shell glands is seen in the Lecithoepitheliata–Prorhynchida,

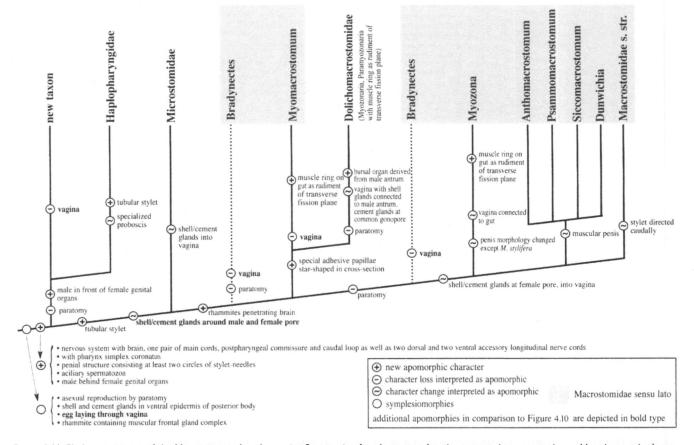

Figure 4.11 Phylogenetic tree of the Macrostomorpha, alternative 2, assuming female pore and vagina present in stem species and lost in certain descendents by virtue of tendency to progenesis.

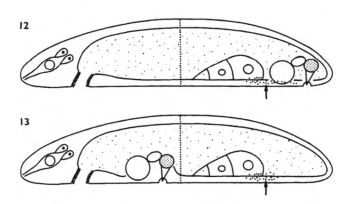

Figure 4.12, 4.13 Schemes of a two-zooid macrostomorphan ancestor similar to paratomizing *Microstomum* and *Myomacrostomum*, illustrating the possibility for reversal of sex organs during reduction of fission. **4.12** Plesiomorphic condition for the Macrostomorpha. **4.13** Reversal of that condition for the Haplopharyngida. Arrows indicate position of shell/cement glands.

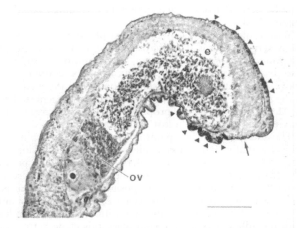

Figure 4.14 Ring of shell glands within the caudal epidermis of *Haplopharynx rostratus*, Rovinj-Istria; slightly parasagittal section through caudal end. e = egg, ov = ovary, arrowheads mark position of shell/cement glands, arrow position of anus. Scale bar: 60 μm.

where a female canal system is present and the female pore lies closer to a midventral position (see Rieger *et al.* 1991: 102 and Figure 31A).

In one of the phylogenetic trees presented here, a female pore and a vagina are assumed to have been absent in the macrostomorphan stem species (Figure 4.10). This necessitates the assumption of several parallel lines of evolution for the female canal system. Alternatively, the female canal system may have regressed, such that egg laying occurred secondarily through rupture of the body wall because of a strong tendency towards progenesis in at least three lines, coupled with the reduction of paratomy (Figure 4.11). More information on the postembryonic development of the genital organs in the Haplopharyngida and in the basal Macrostomida (e.g., Microstomidae) might help to gauge this possibility. The further events of the evolution of the female system may have led then in parallel lines to a complete loss of a female canal (new taxon, *Bradynectes*, *Myomacrostomum*), to a secondary connection to the male antrum (dolichomacrostomids) or, in the case of *Myozona*, to a connection to the gut by a completely new structure: a developing pharynx in the caudal region of a paratomizing ancestor similar in general morphology to *Myomacrostomum*. By this interpretation, the female canal system of *Psammomacrostomum*, *Anthomacrostomum*, *Siccomacrostomum* and *Dunwichia* (see summary of literature in Faubel *et al.* 1994) may be a line of evolution parallel to that of the vagina of *Microstomum* and *Macrostomum*.

In dolichomacrostomids, cement glands are separate from shell glands. They open around the single gonopore, which, according to the interpretation advanced here, would be the original male pore and male antrum of the caudal zooid in an ancestral species.

The copulatory structure and the spermatozoa

With the discovery of the new genus, the ancestral character state of the penial organ of the Macrostomorpha is more easily surmised: that is, it would have had only intracellular needles surrounding the male canal (see also proposals by Doe 1986a,b and Smith *et al.* 1986). The tubular sclerotic stylet found in *Haplopharynx* and the plesiomorphic Macrostomida could be derived from fusion of independent sclerotic needle cells, forming a ring-shaped matrix syncytium that secrets the stylet tube, as suggested by Doe (1982, 1986a,b). The condition in *Haplopharynx quadristimulus*, where four single stylet needles lie outside the tubular stylet, may represent a plesiomorphic condition within the Macrostomorpha. However, we still lack TEM data on accessory spines of *Haplopharynx rostratus*, in which the tubular stylet is not surrounded by accessory spines (see Doe 1986b.)

Needle-like intracellular rods similar in structure to ciliary rootlets occur also to the inside of the tubular stylet in the Macrostomida (for *Macrostomum* and *Paramyozonaria* see Doe 1982; possibly also in *Myozonaria* and *Acanthomacrostomum* and all Karlinginae; see Rieger 1971b, own unpublished observations) but seem to be absent in *Microstomum* (Doe 1982) and Dolichomacrostominae (Rieger 1971c; Brüggemann 1985). Because of their location inside the tubular stylet, these structures have been identified as non-homologous with needle-like intracellular structures located distal to the stylet (Doe 1982, 1986a,b). The intracellular rods could be another autapomorphy of the Macrostomida.

In terms of only copulatory structures, the most parsimonious solution would be to place the new group as the plesiomorphic sister taxon to all other Macrostomorpha. In considering also, however, the character 'fission by paratomy', it is more parsimonious to assume only one event of loss of paratomy for the Haplopharyngida. However, the assumption of two convergent origins for the tubular stylet, one within the Haplopharyngida and one within the Macrostomida then becomes necessary (Figures 4.10 and 4.11).

By what is evidently the plesiomorphic orientation of the male penial organ (pointing rostrally), *Microstomum* and *Myomacrostomum* are clearly set aside from the genus *Macrostomum* (copulatory stylet always points caudally). The genus *Myozona* is also plesiomorphic in this character (see *Myozona stylifera*). A gradual shift toward a caudal orientation

of the penial organ is evident in *Myozona* as well as in the four genera listed with a muscular penis.

On the other hand, the apomorphic character state of the rhammite glands, that is having certain glands penetrate the brain, clearly links *Myomacrostomum* with the dolichomacrostomids and the macrostomids *sensu lato*. Curiously, *Myomacrostomum* forms the plesiomorphic sister group to the dolichomacrostomids by virtue of the apomorphy of the adhesive organs, as noted earlier.

Spermatological evidence also supports the close relationship and thus the monophyly of the Haplopharyngida and the Macrostomida (Rohde and Faubel 1998; own unpublished observations on material supplied by S. Tyler and D. Doe). The 'blunt bristle-like structures' in the lateral cytoplasm of *Haplopharynx* spermatozoa described by Rohde and Faubel (1998) could be precursors of the bristle-like structures of spermatozoa of certain members of the genus *Macrostomum* (see e.g., Ferguson 1939–40 for light microscopy data for many species, and Rohde and Faubel 1998 for TEM literature). However, Watson (1999b) considers a homology of these structures in *Haplopharynx* and *Macrostomum* unlikely. Further information is needed to resolve this issue. I mention it here as another example of possible parallel loss within the Macrostomida: in the dolichomacrostomids (Rohde and Faubel 1998; own unpublished observations), in the genera lacking sclerotic penial structures (*Psammomacrostomum*, *Anthomacrostomum*, *Siccomacrostomum*, and *Dunwichia*; see Faubel *et al.* 1994; own unpublished observations), in various species groups in the Macrostomidae *sensu stricto* (e.g., the *M. hystricinum marinum* species group; Rohde and Faubel 1997; own unpublished observations).

Comparison with the phylogenetic system of the Macrostomorpha proposed by Sopott-Ehlers and Ehlers (1999) (see Figure 4.15)

The system proposed by these authors also places the Haplopharyngida as the plesiomorphic sister group to the Macrostomida. Autapomorphies they propose for the Macrostomorpha are specialized duo-gland adhesive system (Tyler 1976, 1977) and aciliary spermatozoa (Ehlers 1984); those for the Macrostomida are female pore in front of male pore, and special differentiation of penial structures (Doe 1982, 1986a,b; Brüggemann 1985). Except for the autapomorphy 'female pore in front of male pore' for the Macrostomida, the suggested autapomorphies correspond to the phylogenetic tree proposed in this paper. The character 'male behind female genital organs' is interpreted in this chapter as an autapomorphy for the Macrostomorpha (Figures 4.10 and 4.11). Within the Macrostomida, Sopott-Ehlers and Ehlers indicate a basal position of the Microstomidae, but not clearly separated from the Dolichomacrostomidae and *Myozona*. A monophyletic relationship is suggested between the genus *Bradynectes* and the genus *Macrostomum* on the basis of two lateral ledges in the spermatozoon. However, as mentioned above, the 'curved dense structures' (Rohde and Faubel 1998) in *Haplopharynx* might represent an early stage in the evolution of the lateral ledges and bristles in the Macrostomida.

Sopott-Ehlers and Ehlers' (1999) finding of electron-dense 'ledges' in spermatozoa of *Bradynectes* provides new support for the notion that the original spermatozoa of the Macrostomorpha featured such structures. By proposing that a 9 + 1 axoneme and intercentriolar body in the spermatid are autapomorphies of the Rhabditophora, these authors also place these features as plesiomorphies of the Macrostomorpha.

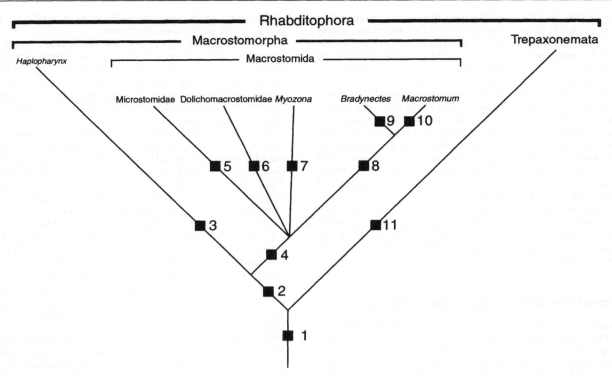

Figure 4.15 Diagram of the basic relationships within the Rhabditophora and the interrelationships of the Macrostomorpha. *Black squares* are sets of hypothesized autapomorphies. Spermatological characteristics are given in italics. The unique 9 + '1' pattern of ciliary axonemes in spermatozoa and an intercentriolar body in spermatids are autapomorphies of the Rhabditophora or of the Trepaxonemata. **1.** Rhabditophora: lamellated rhabdites; duo-gland adhesive systems; protonephridial terminal cells with four or more cilia; pharynx simplex with a prominent nerve ring (see Doe 1981); rhabdomeric photoreceptors with few sensory cells and one mantle (pigment) cell (see Sopott-Ehlers 1996); *biciliated spermatozoon, both cilia inserting at the functional fore-end of the cell*; oocytes with polyphenol-containing eggshell forming granules; female pore. **2.** Macrostomorpha: specialized duo-gland adhesive systems (see Tyler 1976, 1977); *spermatozoon without ciliary axonemes.* **3.** *Haplopharynx*: cranial protrusible proboscis; specialized male organ (see Doe 1986). **4.** Macrostomida: female pore in front of male pore; special differentiation of penial structures (see Brüggemann 1985; Doe 1986); *spermatozoon with two sets of cortical microtubules.* **5.** Microstomidae: gut with preoral blind sac; cranial ciliated sensory pits. **6.** Dolichomacrostomidae: bursal organ, accessory glandular organ, common female and male pore (see Rieger 1971b). **7.** *Myozona: spermatozoon with bone-shaped rods, one cylindrical mitochondrial rod and four sets of microtubules* (see Sopott-Ehlers and Ehlers 1998b). **8.** *(Bradynectes+Macrostomum): spermatozoon with two lateral ledges.* **9.** *Bradynectes*: without female pore; *spermatozoon with membranous lacunae; both sets of microtubules restricted to the posterior region of the cell.* **10.** *Macrostomum*: protonephridia with two-cell terminal weirs showing two rings of interdigitating microvilli (see Kunert 1988; Watson *et al.* 1991); *spermatozoon with two modified lateral ledges = bristles.* **11.** Trepaxonemata: specialized duo-gland adhesive systems with branching gland necks and several papillae on each anchor cell (see Tyler 1976, 1977; Smith *et al.* 1986); with reservation (see Ehlers 1985a): protrusible muscular composite pharynx with pharyngeal cavity (see Rieger *et al.* 1991).

Such an assumption may be correct, but cannot yet be substantiated: no 9 + 1 axoneme has been found in any macrostomorphan. One cannot discount the possiblity that 9 + 2 axonemal structures could be found in spermatozoa of the new taxon. I agree with Watson (1999b), therefore, when she excludes the Macrostomorpha from the Trepaxonemata.

Evidence of recent molecular studies on the phylogeny of the Macrostomorpha (Figure 4.16a,b)

Two molecular phylogenetic studies attempted to elucidate the interrelationships of the Platyhelminthes which also included more than two species of the Macrostomorpha (Littlewood *et al.* 1999a; Litvaitis and Rohde 1999). Littlewood *et al.* (1999a) present, to date, the most representative analysis of complete 18S rDNA sequence data and comparative morphological data which are first analysed separately, and then morphological and DNA evidence in a 'total evidence' approach. These molecular-based studies suggested three alternative evolutionary

hypotheses, depending on which tree-building algorithm was used or whether morphological data were also included in the analyses. In their combined analysis of molecular and morphological characters (their Figure 5) they suggest the positioning of the Haplopharyngida between two clades of macrostomids, the first containing species of the genus *Macrostomum* only, and the second containing microstomids and dolichomacrostomids. The 'total evidence' approach for the within-group relationship of the Macrostomorpha thus would place the Haplopharyngida within the Macrostomorpha, as a monophyletic unit with a dolichomacrostomid and two microstomids; and the genus *Macrostomum* as most ancestral split. This grouping may seem to be in line with the older light histological literature, but seems rather unlikely in the light of new evidence provided in the present study, or of that discussed by Sopott-Ehlers (1999). The topology of the parsimony tree based upon 18S rDNA data (their Figure 3) is equally difficult to connect to a series of recent ultrastructural data. Here, the Haplopharyndida are resolved as a separate lineage as sister group to the Lecithoepitheliata, and outside a clade containing the macrostomids and polycladids. A separation of the

a. Litvaitis and Rohde (1999)

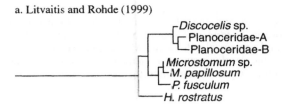

- *Discocelis* sp.
- Planoceridae-A
- Planoceridae-B
- *Microstomum* sp.
- *M. papillosum*
- *P. fusculum*
- *H. rostratus*

b. Littlewood *et al.* (1999a)

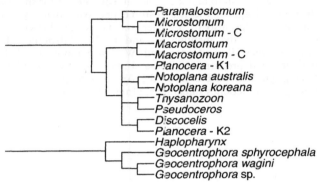

- *Paramalostomum*
- *Microstomum*
- *Microstomum* - C
- *Macrostomum*
- *Macrostomum* - C
- *Planocera* - K1
- *Notoplana australis*
- *Notoplana koreana*
- *Thysanozoon*
- *Pseudoceros*
- *Discocelis*
- *Planocera* - K2
- *Haplopharynx*
- *Geocentrophora sphyrocephala*
- *Geocentrophora wagini*
- *Geocentrophora* sp.

Figure 4.16 Portions of parsimony trees published for the Macrostomorpha based on **A)** partial 28S rRNA by Litvaitis and Rohde (1999), and **B)** complete 18S rRNA by Littlewood et al. (1999a).

Haplopharyngida and Macrostomida is indeed highly unlikely considering the evolution of various morphological characters presented in this chapter. In particular, the occurrence of the pharynx simplex coronatus in *Haplopharynx* and in all macrostomids (see Doe 1981, 1986b) speaks in favour of the monophyly of all Macrostomorpha (Haplopharyngida plus Macrostomida). However, a closer relationship of the Lecithoepitheliata and the Haplopharyngida would appear supported by the suggested evolutionary transitions of two anatomical characters only, the penial structures and the follicular surrounding of accessory cells (see above).

Littlewood *et al.* (1999a) also present the results of a neighbour-joining analysis of the complete 18S rRNA data set based upon the HKY85 maximum likelihood distance model. The neighbour-joining tree (their Figure 4) is similar to the phylogenetic hypothesis presented in this chapter. The Haplopharyngida are placed as most ancestral split and as sister group to two clades of Macrostomida, one comprising *Microstomum* and *Paromalostomum*, and the second two species of the genus *Macrostomum*. The sister group of the *Macrostomum* clade, however, was formed by several polyclads. The placement of the Haplopharyngida as most ancestral split within the Macrostomorpha, followed by two clades of Macrostomida, the first comprised of *Paromalostomum* and *Microstomum* and the second comprised of *Macrostomum*, would be in line with the phylogenetic hypothesis presented in this chapter. In our analysis the first Macrostomid clade is also formed by microstomids and dolichomacrostomids, and the second clade also contains two species of the genus *Macrostomum*. However, the inclusion of the Polycladida as sister-group to the *Macrostomum* clade suggested by the neighbour-joining tree based upon 18S rDNA would place the Macrostomida in paraphyly with respect to the Polycladida.

Litvaitis and Rohde (1999) provide two molecular trees of the Macrostomorpha based on partial 28S rDNA sequences: a neighbour-joining tree (their Figure 2) and a parsimony tree (their Figure 3). Both hypotheses agree with the topology obtained in the neighbour-joining analysis of Littlewood *et al.* (1999a) based upon 18S rDNA in that they also suggest that *Haplopharynx* represents the most ancestral split of a lineage comprising one clade of Macrostomida (*Microstomum* spp. and *Paromalostomum*) and a second comprising the Polycladida. In this analysis the genus *Macrostomum* was not included. A sister group relationship of Macrostomorpha and Polycladida, or even the suggested paraphyly of the Macrostomorpha resulting from the inclusion of the Polycladida, is highly unlikely in relation to the available morphological evidence. Pharynx construction, developmental data as well as body profile point against an equidistant relationship of *Haplopharynx* to the macrostomids and the polyclads. The Polycladida and the Macrostomorpha are – on morphological grounds – two separate, monophyletic assemblages (Karling 1974; Sopott-Ehlers and Ehlers 1999). However, a sister group relationship of the Macrostomorpha and the Polycladida may indeed be supported by our present study, if one considers the new evidence on the evolution of the male scleritic copulatory structures of macrostomorphans (see also Smith *et al.* 1986) and new data on the organization of the ovary of *Haplopharynx rostratus* (see above). In conclusion, molecular phylogenetic analyses provide highly useful insights for the interpretation of alternative morphological transitions in Macrostomorpha. Reciprocal illumination between morphology and molecules will become extremely fruitful, as soon as more phylogenies based upon additional gene segments become available.

Conclusion and suggestions for future work

Thus, we can conclude that:

1. This study supports a basal position of the Macrostomorpha, as is evident in most phylogenetic trees of the Rhabditophora (see Sopott-Ehlers and Ehlers 1999).

2. The Haplopharyngida are the plesiomorphic sister group to the Macrostomida. Haplopharyngids, including the new taxon, may be derived from the macrostomorphan common ancestor by loss of paratomy and the apparent inversion of the male gonad and copulatory structure in relation to the ovary by virtue of that loss. Intracellular needles surrounding the male canal are very likely the plesiomorphous character state for this feature. Similar penial needles, common in the Polycladida, Lecithoepitheliata, and the acoel taxa Paratomellidae and Hofsteniidae are seen as evidence that intracellular needles may have been the original penial structure in the Rhabditophora.

3. A basal position of the taxon Microstomidae in the Macrostomida appears likely.

4. The character 'rhammites penetrating the brain' as well as the lack of pigment-cup ocelli in the new taxon, in the Haplopharyngida, and possibly in the Microstomidae is likely to be a plesiomorphic character in the Rhabditophora and not the result of loss. Such eyes are not known in the Haplopharyngida. Pigmented epithelial eye spots occur in *Microstomum lineare* (see Palmberg *et al.* 1980) and peculiar pigment cup ocelli in other Microstomidae (e.g., the ocelli in *Microstomum dermophthalmum*

Steinböck 1933). The latter have not been studied with TEM.

5. A taxon including the genus *Myozona* and the genera with soft penial structures, such as *Psammomacrostomum*, *Acanthomacrostomum*, *Siccomacrostomum*, and *Dunwichia* might be a new monophyletic subgroup of the Macrostomida.

6. Multiple origins of characters (e.g., vagina), multiple cases of loss of characters (e.g., paratomy, vagina) as well as reversal of character trends (e.g., differentiation of sclerotic stylet needles, their transformation into a tubular sclerotic stylet, subsequent reduction and loss of the sclerotic structure and differentiation of solely muscular penial structures) appear as fundamental processes in the adaptive radiation of the Macrostomorpha.

One point which requires further study in particular is the female canal system in the genera *Haplopharynx* and *Myozona*. A search for homologies of the male penial structures with those of certain taxa of the Acoela (Paratomellidae, Hofsteniidae) could help to identify the sister group within these acoel taxa and the Macrostomida. The absence of a female pore in *Bradynectes* should be corroborated. Finally, the shell and cement glands of the Macrostomorpha and the postembryonic development of the female canal system in the Macrostomida require further investigation.

ACKNOWLEDGEMENTS

I particularly wish to thank S. Tyler and W. Sterrer for sharing their knowledge of the Macrostomorpha over many years. Thanks are due also to S. Tyler and G. Rieger for critical comments on the manuscript, and to W. Salzburger, R. Geschwentner, P. Ladurner and W. Salvenmoser for help with its final preparation. These studies were supported by FWF-grant P13060-Bio from the Austrian Science Foundation, and by Actiones Integradas, grant number 12/99.

Section II
Free-living groups

Chapter 5

The Proseriata

Marco Curini-Galletti

The Proseriata is a diverse and species-rich taxon of small (1–10 mm long), mostly free-living Platyhelminthes. The vast majority of the Proseriata are typical representatives of the marine interstitial mesopsammon, with slender, cylindrical bodies; burrowing species occur in fine sediments, including mud (Reise 1988). A few globose to drop-shaped, mostly epibenthic taxa are known (e.g. species of *Otomesostoma* and *Monotoplana*) (Cannon 1986).

Proseriates are particularly common in coastal habitats, from shallow water to the upper intertidal (Sopott 1972, 1973; Reise 1984, 1988). In some cases, they may be among the dominating organisms of interstitial assemblages, e.g., the Otoplanidae in the high-energy intertidal zone (Remane 1933). Only a few species are known so far from deep water (Westblad 1952). Colonization of fresh water has been marginal: five species are known from Eurasia (including one monotypic family, the Otomesostomidae); one stream dweller occurs on a subantarctic island (Ball and Hay 1977).

Distribution of species appears mostly related to the granulometry of the sediment. In some cases, strict ecological requirements are known. At Roscoff, an intertidal member of the '*Monocelis longiceps*' species complex was only found in sediments among patches of tube-dwelling terebellid polychaetes (personal observation). Similarly, *Polystyliphora filum* and *Coelogynopora faenofurca* are found in the burrows of *Arenicola marina* (Reise 1984; Sopott-Ehlers 1992). In all cases, no trophic relationship with the worms was obvious, and their co-occurrence appears rather related to local modifications of sediment texture. By contrast, most brackish water species are remarkably indifferent to the nature of the substrate. *Monocelis lineata*, one of the most abundant and widespread proseriates, has been found under stones, epibenthic on algae or clumps of mussels, or in sediments ranging from sand to mud (reference in Curini-Galletti and Mura 1998). In the Monocelididae, a correlation between karyotype and breadth of ecological niche was found: species with the basic karyotype occur in low-energy subtidal environments, hypothesized to be the ancestral habitat of the family. Invasions of both high-energy intertidal and stressed brackish habitats were linked with massive karyotype rearrangements, involving chromosome fissions and inversions (Curini-Galletti and Martens 1990).

The narrow ecological niche of many proseriates implies that, generally, the group shows low α-diversity, i.e., in any given locality, each sediment type usually harbours a limited number of species. However, diversity across habitats may be high. The limited data available on the biogeography of proseriates suggest that γ-diversity of the group may be particularly high. In fact, different biogeographical provinces harbour quite distinct proseriate faunas. In Europe, almost the same number of species has been reported from the North Atlantic and the Mediterranean (Figure 5.1). However, only a minor fraction of the species (six, i.e., 7.5% of the total Mediterranean species) is known to be common to the two areas. By comparison, among prosobranch gastropods, a group renowned as good biogeographical descriptors (Briggs 1974), shared species constitute about 18% of the Mediterranean species (Poppe and Goto 1991). Along eastern Australia, the vast majority of the known species appear restricted to one or the other of the marine biogeographical provinces (Solanderian, Peronian, Maugean), which follow from North to South (Curini-Galletti and Cannon 1995, 1996a,b, 1997; Curini-Galletti 1997, 1998; Faubel and Rohde 1998). Furthermore, the so-called 'overlap zone' in southern Queensland and northern New South Wales, where the Solanderian and Peronian provinces overlap, harbours a set of endemic species, in addition to species of the two provinces. This implies that members of the Proseriata may experience physical and climatic barriers to distribution to a finer degree than most macrofaunal taxa, resulting in 'small-scale' biogeography, and ample opportunities for allopatric speciation. Among the few experimental studies, research on the population structure of *Pseudomonocelis ophiocephala* revealed a steep genocline among populations on the two sides of the Otranto channel (between Italy and Albania), about 90 km wide (M. Curini-Galletti 1992, in preparation). The sedentary habit and interstitial way of life of most proseriates are concurrent causes of the poor dispersal power of the group. Furthermore, recent research has highlighted the problem of the presence of sibling species, which appear widespread in groups with few diagnosable characters, such as the Monocelididae with simplex copulatory bulb (M. Curini-Galletti 1993, in preparation). All things considered, the number of species described, which is about 400, is likely to represent only a minor fraction of the actual number of the species of the group. Numbers of species per area closely reflect the distribution of students of the group, with the highest numbers in Europe. Similarly, the very low number of species in the tropical Indo-Pacific probably only reflects lack of research (Figure 5.1).

Feeding ecology

Most members of the Proseriata are carnivorous (Watzin 1983; Martens and Schockaert 1986; Reise 1988). One study on a population of the monocelid *Pseudomonocelis ophiocephala*, with a mean annual density of 14 640 individuals per m^2, revealed an annual intake of 589 000 amphipods (Murina 1981). Gut contents of proseriates often include whole specimens of Copepoda and Nematoda, as well as polychaete setae, and sclerotized pieces of other flatworms (personal observation). Most species appear generalist with respect to prey species. On the other hand, a few taxa present a remarkable degree of specialization. Members of the Archimonocelididae and Nematoplanidae, which prey on cnidarians, store undischarged cnidae in complex cnidosacs, usually arranged dorsally in longitudinal row(s) (Karling 1966a). Dependence on cnidarians as a food source is so tight in the genus *Archimonocelis* that specimens do not survive in cultures on alternative food. At least one species of *Nematoplana* uses sponge microscleres in a similar way, thus exploiting a host food that is unique among Platyhelminthes (M. Curini-Galletti, in preparation). However, a word of caution is necessary: the simple analysis of gut content as a claim for predatory habit may be misleading, as the food might also have been scavenged. Monocelididae in cultures appear unable to prey on juveniles of the isopod *Sphaeroma* sp.pl., although they will willingly feed on freshly crushed specimens (personal observation). A unique case of scavenging has been documented for *Digenobothrium inerme*, only found on fish stranded ashore

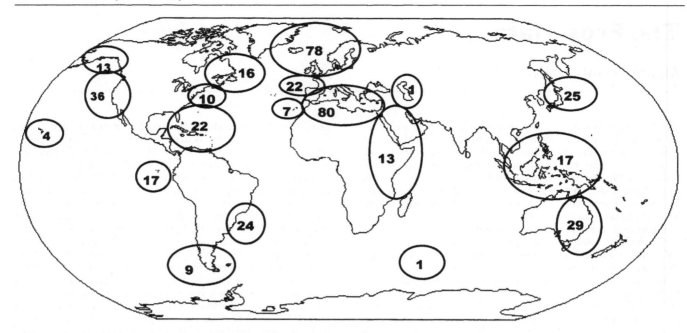

Figure 5.1 Number of species of Proseriata described per geographical area.

(Palombi 1926). Occasionally, predation by members of the Otoplanidae and Monocelididae on harpacticoid copepods and nematodes has been observed in Petri dishes, as have mass attacks on juvenile polychaetes (personal observation). Since escape of prey in nearly sediment-free Petri dishes may be enhanced, paucity of documentation of predation may be misleading. This may offer ground to the observation that predation by members of the Proseriata and other free-living platyhelminth taxa on temporary meiofauna (i.e., macrofaunal larvae) may eventually structure macrofaunal communities (Watzin 1983; Danovaro *et al.* 1995).

A few Proseriata Monocelididae symbiotic on crustaceans are known. They have mostly been found on gills, where they also lay cocoons (Fleming and Burt 1978a,b).

Predation on proseriates has often been observed in Petri dishes, during extraction. Most predators appear to be other flatworms. Convolutid acoels feed on monocelidids, and even on larger proseriates (Curini-Galletti 1998). A kalyptorhynch species has been seen preying on small otoplanids, while a very large Mediterranean graffillid rhabdocoel eagerly preys on *Monocelis lineata*. Furthermore, in the Mediterranean, a large nematode has been observed, taking off bits of tissues from passing monocelidids. A clear escape reaction was triggered in the monocelid when it casually touched this nematode. A similar reaction was not observed when monocelidids came into contact with much smaller bacteriophagous nematodes (personal observation).

Proseriata: the system

Monophyly and outgroup relationships

The Proseriata includes two suborders: Unguiphora and Lithophora. Differences between the two taxa are many and clear-cut (Table 5.1), and may raise the question of the monophyly of the group.

Table 5.1 'State of the art' of the system of the Proseriata, incorporating data and views of Sopott-Ehlers (1973, 1979, 1985, 1991), Cannon (1986), Martens and Schockaert (1988), and Curini-Galletti and Martens (1992). *: main hypothesized autapomorphies. See text for details.

Proseriata Meixner, 1938	
Unguiphora Sopott-Ehlers, 1985	Lithophora Steinböck, 1925
–fam. Nematoplanidae Meixner, 1938	–fam. Archimonocelididae Meixner, 1938
	–fam. Coelogynoporidae, Karling, 1966
	–fam. Monocelididae Hofsten, 1907
	–fam. Monotoplanidae Ax, 1958
	–fam. Otomesostomidae Bresslau, 1933
	–fam. Otoplanidae Hallez, 1892
• statocyst absent	* statocyst present
• pigment cup ocelli	* mantle cell of the rhabdomeric receptors pigment free
* 'multiple' ovaries	• one pair of compact ovaries
* claw-shaped stylet	• sclerotized copulatory structures never claw-shaped
* cocoons with up to nine openings	• cocoons with one opening

Sopott-Ehlers (1985) advocated a monophyletic origin for the Proseriata, based on the lack of lamellated rhabdites in all the approximately 20 species of Lithophora and Unguiphora studied. Ehlers (1985a) added two further characters as autapomorphic for the Proseriata: 1) protonephridial weirs with two rows of cytoplasmic rods, and with the filter region as a compound product of the terminal cell and of the first canal cell; and 2) cranial rootlets of epidermal cilia converging and terminating together at the cranial margins of epidermal cells, which often have a wedge-shaped process (see Bedini and Papi 1974, Figure 24). Autapomorphy of these characters may, however, be debatable.

Lamellated rhabdites are known to be absent in the Lecithoepitheliata, in many Rhabdocoela, including most Dalyellioida, and in the Neodermata (Jondelius and Thollesson 1993), so that the hypothesis of independent losses in Lithophora and Unguiphora should at least be considered.

A two-cell weir in protonephridia is found, besides the Proseriata, in some Macrostomorpha, in the fecampiid *Kronborgia isopodicola*, and in the Neodermata. However, details of morphology differ: in the lithophoran species studied (with the possible exception of the Archimonocelididae, see Ehlers and Sopott-Ehlers 1986) and in *Macrostomum* the terminal cell contributes the external ribs to the weir, whereas it contributes the internal ribs in the weir of Neodermata (Rohde *et al.* 1988a, 1995; Rohde 1990, 1991; but see Ehlers and Sopott-Ehlers 1987). On the other hand, the Unguiphora have been reported to present a single row of ribs (Ehlers and Sopott-Ehlers 1986), together with broad cytoplasmic processes arising from the terminal and the canal cell (Sopott-Ehlers 1985). However, morphological details have never been adequately described. The present state of knowledge thus does not allow us to consider protonephridial ultrastructure as a synapomorphy shared by Lithophora and Unguiphora.

Convergent epidermal ciliary rootlets have only been observed in three species of Otoplanidae (Bedini and Papi 1974), and the character is best considered at the moment as an autapomorphy for that family. A further character observed by Bedini and Papi (1974) and Ehlers and Ehlers (1977) in the Otoplanidae, i.e., the lack of vertical rootlet in epidermal cilia, may be phylogenetically informative, as it occurs only rarely in the Platyhelminthes (in larvae of Neodermata, in *Kronborgia isopodicola*, and in the umagillid *Cleistogamia*) (Rohde *et al.* 1988c; Watson *et al.* 1992b). However, the presence in Lithophora of a small, secondary vertical rootlet has also been reported (Sopott-Ehlers 1979; O. Carcupino and M. Curini-Galletti, in preparaion). No information is available

for the Unguiphora. Present data are therefore far too scanty and conflicting to permit phylogenetic inferences.

Polyphyly of the Proseriata has been championed by Brüggemann (1986b) on the basis of his observations on the formation of sclerotized structures. A modality including several successive steps (intracellular deposition of a framework of microfibrils; envelopment by electron-dense material; final embedding into a fibrous intracellular matrix) was considered plesiomorphic for the Rhabditophora. Such a pathway has been observed in the Lithophora (Ehlers and Ehlers 1980). In the Rhabdocoela, on the other hand, the process involves a synchronous secretion of electron-dense material onto both the inner and the outer membranes of the stylet-forming cell. In the Unguiphora, the distal tip of the stylet originates as in the rhabdocoels, while its basis originates from successive depositions on a rough-form with a tubular framework, as in the basic condition for the Rhabditophora (Brüggemann 1984). Interestingly, a similar intermediate process is found in *Ciliopharyngiella intermedia* ('Typhloplanoida'), where the intracellular rough-form eventually becomes fused to the smooth layers deposited on the membranes of the stylet-forming cells (Brüggemann 1986b). According to Brüggemann (1986b), the deposition on the membrane of the stylet-forming cells is a synapomorphy of the taxa *Ciliopharyngiella*, Unguiphora and Rhabdocoela. Within that group, Unguiphora and Rhabdocoela are considered as sister taxa, based on the reduction of the intracellular rough-form.

Littlewood *et al.* (2000), produced a family-level morphological matrix. The resulting most parsimonious tree had Proseriata paraphyletic, and placed Unguiphora as the sister group of a large assemblage that includes Lithophora (Figure 5.2A). However, a tree with Proseriata monophyletic was only one step longer than the most parsimonious solution (Figure 5.2B).

A larger species-based data set (Appendix 5.1; Figure 5.3B) had Proseriata monophyletic, when *Ciliopharyngiella intermedia* and one 'typhloplanoid' species were used as outgroup; the inclusion, as additional outgroups, of two species of Tricladida Maricola made Proseriata paraphyletic (tree not shown).

If assessment of monophyly of the Proseriata is at present controversial at the morphological level, the situation is no more clear-cut at the molecular level. Phylum-wide molecular systematic assessments failed to support the monophyly of the Proseriata (Littlewood *et al.* 1999a; Litvaitis and Rohde 1999). A denser sampling of the group recovered monophyly, albeit with poor bootstrap support (Littlewood *et al.*, 1999b, 2000). A data set, including 40 species, especially focused on the Proseriata, has been produced by Littlewood *et al.* (2000). The Maximum Parsimony (MP) and Minimum Evolution

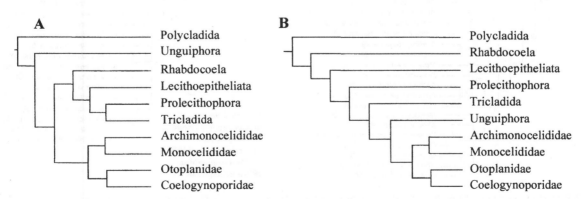

Figure 5.2 Morphology-based relationships of Proseriata (after Littlewood *et al.* 2000). **A)** Most parsimonious solution (length = 27). **B)** A 50% majority rule consensus tree (length = 28), with Proseriata constrained as monophyletic.

(ME) solutions differed remarkably. Seven equally MP trees revealed non-monophyly of the Proseriata, although poorly resolved, whereas the ME tree maintained monophyly, with, however, most clades poorly supported. Strict consensus of the MP solutions (Figure 5.3A) somewhat parallels the family-level morphological results (Figure 5.2A), with the Unguiphora as the outgroup to a cluster containing most Lithophora, Tricladida and Rhabdocoela. However, most Coelogynoporidae (a lithophoran taxon) and one Archimonocelididae (*Calviria solaris*, but see below), cluster with the Unguiphora, although the clade is poorly supported.

Present data, both at the morphological and at the molecular level, appear therefore inconclusive to fully support or reject the monophyly of the Proseriata.

Not surprisingly, assessment of outgroup relationships of Proseriata is similarly tainted with uncertainties. Morphological resemblances between members of the Proseriata and the Tricladida have long been recognized, and the two taxa have in fact been placed together in the order Seriata Bresslau 1933. According to Sopott-Ehlers (1985), the Seriata are a monophyletic taxon, based on two characters: 1) the presence of a tubiform 'pharynx plicatus'; and 2) the presence of follicular gonads (especially vitellaria).

Both characters are not exclusive to the Seriata: follicular gonads are found in representatives of Prolecithophora, Rhabdocoela and Neodermata (Rieger *et al.* 1991), and do not appear particularly phylogenetically informative. A tubiform 'pharynx plicatus' is found in some Polycladida, where, however, it is invariably anteriorly oriented (Prudhoe 1985).

Ultrastructural studies to ascertain the homology of the pharynx type in members of the Tricladida and the Proseriata are lacking. However, light microscopy studies have revealed a complex glandular pattern in the Tricladida (see for example Sluys 1989b), largely absent in the Proseriata. Furthermore, at least the arrangement of muscle layers beneath the outer epithelium of the pharynx appears to be reversed in the Tricladida in comparison with the Proseriata (cf. Kenk 1930 and, for example, Martens and Curini-Galletti 1993).

On morphological grounds, the monophyly of the Seriata has been challenged by Rohde (1994a) who, on the basis of the ultrastructure of protonephridia (but see above), suggested that Proseriata and Neodermata are sister groups.

Molecular systematics has so far failed to support the monophyly of the Seriata (Rohde *et al.* 1995; Carranza *et al.* 1997; Littlewood *et al.* 1999a; Litvaitis and Rohde 1999). However, there are no clear indications to any specific taxon as sister group of the Proseriata. Rather, there are indications that the Proseriata is the sister group to a large assemblage that includes most neoophorans (Rohde 1996; Littlewood *et al.* 1999b, 2000). Bootstrap support, however, appears largely poor, and further studies are necessary before it can be concluded that the Proseriata actually represents an early offshot of the Neoophora. In particular, the set of species sequenced to the present time does not appear to reflect the diversity of the Rhabdocoela, especially of the 'Typhloplanoida', including the problematic family Ciliopharyngiellidae, whose synapomorphies with the Unguiphora + Rhabdocoela have been mentioned above (Brüggemann 1986b). A recent ultrastructural

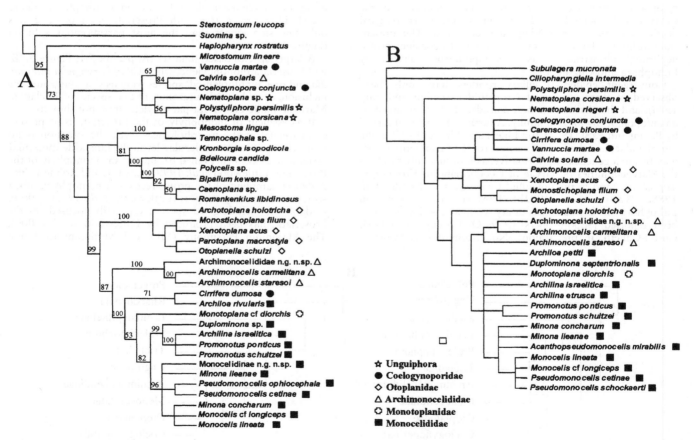

Figure 5.3 Proseriata: in-group relationships. **A)** Molecule-based (after Littlewood *et al.* 2000); Maximum parsimony, strict consensus of seven trees. Bootstrap >50% values are indicated at nodes. **B)** Morphology-based (from the morphological matrix in Appendix 5.1). Strict consensus of 56 equally parsimonious trees (length = 45).

study of the female gonads of *C. intermedia* failed to reveal any unequivocal synapomorphy either with the Proseriata or with the Rhabdocoela + Prolecithophora (Sopott-Ehlers 1997a), and the position of the taxon, superficially very proseriate-like (Ax 1952) is uncertain at present.

Whilst the somewhat unclear situation that currently hinders assessment of both monophyly and of outgroup relationships may be the result of inadequate morphological studies and molecular samplings, the possibility should be mentioned that it may actually reflect a real condition. An early branching both of the Proseriata line (had it ever emerged), into the two taxa Unguiphora and Lithophora, and of the main groups of the Neoophora as a whole, may effectively jeopardize attempts at phylogenetic reconstructions.

In-group relationships

At present, at least seven families are recognized within the Proseriata (Table 5.1), though there have been suggestions that this number may inadequately encompass the diversity of the group (Ax 1994a). Relationships between families on morphological grounds are controversial. Two hypotheses have been formulated:

1. Unguiphora and Lithophora are sister groups. Both are characterized by a number of autapomorphies, the evolution of a statocyst being the main autapomorphy of the latter (Sopott-Ehlers 1985) (Figure 5.4A).

2. Presence of a statocyst is the plesiomorphic condition for the Proseriata, subsequently lost in the Unguiphora. The family Monocelididae is the sister-group to the rest of the Proseriata (the 'Paramonocelida'), a monophyletic group of families with brain capsules and polyploidy-derived karyotype as main synapomorphies. Among the Paramonocelida, two sister groups of sister taxa are recognized: Archimonocelididae + Unguiphora (synapomorphy: morphology and position of epidermal collar receptors), and Otoplanidae + Coelogynoporidae (synapomorphy: presence of a common genital opening) (Martens and Schockaert 1988) (Figure 5.4B).

Littlewood *et al.* (2000) developed a family-level matrix largely based on the characters used by Sopott-Ehlers (1985, and subsequent papers) and Martens and Schockaert (1988). Both the most parsimonious and the monophyletic (one step longer) solutions (Figure 5.2) confirmed the position of the Unguiphora as external to the monophyletic Lithophora, as proposed by Sopott-Ehlers (1985). Among the Lithophora, two sister clades of sister taxa were recognized: Monocelididae + Archimonocelididae, and Otoplanidae + Coelogynoporidae.

Littlewood *et al.* (2000) recently addressed the question of in-group relationships of the Proseriata on a molecular and morphological basis. Both MP and ME trees placed a clade containing Unguiphora + Coelogynoporidae external to the monophyletic group, including all the other lithophorans. Support in the ME tree was very poor throughout. In the strict consensus of MP trees (Figure 5.3A) the clade 'Lithophora excl. Coelogynoporidae' was strongly supported, as were the 'traditional' families Otoplanidae and Archimonocelididae, and a clade including Monocelididae + *Monotoplana* cf. *diorchis* (fam. Monotoplanidae, monotypical) + *Cirrifera dumosa* (Coelogynoporidae) (see Littlewood *et al.* 2000, and below, for the justification of the removal of *Calviria solaris* from the Archimonocelididae, the position of *Monotoplana*, and the anomalous placement of *Cirrifera*). Furthermore, the clade Archimonocelididae + (Monocelididae + *Monotoplana* cf. *diorchis* + *Cirrifera dumosa*) was rather well supported. A major discrepancy between the MP and the family-based trees (Figure 5.2) concerned the removal of the Coelogynoporidae from the rest of the Lithophora. However, the clade Unguiphora + Coelogynoporidae was poorly supported, and a denser sampling is clearly needed in order to assess the position of the Coelogynoporidae. There are, however, strong indications that the clade Coelogynoporidae + Otoplanidae, suggested also by Martens and Schockaert (1988), and by earlier researchers (Ax 1956; Karling 1966b), and the clade Archimonocelididae + Unguiphora (Martens and Schockaert 1988), are not supported by molecular data.

A morphological matrix based on taxa used by Littlewood *et al.* (2000), or congeneric species, in cases where morphological data were missing, has been produced (see Appendix 5.1). Since information from the submicroscopic level is lacking in most cases, emphasis has been placed on information from the light microscopy level. The strict consensus of 56 equally parsimonious trees (length = 45) is shown in Figure 5.3B. Unguiphora clearly emerges as the sister taxon of the monophyletic Lithophora. Three sister groups can be recognized in the Lithophora:

- Coelogynoporidae + *Calviria solaris* (Archimonocelididae);

- Otoplanidae (excl. *Archotoplana holotricha*); and

- a clade including *Archotoplana holothricha* + Archimonocelididae (excl. *Calviria*) + Monocelididae (incl. *Monotoplana diorchis*).

The major discrepancy between the molecular and species-based trees concerns the placement of the Coelogynoporidae, which appears to be the sister group of the Unguiphora in the molecular tree, whereas it clearly belongs in the Lithophora in the morphological tree.

A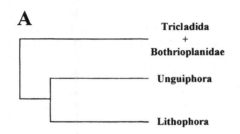

B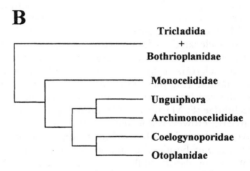

Figure 5.4 Proposed hypotheses of in-group relationships of Proseriata. **A)** From Sopott-Ehlers (1985). **B)** From Martens and Schockaert (1988).

A minor difference concerns the monophyly of the Otoplanidae, which is strongly supported in the molecular tree, but not in the morphological tree. This results from the placement of *Archotoplana holotricha*, which does not cluster with the rest of the Otoplanidae. *Archotoplana* lacks some of the obvious apomorphies of the Otoplanidae (ciliated creeping sole, sensory pits with thick sensory bristles) (Ax 1956). However, it has never been studied ultrastructurally, at which level the Otoplanidae show several autapomorphies (Bedini and Papi 1974; B. Ehlers 1977; Sopott-Ehlers 1993). At the molecular level, it clearly clusters with the rest of the Otoplanidae.

Otherwise, the two trees share numerous similarities, which yield the following phylogenetic and systematic information:

1. Both trees place *Calviria solaris* in a clade including the Coelogynoporidae. At present, the genus belongs in the Archimonocelididae, in the subfam. Calviriinae, as sister group to *Asilomaria ampullata*, formerly attributed to the Monocelididae (Martens and Schockaert 1988; Martens and Curini-Galletti 1993; Karling 1966b). However, it is worth mentioning that *A. ampullata* was placed in the Monocelididae '... only for practical reasons (separate female pore behind the male pore). Anatomically, *Asilomaria* resembles the coelogynoporids in most respects' (Karling 1966b: 502). Both *Calviria* and *Asilomaria* share with most Coelogynoporidae the separate openings of the oviducts in the female gonopore, not merging into a common female duct. Besides these taxa, the feature is only found in a few otoplanids (Ax 1956), where, however, it appears to be a derived condition. There are two interesting implications of the inclusion of the Calviriinae into the Coelogynoporidae: (i) The phylogenetic weight of the monogonoporid condition is overrated (cf. Martens and Schockaert 1988); rather, both trees show this condition to have been independently acquired in the Otoplanidae and in the Coelogynoporidae excl. Calviriinae. (ii) The character 'presence/absence of a (long) female duct' has probably more phylogenetic weight than hitherto assumed (cf. Littlewood *et al.* 2000).

2. There are strong indications that a monotypic family Monotoplanidae should be rejected, as both trees nest *Monotoplana* (cf.) *diorchis* well within the Monocelididae. The numerous autapomorphic features of the taxon (globular shape, two testes, multiple copulatory organs) (Westblad 1952; Ax 1958) have outweighed the synapomorphies shared with the Monocelididae (ciliation, lack of brain capsule, structure of the copulatory bulb) (see also Littlewood *et al.* 2000).

3. A monophyletic clade Monocelididae + Archimonocelididae is supported throughout. Formerly, the Archimonocelididae was variously considered as a subfamily of the Monocelididae (Meixner 1938), as belonging within the Monocelididae Promonotinae (Marcus 1949), or within the Monocelididae Monocelidinae (Karling 1978). The taxon was elevated to family rank by Martens and Schockaert (1988), based on the different modality of formation of the sclerotized structures: intracellular in the Archimonocelididae (the plesiomorphic condition for the Proseriata); derived from a basal lamina differentiation in the Monocelididae (E. Martens 1984). Both taxa appear to be monophyletic. However, one may question the systematic rank given to them. The Monocelididae appears to be a highly derived group. Besides the autamoporphic features of the structure of copulatory bulb and origin of sclerotized structures, other characteristics of the Monocelididae

appear to be linked to their reduced size and interstitial way of life. Among these are the lack of brain capsule and cephalic 'chordoid' gut, present in members of the large Archimonocelididae, which burrow 'head-on' into the substrate. This process of miniaturization might have resulted also in smaller karyotypes, with fewer chromosomes (see Birstein 1991 versus Martens *et al.* 1989), and in a progenetic pathway leading to the loss of the prostatic organ in many genera (Litvaitis *et al.* 1996). One of the implications of the 'crown group' position of the Monocelididae Monocelidinae (a monophyletic taxon in both the molecular and species-based trees) is that the familiar genus *Monocelis*, often taken to exemplify the whole Proseriata, is actually one of the most derived members of the group.

The phylogenetic position of three further taxa is uncertain at present:

- *Otomesostoma auditivum*. A freshwater lithophoran proseriate, placed in the monotypic family Otomesostomidae (Bresslau 1928–33);

- *Digenobothrium inerme*. Only known from the original description of Palombi (1926), it appears to have a 'divisa' copulatory bulb. Considered by Karling (1966b) akin to his *Asilomaria*, but ignored by all subsequent authors.

- *Japanoplana insolita*. A lithophoran, assumed to represent a yet undescribed body-plan for the Proseriata (Ax 1994a). However, its relationships with the Monocelididae should be carefully evaluated (Sopott-Ehlers, personal communication).

None of the above species has been studied ultrastructurally, and for at least the first two, records of morphological studies are very poor.

These cases further emphasize the amount of descriptive morphological work that still needs to be done in the Proseriata. Hopefully, the wealth of new species currently being described from extra-European areas, presenting yet unsuspected combinations of characters, may offer new insights on the evolution of the group, and assist in the construction of a more complete and robust character matrix.

Conclusions

The Proseriata is a species-rich group, widespread in marine habitats, and exhibiting a wide trophic spectrum. While it has long been considered as a monophyletic group, and as a suborder of the Seriata, at present the status and phylogenetic relationships of its members are unclear. In particular:

- Present knowledge, both at the morphological and at the molecular level, appears inconclusive to fully support or to reject the monophyly of the Proseriata. This probably reflects as yet inadequate samplings. However, the possibility of an early splitting of the Unguiphora and the Lithophora should be considered.

- Whilst there are strong suggestions that a taxon Seriata, including Tricladida + Bothrioplanidae + Proseriata, should be rejected, no clear sister-group relationships with any particular neoophoran taxa have emerged so far. However, rejection of the Seriata may have interesting phylogenetic implications, because the morphological characters assumed to support that taxon may be reconsidered as possible apomorphies for the Proseriata. Whilst 'gonads follicular', a widespread character (see above), may be poorly phylogenetically informative, and likely to have evolved independently in taxa with a filiform body-plan, pharynx ultrastructure among taxa of Proseriata should be considered for further research.

- Discrepancy between molecular and morphological analyses on in-group relationships basically concerns the monophyly of the Lithophora, which is robust from the morphological point of view, but not fully supported by molecular analyses – due to one lithophoran group, the Coelogynoporidae, which clusters with the Unguiphora in the molecular tree. A few, particularly problematic taxa exist (including a monotypical

family, the Otomesostomidae). Aspects of morphology are poorly known, and no sequences have yet been obtained; detailed studies of these problematic taxa should be undertaken.

The present blurred scenario of the relationships of the Proseriata may be considered to reflect an 'intermediate state' of knowledge, where some of the certainties of the past have been wiped out by the flow of new molecular information, and a new system has yet to emerge in full detail. Clearly, a denser sampling of taxa and genes for molecular data, and inclusion of additional taxa and characters in morphology-based data sets, as well as deeper insights on character homology, are necessary to resolve the present contradictions.

Appendix 5.1 Morphological character definitions used to reconstruct phylogenetic relationships between Proseriata. See text for further explanation and sources. Except in the case of multistate characters, presence of character described coded as 1; absence coded as 0. Dubious or unknown characters states are coded as '?'. In all analyses, the program used was PAUP 4; tree construction was by heuristic search. All characters were unweighted and unordered.

1. Ciliation. 0) whole body; 1) whole body except tail area; 2) limited to a ventral sole.
2. Statocyst.
3. Lamellated rhabdites.
4. Cnidosacs.
5. Paracnida
6. Sensory ciliated pits, with tactile bristles.
7. Extracellular matrix brain capsule.
8. Pigment cup ocelli.
9. Pharynx tubiformis, of the plicatus type.
10. Follicular testes.
11. Ovoid, muscular copulatory bulb, containing the vesicula seminalis.
12. Vesicula granulorum.
13. Copulatory structure. 0) penis papilla; 1) cirrus; 2) copulatory stylet.
14. Sclerotized structures. 0) absent; 1) intracellular origin; 2) derivation from basal lamina.
15. Prostatoid organ.
16. Ovaria subdivided into smaller follicles, each containing one oocyte.
17. Long common female duct, originated from the prepenial fusion of the two oviducts.
18. Vagina. 0) absent; 1) externa; 2) interna.
19. Common (male + female) genital opening.
20. Bursa. 0) absent; 1) postpenial, derived from, or in close proximity to, the genital atrium or the female antrum; 2) prepenial, interpolated in, or derived from, the common female duct.

Species/character	1	2	3	4	5	6	7	8	9	10	11	12	13	14	15	16	17	18	19	20
Subulagera mucronata	0	0	1	0	0	0	1	0	0	0	0	1	2	1	0	0	0	0	1	1
Ciliopharyngiella intermedia	0	0	0	0	0	0	1	0	0	0	0	1	2	1	0	0	0	0	1	1
Minona concharum	1	1	0	0	0	0	0	0	1	1	1	0	0	2	1	0	1	1	0	2
Minona ileanae	1	1	0	0	0	0	0	0	1	1	1	0	0	2	1	0	1	0	0	2
Pseudomonocelis schockaerti	1	1	0	0	0	0	0	0	1	1	1	0	0	0	0	0	1	1	0	2
Pseudomonocelis cetinae	1	1	0	0	0	0	0	0	1	1	1	0	0	0	0	0	1	1	0	2
Acanthopseudomonocelis mirabilis	1	1	0	0	0	0	0	0	1	1	1	0	0	2	0	0	1	2	0	2
Monocelis lineata	1	1	0	0	0	0	0	0	1	1	1	0	0	0	0	0	1	1	0	2
Monocelis cf longiceps	1	1	0	0	0	0	0	0	1	1	1	0	0	0	0	0	1	1	0	2
Archilina etrusca	1	1	0	0	0	0	0	0	1	1	1	1	1	2	0	0	1	1	0	2
Archilina israelitica	1	1	0	0	0	0	0	0	1	1	1	1	1	2	0	0	1	1	0	2
Promonotus ponticus	1	1	0	0	0	0	0	0	1	1	1	1	1	2	0	0	1	0	0	0
Promonotus schulzei	1	1	0	0	0	0	0	0	1	1	1	1	1	2	0	0	1	0	0	0
Duplominona septentrionalis	1	1	0	0	0	0	0	0	1	1	1	1	1	2	1	0	1	1	0	2
Archiloa petiti	1	1	0	0	0	0	0	0	1	1	1	1	1	2	0	0	1	2	0	2
Monotoplana diorchis	1	1	0	0	0	0	0	0	1	0	1	1	1	2	0	0	1	0	0	2
Archotoplana holotricha	0	1	0	0	0	0	1	0	1	1	0	1	2	1	0	0	1	0	1	1

Continued overleaf

Species/character	1	2	3	4	5	6	7	8	9	10	11	12	13	14	15	16	17	18	19	20
Monostichoplana filum	2	1	0	0	0	1	1	0	1	1	0	1	2	1	0	0	1	0	1	0
Xenotoplana acus	2	1	0	0	0	1	1	0	1	1	0	1	2	1	0	0	0	1	1	0
Otoplanella schulzi	2	1	0	0	0	1	1	0	1	1	0	1	2	1	0	0	1	0	1	0
Parotoplana macrostyla	2	1	0	0	0	1	1	0	1	1	0	1	2	1	0	0	0	0	1	1
Coelogynopora conjuncta	0	1	0	0	1	0	1	0	1	1	0	0	2	1	0	0	0	0	1	1
Carenscoilia biforamen	0	1	0	0	1	0	1	0	1	1	0	0	2	1	0	0	0	0	1	0
Cirrifera dumosa	0	1	0	0	1	0	1	0	1	1	0	1	1	1	0	0	0	0	1	0
Vannuccia martae	0	1	0	0	1	0	1	0	1	1	1	0	1	0	0	0	0	0	1	1
Archimonocelis staresoi	0	1	0	1	0	0	1	0	1	1	0	1	2	1	?	0	1	0	0	2
Archimonocelis carmelitana	0	1	0	1	0	0	1	0	1	1	0	1	2	1	0	0	1	1	0	2
Archimonocelididae n.gen. n.sp.	0	1	0	1	0	0	1	0	1	1	0	1	2	1	0	0	1	1	0	2
Calviria solaris	0	1	0	0	0	0	1	0	1	1	0	0	0	1	?	0	0	0	0	1
Nematoplana riegeri	0	0	0	0	0	0	1	1	1	1	0	1	2	1	0	1	0	0	0	0
Nematoplana corsicana	0	0	0	0	0	0	1	0	1	1	0	1	2	1	0	1	0	0	0	0
Polystyliphora persimilis	0	0	0	0	0	0	1	0	1	1	0	1	2	1	1	1	0	0	0	0

Chapter 6

Molecular taxonomy and phylogeny of the Tricladida

Jaume Baguñà, Salvador Carranza, Jordi Paps, Iñaki Ruiz-Trillo and Marta Riutort

Within the free-living Platyhelminthes, the triclads or planarians are the best known group, partly as a result of being suitable for classroom studies but, largely, because they have been the subject of intensive research concerning the cellular bases of regeneration and pattern formation (for general reviews see Baguñà *et al.* 1994, and Baguñà 1998) and, most recently gene expression (Bayascas *et al.* 1997; Orii *et al.* 1999). The Tricladida Lang, 1884, which is best considered a suborder (Ehlers 1985a), forms, together with the suborder Proseriata Meixner, 1938, the Order Seriata. Autapomorphies for the Seriata are their backwards-directed tubiform and plicate pharynx and the division of testes and vitellaria into serially arranged follicles. Proseriata do not have obvious autapomorphies apart from their lack of lamellate rhabdites (Sopott-Ehlers 1985), whereas Tricladida are characterized by its three-branched intestine and its highly modified embryonic development with the presence of a transitory embryonic pharynx. A family of proseriates, the Bothrioplanidae Hofsten, 1907, was proposed by Sopott-Ehlers (1985) as the actual sister group of the Tricladida forming a taxon N.N. (Bothrioplanida + Tricladida) characterized by the presumed lack of epidermal collar-receptors, a tricladoid intestine, and a crossing-over of muscle layers at the root and the tip of the pharynx. A general phylogenetic scheme of Seriata is summarized in Figure 6.1.

The monophyletic status of the Seriata has been questioned both on morphological and molecular grounds. The main morphological argument against the presumed autapomorphies of Seriata is that, without comparative ultrastructural studies, pharynx types are not useful for phylogenetic studies, whereas the serial arrangement of vitellaria is so general that homology is impossible to test (Sluys 1989a). Moreover, data on the ultrastructure of the excretory system indicates a basal location for Proseriata and, therefore, the paraphyly of Seriata (Rohde 1990). This paraphyly was reinforced from molecular data (18S rDNA sequences; Carranza *et al.* 1997). Neighbour-joining (NJ) distance, maximum-likelihood (ML) and maximum parsimony (MP) trees showed Tricladida clustering either with prolecithophorans or rhabdocoels, whereas proseriates appeared as basal neoophorans close to lecithoepitheliates or Neodermata. A more recent study, combining morphological and molecular characters (Littlewood *et al.* 1999a), confirmed the results of Carranza *et al.* (1997) and suggested a clade made by *Urastoma* Dörler, 1900, Fecampiida (*Kronborgia*, Christensen and Kannerworff, 1964) and *Ichthyophaga* Syriamiatnikova, 1949, as the actual sister group of the Tricladida. The morphological synapomorphies shared by the Bothrioplanida and the Tricladida proposed by Sopott-Ehlers (1985a), have been critically examined by Sluys (1989a). Lack of collar-receptor is a secondary absence and, in the lack of further information, has to be considered a weak character. Instead, the tripartite intestinal system and the muscle crossing-over at the pharynx, the latter not uncommon among other platyhelminths (see references in Sluys 1989a), were considered good synapomorphies.

Within the Tricladida, three infraorders usually have been recognized: Maricola (marine planarians), Paludicola (freshwater planarians) and Terricola (land planarians) (reviewed in Sluys 1989a), to which a new one, the Cavernicola was further added (Sluys 1990). Relationships of these infraorders have been the subject of several morphological-based analyses. Sluys' (1989a) analysis is by far the most detailed and valuable. It supported the monophyly of the Tricladida, the Terricola, the Maricola and the Paludicola, and the existence of a new clade, the Terricola-Paludicola clade. Relationships within the infraorders have only been considered in detail within the Paludicola, in which three families are currently recognized: the Dugesiidae Ball, 1974, the Planariidae Stimpson, 1857, and the Dendrocoelidae Hallez, 1892. The Planariidae and Dendrocoelidae are considered derived groups and form the sister-group of the more primitive Dugesiidae. A radically different view of infraorder relationships of Tricladida emerged recently from molecular studies based on sequence data of the 18S ribosomal genes and the presence of an 18S gene duplication shared by the Terricola and the family Dugesiidae of the Paludicola (Carranza *et al.* 1998a,b). The resulting phylogenetic trees strongly indicated that the Paludicola is paraphyletic since the Terricola and one paludicolan family, the Dugesiidae, share a more recent common ancestor than the dugesiids with the other paludicolans (dendrocoelids and planariids). Therefore, it was suggested that the infraorders Terricola and Paludicola are redundant and should be replaced by a new taxon, the Continenticola (Carranza *et al.* 1998a). A comparison between Sluys' (1989a) and Carranza *et al.*'s (1998a) phylogenetic proposals is depicted in Figure 6.2.

We report here partial sequences of the Cytochrome Oxidase I (COI) gene from 21 taxa of the Terricola and the Paludicola, and new complete 18S rDNA sequences from five species of turbellarians. Published 18S rDNA sequences from the large data set of 18S rDNA sequences available from other Platyhelminthes, namely from the Proseriata, the Prolecithophora and the Rhabdocoela, are also included in the phylogenetic analysis. The aims of this paper are to: 1) Reassess the taxonomic status and the phylogenetic position within the Platyhelminthes of the Proseriata, the Bothrioplanida and the Tricladida; 2) Further test the new triclad phylogeny drawn from molecular data (Carranza *et al.* 1998a,b), in particular the paraphyletic status of the Paludicola and the monophyly of Terricola + Dugesiidae; and 3) Test the monophyletic or paraphyletic status of the Terricola and Dugesiidae and its internal phylogeny. Congruence and conflicts between the morphological and molecular phylogenies of the Tricladida are highlighted.

Materials and methods

The current taxonomic classification of the species used in this study is shown in Table 6.1.

Sequencing of the 18S molecule

High molecular weight DNA was purified according to a modification (Garcia-Fernàndez *et al.* 1993) of the guanidine isothiocyanate method initially described for RNA (Chirgwin *et al.* 1979) from live or ethanol-fixed specimens. The entire length of the 18S rDNA molecule was PCR-amplified applying specific primers and conditions described earlier (Carranza *et al.* 1996, 1998a; Littlewood and Smith 1995). Amplification

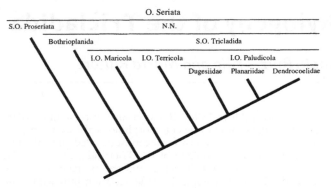

Figure 6.1 Internal phylogenetic relationships of the Seriata after Ehlers (1985), Sopott-Ehlers (1985) and Sluys (1989a), slightly modified. N.N.: unnamed taxon, according to Sopott-Ehlers (1985), formed by the Bothrioplanida and the Tricladida. For the sake of clarity, the Infraorder Cavernicola (Sluys 1990) of the Tricladida is not included.

products were sequenced directly. Sequencing of the clones and the PCR products was performed using an automated sequencer ABI Prism 377, following manufacturer's protocols.

Sequencing of the Cytochrome Oxidase I (COI) molecule

High molecular weight DNA was purified as described for the 18S rDNA gene (see above) from 21 species of Terricola and Paludicola (see Table 6.1). A fragment of approximately 450 nucleotides close to the centre of the cytochrome *c* oxidase subunit I mitochondrial gene was amplified. The primers used (pr-a2 and pr-b2) and the conditions of the PCR reaction were as described in Bessho *et al.* (1992). The PCR products were purified with GenecleanR II kit (BIO 101 Inc.) and directly sequenced using the same primers as for amplification. Cycle sequencing using Dye-labelled terminators (Prism™ Ready Reaction DyeDeoxy™ Terminator Cycle Sequencing Kit) was performed in a DNA Thermal Cycler Perkin-Elmer 480 according to the manufacturer's instructions and run on an automated sequencer ABI 377.

Sequence alignment

18S rDNA sequence data were aligned by hand with the help of a computer editor (GDE 2.2; Smith S.W. *et al.* 1994). Alignment gaps were inserted to account for putative length differences between sequences. A secondary structure model (Gutell *et al.* 1985) was used to optimize alignment of homologous nucleotide positions. Those positions that could not be unambiguously aligned were subsequently excluded resulting in a total of 1483 positions that could be used in the phylogenetic analyses.

CO I sequences were aligned by eye based on the protein sequences. There is a variable region in the middle of the fragment in which some species have one or two extra amino acids; this region has been excluded from the alignment. Given the variability of the third position of the codons it also was excluded from the phylogenetic analyses, resulting in a total of 232 positions that could be used in the subsequent studies.

The full sequence alignments used in these analyses have been deposited with EMBL under accessions ds41997 (18S) and ds42057 (CO I) and are available via anonymous FTP from FTP.EBI.AC.UK under directory pub/databases/embl/align

Phylogenetic analysis

Data sets were analysed using the programs in PHYLIP package v. 3.52 (Felsenstein 1993). A distance matrix was generated from the aligned sequences using the program DNADIST and corrected with the two-parameter method of Kimura (1980). The distances were then converted to phylogenetic trees using the NJ method of Saitou and Nei (1987) provided by the NEIGHBOR program. Bootstrap resampling (Felsenstein 1985) was accomplished with the use of the programs SEQBOOT (1000 replicates)

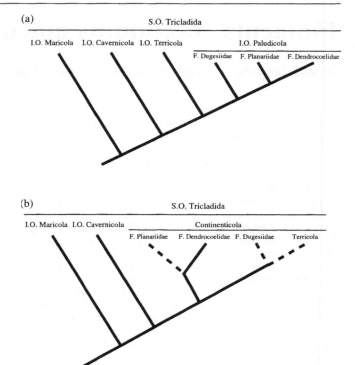

Figure 6.2 a) Phylogenetic relationships of the Tricladida based on morphological characters according to Sluys (1989a); b) Phylogenetic hypothesis for the Tricladida based on molecular characters (18S rDNA sequences and an 18S rDNA duplication event) (Carranza *et al.* 1998a). Dashed lines indicate groups that are not well-supported in the molecular phylogenetic analysis.

and CONSENSE. For the estimation of the maximum-likelihood (ML) trees we used the FastDNAml program v. 1.1.1a (with global rearrangements and reordering of species) (Felsenstein 1981; Olsen *et al.* 1994).

To root the Platyhelminthes internal tree, representatives of lophotrochozoans and ecdysozoans (*sensu* Aguinaldo *et al.* 1997) and from deuterostomates were chosen. For the Terricola + Dugesiidae trees, representatives of families Dendrocoelidae and Planariidae were chosen because they are known to be its sister-group (Carranza *et al.* 1998a).

Results

The phylogenetic position of the Tricladida within the Platyhelminthes

Both NJ and ML methods of phylogenetic reconstruction gave similar results as regards the position of the Tricladida and the Proseriata and the paraphyly of Seriata, only the NJ tree being shown (Figure 6.3). First of all, the Seriata appears as a polyphyletic assemblage, because Proseriata, Bothrioplanida and Tricladida never cluster together. Second, Bothrioplanida never clusters with proseriates and triclads, appearing with poor bootstrap support (NJ tree) as the sister-group of a vast assemblage of neoophorans and in the ML tree as the sister-group of the Neodermata. Third, Proseriata and Tricladida, represented by four or more sequences fall into clear recognizable monophyletic groups, with very high bootstrap support, regardless of the phylogenetic method used. Fourth, although their positions are variable, proseriates fall in both types of trees as a basal clade of neoophorans. And

Table 6.1 Species used in the analysis with classification. Classification of turbellarians following Cannon (1986). 18S sequence (+) and GenBank accession numbers, and COI sequences (*) and GenBank accession numbers (in brackets), are indicated. New 18S rDNA sequences reported in this paper are marked #. Note: The terricolan genus *Artioposthia* is now known as *Arthurdendyus* (Jones and Gerard 1999).

Classification	18S rDNA	COI	Accession number	Classification	18S rDNA	COI	Accession number
Phylum Chordata				**Of uncertain status (see text)**			
Xenopus laevis	+		X04025	Family Urastomidae			
Phylum Echinodermata				*Urastoma cyprinae*	+		U70085
Asterias amurensis	+		D14358	*Ichthyophaga* sp.	+		AJ012512
Phylum Arthropoda				Order Rhabdocoela			
Odiellus troguloides	+		X81441	Dalyelliida			
Scolopendra cingulata	+		U29493	Family Dalyelliidae			
Phylum Annelida				*Microdalyellia rossi*	+		AJ012515
Eisenia foetida	+		X79872	Family Graffillidae			
Lanice conchilega	+		X79873	*Graffilla buccinicola*	+		AJ012521
Phylum Mollusca				Family Pterascolidae			
Acanthopleura japonica	+		X70210	*Pterastericola australis*	+		AJ012518
Nerita albicilla	+		X91971	Family Fecampiidae			
Phylum Platyhelminthes				*Kronborgia isopodicola*	+		AJ012513
Order Catenulida				Temnocephalida			
Family Stenostomidae				Family Temnocephalidae			
Stenostomum leucops	+		U70085	*Temnocephala* sp. #	+		AF051332
Family Catenulidae				*Temnocephala* sp.	+		AJ012520
Suomina sp.	+		L41129	Typhloplanida			
Order Macrostomida				Family Trigonostomidae			
Family Dolichomacrostomidae				*Mariplanella frisia*	+		AJ012514
Paramalostomum fusculum	+		AJ012531	Family Typhloplanidae			
Family Macrostomidae				*Bothromesostoma personatum*	+		M58347
Macrostomum tuba	+		U70080	*Mesocastrada* sp.	+		U70081
Macrostomum hystricinum #	+		AF051329	Kalyptorhynchia			
Family Microstomidae				Family Polycystidae			
Microstomum lineare	+		U70082	*Gyratrix hermaphroditus*	+		AJ012510
Order Polycladida				Schizorhynchia			
Acotylea				Family Schizorhynchidae			
Family Leptoplanidae				*Diascorhynchus rubrus*	+		AJ012508
Notoplana australis	+		AJ228786	Family Karkinorhynchidae			
Notoplana koreana	+		D85097	*Cheliplana* cf. *orthocirra*	+		AJ012507
Family Planoceridae				Order Seriata			
Planocera multitentaculata	+		D17562	Suborder Proseriata			
Family Discocelidae				Family Bothrioplanidae			
Discocelis tigrina	+		U70078	*Bothrioplana semperi* #	+		AF051333
Cotylea				Family Monocelidae			
Family Pseudocerotidae				*Monocelis lineata*	+		U45961
Thysanozoon brochii	+		D85096	*Archiloa rivularis*	+		U70077
Pseudoceros tritriatus	+		AJ228794	Family Otoplanidae			
Order Haplopharyngida				*Otoplana* sp.	+		D85090
Family Haplopharyngidae				*Parotoplana renatae*	+		AJ012517
Haplopharynx rostratus	+		AJ012511	Suborder Tricladida			
Order Lecithoepitheliata				Infraorder Maricola			
Family Prorhynchidae				Family Bdellouridae			
Geocentrophora sp.	+		U70079	*Bdelloura candida*	+		Z99947
Geocentrophora baltica	+		AF065417	Family Procerodidae			
Geocentrophora spyrocephala	+		D85089	*Ectoplana limuli*	+		D85088
Geocentrophora wagini	+		AJ012509	*Procerodes littoralis*	+		Z99950
Order Prolecithophora				Family Uteriporidae			
Combinata				*Uteriporus* sp.	+		AF013148
Proporata				Infraorder Terricola			
Family Pseudostomidae				Family Geoplanidae			
Pseudostomum gracilis	+		AF065423	Subfamily Caenoplaninae			
Reisingeria hexaoculata	+		AF065426	*Artioposthia* sp.		*	(AF178325)
Opisthoporata				*Artioposthia testacea*		*	(AF178305)
Family Cylindrostomidae				*Artioposthia triangulata*	+		AF033038
Cylindrostoma fingalianum #	+		AF051330	*Caenoplana caerulea*	+		AF033040
Cylindrostoma gracilis	+		AF065416	*Australoplana sanguinea*	+		AF033041
Separata				Subfamily Geoplaninae			
Family Plagiostomidae				*Geoplana ladislavi*		*	(AF178313)
Plagiostomum cinctum	+		AF065418	Family Bipaliidae			
Plagiostomum striatum	+		AF065420	*Bipalium adventitium*		*	(AF178306)
Plagiostomum vittatum #	+		AF051331	*Bipalium kewense*	+		AF033039
Plicastoma cuticulata	+		AF065422	*Bipalium* sp.		*	(AF178307)
Vorticeros ijimai	+		D85094	Family Rhynchodemidae			
Family Ulianiniidae				Subfamily Microplaninae			
Ulianinia mollissima	+		AF065427	*Microplana nana*	+	*	AF033042 (AF178317)

Continued overleaf

Table 6.1 Continued.

Classification	18S rDNA	COI	Accession number	Classification	18S rDNA	COI	Accession number
Microplana terrestris		*	(AF178318)	Dugesia ryukyuensis		*	(AF178311)
Subfamily Rhynchodeminae				Neppia montana		*	(AF178319)
Platydemus manokwari		*	(AF178320)	Spathula sp.		*	(AF178324)
Infraorder Paludicola				Girardia tigrina	+	*	AF013157 (AF178316)
Family Planariidae							
Polycelis nigra	+		AF013151	Girardia dorotocephala		*	(AF178314)
Polycelis tenuis	+	*	Z99949 (AF178321)	Girardia anderlani		*	(AF178315)
Crenobia alpina	+		M58345	Neodermata			
Phagocata ullala	+		AF013149	Class 'Monogenea' — incerta sedis			
Phagocata sp.	+		AF013150	Udonella caligorum	+		AJ228796
Family Dendrocoelidae				Class Trematoda			
Dendrocoelum lacteum	+	*	M58346 (AF178312)	Order Strigeida			
				Family Schistosomatidae			
Baikalobia guttata	+		Z99946	Schistosoma mansoni	+		M62652
Family Dugesiidae				Order Echinostomida			
Schmidtea mediterranea	+	*	U31084 (AF178322)	Family Fasciolidae			
				Fasciolopsis buski	+		L06668
Schmidtea polychroa	+	*	AF013152 (AF178323)	Order Plagiorchiida			
				Family Gyliauchenidae			
Cura pinguis	+	*	AF033043 (AF178309)	Gyliauchen sp.	+		L06669
				Class Eucestoda			
Dugesia subtentaculata	+		M58343	Order Cyclophyllidea			
Dugesia etrusca		*	(AF178310)	Family Taeniidae			
Dugesia japonica	+	*	AF013153 (D499166)	Echinococcus granulosus	+		U27015

finally, Tricladida forms a monophyletic group with very high bootstrap support (100%) and appears as the sister-group, with a rather weak support (42%) in NJ trees, to the Prolecithophora. Using both phylogenetic methods, the new clade formed by Tricladida + Prolecithophora, shifts the clade formed by *Urastoma*, Fecampiida (*Kronborgia*) and *Ichthyophaga* (Littlewood *et al.* 1999a) to a more external position forming the sister-group of the Tricladida + Prolecithophora and altogether a new clade with a 79% bootstrap support.

Molecular phylogeny of the Tricladida: the monophyly of the Maricola, the paraphyly of the Paludicola, and evidence for a clade Dugesiidae + Terricola

All dugesiids and all the Terricola sampled so far show two types of 18S ribosomal genes homologous to the type I and type II genes described in the dugesiid *Schmidtea mediterranea* Benazzi *et al.* 1975, by Carranza *et al.* (1996). Using this duplication event and the sequences of either type I or type II, it was found that: 1) the Maricola form a monophyletic primitive group; 2) the Terricola and the Dugesiidae cluster together with high support irrespective of the phylogenetic method used (NJ, MP and ML; Carranza *et al.* 1998a,b); and 3) the other paludicolan families, the Planariidae and Dendrocoelidae form the sister-group of the Terricola + Dugesiidae clade. A NJ tree using only type I 18S rDNA sequences is shown in Figure 6.4. The clade Terricola + Dugesiidae has a 100% bootstrap support, and its sister group a support of 99%. ML trees (not shown) had also both sister-groups very well supported.

Although the topology of the tree drawn from COI sequences is different from that derived from 18S rDNA sequences, it also supports with very high bootstrap support (100%) the clustering of Terricola and Dugesiidae and the sister-group character of the Planariidae and Dendrocoelidae (Figure 6.5).

Are the Terricola and the Dugesiidae monophyletic clades?

In both NJ trees drawn from 18S rDNA sequences (Figures 6.3 and 6.4) the monophyly of the Terricola is not supported or very weakly supported. Likewise, trees derived from COI sequences show monophyly of the Terricola, but with very weak support. A similar situation holds for the Dugesiidae, the anomaly in 18S trees being the clustering of the dugesiid *Girardia tigrina* Girard, 1850 (Figure 6.3) within the Terricola, and in COI trees the early branching of *Spathula* sp. Nurse, 1950 (Figure 6.5). Nevertheless, both sets of trees reproduce with high bootstrap support monophyletic assemblages such as the dugesiid genera *Schmidtea* Ball, 1974, and *Dugesia* Girard, 1850. In addition, the COI tree supports the family Bipaliidae, the subfamily Microplaninae and the genus *Girardia* Ball, 1974. It is worth noting the 'anomalous' position in COI trees of the rhynchodemid *Platydemus manokwari* de Beauchamp, 1962, within the Geoplanidae and of the dugesiid species *Girardia anderlani* Kawakatsu and Hauser, 1983, within the genus *Dugesia* and not with the other *Girardia* species.

Discussion

In his detailed analysis on the phylogenetic relationships of the triclads, Sluys (1989a) stated that research in this group 'is in a state of flux'. The advent of molecular studies, largely based on 18S rDNA sequences, has increased the rate of flux. This reflects the apparent instability of the morphological framework due to increasing difficulties of distinguishing homologies from recurrent convergences or homoplasies but, at the same time, also reflects the inconsistencies of most molecular data often based on incomplete sampling and, so far, on a single (namely 18S rDNA) or a few genes.

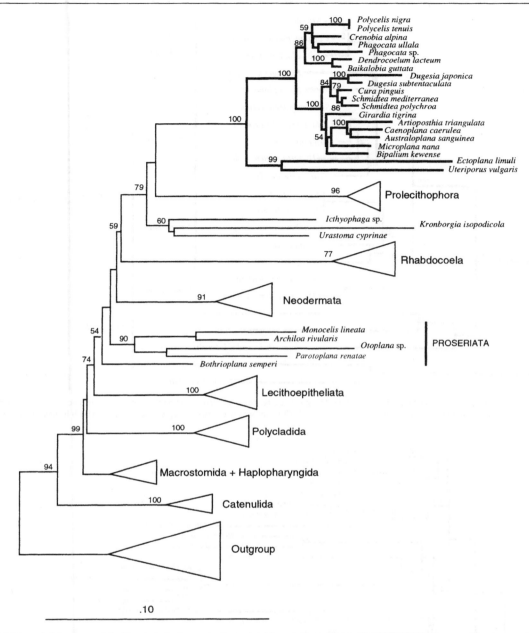

Figure 6.3 Neighbour-joining tree of the Platyhelminthes and other triploblasts (outgroup) based on 18S rDNA data set. Numbers on branches represent bootstrap percentages (n = 1000), only those over 50% being indicated. The tree illustrates the position of the Tricladida (top of the tree, bold lines) and its sister relationship to the Prolecithophora and the paraphyly of the Seriata. For taxa and species names, see Table 6.1. The complete tree with all species names is available on request from the authors. Scale indicates substitutions per position.

The validity of the taxon *Seriata* and the position of the *Tricladida* within the *Platyhelminthes*

The phylogenetic placement of the Tricladida within the Platyhelminthes is an example of these uncertainties. Most taxonomists consider the Tricladida a highly derived group of turbellarians, largely due to the rather complex morphological structure and to the advanced features of embryonic development. The precise phylogenetic position of the Tricladida and its sister-group relationships, however, have not been fully resolved. On the basis of a large set of morphological and embryological characters, Karling (1974) placed the Tricladida close to the Proseriata and not far from Prolecithophora and the

Rhabdocoela, a position also supported by Ehlers (1985a) and by Smith *et al.* (1986). Based on protonephridial structure, Rohde (1990) placed them close to the Polycladida whereas the Proseriata fell close to Neodermata. Finally, using a matrix of 65 equally weighted and unordered morphological characters, the 50% majority-rule consensus solution found also suggested that Tricladida and Proseriata do not constitute a monophylum (Littlewood *et al.* 1999a). The Tricladida was found, however, albeit with very poor support, within a clade together with Polycladida, Macrostomida and Haplopharyngida, whereas Proseriata appeared as the sister group of the Fecampiida. Recoding of some characters to take into account proposed synapomorphies for spermiogenesis between the

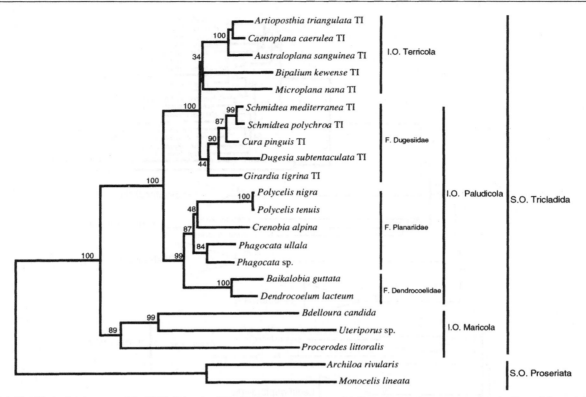

Figure 6.4 Neighbour-joining tree of the Tricladida using 18S rDNA gene sequences with Proseriata as the outgroup. For the Dugesiidae and the Terricola only the type I (TI) 18S rDNA gene sequence (Carranza *et al.* 1998a,b) was used. Bootstrap support (%; n = 1000) indicated above the nodes.

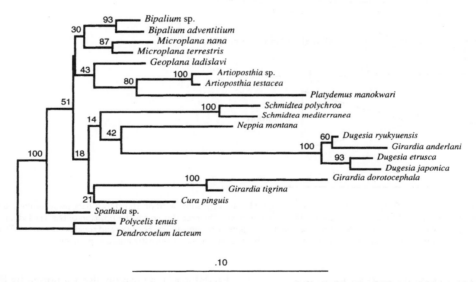

Figure 6.5 Neighbour-joining tree based on partial sequences of the cytochrome oxidase I gene of 21 species of Terricola and Paludicola. Only first and second positions were used for analysis. Numbers on branches represent bootstrap percentages (n = 1000). For taxa and species names, see Table 6.1. Scale indicates substitutions per position.

Fecampiida, *Urastoma* and the Neodermata (Joffe and Kornakova 1998), supported a new tree that, as regards the position of the Tricladida, did not differ substantially to those from previous analyses (Littlewood *et al.* 1999a).

Molecular data based on partial 18S and 28S rDNA sequences have added very little but confusion. Triclads were found either: 1) close to Rhabdocoela (Rohde *et al.* 1995); 2) as a group close to the prolecithophorans (Kuznedelov and

Timoshkin 1995); 3) as a basal platyhelminth group branching early, second only to the acoels (Katayama *et al.* 1996); 4) as forming the sister-group to a clade made up of the Acoela and the rhabdocoel fecampiid *Kronborgia* (Jondelius 1998); 5) as the sister-group to the Acoela (Campos *et al.* 1998); and 6) as the sister-group of the Fecampiidae, both forming a sister-group to some rhabdocoels (Litvaitis and Rohde 1999). Studies based on complete 18S rDNA sequences (Carranza *et al.* 1997; Littlewood *et al.* 1999a; Norén and Jondelius 1999) gave a more consistent position for the Tricladida as they appeared as the sister-group to the Rhabdocoela or to the Prolecithophora (Carranza *et al.* 1997; Norén and Jondelius 1999), or to a clade made by *Urastoma*, Fecampiida (*Kronborgia*) and *Ichthyophaga* (Littlewood *et al.* 1999a). The position of the Proseriata in molecular trees was found to be rather variable. They clustered at the base of the neoophorans (*sensu* Ehlers 1985a) either with the Lecithoepitheliata, the Neodermata or, oddly enough, with the Nemertodermatida. The clustering of Nemertodermatida and Proseriata is, however, unusual and very likely artefactual due to the poor sampling of this group (Carranza *et al.* 1997) and because there are no sound morphological arguments to place them together. When morphological and the 18S rDNA evidence were combined ('total evidence approach'; Littlewood *et al.* 1999a), Tricladida and Proseriata were not monophyletic, i.e., the taxon Seriata is invalid. In addition, Tricladida appeared, as in the molecular trees, as the sister-group to a clade formed by *Urastoma*, Fecampiida (*Kronborgia*) and *Ichthyophaga*.

New sequences of prolecithophorans and rhabdocoels now available from GenBank, added to those available within the large data set of turbellarians, made up 72 full-length sequences of Platyhelminthes (20 from the Tricladida) for analysis. Results show, albeit with a rather weak support (42% bootstrap; Figure 6.3), that prolecithophorans and triclads are monophyletic sister clades. A relationship between both groups had already been advanced by Kuznedelov and Timoshkin (1995) and more recently by Norén and Jondelius (1999). The first study, however, was based on rather short stretches (340–368 bp) of 18S rDNA from a rather restricted sampling of both groups, and lacked suitable outgroups. The second, based on full-length sequences, showed a consistent sister-group relationship between triclads and prolecithophorans and rejected the family Urastomidae as a member of the Prolecithophora. The new position of the prolecithophorans seems to exclude the clade formed by *Urastoma*, Fecampiida (*Kronborgia*) and *Ichthyophaga* as the sister-group of the triclads as suggested by Littlewood *et al.* (1999a). The latter clade and prolecithophorans + triclads form altogether a new clade well supported in NJ (79%) and ML trees, which is worth analysing in the future. The main autapomorphy of prolecithophorans is the aflagellate spermatozoa with no dense bodies and an extensive intracellular membrane system. No synapomorphies are known between prolecithophorans and triclads, though the lack of sclerotized structures, some features of the protonephridial system and the pharynx, and the structure of the female copulatory apparatus are worth exploration. NJ and ML trees also supported the monophyly and the basal position of the Proseriata within the neoophorans, confirming molecular analyses by Carranza *et al.* (1997) and Littlewood *et al.* (1999a). In addition, the Bothrioplanida, here represented by *Bothrioplana semperi*, and considered a proseriate (see Cannon 1986) or the sister-group of the Tricladida (Sopott-Ehlers 1985), was never seen clustering with the Tricladida. Instead it appears as a sister-group, albeit with weak support, to the Neodermata (ML trees), or to a large group of neoophorans (NJ tree; Figure 6.3). Hence, the presumed synapomorphies linking the Bothrioplanida and the Tricladida (a tricladoid intestine and a crossing over of muscle layers at the pharynx; Sopott-Ehlers 1985) may be convergences and should be reassessed. To summarize, available evidence gathered from this as well as from previous molecular analyses, strongly suggests that the Order Seriata is not a valid taxon because the Proseriata and Tricladida are monophyletic unrelated taxa, that the Bothrioplanida is not the sister taxon of the Tricladida, and that the Prolecithophora and Tricladida are likely sister-taxa.

Phylogenetic relationships within the Tricladida: the paraphyly of the Paludicola and the internal phylogeny of the Dugesiidae and the Terricola

The clustering of the Terricola and the paludicolan family Dugesiidae and, therefore, the paraphyly of the Paludicola is mainly based on their sharing of an 18S gene duplication (Carranza *et al.* 1998a,b). That this resulted from a single duplication event in the ancestor of both clades and not from independent duplication events after the split between terricolans and dugesiids was deduced from intra- and interspecific genetic distances found between type I and type II 18S genes (Carranza *et al.* 1998a, 1999). Moreover, using either type I or type II sequences, the Terricola and the Dugesiidae always cluster together with very high bootstrap support in MP and NJ trees (see Figure 6.4 for Type I sequences). In addition to this molecular synapomorphy, a character thought to be an autapomorphy for the dugesiids, the multicellular eye cup, was proposed to be a morphological synapomorphy for terricolans and dugesiids (Carranza *et al.* 1998a).

The phylogenetic tree of paludicolans derived from partial sequences of the cytochrome oxidase c subunit I (COI) gene (Figure 6.5) further reinforces the evidence for a clade Terricola+Dugesiidae, because it shows a 100% bootstrap support for the Planariidae+Dendrocoelidae as the sister-group to the Terricola+Dugesiidae. However, neither 18S gene nor COI gene sequences support the monophyly of either the Terricola and the Dugesiidae rendering them, so far, paraphyletic. In 18S gene trees, the odd placement of *Girardia tigrina* within the Terricola, represents its main drawback. Even so, these trees reproduce the clustering of the genus *Cura* Strand, 1942, and *Schmidtea* Ball, 1974, with the genus *Dugesia* next to them, and *Neppia* Ball, 1974, usually making the external group to the former three. Instead, the genus *Spathula* Nurse, 1950, has, together with *Girardia*, a rather erratic position. The clustering of *Schmidtea* and *Cura* with *Dugesia* contradicts the clustering of *Cura* and *Schmidtea* with *Girardia* (de Vries and Sluys 1991) which was based on a single synapomorphy: the 'angled bursal canal'. This character is only present in *Girardia* (Carranza *et al.* 1998a) and, therefore, has been rejected as a synapomorphy for these three genera (Sluys *et al.* 1998a). In contrast to *Schmidtea* and *Dugesia*, for which more than one species have been sequenced for the 18S gene, only a single species each is so far available for the genera *Girardia* and *Neppia*. This makes the sampling too poor to draw any firm conclusion. Besides, *Neppia* and *Spathula* are poorly defined genera (Sluys 2001, this volume). A similar situation holds for the Terricola, only the clustering of the Geoplanidae (type I gene), represented by three species, being highly supported.

In COI trees, all terricolans and most dugesiids cluster together. However, the very low bootstrap support found does not permit any firm conclusion to be drawn regarding the monophyly of these two clades. Within the Terricola, and despite insufficient sampling, the family Bipaliidae and the subfamily Microplaninae of the Rhynchodemidae appear highly supported. However, the position of the rhynchodemid *Platydemus manokwari* (subfamily Rhynchodeminae) within the Geoplaniidae (also found for type II 18S gene trees; Figure 6.4) suggests that the rhynchodemids are polyphyletic and the geoplaniids are paraphyletic. The family Rhynchodemidae has always been considered an artificial assemblage, with subfamilies Rynchodeminae and Microplaninae being only loosely related (personal communication, Leigh Winsor 1998). The non-clustering of both subfamilies in both 18S and COI gene trees may reflect this situation, and deserves a better sampling and further studies. As regards the Dugesiidae, the clustering of *Dugesia* and *Schmidtea* and, more loosely, *Neppia* is reproduced as it was for the 18S gene tree, but the single *Cura* species, *Cura pinguis*, does not cluster with them. In addition, the single *Spathula* species falls outside the Dugesiidae.

A special mention must be made of the odd placement in COI trees of *Girardia anderlani* within the genus *Dugesia* (Figure 6.5). Taken at face value it may be either an artefact, a misclassified specimen, or that *Girardia anderlani* does not actually belong to *Girardia* but to *Dugesia*. *Girardia anderlani*, described so far from Brazil (Kawakatsu *et al.* 1983), is externally similar to other species belonging to the genus *Girardia* living in South America. However, it differs internally in the presence of dorsal testes (usually ventral in other *Girardia*, though a few also have dorsal testes) and in their large and very asymmetrical penial bulb. Moreover, its chromosome number of $2x = 18$; $n = 9$, differs to those of most *Girardia* species, usually bearing chromosome numbers of $2x = 8$, 16 or 24; $n = 4$ or 8, the only exception being *Dugesia* (*Girardia*) *cubana* Codreanu and Balcesco, 1973, with $2x = 18$. Chromosome numbers of $2x = 18$ have been described for all species of the genus *Dugesia* belonging to the *Dugesia sicula* group (Baguñá *et al.* 1999). Therefore, given the clustering of *Girardia anderlani* within the *Dugesia* species in COI trees, and its specific karyotype, we suggest it may actually belong to the genus *Dugesia*. Otherwise, and considering that the genus *Dugesia* has not been reported from North or South America (Sluys *et al.* 1998a) some misclassification or cross-contamination may have occurred.

Main conclusions and prospects

To summarize, phylogenetic analyses of 18S ribosomal sequences of 72 Platyhelminthes species, including those of 20 species of triclads, together with the phylogenetic analyses of cytochrome oxidase subunit I partial gene sequences from 21 species of triclads show that: 1) the Tricladida is a monophyletic taxon not related to the Proseriata; therefore, validity of the taxon Seriata is rejected; 2) 18S sequence analyses show for both NJ and ML trees a sister-group relationship between the Tricladida and the Order Prolecithophora; 3) a new clade made by Tricladida + Prolecithophora and a clade formed by *Urastoma*, Fecampiida (*Kronborgia*) and *Ichthyophaga* is well supported in both NJ and ML trees deserving further studies; 4) the Bothrioplanida does not appear in any of the analyses as the sister-group of the Tricladida; 5) 18S rDNA sequence analyses, the duplication event involving the 18S ribosomal gene, and COI sequence analyses, strongly support the paraphyly of the Paludicola and the validity of the Terricola + Dugesiidae clade; and 6) 18S rDNA and COI sequences do not validate so far, probably because of insufficient sampling, the monophyly of the Terricola and the Dugesiidae.

Altogether, the new sister-group relationship between the Tricladida and the Prolecithophora, together with 18S rDNA

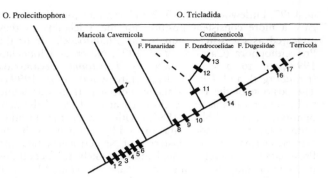

Figure 6.6 A new phylogenetic hypothesis for the Tricladida proposed in the present study. It illustrates the monophyly of the Tricladida and its sister-group relationships to the Order Prolecithophora, the monophyly of the Maricola, the clade formed by the Dugesiidae and the Terricola, and the recently proposed clade, the Continenticola, grouping the present families Dugesiidae, Planariidae, Dendrocoelidae and the Terricola. Selected morphological characters from Sluys (1989a) and the molecular apomorphy (character 15) from Carranza *et al.* (1998a) have been mapped onto the tree, with black prisms referring to derived characters. 1, tricladoid intestine; 2, crossing-over of pharynx muscles; 3, embryology; 4, cerebral position of female gonads; 5, serial arrangement of many nephridiopores; 6, marginal adhesive zone; 7, Haftpapillen in annular zone; 8, loss of Haftpapillen; 9, resorptive vesicles; 10, reduction in number of longitudinal nerve cords; 11, common oviduct opening into atrium; 12, dendrocoelid pharyngeal musculature; 13, anterior adhesive organ; 14, multicellular eye cup with numerous retinal cells; 15, two types of 18S rDNA genes (type I and type II); 16, creeping sole; 17, diploneuran nervous system.

and COI data reported in Carranza *et al.* (1998a,b) and in this work, support a new phylogenetic hypothesis for the Tricladida which is depicted in Figure 6.6.

To further support the conclusions here obtained, and to resolve the contradictions posed, further studies will be needed, including a denser sampling of Tricladida taxa for molecular data, complementary sequences from independent genes, and the broad and thorough morphological database already available for the Tricladida (Sluys 1989a, 1990; Sluys *et al.* 1998a), used together in a combined 'total evidence' approach.

ACKNOWLEDGEMENTS

We are deeply indebted to the following individuals who provided us with fresh and fixed material for this study: Peter Anderson, Mario Benazzi, Marco Curini-Galleti, Eudoxia M. Froehlich, Jacinto Gamo, Boris I. Joffe, Peter M. Johns, Hugh Jones, Masaharu Kawakatsu, Ana Maria Leal Zanchet, Tim Littlewood, Carolina Noreña-Janssen, Carles Novell, Grace Panganiban, Rodrigo Ponce de León, Maria Reuter, Reinhard Rieger, Ernest Schockaert, and Leigh Winsor. Funding was provided by CIRIT grant 1997SGR 00057 to J.B. David Swofford kindly provided a pre-release version of PAUP* (4.0d55).

Chapter 7

Towards a phylogenetic classification and characterization of dugesiid genera (Platyhelminthes, Tricladida, Dugesiidae): A morphological perspective

Ronald Sluys

The classification of triclads (Platyhelminthes, Tricladida) has improved since modern phylogenetic studies, i.e., Hennigian or cladistic analyses, have been used as the basis for taxonomic revisions. In this respect, attention has been paid mostly to: 1) the relationships between the higher taxa within the Tricladida, the sub- or infraorders of current taxonomy (cf. Sluys 1989a; Carranza *et al.* 1998a); 2) the affinities within the suborder of the marine triclads or Maricola (Sluys 1989b); and 3) the higher taxon relationships within the freshwater planarians or Paludicola in general and within the family Dugesiidae Ball, 1974 in particular (cf. Ball 1974a; de Vries and Sluys 1991).

Under the current system (ICZN 1999) and practice of zoological systematics it is customary to provide a diagnosis for each taxon, summarizing its diagnostic, usually morphological and anatomical, characteristics. In principle, such a diagnosis should enable us to determine the extension of a taxon name, i.e., to establish whether a particular species or individual organism fits the diagnosis or definition and thus falls within a certain category. However, it may happen that diagnoses mainly or even exclusively list features that are plesiomorphic on the particular level of generality. For example, Ball (1977a: 3) defined the dugesiid genus *Spathula* Nurse, 1950 as follows: 'Dugesiidae with a rounded, truncate, or spathulate head, with two eyes, or lacking eyes, pigmented or white. With a single seminal vesicle or none at all, without a diaphragm in the ejaculatory duct, with numerous dorsal or ventral testes extending to the tail, without ectal diverticula of the bursal canal, with caudal branches of the oviducts, and in which the oviducts, separately or combined, enter the bursal canal above the zone of the shell glands.' None of the features mentioned in this diagnosis is restricted to species of *Spathula*, but instead can be found in several to many other dugesiids. Thus, apart from the phylogenetic position of the genus within the Tricladida or the Dugesiidae, the question is, 'what is a *Spathula*?' To greater or lesser extent a similar question applies to some of the other dugesiid genera.

In a phylogenetic taxonomy, categorical ranking should reflect the topology of the underlying phylogenetic tree of the taxa. Although ranking provides the structure of the classification, it does not present any information on the diagnostic characteristics of the various monophyletic taxa. Therefore, under a phylogenetic paradigm one would base taxon diagnoses on the postulated (syn)apomorphies of the respective monophyletic groups, as was attempted by Sluys (1989b) for taxa within the Maricola.

However, recently it has been argued that in a truly phylogenetic taxonomy taxon definitions should only refer to phylogenetic tree structure (de Queiroz and Gauthier 1994). Initially, one type of definition was considered that refers to the possession of apomorphic characters in the theoretical ancestor of a clade, but in later studies this apomorphy-based definition of taxon names was judged as unsatisfactory (Bryant 1994; Schander and Thollesson 1995; Lee 1998; Sereno 1999).

Nevertheless, there are three reasons for presently not abandoning the focus on apomorphic features in the formulation of taxon diagnoses or definitions. First, the current code on zoological nomenclature (ICZN 1999) does not provide a framework within which the new proposals for a phylogenetic taxonomy can formally be adopted. Second, the new proposals on the phylogenetic definition of taxon names are presently under full discussion in the literature and all of their implications apparently have not yet been fully explored (cf. Bryant 1997; Dominguez and Wheeler 1997; Sereno 1999). And third, a diagnosis shall remain a necessary part of any current or future taxonomy. Even in the case of fully phylogenetically defined taxon names (i.e., through a node-based or a stem-based definition: cf. de Queiroz and Gauthier 1994; Sereno 1999) we would like to have a diagnosis mentioning the apomorphic characters, thus enabling us to assign a particular organism or species to a certain taxon (see also Pleijel 1999). In such cases, the diagnosis does no longer define the taxon but merely serves to recognize or diagnose individual cases or instances of that particular taxon.

In this context, the present study examines the phylogenetic relationships among the ten genera currently recognized within the Dugesiidae, and thus explores possible diagnostic characteristics of these taxa. Previous analyses (e.g., de Vries and Sluys 1991) usually generalized character state information for each of the genera. However, this practice obscures the facts that a) the taxonomic status of some species is rather uncertain and that upon further reflection they are better assigned to other genera (cf. Sluys 1997), b) that there sometimes is a great deal of variability among species, and c) that some of the genera are poorly defined. As a consequence, genera may not fulfill one of the requirements of supraspecific analysis, viz. being a monophyletic unit, which may thus contribute to errors in inferred relationships (Bininda-Emonds *et al.* 1998).

Therefore, the present study for the first time provides a species-level analysis of phylogenetic relationships within the Dugesiidae. In order not to complicate matters beyond necessity, an exception is made for the genus *Dugesia* Girard, 1850, which is here only included as a single operational taxonomic unit and not with all of its constituent species. Previous studies (de Vries and Sluys 1991; Sluys *et al.* 1998a) have amply demonstrated that members of the genus *Dugesia* share a number of well-defined synapomorphies.

For the first time also, the present study incorporates the unusual monotypic genus *Bopsula* Marcus, 1946. Previous studies (Ball 1974a; de Vries and Sluys 1991; Sluys 1997) have generally omitted this genus from their taxon lists because it was felt that the aberrant characteristics of *Bopsula* should be re-assessed on new material. Unfortunately, no such material has become available since its first description by Marcus (1946). In contrast to Ball (1974a: 376), I have been able to re-examine the two known specimens (one whole mount and one sectioned animal); both specimens are in relatively good histological condition and thus do not suggest that observations on these animals would provide a distorted picture of the species' morphology and anatomy.

Material and methods

A data matrix of potentially phylogenetically informative characters for the included taxa was compiled on the basis of previous analyses, data extracted from the literature, and newly obtained information resulting from extensive specimen examination. The data matrix (Table 7.1) introduces four new names (*G. sphincter*, *C. fortis*, *N. magnibursalis*, and *R. sinuosus*); new species descriptions will be published in a forthcoming paper.

Missing data were coded as '?', and inapplicable conditions were also coded as '?'. The coding of inapplicable or inappropriate character states is a problematic and still largely unresolved issue (for discussions, see Maddison 1993; Waggoner 1996; Rouse and Fauchald 1997; Bininda-Emonds *et al.* 1998). Fortunately, in the present study the number of inapplicable codings are relatively few and mostly limited to one character. Another option for dealing with inapplicable characters consists of assigning to the non-applicable conditions a separate character state (e.g., state 9). Evidently, this practice runs the risk of creating false groupings on the basis of shared non-applicable character states.

Supraspecific taxa principally were coded according to the democratic or majority method: the majority state in a sample of species was taken to represent the higher taxon. In those cases in which the known distribution of two character states in a sample of species approached equality, both states were included and coded as polymorphisms.

Usage of the majority method, notably in the case of the outgroups, was mostly based on practical considerations in that absence of a generally accepted phylogeny prevented the construction of an all-primitive operational taxonomic unit. According to Bininda-Emonds *et al.* (1998), reconstruction of the common ancestor (or groundplan; Yeates 1995) of a group is theoretically the preferred method, while it generally outperforms other methods of supraspecific taxon coding. In contrast, Wiens (1998) found that the majority method generally performed best among the investigated coding methods for higher taxa.

Under the hypothesis of Sluys (1989a) on the higher taxon phylogeny of the Tricladida one would select the Planariidae Stimpson, 1857 plus the Dendrocoelidae Hallez, 1892 as the first, most important outgroup since these two families together constitute the sister-group of the Dugesiidae; supplementary information would be provided by the more distantly related Terricola. However, under the hypothesis of Carranza *et al.* (1998a) the situation is reversed: the Terricola is hypothesized as the sister-group of the Dugesiidae, with the Planariidae + Dendrocoelidae forming a more distantly related monophylum. Nevertheless, for practical purposes it makes little difference since in both cases character states in the land and freshwater planarians have to be considered. Therefore, the present study includes two supraspecific outgroup taxa, one generalizing information on the terrestrial planarians (outgroup T), and another one generalizing character states within the Planariidae and Dendrocoelidae (outgroup PD). Furthermore, in deducing the preferably plesiomorphic states for these outgroup terminals, conditions in the Maricola and Cavernicola were also taken into account, where appropriate.

Phylogenetic analyses were performed with the help of PAUP version 3.1.1 (Swofford 1993), PAUP* version 4.0b2a (Swofford 1998) and MacClade version 3.0.1 (Maddison and Maddison 1992). In these analyses, all characters were treated as unordered, multistate taxa were interpreted as polymorphic, the heuristic search option with random stepwise addition sequence was applied (100 repeats, and HOLD set to 1), with seed number 10 and reference taxon *G. anceps*, as branch swapping option TBR was used, all minimal length trees were kept (MULPARS, MULTREES), zero-length branches were collapsed, the ingroup was made monophyletic, and multiple furcations were treated as soft polytomies. Bootstrap resampling was performed with PAUP* ($n = 1000$, fast heuristic search, with starting trees obtained via stepwise addition with random addition sequence).

Reflections on eye structure

For a long time systematists used to recognize two families within the Paludicola or freshwater planarians, the Planariidae Stimpson, 1857 and the Dendrocoelidae Hallez, 1892, which were defined by Kenk (1930) by the arrangement of the inner muscle layers of the pharynx. In the planariid type of pharynx the longitudinal and circular muscles of the inner zone of the pharynx form two separate layers, whereas in the dendrocoelid pharynx these layers are intermingled (cf. Sluys 1989a, Figure 2B,C). The dendrocoelid type of pharynx has been postulated as the derived condition, supporting the monophyletic status of the Dendrocoelidae (cf. Sluys 1989a), although the situation is slightly more complex in that some dendrocoelids show the planariid type of muscle arrangement (cf. Sluys *et al.* 1998b).

In his study on the phylogeny and taxonomy of freshwater triclads, Ball (1974a) erected the family Dugesiidae for a large group of planarians. In that particular study, Ball failed to find an apomorphic character supporting the presumed monophyletic status of the new family. However, in later publications (Ball 1974b,c) he noted that dugesiid planarians are characterized by an apomorphic eye structure, in that the pigment cup is multicellular and contains numerous retinal cells.

Recently, this presumed good synapomorphy for the dugesiids has been reconsidered in the light of a new phylogenetic hypothesis for the Tricladida. Under the hypothesis of Carranza *et al.* (1998a,b) the multicellular eye cup was postulated as a synapomorphy uniting the Dugesiidae with the Terricola.

Multicellularity of the pigment cup is a very rare condition among platyhelminths, as is the presence of numerous retinal clubs within the eye cup. Therefore, by applying Hennig's auxiliary principle of firstly assuming homology and synapomorphy before hypothesizing convergence, these conditions might be postulated as synapomorphies for the Dugesiidae and the Terricola. However, the situation is rather complex and invites some further reflection upon the morphological differences in eye structure in these taxa, as well as in a few others. Several features have to be considered: (multi)cellularity of the eye cup; number of retinal clubs; orientation of these photoreceptive cells within the eye cup; penetration of the eye cup by the dendrites; and the shape of the retinal clubs.

The observation that the eyes of *Dugesia gonocephala* (Dugès, 1830) consist of a multicellular pigment cup containing very many retinal clubs is usually attributed to Hesse (1897), although earlier Böhmig (1887) and also Jänichen (1896) had already described this situation. Hesse (1897) arranged the planarian eyes under three structural types: the *Planaria torva* type, the *Dendrocoelum lacteum* type, and the *Dugesia gonocephala* type; his three types still function in modern classifications of eye structure (cf. Sluys 1989b).

In support of his hypothesis that the dugesiid eye represents an apomorphy, Ball (1974b: 154) wrote that he had '... examined 19 species of Dugesiidae within the genera *Dugesia*, ... *Schmidtea*, ... *Girardia*, *Neppia*, *Spathula*, and *Cura*, and ... found that all possess eyes of the *Dugesia gonocephala* type ...'. Over the years, I have examined many dugesiid-like planarians and failed to find an exception to the rule that they possess Hesse's *D. gonocephala* type of eye, except in those cases where eyes are absent due to secondary loss. However, I must offer the caveat that in standard histological preparations it is generally difficult to discern whether the pigment cup is multi-nucleated, i.e., multi-cellular, in contrast to the ease with which one can observe the numerous retinal clubs (Sluys 1997). In this context it is noteworthy that only Kishida (1967a, Figures 1d,e; 1967b, Figure 25) and Carpenter *et al.*

Table 7.1 Data matrix for phylogenetic analysis of the Dugesiidae. Generic abbreviations: B = Bopsula, C = Cura, E = Eviella, G = Girardia, N = Neppia, R = Romankenkius, Re = Reynoldsonia, S = Schmidtea, Sp = Spathula. For character codings and further explanation, see text; characters in boxes represent multiple states, e.g. 367 is 3 & 6 & 7.

	1	2	3	4	5	6	7	8	9	10	11	12	13	14	15	16	17	18	19	20	21	22	23	24	26	27	28	30	31	32	33	34	35	36	37	38	39
G. anceps	0	1	0	0	0	1	0	0	0	0	0	1	0	1	1	0	0	0	0	1	1	0	0	0.0	0	0	0	0	0	1	1	0	0	0	0	0	1
G. anderlani	0	1	0	1	1	0	0	0	0	0	0	1	0	1	1	0	0	0	0	1	0	0	0	?	0	1	0	0	0	1	0	0	0	0	0	0	0
G. andina	?	?	0	?	?	?	?	?	?	0	?	0	0	0	?	0	0	0	0	?	?	?	?	?	?	?	?	?	?	?	0	0	0	0	0	0	0
G. antillana	1	2	0	2&3	0&1	0	0	1&2	1	0	1	1	0	0	1	0	0	0	1	0	1	0	0	?	0	1	0	0	0	1	0	0	0	0	0	0	0
G. arimana	1	2	0	1	1	0	0	1&2	2	0	1	1	0	0	0	0	0	0	0	1	0	0	0	?	0	0	0	0	0	0	0	0	0	0	0	0	1
G. arizonensis	1	2	0	0	1	0	0	0	0	4	2	0	0	0	0	0	0	0	0	0	0	?	0	?	1	0	0	0	1	1	0	0	0	0	0	0	0
G. arndti	0	1	0	1&2	1	?	0	0	1	0	?	?	0	0&1 0&1	0&1	0	0	0	0	0	?	?	0	?	0	1	0	0	1	?	0&1	0	0	?	?	0	0&1
G. aurita	1	?	0	1	?	?	?	0	?	0	?	?	0	0	0	0	0	0	0	?	?	?	?	?	0	?	0	?	1	?	?	0	0	0	0	0	0
G. azteca	1	2	0	4	1	?	0	1	0	4	1	?	0	0	0	0	0	0	0	0	0	0	0	?	0	1	0	0	0	?	0	0	0	0	0	0	0
G. barbarae	2	1	0	3	1	?	0	1&2	1	0	1	1	0	0	0	0	0	0	0	0	0	?	0	?	0	0	0	0	1	0	0&1	0	0	0	0	0	0
G. biapertura	0	1	0	3	?	0	0	1&2	2	1	1	1	0	0	0	0	0	0	0	0	0	0	0	?	0	1	0	0	0	0	0	0	0	0	0	0	0
G. bonaerensis	?	?	0	?	?	?	0	0	?	0	?	?	0	0	1	0	0	0	0	0	?	?	0	?	0	1	1	?	1	0	?	0	0	0	0	0	0
G. cameliae	1	2	0	?	1	0	0	0	0	0	0	1	0	0	?	0	0	0	0	1	?	0	0	?	1	0	0	?	0	0	?	0	0	0	0	0	0
G. canai	?	1	0	?	?	?	0	1	0	0	0	?	0	0	0	0	0	0	1	0	?	?	0	?	1	1	0	?	0	0	0	0	0	0	0	0	0
G. chilla	1	2	0	2	0&1	?	?	[012]	0	0	1	0	0	0	0	0	0	0	0	1	1	?	0	?	0	0	1	0	1	0	0	0	0	0	0	0	0
G. cubana	1	2	0	0	1	0	0	0	0	0	1	1	0	0	0	0	0	0	0	0	0	0	0	?	1	0	1	0	1	0	1	0	0	0	0	0	0
G. dimorpha	0	1	0	0	1	0	0	0	0	0	0	0	0	0	0	0	0	0	0	0	1	?	0	?	0	1	1	?	1	0	1	0	0	0	0	0	0
G. dorotocephala	1	2	0	?	?	?	0	0	0	0	?	?	0	0	1	0	0	0	0	1	?	?	0	?	0	1	0	?	0	?	?	0	0	0	0	0	0
G. festae	1	2	0	0	1	0	0	1	0	0	1	0	0	0	0	0	0	0	0	0	0	0	0	?	1	1	0	0	1	1	1	0	0	0	0	0	0
G. guatemalensis	0	1	0	0	0	0	0	0	0	0	0	1	0	0	0	0	0	0	0	1	?	0	0	?	0	1	0	0	1	0	1	0	0	0	0	0	1
G. hypoglauca	1	2	0	0	?	?	?	3	?	?	?	?	0	0	1	0	0	0	?	?	?	?	0	?	0	1	0	?	0	?	0	0	0	0	?	?	0
G. iheringii	?	?	0	?	?	?	?	1	0	?	?	?	0	0	0	0	0	0	0	?	?	?	0	?	0	0	0	?	1	0	0	0	0	0	?	?	0
G. jenkinsae	1	2	0	?	?	?	?	0	0	0	0	?	0	0	0	0	0	0	0	1	?	?	0	?	0	1	0	?	0	?	0	0	0	0	?	?	1
G. longistriata	1	1	0	?	?	?	?	0	0	0	1	?	0	0	0	0	0	0	0	1	1	?	0	?	0	0	0	?	1	?	0	0	0	0	?	?	0
G. mckenziei	1	2	0	?	?	?	?	0	0	0	2	?	0	0	0	0	0	0	0	1	?	?	0	?	0	0	0	?	0	?	0	0	0	0	?	?	1
G. microbursalis	0	1	0	?	1	1	0	0	0	0	1	?	0	1	0	0	0	0	0	0	0	?	0	?	0	0	0	?	0	0	0	0	0	0	0	0	1
G. nonatoi	1	2	0	?	?	0	0	0	?	0	2	?	0	0	0	0	0	0	0	0&1	0&1	?	0	?	0&1	0	0	?	0	1	1	1	0	0	0	0	0
G. paramensis	0	1	0	?	?	?	0	0	0	0	?	?	0	0&1	0&1	0	0	0	0&1	?	?	0	0	?	0&1	0	?	?	?	?	?	0	0	0	?	?	0
G. polyorchis	?	?	0	?	?	?	?	0	?	0	?	?	0	0	?	0	0	0	?	?	?	0	0	?	?	?	?	?	?	?	?	0	?	?	?	?	0
G. rincona	0	1	0	?	1	0	0	0	1	0	1	?	0	1	0	0	0	0	0	0	1	0	0	?	0	1	0	?	1	?	0	0	0	0	0	0	1
G. sanchezi	1	2	0	?	1	0	0	0	0	4	1	?	0	1	?	0	0	0	0	0	?	0	0	?	0	0	1	1	1	?	?	0	0	0	0	0	0
G. schubarti	0	2	0	?	0	0	0	0	0	0	0	?	0	0	0	0	0	0	0	0	0	0	0	?	0	0	0	0	0	1	0&1	0	0	0	0	0	0
G. sphincter	0	?	0	?	0	0	0	0	?	0	1	?	0	0	0	0	0	0	?	?	?	0	0	?	?	?	?	?	?	?	?	0	0	?	?	?	0
G. tahitiensis	0	1	0	?	0&1	0	0	0	1	0	0	?	0	0	0	0	0	0	0	0&1	0&1	0	0	?	0	0	0	?	1	?	?	0	0	0	0	0	0
G. tigrina	0	1	0	0	1	1	0	0	1	0	1	1	0	0&1 0&1	0&1	0	0	0	0&1	0&1	0&1	0	0	0	0&1	0	0	0	1	0&1	1	0	0	0	0	1	0&1
G. typhlomexicana	0	1	1	?	0	?	0	2	1	0	0	?	0	0	0	0	0	?	?	0	1	0	0	0	0	?	?	0	0	0&1	0	0	0	?	?	0	0
G. ururiograndeana	0	1	0	?	1	0	0	1	1	0	2	?	0	0	0	0	0	?	?	0	0	0	0	0	1	0	?	0	0	0	0	0	0	?	?	0	0
B. evelinae	1	2	0	1	1	0	0	1	1	0	2	0	0	0	0	0	0	0	0	0	0	0	0	0	0	0	0	0	0	0	0	1	1	0	0	0	0

Continued overleaf

Table 7.1 Continued.

	1	2	3	4	5	6	7	8	9	10	11	12	13	14	15	16	17	18	19	20	21	22	23	24	25	26	27	28	29	30	31	32	33	34	35	36	37	38	39
C. foremanii	0	1	0	0	0	0	0	1	0	3	1&2	0	0	1	0	1	0	0	0	0	0	0	0	0	0	0	0	1	0	1	1	0	0	0	0	0	0	0	1
C. fortis	?	?	0	0	0	0	0	1	0	3	2	0	0	?	0	0	0	0	0	0	1	1	0	0	0	0	0	1	?	0	0	0	0	0	0	0	0	0	0
C. graffi	?	?	0	0	?	?	?	0	0	2	2	?	?	0	0	0	0	0	0	0	0	0	0	0	0	0	0	1	?	0	0	0	0	0	0	0	0	0	0
C. pinguis	2	0	0	0	0	0	0	0	0	1	0	0	0	0	0	1	0	0	0	0	1	1	1	0	0	0	0	2	1	1	1	0	0	0	0	0	0	0	0
C. evelinae	2	0	0	0	0	0	0	0	1	1	1	1	0	0	0	0	0	0	0&1	0	0	1	1	1	1	0	2	0	1	0	0	0	0	0	0	0	0	1	1
N. falklandica	?	?	0	0	?	0&1	0&1	1&2	0	1	1	0	0	0	0	0	0	0	0&1	0	0	1	1	0	0	0	0	2	?	1	0	0	0	0	0	0	0	0	0
N. jeanneli	0	1	0	0	0	0	1	3	0	1	1	1	0	1	0	0	0	0	1	1	1	1	1	0	1	0	0	0	0	0	0	0	0	0	0	0	0	0	0
N. magnibursalis	?	?	0	0	0	0	0	0	0	0	0	0	0	0	0	0	0	0	0	0	1	1	1	0	0	0	0	0	?	0	0	0	0	0	0	0	0	0	0
N. montana	0	2	0	0	0	1	0	1	3&4	3&4	2	0	0	0	0	0	0	1	0	0	1	1	1	0	0	0	0	0	?	0	0	0	1	0	0	0	0	0	0
N. paeta	0	0	0	0	0	0	0	0	0	0	0	0	0	0	0	0	0	1	0&1	0	1	1	1	0	0	0	0	0	?	0	0	0	0	0	0	0	0	0	0
N. seclusa	3	0	0	?	0	0	0	1	0	0	1&2	?	0	0	0	0	0	0	0&1	0	0	1	1	0	0	0	1	1	?	0	0	0	0	0	0	0	0	0	0
N. tinga	?	1	0	0	0	0	0	0	0	0	0	?	0	0	?	0	0	0	0	0	0	1	1	0	0	0	0	0	?	0	?	0	0	0	0	0	0	0	0
N. wimbimba	0	1	0	0	0	0	0	0	1	1	1	0	1	0	0	0	0	0	0	0	0	?	1	1	1	0	0	0	1	0	0	0	0	0	0	0	0	1	0
E. hynesae	3	0	0	0	?	0	1	0&1	0	0	0	0	1	0	0	0	0	0	0	0	1	1	0	1	0	0	0	0	0	0	0	0	0	0	0	0	0	0	0
R. bilineatus	3	0	0	0	0	0	0	1	4	4	0	0	0	0	0	0	0	0	0	0	0	1	1	1	1	0	0	0	1	0	0	1	0	0	0	0	0	0	2
R. boehmigi	3	0	0	0	?	0	0	1	0	0	1	0	0	0	0	0	0	1	0	0	0	0	1	0	0	0	0	0	1	0	0	0	0	0	0	0	0	0	2
R. glandulosus	3	0	0	3	0	0	0	1	0	0	1	0	0	0	0&1	0	0	0	0	0	0	1	0	0	0	0	0	1	0	0	0	0	0	0	0	0	0	0	2
R. hoernesi	3	0	1	0	0	0	1	1	4	4	1	0	0	0	0	0	0	0	0	0	1	1	0	0	1	0	0	1	?	0	0	0	0	0	0	0	0	0	2
R. kenki	0	1	0	0	0	0	1	1	2	2	1	0	0	0	0	0	0	0	0	0	1	1	0	0	0	0	1	0	1	0	0	0	0	0	0	0	0	0	2
R. libidinosus	3	1	0	0	0	?	0	0	3	3	0	0	0	0	0	0	0	0	0	0	0	0&2	1	1	0&1	0	0	0	0	0	0	?	0	0	0	0	0	0	2
R. michaelseni	?	?	0	?	?	?	?	1	0	0	?	?	?	1	0	0	0	0	0	0	1	0	0	1	1	0	0	0	?	0	?	?	0	0	0	?	0	0	2
R. patagonicus	?	1	0	0	?	0	0	1&2	0	0	0	0	0	1	0	0	0	0	0	0	1	1	1	1	1	?	0	0	?	0	?	0	0	0	0	0	0	0	2
R. pedderensis	3	0	0	0	?	?	0	0	1	1	2	0	1	0	0	0	0	0	0	0	0	1	1	1	1	0	0	0	1	0	?	0	0	0	0	0	0	0	2
R. sinuosus	0	1	0	?	?	?	1	?	?	?	?	?	?	?	?	?	?	0	?	?	?	?	1	1	1	?	0	0	?	0	?	?	0	?	0	?	0	?	2
Sp. agelaea	0	0	1	?	?	?	1	1	?	2	?	?	0	0	0	0	0	0	0	0	0	1	0	0	1	1	1	0	0	0	0	0	0	0	0	0	0	0	0
Sp. alba	4	0	1	?	?	0	0	1	0	0	0	0	1	0	0	0	0	0	0	0	0	0	0	0	0	0	0	1	0	0	0	0	0	0	0	0	0	0	0
Sp. camara	3	0	0	0	0	?	1	0	0	0	0	?	0	0	0	0	0	0	0	0	0	1	0	0	0	0	0	1	?	0	0	0	0	0	0	0	0	0	0
Sp. dittae	3	0	0	0	?	?	0	0	0	0	0	?	0	0	0	0	0	0	0	0	0	1	0	0	0	1	0	0	?	0	0	0	0	0	0	0	0	0	0
Sp. foeni	3	0	0	0	?	0	0	1	0	4	0	0	1	0	0	0	0	0	0	0	0	1	0	0	1	0	0	0	0	0	0	0	0	0	0	0	0	0	0
Sp. fontinalis	4	0	0	0	?	0	1	1&2	0	4	0	0	1	0	0	0	0	0&1	0	0	0	1	1	0	0&1	0	1	1	?	0	0	0	0	0	0	0	0	1	0
Sp. gourbaultae	4	0	1	0	?	?	0	0	0	0	0	1	0	0	0	0	0	0	0	0	0	1	0	0	1	0	0	0	0	0	0	0	0	0	0	0	0	0	0
Sp. limicola	4	0	0	0	0	0	0&1	0	3&4	3&4	0	0	0	0	0	0	1	0	0	0	0	1	0	0	1	0	0	0	?	0	0	0	0	0	0	0	0	0	0
Sp. neara	5	0	0	0	?	0	1	1	0	0	2	0	1	0	0	0	0	0	0	0	0	1	1	0	0	0	0	0	?	0	0	0	0	0	0	0	0	0	0
Sp. ochyra	4	0	0	0	?	0	0	0	0	0	0	0	0	0	0	0	0	0	0	0	0	0	0	0	0	1	0	1	?	0	0	0	0	0	0	0	0	0	0
Sp. schauinslandi	5	0	0	0	0	0	0	0&2	0	0	1	1	1	0	0	0	0	0	0	0	0	1	1	0	0	0	0	0	?	0	0	0	0	0	0	0	0	0	0
Sp. truculenta	5	0	0	0	0	0	1	1	0	0	0	0	1	1	0	0	0	0	0	0	0	1	1	0	1	0	0	0	?	0	0	0	0	0	0	0	0	0	1
Sp. tryssa	3	0	1	?	0	?	1	0	0	0	?	?	1	0	0	0	0	0	0	0	0	1	0	0	0	?	0	0	?	0	0	0	0	0	0	0	0	0	0
S. lugubris	0	0	0	0	0	0	0	1	0	0	0	0	0	0	0	0	0	0	0	0	0	2	0	0	0	0	1	1	?	1	0	0	0	0	0	0	0	0	0
S. mediterranea	0&3	0	0	0	0	0	0	1	0	0	2	1	0	0	0	0	0	0	0	0	0	2	0	0	1	1	1	1	?	1	0	0	0	0	0	0	0	0	0
S. nova	5	0	0	0	0	0	0	0	0	0	1	1	0	0	0	0	0	0	0	0	0	2	0	0	0	0	0	1	?	1	0	0	0	0	0	1	0	0	0
S. polychroa	3	0	0	0	0	?	0	1	0	0	2	1	0	0	0	0	0	1	0	0	0	2	0	0	0	0	0	0	0	1	0	0	0	0	0	0	0	0	0
Re. reynoldsoni	4	0	0	?	0	?	0&1	0&1	0	0	0	?	0	0	0&1	0	1	0	0&1	0	1&2	1	0	0	0	0&1	0	0	0	0	0	0	0	0	0	0	0	0	0
Dugesia	0	1	0	0	0	0&1	1	1	0	12	12	0	0&1	0&1	0	0	0	0	0	0&1	0	1	1	2	0&1	0&1	0	0	?	0	0&1	?	0	0	0	0	0	0	0
outgroup T	367	0	0	1234	0	0	0&1	0&1	1&40&1	1&40&1	40&1	0	0&1	0	0&1	0	0	0	0	0	1	0	0	0	0&10&1	0&10&1	0	0	?	0	0&10&1	0	0	0	0	0	0	0	0&1
outgroup PD	345	0&1	0	0	0	0	0	0	0	0&10&1	0&10&1	0	0	0	0	0	0	0	0	0	0	0	0	0	0	0	0	0	?	0	0	0	0	0	0	0	0	0	0&1

(1974, Figure 1) provided photomicrographs unequivocally showing the multicellular construction of the pigment cup in *Dugesia japonica* Ichikawa and Kawakatsu, 1964 and *Girardia dorotocephala* (Woodworth, 1897), respectively. Furthermore, Kishida (1967a,b) also presented the first schematic reconstructions based on electron microscopic observations (Figures 7.1 and 7.2), later to be followed by Durand and Gourbault (1977, Figure 1) who described the ultrastructure of the eyes in *Cura pinguis* (Weiss, 1909).

It is certainly true that because of its multicellular pigment cup and the presence of numerous retinal clubs (ranging from 25 to 200), the *D. gonocephala* type of eye differs considerably from the eyes usually found in other freshwater triclads, and in marine planarians (cf. Sluys 1989b). However, detailed study

of the literature uncovered a number of statements and observations on multicellularity and multiple retinal clubs in nondugesiid freshwater planarians.

For example, Hesse (1897) already noted that, in contrast to his observations on the unicellular pigment cup of *Dendrocoelum lacteum* (Müller, 1774), the eye cup of the dendrocoelid *Bdellocephala punctata* (Pallas, 1774) consists of numerous small cells. However, in contrast to Hesse (1897), Röhlich and Török (1961) observed characteristic cell borders also within the pigment cup of *D. lacteum* and they cautiously stated that therefore the eye cup can consist of multiple cells; a similar conclusion is suggested by the study of Dyganova and Kiseleva (1988, Figure 1). Actually, Jänichen (1896) had already observed cell borders and multiple nuclei in the pigment cup of *D. lacteum*.

For several members of the dendrocoelid genus *Bdellocephala* de Man, 1875 multicellular eye cups have been described. Zabusova (1929) mentions four to five pigment cup cells and a large number of retinal cells for *Bd. mediobuccalis* Zabusova, 1929 and also discerned a large number of retinal cells and several nuclei in the pigment cup of *Bd. grubiiformis* Zabusova, 1929. Kuchiiwa *et al.* (1991) described six to twelve pigmented eye cup cells for *Bd. brunnea* Ijima and Kaburaki, 1916, with each cup housing 40–50 photoreceptor cells (Figure 7.3). A similar situation may occur in *Baikalobia guttata* (Gerstfeldt, 1858). Although Kiseleva and Dyganova (1988) described only one pigment cell for this species, their reconstruction of the manner in which the dendrites penetrate the eye cup from all sides, suggests that the cup actually consists of several cells (assuming that the dendrites always penetrate *between* cells). A case of all-around penetration was also described for *D. lacteum* (Dyganova and Kiseleva 1988, Figure 1). According to Kuchiiwa *et al.* (1991), penetration of the eye cup by dendrites takes place only at the corneal membrane. In view of the condition in many land planarians, in which the dendrites enter the eye cup from all sides between the pigment cells (see below), it would be very interesting to further explore how common this feature is among freshwater triclads. In this respect it is also interesting that the final structure of the eyes in mature specimens of *B. guttata* results from an ontogenetic trajectory that starts with an eye cup in which a much smaller

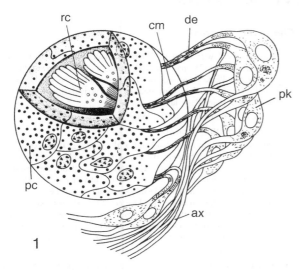

Figure 7.1 Diagram of eye structure in *Dugesia japonica* (after Kishida 1967a). Abbreviations: ax, axon; cm, corneal membrane; de, dendrite; pc, pigment cell; pk, perikaryon; rc, retinal club.

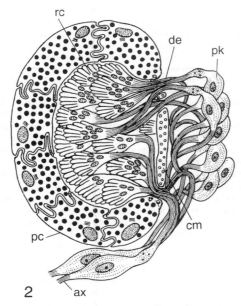

Figure 7.2 Diagram of eye structure in *Dugesia japonica* (after Kishida 1967b). Abbreviations: ax, axon; cm, corneal membrane; de, dendrite; pc, pigment cell; pk, perikaryon; rc, retinal club.

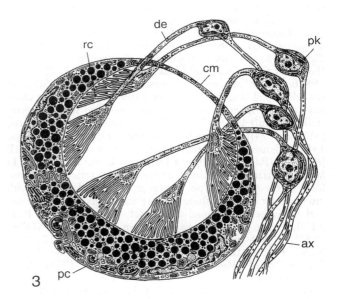

Figure 7.3 Diagram of ultrastructure of the eye in *Bdellocephala brunnea* (after Kuchiiwa *et al.* 1991). Abbreviations as in Figure 7.1.

number of dendrites enter only via the corneal membrane and a few (3–5) retinal clubs face the pigment cup; it is only at later stages that the dendrites increase in number (up to 20), penetrate the eye cup from all sides, and that some of the retinal clubs have a converse orientation (Kiseleva and Dyganova 1988).

With respect to planariid triclads, eye cups consisting of a few pigment cells have been reported for several species of the genus *Phagocata* Leidy, 1847: *Ph. illyrica* (Komárek, 1919) (Beauchamp, 1932), *Ph. coarctata* (Arndt, 1920) (Arndt, 1923), *Ph. olivacea* (Schmidt, 1861) (Beauchamp, 1932), and *Ph. (Atrioplanaria) racovitzai* (Beauchamp, 1928) (Beauchamp, 1932). The number of retinal cells in the eye cups of these species is rather low, ranging from three to ten.

The situation in the land planarians is complex in that there is a large group, the family Bipaliidae, in which the species have multiple marginal eyes formed by a single pigment cell, containing between one and eight retinal cells (Hesse 1897; Graff 1912–17; Shirasawa and Makino 1981), whereas all other terricolans have multicellular eye cups, made up for example of approximately 3300 pigment cells in *Platydemus grandis* (Spencer, 1892) (Graff 1912–17). In the latter group, the dendrites enter the eye cup between the pigment cells and through the corneal membrane, with the result that such eyes contain inversely as well as conversely oriented photoreceptor cells (cf. Sluys 1989b, Figures 22, 24). The number of photoreceptor cells in such eyes varies between one and twenty, in some species being as many as 200 (cf. Sluys 1989b).

The above suggests that multicellular eye cups are not restricted to the Dugesiidae and the Terricola. Furthermore, the number of retinal cells appears to vary much. There is great variability: 1) within the Dugesiidae [e.g., 25–30 in *G. dorotocephala* and *Cura pinguis* (Weiss, 1909) (Carpenter *et al.* 1974; Durand and Gourbault 1977); 150–200 in *D. japonica* (Kishida 1967a), *Schmidtea lugubris* (Schmidt, 1861) (Röhlich and Török, 1961), *Girardia tigrina* (Girard, 1850) (Taliaferro, 1920)]; 2) between dugesiids and other freshwater planarians (see above); and 3) between land planarians (see above). Upon reflection, therefore, statements on the presumed apomorphic presence of 'numerous' photoreceptor cells in eye cups hardly qualify as being of the required accuracy and precision.

There is one other aspect of eye structure that warrants some further discussion, namely the shape of the retinal clubs. Dugesiid planarians have characteristic funnel-shaped retinal clubs (Figures 7.1 and 7.2), whereas those of the dendrocoelids are usually rod- or bottle-shaped (cf. Zabusov 1911, Plate VI, Figures 16, 17 [*Bd. angarensis* (Gerstfeldt, 1858)], 22 [*Sorocelis tigrina = Papilloplana zebra* Kenk, 1974]; Sluys 1989b, Figure 18); in *Bd. brunnea*, however, the clubs are funnel-shaped (Figure 7.3). In planariid planarians the clubs have a more spherical shape and are provided with a shorter stalk (cf. Graff 1912–17, Plate XLV, Figures 1 [*Planaria torva* (Müller, 1774)], 2 [*Polycelis (Ijimia) tenuis* Ijima, 1884]; Sluys, 1989b, Figure 17 [*P. torva*]; Kuchiiwa *et al.* 1991, Figure 1B [*Polycelis (Polycelis) sapporo* Ijima and Kaburaki, 1916]). In land planarians with unicellular eye cups the retinal clubs conform to the *P. torva* type. However, in land planarians with multicellular eye cups there is more variability since their retinal clubs may be funnel-shaped (cf. Graff 1912–17, Plate XLVI, Figure 4 [*Geoplana rufiventris* Schulze and Müller, 1857] or rod- or bottle-shaped (cf. Graff 1912–17, Plate XLVI, Figures 18, 19 [*Platydemus grandis*]; Sluys 1989b, Figures 22, 24 [*Microplana terrestris* (Müller, 1774), *Platydemus grandis*]).

From the above it can be concluded that we need much more comparative data before we can adequately determine the apomorphic or plesiomorphic character conditions with regard to eye structure in planarians. Furthermore, old data (as summarized above) should be checked with more powerful, modern techniques. In that light it remains an open question whether the monophyletic status of the Dugesiidae, or the sister-group relationship between Dugesiidae and Terricola, can be supported by derived eye structural features.

Considering the fact that the monophyletic status of the Dugesiidae is still under discussion in the most recent studies on this topic (Carranza *et al.* 1998a,b), the present study necessarily follows a conservative approach in assuming that the family represents a monophylum, as suggested by earlier studies (Ball 1974a,b,c; Sluys 1989a).

Phylogenetic analysis

Characters selected

1. Head shape: low triangular (0), high triangular (1), bluntly triangular (2), rounded (3), spathulate (4), truncate (5), pointed (6), lunate (7). In the Planariidae and the Dendrocoelidae the head can be spathulate, rounded or truncated. In the Terricola the head shape is rounded, pointed, or lunate.

2. Auricles: absent (0), short (1), long (2).

3. Eyes: present (0), absent (1).

4. Position of mouth: at the hind end of the pharyngeal pocket (0), at 1/5 of the distance between the hind end of the pharyngeal pocket and the root of the pharynx (1), at 1/3 (2), halfway the distance between the hind end and the root of the pharynx (3), at 1/4 (4). In the Planariidae and the Dendrocoelidae, and generally in the Maricola, the mouth opening is located at the posterior end of the pharyngeal cavity. In the Terricola the mouth opening may be close to the hind end of the pharyngeal pocket, but it may also be at the middle of the cavity, or have shifted even more anteriad.

5. Pharynx: unpigmented (0), pigmented (1).

6. Outer pharynx musculature with three layers: absent (0), present (1). The musculature of the pharynx usually consists of an outer, peripheral zone of muscles consisting of two layers, viz. a subepidermal layer of longitudinal muscle, followed by a layer of circular muscle. Some planarians show a third, extra layer of longitudinal muscle entally to the layer of circular muscle, a condition that is not known from the Planariidae or the Dendrocoelidae and that is relatively rare among terricolans.

7. Ciliated pits: absent (0), present (1). Several dugesiids have a pair of deep sensory, ciliated pits, one on either side of the head (Figures 7.4 and 7.5). These pits differ from the much shallower sensory fossae that can be present along the anterior body margin of freshwater planarians; the number of such fossae present varies between species from one to ten on either side (de Vries and Sluys 1991; Sluys 1997). Land planarians also have an anterior sensory border with many sensory pits or 'Sinnesgrübchen' (Dendy 1892; Graff 1912–17; Froehlich 1978). Following Graff (1912-17) and Meixner (1928), I consider these 'Sinnesgrübchen' of the land planarians to be homologous with the auricular sense organs and the sensory fossae in freshwater planarians.

8. Position testes: ventral (0), dorsal (1), extending from dorsal to ventral body surface, i.e., filling the entire dorso-ventral space (2), situated in the middle of the body (3). In the Planariidae the testes usually are ventral; in the Dendrocoelidae they may be either dorsal or ventral, and in the Maricola the follicles are mainly ventral. Therefore, outgroup PD has been scored with ventrally located testes. In the Terricola the testes may be ventral, dorsal, or dorso-ventral (Winsor *et al.* 1998), with the latter condition being relatively rare. In the two supposedly most primitive subfamilies, the Rhynchodeminae and the Microplaninae (Meixner 1928; Marcus 1953), however, the testes follicles are situated at the ventral body surface. It might be argued that character conditions in such basal lineages should override conditions in more advanced clades (Yeates 1995), albeit that basal branches do not necessarily resemble the common ancestor to the greatest extent (Bininda-Emonds *et al.*

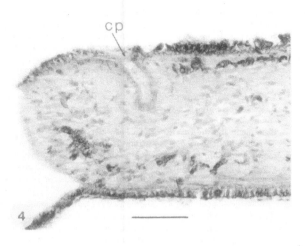

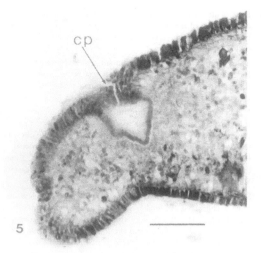

Figure 7.4, 7.5 Anterior ends with ciliated pit. **7.4**. *Spathula ochyra*; **7.5** *Romankenkius kenki*. Abbreviation: cp, ciliated pit. Scale bars: 50 μm.

1998). However, the greatest drawback in applying this method in the present case is that we lack a well-supported phylogeny for the Terricola. Therefore, outgroup T has been scored with ventral and dorsal testes.

9. Interintestinal testes: absent (0), present (1). In some species testis follicles are located not only along the outside of the gut but are situated also between the two posterior main gut branches. This condition is generally absent in the outgroups, but it has been described for the cavernicolan *Mitchellia sarawakana* Kawakatsu and Chapman, 1983 (cf. Sluys 1990).

10. Extension of testes: throughout body length (0), prepharyngeal (1), to about halfway the pharyngeal pocket (2), to the mouth at the posterior end of the pharyngeal pocket (3), to the copulatory apparatus (4).

11. Position of ovary: directly behind the brain (0), at a short distance behind the brain (1), far behind the brain (2). The last-mentioned character state refers to the condition in which the ovaries are located between one-third and one-half of the distance between the brain and the root of the pharynx. In the Planariidae and the Dendrocoelidae the ovaries are situated directly or at a short distance behind the brain. In some terricolans the ovaries may lie just anterior to the pharynx, between the mouth and the gonopore, or alongside the copulatory organs, but the majority condition for the Terricola concerns ovaries that are not too far from the brain (Winsor, *in litt.*). Therefore, the Terricola has been scored with states 0 and 1.

12. Oviduct lined with a nucleate (0) or an infranucleate (1) epithelium.

13. Caudally branched oviducts: absent (0), present (1). In some dugesiids the oviducts show a caudal dichotomy, with one medially directed branch communicating with the female copulatory apparatus and the other branch extending for greater or lesser distance, into the posterior end of the body. Such caudally branched oviducts are generally absent in the outgroups. However, they do occur in the cavernicolan *Rhodax evelinae*, Marcus 1946 (cf. Sluys 1990), and the dendrocoelid *Macrocotyla glandulosa* Hyman, 1956 (Kenk 1975); in *M. lewesi* Kenk, 1975 and *M. hoffmasteri* (Hyman, 1954) short vitelline ducts branch off from the oviducts at a position where *M. glandulosa* shows a true oviducal dichotomy (Kenk 1975). Branched oviducts occur in the terricolan subfamiliy Caenoplaninae, but over all the Terricola the condition is uncommon (Winsor, *in litt.*).

14. Common oviduct communicating with bursal canal: absent (0), present (1). In the Planariidae and the Dendrocoelidae common oviducts are present but are associated with the atrium and not with the bursal canal. In the Maricola the oviducts generally open separately or combined into the bursal canal or the female genital duct. The situation in the Terricola is complex in that the oviducts, separately or combined, may open into the atrium, or into structures that can be considered as possible homologues of the bursal canal, viz. the 'Drüsengang' or glandular duct, the *canalis anonymus*, and

Beauchamp's canal. Therefore, the Terricola has been scored as polymorphic for both character states.

15. Covering epithelium of penial papilla: nucleate (0), infranucleate (1). It is presumed that in the Planariidae and the Dendrocoelidae the penial papillae are generally covered with a nucleate epithelium. In the Terricola, the nucleate covering epithelium of non-eversible penial papillae presumably represents the most common condition (Graff 1899, 1912–17; Winsor *in litt.*).

16. Finger- or thumb-shaped penial papilla: absent (0), present (1).

17. Diaphragm: absent (0), present (1). Refers to the small papilla present in the proximal section of the ejaculatory duct in species of *Dugesia*, usually referred to as the diaphragm (cf. Sluys *et al.* 1998a, Figure 4).

18. Double seminal vesicle: absent (0), present (1). Refers to species of *Schmidtea* Ball, 1974 with two intrabulbar seminal vesicles separated by a relatively long interconnecting duct.

19. Pleated ejaculatory duct: absent (0), present (1). In some dugesiids the ejaculatory duct is not the usual straight tube with a smooth wall but consists of a duct with a pleated wall.

20. Ejaculatory duct: nucleate (0), infranucleate (1). Although occasionally infranucleate ejaculatory ducts do occur, the majority state for the Terricola, Planariidae, and Dendrocoelidae is that the ducts are lined with a nucleate epithelium. In the Maricola the ducts can be nucleate or infranucleate.

21. Penis glands opening through the covering epithelium of the penial papilla: absent (0), present (1). Penis glands usually open into the ejaculatory duct. However, in some dugesiids there are penis glands that open to the exterior through the covering epithelium of the penial papilla (see Leal-Zanchet and Hauser 1999 for a detailed example). It is presumed that the majority state for the outgroups concerns the absence of the last-mentioned type of glands.

22. Musculature of bursal canal: non-reversed (0), reversed (1), mixed (2). In the non-reversed condition the bursal canal is surrounded by a subepithelial layer of circular muscle, followed by a layer of longitudinal muscle. In the reversed condition there is a subepithelial layer of longitudinal muscle, followed by a layer of circular muscle. A mixed layer consists of intermingled circular and longitudinal muscle fibres.

23. Ectal reinforcement: absent (0), present (1), extension (2). In some dugesiids with reversed bursal canal musculature there may be present an extra third layer of longitudinal muscle ectally to the circular muscle layer, thus reinforcing the muscle coat. This ectal reinforcement with longitudinal muscles may be confined to the vaginal area and to the zone around the openings of the oviducts (state 1), but it may also extend much further along the bursal canal for well over half its length, often reaching as far as the copulatory bursa (state 2).

24. Diverticulum of bursal canal: absent (0), present (1). This refers to a diverticulum of the bursal canal located at the most proximal (ventral) section of the canal and receiving the separate or combined openings of the oviducts.

25. Thick circular muscle layer on bursal canal: absent (0), present (1). Refers to the situation that some dugesiids have an exceptionally thick coat of circular muscle on the bursal canal, a condition that is generally absent in the outgroups.

26. Bursal canal: nucleate (0), infranucleate (1). It is important to assess this character on fully mature specimens since the state of nucleation differs between ontogenetic stages (de Vries 1984). In the Maricola (cf. Sluys 1989b) and the Planariidae and the Dendrocoelidae the bursal canal is usually lined with a nucleate epithelium, a condition that also applies to the land planarians (Graff 1899, 1912–17; Winsor *in litt.*; Ogren *in litt.*).

27. Sphincter on bursal canal: absent (0), present (1). A few dugesiids have a characteristic, highly developed muscular sphincter on the bursal canal. This structure is absent in the Terricola and generally lacking in the Planariidae and the Dendrocoelidae (but it has been reported for the planariids *Phagocata kawakatsui* Okugawa, 1956 and *Ph. miyadii* Okugawa, 1939 [= *Ph. vivida* (Ijima and Kaburaki, 1916)? (cf. Kawakatsu *et al.* 1994)].

28. Opening bursal canal into atrium: posterior (0), middle (1), anterior (2). Concerns the point of communication between bursal canal and atrium. Usually, in both of the outgroups and also in the ingroup, the bursal canal opens into the postero-ventral section of the (common) atrium. However, in some dugesiids this point of communication has shifted considerably anteriad in that the bursal canal may open at about halfway along the dorsal male atrium or even at the antero-dorsal section of this atrium.

29. Shell glands opening into the diverticulum of the bursal canal or into the common oviduct: absent (0), present (1). The character can be scored only for those taxa that have a diverticulum (character 24) and/or a common oviduct (character 14) (see also character 39) comparable to those in the dugesiids in question. Therefore, the character has been scored as inapplicable for the Terricola and the Planariidae + Dendrocoelidae; a similar situation applies to the Cavernicola. In the Maricola both character states are found.

30. Atrial fold, tall atrial epithelium, penis bulb musculature extending over atrium: absent (0), present (1). This is a composite feature but it has been judged best presently not to split this complex into several characters, since it is not certain that these separate characters would be independent of each other (see also Figure 7.14). The male atrium, i.e., the section of the atrium closest to the root of the penial papilla, is separated from the common atrium by a fold or constriction. The male atrium is lined with an epithelium that is usually much taller than that of the adjacent common atrium. This tall epithelium of the male atrium is underlain with a thick layer of circular muscle, followed by a well-developed layer of longitudinal muscle that is confluent with the musculature of the penis bulb. This composite feature is generally absent in the outgroups, albeit that strikingly similar conformations occur in species of the planariid (sub)genus *Atrioplanaria* de Beauchamp, 1932.

31. Glandularization of common atrium: absent (0), present (1). In some dugesiids there are abundant glands having their openings into the common atrium. It is presumed that in the majority of the Planariidae and the Dendrocoelidae such glands are absent. In the Terricola differentiation of the atrium into male, female, and common parts is difficult (as is often the case also in paludicolans), but many species have an atrium that is lined with glandular cocoon-forming epithelium (Winsor 1998, *in litt.*). Therefore, the Terricola has been scored for both states.

32. Common atrium: nucleate (0), infranucleate (1). In the Planariidae and the Dendrocoelidae the common atrium is nucleate. The situation in the Terricola is difficult to interpret since epithelia can be both nucleate and infranucleate, depending on what the function is in the manufacturing of the cocoons. In areas that are involved in cocoon production the epithelium is infranucleate (Winsor *in litt.*). Therefore, the Terricola has been scored as polymorphic for both character states.

33. Caudal fold in female or common atrium: absent (0), present (1). In some dugesiids the posterior wall of the female or common atrium (it is not always easy to distinguish between these two parts) shows a characteristic outbulging or fold that as such is absent in the outgroups.

34. Separate male and female gonopores: absent (0), present (1). The derived character (state 1) represents a very rare condition among triclads. With respect to the outgroups, it is known only from: a) a few terricolans: *Digonopyla harmeri* (Graff, 1899) (cf. Fischer 1926), an undescribed species of *Dygonopyla* from Australia (Winsor *in litt.*), a single individual of the species *Othelosoma caffrum* Marcus, 1955 (Marcus 1955, Figure 43), four species of the current genus *Diporodemus* Hyman, 1938 (cf. Ball and Sluys 1990, Figure 7.3.C); b) the cavernicolans *Opisthobursa mexicana* Benazzi, 1972, *O. josephinae* Benazzi, 1975 (cf. Sluys 1990); and c) the marine triclads *Nesion arcticum* Hyman, 1956, *Uteriporus vulgaris* Bergendal, 1890, *U. pacificus* Sluys, 1989, *Tiddles evelinae* Marcus 1963, *Oregoniplana opisthopora* Holmquist and Karling, 1972, and *O. pantherina* Sluys, 1989 (cf. Sluys 1989b). However, the condition with two separate gonopores in these outgroup taxa often is rather different from the situation in *Bopsula*.

35. Penial-atrial duct: absent (0), present (1). The character can easily be gleaned from Figure 7.6.

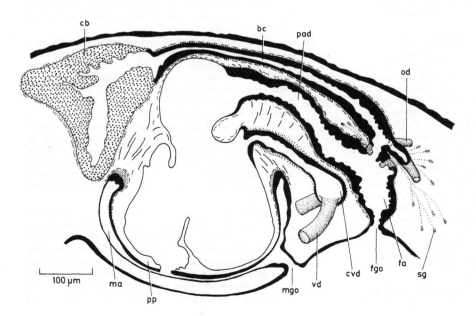

Figure 7.6 Bopsula evelinae. Sagittal reconstruction of copulatory apparatus of the holotype. Abbreviations: bc, bursal canal; cb, copulatory bursa; cvd, common vas deferens; fa, female atrium; fgo, female gonopore; ma, male atrium; mgo, male gonopore; od, oviduct; pad, penial-atrial duct; pp, penial papilla; sg, shell gland; vd, vas deferens.

36. Eversible penial papilla: absent (0), present (1). An eversible penial papilla is rare among freshwater planarians and is confined to several dendrocoelid species (cf. Kenk 1978, Figures 44, 46, 50, 55). Eversible papillae are uncommon among the Terricola (Winsor *in litt.*; cf. Marcus 1951, Figures 153, 154) and the Maricola (cf. Sluys 1989b).

37. Dactylose projections on bursal canal: absent (0), present (1). The presence of these structures seems to be confined to *Reynoldsonia reynoldsoni* Ball, 1974.

38. Fused testes: absent (0), present (1). Generally, planarians have discrete testicular follicles. Among the outgroups, fusion of the follicles to a tube-like system has been reported only for the cavernicolans *Rhodax evelinae* and *Opisthobursa josephinae* (cf. Sluys 1990).

39. Common oviducal opening into bursal canal: absent (0), opening well into bursal canal (1), opening at the most proximal (posteroventral) section of the bursal canal or even into the common atrium (2). This character combines characters 14 (common oviduct) and 24 (diverticulum bursal canal) according to an earlier reinterpretation of these features in species of *Romankenkius* Ball, 1974 representing a feature that is unique among triclads (Sluys 1997).

Results

Minimum length and consensus trees

The data matrix (Table 7.1) was first analysed with PAUP 3.1.1 under combinations of various options: exclusion of characters 14 and 24, or exclusion of character 39; inclusion of both outgroups, making the outgroup taxa either paraphyletic or monophyletic with respect to the ingroup; excluding

one of the outgroup taxa. In each of these runs the program generated the maximum number of 32 700 minimum length trees.

In cases where the effect was studied of either excluding characters 14 and 24, or excluding character 39, with other options kept constant, minimum length trees that excluded the two first-mentioned features in most cases had somewhat higher consistency and retention indices (Table 7.2). This is interpreted as an indication of the situation that character 39 forms indeed a better descriptor of this particular aspect of planarian anatomy. Therefore, in the following most attention shall be paid to generated trees that include character 39 and exclude characters 14 and 24.

In cases where the two outgroups were assigned together to the outgroup taxa box, the program appeared to be unable to keep both outgroups as the most basal members on the phylogenetic tree. It did not matter whether the option to make the outgroups paraphyletic with respect to ingroup, or the option to make the outgroups monophyletic was chosen; in both cases, and with outgroup T as the primary outgroup, the end result was the same, resulting in a 50% majority rule consensus tree in which the species *Cura pinguis* took a basal position between the two outgroups, thus implying that the ingroup is not monophyletic (Figure 7.7 = Run 1; compare also runs 1 and 7 in Table 7.2). This results from the situation that in cases where the tree cannot be rooted such that the ingroup is monophyletic, PAUP will include outgroup taxa to make the ingroup plus the necessary outgroup taxa monophyletic, but the program keeps the primary outgroup always at the root of the tree (Swofford 1993).

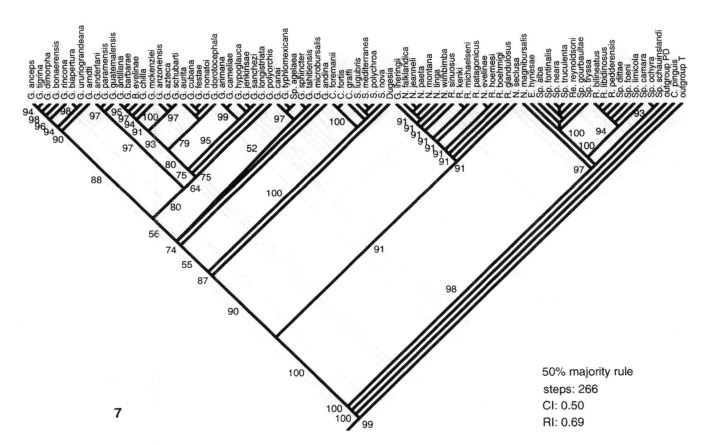

Figure 7.7 A 50% majority-rule consensus tree (Runs 1 and 7; characters 14 and 24 excluded); steps: 266, CI = 0.50, RI = 0.69. Group frequencies of major clades shown on branches; not all frequencies indicated.

Table 7.2 Length and fit measures of minimum length trees (32 700 for each run) generated under various options.

Run	Chars. 14 & 24 excluded	Char. 39 excluded	Outgroups PD & T paraphyl.	Outgroups PD & T monophyl.	Outgroup PD monophyl.	Outgroup T monophyl.	Constraint tree	Length	CI	RI
1	+		+					251	0.506	0.701
2		+	+					253	0.496	0.696
3	+					+		252	0.478	0.683
4		+				+		250	0.482	0.690
5	+				+			242	0.473	0.692
6		+			+			242	0.469	0.692
7	+			+				251	0.504	0.701
8		+		+				253	0.496	0.696
9	+						+	266	0.479	0.675
10		+					+	255	0.496	0.700

There are two reasons for not accepting this result. First, the inclusion of outgroup PD in the ingroup, the Dugesiidae, does not conform to the currently available hypotheses (Carranza *et al.* 1998a, Figure 3; 1998b, Figure 1; 1999, Figure 2) on the higher level phylogenetic relationships within the Tricladida. Second, the present database was not designed to resolve such higher-level relationships but addresses another level in the phylogenetic tree of the Tricladida. Therefore, the data matrix was subsequently analysed under two other options: a) exclusion of one of the two outgroups from the outgroup taxon list; and b) inclusion of a constraint tree, with the outgroup taxa as the two most basal branches of the tree.

With the inclusion of only one of the outgroup taxa in the analysis, the program generated again its maximum of 32 700 trees; the 50% majority rule consensus trees for these runs 3 and 5 (see Table 7.2) are shown in Figures 7.8 and 7.9.

It turned out to be impossible to run analyses with constraint trees under PAUP 3.1.1, and therefore PAUP* was used for enforcing topological constraints. However, searches under PAUP* were limited to a maximum number of 32 700 minimum length trees, in conformity with the limit of PAUP 3.1.1. In these analyses it made no difference whether outgroup PD or T took the most basal position in the constraint tree. The 50% majority rule trees for these runs 9 and 10 (see also Table 7.2) are shown in Figures 7.10 and 7.11.

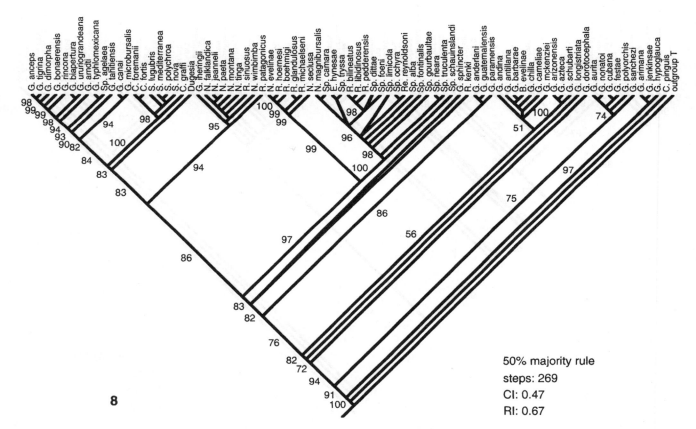

Figure 7.8 A 50% majority-rule consensus tree (Run 3; characters 14 and 24 excluded; outgroup PD excluded); steps: 269, CI = 0.47, RI = 0.67. Group frequencies of major clades shown on branches; not all frequencies indicated.

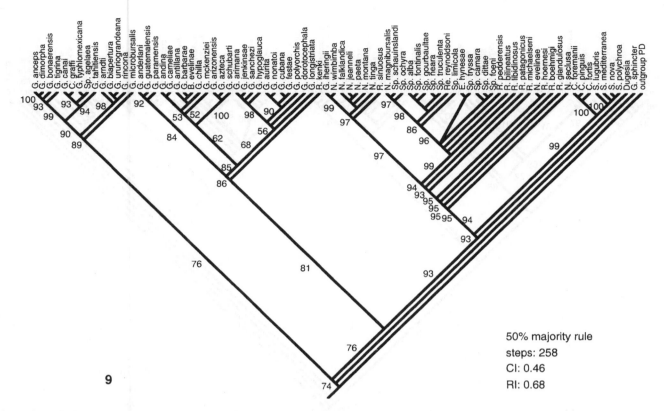

9

50% majority rule
steps: 258
CI: 0.46
RI: 0.68

Figure 7.9 A 50% majority-rule consensus tree (Run 5; characters 14 and 24 excluded; outgroup T excluded); steps: 258, CI = 0.46, RI = 0.68. Group frequencies of major clades shown on branches; not all frequencies indicated.

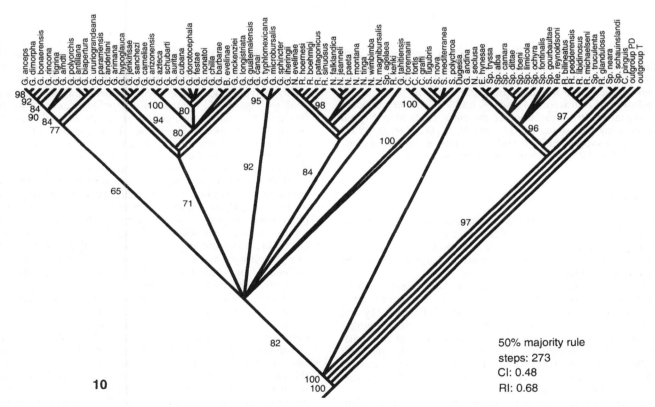

10

50% majority rule
steps: 273
CI: 0.48
RI: 0.68

Figure 7.10 A 50% majority-rule consensus tree (Run 9; characters 14 and 24 excluded; topological constraints enforced); steps: 273, CI = 0.48, RI = 0.68. Group frequencies of major clades shown on branches; not all frequencies indicated.

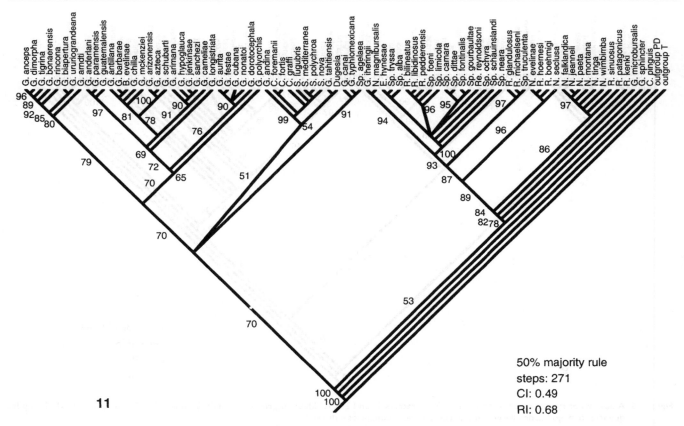

Figure 7.11 A 50% majority-rule consensus tree (Run 10; character 39 excluded; topological constraints enforced); steps: 271, CI = 0.49, RI = 0.68. Group frequencies of major clades shown on branches; not all frequencies indicated.

Bootstrapping of the data matrix led to poor results, with almost all of the clades having a branch support far below 50%. Two clades form an exception, one comprising the four species of *Schmidtea* (bootstrap support 82% under exclusion of characters 14 and 24, and 79% under exclusion of character 39), and the other the clade containing *Neppia jeanneli* (de Beauchamp, 1913) and *N. paeta* (Marcus, 1955) (bootstrap support of 73% and 78% under exclusion of characters 14 and 24, and character 39, respectively).

Tree selection

The five majority rule consensus trees generated under various options (Figures 7.7–7.11), differ considerably in the way that they group species into clades. It is evident that these differences can be traced back to the choice of outgroup(s) and the enforcement of topological constraints.

It is striking that under topologically constrained analyses character 39 appears to perform somewhat less than the combination of characters 14 and 24 (Table 7.2; Figures 7.10 and 7.11), in contrast to most runs without constraints. Although the consistency and retention indices differ not so much between both conditions, the 50% majority rule tree with inclusion of character 39 (and thus exclusion of characters 14 and 24) is much less resolved (Figure 7.10) than the one incorporating character 39 (Figure 7.11).

Inclusion of only one of the outgroup taxa resulted in considerably different consensus trees, as compared with the tree resulting from the first, unconstrained run including both outgroups. The trees differ in many details, but the most conspicuous difference is that in the single-outgroup runs (Figures 7.8 and 7.9) the genus *Girardia* Ball, 1974 is highly polyphyletic. In the unconstrained, multiple outgroup analysis (Figure 7.7) *Girardia* is monophyletic, except for the fact that it includes the monotypic genus *Bopsula* and the poorly known species *Sp. agelaea* Hay and Bal, 1979, and excludes the equally poorly known *G. iheringii* (Böhmig, 1887). In the single-outgroup runs *Bopsula* and *Sp. agelaea* do also settle down in clades with species of *Girardia*.

With these five majority rule consensus trees at hand, the next step is to choose one tree as the best working hypothesis for further analysis of character distribution, comparison with phylogenetic hypotheses resulting from other studies, and for future testing. The choice is based on several criteria. Generally, it is preferable to include more than one outgroup in a phylogenetic analysis, certainly in cases where there still is discussion on the precise sister-group of the ingroup. Therefore, the best hypothesis in the present case should come from analyses incorporating outgroup PD and outgroup T, i.e., the consensus trees depicted in Figures 7.7 and 7.10 (note that the consensus tree in Figure 7.11, although including both outgroups, concerns a different selection of characters).

The constrained analyses were induced by the anomalous basal position of *C. pinguis* between the two outgroups in the unconstrained runs. However, enforcing topological constraints only minimally affected the position of *C. pinguis* by positioning the species as the most basal ingroup taxon. In that respect, the shorter consensus tree (Figure 7.7), resulting from the much shorter minimum length trees generated during the unconstrained run, is to be preferred over the constrained hypothesis (Figure 7.10), ignoring – for the time being (see below) – the anomalous position of *C. pinguis* between the two outgroups. One other reason for choosing the consensus tree in Figure 7.7 as the best working hypothesis is that it is much more resolved than its alternative.

Character tracings

This consensus tree has been explored for features that can be postulated as defining, apomorphic characters for particular clades. This was achieved by tracing the distribution of characters on the tree with the help of MacClade. An indication of possibly useful characters can be found also in the ranges of the fit measures for the characters in all minimum length trees (Table 7.3). I do apprehend the fact that majority rule consensus trees may propose clades that are not supported by other trees that are equally parsimonious. For this reason one might object to using such trees for tracing characters. However, for heuristic purposes I consider the use of majority rule consensus trees for character tracings to be a valid and useful technique (see also Sluys *et al.* 1998a).

On the basis of the present data matrix, the following characters can be considered as unequivocal and strict apomorphies for several taxa: 6; 17 and 23; 18 and 22; 34 and 35; 36 and 37; 38 (Figure 7.12). Character 6 is polymorphic, in that postulated apomorphic state (1) for *N. jeanneli* and *N. paeta* is also present in several species of *Dugesia*. A similar situation applies to character 22(2), where outside of the *Schmidtea* clade the derived condition (2) also occurs in polymorphic

Re. reynoldsoni Ball, 1974 and *R. libidinosus* Sluys and Rohde, 1991.

The character tracings done with MacClade, and also Table 7.3, suggest that there are eight more features with character states that could well be hypothesized as apomorphies for particular clades, albeit that each of these characters exhibits cases of secondary loss or parallelism. These characters will be discussed below; the level of generality at which they are postulated as apomorphies is indicated in Figure 7.12.

A pigmented pharynx (character 5, state 1) is characteristic for a large group of *Girardia's*, including *B. evelinae* Marcus, 1946. In this group there are three species that are polymorphic, in that individuals may have either pigmented or unpigmented pharynges: *G. tigrina* (Girard, 1850), *G. antillana* (Kenk, 1941), and *G. chilla* (Marcus, 1954). Furthermore, *G. anceps* (Kenk, 1930) has an unpigmented pharynx, while information is absent for *G. dimorpha* (Böhmig, 1902), *G. bonaerensis* (Moretto, 1996) *G. biapertura* Sluys, 1997, *G. barbarae* (Mitchell and Kawakatsu, 1973), *G. mckenziei* (Mitchell and Kawakatsu, 1973), *G. aurita* (Kennel, 1888), *G. hypoglauca* (Marcus, 1948), and *G. sanchezi* (Hyman, 1959).

Two clades nested within the previous one can be characterized by the possession of a low triangular head (character 1, state 1) and long auricles (character 2, state 2). The two clades almost overlap in their constituent species, but have apomorphies postulated at slightly different levels of universality. With respect to character 1(1) it is assumed that *G. polyorchis* (Fuhrmann, 1914) (for which information on head shape is lacking) will turn out to have a low triangular head. Concerning character 2(2), information is absent for *G. aurita*, while the feature also occurs in *N. montana* (Nurse, 1950).

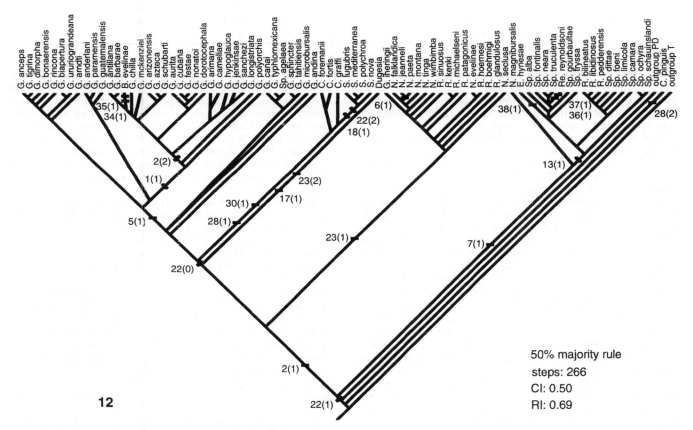

Figure 7.12 A 50% majority-rule consensus tree from Figure 7.7. Postulated apomorphies indicated by black rectangles, accompanied by character number and state (within brackets). For further explanation, see text.

Table 7.3 Length and fit measures for each character over all minimum length trees resulting from Run I (cf. Table 7.2); *, excluded characters.

Character	1	2	3	4	5	6	7	8	9	10
Min. length	16	7	3	14	4	2	5	25	2	14
Max. length	18	8	4	14	6	2	6	26	4	15
CI max.	0.625	0.429	0.333	0.714	1.000	1.000	0.400	0.680	0.500	0.571
CI min.	0.556	0.375	0.250	0.714	0.667	1.000	0.333	0.654	0.250	0.533
RI max.	0.854	0.905	0.667	0.200	1.000	1.000	0.833	0.758	0.889	0.700
RI min.	0.805	0.881	0.500	0.200	0.889	1.000	0.778	0.727	0.667	0.650

Character	11	12	13	14*	15	16	17	18	19	20
Min. length	22	6	3	11	7	3	1	1	7	7
Max. length	23	7	4	12	8	3	1	1	7	8
CI max.	0.364	0.167	0.333	0.364	0.857	0.333	1.000	1.000	0.714	0.714
CI min.	0.348	0.143	0.250	0.333	0.750	0.333	1.000	1.000	0.714	0.625
RI max.	0.500	0.643	0.875	0.562	0.750	0.333	0/0	1.000	0.500	0.714
RI min.	0.464	0.571	0.812	0.500	0.500	0.333	0/0	1.000	0.500	0.571

Character	21	22	23	24*	25	26	27	28	29	30
Min. length	9	9	7	3	9	7	3	4	5	3
Max. length	10	9	7	3	10	7	3	4	5	3
CI max.	0.111	0.444	0.286	0.333	0.444	0.571	0.333	0.500	0.400	0.333
CI min.	0.100	0.444	0.286	0.333	0.400	0.571	0.333	0.500	0.400	0.333
RI max.	0.333	0.839	0.583	0.500	0.615	0.700	0.000	0.750	0.667	0.714
RI min.	0.250	0.839	0.583	0.500	0.538	0.700	0.000	0.750	0.667	0.714

Character	31	32	33	34	35	36	37	38	39
Min. length	9	4	5	1	1	1	1	1	13
Max. length	10	4	5	1	1	1	1	1	14
CI max.	0.222	0.750	0.600	1.000	1.000	1.000	1.000	1.000	0.462
CI min.	0.200	0.750	0.600	1.000	1.000	1.000	1.000	1.000	0.429
RI max.	0.562	0.500	0.500	0/0	0/0	0/0	0/0	0/0	0.588
RI min.	0.500	0.500	0.500	0/0	0/0	0/0	0/0	0/0	0.529

It can be hypothesized that the clade with the long auricles arose within a clade defined by short auricles (character 2, state 1), albeit that this feature shows nine cases of secondary loss and ten species within the clade for which information on auricles is absent (including *G. aurita*).

Ectal reinforcement extending along the bursal canal for a considerable distance (character 23, state 2) is restricted to the genus *Dugesia*. Ectal reinforcement that is confined to the vaginal area (character 23, state 1) is here postulated as an apomorphy for a clade containing mostly species of *Neppia* Ball, 1974 and *Romankenkius* Ball, 1974, but also the species *G. iheringii* (for which information is absent). Within this clade *R. kenki* Ball 1974 and *R. hoernesi* (Weiss, 1909) show the primitive condition (character 23, state 0), while information on the character condition is lacking for five species (including *G. iheringii*). Outside of this clade, character 23(1) occurs in parallel in *Sp. truculenta* Ball, 1977, *R. bilineatus* Ball and Tran, 1979, *R. libidinosus*, and *Sp. ochyra* Ball and Tran, 1979.

Ciliated pits (character 7, state 1) are characteristic for a clade in which one species, *Sp. schauinslandi* (Neppi, 1904), has secondarily returned to the primitive condition without these sensory organs. Pits occur in parallel in the poorly known *Sp. agelaea* (anomalously placed among species of *Girardia*), *N. wimbimba* (Marcus, 1970), and *R. kenki*. Nested within this clade and almost overlapping with it, is a group of species characterized by the synapomorphic presence of caudally branched oviducts (character 13, state 1); one species, *R. bilineatus*, has secondarily reversed to the primitive condition. There is one case of parallelism since *R. sinuosus* also possesses character 13(1).

The apomorphic condition in which the bursal canal communicates with more anteriorly located sections of the atrium (character 28, state 1), as opposed to opening into the postero-ventral part of the atrium, is hypothesized as an apomorphy for a group of *Cura* Strand, 1942 and *Schmidtea* species; there are two cases of parallelism, *G. dimorpha* and *N. evelinae* (Marcus, 1955). Another apomorphic condition from this transformation series, with the opening of the bursal canal shifted even more anteriad (character 28, state 2), is restricted to *C. pinguis*.

The composite character 30(1) is another presumed synapomorphy uniting the current genera *Cura* and *Schmidtea*; the character condition occurs in parallel in the anomalously placed *C. pinguis* and in *N. montana*.

Reversed bursal canal musculature (character 22, state 1) is an apomorphic condition, as compared with the plesiomorphic state (character 22, state 0) in the outgroups, and is here hypothesized as an apomorphy at a basal position in the tree. It is presumed to be secondarily reversed in *R. kenki*, *R. michaelseni* (Böhmig, 1902), and *E. hynesae* Ball, 1977. Furthermore, *Re. reynoldsoni* and *R. libidinosus* are polymorphic, with states 1 and 2 and 0 and 2, respectively; information is absent for *G. iheringii*, *N. wimbimba*, and *R. patagonicus* (Borelli, 1901). It is hypothesized that a reversal to the plesiomorphic condition (0) represents an apomorphy for a clade comprising mostly species of *Girardia*, *Cura* and *Schmidtea*, with a further apomorphic modification to mixed bursal canal musculature (state 2) being expressed in representatives of the last-mentioned genus. Within the clade united by character 22(0), *C. fortis* expresses state (1), while information on character condition is absent for *G. aurita*, *G. tahitiensis* (Gourbault, 1977), *G. microbursalis* (Hyman 1931), *G. andina* (Borelli, 1895), and the poorly known and anomalously placed *Sp. agelaea*.

Discussion

Phylogeny

It is clear from the distribution of the hypothesized apomorphies on the phylogenetic tree (Figure 7.12) that for some major clades unambiguous derived features could not be found. The computer-generated topology of the tree will be retained for further discussion, but strictly speaking some of the clades should be collapsed.

The genus *Dugesia* forms the sister-group of a clade comprising mostly species of *Girardia*, *Cura*, and *Schmidtea*, in contrast to earlier morphological studies (e.g., Sluys 1997) that positioned *Dugesia* as the sister-group of species of *Neppia*. The present result, however, is much more in conformity with molecular analyses that clustered the genera *Schmidtea*, *Cura*, and *Dugesia* (Carranza *et al.* 1998b), sometimes also including *Girardia* (Carranza *et al.* 1998a).

It is evident that on the basis of the present analysis the current taxonomic genera *Neppia*, *Romankenkius*, and *Spathula* do not form monophyletic taxa. Furthermore, it could be argued that the monotypic genera *Reynoldsonia* Ball, 1974 and *Eviella* Ball, 1977 merely represent aberrant species of *Spathula*. In similar vein, it can be argued that monotypic *Bopsula* is an aberrant representative of the genus *Girardia*. Although its copulatory apparatus is highly unusual, *Bopsula*'s external features, that were not depicted before (Figure 7.13), together with its pigmented pharynx, point to a species of *Girardia*. Another aberrant species for the genus, *G. biapertura*, was recently described by Sluys *et al.* (1997), who did not consider it necessary to place this unique species in a genus of its own.

Although tree topology (Figure 7.12) suggests that the current genus *Girardia* is monophyletic, the present database did not suggest an unambiguous apomorphy uniting all of the species. Six species of *Girardia* are placed directly outside of the clade united by the apomorphic presence of a pigmented pharynx and there is no synapomorphy uniting these six species with the rest of the genus. It should be noted that *G. iheringii*, placed in a completely different clade than all other species of *Girardia*, is a poorly known species, which may account for its anomalous position.

Diagnoses

The current genus *Girardia* lacks diagnostic, apomorphic features. However, major clades within the genus can be diagnosed by apomorphies, thus lending some support to the hypothesis implied by current taxonomy, i.e., that *Girardia* represents a natural, monophyletic taxon.

The genus *Dugesia* is a well-diagnosable, monophyletic taxon. The present database suggests the following characters as synapomorphies for species of *Dugesia*: presence of a diaphragm, and extension of the ectal reinforcement along the bursal canal. An earlier study suggested that *Dugesia* is also characterized by oviducts arising from the dorsal surface of the ovaries. This hypothesis should be tested in future studies by incorporating this feature into the data matrix and scoring the

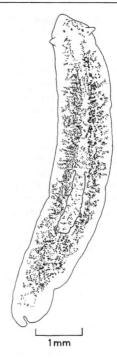

1mm

Figure 7.13 Bopsula evelinae. Whole mount.

various character conditions for all included taxa – something that was not possible in the present analysis.

In a previous study (Sluys 1997) the genus *Cura* was already restricted to the species *C. pinguis*, *C. foremanii* (Girard, 1852), and *C. graffi* (Weiss, 1909). On the basis of the present analysis, the genus lacks defining features and, consequently, does not constitute a monophyletic unit. However, one should realize that *C. graffi* is a poorly known species that was only by default placed in the genus *Cura* (Sluys 1997), and that the anomalous position of *C. pinguis* between the two outgroups is debatable (see below), mostly resulting from limitations of data analysis (see above).

The current genus *Schmidtea* can be diagnosed by two unequivocal apomorphies, the presence of a double seminal vesicle, and a coat of intermingled muscles around the bursal canal – a conclusion that was also reached by earlier studies (de Vries and Sluys 1991; Sluys 1997).

The status of the monotypic genera *Bopsula*, *Eviella*, and *Reynoldsonia* has been already discussed above and the apomorphic characters of the species can be gleaned from Figure 7.12. Previous studies (de Vries and Sluys 1991; Sluys 1997) incorporated the presumed apomorphic presence of a sperm attachment zone in the female genital duct of *E. hynesae*. However, this represents a poorly defined character that does not feature in the current diagnosis of the genus (cf. Ball 1977b: 149), and has been omitted from the present analysis.

In view of the highly dispersed position of their constituent species on the phylogenetic tree (Figure 7.12), it is clear that the current genera *Neppia*, *Spathula*, and *Romankenkius* cannot be diagnosed by uniquely derived characters.

Tree manipulation

It would be premature to draw firm phylogenetic or taxonomic conclusions from the present study. Preliminary trials have already shown that minimal changes in the data matrix may result in large topological changes in the trees generated – a result that is not unexpected in view of the bootstrap results obtained during the present study. However, there is a difference between the accuracy of a phylogenetic hypothesis and the precision of a measure of support, such as the bootstrap. From that perspective, a number of considerations are presented below for manipulating the chosen working hypothesis (Figure 7.12), thus generating a new, 'guestimated' tree that can fruitfully be used in future taxonomic and phylogenetic studies.

First of all, the anomalous position of *C. pinguis* has to be considered. This species exhibits the two highly characteristic taxonomic features that define the clade with four species of *Schmidtea* and three species of *Cura* (including the poorly known *C. graffi*): the composite character 30(1), and a bursal canal communicating with a much more anterior section of the atrium (character 28, states 1 and 2). In point of fact, state (2) of character 28 in *C. pinguis* could well be seen as a more derived condition that evolved from character 28(1). Furthermore, *C. pinguis* possesses a finger- or thumb-shaped penial papilla (character 16, state 1), as do *C. foremanii* and *C. fortis*. These are all well-defined and complex features to which one may wish to assign higher weight on an *a priori* or an *a posteriori* basis. However, differential weighting of characters is a complex and contentious issue and therefore has not been attempted in the present analysis. However, on the basis of the considerations given above, it is predicted that *C. pinguis* is most closely related to *C. foremanii* and *C. fortis*. In this context it is noteworthy that in the bootstrap analysis and in the consensus tree resulting from Run 5 (see Table 7.2 and Figure 7.9), *C. pinguis* grouped with *C. foremanii* and *C. fortis*. Moving the branch of *C. pinguis* to the clade of these two species, and making the resulting clade into a polytomy, enlarges the original tree length (266 steps) by one step only.

A bursal canal communicating with the middle section of the atrium (character 28, state 1) and a finger- or thumb-shaped penial papilla (character 16, state 1) were also scored for *N. evelinae*, suggesting that the species also belongs to the *Schmidtea-Cura* clade in general and to the *Cura* branch in particular. Furthermore, after the above analysis was completed I have had the opportunity to examine the type material of *N. evelinae*. Detailed reconstruction of the copulatory apparatus showed indeed that the conditions of characters 16(1) and 28(1) apply to this species, and that the complex character 30(1) holds true also for *N. evelinae* (Figure 7.14). This suggests strongly that *N. evelinae* is actually a species of *Cura*, as initially envisioned by Marcus (1955). Incorporating *N. evelinae* into the *Cura* clade does not alter the length of the tree.

There are three other characters suggesting some tree manipulation. In a previous study, character 39(2) was considered to be a unique, derived feature for the genus *Romankenkius* (Sluys 1997). Surprisingly, this well-defined and unique character, present in all current species of *Romankenkius*, does not suffice for identifying *Romankenkius* as a monophyletic genus. In view of the uniqueness of the condition one might contemplate giving this feature more weight in an analysis. In addition, most species of *Romankenkius* also possess shell glands opening into this common duct (character 29, state 1), excepting *R. michaelseni* (Böhmig, 1902), and *R. glandulosus* (Kenk, 1930). Character 29(1) is present also in *N. evelinae* and *G. sanchezi*, but it is doubtful whether these conditions are homologous with that in *Romankenkius*.

Most species of *Neppia* have a thick muscle coat on the bursal canal (character 25, state 1), excepting *N. seclusa* (de Beauchamp, 1940); the character condition is also expressed in *G. ururiograndeana* (Kawakatsu, Hauser and Ponce de Léon, 1992), *G. iheringii*, *N. evelinae*, *Sp. fontinalis* Nurse, 1950 (polymorphic), *Sp. gourbaultae* Ball, 1977, *Sp. dittae* Ball and Tran, 1979, *Sp. camara* Ball, 1977, *R. sinuosus*, *R. libidinosus* (polymorphic), and some species of *Dugesia*.

The distribution of these three characters suggests that the genera *Neppia* and *Romankenkius* are closely related and that their constituent species are more closely related to each other than to species from other genera. The tree has been manipulated in order to achieve this representation, with the result that the tree became again seven steps longer.

Although *Sp. agelaea* is a poorly known species, it is certain that it is not a *Girardia* and is most likely a species of *Spathula*; therefore the species has been moved to the *Spathula* clade, thus increasing the tree length with two steps.

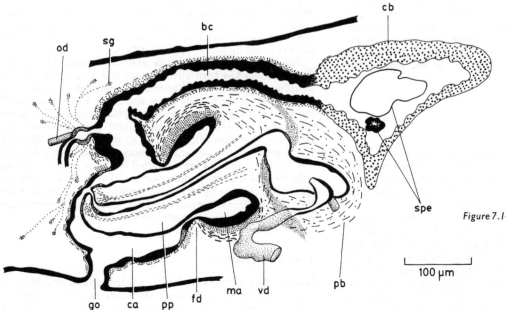

Figure 7.14 *Neppia evelinae*. Lectotype. Sagittal reconstruction of the copulatory apparatus. Abbreviations: bc, bursal canal; ca, common atrium; cb, copulatory bursa; fd, fold; go, gonopore; ma, male atrium; od, oviduct; pb, penis bulb; pp, penial papilla; sg, shell gland; spe, spermatophore; vd, vas deferens.

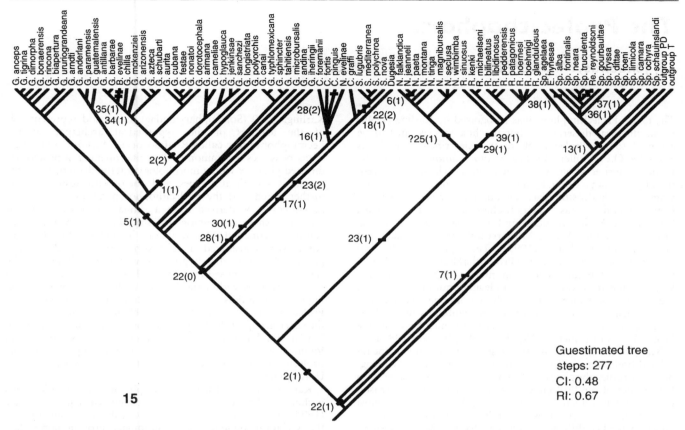

Figure 7.15 'Guestimated' tree. Presumed apomorphies indicated by black rectangles, accompanied by character number and state (within brackets); steps: 277, CI = 0.48; RI = 0.67. For further explanation, see text.

In view of its features, it is considered much more likely that the rather poorly known *G. iheringii* is a species of *Girardia* that is not closely related to the *Neppia* clade. Moving the branch of *G. iheringii* to the base of the *Girardia* clade costs one step.

The resulting 'guestimated' phylogenetic hypothesis (Figure 7.15) is 11 steps longer than the initially selected working hypothesis (Figure 7.12), with only marginally smaller consistency and retention indices.

ACKNOWLEDGEMENTS

I am grateful to the following institutions for loans of material: Museum für Naturkunde der Humboldt-Universität (Berlin), The Natural History Museum (London), Queensland Museum (Brisbane), Royal Ontario Museum (Toronto), Tasmanian Museum and Art Gallery (Hobart), Western Australian Museum (Perth). Drs L. Winsor, R. E. Ogren, and E. M. Froehlich are thanked for sharing their knowledge on character states in the Terricola. Prof. Dr M. Kawakatsu is thanked for providing photocopies of some of the papers cited and for reading the manuscript. Dr D. Swofford kindly made available a pre-release version of PAUP* to the Faculty of Biology of the University of Amsterdam. Dr L. Botosaneanu (Amsterdam) kindly translated parts of some Russian papers. The Linnean Society of London, The Systematics Association, and the British Society of Parasitology are thanked for inviting the author to participate in the symposium on the Interrelationships of the Platyhelminthes.

Chapter 8

The Prolecithophora

Ulf Jondelius, Michael Norén and Jan Hendelberg

The taxon Prolecithophora was recognized by Karling (1940) in his doctoral dissertation, but the first species later to be assigned to the Prolecithophora was described as early as 1788 by O.F. Müller (*Vorticeros auriculatum*). Today, the taxon comprises about 150 known species classified conventionally in about a dozen families: Protomonotresidae, Pseudostomidae, Cylindrostomidae, Scleraulophoridae, Urastomidae, Genostomatidae (as Hypotrichinidae), Baicalarctiidae, Ulianinidae, Multipeniatidae and Plagiostomidae as recognized by Cannon (1986). Most recently the Torgeidae was described by Jondelius (1997). The approximately 150 known species belonging to the Prolecithophora are free-living; though found predominantly in marine habitats, they are known from freshwater. Most prolecithophorans live in soft sediments, sometimes interstitially or among plants. The Urastomidae and Genostomatidae are symbionts of bivalves, crustaceans, or fish, but may not be members of the Prolecithophora (Ehlers 1988; Hyra 1993; Watson 1997a).

Many of the species possess two pairs of eyes and an encapsulated brain (Figure 8.1), and are thus relatively easily recognized as prolecithophorans. In spite of this, there have been few studies of the Prolecithophora in later years. One reason for this may be difficulties in identifying the species, since most of them are quite opaque when studied live in the light microscope. A further difficulty is the absence of sclerotized structures in most of the species.

Karling (1974) stated, while discussing the relationships of 'turbellarian orders' that 'the Prolecithophora for the specialist…appears as a very homogeneous group'. However, he also noted that there were no known synapomorphies for the taxon. Ehlers (1988) suggested the presence of abundantly-folded membrane derivatives in the aflagellar sperm cells of representatives of Pseudostomidae, Protomonotresidae, Cylindrostomidae and Plagiostomidae as a synapomorphy of the Prolecithophora. Through Ehlers's discovery we have a conflict-free synapomorphy for the group. In the following we reserve the use of the name Prolecithophora to flatworms with the sperm structures described above, and worms that form a monophylum with them, excluding taxa with different sperm morphology.

The genus *Ulianinia* was referred to a separate family by Meixner (1938), and later transferred to the Plagiostomidae (as *Plagiostomula*) by Westblad (1956). Karling (1963) resurrected the Ulianinidae as a separate family and considered it intermediate between the Plagiostomidae and Cylindrostomidae.

We studied the sperm morphology at the ultrastructural level in *Ulianinia mollissima* Levinsen, 1879 and *Reisingeria hexaoculatus* Westblad, 1955, to establish the presence or absence of prolecithophoran synapomorphies in the spermatozoa of these taxa, and to provide further data that may prove useful in phylogeny reconstruction.

While the monophyly of the Prolecithophora is established, the sister-group of the taxon has remained obscure. The Tricladida were regarded by Karling (1940) as closely related to the Prolecithophora, but classified separately from the Prolecithophora with the Proseriata for 'morphological and practical reasons'. While discussing the affinities within the Neoophora, Karling proposed *Anthopharynx vaginatus*

Karling, 1940 (Solenopharyngidae) as a good representative of the features of the hypothetical archetype of the Lecithophora (now called Rhabdocoela). *Anthopharynx vaginatus* possesses a common oral–genital opening, a posteriorly directed pharynx with a morphology intermediate between the rosulatus and plicatus types, and dorsorostral testes. Karling further noted that the prolecithophoran taxon Cylindrostominae could be derived from a similar archetype and that they thus share a common ancestor. In this scheme, which could be termed the Lecithophora hypothesis, the Rhabdocoela were considered a subtaxon within the Prolecithophora. However, this hypothesis was not reflected in the classification proposed by Karling (1940; Karling's discussion of relationships within the Prolecithophora is represented in Figure 8.2). Ehlers (1985a) considered the position of the Prolecithophora within the Platyhelminthes uncertain, i.e., the sister-group of the Prolecithophora could not be identified. Norén and Jondelius (1999) analysed 18S rDNA data, and found indications of a sister-group relationship between the Prolecithophora and a taxon consisting of the Tricladida and Urastomidae.

Karling (1940) discussed the phylogenetic relationships within the Prolecithophora, and concurred with Reisinger (1924) in recognizing the groups Combinata, consisting of the Cylindrostomidae, Pseudostomidae, Protomonotresidae, Genostomatidae, Scleraulophoridae and the Urastomidae; and the Separata, consisting of the Baicalarctiidae, Ulianinidae, Plagiostomidae, Multipeniatidae, and later the Torgeidae. Karling's classification was based on the occurrence of separate or combined oral and genital openings. The Combinata was further subdivided into the Proporata, with an anterior, and the Opisthoporata, with a posterior oral–genital opening. This classification is generally considered based on convention rather than on apomorphies (Ehlers 1988; Sluys 1992; Karling and Jondelius 1995; Jondelius 1997). Norén and Jondelius (1999) basing their hypothesis on 18S rDNA, demonstrated the non-monophyly of Separata and proposed that the taxa within the Cylindrostomidae and Ulianinidae be transferred to the Pseudostomidae. In the following, we will use this classification although several 'families' were not represented in their analysis.

In the present study we include 18S rDNA data from four additional prolecithophoran species: *Protomonotresis centrophora* Reisinger, 1923, classified in the Protomonotresidae, which Karling (1940) considered to be a 'link' between the Combinata and Separata, under the name *Archimonotresis limophila*; *Scleraulophorus cephalatus* Karling, 1940, which, together with *Rosmarium* was classified in the Scleraulophoridae by Marcus (1950); *Euxinia baltica* Meixner, 1938, a species without pigmented eyes; *Allostoma neostiliferum* Karling, 1993, which possesses a sclerotized male copulatory organ. In addition, we sequenced 18S rDNA from a rhabdocoel, *Anoplodium stichopi* Bock, 1925 for use as an outgroup taxon and to provide a preliminary test of the 'Lecithophora' hypothesis (Karling 1940).

Some taxa originally classified within the Prolecithophora do not share the morphological synapomorphies proposed by Ehlers (1988). Reisinger (1924) classified *Urastoma* and *Genostoma* (Hypotrichina) within the Cylindrostomidae.

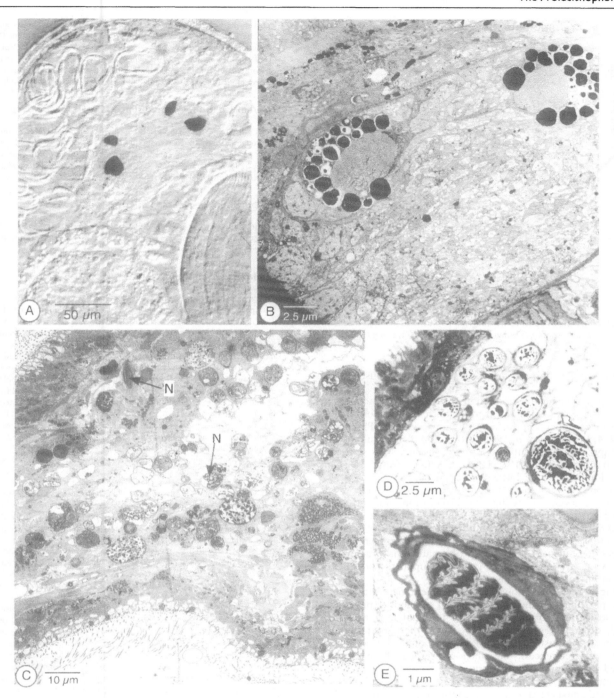

Figure 8.1 Prolecithophoran anatomy. **A**) Light micrograph of anterior end of *Cylindrostoma fingalianum*. Note developing spermatozoa anterior to the two pairs of pigmented eyes. The encapsulated brain is visible below the eyes. Part of the posteriorly directed pharynx is visible in the lower right. **B**) TEM micrograph of the encapsulated brain and one pair of pigmented eyes in *Reisingeria hexaoculatus*. **C**) TEM micrograph of ventral portion of cross-section of *Ulianinia mollissima*. Nematocysts (N) are situated in the parenchyma. **D**) TEM micrograph of nematocysts lodged in the parenchyma of *U. mollissima*. **E**) Nematocyst with coiled thread in parenchyma of *U. mollissima*.

However, Karling (1940) questioned the inclusion of *Urastoma* in the Prolecithophora and later studies of its sperm morphology revealed the presence of two axonemes and the lack of the membrane derivatives characterizing prolecithophorans (Noury-Sraïri *et al.* 1989b; Hyra 1993; Watson 1997a). Regardless of this, *Urastoma* was used as a representative of Prolecithophora in two recent studies of platyhelminth phylogeny using 18S rDNA data (Carranza *et al.* 1997; Littlewood *et al.* 1999a), albeit with some reservations as to its classification in the latter study. In the study by Norén and Jondelius (1999), *Urastoma* grouped with the Tricladida outside the Prolecithophora. Since additional prolecithophoran taxa are included in the present study, among them *P. centrophora* postulated as a basal prolecithophoran by Karling

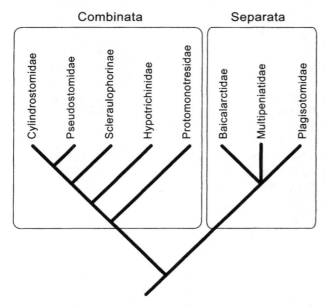

Figure 8.2 Karling's hypothesis. A diagram to illustrate the hypotheses on prolecithophoran phylogeny developed by Karling (1940).

(1940), the position of *Urastoma* is more rigorously tested here.

The phylogenetic information in 18S rDNA sequences has in many cases proven insufficient to establish well-supported phylogenetic hypotheses at the 'inter phylum level' (e.g., Philippe *et al.* 1994; Lipscomb *et al.* 1998). Even deep platyhelminth relationships have been impossible to reconstruct with adequate branch support using 18S rDNA data (e.g., Carranza *et al.* 1997; Jondelius 1998; Littlewood *et al.* 1999a).

The present study deals with branching patterns at a lower taxonomic level; viz. within the Prolecithophora and some related taxa. Therefore rapidly changing sites in the 18S rDNA were expected to contribute more phylogenetic information and less noise. Hence we anticipated alignment of the nucleotide sequences to be less problematic, and branch support, which was measured as jackknife support, to be higher than in phylum-level studies.

Long branch attraction (LBA) is a phenomenon that may cause parsimony to be an inconsistent method of phylogeny reconstruction if branch lengths of terminal taxa are sufficiently different (Felsenstein 1978). While there are no unambiguously demonstrated empirical cases of LBA (see Siddall 1998 with references), the phenomenon has been demonstrated in computer simulations. LBA is sometimes invoked without any supporting evidence when analyses of molecular data yield unexpected results (e.g., Joffe and Kornakova 1998). To eliminate any artifacts resulting from LBA, which by no means has been shown to occur in the present data set, we recoded our data so that all occurring triplet combinations of nucleotides were given a unique code. The number of possible character states thereby increased from four to 64.

Materials and methods

Materials

Specimens were collected from the Gullmaren fjord and from the vicinity of Tjärnö Marine Biological Laboratory on the West Coast of Sweden. Specimens of *Scleraulophorus cephalatus* and *Euxinia baltica* were col-

lected in the Bergen area on the Norwegian West Coast. The specimens were starved for 24 hours and then stored in 70% ethanol until DNA extraction. Nucleotide sequences have been deposited with GenBank (Table 8.1). Published DNA sequences from a selection of other rhabditophoran flatworm taxa were obtained from the SSU rDNA database (van de Peer *et al.* 1997). For specimen and sequence references see Table 8.1.

DNA extraction, PCR amplification, and sequencing

DNA extraction, PCR amplification, and sequencing were according to methods described by Norén and Jondelius (1999). Both strands of the PCR products were sequenced, except for *Scleraulophorus cephalatus* where only one strand could be sequenced in a 500 bp-long segment downstream of the 5F primer.

Preparation for electron microscopy

Methods for electron microscopy were according to Jondelius (1992), except that the sections were studied in a Zeiss CEM 902A transmission electron microscope.

Phylogenetic analysis

Thirty-eight near-complete 18S rDNA sequences were aligned using the multiple alignment program ClustalW version 1.7 (Thompson *et al.* 1994, 1997). Alignment of the 18S rDNA sequences was performed by profile alignment to sequences already aligned by secondary structure from the SSU database (van de Peer *et al.* 1997), using a range of multiple alignment gap open- and gap extension penalties. The results presented here are based on an alignment with the multiple alignment gap open- and gap extension penalties set to 30 and 5, respectively. Similar tree topologies and similar support values were obtained when a number of different profile alignments with multiple alignment gap open- and gap-extension penalties in the ranges 10 to 50 and 1 to 10, respectively, were analysed. Alignment without reference to secondary structure yielded similar results. The full alignment consisted of 2126 sites, of which 882 were parsimony informative.

To control for alignment artefacts, we analysed a 'conservative' alignment where most gapped positions and positions in highly variable regions were excluded. This alignment contained 1616 sites of which 671 were parsimony informative. It could be argued that this approach is not conservative at all: exclusion of data represents *a priori* hypotheses of non-homology of that data. Nevertheless, through comparison with the complete data set, this procedure gave an estimate of the effects of alignment artifacts in our data.

Gaps were treated as missing data in all analyses. Clade support was evaluated through parsimony jack-knifing (Farris *et al.* 1996) performed with the software Xac (Farris). Cut-off frequency was e^{-1}. We performed 1000 replicates per analysis with 10 random addition sequences per replicate and branch swapping was enabled.

To minimize the risk of LBA artefacts, both the full and the conservative alignment were analysed using jack-knifing (1000 replicates) of triplet-coded nucleotides (Farris, pre-release software). All occurring triplet combinations of nucleotides were each given their unique code. This method boosts the number of possible character states from four to 64 at the expense of the number of characters. The full-length triplet-coded data set contained 363 parsimony informative characters. The 'conservative' triplet-coded data set contained 341 parsimony-informative characters.

Results

Electron microscopy

The parenchyma of *Ulianinia mollissima* contained numerous nematocysts (Figure 8.1C,D,E). The nematocysts were always intact; discharged nematocysts were not observed.

The mature spermatozoa in *Reisingeria hexaoculatus* and *U. mollissima* possess the stacked membrane structures reported from other Prolecithophora (Figure 8.3), as well as cortical microtubules (Figure 8.3A,E). Dense bodies were not observed.

Table 8.1 Taxa used in this study, their classification, GenBank accession numbers and original citation.

Species	Classification	GenBank	Reference
Allostoma neostiliferum	**Cylindrostomidae**	AF167420	This study
Anoplodium stichopi	Rhabdocoela	AF167424	This study
Archiloa rivularis	Proseriata	U70077	Carranza *et al.* (1997)
Bipalium sp.	Tricladida	X91402	Mackey *et al.* (1997)
Bothromesostoma sp.	Rhabdocoela	D85098	Katayama *et al.* (1996)
Cylindrostoma fingalianum	**Cylindrostomidae**	AF065415	Norén and Jondelius (1999)
Cylindrostoma gracilis	**Cylindrostomidae**	AF065416	Norén and Jondelius (1999)
Dendrocoelopsis lactea	Tricladida	D85087	Katayama *et al.* (1996)
Dicrocoelium dendriticum	Neodermata	Y11236	Sandoval, H. *et al.* unpublished
Dugesia japonica	Tricladida	D83382, D17560	Katayama *et al.* (1995)
Euxinia baltica	**Cylindrostomidae**	AF167418	This study
Fasciolopsis buski	Neodermata	L06668	Blair, D., Barker, S. C., unpublished
Geocentrophora baltica	Lecithoepitheliata	AF167421	This study
Geocentrophora baltica	Lecithoepitheliata	AF065417	Norén and Jondelius (1999)
Geocentrophora sphyrocephala	Lecithoepitheliata	D85089	Katayama *et al.* (1996)
Macrostomum tuba	Macrostomida	U70080	Carranza *et al.* (1997)
Mesocastrada sp.	Rhabdocoela	U70081	Carranza *et al.* (1997)
Microstomum lineare	Macrostomida	D85092	Katayama *et al.* (1997)
Notoplana koreana	Polycladida	D85097	Katayama *et al.* (1997)
Otoplana sp.	Proseriata	D85090	Katayama *et al.* (1996)
Plagiostomum ochroleucum	**Plagiostomidae**	AF065419	Norén and Jondelius (1999)
Plagiostomum cinctum	**Plagiostomidae**	AF065418	Norén and Jondelius (1999)
Plagiostomum striatum	**Plagiostomidae**	AF065420	Norén and Jondelius (1999)
Plagiostomum vittatum	**Plagiostomidae**	AF065421	Norén and Jondelius (1999)
Planocera multitentaculata	Polycladida	D83383, D17562	Katayama *et al.* (1996)
Protomonotresis centrophorc	**Protomonotresidae**	AF167419	This study
Plicastoma cuticulata	**Plagiostomidae**	AF065422	Norén and Jondelius (1999)
Pseudostomum gracilis	**Pseudostomidae**	AF065423	Norén and Jondelius (1999)
Pseudostomum klostermanni	**Pseudostomidae**	AF06542A	Norén and Jondelius (1999)
Pseudostomum quadrioculatum	**Pseudostomidae**	AF065425	Norén and Jondelius (1999)
Scleraulophorus cephalatus	**Scleraulophoridae**	AF167423	This study
Reisingeria hexaoculata	**Pseudostomidae**	AF065426	Norén and Jondelius (1999)
Thysanozoon brocchii	Polycladida	D85096	Katayama *et al.* (1996)
Ulianinia mollissima	**Ulianinidae**	AF065427	Norén and Jondelius (1999)
Urastoma cyprinae	Urastomidae	AF167422	This study
Urastoma cyprinae	Urastomidae	AF065428	Norén and Jondelius (1999)
Urastoma sp.	Urastomidae	U70085	Carranza *et al.* (1997)
Vorticeros ijimai	**Plagiostomidae**	D85094	Katayama, T. *et al.* unpublished

In spermatozoa of *U. mollissima*, there are two membrane columns (Figure 8.2E,F,G), whereas there is only one in spermatozoa of *R. hexaoculatus*. The membrane columns are in contact with the cell membrane and the outside of the cell in both species (Figure 8.3A,E). The nucleus in cross-sections of *U. mollissima* sperm appears hourglass-shaped (Figure 8.3E).

Phylogenetic analysis

The 60% consensus tree from jack-knife analysis of all nucleotides is shown in Figure 8.4. The Prolecithophora (excluding *Urastoma*) is monophyletic, with Tricladida + *Urastoma* as the sister group. Plagiostomidae is the monophyletic sister group of a taxon consisting of species of *Protomonotresis*, *Allostoma*, *Cylindrostoma*, *Pseudostomum*, *Scleraulophorus*, *Euxinia*, *Ulianinia*, and *Reisingeria*. Norén and Jondelius (1999) previously suggested that these sister-group taxa be referred to the taxon Pseudostomidae. In keeping with this, we will refer to this group as the Pseudostomidae clade. The Rhabdocoela + Neodermata form the sister-group of

Prolecithophora and Tricladida + *Urastoma*. The two most basal neoophoran groups in the analysis are the Proseriata and the Lecithoepitheliata.

In the triplet analysis of the complete alignment (not shown), resolution is lower. The Plagiostomidae clade is highly supported (100%), and so is the Pseudostomidae clade (100%). However, the Prolecithophora is not supported: Plagiostomidae and Pseudostomidae are part of a weakly supported (55%) polytomy with Tricladida and *Urastoma*. The Macrostomida, Proseriata, Neodermata, Rhabdocoela, Polycladida and Lecithoepitheliata are part of an unresolved polytomy.

In the consensus tree resulting from jack-knife analysis of the conservative alignment (Figure 8.5), the Prolecithophora is strongly supported (100%), and consists of the Plagiostomidae and Pseudostomidae clades. Sister-group of the Prolecithophora is a taxon consisting of *Urastoma* and Tricladida.

A similar topology is present in the consensus tree from the triplet analysis of the conservative alignment (Figure 8.6). The Prolecithophora is highly supported, and forms a polytomy with *Urastoma* and Tricladida.

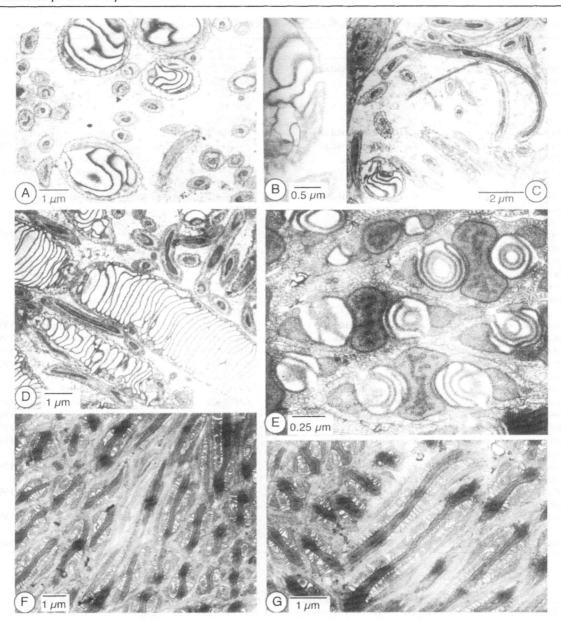

Figure 8.3 TEM micrographs of prolecithophoran sperm cells. **A)** Cross-sectioned sperm cells from seminal vesicle of *Reisingeria hexaoculatus*. Note the large membrane column. **B)** Detail of cross-sectioned sperm cell from seminal vesicle of *Reisingeria hexaoculatus*. Note the connection between the membrane column and outside. **C)**, **D)** Oblique and cross-sections of spermatozoa of *Reisingeria hexaoculatus*. Note the elongate nucleus and large membrane column occupying most of the central sperm cell. **E)** Cross-sectioned spermatozoa from seminal vesicle of *Ulianinia mollissima*. The membrane columns are paired, situated laterally to the hourglass-shaped nucleus. **F**, **G)** Obliquely and longitudinally sectioned sperm cells from the seminal vesicle of *U. mollissima*. Note the elongated, electron-dense nucleus and paired membrane columns.

Discussion

Our TEM micrographs corroborate the apomorphy for the Prolecithophora suggested by Ehlers (1988) (stacked membrane columns in the spermatozoa, absence of axonemes and dense bodies). Stacked membrane columns are present also in *Ulianinia mollissima* and *Reisingeria hexaoculata*, while double columns of stacked membranes, as observed by us in *U. mollissima*, occur in an unidentified *Allostoma* species (Lanfranchi 1998). *Protomonotresis centrophora* appears to possess numerous membrane stacks in the spermatozoa (Ehlers 1988). Our molecular data set included *Allostoma*

neostiliferum, but this species is part of a polytomy within the Pseudostomidae. Hence we cannot determine whether the occurrence of paired membrane columns constitutes an apomorphy for a taxon comprising *Allostoma*, *Ulianinia* and their last common ancestor.

As did Westblad (1956) and Karling (1963), we found nematocysts in the parenchyma of *U. mollissima*. Contrary to the situation in the earlier light microscope studies, none of the nematocysts in our TEM micrographs was discharged.

Triplet coding of the nucleotide data reduces homoplasy through the introduction of 64 character states instead of four (or five if gaps are treated as a fifth character). At the

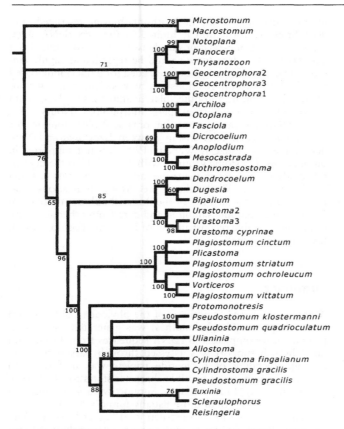

Figure 8.4 Jack-knife analysis of complete nucleotide data set. Majority-rule consensus tree (cut-off 60%) summarizing results of jack-knife analysis of complete nucleotide alignment. Labels indicate jack-knife frequency (obtained using Xac with 1000 replicates, 10 random additions, branch swapping enabled and character deletion frequency e^{-1}).

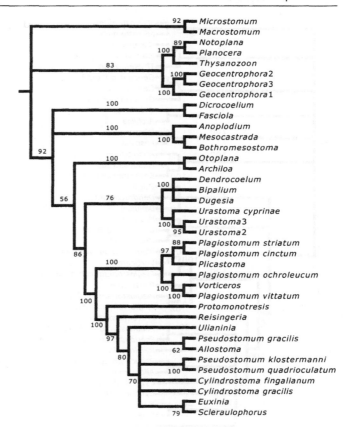

Figure 8.5 Jack-knife analysis of conservative nucleotide data set. Majority-rule consensus tree (cut-off 50%) summarizing results of jack-knife analysis of shortened, 'conservative' nucleotide alignment. Labels indicate jack-knife frequency (obtained using XAC with 1000 replicates, 10 random additions, branch swapping, deletion frequency $= e^{-1}$).

same time, the numbers of characters are reduced, and tree resolution could thus be expected to be diminished. Furthermore, triplet coding is expected to be more sensitive to missing data (gaps in the alignment). In our complete data set, nucleotide jack-knifing yielded 13 supported nodes within the Prolecithophora, but only nine nodes were retained when triplet coding was used. In the conservative alignment, the number of supported nodes within the Prolecithophora was 13 in the nucleotide analysis and 11 in the triplet analysis.

A clear pattern emerges among the prolecithophoran taxa included in our study. There is a basal dichotomy between the Plagiostomidae and Pseudostomidae. Karling (1993) regarded the individual monophyly of the Cylindrostomidae and Pseudostomidae as unsubstantiated by morphological data and referred to the taxa as the 'CP-complex'. Norén and Jondelius (1999) proposed the inclusion of the Cylindrostomidae and Ulianinidae in the Pseudostomidae. This view is further corroborated in the present study: the additional taxa *Euxinia baltica*, *Allostoma neostiliferum* and *Scleraulophorus cephalatus* group in the Pseudostomidae clade. *Ulianinia mollissima* possesses separate mouth and gonopore, but it groups within the Pseudostomidae. The implication of this is that the groups Combinata and Separata are non-monophyletic. *Protomonotresis centrophora* is basal within the Pseudostomidae clade (or its sister-group); a position fully compatible with Karling's views (1940 and Figure 8.1). Within the Plagiostomidae clade, the genus *Plagiostomum* is

non-monophyletic due to the position of *Plicastoma* and *Vorticeros*, and the latter two would have to be included in a monophylum *Plagiostomum*.

Some conclusions can be drawn regarding morphological evolution within the Prolecithophora: an anteriorly directed pharynx appears to be the plesiomorphic condition, and a follicular ovary is the plesiomorphic condition within the Pseudostomidae clade. Pigmented eyes are present in all the Plagiostomidae of our data set, except *Vorticeros auriculatum*, and in the Pseudostomidae clade except *U. mollissima*, *E. baltica* and *S. cephalatus* (the latter two form a monophylum). Thus there are at least three independent reductions of eye pigmentation. The position of *Allostoma neostiliferum* within the Pseudostomidae clade implies that its sclerotized copulatory organ, which is rare in the Prolecithophora, evolved within the Pseudostomidae; it is not a plesiomorphic feature.

The average jack-knife values for the supported nodes within the Prolecithophora were between 84% and 94% in our analyses. The high support values and the robustness with regard to variations in alignment parameters indicate a strong phylogenetic signal in the 18S rDNA data when applied to prolecithophoran phylogeny. Attempts to reconstruct phylogenetic relationships at higher taxonomic levels, i.e., deeper branches, have often been troubled with little branch support, (e.g., Winnepenninckx *et al.* 1995; Winnepenninckx and Backeljau 1996; Carranza *et al.* 1997; Littlewood *et al.* 1999a). As noted by Norén and Jondelius (1999), the 18S

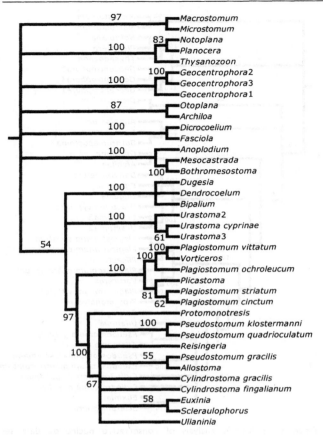

Figure 8.6 Jack-knife analysis of triplet-coded 'conservative' data set. Majority-rule consensus tree (cut-off 50%) summarizing results of jack-knife analysis of shortened, 'conservative' alignment with nucleotide triplets recoded. Labels indicate jack-knife frequency after 1000 replicates (obtained using pre-release software by J.S. Farris).

rDNA gene appears suitable for reconstruction of events on the time-scale of prolecithophoran evolution.

The present cladistic hypothesis of prolecithophoran phylogeny should be viewed as a starting point for future studies. Addition of data from taxa unavailable to us, for example, the endemic taxa of Lake Baikal and the Australian *Torgea karlingi*, would permit better reconstruction of morphological evolution and tracing of biogeographical events within the Prolecithophora. For instance, there are similarities in the anatomy of *Baicalarctia* Friedmann, 1926 and *Torgea*. Are these taxa sister-groups? Is there a whole fauna of similar taxa undescribed in the Australasian region?

Our results indicate a sister-group relationship between the Prolecithophora and the Tricladida + *Urastoma*. However, the sister-group of the Prolecithophora could not be determined with the same level of support as internal prolecithophoran branches. Moreover, our taxonomic sampling of rhabdito-

phoran taxa outside the Prolecithophora was restricted, and deep branch information is limited in the 18S rDNA data. Thus a sister-group relationship between the Prolecithophora and Tricladida + *Urastoma* should be regarded as highly tentative. Data from other genes are needed for a nucleotide-based reconstruction of the relationships between the major groups of rhabditophorans and other platyhelminths. Our DNA data corroborate the results from sperm morphology (Noury-Sraïri *et al.* 1989b) indicating a position outside Prolecithophora for *Urastoma*. A close relationship between *Urastoma* and Neodermata, as suggested by Watson (1997a), is not supported in our analysis. *Urastoma* grouped with Tricladida in the nucleotide analyses, but not in the triplet analyses where its position was unresolved. While there are strong indications that *Urastoma* is not a member of the Prolecithophora, more comprehensive taxonomic sampling and data from other genes are desirable to enable reconstruction of its precise position within the Platyhelminthes.

Summary

A parsimony-based hypothesis of the phylogeny of the taxon Prolecithophora (Platyhelminthes) was generated using complete or near-complete 18S rDNA sequences from 16 ingroup taxa. Branch support was assessed through parsimony jackknifing. Recoding of nucleotide triplets was performed to increase the number of possible character states to 64, thereby eliminating any LBA artifacts. Support values for ingroup clades were higher than those usually found when 18S rDNA sequences are used for reconstruction of higher level phylogenies, e.g., 'inter-phylum' or 'inter-order' relationships. Monophyly of the Prolecithophora and of the family Plagiostomidae was strongly supported. Cylindrostomidae, Pseudostomidae, Scleraulophoridae and Ulianinidae form a monophylum. The taxa Separata, Combinata, and *Plagiostomum* Schmidt, 1852, are shown to be non-monophyletic. *Urastoma* Dörler, 1900 is not part of the Prolecithophora. The results indicate a sister-group relationship between the Prolecithophora and a taxon consisting of Tricladida and *Urastoma*. The results are compared and contrasted with earlier morphology-based classifications. New data on sperm morphology in *Ulianinia mollissima* Levinsen, 1879 and *Reisingeria hexaoculata* Westblad, 1955 provide morphological evidence for their inclusion in the Prolecithophora.

ACKNOWLEDGEMENTS

We thank the staff at the Marine Biological Stations at Kristineberg and Bergen for assistance with the collection and study of live material. Drs J.S. Farris and M. Källersjö allowed us to use their experimental software for nucleotide triplet analyses and Ms I. Holmquist prepared the TEM specimens. The Swedish Natural Science Research Council (NFR) provided financial support for these studies (grant to U.J.).

Section III

Symbionts and parasites

Chapter 9

The Temnocephalida

Lester R.G. Cannon and Boris I. Joffe

'There is no excellent beauty that hath not some strangeness in the proportion'

Francis Bacon

Reported first from Chile as branchiobdellid leeches (Moquin-Tandon 1846), then from the Philippines as monogeneans (Semper 1872), these little ectosymbiotic worms found mainly on freshwater crustaceans were first recognized to have turbellarian affinities by van Beneden (1876). Though formally placed in the Platyhelminthes by Benham (1901), their affinities were uncertain: Haswell (1893), Hett (1925) and Borradaile *et al.* (1948), for example, all assigned them to the trematodes. More recently, while recognizing them as unquestionably related to the rhabdocoel turbellarians, Ehlers (1985a) was unable to provide a clear apomorphy for the group. We now accept the presence of an epidermis made of multiple syncytial plates first reported by Williams (1975) as a uniquely temnocephalan character (see Joffe and Cannon 1998a).

Why has it proven seemingly difficult to agree on the relationships of this group? With approximately 100 known species, they are the biggest taxon of symbiotic turbellarians, and their life style has led to morphological changes in body plan which parallel those of the trematodes, most notably the general, though not exclusive, loss of locomotory cilia and possession in many of a distinct posterior sucker. Interest has focused on them, however, since these characteristics suggest that knowledge of them may lead to a better understanding of the origins of parasitism among the Platyhelminthes.

Here, we summarize the known data about the evolution of the Temnocephalida based on a list of synapomorphies recently proposed by Joffe *et al.* (1998b), but extended to include some putative synapomorphies which need further substantiation. To make the list of synapomorphies clearer, we first provide a brief summary of temnocephalid morphology. Finally, based on the zoogeographical data, we discuss the evolution of the Temnocephalida as a process in time and space.

Morphological overview

Epidermis

Williams (1975) showed with TEM that the presumed 'cuticle' was merely secretory products layered on the epidermis, and later (Williams 1981a) that the epidermis was not only syncytial, but that is was a 'patchwork of several structurally differentiated epithelia', i.e., the epidermis is made up of a mosaic of syncytial epidermal plates.

Joffe (1982) used silver nitrate to stain the mosaic in scutariellids, and Sewell and Cannon (1995) demonstrated the effectiveness of hot fixation for revealing the epidermal sutures with SEM. Subsequently, Joffe *et al.* (1995a,b) studied, using silver nitrate staining, *Didymorchis* and *Diceratocephala* respectively. Analysis of the syncytial mosaic by Joffe and Cannon (1998a) has shown its presence to be an apomorphy for the Temnocephalida. Fusion of syncytia in the evolution of the group appears characteristic of the Temnocephalida (Joffe

and Cannon 1998a). Syncytial plates are known also from within the temnocephalan pharynx where they also demonstrate a clear trend to fusion (Joffe *et al.* 1997b).

Muscles

The presence of anterior tentacles has long been considered a characteristic of temnocephalans, but analysis of the muscle systems reveals that the projections in the Scutarielloidea are more correctly elongate papillae, and similarly *Diceratocephala*, *Decadidymus* and *Actinodactylella* all lack the four characteristic muscle bands found in true tentacles of the remaining Temnocephaloidea. *Didymorchis*, of course, lacks even these extensions and has only an anterior flange.

A posterior adhesive organ is also characteristic of all temnocephalans, but again in the less specialized species it consists merely of an adhesive pad or pads. Muscles controlling this region in the less-derived taxa tend to be weakly arranged and never insert at the adhesive pads themselves. Though *Diceratocephala* has strong posterior muscles, these insert into the walls of the trailing posterior body adjacent to the disc. In more specialized representatives of the group, the muscles are concentrated into an axial group inserted into the base of the disc, and able to retract the disc centrally. The most characteristic facies of a temnocephalan is a worm with a strong pedunculate sucker below, rather than trailing behind the body, and the ability to rotate the body around the attached disc.

Pharynx musculature was analysed by Joffe *et al.* (1997b), who found that the relative position of muscle layers in pharynx walls is similar in *Didymorchis* and other Temnocephalida, but, as far as it is now known from light microscopy, different from that of Dalyelliidae. In *Diceratocephala* and *Decadidymus* the pharynx is spindleform with extremely strong circular muscles in the wall of the pharyngeal lumen. These taxa lack the usually very strong anterior and posterior sphincters found characteristically in all the Temnocephalidae and Actinodactylellidae, with the exception of the subfamily Craspedellinae in which the pharynx is secondarily simplified.

Locomotion

Williams (1981a) characterized temnocephalans as having looping locomotion, which requires the anterior and posterior ends to be capable of independent attachment and release. Scutarielloidea, together with posterior adhesive pads, have a pair of anterior pits. Temnocephaloidea do not all attach anteriorly, e.g., *Didymorchis* and *Diceratocephala* glide on cilia, though the former will squirm and pulsate in a film of water (Sewell and Cannon 1998a). Sewell and Whittington (1995) presented a detailed analysis of locomotion in *Craspedella* and

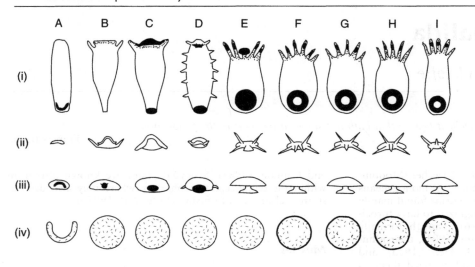

Figure 9.1 Proposed evolutionary series of major genera of Australian Temnocephaloidea. (i) Ventral view showing anterior and posterior adhesive regions and rhabdite distribution (rods); (ii) *en face* view of how anterior is held in life; (iii) *en posterior* view of how adhesive organ is held in life; (iv) posterior adhesive field showing distribution of gland openings and presence of a marginal valve. A, *Didymorchis*, B, *Diceratocephala*, C, *Decadidymus*, D, *Actinodactylella*, E, *Temnomonticellia*, F, *Temnohaswellia*, G, Temnocephala NT, H, *Notodactylus*, I, Craspedellinae. (From Sewell 1998.)

Sewell (1998) provided convincing details of attachment and locomotion in a wide variety of temnocephalans (Figure 9.1).

Protonephridial system

Unlike Dalyelliidae, all the Temnocephalida have at least two commissures linking left and right protonephridia. Both Scutarielloidea and *Didymorchis* have ventro-lateral to lateral excretory pores. In all other temnocephalans the excretory pores are dorsal and generally situated in the trunk syncytium, but in *Temnohaswellia* they may be adjacent to or even enclosed by the post-tentacular syncytium (e.g., *T. comes*) (Joffe and Cannon 1998a). Characteristically in all Neotropical *Temnocephala* each excretory pore lies within each of the paired post-tentacular syncytia (Damborenea and Cannon, in press).

Nervous system and sensory receptors

Joffe and Cannon (1998c) showed in *Craspedella* a reduction in the ventral sensory innervation, but reinforcement of it anteriorly. These characters mirror changes in locomotion in Temnocephalida, from pursuit of prey to sedentary hunting.

The simple adhesive pad of the early temnocephalan has become a true sucker with specialized sensory sensillae (Joffe *et al.* 1998c). Furthermore, worms found externally often flatten the body against the host and lack obvious sensory ornamentation of the surface. *Notodactylus* is most inactive and protected with an extensive armour of dorsal scales and rows of extensible sensory papillae which may push up between them (Jennings *et al.* 1992; Sewell 1998).

Worms living in sheltered sites (branchial chamber) characteristically have little or no pigment and less strongly developed body musculature. Worms which live in regions of rapid movement, such as among the gill lamellae of crayfish (*Actinodactylella* and the Craspedellinae) are further modified and have numerous sensory papillae.

Glands

The most prominent glands in the Temnocephalida, excluding those involved with reproduction, are the rhabdite glands. In *Didymorchis* rhabdites are absent anteriorly, but rhabdite glands open through the adhesive field (Joffe *et al.* 1995a; Sewell and Cannon 1998a). In *Diceratocephala* they are scattered across the anterior lobe, but not on the tentacles (Joffe *et al.* 1995b). In the remaining Temnocephaloidea they occur anteriorly, as well as along the lateral margins of the body (Joffe 1981a,b) but are really abundant only in adhesive regions. In *Dactylocephala* and *Temnomonticellia* all tentacles (including the lateral ones) are adhesive, but in *Temnohaswellia*, and the remaining five-tentacled taxa, ability to adhere (but not rhabdites) is absent from the lateral tentacles: in the Craspedellinae this ability is even concentrated in small pits in the central three tentacles.

So-called postero-lateral glands are found in several members of the Temnocephalidae (including Craspedellinae) (Hett 1925; Cannon 1993; Sewell and Cannon 1998b). Cannon and Watson (1996) have investigated them, but their function remains unclear. They have not been found in other Temnocephalida.

Pigments

Where pigment exists in the Temnocephalida it is generally black (melanin?) and concentrated into small granules acting as a reflective layer for the eyes. It may, however, trail away from the eye region to invade the whole body. In Neotropical *Temnocephala*, however, the whole body sometimes has a pale orange hue, and, although the pigment is concentrated essentially in the eyes, it is red and fugacious (Damborenea and Cannon, in press).

Reproduction

All Temnocephalida have a genitointestinal connection, seen most usually in the Temnocephalidae as a large resorptive vesicle which pushes up, and discharges intermittently, into the gut. All have a single ovary as do the Dalyelliidae. In the Scutarielloidea and *Didymorchis* the vitellarium is in the form of discrete lateral elongated masses as in the majority of rhabdocoels, but elsewhere they form a net-like structure or appear as follicles scattered over the gut. There are a variable number of small seminal receptacles situated at the proximal (internal) end of the female duct, and in some species sclerotic teeth occur in the female canal, though it is doubtful if these structures are significant above the generic level.

Within the male system the testis number varies. A single posterior pair is seen in most rhabdocoels, Scutarielloidea, *Didymorchis* and *Diceratocephala*, but *Decadidymus* has 10 pairs and *Dactylocephala* six. *Actinodactylella* and Temnocephalidae normally have two pairs, though *Achenella* and *Temnocephala brenesi* have only one pair. Changes in the number of testes are due to subdivision of the primary pair of testes.

Temnocephalida have a muscular copulatory bulb. Vasa deferentia unite near the copulatory organ and form a spindle-shaped seminal vesicle. Together with the seminal vesicle, a discrete accessory sac with muscular walls opens into the bulbus in the majority of the Temnocephaloidea, including the Didymorchiidae. Williams (1981a) observed that the Neotropical *Temnocephala* all lacked this sac, which she claimed was present in all Australian temnocephalids, but though generally characteristic it may be incorporated into the bulbus and eventually absent (Sewell and Cannon 1998b). All stages from well-separated to fully reduced are found in the Craspedellinae.

A sclerotic tube with a terminal cirrus is normal for the male armature of the Temnocephalida, but it may be weak or lacking in the Scutarielloidea or lacking a terminal cirrus as in the Caridinicolinae or *Decadidymus* (see Cannon 1991), though some other species have reduced spines, e.g., *Temnocephala dendyi* and *T. santafesiana*. The armature, while of considerable value in recognition of species, may prove of limited value in resolving higher relationships.

The list of synapomorphies (Figure 9.2)

1. **Temnocephalida.** 1.1. Multisyncytial epidermis. 1.2. Syncytial epithelium of the pharynx comprised of several successive belts; a narrow belt-like syncytium harbouring the openings of pharyngeal glands is present in the anterior portion of the pharynx canal, as in the Didymorchiidae, Diceratocephalidae, Scutariellidae. 1.3. Posterior adhesive organ, always including a special syncytium, the adhesive field. 1.4. Two (or more) transverse commissures join the left and right protonephridia so that major protonephridial canals form a ring; at least one pair of smaller internal circles is present in the anterior half of this ring. 1.5. Genitointestinal communication. (Though not unique for the Temnocephalida, the last two characters are not common in the Platyhelminthes and are not known in the Dalyelliidae.) 1.6. The shaft of sperm cells splits at the nuclear end (Watson and Rohde 1995a).

2. **Scutarielloidea.** 2.1. Mouth surrounded by three pairs of sensory papillae and a pair of so-called 'tentacles'. The latter larger than the papillae, but also with a concentration of the sensory endings only at their small terminal zones and lack any parenchymal musculature: they are, therefore, homologous to the other papillae rather than the tentacles of the Temnocephalidae or Diceratocephalidae (see Joffe 1988). 2.2. A pair of small adhesive pits situated on the ventral side at the base of the 'tentacles'. They include specialized glands, a separate paired anterior adhesive syncytium, and are associated with specialized longitudinal muscle fibres on the ventral body wall which insert on the posterior margins of the pits. 2.3. Spermatozoa with a cork-screw anterior region (studied only in a representative of European scutariellids: Iomini *et al.* 1994). 2.4. A terminal anterior mouth (subterminal in all other Temnocephalida and, not less importantly, in the genera of the Dalyelliidae which have a temporary frontal lobe).

3. **Scutariellinae.** 3.1. The stylet of the copulatory organ is reduced. 3.2. The gut is subdivided to an anterior and posterior portions connected with only a very narrow short canal at the level of copulatory organs.

4. **Caridinicolinae.** Current diagnosis for this subfamily is based wholly on plesiomorphic characters (cf. #3). The two genera, *Caridinicola* and *Monodiscus* appear to be monophyletic, but synapomorphies proving their monophyly still need to be found.

5. **Temnocephaloidea.** 5.1. Spiral arrangement of the peripheral microtubules in a specific region of sperm cells (Justine 1991a; Watson and Rohde 1995a). 5.2. The female genital tract forms a special extension, the resorptive vesicle, which participates in resorption of the surplus sperm, bears the genito-intestinal communication (temporary or permanent), and always abuts the posterior wall of the gut or is partially or wholly enclosed in it (Joffe 1981a, 1988). 5.3. Accessory vesicle in male copulatory organ (may be incorporated in the bulbus of copulatory organ in some representatives, e.g., *Temnocephala* species from South America and the Tasmanian genus *Temnomonticellia*). 5.4. The mosaic of the epidermal syncytia follows quite different patterns in the Temnocephaloidea and Scutarielloidea. It does not seem probable that the scutariellid pattern could arise from the temnocephalid one. Therefore this difference additionally indicates early divergence of the Scutarielloidea and Temnocephaloidea.

6. **Didymorchiidae.** 6.1. Epidermis with insunk nuclei (Rohde and Watson 1990; Joffe *et al.* 1995a). 6.2. Modified mitochondria accumulating secretory granules in the epidermis (Rohde and Watson 1990; Joffe *et al.* 1995a).

7. **Diceratocephalidae + Actinodactylellidae + Temnocephalidae.** 7.1. Excretory vesicles present; in all the studied cases, two nuclei have been observed in their walls (see Joffe 1981b for cell composition of the protonephridial system). 7.2. Post-tentacular syncytia involved in ionic regulation (secondarily reduced in *Decadidymus*: Joffe *et al.* 1996b; Joffe and Cannon 1998a). 7.3. The epidermal mosaic consists of only five syncytia: the frontal, post-tentacular (may be paired), trunk, peduncle, and adhesive field syncytia. In a few cases the trunk syncytium fuses with the peduncle or the frontal ones. 7.4. Round posterior terminal adhesive field with highly developed musculature (see Cannon 1991), while it is horseshoe-shaped without a specialized musculature in *Scutariella* and *Didymorchis* (Joffe *et al.* 1995a; Joffe 1982; Joffe and Janashvili 1982). 7.5. The frontal end is supplied with an axial musculature: muscle fibres which emerge from dorsal and ventral longitudinal muscle layers run in antero-transverse and antero-lateral directions to the opposite side of the body. There is no such specialized musculature in Didymorchiidae and Scutariellidae (Joffe 1982; Joffe and Janashvili 1982).

8. **Diceratocephalidae.** 8.1. Lateral portions of the frontal lobe are transformed into a pair of tentacle-like extensions lacking an axial musculature and therefore not to be considered homologous with tentacles of the Temnocephalidae (see Joffe 1982, 1988). 8.2. The frontal syncytium is a very narrow band running along the anterior margin of the frontal lobe, with two lateral dilatations which cover the lateral extensions (in the Didymorchiidae syncytia of the frontal group cover a band along the frontal margin of the body, but the form of this band is not specialized as in the Diceratocephalidae). 8.3. Long spindle-form pharynx with the inner musculature, especially, the circular one, reinforced along all the length of the pharynx (Joffe *et al.* 1997b).

9. **Temnocephalidae + Actinodactylellidae.** 9.1. Pharynx with strong anterior and posterior sphincters (Joffe *et al.* 1997b). An anterior belt-like syncytium in the pharynx has not been observed in any of the studied species (cf. #1.2). 9.2. The post-tentacular syncytium has the form of an

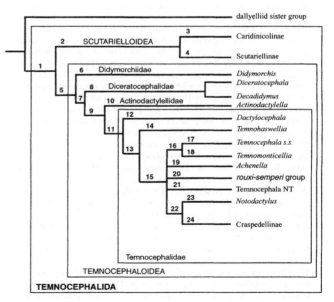

Figure 9.2 Proposed phylogeny of the Temnocephalida (see text for synapomorphies supporting the branches).

unpaired dorsal 'saddle' between the frontal and the trunk syncytia. This condition may be considered a synapomorphy for the two families: *Diceratocephala* has a pair of dorso-lateral syncytia, surrounded by the trunk syncytium, and in *Temnocephala dendyi* an unpaired post-tentacular syncytium is formed from a pair of such syncytia which do not fuse until early post-embryogenesis (Joffe and Cannon 1998a). Some other species of the *Temnocephala dendyi* group, all Neotropical *Temnocephala*, and *Achenella sathonota* have paired post-tentacular syncytia in the adult. By analogy with *T. dendyi,* this condition may be naturally interpreted as a secondary one. **9.3.** The frontal syncytium has a cap-like form, and covers both the ventral and dorsal sides of the anterior end of the body (cf. #8.2).

10. Actinodactylellidae (*Actinodactylella*). 10.1. Sensory projections without axial musculature along the lateral margins of the body. **10.2.** Anterior adhesive field formed by an extension of the trunk syncytium on the ventral side of the body where the rhabdite glands open abundantly (Joffe and Cannon 1998a).

11. Temnocephalidae. 11.1. Anterior lobe is divided into 'true' tentacles with an axial musculature corresponding fully to that of the undivided frontal lobe. **11.2.** A prominent axial bundle of muscle fibres in the posterior adhesive organ.

12. *Dactylocephala*. The only species of this genus, *D. madagascariensis,* is at present not sufficiently known. Though the genus can be diagnosed without problems, no apomorphies which could reliably support its status as the sister-group to the other Temnocephalidae is known. The position of this genus in the tree is, though, reliably indicated by the presence of the apomorphies from #11 and lack of those from #13.

13. The other Temnocephalidae. 13.1. The body with sharp lateral edges marked by a concentration of sensory receptors (Sewell and Cannon 1995) and openings of rhabdite glands (which are less abundant or lacking elsewhere at the posterior half of the body). **13.2.** A true sucker characterized by: 1) insertion of axial muscle fibres only at the central zone of the disk; 2) openings of adhesive glands situated only at the peripheral zone of the disk; and 3) a thin band of epithelium formed by the peduncle syncytium (named the marginal valve by Sewell and Whittington 1995) which covers the margins of the disk. **13.3.** The sucker is situated on a thin peduncle and directed ventrally; there is a short portion of the body ('tail') posterior to the base of the peduncle. **13.4.** The marginal valve and peripheral distribution of the openings of the adhesive glands may not be developed if the sucker has a more or less cylindrical shape (rather than a sucker with a well-developed stalk) as in, e.g., *Achenella* or *Temnomonticellia.* **13.5.** A small number of tentacles is six (*Temnohaswellia*) or five (other genera). A higher number of tentacles is considered plesiomorphic based on the 'oligomerization rule' (Dogiel 1954).

14. *Temnohaswellia*. Six tentacles. No indications are known that the medial tentacle of the temnocephalans with five tentacles was formed by fusion of two medial tentacles of the *Temnohaswellia*. Equally, there is no indication that the even number of tentacles in the *Temnohaswellia* is due to reduction of the unpaired medial tentacle (such a progenitor should have seven tentacles). Origin of the five-tentacle temnocephalids from the six-tentacle ones may be excluded without doubts. Therefore, we tentatively treat the six- and five-tentacle forms as sister-groups. This interpretation is backed by a very realistic scenario that the common progenitor might have a higher (as in *Dactylocephala*) and not fixed number of tentacles (cf. #13), so that states with a single or a pair of medial tentacles might both be present. Independent reduction of the number of tentacles could then lead to states with five and six tentacles.

15. The other genera. Five tentacles (see #14).

16. *Temnocephala* s.s. and *Temnomonticellia*. These two groups are united by incorporation of the accessory sac of the male copulatory organ. Importantly, incorporation and reduction of the accessory sac is characteristic of all representatives of the both groups, and is not related to the size of the body. Incorporation or reduction of the accessory sac is also characteristic of some Australian species of *Temnocephala* and of the genus *Achenella*. Whether they all belong to a unique phylogenetic branch remains unclear. Though close relationships between the Tasmanian and South American forms need confirmation, this assumption looks very sensible from the zoogeographical point of view.

17. *Temnocephala* s.s. Neotropical species of *Temnocephala* proved to form a very well-defined monophyletic taxon (Damborenea and Cannon, in press). **17.1.** The characteristic pattern of the syncytial mosaic which includes only a pair of post-tentacular syncytia enclosing the excretory pores (most probably neoteny) and the adhesive field syncytium, while the

frontal, trunk, and peduncular syncytia are fused. **17.2.** Eyes with red pigment. Though at the first glance this character does not seem convincing, such a state has never been observed in any other temnocephalids or dalyelliids.

18. *Temnomonticellia*. The medial tentacle is transformed into a short bulb.

19. *Achenella*. 19.1. With only a single pair of testes, and these are behind the genital organs. **19.2.** The post-tentacular syncytium is represented by two small plates, insunk into the parenchyma. **19.3.** The frontal and trunk syncytia are fused.

20. *Temnocephala* s.l.: *rouxi-semperi* group of species. The male armament with: (i) a long conical shaft; (ii) a small narrow introvert; and (iii) a small copulatory bulb. (ii) and (iii) relative to the length of the shaft.

21. Other Australian *Temnocephala* species (Temnocephala NT). It remains unclear if the other Australian *Temnocephala* species form a monophyletic clade or a paraphyletic assemblage (Cannon, in preparation).

22. *Notodactylus* + Craspedellinae. 22.1. Sensory papillae on the dorsal side of the body. **22.2.** The peduncular syncytium (unlike that of most five-tentacled temnocephalids) does not extend on to the dorsal side of the body, but occupies all the posterior region of the ventral side. A peduncular syncytium restricted in the trunk to its ventral side is also known in some Australian *Temnocephala* species (i.e., those from the gills of prawns) which may be related to *Notodactylus* and Craspedellinae. On the other hand, monophyly of this group needs confirmation. Sensory papillae in *Notodactylus* might be related to the scales (see below) and not strictly homologous to the papillae of the Craspedellinae.

23. *Notodactylus*. The dorsal side of the body is covered with scales forming a tile-like pattern.

24. Craspedellinae. 24.1. Tentacles with prominent, conical, ciliated papillae arranged in near-regular rows about a central axis. **24.2.** Dorsal surface with one or more transverse ridges bearing raised papillae and with further ridges, also bearing papillae, arranged radially behind the most posterior transverse ridge. **24.3.** Pharynx rudimentary. (The genera included in this taxon were reviewed by Cannon and Sewell (1995) and Sewell and Cannon (1998b), but the phylogenetic relationships among them remain unclear.)

In view of the heuristic value of character matrices especially for generation of computer keys, one of us (see appendix by L.R.G. Cannon) has made such a matrix and also computed a tree based on it. However, this tree should not be considered as a contribution to temnocephalid phylogeny. As a phylogenetic tool, optimal trees have some basic shortcomings (see review and references in Joffe and Kornakova 2001, this volume). In particular, morphological characters obviously have very different phylogenetic values, and the basic postulate behind any method using them for computation of trees is that each branch is supported by *numerous* characters whose values are averaged. The matrix in Appendix 9.1 includes 19 taxa requiring, at the very least, $2(19-1)$ [i.e., 36] characters for description. With only 38 characters in the matrix (most of them binary), the matrix provides an average of only slightly more than one character per branch.

Notes on selected taxa

Scutarielloidea

All the Scutarielloidea are small (they live predominantly on small hosts) and their organization is strongly modified. Formerly, it was considered that the Didymorchiidae are the sister-group to all the other Temnocephalida and that scutarielliids arose from didymorchiid-like ancestors (see, e.g., Bresslau 1933). Though all evidence available now suggests that the Scutarielloidea and Temnocephaloidea are sister-groups, additional convincing synapomorphies for the Temnocephaloidea remain desirable.

At present, the Scutarielloidea are represented by the single family Scutariellidae with two subfamilies, Caridinicolinae Matjasic 1980 and Scutariellinae Baer 1953: the systematics and phylogeny were discussed by Baer (1953) and Matjasic (1990). All characters separating the Scutariellinae and the Caridinicolinae show apomorphic states in the Scutariellinae. Monophyly (rather than paraphyly) of the Caridinicolinae still needs to be confirmed. A third subfamily, Bubalocerinae Matjasic 1980, was characterized primarily by very long tentacles; however, we know of no characters to justify a separate subfamily.

Temnocephalidae and *Temnomonticellia*

There is no doubt that *Dactylocephala* (with 12 tentacles) is the sister-group to the other Temnocephalidae, but the sister-group relations between the five- and six-tentacled representatives of the family need confirmation. The situation is clouded by the uncertainty with *Temnomonticellia* which is endemic to Tasmania. The traditional interpretation of this genus is based on assumption that it has five tentacles and the central one is shortened and bulb-like. This interpretation is supported by the fact that *Temnocephala cita*, the only temnocephalid species not in *Temnomonticellia* which has been found in Tasmania, also has an incorporated accessory sac (Hickman 1967; Joffe 1981a, 1982). Yet theoretically one cannot exclude that the central tentacular bulb was derived from two central tentacles. For example, the rhabdite tracks in the bulb do not fuse, which may or may not be related to the shortness of the bulb. Another specific trait (which might be plesiomorphic or secondary) is adhesion by all the tentacles including the lateral ones while moving. In addition, the sucker of *Temnomonticellia* does not have a marginal valve or a reduced glandular central region. If one assumes that this is a primary state for this genus, one can hypothesize that *Temnomonticellia* branched off from the stem of the Temnocephalidae immediately after *Dactylocephala*. This hypothesis looks over-complicated to one of the authors (B.J.), but obviously deserves more detailed consideration which will be supplied separately (Cannon, in preparation).

Temnocephalidae with five tentacles

Phylogenetic relations within this phylogenetic branch remain the most complex problem of the phylogeny of the Temnocephalida. The largest genus, *Temnocephala*, is obviously a paraphyletic assemblage, but even identification of monophyletic taxa within it – let alone understanding phylogenetic interrelations between them – at present cannot be done reliably.

- *Craspedellinae*: At present, data are insufficient to give a clear indication of the relationships among the genera of the Craspedellinae. However, *Zygopella* and *Craspedella* both come from lowland crayfish (*Cherax*) whereas *Heptacraspedella* and *Gelasinella* are found on *Euastacus*, crayfish associated with wet forests of the south and more mountainous east coast of Australia. These latter are monotypic genera at present and may represent transfer ('host switching') to *Euastacus* in regions of overlap where crayfish of both genera are present. Presumably the Craspedellinae represent a return to the shelter of the branchial chamber and gills from a more exposed environment found on the external exoskeleton. The hypothesis that the branchial chamber is their primary habitat also cannot be excluded at the moment.

- *Neotropical Temnocephala*: Neotropical *Temnocephala* includes *Temnocephala chilensis*, the type species of the genus, and these form a distinct clade, i.e., *Temnocephala s.s.* This has indicated the need for a new name for Australian species now classified as *Temnocephala*, and a new taxon has been suggested by Damborenea and Cannon (in press) in honour of Dr Kim Sewell.

- Temnocephala s.l. (rouxi-semperi group): These range widely in Asia, as well as being from Aru Island off New Guinea and from northern Australia. All Asian specimens have been described as *Temnocephala semperi*, but specimens studied by Merton (1913), Joffe (1981a,b) and Rohde (1966) differ. This is probably a group of species.

- Craniocephala: This genus was established based on presence of three (rather than two) pairs of testes in a worm resembling *Temnocephala s.l.* in all other respects. Such temnocephalans have never been observed subsequently: the validity of this genus must be in doubt.

The position of the Temnocephalida within the Platyhelminthes

The Temnocephalida unquestionably belongs to the Rhabdocoela. Recent 18S rDNA data suggest that the main groups of freshwater rhabdocoels – Typhloplanidae, Dalyelliidae and Temnocephalida – constitute together a monophyletic branch (Littlewood *et al.* 1999a; see also Joffe and

Kornakova 2001, this volume, for qualitative molecular characters corroborating this view). The free-living turbellarians from the family Dalyelliidae and Temnocephalida are united by the presence of: 1) lamellated rhabdites; and 2) a duo-gland adhesive system which is lost in other 'Dalyellioida' (Ehlers and Sopott-Ehlers 1993). Within the Dalyelliidae a group of marine and brackish-water genera – *Halammovortex, Alexlutheria, Thalassovortex, Jensenia, Beauchampiella* – have a male armature more similar to that of the Temnocephalida, than the specialized armature of typical dalyelliid genera. Furthermore, the last three form a widened temporary frontal lobe when alive, as in *Didymorchis*. Significantly, GAIF-positive innervation of this lobe includes an unpaired medial neuron, both in *Thalassovortex* and in the Temnocephalida (Joffe and Cannon 1998b). Another of these genera, *Varsoviella*, is symbiotic: it has a single species living in the branchial chamber of a brackish-water gammarid amphipod from Poland (Gieysztor and Wiszniewski 1947; see also Luther 1955). This fact may indicate the potential of this group for a commensal mode of life rather than specific relationships between *Varsoviella* and Temnocephalida.

On the basis of the organization of the pharyngeal musculature and the protonephridial system, it has been suggested (Joffe 1981a, 1987) that two other forms may be closely related to the Temnocephalida, namely *Pseudograffilla arenicola* and *Anomalocoelus caecus*. *P. arenicola* is a marine 'dalyellioid' turbellarian with an uncertain systematic position. Its possible relationships to the Temnocephalida need to be reconsidered on the basis of other characters and in view of possible close relationships between the Temnocephalida and marine Dalyelliidae. *A. caecus* is an Australian freshwater turbellarian, also with uncertain relationships (Gilbert 1935), which – according to Haswell (1905) – has a protonephridial system very similar to that of the Didymorchiidae and Scutariellidae. If Haswell's data are correct, *Anomalocoelus* should be very closely related to the Temnocephalida.

Biogeography (Table 9.1)

The closest modern-day relatives of the Temnocephalida appear to be the brackish-water members of the Dalyelliidae, and their greatest radiation (in the Temnocephaloidea) is seen on the parastacid crayfish, hosts with presumed marine ancestry, which have radiated in the southern continents, i.e., old Gondwana (Riek 1972).

Two main groups of Temnocephalida can be seen. The 'northern' group, Scutarielloidea, are found along the southern margin of the Eurasian continent from north-eastern Italy and the Balkans through the Caucasus and across Asia reaching Sri Lanka and Japan. The Scutariellinae occur to the west on cave-dwelling (stygean) atyid shrimps, with the single exception of *Scutariella didactyla*. The Caridinicolinae occur from Sri Lanka to Japan on non-stygean atyid shrimps. It is noteworthy that, though the atyid shrimps are common in the southern hemisphere, the Scutariellidae have never been found on them, in particular, in Australia.

The 'southern' group, Temnocephaloidea, are much more diverse and found in Australasia and the Neotropics, but also in Madagascar and into Asia as far west as India and north to southern China. Their hosts are shrimps and crabs, but especially crayfish, though in the Neotropics hosts include molluscs, turtles and insects.

Presumably the Temnocephalida emerged sometime after separation of Gondwana and Laurasia (ca. 350 Ma) and their origin encompassed lands (then neighbouring) now part of North-East Australia, New Guinea, India and Madagascar –

Table 9.1 Temnocephalida and their hosts (parastacid crayfish genera given) in relation to geological time (Ma–millions of years; Ka — thousands of years). *Note:* 1, Those hosts in bold have not been examined. 2, Temnocephala NT is a new taxon for the Australian clade of *Temnocephala.* 3, Those worms in brackets preceded by '?' may prove to be present.

Time	Geography	Hosts	Taxa
> 200 Ma Laurasia + Gondwana			**SCUTARIELLOIDEA**
	Asia	Atyid shrimps (non-Stygean)	*Monodiscus, Caridinicola*
	Asia and Europe	Atyid shrimps (Stygean)	*Bubalocerus, Stygodictyla, Subtelsonia, Scutariella, Troglocaridicola*
~ 160 Ma *Gondwana begins to break up*			
~ 90 Ma Madagascar separates	Madagascar	Astacoides	**TEMNOCEPHALOIDEA** *Dactylocephala, (?Didymorchis)*
> 60 Ma New Zealand separates	**New Zealand**	Paranephrops	*Didymorchis, Temnohaswellia*
> 37 Ma South America separates	South America	*Parastacus, Samastacus* (plus several non-parastacid hosts)	*Didymorchis, Temnocephala* s.s.
~ 5 Ma S.W. Australia climatically climatically separated	S.W. Australia	*Cherax,* **Engaewa**	Temnocephala NT, *Zygopella,* *(?Didymorchis, Actinodactylella)*
~ 20 Ka Tasmania separates	Tasmania	*Astacopsis, Parastacoides,* *Engaeus,* **Geocherax**	*Actinodactylella, Temnomonticellia,* Temnocephala NT, *(?Didymorchis)*
< 10 Ka Cape York separates from New Guinea	E. Australia	*Cherax, Euastacus, Engaeus,* **Geocherax, Gramastacus,** **Tenuibranchiurus** (plus non-parastacid shrimps and crabs)	*Didymorchis, Diceratocephala,* *Decadidymus, Achenella, Actinodactylella,* *Temnohaswellia, Achenella,* Temnocephala NT, *Notodactylus, Craspedella,* *Heptacraspedella, Gelasinella*
< 10 Ka Cape York separates from New Guinea	New Guinea	*Cherax* (plus non-parastacid host *Sesarma*)	*Diceratocephala, Notodactylus,* *Craniocephala (?Didymorchis,* *Craspedella)*
Present Though previously isolated > 120 Ma a second invasion is taking place	S.E. Asia and beyond	(non-parastacid hosts)	Temnocephala NT (*rouxi-semperi* group)

regions where today we find those taxa which diverged early in the group's evolution (Scutariellidae, Didymorchidae, Diceratocephalidae).

Old Gondwana began to break up approximately 160 million years ago when Africa broke free. Remarkably, neither temnocephalids nor parastacids are known from Africa, which suggests that the Temnocephalida did not exist or had not expanded to the African pro-continent by this time. India began moving to its rendezvous with Asia about 100 Ma, and we can possibly date the divergence of the Scutarielloidea and Temnocephaloidea from this time. The divergence of the Scutariellinae and the Caridinicolinae may possibly be dated from the time as India met Laurasia, and the Scutariellinae then evolved in the west. Scutariellinae are found in Europe today, but a Mediterranean distribution with restriction to stygean habitats is characteristic of the remnants of tropical Indian fauna which migrated to the Caucasus and South Europe in the middle Tertiary. Madagascar separated ~90 Ma followed by New Zealand (~60 Ma) and finally the connection via Antarctica between Australia and South America broke ~37 Ma (Talent 1986).

Didymorchis represents the closest to the base plan of the Temnocephaloidea and is found in Australia on both *Euastacus* and *Cherax,* and also in New Zealand on

Paranephrops and South America on *Parastacus.* It has not been found in Madagascar, but its presence there cannot be ruled out. Nor is *Didymorchis* present in India.

Dactylocephala (with 12 tentacles) is the earliest branch of the Temnocephalidae. It is perhaps > 90 Ma, since it is found only in Madagascar on a parastacid *Astacoides* which according to Hamr (1992) is close to the Tasmanian *Astacopsis.* So if the divergence between the Scutarielloidea and Temnocephaloidea dates from the break away of India, presumably all the main events in the evolution of the Temnocephalida (divergence at family level) took place during a short period (ca. 10 Ma) between the respective separation of India and Madagascar.

Tasmanian crayfish are host to *Actinodactylella,* presumed to be earlier derived than the Temnocephalidae, and to *Temnomonticellia,* a member of the five- to six-tentacled Australian temnocephalids, but possibly the earliest of this clade. The New Zealand crayfish *Paranephrops* is close to *Parastacoides* from Tasmania and both form a clade with *Euastacus* and *Astacopsis* (see Crandall *et al.* 1999). New Zealand has both *Didymorchis* and *Temnohaswellia*: the latter has radiated on *Euastacus* the cold water crayfish of forested Eastern Australia which are close relatives of the Tasmanian *Astacopsis.* Clearly *Temnohaswellia* already existed when

New Zealand broke away about 60 Ma, suggesting that *Actinodactylella* must be between 60–90 Ma. *Temnomonticellia* may be either slightly younger than *Actinodactylella* (hypothesis 1), or much younger (< 37 Ma) since it is absent from South America (hypothesis 2).

The absence of an accessory ejaculatory sac probably links Neotropical *Temnocephala* with Tasmanian temnocephalids. Presumably they began to evolve independently after the separation of South America from Australia ~ 40 Ma. Neotropical *Temnocephala* have migrated as far north as Mexico, and though some are found on parastacid crayfish (which according to Riek (1972) are those genera most closely related to the Australian 'land crayfishes', e.g., *Engaeus* and *Geocherax*), most occur on anomuran crabs (*Aegla* spp.) and other crabs as well as a variety of other hosts.

By the time the Antarctic connection between South America and Australia broke ~37 Ma, the phylogenetic branch of the Temnocephalidae with five tentacles was already well diversified. Most of the extant Australian species of *Temnocephala* are presumably more recently derived and have radiated on Australian crayfish (*Cherax and Engaeus* as well as *Euastacus*), though they have extended their hosts to crabs and, like *Temnohaswellia*, to shrimps (Cannon 1993). Temnocephala new taxon species from the *rouxi-semperi* group have radiated, since ~5 Ma, in Asia and are found now in the Philippines and Indo-Malayan region, China, and India. Here it is known as *T. semperi*, though it probably represents a species complex. Association with freshwater crabs has probably facilitated radiation: crayfish have not radiated beyond New Guinea.

Craspedellinae have presumably returned secondarily to the gills and, while radiating in *Cherax* (*Craspedella* and *Zygopella*), have evidently crossed back to *Euastacus* (*Heptacraspedella* and *Gelasinella*). *Zygopella* is found only in the south-west region of Australia, which became climatically isolated ~ 5 Ma (McPhail 1997). Although Tasmania separated from the mainland only about 20 000 years ago, clearly the main host crayfish (*Cherax*) of the Craspedellinae

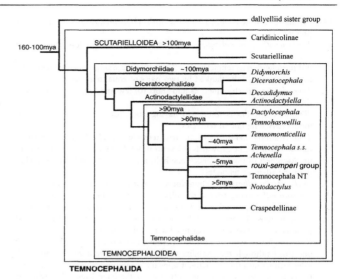

Figure 9.3 Proposed scheme of the evolution of the Temnocephalida integrating phylogenetic, biogeographical and geological data.

must have been isolated from the region geographically by the Great Dividing Range in Australia for much longer, as this group has not established in Tasmania (Hickman 1967; L. Cannon and K. Sewell, unpublished). Thus, *Notodactylus* is presumably > 5 myo, though the Craspedellinae from *Euastacus* must be quite recently evolved. Craspedellinae do occur in New Guinea (B.I. Joffe, unpublished), but the separation of New Guinea from Australia was only about 10 000 years ago. Specialized gill parasites of crayfish, Craspedellinae have not been found in south-east Asia with *T. semperi*.

A proposed evolutionary scenario incorporating phylogenetic and geological data is presented on Figure 9.3.

Appendix 9.1. Character set, cladogram and added notes for 'Temnocephalida' (by Lester R.G. Cannon).

The character set

Several characters derived from electron microscopy of spermatozoa and spermiogenesis have been provided by Rohde and Watson (1995), but have not been included as homologies and are yet to be resolved (Littlewood *et al.* 1999a). Note: '?' is unknown, '–' inappropriate, and '9' species variation.

1. *Locomotory cilia presence:* uniform (0) or reduced (1).
   [Cilia are found uniformly on the surface of *Dalyellia*, are reduced in distribution on *Diceratocephala*, found only on the ventral surface of *Didymorchis* and only about the anterior in *Varsoviella*.]
2. *Locomotory cilia distribution:* reduced partially (0), wholly (1).
   [Tufts of long cilia are present in some species, e.g., *Temnocephala minor* about the excretory pores.]
3. *Dorsal ridges carrying epidermal papillae:* normally absent (0) but present in the Craspedellinae (1).
4. *Epidermal papillae on tentacles:* normally absent (0), but present in Craspedellinae (1). [Those in *Actinodactylella* are possibly not homologous.]
5. *Insunk epidermal nuclei:* normally absent (0) but present in *Didymorchis* (1).
6. *Modified epidermal mitochondria:* normally absent (0) but present in *Didymorchis* (1).
7. *With multiple syncytial epidermal plates:* *Dalyellia* has no syncytial plates (0), but present in all Temnocephalida (1).
   [Joffe and Cannon (1998a) have examined the pattern of epidermal plates in some detail. A posterior adhesive field syncytium is common to all temnocephalans and a general reduction in the number of plates is presumed to be an evolutionary trend. *Dactylocephala* has a unique pattern (Cannon and Sewell, in preparation) in which the frontal syncytium extends back dorsally to reach the dorsal peduncle and the trunk (or peduncular?) syncytium

extends forward ventrally to just posterior of the mouth. This plate has two dorso-lateral extensions which 'capture' the excretory pores. A subtentacular syncytium reminiscent in shape to the 'saddle' or post-tentacular syncytium of other temnocephalans lies between the trunk and frontal syncytia on the ventral surface near the base of the tentacles (see Character 8).]

8. *Post-tentacular syncytium:* absent (0), present (1).
   [The subtentacular syncytium in *Dactylocephala* may not be homologous.]
9. *Post-tentacular syncytium fused:* a fused syncytium (0) or split into 2 (1).
   [Evidence from development and comparative data suggest fusion may not always occur so state (1) may be secondary in some species.]
10. *Peduncular syncytium:* covers posterior part of trunk dorsally and ventrally (0) or confined to ventral (1) absent presumed lost through fusion (2).
11. *Eyespots:* small black (melanized?) granules for the light-reflective layer present (0) or replaced with red fugacious material (1) which disperses on death and is not retained after fixation. [Rarely eyespots are absent, as in *Actinodactylella* and some *Temnocephala* spp.: these are presumed secondarily lost. These pigments may suffuse part or much of the body in many species, most notably those exposed on exterior surfaces, and are less or absent in those species that live in sheltered habitats, e.g., the branchial chamber. Red fugacious material appears limited to the Neotropical *Temnocephala* spp. (Damborenea and Cannon, in press).]
12. *Body compression:* there is no evidence of a lateral ridge (0) or a lateral ridge (or some distinction between dorsal and ventral surfaces) is evident (1).
    [In *Actinodactylella* the lateral borders are served with a row of elongate papillae, *Dactylocephala* has plate suture lines, *Craspedella* has rows of low papillae which mark and delimit dorsal and ventral, and in *Temnocephala chilensis* a distinct lateral border is seen.]

13. *Tentacles*: true tentacles, i.e., those with axial muscles are absent (0) or present (1).
14. *Tentacles equal or subequal*: true tentacles are more or less equal (0) or vary (1).
    [Of those with true tentacles only *Temnomonticellia* has unequal tentacles, having a central lobe flanked by two pairs of equal tentacles].
15. *Tentacles (number)*: numerous and short (0) few and long (1) or long and short (2).
    [*Temnomonticellia* is unique in having one short central pad and four moderately long lateral tentacles].
16. *Tentacles (how held)*: the centre true tentacle(s) held, like the lateral tentacles, down (1), i.e., towards the ventral, while the adjacent tentacles are up. Members of the Craspedellinae hold this central tentacle up (0) higher than the adjacent tentacles.
17. *With anterior adhesion*: the anterior is not able to attach (0) or is adhesive (1).
18. *Anterior adhesion restricted*: the adhesion area may be restricted to pits (1) as in *Scutariella*, an adhesive field (2) as in *Actinodactylella*, the tentacles (3) as in most Temnocephalidae or restricted to the central tentacles (4) such as in *Craspedella*, or scattered across the anterior (0).
19. *Rhabdite tracts*: with two distinct rhabdite tracts per tentacle (1) with only one (0). [Although the rhabdite tracts arc across the anterior with both sides feeding into the tentacles, it appears *Temnomonticellia* has two main tracts to the central lobe, but only one on the lateral tentacles, whereas *Dactylocephala* has paired tracts to all tentacles and all other temnocephalans have only one main tract].
20. *Posterior adhesive field*: the posterior adhesive pad round (1) or crescentic (0). [In Scutariellidae and some *Didymorchis* the crescentic pad may be broken into more than one patch].
21. *Muscles (posterior)*: the whole taxon is characterized by having a posterior adhesive organ. The muscles to this organ are axial inserted in the basal membrane of the adhesive field (1) in *Actinodactylella* (though weak and diffuse) and all the Temnocephalidae, or not axial (0) and inserted in the body wall, usually weak, but strong in *Diceratocephala*.
22. *Protonephridial commissures*: with transverse commissures joining right and left protonephridia to form rings (1), absent (0).
    [*Varsoviella* is reported to have a single commissure joining right and left protonephridia].
23. *Excretory pores*: pores dorso-lateral (1), ventral (0).
    [*Dactylocephala* has the pores almost lateral, rather than dorso-lateral].
24. *Excretory pores (syncytial position)*: pores in post-tentacular syncytium (a) (1), or not (0), i.e in other body plates. In some representatives of *Temnohaswellia* they are adjacent to the post-tentacular syncytium (2).
25. *Pharynx*: with strong anterior and posterior sphincters (1) absent or secondarily lost (0).
26. *Pharynx*: strongly elongate with strongly reinforced ring muscles all along wall of pharynx as in *Diceratocephala* and *Decadidymus* (0); moderately, or hardly elongate; ring muscles either not reinforced or only in part of the wall (1).

27. *Pharynx syncytia*: There are several successive belt-like syncytia which cover the pharynx walls. The pharyngeal glands opening through a narrow anterior belt-like syncytium (1), or otherwise (0).
    [In Scutariellidae, Didymorchidae and Diceratocephalidae, the first case applies, but in *Temnocephala dendyi* they open into the pharyngeal pouch via syncytium P2 and in *Craspedella* the ducts open via the pharyngeal canal. The basis for this is Joffe *et al.* (1997b) who showed progressive change (simplification). Thus a series could be: eight plates *Didymorchis* (0), six plates *Scutariella* and *Diceratocephala* (1), four plates Temnocephala NT (2), three plates *Craspedella* (3)].
28. *Gut*: gut extends the full length of the body so the reproductive structure lie ventral to it (1) or is short and the reproductive structures lie mostly posterior (0).
    [In the Craspedellinae the gut is markedly longer on the right side].
29. *Genitointestinal connection*: a genitointestinal communication exists (1), or not (0).
    [This is most commonly observed as the vesicular resorbens pushing into (and often bursting to release contents) the gut. It is found in all Temnocephalida].
30. *Testes (number)*: two pairs of lateral testes (1) or some other number (0).
    [Scutariellidae, *Didymorchis*, *Diceratocephala*, and *Achenella* have but one pair, as does the aberrant *Temnocephala brenesi* from Costa Rica. *Craniocephala* is reported to have three pairs, but the validity of this (and thus the genus) is in doubt: *Dactylocephala* has six pairs of testes and *Decadidymus* 10 pairs].
31. *Accessory ejaculatory sac*: in many taxa the ejaculatory canal on entering the copulatory bulb reveals an accessory sac of uncertain function (1), but often none is there (0).
    [Since *Dalyellia* lacks an ejaculatory sac this condition is presumably the plesiomorphic state, though the sac is evidently secondarily lost in many cases, e.g., in all Neotropical *Temnocephala* spp.].
32. *Accessory ejaculatory sac*: when present the sac may be free (0) or incorporated (1) [The degree of separation is variable, i.e., attached by a broad or narrow duct].
33. *Vitellaria (distribution)*: vitellaria are discrete in *Didymorchis* (0), or may be scattered over the dorsal, lateral and ventral surfaces of the gut (1).
34. *Sclerotic female duct*: without sclerotized parts of the female duct (0), or with them (1). [Some *Didymorchis* do, some do not, but *Actinodactylella* and all *Temnohaswellia* are characterized by having sclerotic parts of the female duct].
35. *Egg capsules*: egg capsules cemented on the long axis in *Temnomonticellia* (1) or on the short axis as in many species of *Temnocephala* among others (0).
36. *Sucker pedunculate*: not pedunculate (0), with a peduncle (axial muscles always strong) (1).
37. *Gland openings on surface of adhesive area*: even spread of glands (0), restricted to periphery (1).
38. *Marginal valve around sucker*: absent (0), present (1).

The data matrix; 38 characters, 19 taxa

	1 1234567890	2 1234567890	3 1234567890	12345678
Dalyellia	0-0-000---	000---0---	-00-01-000	0-000---
Didymorchis	100-1110--	000---0-00	1000000101	0010000
Temnomonticellia	1100001100	0111201301	1110111111	11101100
Heptacraspedella	1111001101	0110111411	1110011111	0-100111
Temnocephala NT	1100001101	0110101311	1110111111	10100111
Actinodactylella	110-001100	-10---12-1	111011?111	11110000
Dactylocephala	1100001?00	01100?1301	111011?110	0-100000
Temnocephala	1100001112	1110101311	1111111111	01100111
Craspedella	1111001101	0110111411	1110011111	99100111
Notodactylus	1100001101	0110101311	1110111111	10100111
Diceratocephala	1000001110	000---0-10	11000-1101	0100000
Zygopella	1111001101	0110111411	1110011111	11100111
Scutariella	11000010--	000---11-0	010001-010	0-00000
Craniocephala	1100001???	0?101?1?11	1110?1?110	??100???
Achenella	1100001112	0110101311	111011?110	101001??
Decadidymus	11000010-0	00---10-1	011000-110	10100000
Varsoviella	100-00????	-00---??--	-?????0?0	0-00?---
Temnohaswellia	1100001100	0110101311	1112111111	10110111
Gelasinella	1111001101	0110111411	1110011111	10100111

Cladogram: A strict consensus tree of the Temnocephalida (of nine) generated using Hennig86. Trees were generated in Hennig86 (Farris 1988) by running the command sequence 'm* bb* ie*xs w;' three times to get a stable output. This sequence initiates tree generation (m*), utilizes branch breaking (bb*), finds all trees (ie*) and applies successive weighting (xs w) to amplify the signal from characters that are congruent with the cladogram. The taxon sequence was randomly reordered. The outgroup is *Dalyellia* (Fam. Dalyelliidae). The strict consensus tree (of nine produced), generated by running in addition the command 'nelsen', had a length of 403, and indices (to only 2 places) CI = 0.92 and RI = 0.95. Using PAUP 3.1 (Swofford 1993) with the same matrix, a strict consensus tree (from 12 produced) was generated with identical topology, but with a length of 81 and indices CI = 0.741 and RI = 0.790. Resolved nodes in all trees in both analyses were stable and identical save for the grouping of *Varsoviella* and *Scutariella* at the base and ordering of genera within the Craspedellinae.

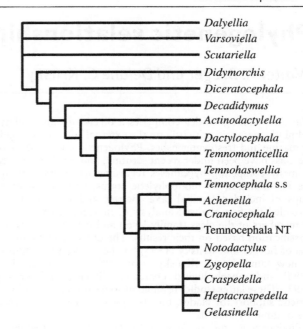

Dalyellia
Varsoviella
Scutariella
Didymorchis
Diceratocephala
Decadidymus
Actinodactylella
Dactylocephala
Temnomonticellia
Temnohaswellia
Temnocephala s.s
Achenella
Craniocephala
Temnocephala NT
Notodactylus
Zygopella
Craspedella
Heptacraspedella
Gelasinella

Chapter 10

Phylogenetic relationships of the Monogenoidea

Walter A. Boeger and Delane C. Kritsky

The class Monogenoidea comprises a diverse group of platy-helminths parasitic primarily on the external surfaces and gills of marine and freshwater fishes. Phylogenetic relationships of the class have been under recent scrutiny, and this has resulted in many competing hypotheses regarding its position within the Neodermata, its monophyletic status, and interrelationships of included familial taxa. Boeger and Kritsky (1993) provided the first cladistic analysis of morphological features of members of the class (family level) and proposed a revised classification based on their results. The phylogenetic hypothesis of Boeger and Kritsky (1993) has been tested by addition of new familial taxa (Kritsky *et al.* 1993; Kritsky and Lim 1995) and morphological characters (Boeger and Kritsky 1997). These latter studies resulted in some alteration of phylogenetic support for clades, but did not significantly challenge the original hypothesis.

Phylogenetic studies using molecular data have frequently suggested that the Monogenoidea is polyphyletic with each of the two basal lineages identified in Boeger and Kritsky's (1993) hypothesis as having independent origins within the Neodermata. Some of these hypotheses are presented in Figure 10.1 (also see Justine 1998a). With the exception of that of Campos *et al.* (1998) (Figure 10.1a), respective topologies of the remaining hypotheses in Figure 10.1 have received little, if any, independent support. The hypothesis of Campos *et al.* (1998), which suggests a monophyletic Monogenoidea, supports the same relationships concluded from morphological studies by Brooks (1989b), Brooks *et al.* (1985b) and Ehlers (1985a,b, 1986). Although gene sequencing and gene expression are very different modalities for examination of phylogenetic relationships and gene expression is poorly understood regarding overt morphological features, Littlewood *et al.* (1999a) suggested monophyly of the Monogenoidea when molecular and morphological data were combined. Justine's (1998a) statement, that all previous phylogenetic hypotheses based on morphological data assumed monophyly of the Monogenoidea, is probably accurate. However, morphological studies have identified some potential synapomorphies for the class (Lambert 1980; Brooks *et al.* 1985b; Ehlers 1985a,b, 1986; Brooks 1989b; Boeger and Kritsky 1993, 1997; Brooks and McLennan 1993a). In the present chapter, the sister-group relationships of the families of Monogenoidea are re-examined based on a new analysis of character evolution, and the monophyly of the class is tested using morphological data. The phylogenetic position of the Udonellidae, recently suggested to be a member of the subclass Polyonchoinea by Rohde and Watson (1993c) and Littlewood *et al.* (1998a), is also examined.

Materials and methods

Homologous series include those of Boeger and Kritsky (1993, 1997) as well as some new series with potential to be informative for testing monophyly of the Monogenoidea; series containing information on sperm morphology and development were updated based on the revised matrix provided by Justine (2001, this volume). Character states were determined from the literature and verified by study of available specimens when possible (see Boeger and Kritsky 1993). Homology of character states was determined according to Remane's criteria on homology as described in Wiley (1981) along with Hennig's Auxiliary Principle (see Brooks and

McLennan 1991). Thus, structural and morphological differences are not considered *a priori* evidence for rejection of a hypothesis of homology of character states. Series in which the derived state represented an autapomorphy for a single ingroup taxon were excluded from the analyses. All homologous series were considered ordered and equally weighted. Determination of the plesiomorphic state was based on outgroup or functional outgroup comparison (Watrous and Wheeler 1981; Maddison *et al.* 1984). Relative plesiomorphy for multistate series was postulated using functional outgroup comparison and tested by running an analysis with all characters unordered using PAUP* 4.0 beta 2 version for Microsoft Windows (Swofford 1998); code assignment of polymorphic characters in specific ingroup taxa was made through prediction of the primitive state (Kornet and Turner 1999). The Urastomidae, Fecampiidae, Trematoda (includes Digenea + Aspidobothrea), and Gyrocotylidea were used as outgroups according to the proposed sister-group relationships of these taxa with the Monogenoidea as suggested by Brooks (1989b), Brooks *et al.* (1985b) and Littlewood *et al.* (1999a). Testing of monophyly of the Monogenoidea was performed by including the Trematoda and Gyrocotylidea as ingroup taxa. The initial hypothesis on the evolutionary relationships of family groups of Monogenoidea and of neodermatan groups was constructed using Hennigian Argumentation (Wiley 1981; Hennig 1966). The resulting hypothesis was tested with PAUP* to determine if it represented a most-parsimonious cladogram. Due to the size of the matrix, most-parsimonious cladograms were determined with the HSEARCH option (heuristic search) of PAUP*, with unlimited MAXTREE. Tree statistics (CI, RI and bootstrap) were determined with the Trematoda and Gyrocotylidea as ingroup taxa and using the command DELETE UNINF of PAUP*; bootstrap support for the respective nodes (replicates = 1000) was determined using MAXTREE = 100 and was considered supportive when ≥50%. Trees presented from bootstrap and consensus analyses are intended as measures of support for clades, and should not be construed to be evolutionary hypotheses for the Monogenoidea. The character analysis is provided in Appendix 10.1; the matrix comprised 66 homologous series (Appendix 10.2).

In the present chapter, taxonomic nomenclature follows that proposed in the revised classification provided by Boeger and Kritsky (1993). The epithet Monogenoidea Bychowsky 1937, is used for the name of the class because: 1) it objectively represents the oldest available name for the taxon at the level of class; 2) it accommodates stability in nomenclature; and 3) it better reflects current hypotheses on the phylogenetic relationships of the taxon. The unconvincing and wistful plea by Wheeler and Chisholm (1995) that 'Monogenea' be used over Monogenoidea, is based on unscientific principles (polling and authoritative directive) and arguably erroneous interpretations of the International Code of Zoological Nomenclature (ICZN) and of the cognizance of previous authors. Polling and authoritarianism aside as irrelevant to the problem (see also Euzet and Prost 1981), two points brought out by Wheeler and Chisholm (1995) require comment. These authors wrongly suggest that we (Boeger and Kritsky 1993) confused the 'letter' of the ICZN with its 'spirit' regarding stability and that nomenclatural stability was threatened with acceptance of Monogenoidea. Indeed, we were fully aware of the importance of these issues when we adopted the name. We recognized that Bychowsky (1937) must have also been concerned because he only changed the ending of the word – a common practice when a categorical change is necessary in classification. Bychowsky's modification of the name does not threaten stability and has no impact on one's understanding of the composition of the taxon (even for non-specialists). On the other hand, utilization of 'Monogenea' for this taxon is not inconsequential. 'Monogenea' was originally proposed as an order of the class Trematoda and was considered the sister-group of the Digenea + Aspidobothrea. Current phylogenetic hypotheses that consider the taxon monophyletic, are supported by several synapomorphies and suggest the taxon to be the sister-group of the class Cestoda. Bychowsky (1937) recognized the latter relationship well before modern methodologies in phylogenetic systematics were available, although an extended period of time was to elapse before general acceptance of his idea. While probably not an issue for phylogeneticists

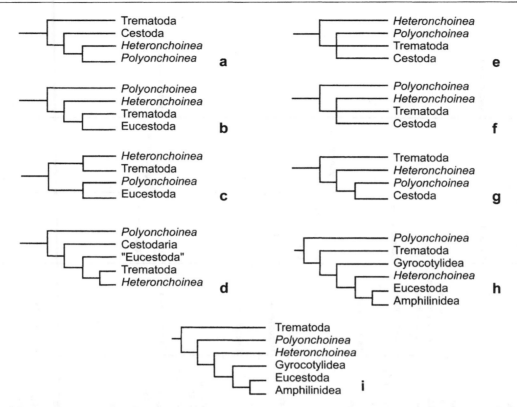

Figure 10.1 Some competing hypotheses on monophyly of the Monogenoidea derived from molecular data (taxonomic nomenclature adapted from Boeger and Kritsky (1993) and based on specific parasite species use in respective analyses). **a.** Hypothesis from Figure 3 of Campos *et al.* (1998). **b.** Hypothesis from Figure 1 of Mollaret *et al.* (1997). **c.** Hypothesis from Figure 3 of Litvaitis *et al.* (1999). **d.** Hypothesis from Figure 5 of Litvaitis *et al.* (1999). **e.** Hypothesis from Figure 1 of Littlewood *et al.* (1998a). **f.** Hypothesis from Figure 2 of Littlewood *et al.* (1998a). **g.** Hypotheses from Figure 3 of Littlewood *et al.* (1999a). **h.** Hypothesis from Figure 3 of Rohde *et al.* (1995). **i.** Hypothesis from Figure 1 (1) of Cunningham *et al.* (1995). Heteronchoinea subclass nov. is proposed in the present paper and comprises the infrasubclasses Polystomatoinea and Oligonchoinea; taxon names for the Cestoda are from Schmidt (1986); taxa in quotes are non-monophyletic in respective hypotheses.

studying the evolutionary relationships of the Platyhelminthes, use of 'Monogenea' does not clearly differentiate between the two alternatives for the non-specialist. The name Monogenoidea, however, does not carry this 'baggage' and is far more stable in its informative content.

In order to justify their preference for 'Monogenea', Wheeler and Chisholm (1995) wishfully argue that classes and orders should be coordinate, and cite several authors with similar feelings. Classes and orders, however, are required categories in Linnaean classification and, as such, cannot be coordinate. In either case, however, coordination does not imply that the change of endings of taxonomic names is improper or unadvisable when a different taxonomic level is required. Indeed, different endings for coordinate names at the familial level (i.e., -inae, -idae) are mandated by the ICZN.

Notes on familial taxa (Monogenoidea)

A total of 53 families of Monogenoidea is included in the analyses. Fifty-two families are those recognized by Boeger and Kritsky (1993, 1997). The Udonellidae was added as an ingroup taxon based on recent accounts from molecular and ultrastructural data that suggest the taxon is aligned within the Polyonchoinea. While the Iagotrematidae (currently comprised of two genera, *Iagotrema* and *Euzetrema*) is apparently polyphyletic, the family is represented in the analyses by *Euzetrema* for which sufficient morphological information is available. Some previously proposed families are not included as ingroup taxa because of questionable taxonomic and/or evolutionary status.

The Microbothriidae was excluded from analyses because it is probably polyphyletic. The family is currently diagnosed by derived characters representing loss of structures (primarily the sclerotized structures in the haptor) that are highly homoplastic among monogenoidean clades and leave no clues on homology. The family has been used as a 'dumping ground' for a variety of monogenoidean species that lack haptoral sclerites for which homologies have not been determined. For example, species of *Dermophthirius* which lack haptoral sclerites (see Benz 1987) are probably members of the Capsalidae based on the general morphology of the reproductive systems. Further, *Anoplocotyle* (= *Anoplodiscus*, see Ogawa and Egusa 1981) and *Enoplocotyle*, previous members of the Microbothriidae (see Yamaguti 1963), have been transferred to other families after character evolution in each genus was determined.

The Ancyrocephalidae, Heterotesiidae, Neocalceostomatidae, Pseudodactylogyridae, and Rhamnocercidae (all Polyonchoinea) are not included as ingroup taxa because of uncertainty of their validity in the current classification system. The Rhamnocercidae is herein considered a subordinate taxon of the Diplectanidae, and the four former families are included in the Dactylogyridae (see Kritsky and Boeger 1989).

The Macrovalvitrematidae and Anchorophoridae are not recognized because of their apparent origins within the Diclidophoridae where each has a separate diclidophorid ancestor (Mamaev 1976). The Bicotylophorinae, recently elevated to family status by Amato (1994), is not included because information on key characters is not available.

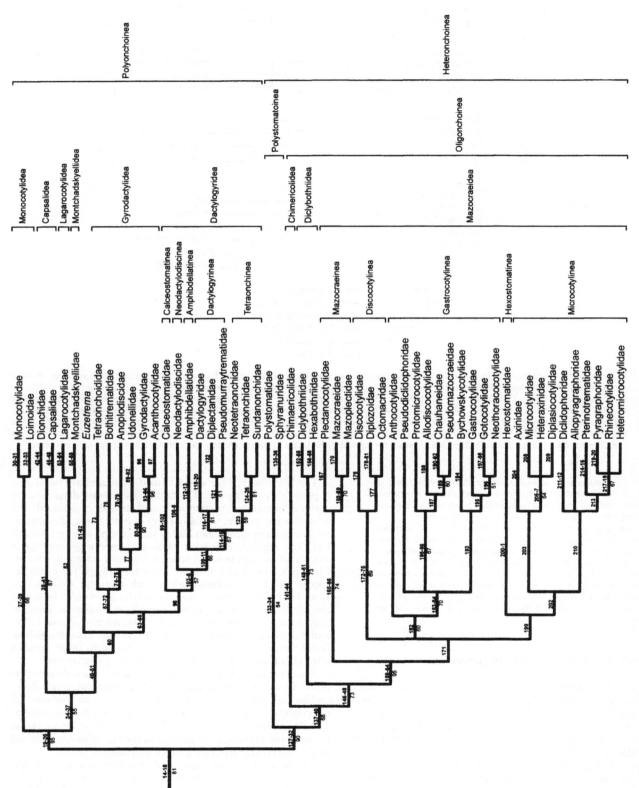

Figure 10.2 Hypothesis for the evolutionary history of 53 families of Monogenoidea based on 66 homologous series of morphological characters. Bold numbers above each branch refer to postulated evolutionary changes as indicated in the character analysis (Appendix 10.1) and character change list (Appendix 10.3). Numbers below each branch refer to bootstrap support for 1000 replicates; only values ≥50% are presented. Tree length = 214; CI = 53%; RI = 87%; all considering the Trematoda and Gyrocotylidea as ingroup taxa.

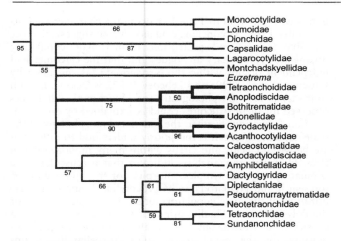

Figure 10.3 Bootstrap showing support for internal clades of the Polyonchoinea, with particular reference to those comprising the Gyrodactylidea. Bootstrap values are percentages of 1000 replicates; only values ⩾50% are presented.

Results and discussion

The phylogenetic hypothesis on the relationships of family groups of Monogenoidea (length = 214; CI = 0.53; RI = 0.87) is presented in Figure 10.2; a list of character-state changes with homoplasies is provided in Appendix 10.3. This hypothesis is represented by 1 of 2899 most-parsimonious cladograms obtained through PAUP* analysis. Bootstrap (Figures 10.2 and 10.3) and strict consensus for all most-parsimonious cladograms (Figure 10.4) provide variable support for the clades of the hypothesis. The present hypothesis is congruent with that provided by Boeger and Kritsky (1993) for ordinal taxa; it lacks congruence at the ordinal level with the latest revision of the Monogenoidea by Boeger and Kritsky (1997) only in the apparent sister-group relationship of the Lagarocotylidea and Montchadskyellidea (see below).

In their classification, Boeger and Kritsky (1993) did not assign a name for the basal clade containing the Polystomatoinea and Oligonchoinea. A name for this major taxon appears desirable because each of the basal clades in the Monogenoidea represents an independent evolutionary lineage

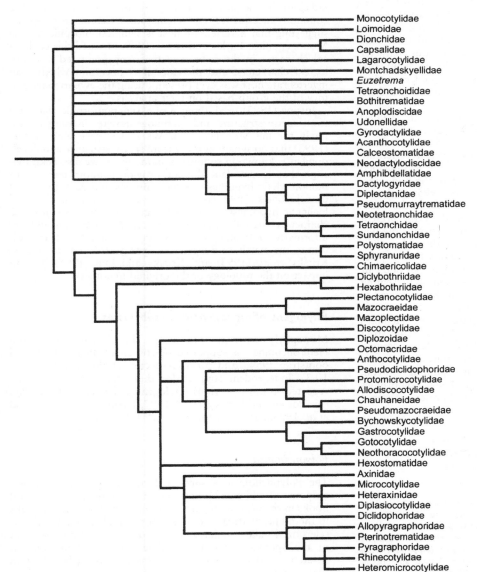

Figure 10.4 Strict consensus tree for sister-group relationships of 53 families of Monogenoidea from the 2899 equally parsimonious solutions obtained with PAUP*.

that coevolved with the Gnathostomata (see Boeger and Kritsky 1997). Thus, Heteronchoinea subclass nov. is proposed for the clade containing the Polystomatoinea and Oligonchoinea. The Polystomatoinea and Oligonchoinea are reduced to infrasubclasses of the Heteronchoinea.

Monophyly of the Monogenoidea

Monophyly of the Monogenoidea has recently been challenged based on molecular data (Figure 10.1) and on sperm morphology and development (Justine 1993). In the present study, the hypothesis for monophyly of the Monogenoidea was tested by including the Trematoda and Gyrocotylidea as ingroup taxa in the data set. Current test results support monophyly of the Monogenoidea (bootstrap = 81%) (Figure 10.5). The resulting sister-group relationships for the Trematoda, Monogenoidea and Gyrocotylidea correspond to those originally proposed by Brooks et al. (1985b), Brooks (1989b), and Ehlers (1985a,b, 1986). Morphological synapomorphies for the Monogenoidea include the larva with three ciliated zones; two pairs of pigmented eyes (larva); two pairs of pigmented eyes (adult), representing retention of number and distribution of larval eyes in the adult; one pair of ventral anchors; and one egg filament. The characters '16 marginal hooks in the larva and adult' may also be synapomorphies for the class. However, our character analysis suggests that these characters could be symplesiomorphic, having developed in the ancestor of the Cerco meromorphae. At the present time, we prefer not to consider them synapomorphies for the class.

In their 'total evidence' approach, Littlewood et al. (1999a) identified four synapomorphies for the Monogenoidea: two pairs (sometimes one) of pigmented eyes in the oncomiracidium, copulatory organ a penis or penis stylet, a vertebrate host only (loss of an invertebrate host?), and a well-defined attachment organ (haptor) not separated from the body parenchyma by a capsule (see Rohde and Watson 1995). Only one of their characters, presence of two pairs of eyes in the oncomiracidium, corresponds to those discovered by the present analysis. Littlewood et al. (1999a) failed to determine the plesiomorphic state for this synapomorphy as indicated in their description of the character ['oncomiracidium with two pairs (sometimes one) of eyes']. Our analysis indicates that one pair of pigmented (fused) eyes in the oncomiracidium (representing loss of one pair) is a secondary development that occurred in the common ancestor of the Mazocraeidea.

Littlewood et al.'s (1999a) clear rejection of Hennig's Auxiliary Principle by their use of binary coding to define strict presence/absence transformations in order to exclude potential homology between character states of the attachment organs (cercomer) and male copulatory organs (among others), resulted in their remaining morphological synapomorphies for the Monogenoidea. Unfortunately, use of presence/absence

in this way may lead to Type 1 errors in the character analysis (see Wiley 1981). Compelling evidence, which suggests that the eversible (cirrus) and protrusible (penis) male copulatory organs are evolutionary analogues and that the cercomer had independent origins within the Neodermata (as suggested by Littlewood et al. 1999a and Lebedev 1987), is lacking. In both characters, the morphological and functional differences observed for these structures could be explained by evolutionary transformation. Further, the functional morphology of the male copulatory organ is too poorly understood to differentiate between a penis and cirrus within many monogenoidean taxa. However, our analysis indicates that a muscular male copulatory organ is symplesiomorphic for the Monogenoidea (present in the Trematoda and Gyrocotylidea), and that sclerotization of the male copulatory organ is synapomorphic for the Polyonchoinea. Finally, origin of the 'cercomer without a capsule' is unknown, but it could also represent a symplesiomorphy for the Neodermata; some 'turbellarian' groups include species with a posterior attachment apparatus that is potentially homologous with the haptor of monogenoideans.

Host, an ecological factor of parasitism, was not utilized for evaluation of monophyly in the present study. Brooks (1989b) indicates that acquisition of a vertebrate host occurred in the common ancestor of the Trematoda + Cercomeromorphae, suggesting that a vertebrate host is symplesiomorphic for the Monogenoidea. Littlewood et al. (1999a) treated this ecological character in the same way that they presented morphological characters, and therefore it is uncertain whether an invertebrate host was ever present during the evolutionary history of the group based on their analysis.

Justine (1993) used absence of a synapomorphic character from sperm structure and spermiogenesis to discount monophyly for the Monogenoidea. While we agree with Justine (1993) that no synapomorphy from sperm structure and development is known in support of monophyly, we also acknowledge that sperm data do not provide support for a nonmonophyletic hypothesis as well. Boeger and Kritsky (1997) used Justine's (1993) sperm data in their revision of the phylogenetic hypothesis for the Monogenoidea. In this study, the characters derived from sperm structure and development were informative in defining clades within the class, but their addition to the matrix of Boeger and Kritsky (1993) as revised by Kritsky et al. (1993) and Kritsky and Lim (1995) had no impact on the topology of the original hypothesis.

Phylogeny of monogenoidean subclasses

Monophyly of both subclasses, the Polyonchoinea and Heteronchoinea, is supported by consensus (Figure 10.4) and bootstrap (95% and 90%, respectively) (Figures 10.2 and 10.3). Independent support for these clades has also been reported in most previous phylogenetic accounts of the class (see Boeger and Kritsky 1993, 1997; Justine 1993; Littlewood et al. 1998a, 1999a; Mollaret et al. 1997; among others).

Polyonchoinea

The Polyonchoinea is supported by eight synapomorphies, only one of which is homoplasious: 14 marginal, two central hooks in the oncomiracidium; 14 marginal, two central hooks in the adult; bilateral osmoregulatory canals fused anteriorly; sclerotized male copulatory organ; dorsoventral microtubules in the spermatozoan absent; intercentriolar body absent; striated rootlets absent; and single testis (shared with Octomacridae + Diplozoidae). The orders Monocotylidea

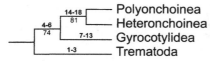

Figure 10.5 Phylogenetic hypothesis of sister-group relationships between major taxa comprising the Neodermata. Numbers above each branch refer to postulated evolutionary changes as indicated in the character analysis (Appendix 10.1) and character change list (Appendix 10.3). Numbers below each branch refer to bootstrap support for 1000 replicates; only values ≥50% are presented.

and Capsalidea are supported (bootstrap values = 66% and 87%, respectively). With the exception of those of the Lagarocotylidea and Montchadskyellidea, the sister-group relationships of the remaining orders comprising the Polyonchoinea are those presented by Boeger and Kritsky (1997), who showed both the Lagarocotylidea and Montchadskyellidea to be unresolved with the clade Dactylogyridea + Gyrodactylidea. In the present hypothesis, the sister-group relationship of the Lagarocotylidea and Montchadskyellidea is supported by only one ambiguous synapomorphy (absence of pigmented eyes in the adult, which also occurs in the Acanthocotylidae, Amphibdellatidae, Gyrodactylidae, Loimoidae, Neodactylodiscidae, and Udonellidae). Bootstrap and consensus do not support the relationship (Figures 10.3 and 10.4).

Monophyly of the order Dactylogyridea and relationships of the suborder Calceostomatinea are not supported by bootstrap; there is bootstrap support for the proposed origins of Neodactylodiscinea (57%), Amphibdellatinea (66%), Dactylogyrinea (61%) and Tetraonchinea (59%). In the Dactylogyridea, only two internal clades, the Sundanonchidae + Tetraonchidae (Tetraonchinea) and the Diplectanidae + Pseudomurraytrematidae (Dactylogyrinea), are reasonably supported (bootstrap values = 81% and 61%, respectively). Nonetheless, strict consensus indicates that sister-group relationships of all families in the Dactylogyridea (except for Calceostomatidae) are constant in equally parsimonious solutions (Figure 10.4).

Mollaret et al. (1997) and Littlewood et al. (1998a, 1999a) using rDNA data and Littlewood et al. (1999a) using 'total evidence' data found the Dactylogyridea to be a basal taxon in the Polyonchoinea (see Figure 5 in Littlewood et al. 1999a), while our hypothesis places the Dactylogyridea as a terminal taxon of the subclass. In a separate analysis using our data with the constraint 'Dactylogyridea basal in Polyonchoinea', the resulting shortest cladograms (length = 230; CI = 0.51; RI = 0.86) provided a significantly less parsimonious explanation than the present hypothesis.

Monophyly of the Gyrodactylidea is supported by six synapomorphies, of which three are non-homoplasious: two seminal vesicles; large vitelline follicles; and hinged hook. The present phylogenetic hypothesis suggests independent sequential origins for the families comprising the order. This configuration differs from that provided by Boeger and Kritsky (1997), in which two clades (Gyrodactylidae + Acanthocotylidae; and Anoplodiscidae + Tetraonchoididae + Bothitrematidae) were suggested. The differences between the two hypotheses resulted when the Udonellidae was included in the analysis. Species in the Udonellidae, Anoplodiscidae and Acanthocotylidae lack anchors, all apparently representing secondary losses. However, independent origins for the gyrodactylidean families as suggested in the present hypothesis may not be the best explanation for evolution of this character because it requires a reversal to presence of ventral anchors in the Gyrodactylidae. Evolutionary loss or absence of structure (in this case, anchors) in monogenoideans are frequent and difficult to evaluate because no visible indication of homology can be ascertained to determine relationship of specific occurrences, i.e., there is no way to determine if absence in a specific family is a result of homologous or homoplasious loss. With Boeger and Kritsky's (1997) configuration, three independent losses of the anchors would be required, increasing tree length by one additional step. Bootstrap values below 50% were obtained for some nodes of the Gyrodactylidea in the present study. However, bootstrap values of 75% and 90%, respectively, provide support for the two clades suggested by Boeger and Kritsky (1997), when the Udonellidae is the sister-group to the Gyrodactylidae + Acanthocotylidae

(Figures 10.3 and 10.4). Phylogenetic relationships within the Gyrodactylidea require further study.

Although Boeger and Kritsky (1993, 1997) showed a sister-group association of the Gyrodactylidea and Dactylogyridea within the Polyonchoinea, doubts of this relationship apparently remain. Shinn et al. (1998) suggest a phylogenetic proximity of the Gyrodactylidae with the Heteronchoinea by asserting that some clusters of body sensilla occurring in species of the two groups are homologous. However, at least some homology suggested by these authors between individual or groups of sensilla in the Gyrodactylidae with those of the Heteronchoinea is suspect. These homologies proposed by Shinn et al. (1998) do not meet Remane's first criterion on homology, i.e., similarity of position, and determination of plesiomorphic and apomorphic states (character evolution) was not accomplished. Thus, the phenetic hypothesis of Shinn et al. (1998) that the Gyrodactylidae is more closely aligned with the Heteronchoinea than the Dactylogyridea is rejected at this time.

Most members of the Gyrodactylidae express many unique characters for the Polyonchoinea, in part including viviparity and absence of a sclerotized male copulatory organ. Although these unique attributes seemed to define clearly the family prior to Harris' (1983) discovery of the oviparous Oogyrodactylus farlowellae from neotropical freshwater fishes, they apparently all represent secondary changes that developed after origin of the family. The phylogenetic reconstruction of the Gyrodactylidae by Boeger et al. (1994) indicates that basal genera of the family include the oviparous species. Further, oviparous species of Gyrodactylidae that possess a copulatory complex comprised of a sclerotized male copulatory organ and accessory piece(s) recently were discovered in Brazil (W. Boeger and D. Kritsky, unpublished). The morphology of the copulatory complex in these species suggests that a sclerotized male copulatory organ with accessory piece is symplesiomorphic for the Gyrodactylidae. Thus, most unique characteristics of the Gyrodactylidae that caused prior confusion regarding the phylogenetic position of the family are secondary modifications of plesiomorphic characters.

The present hypothesis suggests that a new order could be established to accommodate Euzetrema. Bootstrap and strict consensus do not provide any support for this action, however. Further, our results cannot be extended to Iagotrema, which along with Euzetrema currently comprise the Iagotrematidae. We do not propose any changes in the classification of the Monogenoidea to accommodate these genera; as a result, the Iagotrematidae remains incertae sedis.

The phylogenetic position of the Udonellidae within the Platyhelminthes has been controversial since the original description of Udonella caligorum by Johnston (1835). The family has been included in the Monogenoidea by Price (1938), Sproston (1946), and Yamaguti (1963); in the turbellarian group 'Dalyellioidea' by Ehlers (1985a), and as a separate higher taxon within the Platyhelminthes by Ivanov (1952a,b) and Kornakova (1988), among others. Recently, ultrastructural and molecular evidence have suggested that the family has an evolutionary relationship to the Gyrodactylidae (Polyonchoinea) (see Rohde et al. 1989b; Rohde and Watson 1993c; Littlewood et al. 1998a, 1999a; Litvaitis and Rohde 1999). Past confusion concerning the phylogenetic position of the Udonellidae is likely due in part to absences and/or losses of characters in udonellids that were associated with the origin of the group. Our inclusion of the taxon as an ingroup in the analysis of morphological data supports the contention of Littlewood et al. (1998a, 1999a) that the family is a member of the Polyonchoinea with close relationship to the Gyrodactylidae. In the present analysis, the

Udonellidae is placed in the Gyrodactylidea as sister-group to the Gyrodactylidae + Acanthocotylidae. Most of the characters representing absence or loss of structure in the Udonellidae are autapomorphies resulting from secondary character changes, and these features have limited value in determination of phylogenetic affinity. The relationship suggested in the present hypothesis is supported by nine synapomorphies, of which three provide unambiguous evidence for the relationship: presence of a lobate, conspicuous Mehlis gland; absence (loss) of vaginal pore and duct; larva lacking cilia. Bootstrap (90%) and the strict consensus cladogram provide additional support (Figures 10.2 and 10.3).

Udonellids are symbionts of parasitic copepods occurring on the gills of marine fishes. Kearn (1994) indicated that monogenoideans lacking haptoral sclerites are usually limited to the hard surfaces of their hosts where attachment does not depend on the action of sclerites for attachment to host tissue. Littlewood et al. (1998a) provided an apparent teleological explanation for the absence of haptoral sclerites in the Udonellidae by stating, 'lack of hooks in Udonella may be a consequence of its attachment to copepods, whose hard exoskeletons are probably not suitable for penetration by hooks'. Other examples of monogenoideans lacking hooks and attaching to firm surfaces include species of Anoplodiscus (Gyrodactylidea) that perforate the epidermis in order to reach the surface of the 'stratum compactum' of the host's fin (Roubal and Whittington 1990); and some microbothriids that attach themselves to the denticles of elasmobranch scales (Kearn 1965, 1994). Our analysis indicates that loss of hooks occurred in the larvae of udonellids when the family diverged within the Gyrodactylidea and loss of anchors in this family is symplesiomorphic. Contrary to Littlewood et al.'s (1998a) statement, a more likely scenario would be that the ancestor of the Udonellidae lost its hooks which allowed its descendants to attach to the hard surfaces of copepods.

Heteronchoinea

Monophyly of the Heteronchoinea is supported by six synapomorphies, of which four are unambiguous: presence of a genitointestinal canal, two ventrolateral ductus vaginalis, four haptoral suckers associated with hooks, and lateral microtubules in the spermatozoan. Sister-group relationships within the subclass are unchanged from those proposed by Boeger and Kritsky (1993, 1997). The sister-group relationships of the infrasubclasses Polystomatoinea and Oligonchoinea within the Heteronchoinea have been verified in phylogenetic reconstructions using molecular data (Littlewood et al. 1998a, 1999a; Mollaret et al. 1997). Monophyly of the Polystomatoinea (bootstrap = 84%) is supported by two synapomorphies: absence of egg filament (shared with the Lagarocotylidae, Diclybothriidae and Discocotylidae) and presence of three pairs of haptoral suckers with hooks. The monophyly of the Oligonchoinea (bootstrap = 68%) is supported by four synapomorphies, of which three are non-homoplasious: Oncomiracidium with crochet en fléau, haptoral suckers with a mid-sclerite and one pair of lateral sclerites. The only oligonchoineans that do not present all of these characters are the Diclybothriidea, for which the loss of the lateral sclerites is considered a reversal; the absence of a crochet en fléau in the larva of hexabothriids is also considered a secondary event.

Among the orders of Oligonchoinea, the Mazocraeidea is the most diverse. Its monophyly is supported by five unambiguous synapomorphies: Oncomiracidium with one pair of fused pigmented eyes, germarium inverted U-shaped, egg with two polar filaments, oral sucker a buccal organ, and two pairs of lateral sclerites in the haptoral sucker. Both consensus and bootstrap (95%) support the taxon (Figures 10.2 and 10.3). The Mazocraeidea evolved into five clades (suborders), each with variable bootstrap and consensus support. The monophyly of the suborders Mazocraeinea, Gastrocotylinea and Discocotylinea is supported by both bootstrap (74%, 70% and 69%, respectively) and strict consensus.

Although some of the relationships proposed in the present hypothesis are based on single character changes, not supported by bootstrap or by all equally parsimonious solutions, the synapomorphies are considered evolutionarily informative, suggesting that stability of these clades would increase with the incorporation of new characters. The phylogenetic relationship of the clade Discocotylinea + Gastrocotylinea + Hexostomatinea + Microcotylinea is supported by a microcotylid crochet en fléau; the clade Hexostomatinea + Micro-cotylinea is supported by one mid-dorsal ductus vaginalis; and the monophyly of the Microcotylinea is supported by a double inverted, U-shaped germarium.

The only variances in all equally parsimonious cladograms for the sister-group relationships of the families of Mazocraeidea are the relative positions of the Hexostomatidae and Axinidae. The Hexostomatidae may appear in some equally parsimonious cladograms as sister-group of the Discocotylinea, while the Axinidae may occur as an unresolved sister-group of the remaining Microcotylinea. The results of Mollaret et al. (1997) and Littlewood et al. (1999a), although limited to only three and four families, respectively, are compatible with our hypothesis on the Oligonchoinea (Figure 10.2). The reconstruction of Littlewood et al. (1998a), based on complete SSU rDNA, however, is inconsistent with the present hypothesis by the relative phylogenetic position of Plectanocotylidae (represented in their analysis by Plectanocotyle gurnardi). Instead of being a member of Mazocraeinea, as suggested by the sharing of the characters 'posterior plate-like mid-sclerite in the haptoral sucker' and 'two pairs of lateral sclerites (posterior pair with two subunits) in the haptoral sucker', P. gurnardi is shown as an unresolved sister-group of the microcotylinean species Zeuxapta seriolae (Heteraxinidae), Bivagina pagrosomi (Microcotylidae), and Diclidophora denticulata (Diclidophoridae).

Concluding remarks

Characters derived from the haptor, the osmoregulatory system, larval and adult morphology, body sensilla, sperm structure and development, and molecular data have been used independently to construct hypotheses on the phylogeny of the Monogenoidea. However, application of single classes (subsets) of characters has been insufficient to unambiguously resolve phylogenetic relationships within the class. More robust hypotheses would be expected if combined data sets are used. In the present study, morphological information from each of these subsets (excluding molecular data) has been employed for phylogenetic reconstruction. The present analysis demonstrates that characters in each subset are informative in defining internal clades, but none used alone offers comprehensive information. Competing hypotheses developed from molecular data are generally based on limited numbers of species in respective ingroup taxa and probably do not represent total diversity within monogenoidean clades. However, when available morphological and molecular data were combined in an analysis (Littlewood et al. 1999a), a higher level of congruence with the present hypotheses was obtained.

A *priori* assumptions of non-homology of similar characteristics that satisfy some basic criteria of homology in different taxa should be avoided (Hennig 1966; Wiley 1981). Hennig's Auxillary Principle states that homology should be assumed for the analysis, and implies that the cladistic method will indicate the instances of homoplasy (Brooks and McLennan 1991). On the other hand, forced assumptions of non-homology provide minimal means to test hypotheses on analogy, while limiting recognition of homology. For example, when we considered the egg-filament droplets of the Gyrodactylidea and the Capsalidae to be homologous, our analysis indicated that the droplets were homoplastic, having developed independently in the two taxa. If we had considered *a priori* that they were non-homologous, such a result could not have been tested for potential evolutionary association of these characters. A *priori* decisions of non-homology based on strict morphological criteria may lead to erroneous resolution within the analysis because the observed morphological differences could represent sequential evolutionary modifications of homologous traits.

ACKNOWLEDGEMENTS

This study was supported by the Conselho Nacional de Desenvolvimento Científico e Tecnológico (CNPq) and by the Fundação Coordenação de Aperfeiçoamento de Pessoal de Nível Superior (CAPES). We would like to thank Dr J.-L. Justine for sharing his revised matrix on sperm morphology and development prior to its submission for publication in the current text.

Appendix 10.1. Character analysis.

Numbers in parentheses preceding the definition of a character state refer to the coding that state received in the data matrix: (0) represents the plesiomorphic state; (?) denotes missing or inapplicable data; (v) indicates that either (0) or the apomorphic state (1) may apply. Bold numbers in brackets following the definition of a character state refer to respective evolutionary changes depicted in the phylogenetic hypothesis (Figure 10.2).

1. **Larval ciliation (recently hatched).** Plesiomorphy: (0) uniformly ciliated. Apomorphies: (1) with 3 zones of ciliation [**16**]; (2) cilia absent [**88**].

2. **Cephalic collecting ducts of protonephridia (larva).** Plesiomorphy: (0) separate. Apomorphy: (1) fused [**19**]. This series is based upon Euzet *et al.* (1995).

3. **Pigmented eyes (larva).** Plesiomorphy: (000) 1 pair, not fused. Apomorphies: (100) 2 pairs, not fused [**14, 201**]; (200) 1 pair, fused [**159**]; (300) 2 pairs, anterior pair fused [**204**]; (110) 2 pairs, posterior pair fused [**125**]; (??1) absent [**7, 80, 141, 156**]. The coding for absence of eyes in the larva reflects the possibilities that this state may represent either the loss of eyes or of pigmentation from any other state in the series. This series is restricted to pigmented eyes because Kearn (1978) suggested that eyes without pigmentation may have been missed in descriptions of various taxa.

4. **Pigmented eyes (adult).** Plesiomorphy: (00) 1 pair. Apomorphies: (10) 2 pairs [**15**]; (20) 2 pairs, posterior fused [**73, 76, 124**]; (?1) eyes absent [**1, 32, 52, 87, 107, 112, 135, 160**].

5. **Spike sensilla.** Plesiomorphy: (0) absent. Apomorphy: (1) present [**45, 93**].

6. **Mouth.** Plesiomorphy: (0) subterminal [**58**]. Apomorphy: (1) ventral [**37, 153**].

7. **Oral sucker.** Plesiomorphy: (00) circumoral. Apomorphies: (10) present as two buccal organs [**163**]; (?1) absent [**36, 155**]. Rohde and Watson (1996) suggest that the circumoral sucker of some Monogenoidea is not homologous to the buccal organs of Mazocraeidea. However, our analysis supports homology with morphological and ultrastructural differences representing secondary evolutionary changes.

8. **Digitiform projections within the pharynx.** Plesiomorphy: (0) absent [**90**]. Apomorphy: (1) present [**48, 74**].

9. **Number of pharyngeal bulbs.** Plesiomorphy: (0) 1 [**91**]. Apomorphy: (1) 2 [**75**].

10. **Intestine.** Plesiomorphy: (0) bifurcated [**94**]. Apomorphy: (1) single [**67, 123, 178**].

11. **Intestinal diverticula.** Plesiomorphy: (0) absent. Apomorphy: (1) present [**55, 78, 137**].

12. **Number of testes.** Plesiomorphy: (0) more than 2. Apomorphies: (1) 1 [**2, 23, 177**]; (2) 2 [**39**].

13. **Position of testes.** Plesiomorphy: (0) postgermarial. Apomorphy: (1) pregermarial [**8, 185**].

14. **Vas deferens.** Plesiomorphy: (0) not looping left intestinal caecum. Apomorphy: (1) looping left intestinal caecum [**49**]. The character state of *Euzetrema* (Iagotrematidae?) has been coded (?). Although the morphological accounts of Combes (1965), Combes *et al.* (1974), and Timofeeva and Sharpilo (1979) are unclear concerning the state of the vas deferens in species of *Euzetrema*, available illustrations suggest that the vas deferens does loop the left cecum, as it does in all of its sister groups. Confirmation of this state for *Euzetrema* requires further study.

15. **Male copulatory organ.** Plesiomorphy: (0) muscular [**40**]. Apomorphy: (1) sclerotized [**20**].

16. **Accessory piece** (when sclerotized male copulatory organ present). Plesiomorphy: (0) absent. Apomorphy: (1) present [**50**].

17. **Muscular male copulatory organ.** Plesiomorphy: (0) elongate [**154, 187, 197, 219**]. Apomorphy: (1) ovate [**127**].

18. **Spines of male copulatory organ** (when male copulatory organ is muscular). Plesiomorphy: (0) absent [**172, 189, 200**]. Apomorphy: (1) present [**128**].

19. **Number of seminal vesicles.** Plesiomorphy: (0) one vesicle. Apomorphy: (1) two tandem vesicles [**68**].

20. **Genital apertures.** Plesiomorphy: (0) common. Apomorphy: (1) separate [**9, 95**].

21. **Genital pore (when single).** Plesiomorphy: (0) on midline. Apomorphy: (1) lateral [**38, 191**].

22. **Bilateral, armed muscular pads in genital atrium.** Plesiomorphy: (0) absent [**208**]. Apomorphy: (1) present [**203**].

23. **Germarium.** Plesiomorphy: (00) ovate. Apomorphies: (01) lobate [**10, 142**]; (10) elongate, U- shaped [**145, 170**]; (20) elongate, inverted U-shaped [**161**]; (30) elongate, double inverted U- shaped [**202**].

24. **Pathway of oviduct/germarium.** Plesiomorphy: (0) intercaecal [**119**]. Apomorphy: (1) looping right caecum [**30, 56, 103**].

25. **Genitointestinal canal.** Plesiomorphy: (0) absent. Apomorphy: (1) present [**129**].

26. **Mehlis' gland.** Plesiomorphy: (0) follicular, inconspicuous. Apomorphy: (1) lobate, conspicuous [**81**].

27. **Vitelline follicles.** Plesiomorphy: (0) amorphous. Apomorphy: (1) large, well-defined [**69**].

28. **Vagina.** Plesiomorphy: (0000000) true vagina with 1 lateral opening. Apomorphies: (1000000) true vagina with single dorsal opening (Laurer's canal) [**3**]; (0010000) true vagina with 1 midventral opening [**44, 59**]; (0000100) 2 true vaginae with ventrolateral openings [**61, 108**]; (0000001) vagina absent [**82**]; (0100000) ductus vaginalis with bilateral openings [**130**]; (0100010) ductus vaginalis with 1 ventrolateral opening [**186**]; (0101010) ductus vaginalis with 1 midventral opening [**192**]; (0200000) ductus vaginalis with 1 middorsal opening [**195, 199**]; (0300000) ductus vaginalis with 2 dorsal openings [**198, 209**]. This transformation series has been modified to reflect the transformation series proposed by Boeger and Kritsky (1993) with the addition of the state 'true vagina with dorsal opening' to accommodate Laurer's canal of the Trematoda.

29. **Shape of egg.** Plesiomorphy: (0) oval [**83, 114**]. Apomorphy: (1) tetrahedric [**31, 46, 63**].

30. **Egg polar filaments.** Plesiomorphy: (0) polar filament absent [**53, 133, 152, 176**]. Apomorphies: (1) One polar filament [**18, 180, 214**]; (2) Two polar filaments [**162**].

31. **Egg filament droplet.** Plesiomorphy: (0) absent. Apomorphy (1) present [47, 84].

32. **Shape of the haptor (only Monogenoidea).** Plesiomorphy: (0) disc-shaped. Apomorphy: (1) dactylogyrid [104].

33. **Crochet en fléau in oncomiracidium.** Plesiomorphy: (0) absent [157]. Apomorphy: (1) present [138].

34. **Haptoral sucker in oncomiracidum.** Plesiomorphy: (0) absent. Apomorphy: (1) present [173].

35. **Hooks in any stage of development.** Plesiomorphy: (0) absent [92]. Apomorphy: (1) present [4]. Because the primitive number and distribution of hooks in the larva and adult for the Cercomeromorphae presently cannot be determined with confidence, homologous series 35, 36 and 37 are used to recognize that the presence of hooks is shared by the higher taxa of the Cercomeromorphae without forcing an *a priori* assumption on ancestral/descendant relationship between the states 16 or 10 hooks in the larva and adult.

36. **Number and distribution of hooks in larva.** Plesiomorphy: (000) 16 marginal [5, 62]. Apomorphies: (100) 14 marginal and 2 central [21]; (200) 12 marginal and 2 central [100, 117]; (001) 10 marginal [11, 146]; (002) 6 marginal [174]; (003) 4 marginal [181]; (v10) 14 marginal [27, 42, 143].

37. **Number and distribution of hooks in adult.** Plesiomorphy: (00000) 16 marginal [6, 70]. Apomorphies: (1v000) 14 marginal [28, 43, 57, 144]; (01000) 14 marginal + 2 central [22, 97]; (02000) 12 marginal + 2 central [101]; (01100) 10 marginal + 2 central + 4 dorsal [105]; (01200) 8 marginal + 2 central + 4 dorsal [116]; (00010) 1 pair [147]; (00020) 8 in lappets + 2 ventral [216]; (????1) absent [12, 79, 158, 194, 205, 217].

38. **Shape of hooks.** Plesiomorphy: (0) unhinged. Apomorphy: (1) hinged [71].

39. **Anchors in adult.** Plesiomorphy: (00) absent [54, 77, 175, 211]. Apomorphies: (10) 1 pair ventral [17, 72, 96]; (11) present in larva and lost in adult [206, 220]; (20) 2 pairs ventral [60]; (30) 1 ventral, 1 dorsal [109].

40. **Bar in adult (when anchor present).** Plesiomorphy: (0) absent [106]. Apomorphies: (1) 2 ventral [64]; (2) 1 ventral [113]; (3) 1 ventral, 1 dorsal [110]; (4) 1 ventral, 2 dorsal [121, 126].

41. **Number of haptoral suckers.** Plesiomorphy (0000) absent. Apomorphies: (1000) 4 pairs [131]; (2000) 3 pairs [134]; (3000) 1 pair [136]; (1100) many gastrocotylid [190, 193]; (1010) many fire-tongue [218]; (1001) many microcotylid [207]. The term 'haptoral sucker' is used in preference to 'clamp' because it connotes the proposed homology of haptoral suckers lacking sclerites of the Polystomatoinea with those possessing sclerites in the Diclybothriidea, Chimaericolidea and Mazocraeidea. The present analysis suggests the 'clamp' sclerites, that occur in the haptoral suckers of species of Diclybothriidea, Chimaericolidea and Mazocraeidea, are secondary developments in the Heteronchoinea.

42. **Crochet en fléau in adult (when present in oncomiracidium).** Plesiomorphy: (0) present. Apomorphy: (1) absent [188, 196].

43. **Fire-tongue haptoral sucker (haptoral suckers present).** Plesiomorphy: (0) absent. Apomorphy: (1) present [213].

44. **Association between hook and haptoral suckers (haptoral suckers present).** Plesiomorphy: (0) present in adult. Apomorphy: (1) present during ontogeny but absent in adult [148].

45. **Suckers in haptoral appendix (haptoral suckers present).** Plesiomorphy: (0) absent. Apomorphy: (1) present [149].

46. **Mid-sclerite (haptoral suckers present).** Plesiomorphy: (0) absent. Apomorphies: (1) present, flared or truncate [139]; (2) present, terminates in hook [150].

47. **Posterior mid-sclerite (haptoral suckers present).** Plesiomorphy: (00) absent. Apomorphies: (10) rod shaped [210]; (01) present, plate-like [165].

48. **Accessory sclerite (haptoral suckers present).** Plesiomorphy: (0) absent. Apomorphies: (1) parallel to mid-sclerite [182]; (2) oblique to mid-sclerite [183].

49. **Lateral sclerites (haptoral suckers present).** Plesiomorphy: (0) absent [151]. Apomorphies: (1) 1 pair [140]; (2) 2 pairs [164]; (3) 2 pairs, posterior pair with two or more subunits [166, 179, 212].

50. **Anterolateral sclerites (2 pairs of lateral sclerites present).** Plesiomorphy: (0) 1 pair. Apomorphy: (1) fused anteriorly [168].

51. **Distal posterolateral sclerites (when 2 pairs of lateral sclerites present, posterolateral sclerite with 2 subunits).** Plesiomorphy: (0) not fused. Apomorphy: (1) distally fused [169].

52. **Shape of crochet en fléau.** Plesiomorphy: (00) hook-like [215]. Apomorphies: (10) plectanocotylid [167]; (01) microcotylid [171]; (02) gastrocotylid [184].

53. **Number of axonemes during spermiogenesis.** Plesiomorphy: (00) 2 [85]. Apomorphies: (10) 1 + 1 altered [33]; (01) 1 from the beginning of the development [65].

54. **Nucleus in distal region of mature spermatozoon.** Plesiomorphy: (0) distal region with nucleus and other cytoplasmic elements. Apomorphy: (1) distal region with nucleus only [29, 102, 122].

55. **Cytoplasmic middle process and flagella.** Plesiomorphy: (0) separated, then fused during spermiogenesis. Apomorphy: (1) fused from start of spermiogenesis [34].

56. **External ornamentation of cell membrane in zone of differentiation of spermatid.** Plesiomorphy: (0) present [120]. Apomorphy: (1) absent [13, 35].

57. **Number of centrioles in spermatozoon.** Plesiomorphy: (0) 2 [86]. Apomorphy: (1) 1 [66].

58. **Centriole adjunct.** Plesiomorphy: (0) absent. Apomorphy: (1) present [111].

59. **Axoneme structure in mature spermatozoon.** Plesiomorphy: (0) circular [115]. Apomorphy: (1) non-circular [98].

60. **Axonemal b microtubules during spermiogenesis.** Plesiomorphy: (0) complete. Apomorphy: (1) incomplete [99, 118].

61. **Mitochondrion.** Plesiomorphy: (0) normal. Apomorphy: (1) bead-like [41].

62. **Microtubules in zone of differentiation of spermatid.** Plesiomorphy: (0) present [89]. Apomorphy: (1) absent [51].

63. **Dorsoventral microtubules in spermatozoan principal region.** Plesiomorphy: (0) present. Apomorphy: (1) absent [24].

64. **Lateral microtubules in the spermatozoan principal region.** Plesiomorphy: (0) absent. Apomorphy: (1) present [132].

65. **Intercentriolar body.** Plesiomorphy: (0) present. Apomorphy: (1) absent [25].

66. **Striated rootlets.** Plesiomorphy: (0) present. Apomorphy: (1) absent [26].

Appendix 10.2. Morphological data set for the class Monogenoidea. The sequence of characters follows that in the Character Analysis (Appendix 10.1). (0) represents the plesiomorphic state; (?) denotes missing or inapplicable data; (v) indicates that either (0) or the apomorphic state (1) may apply.

	1233344567789	0123456789	01233456788888889	01234566677777899	01111234567789	012233456789	0123456
Sundanonchidae	???1102001?100	1010111??0	00000?0000000000?	?01001???01100030	40000?????????	????01012110	?0?1011
Euzetrema	1?1001001?100	0010?111??0	0000000000001000	1000010001000020	00000?????????	????000??0?0	???10??
Lagarocotylidae	??????101?100	0010111??0	0000000000000000	000001???01000000	?0000?????????	????00011?00	001?011
Udonellidae	2????1?101?100	1010??????	00000001100000010	110000????????00	?0000?????????	????00011000	0001011
Dactylogyridae	111001001?100	0010111??0	0000000000000000	10100120001200030	30000?????????	????01010110	1011011
Neodactylodiscidae	??????101?100	0010111??0	00000100000000100?	?01001???01100020	00000?????????	?????????????	???????
Diplectanidae	111001001?100	0010111??0	0000010000000000	10100120001200030	40000?????????	????01111110	0011011
Pseudomurraytrematidae	1?1001001?100	0010111??0	00000100000000000	10100!?1?012000304	00000?????????	?????????????	???????
Calceostomatidae	1?1001001?100	0010?111??0	0000000000000001	10000120002000020	10000?????????	????01111101	1011011
Gyrodactylidae	2????1?111?111	0010111??1	1?000011000000010	110001???00000110	10000?????????	????00011000	0011011
Monocotylidae	111001000000	0010010??0	00000100000000001	100001v101v000010	00000?????????	????00100000	0001011
Loimoidae	1?????1000000	0010010??0	0000000000000000	100001v101v000010	00000?????????	????10100000	0001011
Dionchidae	1?1001001?100	0020000000	01000000000100000	100001v101v000010	00000?????????	????00011000	0101011
Capsalidae	111001011?110	0020000000	01000000000000001	11000110001000010	00000?????????	????00011000	0101011
Acanthocotylidae	21????1?111?111	0010111??1	1?000001100000010	11000110001000100	?0000?????????	????00011000	0011011

Tetraonchidae	1?1102001?100	1010111??0	00000?00000000000	10100110001100030	40000?????????	????01??????0	???10??
Amphibdellatidae	1?????101?100	0010111??0	00000100000000001	10100110001100030	20000?????????	????01011111	0011011
Bothitrematidae	??????2001?111	1010111???	00000?001???????1	100001???00000110	10000?????????	???????????	???????
Tetraonchoididae	??????001?100	1010111??1	00000?00100000001	100001???00000110	10000?????????	????010?11?0	?0?10??
Polystomatidae	1?10010000000	0000000110	00000010001000000	00000100000000010	02000?0000??0?	????00000000	0000100
Sphyranuridae	1?????1000000	0000000110	00000010001000000	00000???00000010	03000?0000??0?	????00000000	0000100
Chimaericolidae	1???1?000000	0100000110	00001010001000000	100101v101v000010	01000000010001	??0000000000	0000100
Diclybothriidae	1?1001001?100	0100000?0	00010010001000000	00010100100010010	01000001120000	??0000000000	0000100
Hexabothriidae	1???1?000000	0100000110	00010010001000000	10000100??????1010	01000?0?120000	????00000000	0000100
Pterinotrematidae	??????1001000	0100000110	00030010002000000	100101???00020010	01000011011002	000000000000	0000100
Mazocraeidae	10200?1001000	0100000110	00010010001000000	20010100100010010	01000010010103	110000000000	0000100
Hexostomatidae	1?100?1001000	0100000100	00020010002000000	20010100100010010	01000010010002	000100000000	0000100
Plectanocotylidae	1?200?1001000	0100000?10	00020010001000000	20010100100010010	01000010010103	001000000000	0000100
Mazoplectidae	??????1001000	0?00000110	000?00100?????0	2001?1???00010?10	0100000?010103	110000000000	0000100
Discocotylidae	1?200?1001000	0100000100	00020010001000000	00011100200010000	?1000010010002	000100000000	0000100
Diplozoidae	1?200?1001000	11100?0??0	??0200100?????0	10011100300010000	?1000010010003	000100000000	0000100
Diclidophoridae	10????1001000	0100000110	00030010001000000	20010100100010000	?1000000?011003	000100000000	0000100
Anthocotylidae	1?200?1001000	0100000110	0002001000100000?	?0010100100010010	01000010010012	000100000000	0000100
Gastrocotylidae	10200?1001000	0100000110	00020010002000000	20010100100010010	01100001010022	000200000000	0000100
Chauhaneidae	?0200?1001000	0101000000	01020010001010100	200101???00010010	01100001010022	000200000000	0000100
Protomicrocotylidae	??????1001000	0101000110	00020010001000100	2001?1???00010?10	0100000?010022	000200000000	0000100
Gotocotylidae	1?200?1001000	0100000010	00020010003000000	20010100100010010	01100101010022	000200000000	0000100
Microcotylidae	10200?1001000	0100000110	00030010002000000	200101001?????1011	?1001?01010002	000100000000	0000100
Heteraxinidae	10200?1001000	0100000110	00130010002000000	200101001?????1011	?1001?01010002	000100000000	0000100
Allopyragraphoridae	??????1001000	0100000110	00030010002000000	200??1???00010?10	0100000?011002	00??04000000	0000100
Diplasiocotylidae	??200?1001000	0100000110	00130010003000000	200101???????????	?1001?01010002	00??00000000	0000100
Axinidae	10300?1001000	0100000110	00130010002000000	20010100100010010	0100000?010002	000100000000	0000100
Pyragraphoridae	??????1001000	0100000010	00030010002000000	200101001?????1011	01010?11011002	000100000000	0000100
Montchadskyellidae	??????100?104	0110111??0	00000100000100000	1000?1????1v000010	00000?????????	???????????	???????
Pseudodiclidophoridae	??????1001000	0100000110	00020010001000000	2001?1???00010?10	0100000?010022	000200000000	0000100
Neothoracotylidae	??????1001000	0100004110	00020010002000000	200??1???00010?10	01100100?010022	00??00000000	0000100
Bychowskycotylidae	??????1001000	0100004110	0002001000100000?	?00??1?????????1???	?1100?0?010002	00??00000000	0000100
Allodiscocotylidae	??????1001000	0101000010	00020010001000100	200??1???00010?10	0100010?010022	00??00000000	0000100
Rhinecotylidae	10200?1001000	0100000110	000300100??????0	200101001?????1010	01010?11011002	000100000000	0000100
Heteromicrocotylidae	??????1001000	0100000110	00030010002000000	200??1?????????1???	?1010?1?011002	00??00000000	0000100
Octomacridae	??????1001000	0110000100	000200100??????0	2001?1???00010?00	?100000?010002	000100000000	0000100
Pseudomazocraeidae	??????1001000	0101000000	00020010001000100	2001?1???00010010	0100000?010022	000200000000	0000100
Neotetraonchidae	??????1001?100	1010111??0	00000?00000000000?	?010?1???01100030	30000?????????	???????????	???????
Anoplodiscidae	1?1441041?111	1114111??1	04000?40144004001	144001????????1140	?0440?????????	????41011100	0011011
Trematoda	00000?1000000	0010000000	00000000001000000	00?000?????????00	?0004?????????	????00000000	0000000
Gyrocotylidea	00??1??0???00	??01?0000?	1?001000000000000	00?001001?????1000	?0000?????????	????00001000	0000?0
Ancestor	?000000000000	0000000000	00000000000400000	00?000?????????40	?0000?????????	????00000000	0000000

Appendix 10.3. Character Change List.

Postulated evolutionary changes are consecutively numbered in bold as indicated in the phylogenetic hypothesis (Figure 10.2). Respective homoplasies are listed in parentheses; reversals to the plesiomorphic state are identified by an asterisk. Character changes identified by a double asterisk represent either plesiomorphic states for which the ancestor of the state could not be explicitly determined, or autapomorphic states that result from homologous series with a single transformation that occurs in a terminal taxon; these character changes do not add length to the cladogram.

1 (32, 52, 87, 107, 112, 135, 160) pigmented eyes absent in adult; 2 (23, 177) 1 testis; 3** true vagina with single dorsal opening (Laurer's canal); 4 hooks present in any stage of development; 5** (62) 16 marginal hooks in larva; 6** (70) 16 marginal hooks in adult; 7 (80,141,156) pigmented eyes in larva absent; 8 (185) testes pregermarial; 9 (95) genital apertures separate (lateral or midline); 10 (142) germarium lobate; 11 (146) 10 marginal hooks in larva; 12 (79, 158, 194, 205, 217) hooks absent in adult; 13 (35) external ornamentation of cell membrane in zone of differentiation of spermatid absent; 14 (201) 2 pairs, not fused, of pigmented eyes in larva; 15 2 pairs of pigmented eyes in adult; 16 newly hatched larva with 3 ciliated zones; 17 (72, 96) 1 pair of ventral anchor in adult; 18 (180, 214) 1 polar filament in egg; 19 cephalic collecting ducts of protonephridia (larva) fused; 20 male copulatory organ sclerotized; 21 14 marginal + 2 central hooks in larva; 22 (97) 14 marginal +2 central hooks in adult; 23 (2, 177) 1 testis; 24 dorsoventral microtubules in spermatozoon principal region absent; 25 intercentriolar body absent; 26 striated roots absent; 27 (42, 143) 14 hooks marginal (larva); 28 (43, 57, 144) 14 marginal hooks in adult; 29 (102, 122) distal region of mature spermatozoon with nucleus; 30 (56, 103) oviduct/germarium looping right caecum; 31 (46, 63) egg tetrahedric; 32 (1, 52, 87, 107, 112, 135, 160) pigmented eyes absent in adult; 33** 1+1 altered axoneme during spermiogenesis; 34 cytoplasmic middle process and flagella fused from beginning of spermiogenesis; 35 (13) external ornamentation of

cell membrane in zone of differentiation of spermatid absent; 36 (155) oral sucker absent; 37 (153) mouth opening ventral; 38 (191) single genital pore lateral; 39 2 testes; 40* male copulatory organ muscular; 41 mitochondrion of spermatozoa bead-like; 42 (27, 143) 14 hooks marginal (larva); 43 (28, 57, 144) 14 marginal hooks in adult; 44 (59) true vagina with 1 midventral opening; 45 (93) spike sensilla present; 46 (31, 63) egg tetrahedric; 47 (84) egg filament droplet present; 48 (74) digitiform projections within pharynx present; 49 vas deferens looping left intestinal cecum; 50 accessory piece present; 51 microtubules in zone of differentiation of spermatid absent; 52 (1, 32, 87, 107, 112, 135, 160) pigmented eyes absent in adult; 53* (133, 152, 176) polar filament of egg absent; 54* (77, 175, 211) anchor absent adult; 55 (78, 137) intestinal diverticula present; 56 (30, 103) oviduct/germarium looping right caecum; 57 (28, 43, 144) 14 marginal hooks in adult; 58* mouth opening subterminal; 59 (44) true vagina with 1 midventral opening; 60 2 pairs of ventral anchors in adult; 61 (108) 2 true vaginae with 2 ventrolateral openings; 62* (5) 16 marginal hooks in larva; 63 (31, 46) egg tetrahedric; 64 2 ventral bars; 65 1 axoneme from the beginning of development during spermiogenesis; 66 1 centriole in spermatozoon; 67 (123, 178) intestine single; 68 2 tandem seminal vesicles; 69 vitelline follicles large, well defined; 70* (6) 16 marginal hooks in adult; 71 hook hinged; 72 (17, 96) 1 pair of ventral anchor in adult; 73 (76, 124) 2 pairs, posterior pair fused, of pigmented eyes in adult; 74 (48) digitiform projections within pharynx present; **75** 2 pharyngeal bulbs; **76** (73, 124) 2 pairs, posterior pair fused, of pigmented eyes in adult; **77*** (54, 175, 211) anchor absent adult; **78** (55, 137) intestinal diverticula present; **79** (12, 158, 194, 205, 217) hooks absent in adult; **80** (7, 141, 156) pigmented eyes in larva absent; **81** Mehlis' gland lobate, conspicuous; **82** true vagina absent; **83*** (114) egg oval; **84** (47) egg filament droplet present; **85*** 2 axonemes during spermiogenesis; **86*** 2 centrioles in the spermatozoon; **87** (1, 32, 52, 107, 112, 135, 160) pigmented eyes absent in adult; **88** newly hatched larva lacking cilia; **89*** microtubules in zone of differentiation of spermatid present; **90*** digitiform projections within pharynx absent; **91*** 1 pharyngeal bulb; **92*** hooks absent from all stages of development; **93** (45) spike sensilla present; **94*** intestine bifurcated; **95** (9) genital apertures separate (lateral or midline); **96** (17, 72) 1 pair of ventral anchor in adult; **97** (22) 14 marginal + 2 central hooks in adult; **98** axoneme structure in mature spermatozoon non-circular; **99** (118) axonemal b microtubules during spermiogenesis incomplete; **100** (117) 12 marginal + 2 central hooks in larva; **101** 12 marginal + 2 central hooks in adult; **102** (29, 122) distal region of mature spermatozoon with nucleus; **103** (30, 56) oviduct/germarium looping right caecum; **104** haptor dactylogyrid; **105** 10 marginal + 2 central + 4 dorsal hooks in adult; **106*** bars absent; **107** (1, 32, 52, 87, 112, 135, 160) pigmented eyes absent in adult; **108** (61) 2 true vaginae with 2 ventrolateral openings; **109** 1 pair ventral, 1 pair dorsal anchors in adult; **110** 1 ventral, 1 dorsal bar; **111** centriole adjunct present; **112** (1, 32, 52, 87, 107, 135, 160) pigmented eyes absent in adult; **113** 1 ventral bar; **114*** (83) egg oval; **115*** axoneme structure in mature spermatozoon circular; **116** 8 marginal + 2 central + 4 dorsal hooks in adult; **117** (100) 12 marginal + 2 central hooks in larva; **118** (99) axonemal b microtubules during spermiogenesis incomplete; **119*** oviduct/germarium intercaecal; **120*** external ornamentation of cell membrane in zone of differentiation of spermatid present; **121** (126) 1 ventral, 2 dorsal bars; **122** (29, 102) distal region of mature spermatozoon with nucleus; **123** (67, 178) intestine single; **124** (73, 76) 2 pairs, posterior pair fused, of pigmented eyes in adult; **125** 2 pairs, posterior pair fused, of pigmented eyes in larva; **126** (121) 1 ventral, 2 dorsal bars; **127** muscular male copulatory organ ovate; **128** spines of male copulatory organ present; **129** genitointestinal canal present; **130** ductus vaginalis with bilateral openings; **131** 4 pairs of haptoral suckers; **132** lateral microtubules in the spermatozoon's principal region present; **133*** (53, 152, 176) polar filament of egg absent; **134** 3 pairs of haptoral suckers; **135** (1, 32, 52, 87, 107, 112, 160) pigmented eyes absent in adult; **136** 1 pair of haptoral suckers; **137** (55, 78) intestinal diverticula present; **138** crochet en fléau in oncomiracidium present; **139** mid-sclerite of haptoral sucker flared or truncate; **140** 1 pair of lateral sclerites of haptoral suckers; **141** (7, 80, 156) pigmented eyes in larva absent; **142** (10) germarium lobate; **143** (27, 42) 14 hooks marginal (larva); **144** (28, 43, 57) 14 marginal hooks in adult; **145** (170) germarium elongate, U-shaped; **146** (11) 10 marginal hooks in larva; **147** 1 pair of hooks in adult; **148** absence of association of hook and haptoral suckers in adult; **149** suckers in haptoral appendix present; **150** mid-sclerite of haptoral sucker terminates in hook; **151*** lateral sclerites

of haptoral suckers absent; **152*** (53, 133, 176) polar filament of egg absent; **153** (37) mouth opening ventral; **154*** (187, 197, 219) muscular male copulatory organ elongate; **155** (36) oral sucker absent; **156** (7,80,141) pigmented eyes in larva absent; **157*** crochet en fléau in oncomiracidium absent; **158** (12, 79, 194, 205, 217) hooks absent in adult; **159** 1 pair of fused pigmented eyes in larva; **160** (1, 32, 52, 87, 107, 112, 135) pigmented eyes absent in adult; **161** germarium elongate, inverted U-shaped; **162** 2 polar filaments in egg; **163** oral sucker as 2 buccal organs; **164** 2 pairs of lateral sclerites in haptoral suckers; **165** posterior mid-sclerite of haptoral sucker plate-like; **166** (179, 212) 2 pairs of lateral sclerites in haptoral suckers, posterior pair with 2 subunits or more; **167**** crochet en fléau plectanocotylid; **168** anterolateral sclerites of haptoral sucker fused anteriorly; **169** distal posterolateral sclerites of haptoral sucker distally fused; **170** (145) germarium elongate, U-shaped; **171** crochet en fléau microcotylid; **172*** (189, 200) spines of male copulatory organ absent; **173** haptoral sucker in oncomiracidium present; **174** 6 marginal hooks in larva; **175*** (54, 77, 211) anchor absent adult; **176*** (53, 133, 152) polar filament of egg absent; **177** (2, 23) 1 testis; **178** (67, 123) intestine single; **179** (166, 212) 2 pairs of lateral sclerites in haptoral suckers, posterior pair with 2 subunits or more; **180** (18, 214) 1 polar filament in egg; **181** 4 marginal hooks in larva; **182** accessory sclerite parallel to mid-sclerite; **183** accessory sclerite oblique to mid-sclerite; **184** crochet en fléau gastrocotylid; **185** (8) testes pregermarial; **186** ductus vaginalis with 1 ventrolateral opening; **187*** (154, 197, 219) muscular male copulatory organ elongate; **188** (196) crochet en fléau absent in adult; **189*** (172, 200) spines of male copulatory organ absent; **190** (193) many gastrocotylid haptoral suckers; **191** (38) single genital pore lateral; **192**** ductus vaginalis with 1 midventral opening; **193** (190) many gastrocotylid haptoral suckers; **194** (12, 79, 158, 205, 217) hooks absent in adult; **195** (199) ductus vaginalis with 1 middorsal opening; **196** (188) crochet en fléau absent in adult; **197*** (154, 187, 219) muscular male copulatory organ elongate; **198** (209) ductus vaginalis with 2 dorsal openings; **199** (195) ductus vaginalis with 1 middorsal opening; **200*** (172, 189) spines of male copulatory organ absent; **201** (14) 2 pairs, not fused, of pigmented eyes in larva; **202** germarium elongate, double inverted U-shaped; **203** bilateral, armed muscular pads in the genital atrium present; **204** 2 pairs, anterior pair fused, of pigmented eyes in larva; **205** (12, 79, 158, 194, 217) hooks absent in adult; **206** (220) anchor absent in adult, present in larva; **207** many microcotylid haptoral suckers; **208*** bilateral, armed muscular pads in the genital atrium absent; **209** (198) ductus vaginalis with 2 dorsal openings; **210** posterior mid-sclerite of haptoral sucker rod shaped; **211*** (54, 77, 175) anchor absent adult; **212** (166, 179) 2 pairs of lateral sclerites of haptoral suckers, posterior pair with 2 subunits or more; **213** fire-tongue haptoral sucker present; **214** (18, 180) 1 polar filament in egg; **215*** crochet en fléau hook-like; **216** 8 in lappets + 2 ventral hooks in adult; **217** (12, 79, 158, 194, 205) hooks absent in adult; **218** many fire-tongue haptoral suckers; **219*** (154, 187, 197) muscular male copulatory organ elongate; **220** (206) anchor absent in adult, present in larva.

Chapter 11

The Gyrocotylidea, Amphilinidea and the early evolution of Cestoda

Willi E.R. Xylander

The systematics of the Cestoda has been disputed in recent decades. In particular, the position of the monozoic taxa (Gyrocotylidea, Amphilinidea and Caryophyllidea) has been controversial (e.g., Arme and Pappas 1983). During recent years, however, after comprehensive ultrastructural and genetic investigations, a consensus phylogenetic system has been established for the cestodes. The Cestoda are monophyletic, characterized by at least eight autapomorphies. The Cestoda comprises the Gyrocotylidea, the Amphilinidea and the Cestoidea *sensu* Ehlers (1984, 1985a). The Gyrocotylidea is the sister-group of the other two taxa, which together form the Nephroposticophora Ehlers 1984. The Gyrocotylidea – including the genus *Gyrocotyloides* – are considered monophyletic. Within the Nephroposticophora, the Amphilinidea which are well-characterized by a number of autapomorphies and the Cestoidea (synonymous to the so-called Eucestoda of Hoberg *et al.* 1999b: all other Cestoda characterized by a six-hooked primary larva) are sister-groups. The Cestoidea (in this chapter this suffix will be used according to the arguments for a differentiation between Cestoidea, Eucestoda and Caryophyllidea given by Ehlers 1985a) share at least six autapomorphies.

The characters are presented and discussed in relation to their evolution within the Cestoda and the other groups of the Neodermata Ehlers 1984. The phylogenetic relationships within the Cestoda presented here correspond with those based on recent molecular investigations (e.g., Rohde *et al.* 1993b; Littlewood *et al.* 1999a).

The autapomorphies of the Cestoda

All stages are without an intestine

The Cestoda (including the monozoic taxa) lack a gut in all stages of their life cycle. In comparison to the other Neodermata (Digenea, Aspidogastrea, Monogenea) where, at least in the reproductive stage, a gut normally occurs, the complete lack of an intestine in all stages of Cestoda is an autapomorphy (see Ehlers 1985a; Xylander 1996). The invagination at the anterior end (the so-called 'sucker' or 'apical pit'; see Figures 11.1 and 11.4E) shows some similarities to prepharyngeal regions of other Neodermata. Such an anterior pit is lacking in all primary larval stages (lycophore, coracidium and oncosphere; see Figure 11.3) and is formed during the development within the first intermediate host; in the parasitic postlarval stages it is completely covered by neodermis (see Figure 11.4E).

Neodermis with a distinct type of microvilli

Cestodes have, in their gut-living parasitic stages, typical microvilli which differ from those of other Neodermata in their ultrastructure and arrangement. They are slender, regular in shape, and contain an electron-dense hollow cylinder which

has translucent cytoplasm peripherally and centrally (see Figure 11.4A,C,D). Many of these microvilli (or microtriches, see below) also have an electron-dense cap (see Figure 11.4D).

Microvilli with an electron-dense cap (which may differ in size) are found in the anterior and mid-body region of Gyrocotylidea and most body regions of stages of Cestoidea living in the intestine (e.g., Morseth 1966; Bråten 1968; Lyons 1969a; Featherston 1972; McVicar 1972; Lumsden and Specian 1980; Xylander 1986a). These microvilli are typical for the neodermis of tapeworms, and are not found in any other group of parasitic Platyhelminthes; they constitute an apomorphy of the Cestoda (Ehlers 1985a; Xylander 1986a). Within the Cestoidea it has been considered that the evolution of such (regularly shaped) microvilli arose because of the necessity to take up nutrients exclusively through the body surface (due to the lack of a gut). However, some arguments contradict this widely accepted view, indicating that they may be involved in the maintenance of a pH-microclimate at the body surface of the worm which inhibits digestion by host enzymes by significantly decreasing the pH below the pH-optimum of the proteases and other enzymes in the gut of their host (see Uglem and Just 1983).

The first canal cell of protonephridia is without a cell gap and desmosome

The first canal cell of protonephridia in all cestodes lacks a cell gap and a desmosome as seen from TEM (see Figure 11.4B). This cell forms a hollow cylinder distally to the filtration area (Howells 1969; Wilson and Webster 1974; Swiderski *et al.* 1975; Lumsden and Specian 1980; Rohde and Georgi 1983; Xylander 1987a,b, 1992b; McCullough and Fairweather 1991). In the other groups of the Neodermata (with very few exceptions in the monopisthocotylean monogeneans which most probably evolved independently from the conditions in the Cestoda) and most free-living Platyhelminthes, those peripheral parts of the first canal cell forming the ductule bend towards each other and these cell margins are interconnected by a desmosome (for literature see Xylander 1987a, 1992b). This type of first canal cell represents the plesiomorphic condition, whereas the structure found in the Cestoda is derived.

The protonephridial system of postlarval stages is reticulate

The protonephridial system of the Gyrocotylidea, Amphilinidea, early postlarval stages of the Cestoidea and the anteriormost part of the protonephridial system of the gut living stages of Cestoidea is reticulate (Figure 11.2), building numerous anastomoses in an irregular arrangement (Hein 1904; Fuhrmann 1931; Malmberg 1971, 1974; Lindroos and Gardberg 1982; Ehlers 1985a; Gibson *et al.* 1987; Xylander 1992b, 1996). During proglottid formation in the Cestoidea a bilateral

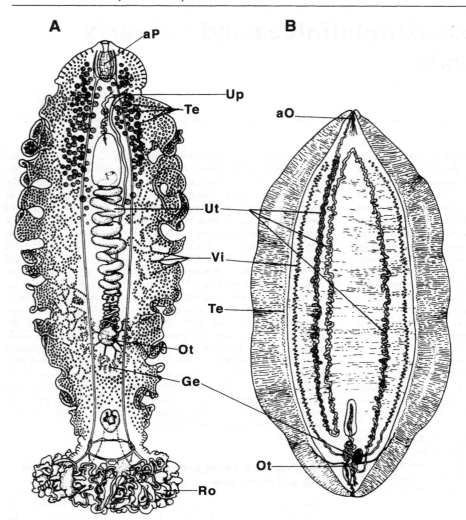

Figure 11.1 Representatives of the Gyrocotylidea and Amphilinidea. **A)** *Gyrocotyle fimbriata* (after Lynch 1945 modified from Xylander 1996). **B)** *Nesolecithus africanus* (modified after Dönges and Harder 1966 from Xylander 1996). Abbreviations: aO, apical organ; aP, apical pit; Ge, germarium; Ot, ootype; Ro, rosette organ; Te, testes; Ut, uterus; Up, uterine pore; Vi, vitellarium.

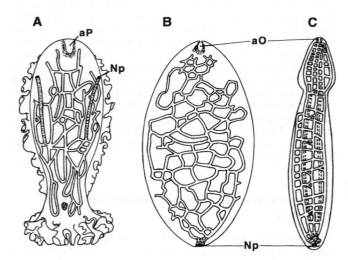

Figure 11.2 Protonephridial system of adult Gyrocotylidea (after Malmberg 1974) and Amphilinidea (after Hein 1904) and the plerocercoid of *Diphyllobothrium dendriticum* (after Lindroos and Gardberg 1982; modified after Xylander 1992b). Abbreviations: aO, apical organ; aP, apical pit; NP, nephridiopore.

symmetrical system is formed secondarily which is oriented more or less alongside the main strands. This evolutionary innovation is, however, a character developed within the Cestoidea (see below). In the other taxa of the Neodermata the protonephridial system is bilateral and more or less symmetrical.

Cell bodies of the protonephridial canal cells are located under a basal lamina

The cell bodies of the canal cells containing the nuclei are located beneath the well-elaborated basal lamina in postlarval stages of all cestode taxa investigated (Bugge 1902; Lindroos 1983; Lumsden and Hildreth 1983; Xylander 1987b, 1992b; McCullough and Fairweather 1991). The canal tissue appears to be syncytial at least in most parts of the system. High numbers of vesicles and grana in the canal tissue indicate a high metabolic turnover and a transport towards the lumen (luminad) and/or to the surrounding tissues (contraluminad) (see Xylander 1992b).

The formation of a reticulate protonephridial system during development of postlarval stages and the 'translocation' of nuclei to the surrounding tissues are considered to be apomorphic relative to the correponding structures of larval stages of the Cestoda, the other taxa of Neodermata (e.g., Krupa *et*

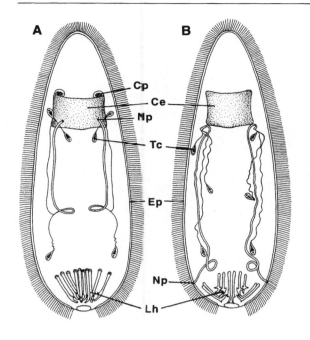

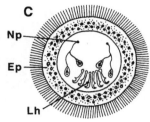

Figure 11.3 Ciliated larvae of the Cestoda showing the epidermis, larval hooks, the protonephridial system, the cerebrum and ciliary photoreceptors. **A**) Lycophore of *Gyrocotyle urna*. **B**) Lycophore of *Austramphilina elongata*. **C**) Coracidium (from different authors, modified after Xylander 1996). Abbreviations: Ce, cerebrum; Cp, ciliary photoreceptor; Ep, ciliated epidermis; Lh, larval hooks; Np, nephridiopore; Tc, terminal cell.

al. 1969; Ebrahimzadeh and Kraft 1971), and the free-living Platyhelminthes. These changes may indicate that this system takes over functions in nutrient transport as a consequence of an increase in body size and the simultaneous reduction of the 'classical distributive system of Platyhelminthes': the gut (Xylander 1987b; see Xylander 1992b for further discussion).

The larval epidermis is syncytial (without interspaced neodermal areas)

The epidermis of all ciliated cestode larvae (lycophore, coracidium) is syncytial. Neodermal tissue does not reach the body surface prior to the infection of the intermediate host. It is either located between the epidermis and the basal lamina or attached to the epidermis by 'synaptic-like' connections (Grammeltvedt 1973; Lumsden *et al.* 1974; Rohde and Georgi 1983; Xylander 1986b, 1987c).

In all other neodermatans the ciliated larvae have an epidermis which is interrupted by neodermal ridges (miracidia; Wilson 1969a; Southgate 1970; Meuleman *et al.* 1978; Pan 1980; Eklu-Natey *et al.* 1985) or larger regions which are considered to originate by fusion of the epidermal and neodermal cells (oncomiracidia; Llewellyn 1963; Lyons 1973b; Fournier 1979) or their origin is unknown (cotylocidia, Rohde 1972; Fredericksen 1978; see Xylander 1987c for a comparison of the epidermal ultrastructure in the Neodermata). The definitive body covering in these larvae reaches the body surface. This situation is considered to be plesiomorphic for the Neodermata. The syncytial, complete body covering of the Cestoda therefore appears to be derived. An epidermis completely covering the body, uninterrupted by neodermal tissue, as found in the lycophores and coracidia, may be a consequence of its syncytial nature.

Only ten larval hooks

The number of larval hooks in lycophores is ten, while in coracidia and oncospheres it is six (Figure 11.3). These hooks in cestode larvae are considered homologous to those of the monogeneans (see Ehlers 1985a for references), as indicated by their histochemistry, morphology and original function (attachment in the Monogenea and Gyrocotylidea). The Monogenea and Cestoda therefore are regarded as sister-groups, both constituting the Cercomeromorpha Bychowsky 1937 (see Ehlers 1984, 1985a). Relative to the Monogenea, which originally have 16 hooks (see Boeger and Kritsky 2001, this volume), their number is presumably reduced in the Cestoda (see Llewellyn 1963; Ehlers 1985a; Xylander 1991; but see Malmberg 1979 for a conflicting hypothesis).

This reduction constitutes an autapomorphy for the Cestoda (see Ehlers 1985a; Xylander 1991) which may have evolved due to the fact that the hooks did not play an important role for the contact between host and parasite as the period of attachment of larvae in cestodes is short (from the definitive contact to the intermediate host until penetration; this is much shorter than in Monogenea) and even in the Gyrocotylidea – which still attach themselves with their posterior end – the hooks are of minor (if any) importance; thus the number of hooks could be reduced without evolutionary disadvantage. Taking into account the presumed homology of the hooks of the Cercomeromorpha and the higher number of hooks in the Monogenea as the plesiomorphic condition, there is no evidence that the number of ten represents a synapomorphy of the Amphilinidea and Gyrocotylidea.

Large body dimensions

The vast majority of cestodes are rather large: Mature Gyrocotylidea are 3 to 30 cm in length, Amphilinidea 4 to 30 cm and most Cestoda range between 5 and 400 cm (up to 20 m in *Diphyllobothrium latum*). When comparing the more primitive taxa of Cestoda and Eucestoda (e.g., Gyrocotylidea, Amphilinidea, Caryophyllidea, *Ligula*, many pseudophyllideans) these taxa are significantly larger than most Digenea, Aspidogastrea and Monogenea.

Larger body size therefore seems to be a specific derived character – an autapomorphy – of the Cestoda.

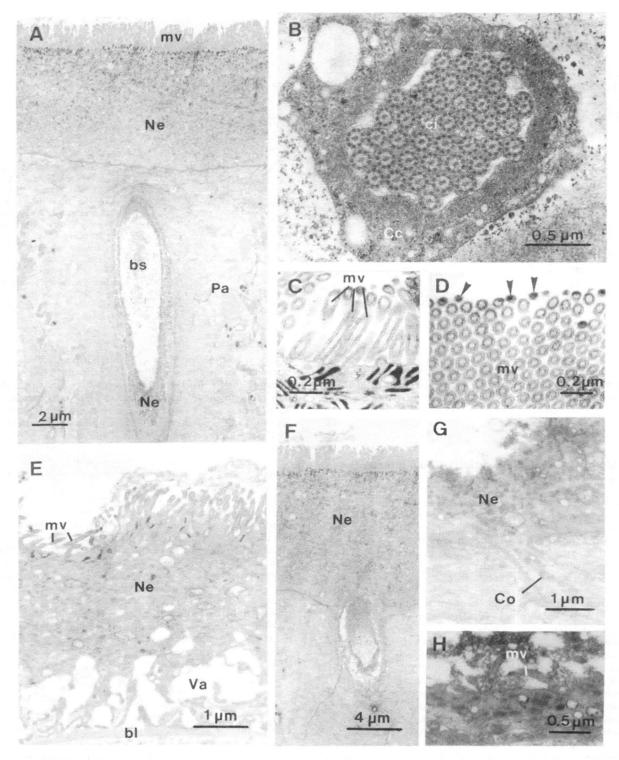

Figure 11.4 TEM-micrographs from *Gyrocotyle urna*. **A**) Neodermis of a immature specimen from the host *Chimaera monstrosa* with a 'body spine' located in the subneodermal parenchyma, and typical cestode microvilli. **B**) Terminal protonephridial cilia and first canal cell surrounding the ciliary bulb. Note that no cell gap closed by a desmosome occurs in the first canal cell. **C**) Neodermal microvilli of the body surface in longitudinal section. Note the central electron-dense hollow cylinder. **D**) Neodermal microvilli of the body surface in oblique section. Note the electron-dense tip (arrowheads). **E**) Neodermis of the apical pit of a 4 mm-long *Gyrocotyle*. Note the shorter microvilli and a high number of vacuoles on the contraluminal side of the neodermis close to the basal lamina. **F**) Body spine surrounded with neodermal tissue fusing with the surface neodermis. **G**) Neodermis surrounding a body spine deep in the parenchyma showing electron-dense bodies and a connection to the cell body. **H**) Microvilli of the neodermis surrounding a body spine. Abbreviations: bl, basal lamina; bs, body spine; Cc, first canal cell; ci, protonephridial cilia; Co, connection between neodermis and cell body; mv, microvilli; Ne, neodermis; Pa, parenchyma; Va, vacuole.

Formation of an apical pit during the development in the first host

An anterior pit ('apical pit') is found in all postlarval stages of Gyrocotylidea and Amphilinidea as well as procercoids and plerocercoids and may persist thoughout life; this cavity is surrounded by musculature elaborated to a different extent in the various taxa. Glands may open into it if subsequent stages use this organ for penetration of the gut or body wall of another host (e.g., late procercoids, plerocercoids and Amphilinidea; see Fuhrmann 1931; Kuperman and Davydov 1982; own observations, Figure 11.5B,C); it is then called an 'apical organ'.

This cavity is completely covered by a neodermis in a 3 mm-long postlarva of Gyrocotyle urna (own unpublished observations), which bears neodermal microvilli and shows large vesicles in the cytoplasm close to the basal lamina (Figure 11.4E). As on several occasions material from the host intestine had been observed to be sucked into this cavity, these large vesicles may result from nutrient uptake. In Amphilinidea, microvilli, which look like typical cestode neodermal microvilli and not like the shorter microvilli of the body surface of Amphilinidea, have been found in this pit (Figure 11.5C; see also Davydov and Kuperman 1993).

The apical pit (or in its more differentiated form, the apical organ) found in many (so-called basal) groups of the Cestoda (see Caira et al. 2001, this volume) must be considered homologous. At least its kind of formation in the first intermediate host is a specific character of this group.

The autapomorphies of the Gyrocotylidea

Epidermis of lycophora larva without nuclei

The epidermis of the free-swimming lycophore in Gyrocotyle urna does not have nuclei (Xylander 1987c). This also seems to be the case in the lycophore of Gyrocotyle fimbriata (see Simmons 1974).

In the other neodermatans the epidermis of their ciliated larvae contains nuclei (e.g., coracidia: Grammeltvedt 1973; Lumsden et al. 1974; lycophora of Amphilinidea: Rohde and Georgi 1983; miracidia: Wilson 1969a; Southgate 1970; Meuleman et al. 1978; Pan 1980; oncomiracidia: Lyons 1973b; Fournier 1979; cotylocidia: Rohde 1972; Fredericksen 1978). The anucleate state of the syncytial epidermis in Gyrocotyle has to be regarded as an autapomorphy of this group (see Xylander 1986b, 1987c). However, further investigations on lycophores of other Gyrocotyle species are necessary to prove this hypothesis. For oncomiracidia, a loss of nuclei by discharge from epidermal cells during embryogenesis has been described (Lyons 1973b; Ehlers 1985a). Whether the development of the epidermis in Gyrocotyle follows this pattern is still unknown.

Parasites of Holocephali

Species of the genus Gyrocotyle and Gyrocotyloides nybelini have only been found in Holocephali. Other taxa are not parasitized. Most probably, the earliest tapeworms were parasitizing several taxa of primitive fish (e.g., holocephalans, sharks, oestoglossid teleosteans), as indicated from the taxa parasitized by the Amphilinidea and also some primitive Cestoidea. During or after evolutionary separation of the Gyrocotylidea and the Nephroposticophora (most probably during the evolution of recent Gyrocotylidea) the Gyrocotylidea became restricted to the Holocephali.

Existence of parenchymatic postlarvae

Postlarvae are frequently found in the parenchyma of various species of Gyrocotyle. These embedded worms show numerous characters of postlarval gyrocotylideans: ten typically shaped hooks, anterior pit, neodermis, lack of cilia (see Fuhrmann 1931; Malmberg 1986). They, most probably, belong to the same species as their hosts, as far as ascertained from the characters developed in specimens that are only a few millimetres long. These specimens have been described as 'parenchymatic postlarvae'. The number of such embedded specimens is higher in smaller gyrocotylideans than in larger ones (Halvorsen and Williams 1968).

The function of these larvae is unclear. It seems possible that the postlarvae are 'absorbed' neodermal stages of smaller gyrocotylids taken up by a larger specimen involving an unknown mechanism (see Lynch 1945; Dienske 1968; Simmons and Laurie 1972). This uptake may restrict the number of parasites per host to two, the number most frequently encountered (Halvorsen and Williams 1967, 1968; Dienske 1968; Simmons and Laurie 1972; Xylander 1989a); this number is 'adjusted' during development, with larger hosts usually harbouring a total of two larger parasites whereas smaller hosts have greater numbers.

A phenomenon comparable to the parenchymatic postlarvae (whatever its function may be) has not been described from any other cestode, and must be regarded as an autapomorphy of the Gyrocotylidea.

Neodermal spines of typical shape

In Gyrocotyle, multilayered spines have been described in the outer body regions (especially at the anterior body end). They have a species-specific shape, number and location, and are used as characters for species description and determination (Lynch 1945; van der Land and Dienske 1968; van der Land and Templeman 1968). Our own TEM investigations have shown that these spines are multilayered and located in pockets within the neodermis directed proximally. These pockets penetrate the basal lamina extending into the parenchyma (Figure 11.4A,F,G), but connections of the pockets to the outside could not be detected. The neodermal pockets occasionally form microvilli that extend into the lumen in which the spines are located (Figure 11.4H). These proximad tissue strands containing the spines, however, did not bear nuclei in any of the cases investigated.

The neodermal spines of the Gyrocotylidea (Figure 11.4A) show some similarity with the calcareous corpuscles of the Amphilinidea (Figure 11.5E) and Cestoidea (e.g., McCullough and Fairweather 1987; Pawlowski et al. 1988): These increase in number during maturation and have a multi-layered structure. Calcareous corpuscles in the Cestoda are usually spherical and considered to function as excretory deposits. Such corpuscles are lacking in Gyrocotyle. The neodermal spines in this taxon may have a function similar to that of the corpuscles. In any case, in view of their extraordinary shape, the neodermal spines must be regarded as an autapomorphy of the Gyrocotylidea.

No intraepithelial multiciliary sensory structures

In the Gyrocotylidea all sensory structures described from the body surface (from the epidermis of the lycophora and the neodermis of the postlarval stages) are monociliary (Lyons 1969a; Allison 1980; Xylander 1986a, 1987d, 1992a). The

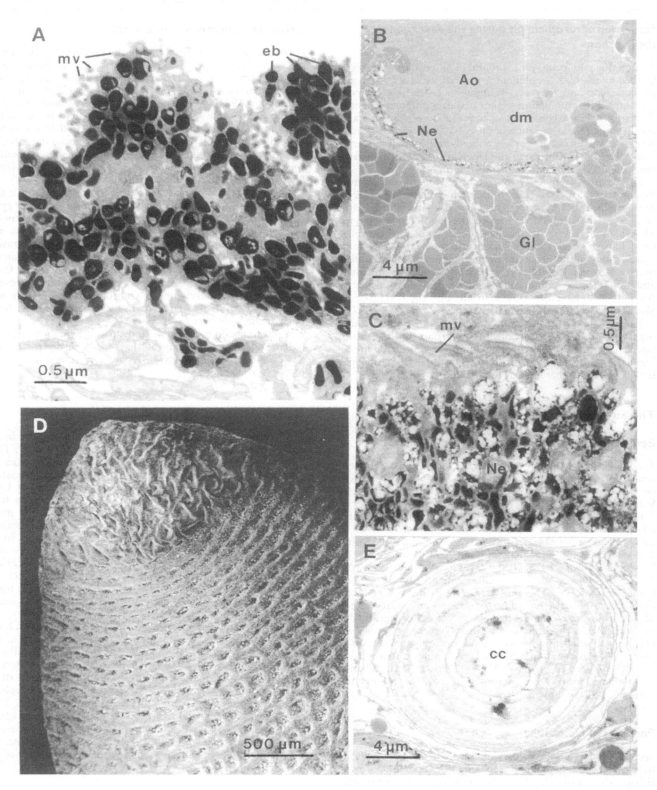

Figure 11.5 TEM-micrographs from mature specimens and a SEM-micrograph from an immature specimen of *Amphilina foliacea*. **A)** Surface neodermis showing numerous electron-dense bodies and microvilli. **B)** Section through the apical organ with a thin neodermis and numerous glands opening into the lumen (these glands have obviously discharged the moderately electron-dense material). **C)** Neodermis of the apical organ in higher magnification showing few long and slender microvilli. **D)** Posterior end of a ca. 10 mm-long specimen of *Amphilina foliacea* showing a surface folding resembling the (simple) types of rosette organs found in some Gyrocotylids. **E)** Lamellate structure of a cross-sectioned 'calcareous corpuscles'. Abbreviations: Ao, apical organ; cc, calcareous corpuscle; dm, glandular material discharged into the apical organ; eb, neodermal electron-dense bodies; Gl, glands of the apical organ; mv, microvilli; Ne, neodermis.

only multiciliary sensory structure reported so far is a presumptive photoreceptor of the lycophora of *Gyrocotyle urna*, which is located subepithelially at the anterior end of the cerebrum (Xylander 1984, 1987d).

In larval as in postlarval stages of all other parasitic Platyhelminthes multiciliary intraintegumental sensory structures have been found (e.g., miracidia: Wilson 1970; Pan 1980; postlarval Digenea: Zdárská *et al.* 1988; Czubaj and Niewiadomska 1996; oncomiracidia: Lyons 1972, 1973a; cotylocidia: Rohde 1972; postlarval Aspidogastrea: Rohde 1972; lycophora of Amphilinidea: Rohde and Garlick 1985a; Rohde *et al.* 1986). Taking into account the intensity of investigations, it seems unlikely that multiciliary intraintegumental receptors could have been overlooked in those species of *Gyrocotyle* investigated. Therefore, the lack of epithelial multiciliary sensory structures can currently be considered an autapomorphy of the Gyrocotylidea.

The autapomorphies of the Nephroposticophora (Amphilinidea + Cestoidea)

Unpaired excretory pore at the posterior end in postlarval stages

In the Amphilinidea and Cestoidea the nephridiopore (= excretory pore) is unpaired and originally located at the posterior end of the body (Hein 1904; Malmberg 1974; Lindroos and Gardberg 1982; Lindroos 1983; Gibson *et al.* 1987; Xylander 1987b, 1992b, 1996; see Figures 11.2 and 11.5D). In the Cestoidea with proglottids this original nephridiopore is lost when discarding the first proglottid. Simultaneously, twin secondary excretory pores are formed from the terminal parts of the nephridioducts.

In most other parasitic Platyhelminthes there is a pair of nephridiopores (e.g., Gyrocotylidea: Malmberg 1974; Xylander 1987a,b, 1992b; see Figure 11.2; miracidia of the Digenea: e.g., Pan 1980; cercaria of the Digenea: Rees 1977; Monogenea: Malmberg 1974; Aspidogastrea: Rohde 1972). The unpaired excretory pore of the postcercarial stages in Digenea is formed from distal parts of the nephridioduct after discarding the cercarial tail on which the paired nephropori of the cercaria are located. The 'cercomer', the body part containing the larval hooks, is still present in the Amphilinidea. However, it is lost prior to proglottid formation in those Cestoidea with proglottids. Therefore, the secondary unpaired nephridioducts in adult Digenea and Nephroposticophora have most likely evolved independently.

How the paired excretory pores in lycophores and coracidia (see Figure 11.3) are transformed to an unpaired nephridiopore in postlarval amphilinideans and the metacestode stages of the cestoideans is unknown.

Larger nephridioducts unciliated

In the Nephroposticophora the larger excretory ducts lack a non-terminal ciliation (= lateral flames) (see Xylander 1987b, 1992b). This ciliation is found in larger nephridioducts of other Platyhelminthes, bringing about the transport of the excretory fluid towards the nephridiopore. So in Nephroposticophora the transport of excretory fluids must be ascertained by another system. The well-developed musculature on the contraluminal side of the thick basal lamina surrounding the ducts indicates that the fluid transport is brought about by muscle action.

In other parasitic platyhelminths there is a constant beating of the lateral flames of the excretory ducts (Gyrocotylidea: Fuhrmann 1931; Xylander 1987b, 1992b; Digenea: Bugge 1902; Monogenea: Rohde 1973a, 1975a; Aspidogastrea: Rohde 1972). The reduction of this ciliation is an autapomorphy of the Nephroposticophora (Xylander 1987a,b, 1992b, 1996). Xylander (1987b, 1992b) considered the reduction of nephridioduct ciliation to be related to a presumed additional function of the protonephridial system in nutrient transport: cilia that move without neuronal control have been reduced, and movement of the nephridial fluids is controlled by the musculature surrounding the larger ductules. Thus, the presumed transfer of nutrients via the reticulate protonephridial system may become directed, controllable and more effective (see Xylander and Bartolomaeus 1995).

The autapomorphies of the Amphilinidea

Adult stages are coelomic parasites

All Amphilinidea become mature within the coelomic cavity of their host. No species of this group has ever been found in the intestine of a vertebrate host (although Janicki (1930) speculated about a 'reptile host' involved in the life cycle of the basic character pattern of Amphilinidea).

Gyrocotylidea and Cestoidea usually become mature in the intestine of their vertebrate host. Many 'primitive' Eucestoda (e.g., Liguloidea, Pseudophyllidea) have a second intermediate host in which they live in the coelomic cavity (usually fishes); some species show maturation of the reproductive organs (in few cases even egg production) already in the second intermediate host, or even in the primary invertebrate host. Facultative maturity in this host has also been described. Nevertheless, for the ground pattern of Cestoidea and Gyrocotylidea a vertebrate host must be considered, the intestine of which is the habitat of the mature stage. The development of coelomic parasitism in the Amphilinidea is an autapomorphy of this taxon most probably arisen from the omission of the definitive host (which more likely has been a larger fish than a sauropsid).

As peristaltic movement is lacking in the coelom, the posterior attachment organ was no longer necessary and lost its function. It is, however, still present as a 'rudimentary rosette organ' at the posterior end of *Amphilina* (Figure 11.5D; see also Xylander 1996). The rosette organ, therefore, most probably does not represent an autapomorphy of the Gyrocotylidea, as presumed by several authors (Brooks *et al.* 1985b; Bandoni and Brooks 1987; Brooks 1989a; Littlewood *et al.* 1999a) but was developed in the stem lineage of the Cestoda.

The most primitive species of the Amphilinidea use the abdominal pores of their hosts for deposition of their eggs. More derived species penetrate the body wall or throat of their hosts with their apical organ which has a strong musculature and numerous glands (see Janicki 1930; Dönges and Harder 1966; Xylander 1986a; Rohde and Watson 1989; Davydov and Kuperman 1993).

Neodermal microvilli short and stubby

The neodermal microvilli of all amphilinidean species investigated are regularly shaped and distributed, but differ from the typical microvilli of Cestoidea in being short and stubby, lacking an electron-dense tip (Lyons 1977; Xylander 1986a, 1996; Davydov and Kuperman 1993; Figure 11.5A). Only in the apical organ ('haptor', Figure 11.5B,C) have microvilli been found, which are long and slender (Figure 11.5C) and resemble cestoidean microvilli or microtriches (see Davydov and Kuperman 1993 for another type of microvilli).

The body microvilli of the Amphilinidea correspond to the microvilli of the Gyrocotylidea and Cestoidea regarding their regular arrangement. Such regularly arranged microvilli are lacking in the other Neodermata. However, they differ from the microvilli of other cestodes in their length. The view that long microvilli belong to the ground pattern of the Cestoda is supported by the occurrence of typical cestode microvilli in the anterior body invagination. The fact that long microvilli are lacking in these parasites of the coelomic cavity indicates that one main function of the typical cestode microvillar surface of gut living stages is the protection against digestion.

Uterus tripartite

The uterus of all amphilinideans has two ascending (anteriad) and one descending (posteriad) parts (Xylander 1988a; see Figure 11.1B). The eggs fertilized in the ootype mature on their way to the anterior end of the uterus.

Such a uterine structure is unique within the Neodermata, and represents an autapomorphy of the Amphilinidea (Xylander 1986a, 1998).

Uterine pore at the anterior end

The uterus of all amphilinideans terminates close to or in the apical organ (e.g., Janicki 1908; Dönges and Harder 1966; Dubinina 1974; Xylander 1988a; Figure 11.1B). This organ, which is equipped with a strong musculature, is projected into the coelomoduct of the host, penetrates the wall of the body cavity (in the few species with a host without a coelomoduct) and the Amphilinidea deposit their eggs (Dönges and Harder 1966) by exposing their body tip to the water outside the host or into the tracheal lumen or cloacae of its turtle host in *Austramphilina elongata*.

Such a translocation of the uterine pore to the very anterior tip of the body appears to be an adaptation to a life cycle in which the definitive host has been lost and maturity is reached in the coelomic cavity of the second (originally intermediate) host (see Janicki 1928) (in the presumed gut-living predecessors of the recent Amphilinidea, eggs were most probably set free with the faeces of the hosts). The use of the coelomoducts of the hosts requires an organ which is able to project into this duct, which usually has a small lumen. Thus, the apical organ with its extraordinary strong musculature may have evolved as an adaptation of the amphilinideans to the change from an intestinal to a coelomic parasite. As this organ had become available, a few species could change to fish hosts without coelomic ducts. These amphilinidean species use this organ to penetrate the body wall for oviposition (see Dönges and Harder 1966) while the rest of the body remains sheltered within the host. For this kind of oviposition the uterine pore had to be translocated to the anterior tip of the body.

The autapomorphies of the Cestoidea

Neodermis with typical microtriches

In the Cestoidea the electron-dense caps of the microvilli of some body regions may be enlarged forming a characteristic spine. This type of microvillus is called a microtriche.

In no other cestode group is there such an enlargement of the electron-dense microvillar caps, although small caps can already be found in the Gyrocotylidea (Figure 11.4C,D; see also Xylander 1986a, 1996). Microtriches are an autapomorphy of the Cestoidea.

Spermatozoa without mitochondria

In all Cestoidea, spermatozoa lack mitochondria (e.g., Featherston 1971; Mokthar-Maamouri and Swiderski 1976; Swiderski and Eklu-Natey 1978; Swiderski and Mokthar-Maamouri 1980; Euzet et al. 1981; Swiderski 1981; Davis and Roberts 1983; MacKinnon and Burt 1984, 1985; Justine 1986, 1995, 1998b). In the ground pattern, two incorporated axonemata with a $9 \times 2 + '1'$-pattern, two rows of microtubules between the axonemes and an elongated nucleus are present.

In all other taxa of the Neodermata, the Gyrocotylidea (Xylander 1989b), Amphilinidea (Rohde and Watson 1986; Xylander 1986b, 1993, 1996), Monogenea (Halton and Hardcastle 1976; Justine 1983; Justine and Mattei 1981, 1983c, 1984b, 1986, 1987; Justine et al. 1985a; Xylander 1988b), Digenea (e.g., Rees 1979; Daddow and Jamieson 1983; Hendow and James 1988) and Aspidogastrea (Rohde 1972), mitochondria (or a single mitochondrion) can be found within the spermatozoa. The lack of mitochondria is an autapomorphy of the Cestoidea (Ehlers 1985a; Xylander 1986a, 1989b, 1993, 1996).

First larval stage without sensory structures and cerebrum ('passive larva')

Oncospheres and coracidia, the first stages of Cestoidea following embryogenesis, lack all kinds of sensory structures and have only very few nerve cells (Fairweather and Threadgold 1981b; Swiderski 1983) which do not constitute a cerebrum as in the lycophorae (Figure 11.3) and other ciliated larvae of the Neodermata. Coracidia and oncospheres are 'passive larvae' which do not search for their host actively, but have to 'wait' until taken up orally by their first host to continue the life cycle.

In the ground pattern of all other groups of Neodermata, ciliated larvae are equipped with a heterogeneous set of sensory structures and search actively for their host (miracidia: see Wilson 1970; Pan 1980; oncomiracidia: see Lyons 1969b,c, 1972, 1973b; lycophorae: Xylander 1984, 1987d, 1989a; Rohde and Garlick 1985a,b,c; Rohde et al. 1986). The few cases of 'passive host finding' in a few taxa of the Neodermata other than Cestoidea (e.g., the digenean *Dicrocoelium dendriticum*) have secondarily evolved from an active host-finding behaviour. The passive mode in Cestoidea is an autapomorphy of this group.

Reduction of several tissues and organs in the primary larval stage (coracidium)

Compared with the situation found in lycophores, miracidia, oncomiracidia and cotylocidia, many organs of the coracidia (and oncospheres) are retarded or reduced. Coracidia and oncospheres have a maximum of four glands, two protonephridial terminal organs and only six hooks (see Fairweather and Threadgold 1981b; Kuperman and Davydow 1982; Chew 1983; Swiderski 1983; Xylander 1986a, 1987a, 1990, 1996; Figure 11.3C).

In the lycophores of *Gyrocotyle urna*, eight larval glands belonging to four different types were described at the ultrastructural level; all these glands terminated at the very anterior tip of the larvae (Xylander 1990). Rohde (1986b, 1987b), Rohde and Georgi (1983) and Rohde and Watson (1989) described a much greater number of glands belonging to three types for the lycophora of *Austramphilina elongata*. The miracidia and the oncomiracidia also have a large number of different glands (e.g., Wilson 1971; Rohde 1975a; Pan 1980). The reduction of the number and types of glands in larvae of Cestoidea is an autapomorphy of this group.

In lycophores, the number of protonephridial terminal organs is generally six (Malmberg 1974; Rohde and Georgi 1983; Xylander 1986b, 1987a; Figures 11.3A,B). In oncomiracidia there often is a number of eight (Malmberg 1974), and in miracidia one or two pairs of terminal cells have been described (Wilson 1969b; Pan 1980; Smyth and Halton 1983). The retardation of the number of protonephridial terminal organs in the first larvae of Cestoidea is an autapomorphy of this taxon.

The number of hooks in the ground pattern of Cestoda is 10 (see above; Figure 11.3A,B); however, this number has only been retained in the lycophores (see Xylander 1991). Within the stem lineage of the Cestoidea the number is reduced to six (Figure 11.3C), and the mode of action is now mainly (or exclusively) that of a raptor tearing the host's tissue in order to enable the penetration of the intestine of the intermediate host and to enter its coelom. The reduction in number and the change in the mode of action are autapomorphies of the Cestoidea (for further information and a comprehesive discussion, see Xylander 1991).

Cercomer is discharged during larval development

The cercomer is that part of the body of cestodes which contains the larval hooks. It becomes a useless appendage during the development of the earliest metacestode stages in Cestoidea and is mostly lost.

However, in Gyrocotylidea and Amphilinidea, these hooks are incorporated and can be found in the body throughout the entire life span (although they are without function). The shedding of the cercomer in the Cestoidea during metacestode development is an autapomorphy of the Cestoidea.

Conclusions

The Cestoda represent a monophyletic taxon, including the Gyrocotylidea and Amphilinidea, as demonstrated by a great number of common derived characters. Within the Cestoda, the Gyrocotylidea and Nephroposticophora are sister-groups, the first characterized by for example the lack of multiciliary sensory receptors, the occurrence of parenchymatic postlarvae, and a larval epidermis lacking nuclei. The Nephroposticophora share an unpaired nephridiopore at the caudal end of the body and reduction of the non-terminal ciliation in the protonephridial ducts. Within the Nephroposticophora, the Amphilinidea and the Cestoidea have the rank of sister-groups, each of which is well-characterized by autapomorphies.

ACKNOWLEDGEMENTS

I would like to thank my colleague Ulrich Ehlers for his comprehensive support during my first approach to the phylogeny, evolution and comparative ultrastructural research on parasitic Platyhelminthes. I furthermore would like to thank my former supervisor Peter Ax for continuous inspiration. Andrea Hanel, Rainer Stephan, Wolfgang Junius, Klaus Rohde, Nikki Watson, Eric Hoberg and Herbert Boyle helped with the manuscript and the figures.

Chapter 12

Phylogeny among orders of the Eucestoda (Cercomeromorphae): Integrating morphology, molecules and total evidence

Eric P. Hoberg, Jean Mariaux and Daniel R. Brooks

Advances in our understanding of genealogy among the orders of the Eucestoda have been based on independent approaches linked to comparative morphology and analysis of molecular sequence data, particularly from 18S rDNA. Parsimony analyses of molecular or morphological databases have yielded largely concordant trees supporting monophyly for the Eucestoda: 1) monozoic Caryophyllidea are basal; 2) difossate and segmented forms such as Pseudophyllidea are the sister for the remaining orders; and 3) tetrafossates including the paraphyletic Tetraphyllidea, Proteocephalidea, Nippotaeniidea, Tetrabothriidea, Mesocestoidata, and Cyclophyllidea are highly derived. Hypotheses by Hoberg *et al.* (1997c) and Mariaux (1998) differed in placement of the Diphyllidea and the Trypanorhyncha. A 'total evidence' approach now combines substantial components of currently available morphological and molecular data and additional taxa to further examine the putative relationships among the Eucestoda using two complementary strategies: 1) a top-down analysis employing a 'consensus' or reduced matrix for molecular and morphological data for the ingroup across 11 ordinal-level taxa, three families of Tetraphyllidea and two families of Pseudophyllidea; and 2) a bottom-up analysis employing character data for individual, representative genera and species including 48 ingroup taxa across 16 ordinal and family-level taxa in a 'comprehensive' matrix. Parsimony analysis of the consensus matrix resulted in two most-parsimonious trees (MPTs) (CI = 0.671, RC = 0.378) largely similar to general structure outlined by Hoberg *et al.* (1997c) and Mariaux (1998). Analysis of the comprehensive matrix resulted in 48 equal-length trees (CI = 0.484, RC = 0.378) congruent to the MPTs derived from analysis of the consensus matrix in diagnosing the orders and putative relationships of the eucestodes. Results overall contrasted minimally with Hoberg *et al.* (1997c) or Mariaux (1998). Comparative data from morphology, ontogeny and ultrastructure are validated; a complementary nature is emphasized for: 1) morphological and molecular characters; and 2) top-down versus bottom-up approaches. Phylogenetic resolution among the Eucestoda will lead to development of model systems for evolutionary biology, cospeciation analysis and historical biogeography.

A history and background for tapeworm phylogenetics

Historically, the phylogeny for the tapeworms has been problematic and unresolved with numerous competing hypotheses having been presented over the past century (e.g., Loennberg 1897; Fuhrmann 1931; Skrjabin 1940; Baer 1950; Spasskii 1951, 1958; Euzet 1959, 1974; Freeman 1973; Dubinina 1980; Euzet *et al.* 1981; Brooks *et al.* 1991). Although the morphological limits for most orders are apparent (e.g., Wardle and McLeod 1952; Yamaguti 1959; Schmidt 1986;

Spasskii 1992; Khalil *et al.* 1994) there has been disagreement over the validity and rank of certain taxa. The contentious nature of these hypotheses has been driven by a variety of issues: 1) concepts for relationships were often based on assessments of single characters or structural/ontogenetic attributes; 2) assumptions about putative coevolutionary linkages among tapeworms and their hosts strongly influenced concepts for parasite phylogeny (e.g., Wardle and McLeod 1952; Brooks *et al.* 1991; Klassen 1992); 3) philosophical differences over the evolutionary process led to divergent interpretations of relationship; 4) adequacy and applicability varied for a diversity of classes of characters as indicators of relationship (e.g., morphology, ontogeny, etc.); 5) interpretations for homology were contradictory; and 6) the methods used to develop and assess competing phylogenetic hypotheses were often not directly comparable. Although recent diagnostic keys have provided new critical data and interpretation extending to generic-level taxa, there has been no general attempt to reflect evolutionary history (Schmidt 1986; Khalil *et al.* 1994). Such concerns led to recognition that a standardized approach, emphasizing cladistic methodology (Hennig 1950, 1966; Wiley 1981; Wiley *et al.* 1991) was requisite for resolution and formulation of a synoptic understanding of the history and genealogical relationships for the Eucestoda (Hoberg *et al.* 1997b).

Phylogenetic studies of cestodes, first initiated in the late 1970s (e.g., Brooks 1978), have been limited in taxonomic scope to families, genera and species (summarized in Brooks and McLennan 1993a) and higher-level systematics was not addressed in detail (Mariaux 1996). Thus, phylogenetic reconstruction until recently had evaluated intraordinal relationships for some Proteocephalidea (Brooks 1978, 1993), Tetraphyllidea (Brooks *et al.* 1981; Brooks 1992), Tetrabothriidea (Hoberg 1989, 1995; Hoberg and Adams 1992), and Cyclophyllidea (Hoberg 1986, 1992; Moore and Brooks 1987). Brooks *et al.* (1991) and Brooks and McLennan (1993a) were the first to apply cladistic methods to develop a synoptic hypothesis for the phylogeny of the major lineages and orders of the Eucestoda.

Over the years since 1996, the focus on phylogenetic studies among the Eucestoda has increased dramatically (Hoberg *et al.* 1997b). New working hypotheses for relationships of the orders (Figure 12.1) were developed based on comparative morphology (Hoberg *et al.* 1997c, 1999b) and molecular sequence data (Mariaux 1998; Olson and Caira 1999). Justine (1998b) evaluated an extensive literature and examined the utility of spermatozoon ultrastructure for phylogenetic reconstruction. Concurrently, Hoberg *et al.* (1997c, 1999b) outlined, summarized and compared prior explicit concepts for phylogeny. Mariaux and Olson (2001, this volume) summarized progress in molecular systematics studies of tapeworms.

Phylogenetic reconstruction has now been conducted at the intraordinal level to examine relationships within eight of

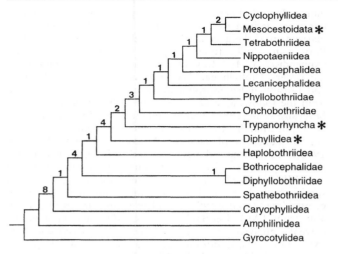

Figure 12.1 Phylogeny for the Eucestoda based on comparative morphology. Shown is one of two MPTs (length = 159; CI = 0.803; RC = 0.645) that is largely congruent with prior hypotheses (Hoberg *et al.* 1997c, 1999b); an alternative hypothesis places the Diphyllobothriidae basal to the Bothriocephalidae and remaining eucestodes. Contrasting with Hoberg *et al.* (1997c), this hypothesis now includes the Mesocestoidata, Bothriocephalidae and Diphyllobothriidae as independent taxa; Litobothriidae is not included here, as data are unknown for 22 of 51 characters (spermatozoa, spermiogenesis, ontogeny and eggs). Departures from the topology reported by Mariaux (1998) based on 18S rDNA are denoted by asterisks. Reconciliation of the morphological hypothesis with that from molecular analysis requires 18 additional steps in the former tree (length = 178; CI = 0.72; RC = 0.50). Bremer decay values are shown above each node.

12–13+ currently recognized orders: Tetrabothriidea (Hoberg 1989; Hoberg and Adams 1992); Pseudophyllidea (Bray *et al.* 1999); Trypanorhyncha (Beveridge *et al.* 1999); Tetraphyllidea and Lecanicephalidea (Caira *et al.* 1999; Olson and Caira 1999) where recognition of the ordinal level diversity within this group is now being highlighted; Proteocephalidea (Rego *et al.* 1998; Zehnder and Mariaux 1999); Diphyllidea (Ivanov and Hoberg 1999); and Cyclophyllidea (Hoberg *et al.* 1999a; von Nickisch-Rosenegk *et al.* 1999). Studies within the Spathebothriidea, Caryophyllidea and Nippotaeniidea, continue to be lacking. These represent a suite of testable hypotheses for in-depth analysis of the phylogeny for the Eucestoda.

Independent analysis of molecular and morphological data using parsimony has yielded largely congruent trees supporting monophyly for the Eucestoda: 1) monozoic Caryophyllidea are basal; 2) difossate and segmented forms such as Pseudophyllidea are the sister for the remaining orders; and 3) tetrafossates including the Tetraphyllidea, Proteocephalidea, Nippotaeniidea, Tetrabothriidea, Mesocestoidata, and Cyclophyllidea are highly derived (Figure 12.1). Hypotheses by Hoberg *et al.* (1997c) and Mariaux (1998) were largely congruent and differed primarily in placement of the Diphyllidea and the Trypanorhyncha and whether or not the Mesocestoidata should be subsumed within the Cyclophyllidea. Alternatively, Olson and Caira (1999) found the Spathebothriidea to be basal, a possible separation between lineages containing difossate versus tetrafossate taxa, and a variable position for the Trypanorhyncha and Diphyllidea. It is now clear that the Tetraphyllidea is paraphyletic (Euzet *et al.* 1981; Hoberg *et al.* 1997c; Caira *et al.* 1999; Olson and Caira 1999). There is also some indication that Mesocestoidata and

Litobothriidae should be resurrected at the ordinal level (see Mariaux 1998; Hoberg *et al.* 1999a; Miquel *et al.* 1999; Olson and Caira 1999). Apparent resolution over the relationships of the Tetrabothriidea as the putative sister of the Cyclophyllidea or of the Mesocestoidata + Cyclophyllidea (Hoberg *et al.* 1997c, 1999a; Mariaux 1998; Olson and Caira 1999) contrasts with a diversity of opinions over the past 100 years (Baer 1954; Spasskii 1958, 1992; Temirova and Skrjabin 1978; Galkin 1987, 1996; Hoberg 1987, 1994).

The current studies outlined in this chapter are designed to further elucidate the higher-level hypotheses for the Eucestoda. The most recent consensus (Khalil *et al.* 1994) recognized 12 orders of the Eucestoda, and 9–19 orders have been distinguished by taxonomic consensus in the past (Mariaux 1996). It is suggested that ordinal-level diversity may reside between these extremes. A 'total evidence' approach (Kluge 1989; de Queiroz *et al.* 1995; Sanderson *et al.* 1998) combining the currently available morphological and molecular databases (18S rDNA only) was used to examine the putative relationships among the Eucestoda using two complementary strategies: 1) a top-down analysis employing a 'consensus' or reduced matrix for molecular and morphological data for 16 ordinal and family-level taxa of the ingroup; and 2) a bottom-up analysis employing character data for individual, representative genera and species including 48 in-group taxa in a 'comprehensive' matrix. The limiting factor for inclusion of taxa was based on the availability of molecular sequence data which are lacking for inclusive families within the paraphyletic Tetraphyllidea (see Euzet 1994). The current analyses extend those of Hoberg *et al.* (1997c, 1999b) and Mariaux (1998) with the addition of members of previously unrepresented families and orders and address: 1) examination of the monophyly for orders of the Eucestoda; 2) development of a more robust hypothesis for relationships among the major taxa based on these data; and 3) definition of a clear working hypothesis for phylogeny that will be useful in delineating areas of future research linked to systematics, classification, and historical ecological research in coevolution, biogeography and ecology of the tapeworms.

Foundations for the analyses: taxa and characters

Relationships among 16 putative orders and several families of the Eucestoda were evaluated based on data derived from comparative morphology and molecular sequences of 18S rDNA. Taxa and characters included those in Hoberg *et al.* (1997c) and Mariaux (1998) with sequences for two additional taxa from Olson and Caira (1999); details of outgroups and polarity argumentation are presented in these papers. Mesocestoidata is treated here as an independent taxon based on recent studies by Mariaux (1998), Hoberg *et al.* (1999a) and Miquel *et al.* (1999). Tetraphyllidea is deconstructed to represent the Onchobothriidae, Phyllobothriidae and Litobothriidae. Pseudophyllidea is represented by two putative lineages encompassing the Diphyllobothriidae and Bothriocephalidae, with the latter including Triaenophoridae, Echinophallidae, Philobythiidae and Cephalochlamididae (consistent with Bray *et al.* 1999). Characters for spermatozoons and spermiogenesis are included for the Mesocestoidata (Miquel *et al.* 1999) and Lecanicephalidea (personal communication J.-L. Justine and L. Euzet). Additionally, molecular sequences from species of Haplobothriidea (*Haplobothrium globuliforme*), Litobothriidae (*Litobothrium janovyi*), Lecanicephalidea (*Eniochobothrium gracile*), and Onchobothriidae (*Calliobothrium* sp.), kindly provided by P. D. Olson, were included in the current analysis, otherwise data and sequence alignments are consistent with Mariaux (1998).

'Total evidence' or combined analysis

Molecular and morphological data were combined in the current analysis in accordance with rationale established for 'total evidence' and phylogenetic reconstruction among other taxa (e.g., Kluge 1989; de Queiroz et al. 1995; Huelsenbeck et al. 1996; Siddall 1997; Blair et al. 1998; Sanderson et al. 1998; Littlewood et al. 1999a). Because the original data did not contain the same taxa, and the numbers of taxa were relatively limited, we initially used a 'supermatrix' approach in which the matrices were combined (Kluge and Wolf 1993; Sanderson et al. 1998). Additionally, separate data were compared for heterogeneity according to the Partition Homogeneity Test (PHT) as implemented in PAUP* (Swofford 1998).

Consensus matrix

The 'consensus' or 'reduced' matrix was designed to address relationships among putative orders in a top-down analysis where supraspecific taxa are used as terminals (see for example discussion in Bininda-Emonds et al. 1998). There were 16 ingroup taxa and two outgroups (consistent with Hoberg et al. 1997c, plus the Mesocestoidata, Litobothriidae, Bothriocephalidae and Diphyllobothriidae). The Tetraphyllidea, recognized to be paraphyletic (e.g., Hoberg et al. 1997c; Caira et al. 1999) was deconstructed to the family level to represent constituent taxa. Explicitly, the current analysis does not address the relationships for the diversity of minor 'tetraphyllidean' families (e.g., Cathetocephalidae, Chimaerocestidae, Disculicipitidae, Prosobothriidae and Dioecotaeniidae), nor the monophyly of these inclusive groups. Included were 51 binary and multistate characters from comparative morphology (Appendix 12.1) (consistent with Hoberg et al. 1997c, 1999b), and 1102 aligned nucleotide sites representing bases from partial sequences of 18S rDNA (consistent with Mariaux 1998; Olson and Caira 1999); 144 characters were informative for parsimony analysis, respectively, 100 molecular and 44 morphological.

Multistate taxa were coded as polymorphic for both morphological and molecular data, where families, genera or species possessed alternative character states. The potential influence of coding for polymorphism in multistate taxa, was considered (see Maddison and Maddison 1992; Swofford 1993). Estimation of ancestral states based on a prior phylogeny such as that outlined by Yeates (1995) or Bininda Emonds et al. (1998) was not applied to supraspecific taxa in this analysis. Although this method for coding would eliminate polymorphism by explicitly recognizing the ancestral state for each supraspecific taxon, decisions could not be based on application of a consistent convention.

Comprehensive matrix

Alternatively, a 'comprehensive' matrix included data for representative species or exemplars for 48 in-group taxa across eucestode ordinal diversity, and two outgroups in a bottom-up analysis. Generic and species-level taxa are consistent with Mariaux (1998) and Olson and Caira (1999) and a complete list of taxa is included in the former study. Ordinal-level taxa were deconstructed to the species level with respect to

morphological characters; 113 binary and multistate characters were derived from comparative morphology (data summarized in Hoberg et al. 1997c, 1999a; Rego et al. 1998; Bray et al. 1999); autapomorphies were generally excluded (Appendices 12.1, 12.2). Due to limited representation for the orders Diphyllidea, Trypanorhyncha, Lecanicephalidea, and Tetraphyllidea, morphological data for species and genera from the following studies were not included in this matrix: Beveridge et al. (1999); Ivanov and Hoberg (1999); Caira et al. (1999). A total of 1102 nucleotide sites represented bases from partial sequences of 18S rDNA (consistent with Mariaux 1998); 1215 total characters, and 271 informative for parsimony analysis. Sequence alignments have been deposited in EMBL/GenBank and a complete list of genera and species and their GenBank accession numbers are documented (see Mariaux 1998; complete alignments are also available from J.M.). GenBank data for additional taxa in the current analysis are as follows: *Calliobothrium* sp. (AF124469); *Litobothrium janovyi* (AF124468); *Haplobothrium globuliforme* (AF124458); *Eniochobothrium gracile* (AF124465).

Parsimony analysis

The two contrasting matrices allowed examination of the influence of character coding and different strategies of analysis (e.g., top-down employing supraspecific taxa versus bottom-up employing a series of genera and species as representatives of higher taxa) on tree structure and stability along with recovery and diagnosis of higher taxa. We examined the issue of maintenance of position for terminal supraspecific taxa in a cladogram with respect to a solution involving 'all' species (Bininda-Emonds et al. 1998; Wiens 1998).

Analysis of the matrices, written with MacClade 3.05 (Maddison and Maddison 1992), was conducted using PAUP 3.1.1, and PAUP* 4.0 (Swofford 1993, 1998). Analyses were done initially in a heuristic search mode (HS), with step-wise addition = simple and branch swapping either by nearest neighbour interchanges (NNI) or by tree bisection-reconnection (TBR); results were confirmed with branch and bound (B&B) for the consensus matrix. In HS and B&B, multistate characters were unordered; character weights were not applied; multistate taxa were designated as polymorphic; gaps in sequence alignments were treated as missing data, however, handling these as a fifth base had no influence on results; optimization was by ACCTRAN. Results are shown as a phylogenetic tree(s) with associated statistics, including the consistency index (CI), and rescaled consistency index (RC), or as strict consensus trees as defined by Swofford (1993). As implemented in PAUP* (Swofford 1998), resulting hypotheses were further evaluated via bootstrap and jack-knife resampling (Farris et al. 1996). Decay or Bremer-support indices (Bremer 1994) were calculated using AutoDecay 3.0.3 (distributed by the authors, T. Eriksson and N. Wikström 1995).

Host–parasite coevolution

Putative coevolutionary associations of definitive hosts and eucestode taxa were examined by mapping extant vertebrate taxa (e.g., Chondrichthyes, Holocephali, Actinopterygii, Teleostei, Amphibia, Chelonia, Mammalia, Lepidosauria, and Aves) onto the parasite tree. A host-matrix (not shown) was written with MacClade 3.05 and hosts as characters were optimized onto the phylogeny of the Eucestoda (Maddison and Maddison 1992).

Total evidence and a phylogeny for the Eucestoda

Data set heterogeneity

The PHT revealed that the partitions (molecular versus morphological) were not homogeneous ($P < 0.01$). We were aware, however, of areas of conflict in these data, particularly in the placement of the Diphyllidea. With the Diphyllidea excluded, homogeneity was indicated with values dependent on the matrix and conditions of analysis ($0.08 > P > 0.02$).

Phylogeny of the Eucestoda

Parsimony analysis of the consensus matrix (Figure 12.2) resulted in two most-parsimonious trees (MPT) (excluding uninformative characters: CI = 0.671, RC = 0.378); monophyly for the Eucestoda was strongly supported. The MPTs were fully resolved except for a polytomy linking the Haplobothriidea with the Diphyllobothriidae and Bothriocephalidae. Trees were largely similar in general structure to those outlined by Hoberg *et al.* (1997c) or Mariaux (1998) but contrasted as follows: 1) Haplobothriidea + Pseudophyllidea as sister groups; 2) Trypanorhyncha placed basal to the tetrafossates (consistent with Hoberg *et al.* 1997c); 3) Litobothriidae and Lecanicephalidea as groups basal to the 'Tetraphyllidea' + remaining tetrafossates; and 4) Diphyllidea as the sister-taxon of the Proteocephalidea (consistent with Mariaux 1998). Bootstrap and jack-knife values were generally within a range of 70–100% except for three nodes (Figure 12.2).

Parsimony analysis of the comprehensive matrix resulted in 48 equal-length trees (CI = 0.484, RC = 0.378). The strict consensus was fully resolved (except for a polytomy among crown groups in the Cyclophyllidea) and congruent to the MPTs derived from analysis of the reduced matrix (Figure 12.3); the tree is consistent with phylogenetic diagnoses of a minimum of 16 orders. Bootstrap and jack-knife resampling revealed equivocal support in two sectors of the tree: 1) in relationships of the relatively basal Haplobothriidea + Pseudophyllidea (including Bothriocephalidae and Diphyllobothriidae); and 2) within the Cyclophyllidea.

Based on separate analyses (e.g., Hoberg *et al.* 1997c; Mariaux 1998) and the results of these combined, total evidence, analyses (Figures 12.1–12.3) the following observations are supported: 1) monophyly for Eucestoda; 2) a basal position for the monozoic Caryophyllidea; 3) Pseudophyllidea and Haplobothriidea as basal, polyzoic taxa with separation of the Diphyllobothriidae from the Bothriocephalidae and remaining pseudophyllideans; 4) Trypanorhyncha as the sister for the Litobothriidae and the tetrafossates; 5) Tetraphyllidean paraphyly; and 6) unequivocal support for the relationships and placement of the Nippotaeniidea + Tetrabothriidea + Mesocestoidata + Cyclophyllidea. In contrast to Hoberg *et al.* (1997c), the Pseudophyllidea + Haplobothriidea are sister-groups, the Lecanicephalidea are the sister-group for the remaining tetrafossates, and the Diphyllidea are the putative sister of the Proteocephalidea. In contrast to Mariaux (1998), the Trypanorhyncha are postulated as the sister of the Litobothriidae + tetrafossate tapeworms and Mesocestoidata are the sister of the Cyclophyllidea. Haplobothriidea, Litobothriidae, Lecanicephalidea, and Onchobothriidae had

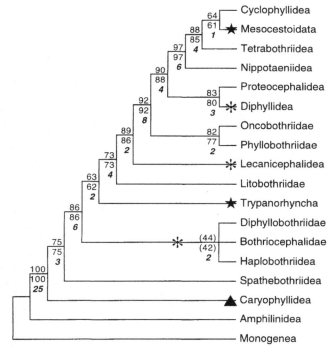

Figure 12.2 Phylogeny for the Eucestoda based on the 'consensus' matrix. Shown is the strict consensus of two MPTs resulting from analysis of total evidence combining morphological and molecular databases (length = 578; CI = 0.671; RC = 0.378). Incongruence with the morphologically based hypothesis (Figure 12.1; and Hoberg *et al.* 1997c, 1999b) is indicated by asterisks and is evident in the following areas: 1) placement of the Pseudophyllidea basal to the Haplobothriidea; 2) Diphyllidea as the sister of Trypanorhyncha + remaining eucestodes; and 3) Lecanicephalidea as the sister of the Proteocephalidea + remaining eucestodes. Departures from the molecular-based hypothesis (Mariaux 1998) are indicated by stars and include: 1) Trypanorhyncha as the sister of the Pseudophyllidea + remaining eucestodes; and 2) Mesocestoidata as the sister of the Tetrabothriidea + Cyclophyllidea. A major contrast with respect to the hypothesis by Olson and Caira (1999) is the alternative placement of the Spathebothriidea and Caryophyllidea as indicated by a triangle. Indices shown on the strict consensus tree: 1) bootstrap, above branches (1000 replicates with 10 repeats each, values < 50% shown in brackets); 2) jack-knife, below branches (1000 replicates with 10 repeats, HS; Jac emulation < 50% in brackets); 3) Bremer decay indices below branches, in bold italics.

not been represented in the original analysis of sequences from 18S rDNA.

Host–parasite relationships

Hosts were mapped onto a tree that summarizes the putative relationships for eucestodes based on analysis of the comprehensive matrix (Figure 12.4). A complex history involving cospeciation, colonization and extinction is indicated by this hypothesis. Actinopterygians are postulated as basal or ancestral hosts for eucestodes. Patterns of occurrence for cestodes are indicative of episodes involving serial colonization followed by rapid and potentially explosive radiations in fishes (e.g., neoselachians and teleosts), tetrapods, and amniotes (lineages leading to extant amphibians, mammals, chelonians, lepidosaurians, and birds).

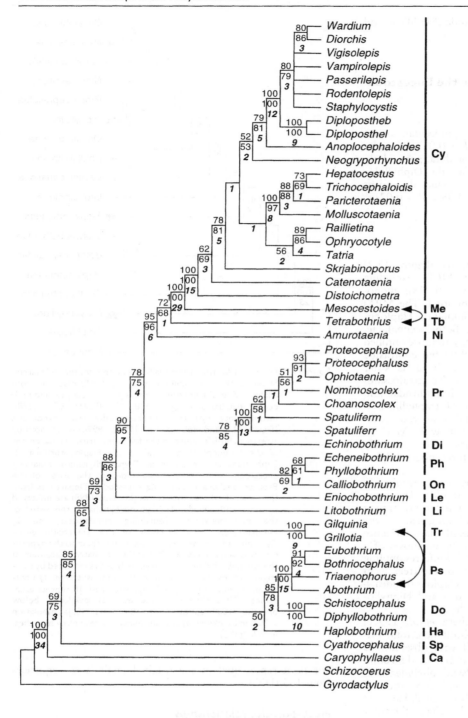

Figure 12.3 Phylogeny for the Eucestoda, based on total evidence and a 'comprehensive' matrix. The strict consensus is based on 48 MPTs (length = 1124; CI = 0.484; RC = 0.376). Note general congruence with relationships among ordinal level taxa in the 'consensus'/total evidence analysis (Figure 12.2). Bi-directional arrows show alternative placement of Pseudophyllidea/Trypanorhyncha and Tetrabothriidea/Mesocestoidata based on the comprehensive dataset for 18S rDNA (Mariaux 1998); departures in placement of Diphyllidea and Lecanicephalidea relative to the morphological hypothesis (Figure 12.1; Hoberg et al. 1997c) have been indicated in Figure 12.2. At a minimum, 16 orders may be diagnosed, with a clear separation of the Diphyllobothriidae from the Bothriocephalidae and Triaenophoridae in the Pseudophyllidea, and paraphyly for the Tetraphyllidea. Indices include: 1) bootstrap, above branches (100 replicates with 10 repeats each, HS; values < 50% shown in brackets); 2) jack-knife, below branches (100 replicates with 10 repeats each, HS; Jac emulation < 50% in brackets); 3) Bremer decay indices in bold italics. Family and ordinal-level taxa, in phylogenetic order, are indicated as follows: Ca = Caryophyllidea; Sp = Spathebothriidea; Ha = Haplobothriidea; Do = Diphyllobothriidae; Ps = Pseudophyllidea (including: Bothriocephalidae, Triaenophoridae); Tr = Trypanorhyncha; Li = Litobothriidae; Le = Lecanicephalidea; On = Onchobothriidae; Ph = Phyllobothriidae; Di = Diphyllidea; Pr = Proteocephalidea; Ni = Nippotaeniidea; Tb = Tetrabothriidea; Me = Mesocestoidata; Cy = Cyclophyllidea.

Interpreting the phylogeny of the Eucestoda

Monophyly, characters and ordinal level relationships

Monophyly for the Eucestoda is independently corroborated based on molecular, comparative morphological, ontogenetic and ultrastructural characters (Ehlers 1984, 1985a,b, 1986; Brooks *et al.* 1985b, 1991; Brooks 1989a,b; Justine 1991a, 1998b; Brooks and McLennan 1993a; Hoberg *et al.* 1997c, 1999b; Mariaux 1998; Littlewood *et al.* 1999a). Phylogenetic trees resulting from separate (Hoberg *et al.* 1997c; Mariaux 1998) or combined analysis and from either a consensus

or comprehensive matrix were well resolved and diagnosed largely congruent relationships for major taxa (Figures 12.1–12.3).

Given the results emanating from molecular data only (e.g., Olson and Caira 1999), we cannot exclude a fundamentally different scenario for the evolution of the eucestodes. Several lines of information (based on different weighting schemes in Mariaux 1998) from the current analysis and that of Olson and Caira (1999) would support diagnosis of largely difossate and tetrafossate clades (see also Mariaux and Olson 2001, this volume). A further contrast with the current analysis is the basal placement of the Spathebothriidea resulting from

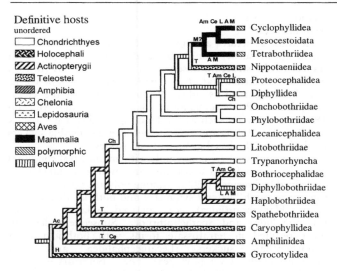

Definitive hosts
unordered
☐ Chondrichthyes
▨ Holocephali
▨ Actinopterygii
▨ Teleostei
▨ Amphibia
▨ Chelonia
▨ Lepidosauria
▨ Aves
■ Mammalia
▨ polymorphic
▨ equivocal

Cyclophyllidea
Mesocestoidata
Tetrabothriidea
Nippotaeniidea
Proteocephalidea
Diphyllidea
Onchobothriidae
Phylobothriidae
Lecanicephalidea
Litobothriidae
Trypanorhyncha
Bothriocephalidae
Diphyllobothriidae
Haplobothriidae
Spathebothriidae
Caryophyllidea
Amphilinidea
Gyrocotylidea

Figure 12.4 Phylogenetic hypothesis for the Eucestoda based on total evidence, showing the distribution of major vertebrate definitive hosts. Host taxa are mapped and optimized on the tree representing the 'consensus' analysis, with further resolution of relationships summarized from Figures 12.1–12.3, with MacClade 3.05 (Maddison and Maddison 1992). Host distributions are indicative of an early association with basal actinopterygian fishes, multiple colonization events of basal teleosts, and a secondary colonization of and radiation in chondrichthyans; colonization also accounts for the distribution of the tetrabothriideans in marine mammals and seabirds (Hoberg et al. 1997c, 1999a,b). Although the root of the subclade for Tetrabothriidea, Mesocestoidata + Cyclophyllidea optimizes to mammalian hosts, the associations of these groups with tetrapods appear to be older than implied and may reflect radiations in now extinct host groups (Hoberg et al. 1999a,b). The broader aspects of the putative history for these associations is not refuted by alternative placement of the Lecanicephalidea or Diphyllidea and inclusion of the Litobothriidae (Hoberg et al. 1999b). Host taxa are indicated as follows: A = Aves; Ac = Actinopterygii; Am = Amphibia; Ce = Chelonia; Ch = Chondrichthyes; H = Holocephali; L = Lepidosauria; M = Mammalia; T = Teleostei.

analysis of *Elongation factor* -1α and 18S rDNA (Olson and Caira 1999). Some aspects of these hypotheses appear weakly corroborated, however, and may be contradicted by both comparative morphology and total evidence. Additionally, the placement of the Diphyllidea with the Proteocephalidea in the current analysis and that of Mariaux (1998) seem ambiguous, given that this topology has not been observed in any other phylogenetic studies of molecular or morphological data.

Prior hypotheses for relationships among orders of the eucestodes represent two basic patterns (reviewed in Hoberg et al. 1997c; Mariaux 1998). Loennberg (1897) and Baer (1950) proposed a dichotomous and hierarchical phylogeny. In contrast, major lineages, diagnosing separate difossate and tetrafossate groups derived from a common ancestor, were defined by Fuhrmann (1931), Euzet (1959, 1974), Dubinina (1980), Euzet et al. (1981), and Olson and Caira (1999). Among difossate or tetrafossate taxa Freeman (1973) recognized five distinct lineages. Brooks et al. (1991), and Brooks and McLennan (1993a) defined a difossate lineage and a tetrafossate lineage with three coordinate sister-groups. Based on either separate or combined analyses, however, there is no apparent support for the diagnosis of distinct clades or lineages for difossate versus tetrafossate tapeworms. Detailed comparisons among this range of hypotheses based either on

comparative morphology or sequence data have been considered previously (Hoberg et al. 1997c, 1999b; Mariaux 1998; Mariaux and Olson 2001, this volume).

The following areas of concordance are notable with respect to the current and prior hypotheses: 1) The caryophyllids are basal with monozooy postulated as ancestral (Llewellyn 1965; Mackiewicz 1972; Dubinina 1980; Ehlers 1986; Hoberg et al. 1997c; Mariaux 1998); 2) difossate forms are primitive and basal to the tetrafossate groups of the higher tapeworms, and the difossate forms (e.g., pseudophyllideans) are the sister-group of the strongly polyzoic tapeworms (Brooks et al. 1991; Hoberg et al. 1997c); 3) Pseudophyllidea contains two distinct lineages represented by Diphyllobothriidae and Bothriocephalidae (Bray et al. 1999); 4) the Pseudophyllidea and Haplobothriidea are closely related (Olson and Caira 1999); 5) Trypanorhyncha are the putative sister for the Litobothriidae + higher tetrafossates; 6) higher tapeworms or tetrafossates, e.g., lecanicephalideans, onchobothriids, phyllobothriids, proteocephalideans (+ diphyllideans), nippotaeniideans, tetrabothriideans and mesocestoidatans + cyclophyllideans, are closely related or potentially coordinate groups (Loennberg 1897; Fuhrmann 1931; Baer 1950; Spasskii 1958, 1992; Euzet 1959; Freeman 1973; Dubinina 1980; Brooks et al. 1991; Brooks and McLennan 1993a; Hoberg et al. 1997c; Mariaux 1998; Olson and Caira 1999); 7) the tetraphyllideans do not constitute a natural group but are paraphyletic (Loennberg 1897; Euzet et al. 1981; Hoberg et al. 1997c, 1999b; Caira et al. 1999); 8) nippotaeniideans are derived within the higher tapeworms (Dubinina 1980; Hoberg et al. 1997c; Mariaux 1998; Olson and Caira 1999); and 9) the tetrabothriideans are the putative sister-group of the cyclophyllideans (Dubinina 1980; Hoberg et al. 1997c, 1999a; Olson and Caira 1999) or the mesocestoidatans + cyclophyllideans (Mariaux 1998). The status of the Mesocestoidata as an independent order, or as the basal member of the Cyclophyllidea may require further examination (see Mariaux 1998; Miquel et al. 1999; Hoberg et al. 1999a), although unequivocal separation from tetrabothriideans and cyclophyllideans is apparent (Figures 12.1–12.3).

The current study has minimal bearing on resolving the placement of the minor constituent taxa of the paraphyletic Tetraphyllidea (see Hoberg et al. 1997c; Caira et al. 1999). Although morphological characters have been studied for a diversity of species and genera among these taxa (e.g., Caira et al. 1999), a completely comparable molecular database is currently under development (Olson et al. 1999). Relationships of these taxa could have a bearing on understanding broader issues within the Eucestodes, if (1) they do not represent a domain in the tree or a series of coordinate taxa; or (2) other groups such as the Proteocephalidea and Lecanicephalidea are found to be paraphyletic- or subsumed within one or another of the 'tetraphyllidean' groups (see Caira et al. 1999; Olson and Caira 1999). Evidence presented from analyses of total evidence is compatible with the former contention in postulating a close relationship for 'tetraphyllideans', the Litobothriidae and Lecanicephalidea. Proteocephalideans have been regarded as monophyletic in most recent analyses (Rego et al. 1998; Zehnder and Mariaux 1999) although this requires more complete examination. Paraphyly for the Proteocephalidea, however, would in part be consistent with the systematics proposed by Spasskii (1958) in which it is placed as a suborder in the Tetraphyllidea with Phyllobothriata, Tetrabothriata, and Nippotaeniata.

Comparative data from morphology, ontogeny and ultrastructure are validated; a complementary nature is emphasized for: 1) morphological and molecular characters; and 2) top-down versus bottom-up approaches (see Yeates 1995; Bininda-Emonds *et al.* 1998; Wiens 1998). Recognition of a robust phylogenetic signal and support for these hypotheses is indicated, based on both consensus and comprehensive data, within the context of a total evidence approach (Kluge 1989; de Queiroz *et al.* 1995; Sanderson *et al.* 1998). This is consistent with the contention of Bininda-Emonds *et al.* (1998) that analysis of supraspecific taxa as terminals can be phylogenetically informative. Additionally, this suggests that in studies of the Platyhelminthes, there is no compelling reason to exclude certain classes of morphological characters – and in essence select only those that give solutions (or resolution) congruent to that based on molecular sequence data (e.g., see arguments in de Queiroz *et al.* 1995).

Trends in character evolution

Major conclusions about morphological characters and their evolution (origin of a segmented strobila; ovarian, vitelline and uterine development; eggs; spermatozoa and spermiogenesis) have not dramatically changed from those presented in Hoberg *et al.* (1997c) and Mariaux (1998). Certain points, particularly those related to the origin of polyzooy and the structure of ancestral cestodes can be reiterated.

Based on the monozoic form of the Amphilinidea, the putative sister-group of the tapeworms and the basal position of the Caryophyllidea, an unsegmented strobila is postulated as plesiomorphic based on the analyses herein. A monozoic strobila as ancestral for the eucestodes is consistent with hypotheses presented by Llewellyn (1965), Rees (1969), Mackiewicz (1972), and Ehlers (1984, 1985a,b, 1986). A close relationship for the caryophyllids and spathebothriids may also be suggested by karyological characters (Petkeviciute 1996). Thus, the suggestion that these groups are subsumed within the Pseudophyllidea and that monozooy or the tendency toward monozooy is secondary in these orders (e.g., Baer 1950; Freeman 1973; Brooks *et al.* 1991; Brooks and McLennan 1993a) is not supported (see discussion in Brooks *et al.* 1991). This may refute the contention by Wardle and McLeod (1952) for convergence of the polyzoic strobila in difossate and tetrafossate lineages. Polyzooy appears to have been derived once among the eucestodes, initially in the common ancestor of the Pseudophyllidea + strobilate tapeworms.

We should note, however, that Olson and Caira (1999) have proposed a basal position for the Spathebothriidea, a contention in part supported by their molecular analyses of two independent gene loci and by a shared deletion in the *Ef*-1α gene among their spathebothriidean-exemplar and platyhelminth outgroups. Within the context of phylogeny for the outgroups, cestodarians and eucestodes, however, absence of the intron in *Ef*-1α represents the plesiomorphic condition and thus is equivocal in diagnosing support for a basal position for the Spathebothriidea. In essence, the presence of the intron is synapomorphic for a restricted group of eucestodes, but as a consequence absence may not be phylogenetically informative with respect to the Spathebothriidea.

If the spathebothriideans are the basal lineage of the tapeworms a single origin of polyzooy among the eucestodes remains supported. The monozoic condition in the caryophyllideans, as a component of the pseudophyllidean/ haplobothriidean lineage, would be consistent with a single reversal to monozooy in the former group (Olson and Caira 1999). Results outlined herein, however, and those from independent analyses (Hoberg *et al.* 1997c; Mariaux 1998) support a basal position for the caryophyllideans that is bolstered by strong Bremer and bootstrap indices, and fail to diagnose a

relationship between the pseudophyllidean and caryophyllidean lineages.

Holdfast structures among the eucestodes are diverse. The single apical sucker constitutes a deep plesiomorphy with respect to the structure observed in the amphilinids (Brooks *et al.* 1991; Littlewood *et al.* 1999a). Ancestral tapeworms have been regarded as either tetrafossate, such as the Tetraphyllidea or Proteocephalidea (e.g., Wardle and McLeod 1952), or difossate such as the Pseudophyllidea (e.g., Joyeux and Baer 1961). It is apparent, however that definable suckers are lacking in the basal groups of eucestodes.

Current analyses are compatible with an ancestral condition being represented by a unifossate (and monozoic) tapeworm, as defined by Freeman (1973). Among polyzoic cestodes, difossate forms, with bilateral symmetry and bothria (Pseudophyllidea, Haplobothriidea) are derived relative to the Caryophyllidea and Spathebothriidea. Subsequent development resulted in tetrafossate forms with bothridia (Trypanorhyncha), suckers (Lecanicephalidea and later Proteocephalidea–Cyclophyllidea) or bothridia + suckers (Onchobothriidae, Phyllobothriidae). Relative placement of the Trypanorhyncha would suggest that bothridia are plesiomorphic among the tetrafossate tapeworms, and that suckers are derived from bothridial margins, in contrast to Brooks *et al.* (1991). Difossate forms with bothridia such as the Diphyllidea, the putative sister of the Proteocephalidea remain enigmatic (Ivanov and Hoberg 1999). Tetraphyllidean and tetrabothriidean bothridia, or bothridia-like suckers with auricular appendages, are not homologous, and more definitive studies of the structure of tetrabothriidean 'bothridia' are required (Rees 1956; Andersen and Lysfjord 1982; Caira *et al.* 1999). Current placement of the Nippotaeniidea is consistent with paedomorphic development for the apical sucker, a character that may show ontogenetic convergence in the Litobothriidae.

Continued resolution of relationships at the ordinal and intraordinal levels for eucestodes is dependent on exploitation of new and poorly studied classes of characters (Mariaux 1996; Hoberg *et al.* 1997b,c; Caira *et al.* 1999; Olson and Caira 1999). A wealth of data remains to be revealed and evaluated within the context of homology for the ultrastructure of organ systems, structure and ontogeny of complex organs such as the uterus and rostellum (e.g., among the Cyclophyllideans), life history and larval stages, spermatozoons and spermiogenesis, and karyology. Integration with a rapidly expanding database for molecular sequences from multiple gene systems may dramatically contribute to confidence in resolving and reconstructing the evolutionary history for the eucestodes.

Parasite–host coevolution and systematics

Hypotheses for the phylogeny of the eucestodes have been strongly tied to concepts of host–parasite specificity as an indicator of extended and extensive coevolutionary histories of these assemblages (e.g., Fuhrmann 1931; Wardle and McLeod 1952; Dubinina 1980; discussed in Brooks *et al.* 1991) where extant representatives of primitive fishes, e.g., chondrichthyans, were considered to be hosts for the most ancestral taxa of tapeworms (Loennberg 1897). Results of the analyses currently presented, which place the Caryophyllidea, Spathebothriidea, Pseudophyllidea, and Haplobothriidea as basal to the 'higher tapeworms' fail to support this contention.

Examination of host distribution within the context of the current phylogeny for the eucestodes suggests a basal association with actinopterygian fishes and independent colonization(s) and secondary radiation among chondrichthyans, teleosts, and tetrapods (Figure 12.4) (Hoberg *et al.* 1999a,b). This emphasizes the putative great age for origin of the eucestodes, and a temporally deep association with their hosts extending perhaps to the Devonian, ≥ 350–420 Ma. Hypotheses for deep age of the eucestodes and the relative contributions of cospeciation, host-switching and extinction as determinants of the structure of the contemporary fauna were initially outlined in studies of the higher phylogeny for eucestodes and the intraordinal relationships of cyclophyllideans (Hoberg *et al.* 1997c, 1999a,b). Independent estimates for the age of divergence for disparate segments of the parasite tree(s) were based on an integration of parasite and host phylogenies, host-distribution, biogeography for host–parasite assemblages, and historical geological data (Hoberg 1997). Archaic radiations previously postulated for the Monogenea, Digenea, Gyrocotylidea and Amphilinidea are consistent with these temporal estimates (e.g., Brooks 1989b; Brooks and McLennan 1993a; Boeger and Kritsky 1997). Conceptually, this theoretical framework has recently received corroboration through investigations on the broader history of the Platyhelminthes (Littlewood *et al.* 1999b).

Total evidence and concordant relationships

There is a general concordance in the relationships observed in trees derived from separate analysis of morphological and molecular data (Hoberg *et al.* 1997c; Mariaux 1998; Olson and Caira 1999; Mariaux and Olson 2001, this volume), and in those from a total evidence approach (Figures 12.1–12.3). Recognition of discrete lineages or clades for difossate and tetrafossate taxa is not generally supported, although such has been demonstrated in analyses of some molecular data (Olson and Caira 1999).

Principle conclusions derived from separate and combined analyses are the following: 1) Caryophyllids are basal within the Eucestoda and monozooy is ancestral; 2) Pseudophyllidea includes two distinct lineages represented by the Diphyllobothriidae and the Bothriocephalidae, Triaenophoridae and other families; 3) Pseudophyllidea + Haplobothriidea are the sister-group of the strongly polyzoic tapeworms and difossate forms are relatively primitive; 4) Trypanorhyncha is a relatively basal taxon; 5) Litobothriidae is the sister-group for the

tetrafossates; 6) tetrafossates or higher tapeworms (Lecanicephalidea, 'Tetraphyllidea', Proteocephalidea + Diphyllidea, Nippotaeniidea, Tetrabothriidea, Mesocestoidata and Cyclophyllidea) are strongly diagnosed; 7) Tetraphyllidea is paraphyletic; 8) Nippotaeniidea is highly derived; and 9) Tetrabothriidea is a distinct order and the sister-group of the Mesocestoidata + Cyclophyllidea. Also emphasized is the putative deep age for origin and initial radiation of the eucestodes in the Paleozoic. The generality of these conclusions appears robust, based on the data available.

Phylogenetic studies of the Eucestoda are integral to biodiversity assessment and refinement of historical research programmes linking ecology and biogeography (e.g., Brooks and McLennan 1991, 1993a; Hoberg 1997; Hoberg *et al.* 1997a,b,c). Survey and inventory in biologically critical habitats is required in formulating a synoptic view of the history, structure and evolution of global biodiversity (Brooks and Hoberg 2000). Elucidation of coevolutionary and biogeographic histories for host–parasite assemblages involving cestodes, lends relevant information to programmes in conservation biology. Predictions about the age and duration of specific faunal associations (e.g., Hoberg *et al.* 1999a,b), recognition of regions of endemism and evolutionary 'hot spots' and the historical structure of ecosystems and communities emanate directly from phylogenetic studies of tapeworms (Hoberg 1997). Our phylogenetic knowledge of cestodes is an integrative framework to be applied to questions of the origin, maintenance and distribution of organismal diversity.

ACKNOWLEDGEMENTS

Seminal work on the higher phylogeny of the tapeworms was conducted with Jean-Lou Justine, Peter Weekes, Rod Bray, and Arlene Jones during and subsequent to the 2nd International Workshop for Tapeworm Systematics at Lincoln, Nebraska in September 1996. The authors further wish to thank Boyko Georgiev, Alain de Chambrier, Ronald A. Campbell, Robert Rausch, Louis Euzet, Vladimir Gulyaev, Vasyl Tkach, Veronica Ivanov, and Claude Vaucher for critical discussions that contributed to ideas presented in this chapter. Peter Olson is acknowledged for sharing preprints and unreleased sequence data. Jean Mariaux was supported by the Swiss National Foundation for Scientific Research, Grant No. 3100-037548.93.

Appendix 12.1. Morphological characters for analysis using supraspecific taxa as terminals, and the foundation for the 'consensus' matrix. Character descriptions, justification, polarity, and numbering are consistent with higher-level analyses among the Eucestoda (Hoberg *et al.* 1997c, 1999a).

Orders of the Eucestoda. 51 characters (characters 160–210 for consensus matrix; 1103–1153 for comprehensive matrix). Corresponding to Hoberg *et al.* (1997c, 1999a). Numbers correspond to those for characters in original papers. Polarity is discussed and defined in Hoberg *et al.* (1997c, 1999b). Morphological characters were polarized by taxonomic outgroup criteria (Watrous and Wheeler 1981; Maddison *et al.* 1984; Wiley *et al.* 1991), with reference to the Monogenea, Gyrocotylidea and the Amphilinidea, groups recognized as basal to the Eucestoda (Ehlers 1984, 1985a,b, 1986; Brooks *et al.* 1985b; Brooks 1989a,b; Brooks *et al.* 1991; Justine 1991a; Brooks and McLennan 1993a).

1. UTERUS (structure at initial stage of development). 4 states. 0 = tubular, coiled, sinuous; 1 = tubular, straight; 2 = granular or cellular Anlagen, which expands to form tubular and saccate structure; 3 = saccate, bilateral.

2. UTERUS (structure when gravid). 4 states. 0 = tubular, including straight and coiled forms; 1 = tubular, with anterior saccate expansion; 2 = saccate, without bilateral expansion; 3 = saccate, bilateral, usually with distinct

diverticula, and including modifications from saccate condition. Seen in the Cyclophyllidea (including, reticulate, paruterine organs, capsules).

3. UTERINE PORE (structure). 4 states. 0 = permanent, ventral; 1 = dehiscence, ventral; 2 = dehiscence, dorsal; 3 = absent.

4. UTERUS (position). 2 states. 0 = ventral; 1 = dorsal.

5. GENITAL PORE (position). 3 states. 0 = marginal and separate; 1 = median; 2 = marginal and single.

6. GENITAL PORES (position and fusion). 3 states. 0 = male and female pores separate; 1 = separate or fused but opening in common atrium; 2 = fused.

7. ANTERIOR HOLDFAST (orientation and structure in adult). 3 states. 0 = apical sucker, single; 1 = bilateral, difossate; 2 = bilateral, tetrafossate.

8. ANTERIOR HOLDFAST (adult, structure of adhesive organs). 4 states. ? = not applicable as character; 0 = bothria; 1 = bothridia only; 2 = bothridia + suckers; 3 = suckers only; 4 = bothridia-like acetabulae with auricular appendages.

9. APICAL HOLDFAST (myzorhynchus). 2 states. 0 = absent; 1 = present.
10. ROSTELLUM. 2 states. 0 = absent; 1 = present and either non-retractable or retractable.
11. BOTHRIDIAL MARGINS (adult scolex). 3 states. 0 = absent; 1 = flimsy; 2 = rigid.
12. SCOLEX TENTACLES. 2 states. 0 = absent; 1 = present.
13. SCOLEX HOOKS (adult). 3 states. 0 = absent; 1 = present, on bothridia; 2 = present, on region anterior to the bothridia or bothria.
14. APICAL HOOKS (adult). 2 states. 0 = absent; 1 = spines, or hooks present.
15. HOOKS (structure of true hooks with base, handle and blade). 4 states. 0 = absent; 1 = 'Diphyllidean' spines and prongs; 2 = 'Tetraphyllidean/Trypanorhynchan', prong-like; 3 = taenioid.
16. SCOLEX (ontogeny of adult form). 2 states. 0 = adult form in intermediate host with limited differentiation in definitive host; 1 = adult form in definitive host only.
17. SCOLEX (structure at base). 2 states. 0 = lacking pedunculate form; 1 = pedunculate.
18. LONGITUDINAL MUSCULATURE (adult). 2 states. 0 = weakly developed; 1 = well developed, with clear demarcation of the cortical and medullary regions.
19. EGG DEVELOPMENT. 2 states. 0 = polylecithal; 1 = oligolecithal.
20. EGG (operculum). 2 states. 0 = present; 1 = absent.
21. EGG (quinone tanning). 2 states. 0 = present; 1 = absent.
22. EMBRYONIC MEMBRANES. 2 states. 0 = single membrane formed by embryo; 1 = 2 membranes with embryophore.
23. PROTONEPHRIDIUM (in hexacanth). 2 states. 0 = present; 1 = absent.
24. EMBRYOPHORE. 3 states. 0 = ciliated, developed as a coracidium; 1 = nucleated, non-rigid; 2 = rigid.
25. EMBRYOGENESIS. 2 states. 0 = not delayed; 1 = delayed.
26. HEXACANTH. 2 states. 0 = hexacanth oncosphere absent; 1 = present.
27. OVARY (structure in transverse section). 2 states. 0 = flat, lobate, globular; 1 = 'X'-shaped.
28. VITELLARIUM (position). 3 states. 0 = cortical; 1 = medullary; 2 = paramuscular.
29. VITELLARIUM (distribution, and form). 3 states. 0 = lateral, follicular; 1 = median, preovarian, globular; 2 = median, postovarian, globular.
30. VITELLARIUM (structure and disposition of ducts). 3 states. 0 = follicular, with multiple ducts; 1 = bilobed, globular, connected by two ducts, narrow isthmus and reservoir; 2 = compact, globular, with single duct.

31. STROBILA (external segmentation). 2 states. 0 = absent (monozoic); 1 = present (polyzoic).
32. STROBILA (internal segmentation). 2 states. 0 = absent; 1 = present.
33. STROBILA (segment structure). 3 states. 0 = monozoic; 1 = acraspedote; 2 = craspedote.
34. TESTES (position). 2 states. 0 = cortical; 1 = medullary.
35. CERCOID (ontogeny and structure). 2 states. 0 = larval scolex not invaginated (e.g. procercoids with bothria); 1 = larval scolex invaginated (cysticercoid, cysticercus, some cercoscolex).
36. CERCOID (ontogeny). 2 states. 0 = cercomer not forming cyst (acystic); 1 = cercomer forms cyst (cystic).
37. METACESTODE (apical). 2 states. 0 = apical sucker present; 1 = apical sucker absent.
38. METACERCOID (form). 6 states. 0 = absent; 1 = plerocercoid (bothriate); 2 = plerocercus (acetabulate type); 3 = caudate adults; 4 = neoplerocercus; 5 = acetabulo- or glandulo-procercoid.
39. SPERM (mitochondrion). 2 states. 0 = present; 1 = absent.
40. SPERM (crested-body). 2 states. 0 = absent; 1 = present.
41. SPERM (peripheral microtubules). 2 states. 0 = parallel; 1 = twisted.
42. SPERMIOGENESIS. 4 states. 0 = flagellar rotation present, with 2 axonemes; 1 = Caryophyllidean type, single axoneme, with flagellar rotation and proximodistal fusion; 2 = single axoneme, flagellar rotation absent, with proximodistal fusion; 3 = single axoneme, without flagellar rotation or proximodistal fusion.
43. SPERM (periaxonemal sheath). 2 states. 0 = absent; 1 = present.
44. SPERM (formation and number of axonemes). 3 states. 0 = 2 axonemes; 1 = 1 axoneme, one free flagellum and flagellar bud + rootlet; 2 = 1 axoneme lacking rootlet structure.
45. ONTOGENY OF SCOLEX. 2 states. 0 = lacking heterochrony; 1 = with sequential heterochrony.
46. APICAL ORGAN (in adult). 3 states. 0 = absent; 1 = present as pit-like glandular structure at apex of scolex; 2 = present as a muscular cushion.
47. SPERM (striated root associated with each centriole). 2 states. 0 = present; 1 = absent.
48. SPERM (intercentriolar body). 2 states. 0 = present; 1 = absent.
49. SPERM (peripheral microtubules). 2 states. 0 = single arrangement of microtubules; 1 = dual arrangement of microtubules.
50. MICROTRICHES (scolex). 2 states. 0 = palmate structure absent; 1 = palmate structure present.
51. LARVAL CYST (structure). 2 states. 0 = lacking well-defined cyst; 1 = with well-defined cyst.

'Consensus' or reduced matrix for morphological characters (characters 1–51*).
New data represented here and for Appendix 12.2: 1) Pseudophyllidea are represented by constituent taxa as follows: Diphyllobothriidae and Bothriocephalidae (including morphological data for Triaenophoridae, Echinophallidae, Philobythiidae and Cephalochlamydidae); 2) Tetraphyllidea are represented by constituent taxa as follows: Onchobothriidae, Phyllobothriidae, and Litobothriidae ('minor families', Disculicipitidae, Prosobothriidae, Dioecotaeniidae, Chimaerocestidae, and Cathetocephalidae are not represented); 3) characters. 39–44 and 47–48 (sperm and spermiogenesis) for *Mesocestoides* based on Miquel *et al.* (1998) and Lecanicephalidea based on J.-L. Justine and L. Euzet (personal communication 1999); 4) all other characters for Mesocestoidata, which was originally subsumed within Cyclophyllidea (Hoberg *et al.* 1997c, 1999a); 5) character 47, striated root, was originally coded incorrectly for the Tetrabothriidea in Hoberg *et al.* (1997c). Codes: [n] = polymorphism; '?' = missing/unknown/inapplicable.

	1	2	3	4	5	6	7	8	9	10	11	12	13	14	15	16	17	18	19	20	21	22	23	24	25
Monogenea	0	0	0	0	0	0	0	?	0	0	0	0	0	0	0	0	0	0	0	0	0	0	0	0	0
Amphilinidea	0	0	0	0	0	0	0	?	0	0	0	0	0	0	0	0	0	0	1	0	0	0	0	0	0
Caryophyllidea	0	0	0	0	1	[01]	1	0	0	0	0	0	0	0	0	0	0	0	0	0	0	0	0	[01]	0
Spathebothriidea	0	0	0	0	1	0	1	?	0	0	0	0	0	0	0	0	0	0	0	0	0	0	0	1	0
Diphyllobothriidae	0	0	0	0	1	[01]	1	0	0	0	0	0	0	0	0	0	0	0	0	0	0	0	0	0	1
Bothriocephalidae	0	1	0	0	[12]	1	1	0	0	0	0	0	[02]	[01]	0	0	0	0	0	[01]	0	0	0	[01]	1
Haplobothriidea	0	1	1	0	1	1	1	0	0	0	0	1	0	0	0	0	0	0	0	0	0	0	0	0	0
Diphyllidea	1	1	3	0	1	1	1	1	0	0	1	0	2	1	1	0	1	0	0	1	?	?	?	1	0
Trypanorhyncha	1	2	[13]	0	2	2	2	1	0	0	2	1	0	0	2	0	1	0	0	[01]	0	0	1	0	1
Onchobothriidae	1	2	1	0	2	2	2	2	0	0	2	0	1	0	2	1	1	0	1	1	1	1	1	1	0
Phyllobothriidae	1	2	1	0	2	2	2	[12]	1	0	[12]	0	0	0	0	0	0	1	1	1	1	1	1	1	0
Litobothriidae	1	2	?	0	2	1	0	?	0	0	0	0	0	0	0	?	0	0	1	1	?	?	?	?	?
Lecanicephalidea	1	2	1	0	2	2	2	3	1	0	0	0	0	[01]	0	0	0	0	1	1	1	1	1	1	0
Proteocephalidea	2	3	1	0	2	2	2	3	0	[01]	0	0	0	[01]	[03]	0	0	[01]	1	1	1	1	1	[12]	[01]
Nippotaeniidea	1	2	3	0	2	2	2	0	0	0	0	0	0	0	0	0	1	1	1	1	1	1	1	1	0
Tetrabothriidea	2	2	2	1	2	2	2	4	0	0	0	0	0	0	1	0	1	1	1	1	1	1	1	1	0
Cyclophyllidea	3	3	3	[01]	2	2	2	3	0	[01]	0	0	0	[01]	[03]	[01]	0	1	1	1	1	1	1	2	0
Mesocestoidata	[13]	3	3	0	1	2	2	3	0	0	0	0	0	0	0	0	0	1	1	1	1	1	1	2	0

	26	27	28	29	30	31	32	33	34	35	36	37	38	39	40	41	42	43	44	45	46	47	48	49	50	51
Monogenea	0	0	0	0	0	0	0	0	0	0	0	0	0	0	0	0	0	0	0	0	0	0	0	0	0	0
Amphilinidea	0	0	0	0	0	0	0	0	0	0	0	0	0	0	0	0	0	0	0	0	0	0	0	0	0	0
Caryophyllidea	1	0	[01]	0	0	0	0	0	0	0	0	1	3	1	1	0	1	0	1	0	0	0	0	0	0	0
Spathebothriidea	1	0	0	0	0	0	1	0	0	0	0	1	3	1	?	0	?	0	?	0	0	?	?	?	0	0
Diphyllobothriidae	1	0	0	0	0	1	1	1	1	0	0	1	1	1	1	0	0	0	0	0	0	0	0	0	0	0
Bothriocephalidae	1	0	[01]	[02]	[02]	1	1	[12]	1	0	0	1	1	1	1	0	0	0	0	0	0	0	0	0	0	0
Haplobothriidea	1	0	1	0	0	1	1	1	1	0	0	1	1	1	1	0	0	0	0	0	0	0	0	0	0	0
Diphyllidea	1	0	0	0	0	1	1	1	1	0	?	?	?	1	1	0	[01]	0	[01]	0	0	0	0	0	1	1
Trypanorhyncha	1	[01]	0	0	0	1	1	[12]	1	0	0	1	4	1	1	0	0	?	0	0	0	0	0	0	1	1
Onchobothriidae	1	[01]	1	0	0	1	1	1	1	0	0	0	5	1	1	0	0	0	0	1	1	0	0	1	0	0
Phyllobothriidae	1	[01]	1	0	0	1	1	[12]	1	0	0	1	5	1	1	0	1	0	1	0	1	0	0	1	0	0
Litobothriidae	1	[01]	1	0	0	1	1	2	1	?	?	?	?	?	?	?	?	?	1	?	?	0	?	?	0	?
Lecanicephalidea	1	0	1	0	0	1	1	1	1	0	0	1	5	1	1	0	1	0	1	0	0	0	0	1	0	0
Proteocephalidea	1	0	[012]	0	0	1	1	[12]	[01]	[01]	0	[01]	2	1	1	[01]	[01]	0	[01]	0	[01]	0	[01]	0	0	0
Nippotaeniidea	1	0	1	1	1	1	1	1	1	0	0	0	2	1	?	0	?	0	1	0	0	?	?	?	0	0
Tetrabothriidea	1	0	1	1	[02]	1	1	2	1	0	0	0	5	1	1	1	1	1	1	1	0	1	0	0	0	0
Cyclophyllidea	1	0	1	2	2	1	1	2	1	1	[01]	[01]	2	1	1	1	[23]	1	2	[01]	0	1	1	0	0	0
Mesocestoidata	1	0	1	2	2	1	1	2	1	1	0	1	2	1	1	0	1	?	1	0	1	0	0	0	0	0

Appendix 12.2. Morphological characters for analysis using generic and species level taxa as terminals, and the foundation for the 'comprehensive' matrix. Character descriptions, justification and polarity, and numbering are consistent with: 1) higher-level analyses among the Eucestoda (Hoberg *et al.* 1997c, 1999b); 2) Cyclophyllidea (Hoberg *et al.* 1999a); 3) Proteocephalidea (Rego *et al.* 1998); and 4) Pseudophyllidea (Bray *et al.* 1999).

Order Cyclophyllidea. 35 characters (characters 1154–1188 for comprehensive matrix). Corresponding to Hoberg *et al.* (1999a).

1. STROBILA (shape). 2 states. 0 = flattened, ribbon-like; 1 = cylindrical.
2. STROBILA (sexual development of proglottid). 2 states. 0 = hermaphroditic; 1 = dioecious.
3. STROBILA (segmentation). 2 states. 0 = strongly defined; 1 = absent or poorly defined.
4. STROBILA (segment structure). 2 states. 0 = acraspedote; 1 = strongly craspedote.
5. GENITAL PORE (position). 2 states. 0 = marginal; 1 = ventral, median.
6. ACCESSORY REPRODUCTIVE DUCTS. 2 states. 0 = absent; 1 = present.
7. TESTES (number). 2 states. 0 = numerous; 1 = generally not > 2 or 3.
8. SEMINAL VESICLE (internal). 2 states. 0 = absent; 1 = present.
9. SEMINAL VESICLE (external). 2 states. 0 = absent; 1 = present.
10. VAGINA. 2 states. 0 = present; 1 = absent.
11. GENITALIA. 2 states. 0 = single; 1 = double.
12. PROGLOTTID (ontogeny). 2 states. 0 = protandrous; 1 = proterogynous.
13. OVARY (position and form). 2 states. 0 = positioned in far posterior of proglottid, strongly bilobed and compact; 1 = not positioned in far posterior, and usually multilobate.
14. UTERUS (position relative to female organs). 0 = ventral; 1 = dorsal.
15. UTERUS (structure). 6 states. 0 = longitudinal sac with diverticula; 1 = tubular; 2 = initially transverse tubular, later saccate expanding to fill segment; 3 = saccate; 4 = reticulate; 5 = labyrinthine.
16. UTERUS (persistence). 2 states. 0 = persistent; 1 = ephemeral or transient.
17. PARUTERINE ORGAN. 2 states. 0 = absent; 1 = present.
18. EGGS (structure, striated embryophore). 2 states. 0 = embryophore not striated; 1 = embryophore striated.
19. EGGS (pyriform apparatus). 2 states. 0 = absent; 1 = present.
20. EMBRYOPHORE (polar processes). 2 states. 0 = absent; 1 = present.
21. EGGS (thin-walled capsules). 2 states. 0 = absent; 1 = present.
22. EGGS (fibrous capsules). 2 states. 0 = absent; 1 = present.
23. ROSTELLUM (structure). 5 states. 0 = absent; 1 = sucker-like; 2 = taeniid; 3 = sac-like; 4 = davaineid.
24. ROSTELLAR SHEATH. 2 states. 0 = absent; 1 = present.
25. ROSTELLAR HOOKS (presence). 2 states. 0 = absent; 1 = present.
26. ROSTELLAR HOOKS (number of rows). 4 states. 0 = absent; 1 = 1 row; 2 = 2 rows; 3 > 2 rows.
27. ROSTELLAR HOOKS (hammer form). 2 states. 0 = hammer-shaped hooks absent; 1 = hammer-shaped hooks present.
28. ROSTELLAR HOOKS (epiphyseal form). 2 states. 0 = hooks lacking well-demarcated epiphysis between handle and blade; 1 = hooks with epiphysis.

29. SUCKERS (armature). 2 states. 0 = consistently absent; 1 = present.
30. SCOLEX (structure during ontogeny). 3 states. 0 = retracted; 1 = invaginated; 2 = neither retracted nor invaginated.
31. LIFE CYCLE (number of hosts). 2 states. 0 = 3 hosts; 1 = 2 hosts.
32. PRIMARY LACUNA (larval ontogeny). 2 states. 0 = absent; 1 = present.
33. CERCOMER (larval ontogeny). 2 states. 0 = present (caudate); 1 = absent (acaudate).
34. METACESTODE (apical structure). 2 states. 0 = apical sucker present during development; 1 = apical sucker always absent.
35. METACESTODE ('cuticular hairs'). 2 states. 0 = metacestode with prominent 'cuticular hairs' (hair-like microtriches?); 1 = fibrous layer surrounding cysticercoid.

Order Proteocephalidea. 16 characters (characters 1189–1204 for comprehensive matrix). Corresponding to Rego *et al.* (1998).

3. OVARY (position). 3 states. 0 = medullary; 1 = initially medullary, developing cortically; 2 = cortical.
5. TESTICULAR FIELDS (position). 3 states. 0 = single field; 1 = two fields confluent in anterior; 2 = two distinctly separate fields.
6. UTERUS (position). 3 states. 0 = medullary; 1 = cortical; 2 = initially cortical developing medullary.
8. EGG (structure). 3 states. 0 = spherical to oval, external hyaline membrane present; 1 = internal polar circle-like structures present; 2 = polar filaments present.
10. VAGINAL SPHINCTER. 2 states. 0 = present; 1 = absent.
11. VAGINA (position relative to cirrus sac). 3 states. 0 = anterior; 1 = posterior; 2 = alternating.
12. GENITAL PORE. 2 states. 0 = alternating irregularly; 1 = unilateral.
15. TEGUMENTAL WRINKLES (transverse). 2 states. 0 = absent; 1 = present.
16. METASCOLEX. 2 states. 0 = absent; 1 = present.
18. SUCKERS. 2 states. 0 = spherical; 1 = alternative structures.
19. SUCKERS (appendages, auricular, papilla-like). 2 states. 0 = absent; 1 = present.
20. SUCKERS (distal sphincter). 2 states. 0 = present; 1 = absent.
23. LONGITUDINAL MUSCULATURE (arrangement). 2 states. 0 = isolated fibres; 1 = fibres organized in discrete bundles.
24. CIRRUS (spination). 2 states. 0 = present; 1 = absent.
25. SUCKERS (spination). 2 states. 0 = absent; 1 = present.
27. TEGUMENTAL WRINKLES (longitudinal). 2 states. 0 = absent; 1 = present.

Appendix 12.2 Continued.

Order Pseudophyllidea including Bothriocephalidae, Triaenophoridae and Diphyllobothriidae.

11 characters (characters 1205–1215 for comprehensive matrix). Corresponding to Bray *et al.* (1999). Numbers correspond to those for characters in original papers. Polarity is discussed and defined in Rego *et al.* (1998), Bray *et al.* (1999), and Hoberg *et al.* (1997c, 1999a,b).

3.   SEGMENTATION (external). 3 states. 0 = absent; 1 = incomplete; 2 = present.
7.   GENITAL PORE. 3 states. 0 = single; 1 = double; 2 = multiple.
8.   GENITAL PORE (when median). 2 states. 0 = dorsal; 1 = ventral.
15.   BOTHRIA (shape). 2 states. 0 = elongate slit; 1 = tiny slit.
21.   APICAL DISC. 2 states. 0 = absent; 1 = present.
23.   APICAL DISC (armature). 2 states. 0 = unarmed; 1 = armed.

27.   OSMOREGULATORY CANALS. 2 states. 0 = four; 1 = many.
28.   OSMOREGULATORY CANALS. 2 states. 0 = medullary; 1 = cortical.
37.   VAGINA (sphincter). 2 states. 0 = absent; 1 = present.
39.   UTERUS (shape, structure). 4 states. 0 = sigmoid, sinuous; 1 = saccular; 2 = diverticulate; 3 = Y-shaped.
40.   UTERINE PORE. 2 states. 0 = median; 1 = submedian.

'Comprehensive' matrix for morphological characters. Characters 1–113. Complete listing of species-level taxa is found in Mariaux (1998). Uniformity in characters for spermatozoa and spermiogenesis is assumed within families; data for spermatozoa and spermiogenesis in Nematotaeniidae (*Distoichometra*) are for *Nematotaenia chantale* summarized in Miquel *et al.* (1999). Codes: [n] = polymorphism; '?' = missing/unknown; '9' = inapplicable character.

	1	2	3	4	5	6	7	8	9	10	11	12	13	14	15	16	17	18	19	20	21	22	23	24	25
Gyrodactylus	0	0	0	0	0	0	0	?	0	0	0	0	0	0	0	0	0	0	0	0	0	0	0	0	0
Schizochoerus	0	0	0	0	0	0	?	0	0	0	0	0	0	0	0	0	0	0	1	0	0	0	0	0
Caryophyllaeus	0	0	0	0	1	[01]	1	0	0	0	0	0	0	0	0	0	0	0	0	0	0	0	0	[01]	0
Cyathocephalus	0	0	0	0	1	0	1	?	0	0	0	0	0	0	0	0	0	0	0	0	0	0	0	1	0
Molluscotaenia	3	3	3	0	2	2	2	3	0	1	0	0	0	1	3	0	0	1	1	1	1	1	1	2	0
Hepatocestus	3	3	3	0	2	2	2	3	0	1	0	0	0	1	3	0	0	1	1	1	1	1	1	2	0
Neogryporhynchus	3	3	3	0	2	2	2	3	0	1	0	0	0	1	3	0	0	1	1	1	1	1	1	2	0
Paricterotaenia	3	3	3	0	2	2	2	3	0	1	0	0	0	1	3	0	0	1	1	1	1	1	1	2	0
Trichocephaloidis	3	3	3	0	2	2	2	3	0	1	0	0	0	1	3	0	0	1	1	1	1	1	1	2	0
Wardium	3	3	3	0	2	2	2	3	0	1	0	0	0	1	3	0	0	1	1	1	1	1	1	2	0
Vigisolepis	3	3	3	0	2	2	2	3	0	1	0	0	0	1	3	0	0	1	1	1	1	1	1	2	0
Vampirolepis	3	3	3	0	2	2	2	3	0	1	0	0	0	1	3	0	0	1	1	1	1	1	1	2	0
Diplopostheb	3	3	3	0	2	2	2	3	0	1	0	0	0	1	3	0	0	1	1	1	1	1	1	2	0
Diploposthel	3	3	3	0	2	2	2	3	0	1	0	0	0	1	3	0	0	1	1	1	1	1	1	2	0
Diorchis	3	3	3	0	2	2	2	3	0	1	0	0	0	1	3	0	0	1	1	1	1	1	1	2	0
Passerilepis	3	3	3	0	2	2	2	3	0	1	0	0	0	1	3	0	0	1	1	1	1	1	1	2	0
Rodentolepis	3	3	3	0	2	2	2	3	0	1	0	0	0	1	3	0	0	1	1	1	1	1	1	2	0
Staphylocystis	3	3	3	0	2	2	2	3	0	1	0	0	0	1	3	0	0	1	1	1	1	1	1	2	0
Anoplocephaloides	3	3	3	1	2	2	2	3	0	0	0	0	0	0	0	0	0	1	1	1	1	1	1	2	0
Catenotaenia	3	3	3	0	2	2	2	3	0	0	0	0	0	0	1	0	0	1	1	1	1	1	1	2	0
Tatria	3	3	3	0	2	2	2	3	0	1	0	0	0	1	3	0	0	1	1	1	1	1	1	2	0
Raillietina	3	3	3	0	2	2	2	3	0	1	0	0	0	1	3	0	0	1	1	1	1	1	1	2	0
Ophryocotyle	3	3	3	0	2	2	2	3	0	1	0	0	0	1	3	0	0	1	1	1	1	1	1	2	0
Skrjabinoporus	3	3	3	1	2	2	2	3	0	1	0	0	0	1	3	0	0	1	1	1	1	1	1	2	0
Mesocestoides	1	3	3	0	1	2	2	3	0	0	0	0	0	0	0	0	0	1	1	1	1	1	1	2	0
Distoichometra	3	3	3	?	2	2	2	3	0	0	0	0	0	0	0	0	0	1	1	1	1	1	1	2	0
Choanoscolex	2	3	1	0	2	2	2	3	0	0	0	0	0	0	0	0	0	1	1	1	1	1	1	1	1
Spatuliferm	2	3	1	0	2	2	2	3	0	0	0	0	0	0	0	0	0	1	1	1	1	1	1	?	1
Nomimoscolex	2	3	1	0	2	2	2	3	0	0	0	0	0	0	0	0	0	1	1	1	1	1	1	1	1
Spatulifer	2	3	1	0	2	2	2	3	0	0	0	0	0	0	0	0	0	1	1	1	1	1	1	?	0
Proteocephalusp	2	3	1	0	2	2	2	3	0	0	0	0	0	0	0	0	0	1	1	1	1	1	1	1	0
Proteocephaluss	2	3	1	0	2	2	2	3	0	0	0	0	0	0	0	0	0	1	1	1	1	1	1	1	0
Ophiotaenia	2	3	1	0	2	2	2	3	0	0	0	0	0	0	0	0	0	1	1	1	1	1	1	2	0
Eubothrium	0	1	0	0	2	2	1	0	0	0	0	0	0	0	0	0	0	0	0	0	0	0	0	0	1
Triaenophorus	0	1	0	0	2	2	1	0	0	0	0	0	2	0	0	0	0	0	0	0	0	0	0	0	1
Abothrium	0	1	0	0	2	2	1	0	0	0	0	0	0	0	0	0	0	0	0	0	0	0	0	1	1
Bothriocephalus	0	1	0	0	1	2	1	0	0	0	0	0	0	0	0	0	0	0	0	0	0	0	0	0	1
Schistocephalus	0	0	0	0	1	2	1	0	0	0	0	0	0	0	0	0	0	0	0	0	0	0	0	0	1
Diphyllobothrium	0	0	0	0	1	2	1	0	0	0	0	0	0	0	0	0	0	0	0	0	0	0	0	0	1
Tetrabothrius	2	2	2	1	2	2	2	4	0	0	0	0	0	0	1	0	1	1	1	1	1	1	1	1	0
Echeneibothrium	1	2	1	0	2	2	2	1	1	0	1	0	0	0	0	0	0	1	1	1	1	1	1	1	0
Phyllobothrium	1	2	1	0	2	2	2	0	0	0	1	0	0	0	0	0	0	1	1	1	1	1	1	1	0
Calliobothrium	1	2	1	0	2	2	2	0	0	2	0	1	0	2	1	1	0	1	1	1	1	1	1	1	0
Litobothrium	1	2	?	0	2	1	0	9	0	0	0	0	0	0	?	0	0	1	1	?	?	?	?	?	?
Echinobothrium	1	1	3	0	1	1	1	1	0	0	1	0	2	1	1	0	0	0	1	?	?	?	1	0	
Gilquinia	1	2	3	0	2	2	1	0	0	2	1	0	0	2	0	1	0	0	[01]	0	0	1	1	1	
Grillotia	1	2	1	0	2	2	1	0	0	2	1	0	0	2	0	1	0	0	[01]	0	0	1	0	1	
Amurotaenia	1	2	3	0	2	2	0	9	0	0	0	0	0	0	0	0	0	1	1	1	1	1	1	0	
Eniochobothrium	1	2	1	0	2	2	2	3	1	0	0	0	0	[01]	0	0	0	1	1	1	1	1	1	0	
Haplobothrium	0	1	1	0	1	1	1	0	0	0	0	1	0	0	0	0	0	0	0	0	0	0	0	0	

	26	27	28	29	30	31	32	33	34	35	36	37	38	39	40	41	42	43	44	45	46	47	48	49	50
Gyrodactylus	0	0	0	0	0	0	0	0	0	0	0	0	0	0	0	0	0	0	0	0	0	0	0	0	0
Schizochoerus	0	0	0	0	0	0	0	0	0	0	0	0	0	0	0	0	0	0	0	0	0	0	0	0	0
Caryophyllaeus	1	0	[01]	0	0	0	0	0	0	0	0	1	3	1	1	0	1	0	1	0	0	0	0	0	0
Cyathocephalus	1	0	0	0	0	0	1	0	0	0	0	1	3	1	?	0	?	0	?	0	0	?	?	?	0
Molluscotaenia	1	0	1	2	2	1	1	2	1	1	0	1	2	?	?	?	?	?	?	0	0	?	?	?	0
Hepatocestus	1	0	1	2	2	1	1	2	1	1	0	1	2	?	?	?	?	?	?	0	0	?	?	?	0
Neogryporhynchus	1	0	1	2	2	1	1	2	1	1	0	1	2	?	?	?	?	?	?	0	0	?	?	?	0
Paricterotaenia	1	0	1	2	2	1	1	2	1	1	0	1	2	?	?	?	?	?	?	0	0	?	?	?	0
Trichocephaloidis	1	0	1	2	2	1	1	2	1	1	0	1	2	?	?	?	?	?	?	0	0	?	?	?	0
Wardium	1	0	1	2	2	1	1	2	1	1	1	1	2	1	1	1	3	0	2	0	0	1	1	0	0
Vigisolepis	1	0	1	2	2	1	1	2	1	1	1	1	2	1	1	1	3	0	2	0	0	1	1	0	0
Vampirolepis	1	0	1	2	2	1	1	2	1	1	1	1	2	1	1	1	3	0	2	0	0	1	1	0	0
Diplopostheb	1	0	1	2	2	1	1	2	1	1	1	1	2	1	1	1	3	0	2	0	0	1	1	0	0
Diploposthel	1	0	1	2	2	1	1	2	1	1	1	1	2	1	1	1	3	0	2	0	0	1	1	0	0
Diorchis	1	0	1	2	2	1	1	2	1	1	1	1	2	1	1	1	3	0	2	0	0	1	1	0	0
Passerilepis	1	0	1	2	2	1	1	2	1	1	1	1	2	1	1	1	3	0	2	0	0	1	1	0	0
Rodentolepis	1	0	1	2	2	1	1	2	1	1	1	1	2	1	1	1	3	0	2	0	0	1	1	0	0
Staphylocystis	1	0	1	2	2	1	1	2	1	1	1	1	2	1	1	1	3	0	2	0	0	1	1	0	0
Anoplocephaloides	1	0	1	2	2	1	1	2	1	1	0	1	2	1	1	1	3	1	2	0	0	1	1	0	0
Catenotaenia	1	0	1	2	2	1	1	1	1	1	0	0	2	1	1	1	2	1	2	1	0	1	?	0	0
Tatria	1	0	1	2	2	1	1	2	1	1	1	1	2	?	?	?	?	?	?	0	0	?	?	?	0
Raillietina	1	0	1	2	2	1	1	2	1	1	1	1	2	1	1	1	2	1	2	0	0	1	1	0	0
Ophryocotyle	1	0	1	2	2	1	1	2	1	1	1	1	2	1	1	1	2	1	2	0	0	1	1	0	0
Skrjabinoporus	1	0	1	2	2	1	1	2	1	1	0	1	2	?	?	?	?	?	?	0	0	?	?	?	0
Mesocestoides	1	0	1	2	2	1	1	2	1	1	0	1	2	1	1	0	1	?	1	0	0	0	0	0	0
Distoichometra	1	0	1	2	2	1	1	1	1	1	0	1	2	1	1	1	2	?	2	0	0	1	1	0	0
Choanoscolex	1	0	0	0	0	1	1	1	0	?	0	?	2	1	1	?	?	0	?	0	0	0	?	0	0
Spatuliferm	1	0	0	0	0	1	1	1	0	0	0	?	2	1	1	?	?	0	?	0	0	0	?	0	0
Nomimoscolex	1	0	0	0	0	1	1	2	1	?	0	?	2	1	1	0	0	0	0	0	1	0	0	0	0
Spatulifer	1	0	0	0	0	1	1	1	1	0	0	?	2	1	1	?	?	0	?	0	0	0	?	0	0
Proteocephalusp	1	0	2	0	0	1	1	1	1	[01]	0	[01]	2	1	1	0	0	0	0	0	[012]	0	0	0	0
Proteocephaluss	1	0	2	0	0	1	1	1	1	[01]	0	[01]	2	1	1	0	0	0	0	0	[012]	0	0	0	0
Ophiotaenia	1	0	2	0	0	1	1	1	1	1	0	0	2	1	1	?	?	0	?	0	[012]	0	?	0	0
Eubothrium	1	0	0	0	0	1	1	1	1	0	0	1	1	1	1	1	0	0	0	0	0	0	0	0	0
Triaenophorus	1	0	0	0	0	1	1	1	1	0	0	1	1	1	1	1	0	0	0	0	0	0	0	0	0
Abothrium	1	0	0	0	0	1	1	1	1	0	0	1	1	1	1	1	0	0	0	0	0	0	0	0	0
Bothriocephalus	1	0	0	0	0	1	1	1	1	0	0	1	1	1	1	1	0	0	0	0	0	0	0	0	0
Schistocephalus	1	0	0	0	0	1	1	2	1	0	0	1	1	1	1	1	0	0	0	0	0	0	0	0	0
Diphyllobothrium	1	0	0	0	0	1	1	2	1	0	0	1	1	1	1	1	0	0	0	0	0	0	0	0	0
Tetrabothrius	1	0	1	1	2	1	1	2	1	0	0	0	5	1	1	1	1	1	1	1	1	0	1	0	0
Echeneibothrium	1	1	1	0	0	1	1	[12]	1	0	0	[01]	5	1	1	0	1	0	1	0	1	0	0	1	0
Phyllobothrium	1	[01]	1	0	0	1	1	[12]	1	0	0	[01]	5	1	1	0	1	0	1	0	1	0	0	1	0
Calliobothrium	1	0	1	0	0	1	1	2	1	0	0	0	5	1	1	0	0	0	0	1	1	0	0	1	0
Litobothrium	1	[01]	1	0	0	1	1	2	1	?	?	?	?	?	?	?	?	?	?	0	?	?	?	?	0
Echinobothrium	1	0	0	0	0	1	1	1	1	0	?	1	?	1	1	0	[01]	0	[01]	0	0	0	0	0	1
Gilquinia	1	0	0	0	0	1	1	1	1	0	0	1	4	1	?	0	0	?	0	0	0	0	0	0	1
Grillotia	1	0	0	0	0	1	1	1	1	0	0	1	4	1	?	0	0	?	0	0	0	0	0	0	1
Amurotaenia	1	0	1	1	1	1	1	1	1	0	0	0	2	1	?	0	?	0	1	0	0	?	?	?	0
Eniochobothrium	1	0	1	0	0	1	1	1	1	0	0	1	5	1	1	0	1	0	1	0	0	0	0	1	0
Haplobothrium	1	0	1	0	0	1	1	1	1	0	0	1	1	1	1	?	0	0	0	0	0	0	0	0	0

	51	52	53	54	55	56	57	58	59	60	61	62	63	64	65	66	67	68	69	70	71	72	73	74	75
Gyrodactylus	0	0	0	9	9	0	0	9	0	0	0	0	9	9	0	9	0	9	9	9	9	9	9	9	9
Schizochoerus	0	0	0	9	9	0	0	9	0	0	0	0	9	9	0	9	0	9	9	9	9	9	9	9	9
Caryophyllaeus	0	9	9	9	9	9	9	9	9	9	9	9	9	9	9	9	9	9	9	9	9	9	9	9	9
Cyathocephalus	0	9	9	9	9	9	9	9	9	9	9	9	9	9	9	9	9	9	9	9	9	9	9	9	9
Molluscotaenia	0	0	0	0	1	0	0	0	0	0	0	0	0	1	0	3	0	0	0	0	0	0	0	3	1
Hepatocestus	0	0	0	0	1	0	0	0	0	0	0	0	0	1	0	5	0	0	0	0	0	0	0	3	1
Neogryporhynchus	0	0	0	0	1	0	0	1	0	0	0	0	0	1	0	3	0	0	0	0	0	0	0	3	0
Paricterotaenia	0	0	0	1	0	0	0	0	0	0	0	0	0	1	0	3	0	0	0	0	0	0	0	3	1
Trichocephaloidis	0	0	0	0	1	0	0	[01]	0	0	0	0	0	1	0	3	0	0	0	0	0	0	0	3	1
Wardium	0	0	0	0	1	0	0	1	1	1	0	0	0	1	0	3	0	0	0	0	[01]	0	0	3	1
Vigisolepis	0	0	0	0	1	0	0	1	1	1	0	0	0	1	0	3	0	0	0	0	0	0	0	3	1
Vampirolepis	0	0	0	0	1	0	0	1	1	1	0	0	0	1	0	2	0	0	0	0	0	0	0	3	1
Diplopostheb	0	0	0	0	1	0	0	1	1	1	0	0	0	1	0	2	0	0	0	0	0	0	0	3	1
Diploposthel	0	0	0	0	1	0	0	1	1	1	0	0	0	1	0	2	0	0	0	0	0	0	0	3	1
Diorchis	0	0	0	0	1	0	0	1	1	1	0	0	0	1	0	3	0	0	0	0	1	0	0	3	1
Passerilepis	0	0	0	0	1	0	0	1	1	1	0	0	0	1	0	3	0	0	0	0	[01]	0	0	3	1
Rodentolepis	0	0	0	0	1	0	0	1	1	1	0	0	0	1	0	5	0	0	0	0	0	0	0	3	1
Staphylocystis	0	0	0	0	1	0	0	1	1	1	0	0	0	1	0	3	0	0	0	0	0	0	0	3	1
Anoplocephaloides	0	0	0	1	0	0	0	0	1	1	0	0	0	1	1	2	0	0	0	1	0	0	0	0	0
Catenotaenia	0	0	0	0	0	0	0	0	0	0	0	0	0	1	0	0	0	0	0	0	0	0	0	0	0
Tatria	0	0	0	0	1	0	1	0	1	1	1	0	0	1	0	3	0	0	0	0	[01]	0	0	3	0
Raillietina	0	0	0	0	1	0	0	0	0	0	0	0	0	1	0	3	1	0	0	0	0	0	1	4	0
Ophryocotyle	0	0	0	0	1	0	0	0	0	0	0	0	0	1	0	3	0	0	0	0	0	0	0	4	0
Skrjabinoporus	0	0	0	0	1	0	0	0	0	0	0	0	0	1	1	3	1	0	0	0	0	0	0	1	0
Mesocestoides	0	0	0	0	0	1	0	0	0	0	0	0	0	0	0	1	1	1	0	0	0	0	0	0	0
Distoichometra	0	1	0	1	0	0	0	1	0	0	0	0	0	1	?	3	1	1	0	0	0	0	0	0	0
Choanoscolex	0	9	9	9	9	9	9	9	9	9	9	9	9	9	9	9	9	9	9	9	9	9	9	9	9
Spatuliferm	0	9	9	9	9	9	9	9	9	9	9	9	9	9	9	9	9	9	9	9	9	9	9	9	9
Nomimoscolex	0	9	9	9	9	9	9	9	9	9	9	9	9	9	9	9	9	9	9	9	9	9	9	9	9
Spatulifer	0	9	9	9	9	9	9	9	9	9	9	9	9	9	9	9	9	9	9	9	9	9	9	9	9
Proteocephalusp	0	9	9	9	9	9	9	9	9	9	9	9	9	9	9	9	9	9	9	9	9	9	9	9	9
Proteocephaluss	0	9	9	9	9	9	9	9	9	9	9	9	9	9	9	9	9	9	9	9	9	9	9	9	9
Ophiotaenia	0	9	9	9	9	9	9	9	9	9	9	9	9	9	9	9	9	9	9	9	9	9	9	9	9
Eubothrium	0	9	9	9	9	9	9	9	9	9	9	9	9	9	9	9	9	9	9	9	9	9	9	9	9
Triaenophorus	0	9	9	9	9	9	9	9	9	9	9	9	9	9	9	9	9	9	9	9	9	9	9	9	9
Abothrium	0	9	9	9	9	9	9	9	9	9	9	9	9	9	9	9	9	9	9	9	9	9	9	9	9
Bothriocephalus	0	9	9	9	9	9	9	9	9	9	9	9	9	9	9	9	9	9	9	9	9	9	9	9	9
Schistocephalus	0	9	9	9	9	9	9	9	9	9	9	9	9	9	9	9	9	9	9	9	9	9	9	9	9
Diphyllobothrium	0	9	9	9	9	9	9	9	9	9	9	9	9	9	9	9	9	9	9	9	9	9	9	9	9
Tetrabothrius	0	9	9	9	9	9	9	9	9	9	9	9	9	9	9	9	9	9	9	9	9	9	9	9	9
Echeneibothrium	0	9	9	9	9	9	9	9	9	9	9	9	9	9	9	9	9	9	9	9	9	9	9	9	9
Phyllobothrium	0	9	9	9	9	9	9	9	9	9	9	9	9	9	9	9	9	9	9	9	9	9	9	9	9
Calliobothrium	0	9	9	9	9	9	9	9	9	9	9	9	9	9	9	9	9	9	9	9	9	9	9	9	9
Litobothrium	?	9	9	9	9	9	9	9	9	9	9	9	9	9	9	9	9	9	9	9	9	9	9	9	9
Echinobothrium	1	9	9	9	9	9	9	9	9	9	9	9	9	9	9	9	9	9	9	9	9	9	9	9	9
Gilquinia	1	9	9	9	9	9	9	9	9	9	9	9	9	9	9	9	9	9	9	9	9	9	9	9	9
Grillotia	1	9	9	9	9	9	9	9	9	9	9	9	9	9	9	9	9	9	9	9	9	9	9	9	9
Amurotaenia	0	9	9	9	9	9	9	9	9	9	9	9	9	9	9	9	9	9	9	9	9	9	9	9	9
Eniochobothrium	0	9	9	9	9	9	9	9	9	9	9	9	9	9	9	9	9	9	9	9	9	9	9	9	9
Haplobothrium	0	9	9	9	9	9	9	9	9	9	9	9	9	9	9	9	9	9	9	9	9	9	9	9	9

	76	77	78	79	80	81	82	83	84	85	86	87	88	89	90	91	92	93	94	95	96	97	98	99	100
Gyrodactylus	9	9	9	9	9	9	9	9	9	9	9	9	9	9	9	9	9	9	9	9	9	9	9	9	9
Schizochoerus	9	9	9	9	9	9	9	9	9	9	9	9	9	9	9	9	9	9	9	9	9	9	9	9	9
Caryophyllaeus	9	9	9	9	9	9	9	9	9	9	9	9	9	9	9	9	9	9	9	9	9	9	9	9	9
Cyathocephalus	9	9	9	9	9	9	9	9	9	9	9	9	9	9	9	9	9	9	9	9	9	9	9	9	9
Molluscotaenia	1	2	0	0	0	0	1	1	1	1	1	9	9	9	9	9	9	9	9	9	9	9	9	9	9
Hepatocestus	1	2	0	0	0	0	1	1	1	1	1	9	9	9	9	9	9	9	9	9	9	9	9	9	9
Neogryporhynchus	1	2	0	0	0	0	0	1	0	1	?	9	9	9	9	9	9	9	9	9	9	9	9	9	9
Paricterotaenia	1	1	0	0	0	0	1	1	1	1	1	9	9	9	9	9	9	9	9	9	9	9	9	9	9
Trichocephaloidis	1	1	0	0	0	0	1	1	1	1	1	9	9	9	9	9	9	9	9	9	9	9	9	9	9
Wardium	1	1	0	0	0	0	1	1	0	1	1	9	9	9	9	9	9	9	9	9	9	9	9	9	9
Vigisolepis	1	1	0	0	0	0	1	1	0	1	1	9	9	9	9	9	9	9	9	9	9	9	9	9	9
Vampirolepis	1	1	0	0	0	0	1	1	0	1	1	9	9	9	9	9	9	9	9	9	9	9	9	9	9
Diplopostheb	1	1	0	0	0	0	1	1	0	1	1	9	9	9	9	9	9	9	9	9	9	9	9	9	9
Diploposthel	1	1	0	0	0	0	1	1	0	1	1	9	9	9	9	9	9	9	9	9	9	9	9	9	9
Diorchis	1	1	0	0	[01]	0	1	1	0	1	1	9	9	9	9	9	9	9	9	9	9	9	9	9	9
Passerilepis	1	1	0	0	0	0	1	1	0	1	1	9	9	9	9	9	9	9	9	9	9	9	9	9	9
Rodentolepis	1	1	0	0	0	0	1	1	0	1	1	9	9	9	9	9	9	9	9	9	9	9	9	9	9
Staphylocystis	1	1	0	0	0	0	1	1	0	1	1	9	9	9	9	9	9	9	9	9	9	9	9	9	9
Anoplocephaloides	0	0	0	0	0	0	1	1	0	1	1	9	9	9	9	9	9	9	9	9	9	9	9	9	9
Catenotaenia	0	0	0	0	0	0	1	1	0	0	0	9	9	9	9	9	9	9	9	9	9	9	9	9	9
Tatria	1	1	0	0	[01]	0	1	1	0	1	1	9	9	9	9	9	9	9	9	9	9	9	9	9	9
Raillietina	1	2	1	0	1	0	1	1	0	1	1	9	9	9	9	9	9	9	9	9	9	9	9	9	9
Ophryocotyle	1	2	1	0	1	?	1	?	?	1	1	9	9	9	9	9	9	9	9	9	9	9	9	9	9
Skrjabinoporus	1	[12]	0	1	0	?	1	?	?	1	0	9	9	9	9	9	9	9	9	9	9	9	9	9	9
Mesocestoides	0	0	0	0	0	0	0	0	0	1	0	9	9	9	9	9	9	9	9	9	9	9	9	9	9
Distoichometra	0	0	0	0	0	?	?	0	1	1	?	9	9	9	9	9	9	9	9	9	9	9	9	9	9
Choanoscolex	9	9	9	9	9	9	9	9	9	9	9	2	0	1	2	0	1	0	1	1	1	0	0	0	1
Spatuliferm	9	9	9	9	9	9	9	9	9	9	9	1	0	0	0	1	2	0	1	1	0	0	0	0	1
Nomimoscolex	9	9	9	9	9	9	9	9	9	9	9	1	2	2	0	0	2	1	1	1	1	1	1	1	0
Spatulifer	9	9	9	9	9	9	9	9	9	9	9	1	0	0	0	1	2	0	1	1	0	0	0	0	1
Proteocephalusp	9	9	9	9	9	9	9	9	9	9	9	0	1	0	[01]	0	2	0	0	0	1	0	0	1	1
Proteocephaluss	9	9	9	9	9	9	9	9	9	9	9	0	1	0	[01]	0	2	0	0	0	1	0	0	1	1
Ophiotaenia	9	9	9	9	9	9	9	9	9	9	9	0	1	0	[01]	0	2	0	0	0	1	0	0	1	1
Eubothrium	9	9	9	9	9	9	9	9	9	9	9	9	9	9	9	9	9	9	9	9	9	9	9	9	9
Triaenophorus	9	9	9	9	9	9	9	9	9	9	9	9	9	9	9	9	9	9	9	9	9	9	9	9	9
Abothrium	9	9	9	9	9	9	9	9	9	9	9	9	9	9	9	9	9	9	9	9	9	9	9	9	9
Bothriocephalus	9	9	9	9	9	9	9	9	9	9	9	9	9	9	9	9	9	9	9	9	9	9	9	9	9
Schistocephalus	9	9	9	9	9	9	9	9	9	9	9	9	9	9	9	9	9	9	9	9	9	9	9	9	9
Diphyllobothrium	9	9	9	9	9	9	9	9	9	9	9	9	9	9	9	9	9	9	9	9	9	9	9	9	9
Tetrabothrius	9	9	9	9	9	9	9	9	9	9	9	9	9	9	9	9	9	9	9	9	9	9	9	9	9
Echeneibothrium	9	9	9	9	9	9	9	9	9	9	9	9	9	9	9	9	9	9	9	9	9	9	9	9	9
Phyllobothrium	9	9	9	9	9	9	9	9	9	9	9	9	9	9	9	9	9	9	9	9	9	9	9	9	9
Calliobothrium	9	9	9	9	9	9	9	9	9	9	9	9	9	9	9	9	9	9	9	9	9	9	9	9	9
Litobothrium	9	9	9	9	9	9	9	9	9	9	9	9	9	9	9	9	9	9	9	9	9	9	9	9	9
Echinobothrium	9	9	9	9	9	9	9	9	9	9	9	9	9	9	9	9	9	9	9	9	9	9	9	9	9
Gilquinia	9	9	9	9	9	9	9	9	9	9	9	9	9	9	9	9	9	9	9	9	9	9	9	9	9
Grillotia	9	9	9	9	9	9	9	9	9	9	9	9	9	9	9	9	9	9	9	9	9	9	9	9	9
Amurotaenia	9	9	9	9	9	9	9	9	9	9	9	9	9	9	9	9	9	9	9	9	9	9	9	9	9
Eniochobothrium	9	9	9	9	9	9	9	9	9	9	9	9	9	9	9	9	9	9	9	9	9	9	9	9	9
Haplobothrium	9	9	9	9	9	9	9	9	9	9	9	9	9	9	9	9	9	9	9	9	9	9	9	9	9

	101	102	103	104	105	106	107	108	109	110	111	112	113
Gyrodactylus	9	9	9	9	9	9	9	9	9	9	9	9	9
Schizochoerus	9	9	9	9	9	9	9	9	9	9	9	9	9
Caryophyllaeus	9	9	9	9	9	9	9	9	9	9	9	9	9
Cyathocephalus	9	9	9	9	9	9	9	9	9	9	9	9	9
Molluscotaenia	9	9	9	9	9	9	9	9	9	9	9	9	9
Hepatocestus	9	9	9	9	9	9	9	9	9	9	9	9	9
Neogryporhynchus	9	9	9	9	9	9	9	9	9	9	9	9	9
Paricterotaenia	9	9	9	9	9	9	9	9	9	9	9	9	9
Trichocephaloidis	9	9	9	9	9	9	9	9	9	9	9	9	9
Wardium	9	9	9	9	9	9	9	9	9	9	9	9	9
Vigisolepis	9	9	9	9	9	9	9	9	9	9	9	9	9
Vampirolepis	9	9	9	9	9	9	9	9	9	9	9	9	9
Diplopostheb	9	9	9	9	9	9	9	9	9	9	9	9	9
Diploposthel	9	9	9	9	9	9	9	9	9	9	9	9	9
Diorchis	9	9	9	9	9	9	9	9	9	9	9	9	9
Passerilepis	9	9	9	9	9	9	9	9	9	9	9	9	9
Rodentolepis	9	9	9	9	9	9	9	9	9	9	9	9	9
Staphylocystis	9	9	9	9	9	9	9	9	9	9	9	9	9
Anoplocephaloides	9	9	9	9	9	9	9	9	9	9	9	9	9
Catenotaenia	9	9	9	9	9	9	9	9	9	9	9	9	9
Tatria	9	9	9	9	9	9	9	9	9	9	9	9	9
Raillietina	9	9	9	9	9	9	9	9	9	9	9	9	9
Ophryocotyle	9	9	9	9	9	9	9	9	9	9	9	9	9
Skrjabinoporus	9	9	9	9	9	9	9	9	9	9	9	9	9
Mesocestoides	9	9	9	9	9	9	9	9	9	9	9	9	9
Distoichometra	9	9	9	9	9	9	9	9	9	9	9	9	9
Choanoscolex	0	1	9	9	9	9	9	9	9	9	9	9	9
Spatuliferm	0	0	9	9	9	9	9	9	9	9	9	9	9
Nomimoscolex	1	1	9	9	9	9	9	9	9	9	9	9	9
Spatulifer	0	0	9	9	9	9	9	9	9	9	9	9	9
Proteocephalusp	1	1	9	9	9	9	9	9	9	9	9	9	9
Proteocephaluss	1	1	9	9	9	9	9	9	9	9	9	9	9
Ophiotaenia	1	1	9	9	9	9	9	9	9	9	9	9	9
Eubothrium	9	9	0	2	9	0	1	1	?	?	1	1	1
Triaenophorus	9	9	0	2	9	0	1	1	?	?	1	1	1
Abothrium	9	9	0	2	9	0	1	1	?	?	1	1	1
Bothriocephalus	9	9	2	0	0	0	1	0	0	0	0	1	0
Schistocephalus	9	9	2	0	1	1	0	9	1	0	0	0	0
Diphyllobothrium	9	9	2	0	1	0	0	9	1	1	1	0	0
Tetrabothrius	9	9	9	9	9	9	9	9	9	9	9	9	9
Echeneibothrium	9	9	9	9	9	9	9	9	9	9	9	9	9
Phyllobothrium	9	9	9	9	9	9	9	9	9	9	9	9	9
Calliobothrium	9	9	9	9	9	9	9	9	9	9	9	9	9
Litobothrium	9	9	9	9	9	9	9	9	9	9	9	9	9
Echinobothrium	9	9	9	9	9	9	9	9	9	9	9	9	9
Gilquinia	9	9	9	9	9	9	9	9	9	9	9	9	9
Grillotia	9	9	9	9	9	9	9	9	9	9	9	9	9
Amurotaenia	9	9	9	9	9	9	9	9	9	9	9	9	9
Eniochobothrium	9	9	9	9	9	9	9	9	9	9	9	9	9
Haplobothrium	9	9	9	9	9	9	9	9	9	9	9	9	9

Chapter 13

Cestode systematics in the molecular era

Jean Mariaux and Peter D. Olson

Both cladistic methodology and molecular data were slow to be implemented in cestode systematics, and have been applied inequitably among the major lineages within the class (Mariaux 1996). With regard to molecular techniques, enzyme-based works were first applied to cestodes around 1980 (e.g., Le Riche and Sewell 1978; Renaud *et al.* 1983) and works based on DNA sequences first appeared in the 1990s (e.g., Bowles *et al.* 1992). In recent years, however, molecular systematics studies on cestodes have been accumulating with increasing speed. The listing of such works in a separate 'Molecular and biochemical taxonomy' section of *Helminthological Abstracts* since 1990 illustrates this trend. Here, we examine the diversity of methodological approaches used and the taxonomic groups that have been investigated, review the main progress attained with such methods, and propose directions for future work. Although not exhaustive, we have attempted to summarize the majority of systematic works on cestodes utilizing molecular data.

Systematic applications within the Cestoda

Molecular systematic works on tapeworms range from intraspecific questions to the interrelationships of the major lineages, although few orders in the class have been examined across all taxonomic levels. We review these works briefly in categories that broadly represent the taxonomic focus of the studies: *species and intraspecific* (species circumscription, subspecific and strain definition, and population structure); *intraordinal* (familial and generic circumscription and interrelationships); and *interordinal* (ordinal circumscription and interrelationships of the major lineages within the class). An overview of most of the published taxa and gene systems reviewed herein is shown in Table 13.1.

Species and intraspecific applications

Most molecular studies aimed at the specific or intraspecific level concern primarily parasites of economic or medical importance belonging to the order Cyclophyllidea and, to a lesser extent, the orders Proteocephalidea and Pseudophyllidea. Although many molecular studies of cyclophyllidean taxa employ phylogenetic methodology, their focus is often pathogen identification, novel detection methods or chemotherapeutics, and thus systematic aspects are generally marginal or not considered. From a systematic perspective, however, these works have led to the recognition of many subspecific entities ('strains') whose systematic and taxonomic status are difficult to define (Thompson and Lymbery 1990). Moreover, the relationships among the multiple genotypes detected are not always clear. Such variation has only begun to be evaluated in a phylogenetic framework (Bowles *et al.* 1995a; Thompson *et al.* 1995), but clearly the fine differences revealed by molecular markers can greatly complicate the building of traditional classification schemes (Thompson 1998). This is even more so the case when considering the problems linked to the definition of species in herma-phroditic, and potentially self-fertilizing, organisms (Bray 1991; Thompson and Lymbery 1996).

Order Cyclophyllidea

With more than 50% of known tapeworm genera belonging to the Cyclophyllidea (Khalil *et al.* 1994), it is by far the most species-rich order of cestodes. It is also the primary order containing almost exclusively parasites of homeotherm tetrapods and several genera of significant economic or medical importance (e.g., *Echinococcus*). Within the Cyclophyllidea, no detailed DNA study has been conducted at the species level apart from works on *Echinococcus* and *Taenia* spp., both members of the family Taeniidae (e.g., Bowles *et al.* 1995a; Thompson *et al.* 1995; Kedra *et al.* 1999), and more recently on *Mesocestoides* spp. (Crosbie *et al.* 2000). In the latter case, the small subunit (SSU), and especially ITS2, ribosomal DNA genes have allowed for a detailed analysis of the genus and a potential definition of subspecific evolutionary lineages within it. Regarding the genus *Taenia*, molecular characters have been widely used to differentiate *T. asiatica* from *T. saginata* and to assess relationships among members of the genus (Zarlenga *et al.* 1991; Gasser and Chilton 1995). Although controversy still exists, a number of molecular markers appear to be decisive in establishing a subspecific status of *T. asiatica* within *T. saginata* (Bowles and McManus 1994; Zarlenga and George 1995). In a related field, *Taenia* molecular phylogenies (based on CO1 and nuclear large subunit (LSU) rDNA) have already been exploited to infer evolutionary scenarios of colonization of human beings by these parasites (de Queiroz and Alkire 1998). As for the genus *Echinococcus*, the existence of strains, especially within *E. granulosus*, has been extensively studied, including the sequencing of a variety of genes (for a review see Bowles and McManus 1993a). The situation is complicated, however, and a recent study on pig and human isolates of *E. granulosus* in Eastern Europe by Kedra *et al.* (1999) found no differences among isolates using the ND1 gene. The ITS-1 gene, on the other hand, showed differences even among clones of individual isolates, suggesting that polymorphism of the rDNA copies makes the ITS-1 an unreliable molecular marker. A similar situation exists for *E. multilocularis*, and cases where molecular techniques have allowed for the differentiation of unique strains have also been documented in this species complex (Rinder *et al.* 1998). These studies have allowed for the definition of novel molecular probes (e.g., Bowles *et al.* 1994; Gasser and Chilton 1995) that may greatly benefit studies concerning epidemiology.

A few studies, most of them based on isoenzymes, have looked into the systematics of species of less medical or economic importance: Dixon and Arai (1985, 1989), Novak *et al.* (1989) and Okamoto *et al.* (1997) examined interrelationships within the genus *Hymenolepis*; Okamoto *et al.* (1995) examined intraspecific variation within *Taenia taeniaeformis*; Baverstock *et al.* (1985) revealed an unsuspected diversity of *Progamotaenia* spp. (Anoplocephalidae) in marsupials; and Bâ *et al.* (1994) conducted a similar, although more limited, study on the genera *Avitellina* and *Thysaniezia* (also Anoplocephalidae) in African cattle.

Table 13.1 Summary of cestode systematic studies employing molecular data.

Level of inference Genome	Gene locus	Taxon* (No. of representative species/isolates)	N†	Method of analysis‡	Sequence length (regions)	Reference
Inter/intraspecific						
Genomic						
		Cyc: *Taenia crassiceps, T. saginata, T. solium,* 'Asian taenia'	4	RFLP (14 enzymes)		Zarlenga and George 1995
		Pse: *Bothriocephalus funiculus (72)*	1	RAPD (65 primers)		Verneau et al. 1995
		Pro: *Proteocephalus exiguus (20)*	1	RAPD (8 primers)		Král'ová and Spakulová 1996
		Pro: *Proteocephalus exiguus, P. percae (6)*	2	RAPD (8 primers)		Král'ová 1996
		Pro: *Proteocephalus percae (141)*	1	RAPD (4 primers)		Hanzelová et al. 1999
		Pse: *Bothriocephalus andresi, B. barbatus, B. clavibothrium, B. claviceps, B. funiculus, B. gregarius, B. scorpii*	7	DNA/DNA Hybrid	300–500	Verneau et al. 1997a,b
Nuclear						
	SSU rDNA	**Pse:** *Spirometra erinaceieuropaei*	1	Seq	(complete)	Liu et al. 1997
		Pse: *Bothriocephalus andresi, B. barbatus, B. clavibothrium, B. claviceps, B. funiculus, B. gregarius, B. scorpii*	7	Seq	650 (3' end)	Verneau et al. 1997b
		Cyc: *Mesocestoides sp. (15)*	1	Seq	1050 (5' and 3' ends)	Crosbie et al. 2000
		Tet: *Acanthobothrium* spp. (2), *Anthocephalum* spp. (3), *Calliobothrium* spp. (2), *Duplicibothrium* spp. (3), *Litobothrium* spp. (2), *Rhinebothrium* spp. (3)	28	Seq	1500 (V4, V7–9)	Olson et al. 1999
	LSU rDNA	**Cyc:** *Taenia saginata, T. solium, T. multiceps, T. ovis, T. hydatigena, T. pisiformis, T. crassiceps, T. taeniaformis,* 'Asian taenia', *Echinococcus granulosus*	10	Seq	~300 (D1)	Bowles and McManus 1994
		Pro: *Peltidocotyle, Othinoscolex, Crepidobothrium, Monticellia* spp.	4	Seq	1200 (D1–D3)	Zehnder and Mariaux 1999
		Pse: *Spirometra erinacei, S. mansonoides*	2	RFLP	~300 (D1)	Lee et al. 1997
	5.8S LSU rDNA	**Pro:** *Peltidocotyle, Othinoscolex, Crepidobothrium, Monticellia* spp.	4	Seq	(complete)	Zehnder and de Chambrier 2000
	5.8S, ITS-2, LSU rDNA	**Pro:** Monticelliidae *Nomimoscolex* (9), other monticellids (7), Proteocephalidae (2)	18	Seq	1740	Zehnder et al. 2000
	ITS-1 rDNA	**Cyc:** *Echinococcus granulosus (9), E. multilocularis (2), E. vogeli (2), E. oligarthrus (1)*	4	Seq	(complete)	Bowles et al. 1995a
		Cyc: *Echinococcus granulosus (5)*	1	Seq	(complete)	
		Pse: *Bothriocephalus andresi, B. barbatus, B. clavibothrium, B. claviceps, B. funiculus, B. gregarius, B. scorpii*	7	Seq	(complete)	Verneau et al. 1997b
		Pse: *Spirometra erinacei, S. mansonoides*	2	RFLP	(complete)	Lee et al. 1997
	ITS-2 rDNA	**Cyc:** *Echinococcus granulosus (11), E. multilocularis (5), Taenia hydatigena (5), T. ovis (4), T. pisiformis (5), T. multiceps (2), T. serialis (4)*	7	RFLP	(complete)	Gasser and Chilton 1995
		Cyc: *Hymenolepis diminuta (2), H. microstoma, H. nana (2)*	3	Seq	(complete)	Okamoto et al. 1997
		Cyc: *Mesocestoides sp. (15)*	1	Seq	(complete)	Crosbie et al. 2000
		Pro: *Peltidocotyle, Othinoscolex, Crepidobothrium, Monticellia* spp.	4	Seq	(complete)	Zehnder and de Chambrier 2000
Mitochondrial						
	LSU rDNA	**Pro:** *Peltidocotyle, Othinoscolex, Crepidobothrium, Monticellia* spp.	4	Seq	440 (partial)	Zehnder and de Chambrier 2000
	Cytochrome c oxidase I (COI)	**Cyc:** *Echinococcus granulosus (53), E. multilocularis (4), E. vogeli (2), E. oligarthrus (1)*	4	Seq	366 (partial)	Bowles et al. 1995a
		Cyc: *Echinococcus granulosus, E. multilocularis, Taenia*	9	Seq	391 (partial)	Okamoto et al. 1995
		Cyc: *Hymenolepis diminuta (2), H. microstoma, H. nana (2)*	3	Seq	391 (partial)	Okamoto et al. 1997
		Cyc: *Taenia saginata, T. solium, T. multiceps, T. ovis, T. hydatigena, T. pisiformis, T. crassiceps, T. taeniaformis,* 'Asian taenia', *Echinococcus granulosus*	10	Seq	~400	Bowles and McManus 1994
		Pse: *Spirometra erinacei, S. mansonoides*	2	RFLP	~400	Lee et al. 1997
	NADH dehydrogenase I (ND1)	**Cyc:** *Echinococcus granulosus (63), E. multilocularis (4), E. vogeli (2), E. oligarthrus (1)*	4	Seq	471 (partial)	Bowles et al. 1995a
		Cyc: *Echinococcus granulosus (61)*	1	Seq	~450 (partial)	Kedra et al. 1999
	NADH dehydrogenase 3 (ND3)	**Pse:** *Diphyllobothrium nihokaiense (2), Sparganum erinacei (2), S. proliferum (2)*	3	Seq	(complete)	Kokaze et al. 1997
Intraordinal						
Nuclear						
	SSU rDNA	**Cyc:** Amabiliidae (1), Anoplocephalidae (1), Catenotaeniidae (1), Davaineidae (2), Dilepididae (5), Hymenolepididae (9), Mesocestoididae (1), Metadilepididae (1), Nematotaeniidae (1)	22	Seq	1300 (V1–2, V5, V8–9)	Mariaux 1998
		Tet: Dioecotaeniidae (1), Litobothriidae (2), Onchobothriidae (6), Phyllobothriidae: Echeneibothriinae (1), Phyllobothriinae (8), Rhinebothriinae (7); Prosobothriidae (1), **Lec** (2)	28	Seq	1500 (V4, V7–9)	Olson et al. 1999

Table 13.1 Continued.

Level of inference Genome	Gene locus	Taxon* (No. of representative species/isolates)	N†	Method of analysis‡	Sequence length (regions)	Reference
	LSU rDNA	**Pro:** Monticelliidae: Endorchiinae (1), Monticelliinae (1), Peltidocotylinae (4), Rudolphiellinae (2), Zygobothriinae (17); Proteocephalidae: Acanthotaeniinae (1), Corallobothriinae (1), Gangesiinae (2), Proteocephalinae (24)	53	Seq	1149 (D1–D3)	Zehnder and Mariaux 1999
Mitochondrial	SSU rDNA	**Cyc:** Anoplocephalidae (2), Catenotaeniidae (2), Dilepididae (1), Dipylidiidae (1), Hymenolepididae (3), Mesocestoididae (3), Taeniidae (6), **Pse** (2)	20	Seq	314 (partial)	von Nickisch-Rosenegk et al. 1999
	LSU rDNA	**Pro:** Monticelliidae: Endorchiinae (1), Monticelliinae (1), Peltidocotylinae (4), Rudolphiellinae (2), Zygobothriinae (17); Proteocephalidae: Acanthotaeniinae (1), Corallobothriinae (1), Gangesiinae (2), Proteocephalinae (24)	53	Seq	437 (partial)	Zehnder and Mariaux 1999
Interordinal *Nuclear*	SSU rDNA	**Amp** (1), **Car** (1), **Cyc** (22), **Dip** (1), **Nip** (1), **Pro** (7), **Pse** (6), **Spa** (1), **Teb** (1), **Tet** (3), **Try** (2)	46	Seq	1300 (V1–2, V5, V8–9)	Mariaux 1998
		Amp (1), **Car** (1), **Cyc** (1), **Dip** (2), **Gyr** (1), **Hap** (1), **Lec** (2), **Nip** (1), **Pro** (1), **Pse** (2), **Spa** (1), **Teb** (1), **Tet** (6), **Try** (2)	23	Seq	(complete)	Olson and Caira 1999
		Amp, Car, Cyc, Dip, Gyr, Hap, Lec, Nip, Pro, Pse, Spa, Teb, Tet, Try	71	Seq	(complete)	Olson et al., in press
	Elongation factor 1-α (Ef-1α)	**Amp** (1), **Car** (1), **Cyc** (1), **Dip** (2), **Hap** (1), **Lec** (2), **Nip** (1), **Pro** (1), **Pse** (2), **Spa** (1), **Teb** (1), **Tet** (6), **Try** (2)	21	Seq	825 (partial)	Olson and Caira 1999
	LSU rDNA	**Amp, Car, Cyc, Dip, Gyr, Hap, Lec, Nip, Pro, Pse, Spa, Teb, Tet, Try**	71	Seq	1400 (D1–D3)	Olson et al., in press
Class *Nuclear*	SSU rDNA	**Amp** (1), **Gyr** (1), **Pse** (1), **Tet** (1)	4	Seq	469 (4 separate regions)	Baverstock et al. 1991
		Amp (1), **Gyr** (1), **Tet** (1)	3	Seq	580 (5′ end)	Rohde et al. 1993b
		Amp (1), **Cyc** (3), **Gyr** (1), **Pse** (1), **Tet** (1)	7	Seq	(complete and partial)	Campos et al. 1998
		Cyc (1), **Gyr** (1), **Pro** (1), **Pse** (3), **Try** (1)	7	Seq	(complete)	Littlewood et al. 1999a

* Classification follows Khalil et al. (1994). **Amp**, Amphilinidea; **Car**, Caryophyllidea; **Cyc**, Cyclophyllidea; **Dip**, Diphyllidea; **Gyr**, Gyrocotylidea; **Hap**, Haplobothriidea; **Lec**, Lecanicephalidea; **Nip**, Nippotaeniidea; **Pro**, Proteocephalidea; **Pse**, Pseudophyllidea; **Spa**, Spathebothriidea; **Teb**, Tetrabothriidea; **Tet**, Tetraphyllidea; **Try**, Trypanorhyncha.
† Total number of species in analysis.
‡ Seq, phylogenetic and/or phenetic analysis using nucleotide sequences; DNA/DNA Hybrid, temperature-dependent hybridization of genomic DNA anaylsis; RAPD, randomly amplified polymorphic DNA analysis; RFLP, restriction-fragment length polymorphism analysis.

Order Proteocephalidea

The proteocephalideans are a relatively large order of tapeworms with freshwater fishes, reptiles and amphibians as definitive hosts (Rego 1994). The fact that they are of little consequence to human affairs, as are most cestodes, probably accounts for the paucity of studies in comparison to members of the Cyclophyllidea. It is becoming clear, however, that an understanding of their role in tapeworm evolution is important for understanding the history of the class itself, especially that of the higher tetrafossate orders (Caira *et al.* 1999; Olson and Caira 1999; Olson *et al.* 1999). Only a few molecular works have dealt with proteocephalideans at the species level, and are mainly on Palaearctic species in the genus *Proteocephalus*. Snábel *et al.* (1994) used isoenzymes to show the possible conspecificity of two previously well-established taxa, and Snábel *et al.* (1996) also used isoenzymes for examining their genetic variation and mode of fertilization. Similarly, Král'ová (1996), Král'ová and Spakulová (1996), and Hanzelová *et al.* (1999) used randomly amplified polymorphic DNA analysis (RAPD) and isoenzymes markers to distinguish between closely related *Proteocephalus* spp. or populations and revealed intraspecific variation not observed with classical methods. Zehnder, in a series of works dealing with the interrelationships of the order, examined a number of genera and found sequence data useful to resolve or complement analyses not only within the genus *Proteocephalus* (Zehnder and Mariaux 1999), but also within *Peltidocotyle* and other closely related genera (Zehnder and de Chambrier, 2000), and within the genus *Nomimoscolex* (Zehnder *et al.* 2000). These contributions aimed at lower systematic levels help to circumscribe monophyletic taxa in the order and can thus guide the reconstruction of the higher systematic structure of the group.

Order Pseudophyllidea

The Pseudophyllidea is a cosmopolitan group of tapeworms especially abundant in fishes, and less frequently found in other vertebrate groups (Bray *et al.* 1994a). They have been little studied using molecular techniques apart from members of the family Diphyllobothriidae, of which a few species are occasional parasites of human beings. Unlike all other pseudophyllidean cestodes, the diphyllobothriids parasitize mammals and are likely to represent a distinct lineage of equal taxonomic rank (Mariaux 1998; Hoberg *et al.* 2001, this volume). Only a few pseudophyllidean genera or groups of species have been studied to date. Matsuura *et al.* (1992) used RFLP analysis to differentiate *Diphyllobothrium latum* and *D. nihonkaiense*. Renaud *et al.* (1983, 1986), followed by Verneau *et al.* (1995, 1997a,b) worked on *Bothriocephalus* species complexes and employed a wide array of techniques on this model system including isoenzymes, RAPD, sequencing and DNA–DNA hybridization (the only use of the latter technique in tapeworms in so far as we know). These authors have described several cryptic species, studied their evolutionary relationships and evaluated the host–parasite evolution of the group. In addition, their DNA–DNA hybridization studies allowed for a rough assessment of the times of divergence within the genus (Verneau *et al.* 1997b). This series of works makes *Bothriocephalus* one of the best understood non-medically important genera of tapeworms with regard to host specificity, hidden diversity and fine-scale evolutionary relationships.

Within the Diphyllobothriidae, most studies have been aimed at using molecular markers to identify larval forms. For example, deVos and Dick (1989) used isoenzymes to associate larvae and adult species of *Diphyllobothrium*, and Kokaze *et al.* (1997) used mitochondrial sequences to probe sparganum larvae against potential adult forms. In the same group of organisms, Afanas'ev and Tseitlin (1993) looked into the population structure of *Triaenophorus nodulosus* with enzymes, Liu *et al.* (1997) determined the primary sequence and possible secondary structure of the SSU gene of *Spirometra erinaceieuropaei*, and Lee *et al.* (1997) used RFLP analysis to compare *S. erinacei* and *S. mansonoides*.

Summary

On the whole, molecular systematic studies addressing species and subspecific levels in cestodes are still restricted to very few, usually medically or economically important taxa. Other studies are the result of a few specialists who have been able to incorporate molecular approaches to a long-studied model organism. A consequence of this situation is that species whose genetic structure is well understood represent taxa found in a limited diversity of hosts (i.e., human and domesticated animals). Clearly, a far greater diversity of taxonomic studies is needed before we can characterize the level of molecular variability typical of tapeworm populations and species. However, the ability to detect molecular variation at the population level, and even within individuals, makes delimiting species boundaries based on molecules both a subjective and complicated task.

Intraordinal applications

Molecular works dealing with suborders, families and genera of tapeworms are scarce. At present, they concern almost exclusively the three large tetrafossate orders: Cyclophyllidea, Proteocephalidea and Tetraphyllidea.

Order Cyclophyllidea

To date, only von Nickisch-Rosenegk *et al.* (1999) have addressed specifically intraordinal relationships within the Cyclophyllidea using molecular data, but their preliminary work considered members of only eight families and was based on a rather short partial sequence of the mitochondrial SSU gene (314 bp). Their results supported the monophyly of most families (with the notable and somewhat surprising exception of the Hymenolepididae), placed the Cateno-taeniidae as the most basal family in the order (a result congruent with the recent total evidence analysis of Hoberg *et al.* (2001, this volume), and positioned the Mesocestoididae within the derived branches of the tree (although its position

was subject to the type of analysis performed). The position of the Mesocestoididae, as well as other systematic implications of their analysis, contradict previous hypotheses and should be taken with caution given the very low support shown for most interfamilial nodes, and the likely saturation of the mitochondrial SSU gene in their dataset. Mariaux (1998) examined relationships of nine cyclophyllidean families represented by 22 taxa in the context of a higher systematics study based on 1.2 kb of the nuclear SSU gene. His work showed most presently recognized families to be monophyletic, and supported the recognition of the family Gryporhynchidae. Although not all interfamilial relationships were resolved, strong support was found for the exclusion of the Mesocestoididae from the order, and for a sister-group relationship between Hymenolepididae and Anoplocephalidae. These findings confirm previously proposed hypotheses (for a recent review see Hoberg *et al.* 1999a), and in this case, molecular data merely helped to choose between ambiguous support based on different subsets of morphological data.

Order Proteocephalidea

Mariaux (1998) examined the relationships of a small number of taxa with partial SSU sequences, but no resolution was achieved by strict or even 50% majority rule consensus of the trees resulting from analysis by parsimony. Later, Zehnder and Mariaux (1999) examined the interrelationships of the Proteocephalidea specifically in a study including over 50 taxa using both nuclear and mitochondrial LSU rDNA. These preliminary works allowed for the first evaluation of the current systematic structure of the Proteocephalidea (Rego 1994), and suggested that fundamental changes are necessary (Rego *et al.* 1998). Unlike the Cyclophyllidea, the backbone of the proteocephalidean classification (i.e., its partition into two widely accepted families: Monticellidae and Proteocephalidae), and monophyly of long-accepted taxa (including the type genus, *Proteocephalus*) appear to be artificial (Zehnder and Mariaux 1999; Zehnder *et al.* 2000). A complete reconstruction of the group's phylogeny from the bottom up is needed in order to provide a phylogenetic basis for the higher classification of the group. Molecular characters have justified this approach by demonstrating the weakness of present hypotheses and the unreliability of some key morphological characters for defining natural familial and subfamilial taxa (e.g., the position of reproductive organs relative to the medulla).

Order Tetraphyllidea

The Tetraphyllidea is the third largest order in the class, and the largest group of tapeworms of elasmobranchs (Khalil *et al.* 1994). Historically, the composition of the order has been unstable and difficult to define, and recent analyses based on morphology (Caira *et al.* 1999; Caira *et al.* 2001, this volume) as well as molecules (Mariaux 1998; Olson and Caira 1999; Olson *et al.* 1999) have shown the order to represent a paraphyletic assemblage. Perhaps more so than the Proteocephalidea, the pivotal position of the Tetraphyllidea in the evolution of the higher tetrafossate lineages makes monophyletic circumscription of the order and a better understanding of their interrelationships important in the larger context of tapeworm evolution. The strict host specificity common among elasmobranch-hosted tapeworms (Caira 1990) also makes the study of their interrelationships interesting from the perspective of host–parasite coevolution (see e.g., Olson *et al.* 1999). The potential for such tapeworm groups to be used as model systems for coevolutionary studies has been vastly under-utilized.

To date, only Olson *et al.* (1999) have examined the intraordinal relationships of tetraphyllideans using molecular data. Based on partial SSU sequences, this preliminary investigation included 26 species representing four families in the order (Table 13.1), but focused largely on phyllobothriid parasites of rays in the subfamilies Phyllobothriinae and Rhinebothriinae. In addition, representatives of the enigmatic family Litobothriidae, and the order Lecanicephalidea were included in order to test their positions relative to the Tetraphyllidea as currently defined (Euzet 1994). Comparison of the V4 and V7–9 variable regions of the molecule showed a higher level of divergence than would be expected among closely related genera (Hillis and Dixon 1991). The primary implication of the phylogenetic study was that the interrelationships reflected common host associations among the taxa better than their current classification. For example, the family 'Phyllobothriidae' was found to be paraphyletic, with those species infecting sharks more closely related to onchobothriid and prosobothriid tetraphyllideans of sharks than to the other phyllobothriid species included in the analysis, all of which parasitize rays. Other common host associations found were a clade of primarily diamond ray (*Dasyatis* spp.) parasites and a clade of cownose ray (*Rhinoptera* spp.) parasites; both of which contain members from different taxonomic groups. This work further illustrates the need for taxonomic revision of the order, although a much denser sampling of taxa is needed before revision based on molecular results could be justified.

Summary

Whereas paraphyly of an order may be indicated using a limited number of representative taxa, the internal structure of an order generally requires a much broader sampling to evaluate. Relative to other taxonomic levels, intraordinal classifications of tapeworms are the most unstable, and considerable work remains to be done in order to erect classification schemes that more accurately reflect phylogenetic history. It is clear from the few works to date that molecular data are informative for evaluating the putative morphological homologies used to define intraordinal clades. Unfortunately, this level of phylogenetic inference has thus far received the least attention from molecular cestodologists.

Interordinal applications

Defining the major lineages of tapeworms and determining their interrelationships have been largely intractable problems. As reviewed by Hoberg *et al.* (1997c, 1999a), no general consensus is found among the various hypotheses proposed over the course of the previous century, and many of the minor orders (e.g., Haplobothriidea, Lecanicephalidea, Spathebothriidea) have not been universally recognized until the recent publication of the classification adopted by Khalil *et al.* (1994). As a result, answers to long-standing questions regarding the evolution of the class have persisted without general agreement or decisive evidence. The application of molecular data to the problem was hoped to finally overcome the difficulties associated with other classes of data.

Only two works to date have investigated ordinal-level interrelationships using molecular data (Figure 13.1). Mariaux (1998) analysed partial SSU sequences of 46 species (including 23 cyclophyllideans) representing all currently recognized orders (Khalil *et al.* 1994) except the Haplobothriidea, Lecanicephalidea and Gyrocotylidea, and Olson and Caira (1999) analysed complete SSU and partial *elongation factor 1-α* (Ef-1α) sequences of 23 taxa representing the 14 orders recognized in Khalil *et al.* (1994). With the exceptions of the amphilinidean, *Schizochoerus liguloideus*, and the nippotaeniidean, *Amurotaenia decidua*, no overlap of exemplar species is found between the two studies, which is interesting in that it helps to evaluate the effect of species representation on the ordinal-level interpretation of the phylogenetic estimates. However, differences in ordinal representation between the studies confound direct comparison of the results.

Both studies supported the monophyly of the Eucestoda and of a clade including the tetrafossate orders Cyclophyllidea, Proteocephalidea, Tetrabothriidea and Tetraphyllidea, as well as the monofossate order Nippotaeniidea. Basal to the 'tetrafossate' clade were the remaining monofossate and difossate orders, although their relative positions differed between the studies. Both studies also showed the order Tetraphyllidea to be paraphyletic. Unfortunately, neither study resolved strongly all interordinal relationships, and this was particularly true of the basal difossate groups (e.g., Diphyllidea, Trypanorhyncha and Pseudophyllidea). A lack of character support appears to be largely responsible for differences in the results of the studies, and between the SSU and Ef-1α genes, whereas strongly supported clades showed congruence between the studies. For example, both studies showed the orders Nippotaeniidea and Tetrabothriidea to form a derived clade together with the Cyclophyllidea, whereas previous hypotheses based on morphology have been either conflicting or ambiguous in the placement of these taxa, with the exception of the most recent cladistic studies by Hoberg *et al.* (1997c, 1999a, 2001, this volume), in which this clade was also recovered.

Other strongly supported results were unique to the individual studies due to differences in species representation. The position of the enigmatic species *Haplobothrium globuliforme* was shown by Olson and Caira (1999) to be closest to the Diphyllobothriidae, thus refuting a potential homology between the tentacular structures of *H. globuliforme* and of the trypanorhynchs. Their study also supported recognition of the Litobothriidae as a lineage separate from the order Tetraphyllidea, in which the group is currently recognized as a family (Euzet 1994). Mariaux (1998) showed paraphyly of the order Pseudophyllidea, with members of the Diphyllobothriidae forming a lineage separate from the other pseudophyllidean taxa in his analysis. Similarly, the family Mesocestoididae was shown to be distinct from other members of the order Cyclophyllidea. The total evidence analyses of Hoberg *et al.* (2001, this volume) and mitochondrial SSU data (von Nickisch-Rosenegk *et al.* 1999), however, do not support exclusion of the Mesocestoididae from the Cyclophyllidea.

The positions of the cestodarian orders, Amphilinidea and Gyrocotylidea, relative to the eucestodes was examined by Olson and Caira (1999) using SSU data. Their results supported, albeit weakly, a closer relationship of the Gyrocotylidea, rather than the Amphilinidea, to the Eucestoda. This result is in contrast to previously proposed hypotheses regarding the positions of these basal orders. Both groups show a high degree of divergence of their SSU gene relative to those in eucestodes, and indeed to most other rhabditophoran platyhelminths (Littlewood and Olson 2001, this volume). The amphilinideans in particular possess extremely large expansion regions (see Appendix A in Olson and Caira 1999; Littlewood and Olson 2001, this volume), and in some taxa are missing highly conserved stem regions found in all other platyhelminth, if not all eukaryote SSU genes (Neefs *et al.*

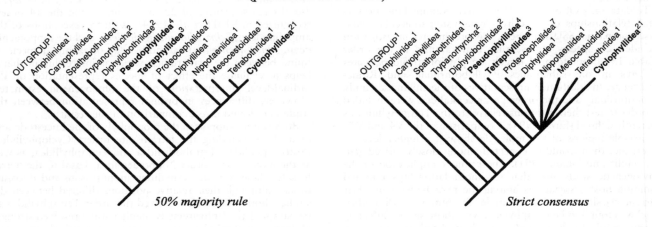

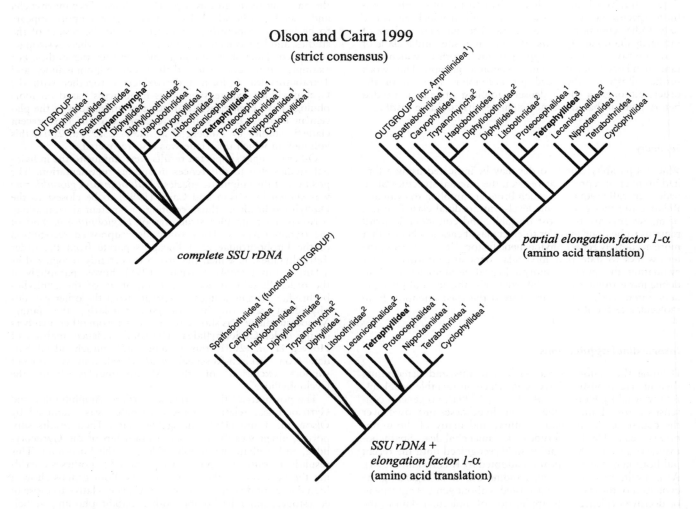

Figure 13.1 Higher-level interrelationships of tapeworms based on parsimony analyses of molecular data. Classifications follow Khalil *et al*. (1994). Bold font indicates paraphyly of the taxonomic groups. Superscripts denote the number of taxa in the analyses representing each 'ordinal' lineage. See citations for levels of nodal support and the specific exemplar species. Note: Mariaux's analysis (1998) supported the Diphyllobothriidae (Pseudophyllidea) and Mesocestoididae (Cyclophyllidea) as independent lineages, and Olson and Caira's analyses (1999) supported the Litobothriidae (Tetraphyllidea) as an independent lineage. See Hoberg *et al*. (2001, this volume) and Olson *et al*. (in press) for analyses of SSU rDNA combined with morphological data.

1990, 1993). Although both groups represent 'long branches', from a phenetic perspective, the gyrocotylidean SSU sequences show a greater overall similarity to those of eucestodes (Olson and Caira 1999). However, when analysing only the regions of the gene conserved enough to be alignable across all platyhelminth taxa, the traditional arrangement of the amphilinideans as the sister-group to the Eucestoda, with the Gyrocotylidea forming the sister-group to the Amphilinidea + Eucestoda is found (Littlewood and Olson 2001, this volume). This suggests that the analysis of Olson and Caira (1999) was subject to problems associated with long branches (Felsenstein 1978; Siddall 1998).

An important difference between the results of Mariaux (1998) and of Olson and Caira (1999) was the determination of the most basal eucestode order. A basal position of the Spathebothriidea was supported in Olson and Caira (1999) by two independent gene loci, and by the shared absence of an intron within the Ef-1α gene among the spathebothriidean, amphilinidean and outgroup taxa, but otherwise universally observed among the Eucestoda. A basal position of the Caryophyllidea, however, was found by Mariaux (1998), and is supported by morphological and total evidence analyses (Hoberg *et al.* 1997c, 2001, this volume). In either case, members of both orders are non-segmented worms, suggesting that external segmentation within the Eucestoda is a derived condition. The overall pattern of tapeworm evolution differed between the two studies as well. Most previous hypotheses have suggested either a hierarchical (Hoberg *et al.* 1997c) or diphyletic (Euzet 1959; Euzet *et al.* 1981) pattern; the latter being split among difossate and tetrafossate lineages. Mariaux's (1998) results support a hierarchical pattern, whereas Olson and Caira's analyses (1999) generally showed a somewhat more diphyletic pattern (Figure 13.1).

Summary

The use of molecular data for evaluating higher-level relationships within the Cestoda has resulted in a better understanding of their evolution, although the positions of a number of taxa remain unclear. If the inability to resolve these relationships represents a 'hard polytomy', then a literal interpretation suggests that the diversification of the basal orders was too rapid to allow sufficient time for genetic mutations to accumulate. Alternatively, if the situation represents a 'soft polytomy', then a gene conserved enough to maintain such ancient mutations must be sampled – but has yet to be identified. Neither explanation can be discounted at present, but congruence in the *lack* of resolution from SSU and Ef-1α data for the nodes subtending the difossate and tetrafossate lineages points toward explosive radiations of these potential clades. It is interesting how the molecular data reflect morphology: for example, the phylogenetic position of orders such as the Diphyllidea have been equally problematic to define using traditional means, and with regard to the Amphilinidea and Gyrocotylidea, their ribosomal sequences, like their morphology, are as distinct from one another as is either to those of the eucestodes.

Overview

Key taxa

Although much work needs to be done at all taxonomic levels within the Cestoda, current SSU data are likely to be sufficient to determine the affinities of taxa listed as *incertae sedis* (e.g., in Khalil *et al.* 1994), or, more generally, of any unknown taxon, to at least the ordinal level, and to the familial level in many cases. This in turn will allow for a better understanding of the limits and extremes of morphological variability within lineages, and may thus help to determine homology among disparate morphological features. In this regard, it is particularly important to include type-species as often as possible in molecular studies so that there is a more direct correspondence between molecular characterizations and the typological nature of the zoological nomenclature.

Amphilinidean and gyrocotylidean taxa would be interesting to study in greater detail both from the perspectives of organismal, as well as genetic evolution. Why they possess such extremely divergent SSU sequences, and how their unique sequences deviate from the 'universal' eukaryotic models of ribosomal secondary structure (de Rijk *et al.* 1992; Neefs *et al.* 1993) could hold new insights into the molecular evolution of parasitism.

With regard to higher-level relationships, molecular data indicate that the Tetraphyllidea represents not only a paraphyletic group, but may be key in understanding the origins and diversification of the higher tetrafossate forms. Its precise relationship with the Proteocephalidea, as well as the understanding of the evolution of the latter order is fundamental in retracing the colonization of higher vertebrates by cestodes. It will be thus important to circumscribe monophyletic groups of 'tetraphyllidean' worms and examine the phylogenetic positions of these separate lineages in relation to other closely related groups, such as the Lecanicephalidea and Litobothriidea.

Gene systems

It is safe to say that ribosomal data will continue to play the primary role in the molecular systematics of tapeworms, if only because the database of these genes (particularly the SSU) is considerably larger than for any other locus. The high copy number of the ribosomal array also facilitates the ease with which the genes may be amplified, even from degraded or marginal quantities of genomic DNA. Furthermore, new and promising candidate genes have yet to be demonstrated to be more useful for examining distantly related taxa. Ef-1α, a nuclear protein coding gene involved in the translation of mRNA, was determined by Friedlander *et al.* (1992) to be slowly evolving, and thus one such promising candidate. However, its utility in cestode systematics appears neither better nor worse than that of the SSU gene (Olson and Caira 1999), and technical difficulties in determining Ef-1α sequences make its use less practical. On the other hand, faster-evolving genes (especially those of the mitochondrial genome) appear to be either too variable for analysing even intraordinal relationships, or difficult to amplify, and thus the choice of good candidates at least for the higher systematics of the group is presently limited.

Problems

Besides the difficulty of obtaining fresh material suitable for genetic analysis (Mariaux 1996), many problems inherent to the analysis of molecular data have been well documented (see Felsenstein 1978; Simon *et al.* 1994; Nadler 1995; Moritz and Hillis 1996; Siddall 1998), and Nadler (1995) rightfully notes that 'the youthful optimism once characteristic of the discipline ... has been replaced by a more balanced perspective that recognizes the advantages and limitations of nucleotide sequence data for inferring organismal relationships'. Some problems may be recognized and avoided (e.g., long branches) or at least partially overcome via computational methods (e.g. among-site rate variation and among-taxon nucleotide heterogeneity). Occasionally, however, the problems may be more elusive to detect. For example, it has been recently shown that the SSU molecule may be present in different copies among some platyhelminths, thus posing problems of orthology (see e.g., Carranza *et al.* 1996).

Isoenzyme-based studies tend to be less numerous given the ease and efficiency by which nucleotide sequences can now be generated. It should be emphasized, however, that these techniques, especially at lower taxonomic levels, may answer questions regarding species definition rapidly and inexpensively (e.g., Renaud *et al.* 1983; deVos and Dick 1989; Zehnder *et al.* 2000). Although such data are not always well suited for inferring phylogenetic history and are plagued with a number of problems (recently reviewed by Beveridge 1998), their use for such applications remains possible in certain cases (Andrews and Chilton 1999; Buth and Murphy 1999).

Utility of the genetic database beyond systematics

In addition to providing characters useful for phylogenetic inference, the characterization of cestode gene sequences provides a highly useful reference database for diagnostic purposes; the utility of which grows considerably with each new species submitted for public access (e.g., to GenBank/EMBL). Novel uses of the data will surely come to light, but perhaps the most obvious diagnostic applications are in the identification of larval forms and unrecognized or aberrant pathogens. The latter approach was used to help determine the identity of a lethal pathogen in an AIDS patient by Santamaría-Fríes *et al.* (1996), and although the database was then too limited to provide an exact identity, the results indicated that the culprit was a hymenolepidid species – limiting greatly the sequencing effort required in a follow-up study (P.D.O., unpublished data).

In the coming years, we may expect to see this approach used more widely for elucidating complex life cycles, and thereby reducing the need to maintain intermediate and definitive hosts in the laboratory. Life cycles for some tapeworm orders have yet to be determined for even a single representative species, and because a significant portion of tapeworm diversity is found in elasmobranch hosts, this approach will be especially valuable where the maintenance of such marine species can be too impractical to justify the effort. To date, this approach has been used in cestodes only for identifying diphyllobothriid larvae (Kokaze *et al.* 1997); however, it has been used successfully to identify digenean larvae recovered from intermediate hosts (Schulenburg and Wägele 1998; Jousson *et al.* 1999). Its full potential for evaluating host utilization and specificity has not yet begun to be realized. Other uses include parasite-specific PCR detection of larvae in commercially important food species, fluorescent probes for *in situ* detection, parasite localization in tissue sections, etc. Indeed, the ability to identify simply tapeworm species of any ontogenetic stage using molecular markers will allow workers to concentrate on questions that have been previously intractable.

ACKNOWLEDGEMENTS

We thank I. Beveridge, J. Caira, E. Hoberg, T. Ruhnke and T. Scholz for discussions and ideas regarding tapeworm systematics and evolution. Marc Zehnder kindly made preprints and unpublished results available. Thanks to D. Blair and B. Georgiev for helpful comments and criticisms of the text. J.M. was supported by the Swiss National Foundation for Scientific Research (grant No. 3100-037548.93), and P.D.O. by a Marshall-Sherfield Fellowship (Marshall Aid Commemoration Commission, UK) during the writing of this manuscript. We are grateful to D.T.J. Littlewood for facilitating this collaboration.

Chapter 14

Interrelationships among tetraphyllidean and lecanicephalidean cestodes

Janine N. Caira, Kirsten Jensen and Claire J. Healy

The cestodes are among the most diverse of the major platy-helminth groups. The adult stages of these obligate parasites tend to exhibit a fairly high degree of host specificity. This specificity is reflected to some extent in the current ordinal level organization of the class (see Khalil *et al.* 1994); the adults of most tapeworm orders parasitize members of one or two classes of vertebrates. That is not to say that ordinal membership is based entirely on adult host associations, because many orders are also defined by conspicuous, unique morphological features. Recent phylogenetic work, however, has begun to call into question the monophyly of at least some of the 14 currently recognized orders of cestodes (see Brooks *et al.* 1991; Hoberg *et al.* 1997c, 1999b; Mariaux 1998; Caira *et al.* 1999; Olson and Caira 1999). Particularly pivotal in these discussions are two of the tapeworm orders that parasitize elasmobranchs (the Tetraphyllidea and Lecanicephalidea) and one that commonly parasitizes freshwater bony fishes (the Proteocephalidea) because preliminary evidence suggests that at least some species in these orders have closest relatives that are currently members of one or more other cestode orders. If this is the case, the classification of cestodes cannot be revised to reflect monophyly until the relationships among and within these groups are more completely understood and a stable phylogenetic hypothesis has been established. One of the primary objectives of the present study was to further explore the relationships of these groups, concentrating on species in the two orders that parasitize elasmobranchs.

At present, the Tetraphyllidea is considered to include approximately 62 genera and 800 species, while the Lecanicephalidea includes approximately ten genera and 38 species. As adults, all species in both orders parasitize the spiral intestines of chondrichthyan fishes, primarily elasmobranchs. Tetraphyllidean cestodes are fairly widely distributed among chondrichthyans; species are known from all of the major groups of chondrichthyan fishes that have been examined for internal parasites. Lecanicephalidean cestodes are primarily parasites of rays. Among tapeworms, these two groups appear to be remarkably host-specific; in most cases, each species is known to parasitize only a single species of host.

The Tetraphyllidea and Lecanicephalidea are also among the most morphologically diverse of the cestode orders. Collectively these groups show interesting variation in segment morphology, but the variation in scolex morphology they exhibit (Figures 14.1–14.68) exceeds that of any of the other tapeworm orders. The majority of the members of these two orders have a scolex that is tetrafossate in form. Curiously, much of the variation in scolex morphology seems attributable to basic modifications of this general tetrafossate scolex condition. For example, the four acetabula can be completely sessile (Figure 14.62), free along one or more margins (Figure 14.1), pedicel-late (Figure 14.50), or stalked (Figure 14.41). Acetabula can be unloculate (Figure 14.25), marginally loculate (Figure 14.4), facially loculate (Figure 14.22), or both facially and marginally loculate (Figure 14.10). Acetabula can have one or more free margins that are folded (Figure 14.47), reflexed and fused (Figure 14.24), flat (Figure 14.31) or cup-shaped (Figure

14.45). Acetabula can bear a specialized anterior region (Figure 14.1) or not (Figure 14.54). If present, the specialized anterior region can be in the form of a sucker (Figure 14.24) or a loculus (Figure 14.29), or in the form of a thickened loculus (a pad) that bears its own sucker (Figure 14.1). Acetabula can bear hooks (Figure 14.34), or not (Figure 14.29); if acetabular hooks are present, they can consist of a single pair of hooks that are unipronged (Figure 14.46), bipronged (Figure 14.49), or tripronged (Figure 14.35), or they can be arranged in two pairs (Figure 14.7). In adult worms, the apex of the scolex proper can be completely unmodified (Figure 14.48) or can be conspicuously modified (Figure 14.56), and/or can bear an apical organ (Figures 14.56 and 14.61). The apical organ can be minute (Figure 14.60) or extensive (Figures 14.55–14.57), divided into tentacles (Figure 14.62) or undivided (Figure 14.61). Particularly intriguing, however, are the few groups that, despite their possession of segments that are generally morphologically consistent with some of the above taxa, exhibit a scolex that bears little or no evidence of the tetra-fossate condition (Figures 14.12, 14.21 and 14.27).

These interpretations of lecanicephalidean and tetraphylli-dean scolex morphology, however, rest on a number of assumptions about scolex homologies that have not been suffi-ciently investigated. Examples of some of the more significant assumptions are:

1. Scolex suckers (Figure 14.62) and bothridia (Figure 14.1) are homologous (thus the use of the collective term acet-abulum).

2. All acetabular hooks are homologous, but these hooks are not homologous to the apical organ hooks of cyclophylli-deans (Figure 14.69) or diphyllideans (Figures 14.71 and 14.72), or the tentacular hooks of trypanorhynchs (Figures 14.77–14.82) or the bothrial hooks of pseudo-phyllidean groups such as *Triaenophorus* Rudolphi, 1793.

3. Bothridial (acetabular) specialized anterior region suckers (Figure 14.24) are homologous to bothridial (acetabular) specialized anterior region loculi (Figure 14.29).

4. Apical organs are homologous across all tapeworm groups (e.g., Figures 14.55–14.57, 14.60–14.62, 14.64, 14.69–14.71).

5. All tapeworms in which segmentation does not begin immediately behind the scolex proper possess a cephalic peduncle (sometimes referred to as the neck); this pedun-cular region is considered to be part of the scolex and is further considered to be homologous among all tapeworm groups, regardless of whether the boundary between the peduncle and the strobila is conspicuous or inconspicuous.

Comparative morphological studies including members of as great a diversity of eucestodes as possible, beyond those exam-ined here, would provide useful data for evaluation of the appropriateness of these and many additional assumptions of morphological homology among these groups. A well-substan-tiated phylogenetic tree for the group would assist with evalua-tion of the degree of homoplasy in these characters under the current assumptions of homology and therefore would also allow evaluation of the hypotheses of homology themselves, for it is possible that the brilliance of the morphological diversity of the Tetraphyllidea and Lecanicephalidea is merely an artifact of the non-monophyly of the groups.

The phylogenetic relationships of the Tetraphyllidea and Lecanicephalidea have received some recent attention but, nonetheless, remain poorly understood. Several analyses have focused on relationships within particular genera (e.g., Brooks 1992; Brooks and McLennan 1993; Nasin et al. 1997). Selected tetraphyllideans and lecanicephalideans have been included in phylogenetic analyses investigating broader (ordinal) cestode relationships (e.g., Mariaux 1998; Olson and Caira 1999). The relationships among members of 15 genera were recently investigated by Olson et al. (1999) based on analysis of sequence data from the 18S ribosomal DNA gene. In addition, Caira et al. (1999) recently attempted the first analysis with broad generic representation in a study based on morphological data for one species in each of 48 tetraphyllidean and eight lecanicephalidean genera. Differences in taxon sampling among these various studies impede direct comparison of the results. Overall, there is only moderate congruence in the placement of specific taxa among the analyses in which such comparisons can be made. However, in general, these investigations all provide evidence questioning the monophyly of the Tetraphyllidea and/or the Lecanicephalidea and Proteocephalidea.

The present study is an extension of the investigation of Caira et al. (1999). It seemed that the monophyly of the Tetraphyllidea and Lecanicephalidea, in particular their intergeneric relationships and the many questions of character homology raised in that analysis, could most effectively be examined through the collection of data on additional species. Thus, in the present study, a replicated exemplar approach was employed. A minimum of two species of as many of the non-monotypic tetraphyllidean and lecanicephalidean genera as possible was examined. Given the conflicting hypotheses regarding the interordinal relationships of cestodes, the taxonomic constituency of the outgroup was expanded to include one or more representatives of six, rather than three, other orders of tapeworms.

Analysis of this suite of taxa also facilitated the investigation of several intriguing methodological issues. The effect of taxon sampling on the results of phylogenetic analyses has recently been explored fairly extensively both in simulation studies (e.g., Hillis 1996, 1998; Graybeal 1998; Rannala et al. 1998) and empirical studies using molecular data (e.g., Halanych 1998; Poe 1998), but it has received little attention in empirical studies using morphological data. In the present study, the effect of taxon sampling was investigated by comparison of trees generated using the exemplar approach (i.e., from analysis of a single species in each of the target genera; see Caira et al. 1999) with trees generated using the replicated exemplar approach (i.e., from analysis of several species in most of the non-monotypic target genera). In addition, to a limited degree the effect of number of characters was investigated by comparison of the results of the analyses of 63 taxa coded for 113 characters by Caira et al. (1999), with analysis of the same 63 taxa for 148 characters.

The effect of coding inapplicable characters as unknown or as a separate character state was also investigated. The theoretical importance of the distinction among polymorphic, unknown and inapplicable characters in phylogenetic analyses has been recognized (e.g., Maddison 1993), but the effect has not been extensively explored using empirical data. When this issue arises in molecular analyses, the volume of data often allows investigators to exclude the inapplicable regions and thereby avoid the problems associated with coding these regions for inclusion in the analysis. Elimination of inapplicable characters, however, is a method not so easily justified for

---

(See the following four pages)

Figures 14.1–14.20 Scanning electron micrographs of scoleces. 1. Acanthobothrium parviuncinatum. 2. Acanthobothroides thorsoni. 3. Anthobothrium c. f. cornucopia, only one of four acetabula shown. 4. Anthocephalum duszynskii. 5. Balanobothrium n. sp. 6. Bibursibothrium gouldeni. 7. Biloculuncus pritchardae. 8. Calliobothrium evani. 9. Calyptrobothrium sp. 10. Cardiobothrium beveridgei. 11. Carpobothrium n. sp. 12. Cathetocephalus sp. 13. Caulobothrium n. sp. 1. 14. Ceratobothrium xanthocephalum. 15. Chimaerocestos n. sp. 16. Clistobothrium carcharodoni. 17. Crossobothrium laciniatum. 18. Dicranobothrium spinulifera. 19. Dinobothrium sp. 20. Dioecotaenia sp. Abbreviations: A, acetabulum; AH, acetabular hook; FL, facial loculus; ML, marginal loculus; S, sucker; SAR, specialized anterior region of acetabulum.

Figures 14.21–14.42 Scanning electron micrographs of scoleces. 21. Disculiceps sp., note that this species of Disculiceps was not included in the analysis. 22. Duplicibothrium n. sp. 2. 23. Echeneibothrium sp. 2. 24. Flexibothrium ruhnkei. 25. Gastrolecithus planus. 26. Glyphobothrium zwerneri. 27. Litobothrium amplifica. 28. Marsupiobothrium gobelinus. 29. Monorygma sp. 30. Myzocephalus sp. 31. Orygmatobothrium sp. 2. 32. Pachybothrium hutsoni. 33. Paraorygmatobothrium sp., note that this species of Paraorygmatobothrium was not included in the analysis. 34. Pedibothrium longispine. 35. Phoreiobothrium n. sp. 3. 36. Phyllobothrium sp., note that this species of Phyllobothrium was not included in the analysis. 37. Pinguicollum pinguicollum. 38. Platybothrium parvum. 39. Potamotrygonocestus magdalenensis. 40. Prosobothrium sp. 41. Pseudanthobothrium n. sp. 1. 42. Rhinebothrium urobatidium. Abbreviations: A, acetabulum; AH, acetabular hook; AMSP, apical modification of scolex proper; FL, facial loculus; SAR, specialized anterior region of acetabulum; S, stalk.

Figures 14.43–14.62 Scanning electron micrographs of scoleces. 43. Rhinebothroides moralarai. 44. Rhodobothrium sp. 45. Scyphophyllidium sp. 46. Spiniloculus n. sp. 47. Thysanocephalum sp. 48. Trilocularia acanthiaevulgaris. 49. Uncibilocularis sp. 50. Yorkeria n. sp. 51. Zyxibothrium kamienae. 52. new genus 1 n. sp. 53. Adelobothrium sp. 54. Anteropora sp. 55. Cephalobothrium n. sp. 1. 56. Cephalobothrium n. sp. 2. 57. Corrugatocephalum ouei. 58. 'Discobothrium' n. sp. 59. Eniochobothrium sp. 60. Hornellobothrium n. sp. 1. 61. Lecanicephalum sp. 62. Polypocephalus n. sp. 3. Abbreviations: A, acetabulum; AH, acetabular hook; AMSP, apical modification of scolex proper; AO, apical organ; FAR, folded acetabular region; P, pedicel; UAR, unfolded acetabular region.

Figures 14.63–14.82 Scanning electron micrographs of scoleces. 63. Tetragonocephalum sp. 64. Tylocephalum n. sp. 65. new genus 2 n. sp. 66. new genus 3 n. sp. 67. new genus 4 n. sp. 68. new genus 5 n. sp. 69. Echinococcus granulosus. 70. Ditrachybothridium macrocephalum. 71. Echinobothrium n. sp. 72. Macrobothridium sp. 73. Monticellia lenha. 74. Proteocephalus perplexus. 75. Diphyllobothrium cordatum. 76. Tetrabothrius cylindraceus. 77. Floriceps minacanthus. 78. Grillotia similis. 79. Gymnorhynchus isuri. 80. Mixodigma leptaleum. 81. Otobothrium n. sp. 82. Tentacularia sp. Abbreviations: AO, apical organ; AOH, apical organ hook; AMSP, apical modification of scolex proper; LH, lateral hooklets; TH, tentacular hook.

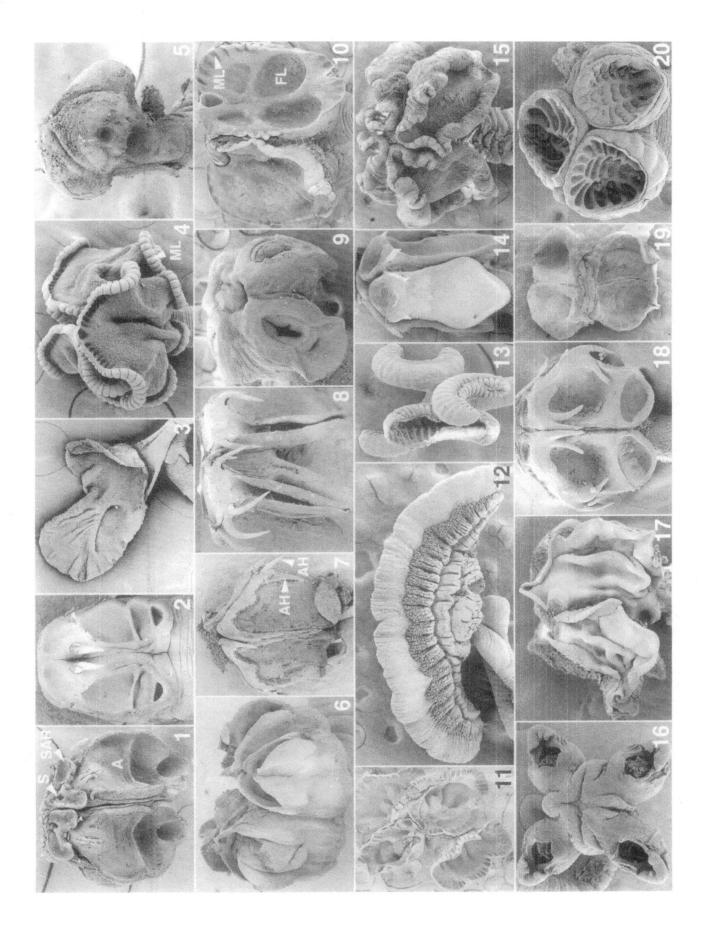

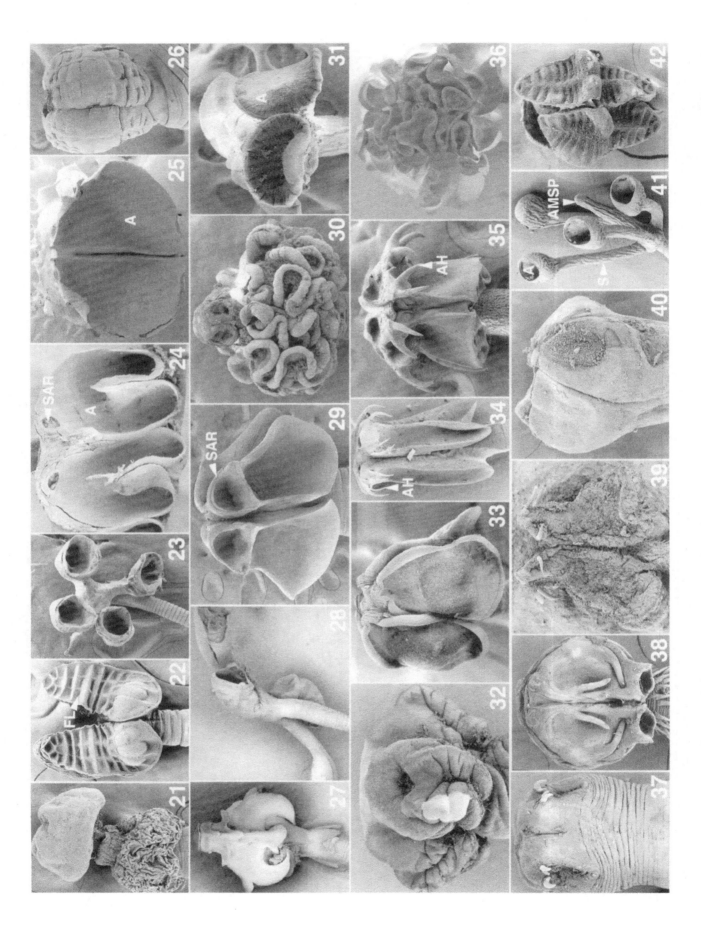

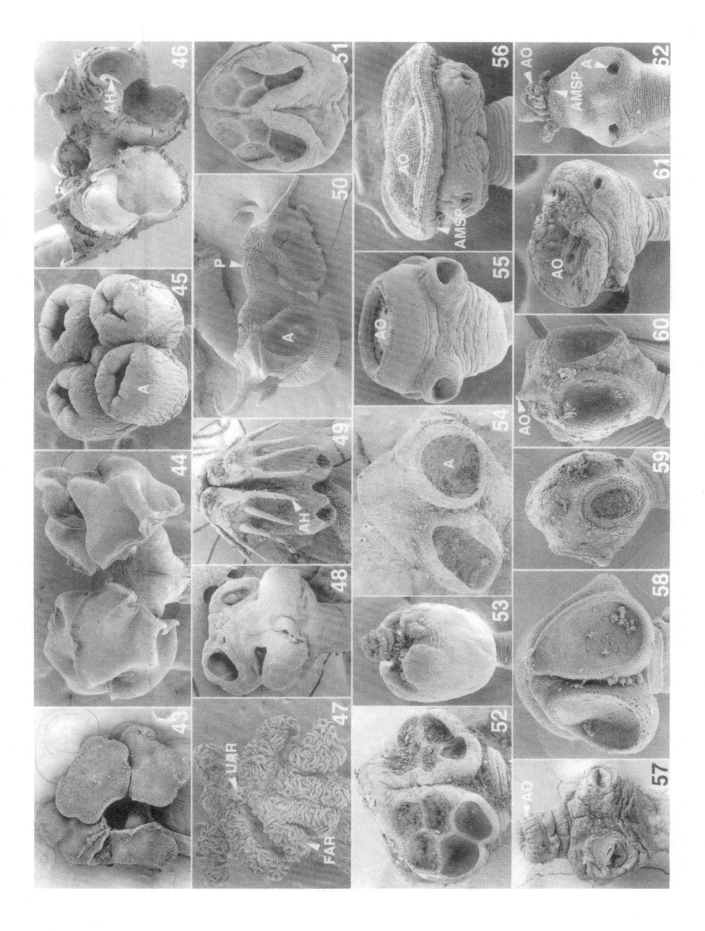

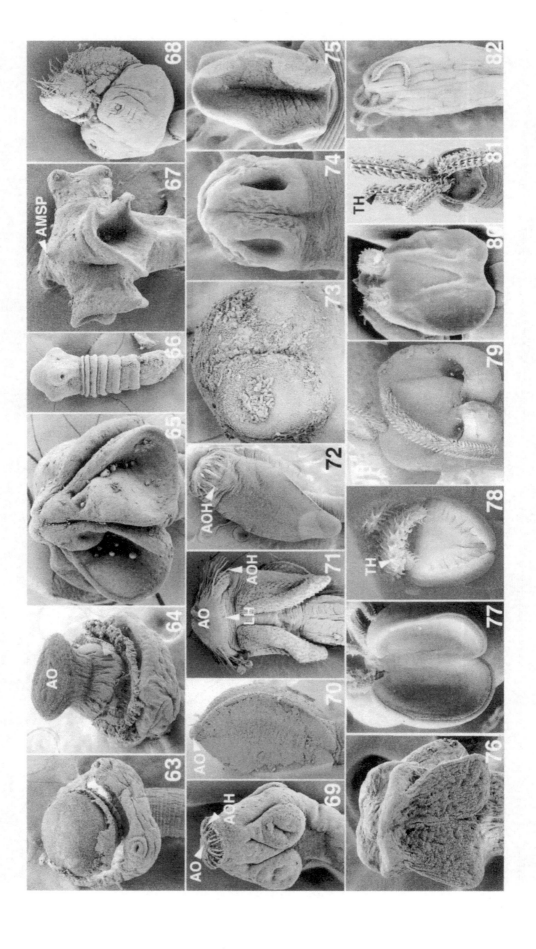

the more limited data sets of morphological analyses. The data matrix generated in this study consisted of morphological characters coded across relatively distantly related taxa and provided a useful opportunity to explore the effect of two possible codings of inapplicable characters on tree topology.

Materials and methods

The present investigation involved a total of 127 species: the 63 used by Caira et al. (1999) and 64 that were new to this study. Although it seems clear that the number and choice of specific exemplars can dramatically affect the results of an analysis (e.g., see Yeates 1995), the choice of the additional exemplars in this study rested to a large extent on the availability of new material. In addition, a special effort was made to include representatives of tetraphyllidean and lecanicephalidean genera not examined by Caira et al. (1999). The 127 species used in the present study are identified in Appendix 14.1, along with information on the source of the character data used to code the 64 species added to the study. In addition, source information for specimens of the 63 species examined by Caira et al. (1999), beyond those used in that study, but examined here, is provided. In several cases, character information was obtained, at least in part, from the literature; whenever possible, type material of such species was also examined. Museum abbreviations used for the sources of this material are as follows: LRP, Larry R. Penner Parasitology Collection, Department of Ecology & Evolutionary Biology, University of Connecticut, Storrs, Connecticut, U.S.A.; MPM, Meguro Parasitological Museum, Toyko, Japan; USNPC, U.S. National Parasite Collection, Beltsville, Maryland, U.S.A. Readers are referred to Table I of Caira et al. (1999) for the source information for the material of the 63 taxa examined for that study.

Material of most of the species new to this study came from collections made within the last five years from hosts (primarily elasmobranchs) from a variety of different geographic localities throughout the world. Specimens were prepared as whole mounts and histological sections, as well as for examination with the scanning electron microscope following the procedures given in Caira et al. (1999).

Full phylogenetic analyses utilized an ingroup consisting of 88 species of 50 tetraphyllidean genera (including four genera new to this study: *Myzocephalus* Shipley and Hornell, 1906, *Scyphophyllidum* Woodland, 1927, *Uncibilocularis* Southwell, 1925 and a genus new to science, new genus 1) and 21 species of 15 lecanicephalidean genera (including seven genera new to this study: *Adelobothrium* Shipley, 1900, *Tetragonocephalum* Shipley, 1900, *Cephalobothrium* Shipley and Hornell, 1906, and four genera new to science, new genus 2, new genus 3, new genus 4 and new genus 5). The outgroup consisted of a total of 18 species: four species in two proteocephalidean genera, four species in three diphyllidean genera, six species in six trypanorhynchan genera, two species belonging to one tetrabothriid genus, one cyclophyllidean and one pseudophyllidean species. The unusual morphologies exhibited by the five genera new to science (Figures 14.52, 14.65–14.68) made their inclusion in the analysis an attractive prospect, despite the fact that they have not yet formally been named.

As might be expected, the addition of taxa representing orders and genera not included in the analyses of Caira et al. (1999), as well as the addition of more than a single species in each genus, provided the opportunity for (and in some cases necessitated) the articulation of additional characters. As a consequence, 37 characters were added to the 120 characters of Caira et al. (1999). In addition, the interpretation of several of the original characters was revised following examination of additional material. It should be noted however, that this character list does not do justice to the morphological variation found in the outgroup taxa as they were not the focus of this analysis. The 157 characters and their character states are listed in Appendix 14.2. Characters have been renumbered consecutively for this study, but the original character numbers from Caira et al. (1999) are provided in parentheses for characters used in both studies. Daggers following parenthetical character numbers indicate that the description or interpretation of one or more of the character states has changed in some way from that used by Caira et al. (1999). In addition, the conditions under which the inapplicable character coding ('9') was invoked are given for each character new to this study. Detailed discussion of each of the new characters is not possible due to space

limitations; however, several characters that are worthy of special explanation are described below.

Hooks on apical organ

In the analyses of Caira et al. (1999), the hooks on the scolex of diphyllideans of the genera *Echinobothrium* van Beneden, 1849 (Figure 14.71) and *Macrobothridium* Khalil and Abdul-Salam, 1989 (Figure 14.72) were considered to be associated with the apical modification of the scolex proper. However, re-examination of the initial material, following the discussions of apical organs that took place at the 3rd International Workshop for Systematic Zoology of Tapeworms in Bulgaria in August of 1999, now leads us to believe that the hooks of diphyllideans are more appropriately considered to be associated with the apical organ. Of primary importance in this re-interpretation is the fact that the region of the scolex that bears the hooks in such taxa is separated from the remainder of the scolex by a discrete boundary in the form of a distinct membranous layer of tissue (Figure 14.83). In this analysis, the diphyllideans, like the cyclophyllidean, have been coded as possessing hooks on the apical organ. Two characters have been included to describe the difference between the apical organ hooks of the diphyllideans and those of the cyclophyllidean, specifically Character 17 (apical organ hook arrangement) and Character 18 (apical organ hook shape). In addition, Character 19 (discrete muscle bundles associated with apical organ hook bases) and Character 20 (lateral hooklets on apical organ), illustrated in Figures 14.83 and 14.71, respectively, were added to the analysis.

Position of attachment of scolex muscle bundles to apical organ

We are far from adequately understanding the intricacies of the musculature associated with the scolex of most tapeworms, and none of this musculature was included in the characters recognized by Caira et al. (1999). However, as a minimum, a character coding for the position of attachment of conspicuous muscle bundles associated with the apical organ (Character 28) was included in the present analysis. Examination of longitudinal sections through the scolex and whole mounts of a wide diversity of species suggests that, if present, these muscle bundles are either broadly attached throughout the entire basal margin of the apical organ (Figure 14.84) or have limited attachment points either to just the peripheral (Figure 14.83) or, rarely, just the central regions of the apical organ. Alternatively, the apical organs of some taxa, such as members of *Litobothrium* Dailey, 1969, for example, appear to lack this musculature entirely (Figure 14.85).

Scolex folding

Several groups of tapeworms exhibit folding of part, or all, of the surface of the scolex. Particularly notable are members of the genera *Thysanocephalum* Linton, 1891, *Phyllobothrium* van Beneden, 1849 and *Disculiceps* Joyeux and Baer, 1936 (Figures 14.47, 14.36 and 14.21, respectively). Caira et al. (1999) included only a single character coding for acetabular folding. In that study, species in the former two genera were coded as possessing this feature. *Disculiceps galapagoensis* was coded as lacking this feature because cross-sections through its folded region provided no evidence of either the tetrafossate condition or the classic bounding membrane or radial muscle fibres typical of an acetabulum; thus, this region was considered to be non-acetabular and the folding seen in *Disculiceps galapagoensis* went unrecognized.

The addition of *Myzocephalus* sp. (Figure 14.30) to the present study led us to re-evaluate the homology of the folded portions of the scolex among the study species. Cross-sections through the extensively folded region of the scolex of *Myzocephalus* sp. revealed this region to be tetrafossate (Figure 14.86); however, the classic delimiting membrane and radial musculature typical of an acetabulum are not well developed in this region. Interpretation was further complicated by the fact that, unlike the condition seen in *Thysanocephalum* sp. (Figure 14.47), in which the elaborately folded region is continuous with the unfolded anterior region of the acetabulum, the extensively folded region seen in *Myzocephalus* sp. is conspicuously separated from the anterior unfolded acetabular region of the scolex (Figure 14.87). Thus, the homologies between the extensively folded region of *Myzocephalus* sp., and those of *Thysanocephalum* sp. and *Disculiceps galapagoensis* are unclear. To accommodate the condition seen in *Disculiceps* in the present study, we added a character for

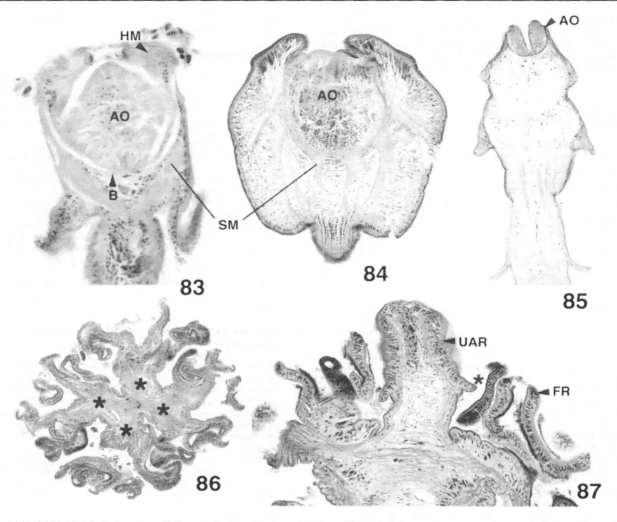

Figures 14.83–14.87 Histological sections. 83. Longitudinal section through scolex of *Echinobothrium* n. sp., note muscles attached to periphery of apical organ only. 84. Longitudinal section through scolex of *Cephalobothrium* n. sp. 1, note muscles attached throughout entire boundary of apical organ. 85. Longitudinal section through scolex of *Litobothrium amplifica*, note absence of muscles attached to apical organ. 86. Cross-section through posterior folded region of scolex of *Myzocephalus* sp., * indicates tetrafossate condition of folded region of scolex. 87. Longitudinal section through scolex of *Myzocephalus* sp., * indicates posterior margin of unfolded acetabular region of scolex. Abbreviations: AO, apical organ; B, boundary of apical organ; FR, folded region; HM, muscles of apical organ hooks; SM, scolex muscles; UAR, unfolded acetabular region.

peduncular folding (Character 68). However, the ambiguity of the data for the folded region of *Myzocephalus* sp. left us with no way to code this taxon with any confidence for either Character 61 or 68. Thus, we coded *Myzocephalus* sp. as unknown ('?') for both Characters 61 and 68, as well as for several characters associated with acetabular morphology. It seemed that the position of *Myzocephalus* sp. in the tree resulting from analysis of other characters might provide some insight as to the most appropriate interpretation of the homology of the folded region of the scolex between this and the other tetraphyllidean taxa.

Cortex versus medulla

Traditionally, the acoelomate body of tapeworms has been considered to consist of an outer cortical region and an inner medullary region (e.g., see Hyman 1951; Joyeux and Baer 1961; etc.). The boundary between these two regions is often marked by a layer of circular muscle fibres (e.g., Figure 14.88). In taxa in which this is the case, there is little disagreement about the extent of the cortex or the medulla. However, some groups lack a conspicuous circular muscle layer. In such groups, the boundary between cortex and medulla is not so clearly defined. The position of the longitudinal muscle bundles, when present, has also been used to assist

with the distinction between cortex and medulla. The longitudinal muscle bundles have generally been considered to be cortical in position when circular muscle fibres are present (e.g., Figure 14.88), or as the actual indicators of the boundary between the two regions when circular muscle fibres are absent (Figure 14.91) (e.g., see Rego 1994). Thus, it seems that two different criteria have been used to recognize the boundary between the cortex and the medulla, depending on whether circular muscle fibres are present or absent. The characterization of the cortex and medulla has importance beyond just this single character because several other taxonomically important characters are considered to depend on it. For example, the position of the vitelline follicles relative to these two regions has been used to distinguish between the two families of proteocephalideans (e.g., see Rego 1994).

Characters describing or relating to the configuration of the cortex and medulla were totally lacking from Caira *et al.* (1999). The articulation of characters describing these features was explored in the present study. Comparison of cross-sections through mature segments of the 127 study taxa initially presented numerous difficulties with consistently coding the distinction between cortex and medulla and the suite of related characters. Particularly problematic were taxa, such as many of the tetraphyllideans, that lack both circular muscle fibres and discrete longitudinal

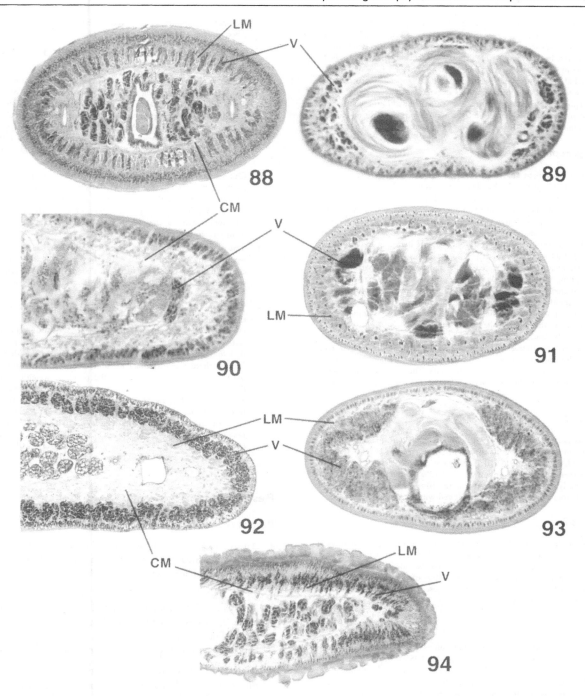

Figures 14.88–14.94 Segment cross-sections. **88.** *Grillotia similis.* **89.** '*Discobothrium*' n. sp. **90.** *Proteocephalus exiguus.* **91.** *Echeneibothrium* sp. 1. **92.** *Gymnorhynchus isuri.* **93.** *Rhodobothrium* sp. **94.** *Thysanocephalum* sp. Abbreviations: CM, circular muscle fibres; LM, longitudinal muscle bundles; V, vitelline follicle.

muscle bundles (Figure 14.89). In such cases, it was questionable as to whether the cortex should be considered to be reduced, or possibly lacking. Also problematic was the classic description of vitellaria as medullary or cortical because vitellaria were found to occupy a wide variety of positions relative to the longitudinal muscle bundles. The most objective coding seemed to be to code for the presence or absence of the features used to recognize these regions, rather than for the regions themselves. Thus, we added the following three characters to this study: Character 109 (Discrete circular muscle fibres) which were considered to be either absent (Figure 14.89) or present (Figure 14.88); Character 110 (Discrete

longitudinal muscle bundles) also considered to be absent (Figure 14.90) or present (Figure 14.92); and Character 146 (Vitelline element position relative to longitudinal muscle bundles) with states describing each of the five positions seen among the study taxa (Figures 14.88 and 14.91–14.94). These characters allowed us to code for all four possible combinations of circular muscle fibres and longitudinal muscle bundles; specifically: both circular and longitudinal muscles present (Figure 14.88), circular muscles present but longitudinal muscles absent (Figure 14.90), circular muscles absent but longitudinal muscles present (Figure 14.91), and both circular and longitudinal muscles absent (Figure 14.89).

The complete data matrix (Appendix 14.3) consisted of 19 939 character/taxon combinations (157 characters for 127 taxa), 463 (2.32%) of which were unknown at the time of analysis and were thus coded with '?' in the matrix. In analyses in which the 10% exclusion rule was applied, characters in which states were unknown in greater than 10% of the taxa were excluded from the analysis. All characters were run unordered and equally weighted. In the four instances in which taxa were found to be polymorphic for a character, both states were included in the matrix. To ensure consistency in coding between the initial 63 taxa and the additional 64 taxa, material of all 127 species was examined for all 157 characters.

Phylogenetic analysis

All of the phylogenetic analyses were conducted using NONA version 1.6 (Goloboff 1997) using 'the Ratchet' search routine (Nixon 1999). The relatively large number of taxa necessitated the use of Heuristic search algorithms, with the likelihood of finding all shortest, equally parsimonious trees resting to a large extent on how aggressively tree space was explored. For each analysis, 'the Ratchet' was run four to 15 different times with 100 to 5000 iterations, saving one tree per iteration, for a total of at least 20 000 iterations. Strict (Nelson) consensus trees were used to summarize relationships indicated by the shortest, equally parsimonious trees resulting from each analysis. The number of equally parsimonious trees presented for each analysis is of unambiguously supported trees only, and should be considered to be a minimum. Support values, similar to Bremer support, were determined by importing the 20 000 or more trees resulting from the NONA iterations into PAUP* version 4.0.0.d64 (Swofford 1998) and using the 'Filter trees' option to examine nodal support for consensus trees of trees one or more steps longer than the shortest tree.

The following five analyses were conducted using different combinations of characters, taxa and coding strategies. The strategy used to compare the results of the different analyses follows the descriptions of the five analyses.

Analysis 1

The 63 taxa from Caira *et al.* (1999) with the same ingroup and outgroup constitution as used in that analysis, and 148 characters. Characters 84, 85, 91, 93, 100, 105, 106, 107, 157 were excluded from the analysis because their states were unknown in > 10% of the 127 taxa of the present study (i.e., they conformed with the 10% exclusion rule described above). Inapplicable characters were coded with '9'.

Analysis 2

The 63 taxa from Caira *et al.* (1999) with the same ingroup and outgroup constitution as used in that analysis, and 147 characters. Characters 84, 85, 91, 92, 93, 100, 105, 106, 107, 157 were excluded from the analysis because their states were unknown in > 10% of the 63 taxa (i.e., they conformed with the 10% exclusion rule described above). Inapplicable characters were coded with '9'.

Analysis 3

All 127 taxa with the ingroup and outgroup as described above, and 148 characters. Characters 84, 85, 91, 93, 100, 105, 106, 107, 157 were excluded from the analysis because their states were unknown in > 10% of the 127 taxa (i.e., they conformed with the 10% exclusion rule described above). Inapplicable characters were coded with the character state '9'. Nodal support values were calculated for this analysis as described above.

Analysis 4

All 127 taxa with the ingroup and outgroup as described above, and all 157 characters, and with inapplicable characters coded with the character state '9'.

Analysis 5

All 127 taxa with the ingroup and outgroup as described above, and all 157 characters, and with inapplicable characters coded as unknown ('?'). It should be noted that this coding strategy resulted in unknown codings for 7947 (39.9%) of the character/taxon combinations in the data

matrix. Nodal support values were calculated for this analysis as described above.

The results of these analyses were compared as described below to examine the following effects:

- **Taxon sampling:** The effect of taxon sampling on tree topology was investigated by comparing trees generated from analysis of the original 63 taxa of Caira *et al.* (1999) and the 127 taxa of the present study using the same set of characters. Thus, the results of Analysis 1 were compared to those of Analysis 3. The results of Analysis 3 were also compared to those of Analysis 2 in case character differences resulting from differences in the group to which the 10% exclusion rule was applied affected the results of the analysis.

- **Character sampling:** The effect of the changes in the character list (both addition and alteration of characters) was investigated by comparing trees generated for the same set of taxa (the 63 of Caira *et al.* 1999) using different sets of characters. Thus, the tree resulting from the Comprehensive analysis in Caira *et al.* (1999, figure 87) was compared to the results of Analysis 2.

- **Ten per cent exclusion rule:** The effect of removing characters that were unknown in greater than 10% of the taxa was investigated by comparing trees generated for the 127 taxa based on matrices that included and excluded such characters. Thus, the results of Analysis 3 were compared to those of Analysis 4.

- **Treatment of inapplicable characters:** The issue of how best to handle the coding of inapplicable characters is not yet resolved. Caira *et al.* (1999) chose to code inapplicable characters with a separate character state ('9') rather than as unknown ('?') for these characters. They felt that the advantages of preserving information on the absence of a feature, thereby constraining possible character-state optimizations, outweighed the disadvantages of the potential weighting effects of this coding, in other words that taxa are coded for lacking a structure as well as for lacking all of the characters coding for the various features of that structure. The effects of these two coding strategies were investigated by comparing trees generated using the same set of taxa and characters, in which inapplicable characters were treated as either unknown or as a separate character state. Thus, the results of Analysis 4 were compared to those from Analysis 5.

Results

Results of analyses were as follows:

- **Analysis 1:** This analysis resulted in 396 most-parsimonious trees, each 860 steps long with a CI of 0.344 and an RI of 0.64. The strict consensus of these trees is illustrated in Figure 14.95.

- **Analysis 2:** This analysis resulted in 453 most-parsimonious trees, each 856 steps long with a CI of 0.342 and an RI of 0.64. The strict consensus of these trees was identical in topology to that resulting from Analysis 1 (see Figure 14.95).

- **Analysis 3:** This analysis resulted in 4491 most-parsimonious trees, each 1329 steps long with a CI of 0.256 and an RI of 0.727. The strict consensus of these trees is illustrated in Figure 14.96. Nodal support values are given for each node on the branch leading to the node.

- **Analysis 4:** This analysis resulted in 3497 most-parsimonious trees, each 1394 steps long with a CI of 0.26 and an RI of 0.723. The strict consensus of these trees is illustrated in Figure 14.97.

- **Analysis 5:** This analysis resulted in 22 most-parsimonious trees, each 1021 steps long with a CI of 0.245 and an RI of 0.606. The strict consensus of these trees is illustrated in Figure 14.98. Nodal support values are given for each node on the branch leading to the node.

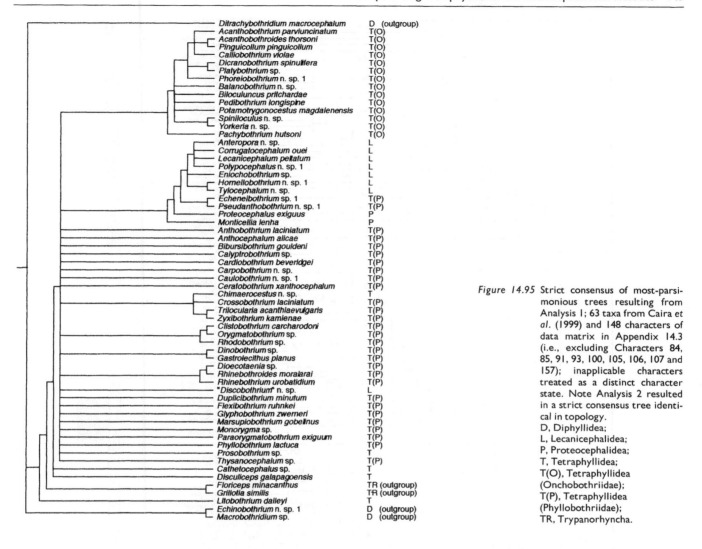

Figure 14.95 Strict consensus of most-parsimonious trees resulting from Analysis 1; 63 taxa from Caira et al. (1999) and 148 characters of data matrix in Appendix 14.3 (i.e., excluding Characters 84, 85, 91, 93, 100, 105, 106, 107 and 157); inapplicable characters treated as a distinct character state. Note Analysis 2 resulted in a strict consensus tree identical in topology. D, Diphyllidea; L, Lecanicephalidea; P, Proteocephalidea; T, Tetraphyllidea; T(O), Tetraphyllidea (Onchobothriidae); T(P), Tetraphyllidea (Phyllobothriidae); TR, Trypanorhyncha.

Discussion

Taxon sampling

The analysis of 127 taxa (Analysis 3) resulted in a consensus tree that was generally congruent with, but substantially more resolved than, the trees resulting from either of the analyses of 63 taxa (Analyses 1 and 2). The analyses of the smaller number of taxa supported only an onchobothriid clade, a lecanicephalidean clade and four minor phyllobothriid clades, whereas all but seven of the nodes in the tree resulting from analysis of the 127 taxa were fully resolved. This increase in resolution is generally consistent with the simulations of Hillis (1996). Of course, the accuracy of the groupings represented in the more resolved tree remains to be determined.

Character sampling

The consensus tree resulting from Analysis 1 (see Figure 14.95) differed somewhat in topology from the tree resulting from the Comprehensive analysis conducted by Caira et al. (1999; figure 87). Only two of the four major groups seen in that study were supported by analysis of the more extensive character data in the present study. These consisted of a clade

including the majority of the lecanicephalideans and a clade consisting of the onchobothriids. Minor differences in the relationships among the members of both clades were seen between the two analyses. For example, the outgroup species *Proteocephalus exiguus* was a member of the lecanicephalidean clade in Analysis 2 (see Figure 14.95), but not in that of Caira et al. (1999). Neither of the two major phyllobothriid clades seen in Caira et al. (1999) were supported by the more extensive character data, although a small group consisting of four phyllobothriids was present. Clearly, the additional character data alone had some effect on the relationships recovered.

Ten per cent exclusion rule

The analyses of the 127 taxa using all characters (Analysis 4) and the 127 taxa using all but the nine characters that were unknown in > 10% of the taxa (Analysis 3) resulted in trees with fairly congruent, but not identical topologies. The former analysis resulted in a consensus tree that was much less resolved than that resulting from the latter analysis. The former tree provided no resolution of the relationships among the unloculate onchobothriid groups (e.g., *Pachybothrium*, *Pedibothrium*, *Potamotrygonocestus*) and little resolution for

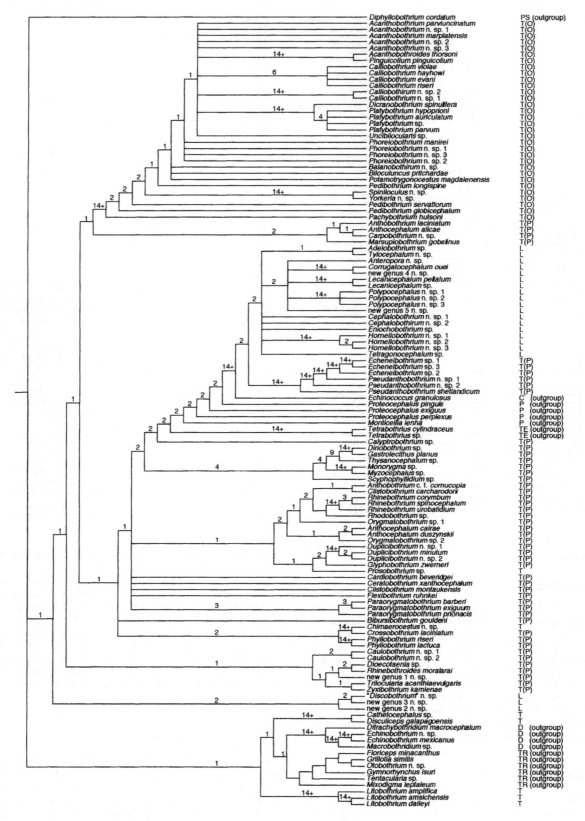

Figure 14.96 Strict consensus of most-parsimonious trees resulting from Analysis 3; 127 taxa and 148 characters of data matrix in Appendix 14.3 (i.e., excluding characters 84, 85, 91, 93, 100, 105, 106, 107 and 157); inapplicable characters treated as a distinct character state. Nodal support values are shown next to each node. C, Cyclophyllidea; D, Diphyllidea; L, Lecanicephalidea; P, Proteocephalidea; PS, Pseudophyllidea; T, Tetraphyllidea; TE, Tetrabothriidea; T(O), Tetraphyllidea (Onchobothriidae); T(P), Tetraphyllidea (Phyllobothriidae); TR, Trypanorhyncha.

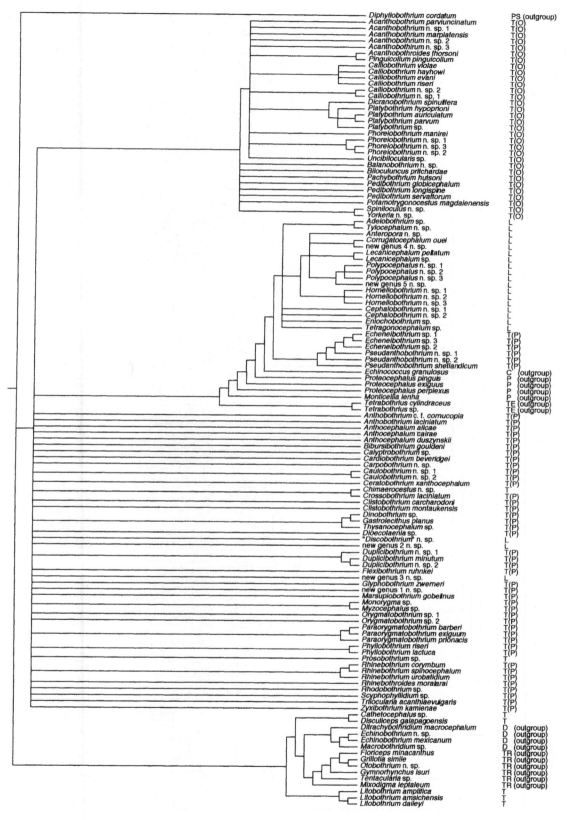

Figure 14.97 Strict consensus of most-parsimonious trees resulting from Analysis 4; 127 taxa and 157 characters of data matrix in Appendix 14.3; inapplicable characters treated as a distinct character state. C, Cyclophyllidea; D, Diphyllidea; L, Lecanicephalidea; P, Proteocephalidea; PS, Pseudophyllidea; T, Tetraphyllidea; TE, Tetrabothriidea; T(O), Tetraphyllidea (Onchobothriidae); T(P), Tetraphyllidea (Phyllobothriidae); TR, Trypanorhyncha.

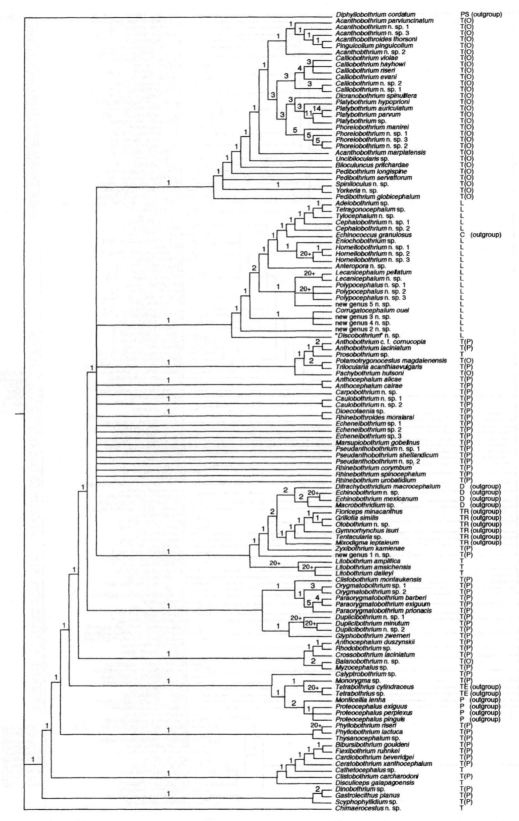

Figure 14.98 Strict consensus of most-parsimonious trees resulting from Analysis 5; 127 taxa and 157 characters of data matrix in Appendix 14.3; inapplicable characters treated as unknown characters. Nodal support values are shown next to each node. C, Cyclophyllidea; D, Diphyllidea; L, Lecanicephalidea; P, Proteocephalidea; PS, Pseudophyllidea; T, Tetraphyllidea; TE, Tetrabothriidea; T(O), Tetraphyllidea (Onchobothriidae); T(P), Tetraphyllidea (Phyllobothriidae); TR, Trypanorhyncha.

most of the phyllobothriid groups. This analysis did, however, provide greater resolution within the lecanicephalidean clade.

Treatment of inapplicable characters

The shortest trees resulting from treating inapplicable characters as unknown (Analysis 5) were 373 steps shorter and comparatively more resolved than the trees resulting from analysis in which the inapplicable characters were treated as a separate character state (Analysis 4). However, nodal support values suggest that almost 70% of the nodes on the tree from Analysis 5 are not supported on trees one step longer. The trees share some points of congruence. An onchobothriid and a lecanicephalidean clade are present on both trees, although the membership of the clades differs somewhat between the two trees. For example, in the tree resulting from Analysis 5, the lecanicephalidean clade includes one of the new genera (new genus 3), but in the tree from Analysis 4, this taxon was located well outside of the lecanicephalideans, among the tetraphyllideans. The six species of *Pseudanthobothrium* Baer, 1956 and *Echeneibothrium* van Beneden, 1849 do not group with the lecanicephalideans in Analysis 5; rather, they are part of an unresolved polytomy with the lecanicephalideans and many of the other phyllobothriid groups. The proteocephalidean species also do not group with the lecanicephalideans in Analysis 5, rather they are among the most basal taxa on the tree.

One of the most surprising results of Analysis 5 was that the monophyly of the onchobothriids was not supported. Although an onchobothriid clade is present on the tree, this group does not include all of the hooked taxa. *Potamotrygonocestus magdalenensis* is grouped as the sister taxon to *Trilocularia acanthiaevulgaris*, and *Balanobothrium* sp. is found even more basally as the sister taxon to *Myzocephalus* sp.

However, perhaps the most unexpected feature of the tree resulting from treatment of the inapplicable characters as unknown, was that it supported a monophyletic group consisting of a number of phyllobothriids and all of the diphyllidean and trypanorhynchan outgroup species. In fact, *Diphyllobothrium cordatum* was the only outgroup species that remained outside of all of the ingroup species in this tree. Again, it should be noted that the nodal support values suggest that these nodes are not well supported and thus this topology should be considered with some reservation.

Relationships among Tetraphyllidea and Lecanicephalidea

The trees resulting from the three different analyses of all 127 taxa differed in topology and/or resolution. Because of our reservations regarding the treatment of inapplicable characters, we have most confidence in the topology of trees resulting from Analyses 3 and 4. In light of the discussion of Wiens (1998) on the benefits of a complete sampling of taxa over a greater sampling of characters, we have deferred to the tree resulting from analysis of a greater percentage of known data (Analysis 3) for discussion of the relationships among the Tetraphyllidea and Lecanicephalidea.

In addition to fairly well-resolved onchobothriid and lecanicephalidean clades, the tree resulting from Analysis 3 supports six clades consisting of species of three or more phyllobothriid genera. Nodal support values suggest that the onchobothriid and lecanicephalidean clades are well supported, but that there is relatively little support for most of the phyllobothriid clades,

only one of which remains in trees greater than two steps longer. This group, consisting of *Scyphophyllidium* sp., *Myzocephalus* sp., *Monorygma* sp., *Thysanocephalum* sp., *Gastrolecithus planus*, and *Dinobothrium* sp., is supported in trees up to four steps longer, and the group consisting of the latter three taxa is supported in trees up to nine steps longer. The grouping of *Myzocephalus* sp. close to *Thysanocephalum* sp. suggests that it may be more appropriate to consider the folded region of the scolex of *Myzocephalus* sp. as acetabular rather than peduncular in origin.

Relationships among the onchobothriids were fairly well resolved, but it should be noted that none of these nodes is especially well supported and that these relationships differ somewhat from those suggested in the analysis of Caira et al. (1999). The most basal onchobothriids have unloculate acetabula and at least the most basal taxon has unipronged hooks. Multipronged hooks are derived, as is the existence of two rather than a single pair of hooks. The more basal placement of *Biloculuncus pritchardae* relative to the six species of *Calliobothrium* van Beneden, 1850 suggests that bipronged hooks may have split more than once to form two independent pairs of hooks.

This analysis supports the monophyly of each of the following ingroup genera: *Platybothrium* Linton, 1891, *Polypocephalus* Braun, 1878, *Hornellobothrium* Shipley and Hornell, 1906, *Echeneibothrium*, *Rhinebothrium* Linton, 1891, *Duplicibothrium* Williams and Campbell, 1978, *Paraorygmatobothrium* Ruhnke, 1994, *Phyllobothrium*, and *Litobothrium*. However, caution should be exercised regarding statements of generic monophyly in analyses such as this because inclusion of additional members of each genus could significantly alter such statements. Even in this analysis, the nodes of some of these groups are not well supported. The analysis failed to provide evidence either in support of or against the monophyly of *Acanthobothrium* van Beneden, 1849, *Calliobothrium*, *Phoreiobothrium* Linton, 1889 and *Cephalobothrium*, as the relationships within each of these groups were not sufficiently resolved. Perhaps most interesting, this analysis calls into question the monophyly of each of the following ingroup genera: *Pedibothrium*, *Anthocephalum* Linton, 1890, *Anthobothrium* van Beneden, 1849, *Clistobothrium* Dailey and Vogelbein, 1990, *Pseudanthobothrium* and *Orygmatobothrium* Diesing, 1863.

With respect to the monophyly of the Lecanicephalidea, the results of this analysis are generally consistent with those of Caira et al. (1999) in that limited support for the monophyly of the order is provided. The results of the analysis suggest that two of the new genera (new genera 2 and 3) should not necessarily be considered as lecanicephalideans. Despite their conspicuously lecanicephalidean segment morphology, these two taxa are part of a lineage that is sister to most of the other lecanicephalidean and tetraphyllidean species included in the analysis. It is likely that the lack of both an apical modification of the scolex proper and an apical organ from these taxa had a major influence on their position in the tree. Also consistent with the results of Caira et al. (1999), the species of *Echeneibothrium* and *Pseudanthobothrium* grouped as close relatives of the lecanicephalideans, as did the outgroup proteocephalidean species *Monticellia lenha*.

In this study, *Cathetocephalus* sp., *Disculiceps galapagoensis* and the three litobothriids all grouped with the majority of the outgroup species rather than with members of the ingroup. If our interpretation of the unusual (non-tetrafossate) scoleces of each of these groups is correct, it is possible that this is an appropriate placement for these taxa, i.e., that they are basal groups. However, we remain uncertain of the homologies of the various features of the scolex either among these groups or relative to the other tetraphyllidean and lecanicephalidean species. This basal position may be an artifact of the misinterpretation of the scolex morphology in one or more of these groups.

To date, both Olson and Caira (1999) and Olson *et al.* (1999) have conducted molecular studies that included at least modest representation of the lecanicephalidean and tetraphyllidean genera and which, thus, might provide independent evaluation of our results regarding some of the more specific sets of relationships. Unfortunately, the relationships suggested by these analyses differ with respect to one another. For example, the two lecanicephalidean species included are relatively basal (Olson *et al.* 1999) or relatively derived (Olson and Caira 1999). The included onchobothriids are monophyletic (Olson and Caira 1999) or not (Olson *et al.* 1999), depending on the gene used as the source of the data and the specific taxa included in the analysis. The results of both studies also differ somewhat from the relationships suggested by our data. Regardless, it should be noted that none of the deeper branches in any of these trees or the trees presented in the present study is supported by greater than one or two steps. Thus, all of these trees should be considered to be nothing more than useful approximations of the relationships among these taxa. This system seems primed for a total evidence approach combining morphological and molecular data.

The general results of this analysis are consistent with much previous work (Hoberg *et al.* 1997c; Mariaux 1998; Caira *et al.* 1999; Olson and Caira 1999) in that they support the monophyly of a group consisting of tetrafossate species, but question the monophyly of the Tetraphyllidea with respect to the other members of that broader tetrafossate group. Caira *et al.* (1999) included two tetrafossate (proteocephalidean) species as members of the outgroup in their analysis, yet both species consistently grouped among the tetraphyllideans. In the present study, the outgroup was expanded beyond these two taxa to include one or more representatives from two additional tetrafossate orders, as well as two more proteocephalidean species. Without exception, all of these species grouped among the tetraphyllidean and/or lecanicephalidean taxa. It is especially interesting that one of the most stable taxa in the analysis was the cyclophyllidean outgroup species *Echinococcus granulosus*, which consistently grouped deep within the ingroup near the lecanicephalidean clade.

Our data suggest that it is possible that several different lineages of tapeworms have closest relatives that parasitize elasmobranchs. It seems that the proteocephalideans represent one (or more?) lineage that has invaded freshwater fish from a basal association with elasmobranchs. There is very preliminary support to suggest that this may also be true independently for the tetrabothriideans and the cyclophyllideans, but this result is in conflict with the relationships suggested by the recent molecular analyses that included sufficient taxon representation to allow this question to be even superficially addressed. Both of these studies (Mariaux 1998; Olson and Caira 1999) support a sister taxon relationship between the tetrabothriideans and cyclophyllideans. It is clear that analysis of this issue would benefit greatly from the inclusion of a significant number of species (and characters) representing the other tetrafossate orders; most interesting would be members of the orders Proteocephalidea and Cyclophyllidea, but additional tetrabothriidean representation would also be useful. Clearly, this type of expanded analysis is necessary if the issue of the monophyly of the Tetraphyllidea is to be convincingly resolved.

Our results indicate that the replicated exemplar approach is preferable to the single exemplar approach because addition of taxa led to greater resolution, even though both analyses were based on essentially the same set of characters. Admittedly, however, our selections of additional exemplars were based to a large extent on availability of material, rather than on any of the useful strategies for multiple exemplar selection provided by Yeates (1995). At least some of the exemplars were chosen from within monophyletic groups (i.e. congeners) and thus, based on the simulations of Rannala *et al.* (1998), may have increased the accuracy of the tree, but this is difficult to assess with empirical data.

We remain convinced that it is preferable to code inapplicable characters with a discrete character state rather than as unknown characters because of the resolution that was achieved with the latter coding. The difference in tree topology between these two coding schemes was dramatic both in terms of resolution and supported groupings. That the tree with the less traditional, but better resolved topology (resulting from the '?' coding), was based on a matrix in which almost 40% of the character/taxon combinations were coded as unknown, leads us to have less confidence in the topology of this tree than in the tree resulting from the alternate coding strategy. That is not to say that the potential problems associated with coding inapplicable characters with a distinct character state are trivial, but we find this method to be preferable given the limitations of all currently available analysis programs.

Given the lack of consensus among trees resulting from analyses using different parameters, it cannot be said that the relationships among the lecanicephalidean and tetraphyllidean groups are well understood at this point in time. Support is accumulating for several broad generalizations such as the monophyly of the Lecanicephalidea and the non-monophyly of the Tetraphyllidea, but the intergeneric relationships are far from clear. Especially unclear are the interrelationships among the phyllobothriid genera, but non-monophyly of the onchobothriids is also an intriguing idea that certainly deserves additional investigation. It seems wise to refrain from discussions of the coevolutionary aspects of this remarkably host specific system until a more stable (and possibly accurate) hypothesis of the interrelationships of these groups is obtained.

ACKNOWLEDGEMENTS

We are grateful to the following individuals for providing us with fresh fixed material for this study: P. Olson (*Proteocephalus pinguis*, and *P. perplexus*), T. Osipa (new genus 1), R. Rausch (*Echinococcus granulosus* and *Diphyllobothrium cordatum*), C. Richmond (*Otobothrium* n. sp.), N. Riser (*Scyphophyllidium* sp.), T. Ruhnke (*Duplicibo-thrium* n. sp. 1 and 2) and G. Tyler (*Echinobothrium mexicanum*). In addition, we thank J. R. Lichtenfels of the U. S. National Parasite Collection, Beltsville, Maryland, U.S.A. and S. Kamegai of the Meguro Parasitological Museum, Tokyo, Japan for lending specimens. We found the discussions of apical structures that occurred at the 3rd International Workshop for Systematic Zoology of Tapeworms, held in Sofia, Bulgaria by B. Georgiev, to be of much assistance with the interpretation of scolex features. We thank E. Hoberg and T. Scholz for assistance with identification of some of the tetrabothriidean and proteocephalidean specimens, respectively. We are especially grateful to M. Cantino and J. Romanow, of the University of Connecticut Electron Microscopy Facility, for their patience, tolerance and expert technical assistance during the numerous hours we spent in that facility conducting the SEM for this study. K. Nixon provided invaluable assistance and advice greatly facilitating our use of NONA for all phylogenetic analyses.

Appendix 14.1 Material and literature used to code for each taxon in the data matrix. Taxa are listed alphabetically within each order.

SPECIES (ex. HOST) used for coding	LOCALITY of material used for coding	SPECIMENS*† used for coding	LITERATURE used for coding
TETRAPHYLLIDEA			
Acanthobothrium marplatensis Ivanov and Campbell, 1998 ex. *Rioraja castelnaui* (Ribeiro)	Mar del Plata, Buenos Aires, Argentina	1 w.m. (paratype, USNPC 87475)	Ivanov and Campbell 1998
Acanthobothrium n. sp. 1 ex. *Heterodontus francisci* (Girard)	Puertecitos, Baja California, México	2 w.m.; 2 seg. c.s.; 3 spcm. SEM	
Acanthobothrium n. sp. 2 ex. *Urobatis maculatus* Garman	Santa Rosalia, Baja California, México	3 w.m.; 2 seg. c.s.; 1 spcm. SEM	
Acanthobothrium n. sp. 3 ex. *Urobatis halleri* (Cooper)	Bahia de Los Angeles, Baja California, México	3 w.m.; 4 seg. c.s.; 1 spcm. SEM	
Acanthobothrium parviuncinatum Young, 1954 (see Caira *et al.* 1999)			
Acanthobothroides thorsoni Brooks, 1977 (see Caira *et al.* 1999)			
Anthobothrium c.f. *cornucopia* ex. *Galeorhinus galeus* (Linnaeus)	San Remo, Victoria, Australia	3 w.m.; 3 seg. c.s.; 1 spcm. SEM	
Anthobothrium laciniatum Linton, 1890 (see Caira *et al.* 1999)			
Anthocephalum alicae Ruhnke, 1994 (see Caira *et al.* 1999)		6 w.m.; 1 sc. c.s.; 1 sc. l.s.; 1 spcm. SEM	
Anthocephalum cairae Ruhnke, 1994 ex. *Dasyatis americana* Hildebrand and Schroeder	Florida Bay, Florida Keys, USA; Bimini Islands, Bahamas	None	Ruhnke 1994a
Anthocephalum duszynskii Ruhnke, 1994 ex. *Urobatis halleri* (Cooper) and *Urobatis maculatus* Garman	Cholla Bay and Isla San Esteban, Baja California, México	1 w.m.	Ruhnke 1994a
Balanobothrium n. sp. (see Caira *et al.* 1999)			
Bibursibothrium gouldeni McKenzie and Caira, 1998 (see Caira *et al.* 1999)		4 w.m.	
Biloculuncus pritchardae (Caira and Ruhnke, 1990) Nasin, Caira and Euzet, 1997 (see Caira *et al.* 1999)			
Calliobothrium evani Caira, 1985 ex. *Mustelus lunulatus* Jordan and Gilbert	Santa Rosalia, Baja California, México	1 w.m.; 1 seg. c.s.; 1 spcm. SEM	Caira 1985 (for hooks); Nasin *et al.* 1997
Calliobothrium hayhowi Nasin, Caira and Euzet, 1997 ex. *Mustelus antarcticus* Günther	Port Philip Bay, Port Albert and San Remo, Victoria, Australia	4 w.m.	Nasin *et al.* 1997
Calliobothrium n. sp. 1 ex. *Mustelus canis* (Mitchill)	Long Island Sound, Connecticut, USA	2 w.m.; 2 seg. c.s.; 1 spcm. SEM	
Calliobothrium n. sp. 2 ex. *Mustelus antarcticus* Günther	San Remo, Victoria, Australia	2 w.m.; 5 seg. c.s.; 1 spcm. SEM	
Calliobothrium riseri Nasin, Caira and Euzet, 1997 ex. *Mustelus henlei* Gill	Puertecitos and Santa Rosalia, Baja California, México	None	Nasin *et al.* 1997
Calliobothrium violae Nasin, Caira and Euzet, 1997 (see Caira *et al.* 1999)		1 w.m.; 3 seg. c.s. TEM	
Calyptrobothrium sp. (non *Calyptrobothrium sensu* Euzet, 1994) (see Caira *et al.* 1999)			
Cardiobothrium beveridgei McKenzie and Caira, 1998 (see Caira *et al.* 1999)			
Carpobothrium n. sp. (see Caira *et al.* 1999)			
Cathetocephalus sp. (see Caira *et al.* 1999)			
Caulobothrium n. sp. 1 (see Caira *et al.* 1999)			
Caulobothrium n. sp. 2 ex. *Myliobatis californicus* Gill	Bahia de Los Angeles, Baja California, México	4 w.m.; 2 seg. c.s.; 1 spcm. SEM	
Ceratobothrium xanthocephalum Monticelli, 1892 (see Caira *et al.* 1999)		1 w.m.	
Chimaerocestos n. sp. (see Caira *et al.* 1999)			
Clistobothrium carcharodoni Dailey and Vogelbein, 1990 (see Caira *et al.* 1999)		1 w.m.	
Clistobothrium montaukensis Ruhnke, 1993 ex. *Isurus oxyrinchus* Rafinesque	Atlantic Ocean off Montauk, New York, USA	2 w.m.; 3 seg. c.s.; 3 sc. l.s.; 1 spcm. SEM	Ruhnke 1993
Crossobothrium laciniatum Linton, 1889 (see Caira *et al.* 1999)			
Dicranobothrium spinulifera (Southwell, 1912) Euzet, 1953 (see Caira *et al.* 1999)			
Dinobothrium sp. (see Caira *et al.* 1999)			
Dioecotaenia sp. (see Caira *et al.* 1999)			
Disculiceps galapagoensis Nock and Caira, 1988 (see Caira *et al.* 1999)			
Duplicibothrium minutum Williams and Campbell, 1978 (see Caira *et al.* 1999)		5 w.m.	
Duplicibothrium n. sp. 1 ex. *Rhinoptera steindachneri* Everman and Jenkins	Puertecitos, Baja California, México	2 w.m.; 1 spcm. SEM	Ruhnke *et al.* 2000
Duplicibothrium n. sp. 2 ex. *Rhinoptera steindachneri* Everman and Jenkins	Puertecitos, Baja California, México	4 w.m. 1 spcm. SEM	Ruhnke *et al.* 2000
Echeneibothrium sp. 1 (see Caira *et al.* 1999 as *Echeneibothrium* sp.)			
Echeneibothrium sp. 2 ex. *Raja fyllae* Lütken	Faroe Islands, North Sea	4 w.m.; 2 seg. c.s.; 1 spcm. SEM	
Echeneibothrium sp. 3 ex. *Raja erinacea* Mitchill	Long Island Sound, USA	2 w.m.; 1 seg. c.s.; 1 spcm. SEM	
Flexibothrium ruhnkei McKenzie and Caira, 1998 (see Caira *et al.* 1999)			
Gastrolecithus planus (Linton, 1922) Yamaguti, 1952 (see Caira *et al.* 1999)			
Glyphobothrium zwerneri Williams and Campbell, 1977 (see Caira *et al.* 1999)			
Litobothrium amplifica (Kurochkin and Slankis, 1973) Euzet 1994 ex. *Alopias pelagicus* Nakamura	Bahia de Los Angeles, Baja California, México	3 w.m.; 2 seg. c.s.; 2 spcm. SEM	
Litobothrium amsichensis Caira and Runkle, 1993 ex. *Mitsukurina owstoni* Jordan	ESE of Ulladulla, New South Wales, Australia	3 spcm. SEM	Caira and Runkle 1993
Litobothrium daileyi Kurochkin and Slankis, 1973 (see Caira *et al.* 1999)			
Marsupiobothrium gobelinus Caira and Runkle, 1993 (see Caira *et al.* 1999)			
Monorygma sp. (see Caira *et al.* 1999)			

Appendix 14.1 Continued.

SPECIES (ex. HOST) used for coding	LOCALITY of material used for coding	SPECIMENS*† used for coding	LITERATURE used for coding
Myzocephalus sp. ex. *Aetobatus narinari* (Euphrasen)	Lee Point, Darwin, Northern Territory, Australia	7 w.m.; 3 seg. c.s.; 2 sc. c.s.; 1 sc. l.s.; 1 spcm. SEM	
Orygmatobothrium sp. 1 (see Caira *et al.* 1999)			
Orygmatobothrium sp. 2 ex. *Mustelus manazo* Bleeker	Prefecture Aomori, Honshu, Aomori, Mutsu Bay, Japan	3 w.m.; 2 seg. c.s.; 1 spcm. SEM	
Pachybothrium hutsoni (Southwell, 1911) Baer and Euzet, 1962 (see Caira *et al.* 1999)			
Paraorygmatobothrium barberi Ruhnke, 1994 ex. *Triakis semifasciata* Girard	Hermosa Beach Pier, Hermosa Beach, California, USA	None	Ruhnke 1994b
Paraorygmatobothrium exiguum (Yamaguti, 1935) Ruhnke, 1994 (see Caira *et al.* 1999)			
Paraorygmatobothrium prionacis (Yamaguti, 1934) Ruhnke, 1994 ex. *Prionace glauca* (Linnaeus)	Japanese coastal waters, Japan Atlantic Ocean off Montauk, New York and South Yarmouth, Massachussets, USA	None	Ruhnke 1994b
Pedibothrium globicephalum Linton, 1908 ex. *Ginglymostoma cirratum* (Bonnaterre)	Dry Tortuga Islands, Florida, USA	3 w.m.; 1 spcm. SEM	Caira 1992
Pedibothrium longispine Linton, 1909 (see Caira *et al.* 1999)			
Pedibothrium servattorum Caira, 1992 ex. *Ginglymostoma cirratum* (Bonnaterre)	Bimini Islands, Bahamas	None	Caira 1992
Phoreiobothrium manirei Caira, Healy and Swanson, 1996 ex. *Sphyrna mokarran* (Rüppell)	Pine Island Sound, Florida, USA	1 w.m.; 2 sc. c.s.; 4 sc. and seg. l.s.; 1 spcm. SEM	Caira *et al.* 1996
Phoreiobothrium n. sp. 1 (see Caira *et al.* 1999)			
Phoreiobothrium n. sp. 2 ex. *Sphyrna lewini* (Griffith and Smith)	Woods Hole, Massachusetts, USA	4 w.m.; 1 seg. c.s.; 1 spcm. SEM	
Phoreiobothrium n. sp. 3 ex. *Carcharhinus leucas* (Müller and Henle (ex.Valenciennes))	Bahia de Los Angeles, Baja California, México	4 w.m.; 1 seg. c.s.; 1 spcm. SEM	
Phyllobothrium lactuca van Beneden, 1849 (see Caira *et al.* 1999)			
Phyllobothrium riseri Ruhnke, 1996 ex. *Triakis semifasciata* Girard	Monterey Bay, California, USA	None 3 seg. c.s.	Ruhnke 1996
Pinguicollum pinguicollum (Sleggs, 1927) Riser, 1955 (see Caira *et al.* 1999)			
Platybothrium auriculatum Yamaguti, 1952 ex. *Prionace glauca* (Linnaeus)	Pacific coast of Mie Prefecture, Japan La Paz, Baja California, México Atlantic Ocean off Montauk, New York, USA	1 w.m. (holotype, MPM 22812); 1 spcm. SEM 1 w.m.; 2 seg. c.s.	
Platybothrium hypoprioni Potter, 1937 ex. *Negaprion brevirostris* (Poey)	Dry Tortuga Islands, Florida, USA Bimini Islands, Bahamas	4 w.m. (2 cotypes, USNPC 9062); 1 seg. c.s.; 1 spcm. SEM	
Platybothrium parvum Linton, 1901 ex. *Sphyrna lewini* (Griffith and Smith) and *Sphyrna zygaena* (Linnaeus)	Bahia de Los Angeles and San Jose del Cabo, Baja California, México	4 w.m.; 1 seg. c.s.; 1 spcm. SEM	
Platybothrium sp. (see Caira *et al.* 1999 as *Platybothrium cervinum* Linton, 1890)	Woods Hole, Massachusetts, USA	2 w.m. (USNPC 7690)	
Potamotrygonocestus magdalenensis Brooks and Thorson, 1976 (see Caira *et al.* 1999)			
Prosobothrium sp. (see Caira *et al.* 1999)			
Pseudanthobothrium n. sp. 1 ex. *Raja erinacea* (see Caira *et al.* 1999)			
Pseudanthobothrium n. sp. 2 ex. *Raja senta* Garman	Grand Banks, Newfoundland, Canada	3 w.m.; 2 seg. c.s.	
Pseudanthobothrium shetlandicum Wojciechowska, 1990 ex. *Bathyraja maccaini* Springer	Elephant Shelf, South Shetlands	10 w.m.; 9 seg. c.s.; 2 seg. l.s.; 3 sc. c.s.; 13 sc. l.s.; 1 spcm. SEM	
Rhinebothrium corymbum Campbell, 1975 ex. *Dasyatis americana* Hildebrand and Schroeder	Gulf of Mexico off Florida, USA	6 w.m.; 5 seg. c.s.; 1 seg. l.s.; 1 sc. c.s.; 2 sc. l.s.; 1 spcm. SEM	Campbell 1975
Rhinebothrium spinocephalum Campbell, 1970 ex. *Dasyatis americana* Hildebrand and Schroeder	Gulf of Mexico off Florida, USA	2 w.m.; 1 seg. c.s.; 1 spcm. SEM	Campbell 1970
Rhinebothrium urobatidium (see Caira *et al.* 1999)			
Rhinebothroides moralarai (Brooks and Thorson, 1976) Mayes, Brooks and Thorson, 1981 (see Caira *et al.* 1999)			
Rhodobothrium sp. (see Caira *et al.* 1999)			
Scyphophyllidium sp. ex. *Galeorhinus zyopterus* Jordan and Gilbert	Monterey Bay, California, USA	4 w.m.; 3 seg. c.s.; 2 sc. l.s.	
Spiniloculus n. sp. (see Caira *et al.* 1999)			
Thysanocephalum sp. (see Caira *et al.* 1999)			
Trilocularia acanthiaevulgaris (Olsson, 1867) (see Caira *et al.* 1999)			
Uncibilocularis sp. ex. *Pristis zijsron* Bleeker	Buffalo Creek, Darwin, Northern Territory, Australia	4 w.m.; 6 seg. c.s.; 1 spcm. SEM	
Yorkeria n. sp. (see Caira *et al.* 1999)			
Zyxibothrium kamienae Hayden and Campbell, 1981 (see Caira *et al.* 1999)			
new genus 1 n. sp. ex. *Parascyllium collare* Ramsay and Ogilby	Sydney, New South Wales, Australia	2 w.m.; 1 seg. sec.; 1 spcm. SEM	
LECANICEPHALIDEA			
Adelobothrium sp. ex. *Aetobatus narinari* (Euphrasen)	Lee Point, Darwin, Northern Territory, Australia	2 w.m.; 1 seg. c.s.; 1 seg. l.s.; 2 sc. l.s.; 1 spcm.	
Anteropora n. sp. (see Caira *et al.* 1999)			
Cephalobothrium n. sp. 1 ex. *Aetobatus narinari* (Euphrasen)	Bangsaray, Gulf of Thailand	4 w.m.; 8 seg. c.s.; 2 sc. l.s.; 2 spcm. SEM	
Cephalobothrium n. sp. 2 ex. *Aetobatus narinari* (Euphrasen)	Bangsaray, Gulf of Thailand	4 w.m.; 1 w.w. l.s.; 1 spcm. SEM	
Corrugatocephalum ouei Caira, Jensen and Yamane, 1997 (see Caira *et al.* 1999)			
'Discobothrium' n. sp. (see Caira *et al.* 1999)			
Eniochobothrium sp. (see Caira *et al.* 1999)			
Hornellobothrium n. sp. 1 (see Caira *et al.* 1999)			

Appendix 14.1 Continued.

SPECIES (ex. HOST) used for coding	LOCALITY of material used for coding	SPECIMENS*† used for coding	LITERATURE used for coding
Hornellobothrium n. sp. 2 ex. *Aetobatus narinari* (Euphrasen)	Lee Point, Darwin, Northern Territory, Australia	5 w.m.; 3 seg. c.s.; 2 w.w. l.s.; 2 spcm. SEM	
Hornellobothrium n. sp. 3 ex. *Aetobatus narinari* (Euphrasen)	Lee Point, Darwin, Northern Territory, Australia	2 w.m.; 1 w.w. c.s.; 1 w.w. l.s.; 1 spcm. SEM	
Lecanicephalum peltatum Linton, 1890 ex. *Dasyatis centroura* (Mitchill) (see Caira *et al.* 1999)			
Lecanicephalum sp. ex. *Dasyatis centrura* (sic) (Mitchill)	Atlantic Ocean off North Carolina, USA	4 w.m.; 4 w.w. c.s.; 2 w.w. l.s.; 2 sc. l.s.; 5 spcm. SEM	
Polypocephalus n. sp. 1 (see Caira *et al.* 1999)			
Polypocephalus n. sp. 2 ex. *Rhinoptera* sp.	Fog Bay, Northern Territory, Australia	4 w.m.; 4 seg. c.s.; 2 w.w. l.s.; 2 spcm. SEM	
Polypocephalus n. sp. 3 ex. *Dasyatis centroura* (Mitchill)	Atlantic Ocean off North Carolina, USA	3 w.m.; 1 seg. c.s. and sc. c.s.; 2 sc. l.s. and seg. c.s.	
Tetragonocephalum sp. ex. *Himantura* sp.	Lee Point, Darwin, Northern Territory, Australia	5 w.m.; 1 seg. c.s.; 1 seg. l.s.; 1 sc. l.s.; 1 spcm. SEM	
Tylocephalum n. sp. (see Caira *et al.* 1999)			
new genus 2 n. sp. ex. *Squatina california* Ayres	Santa Rosalia, Baja California, México	6 w.m.; 2 seg. c.s.; 1 sc. l.s.; 1 spcm. SEM	
new genus 3 n. sp. ex. *Mobula japonica* (Eschmeyer and Herald)	Punta Arenas, Baja California, México	4 w.m.; 1 w.w. c.s.; 3 w.w. l.s.; 2 spcm. SEM	
new genus 4 n. sp. ex. *Mobula japonica* (Eschmeyer and Herald)	Punta Arenas, Baja California, México	4 w.m.; 1 w.w. c.s.; 2 w.w. l.s.; 1 sc. l.s.; 2 spcm. SEM	
new genus 5 n. sp. ex. *Pastinachus sephen* Forsskål	Buffalo Creek, Darwin, Northern Territory, Australia	3 w.m.; 1 seg. c.s.; 1 sc. l.s.; 1 spcm. SEM	
CYCLOPHYLLIDEA			
Echinococcus granulosus (Batsch, 1786) Rudolphi 1801 ex. *Canis familiaris* Linnaeus	Argentina	3 w.m.; 4 w.w. c.s.; 3 spcm. SEM	
DIPHYLLIDEA			
Ditrachybothridium macrocephalum Rees 1959 (see Caira *et al.* 1999)			
Echinobothrium mexicanum Tyler and Caira 1999 ex. *Myliobatis longirostris* Applegate and Fitch	Loreto, Baja California, México	2 w.m.; 1 seg. c.s.; 1 spcm. SEM	Tyler and Caira 1999
Echinobothrium n. sp. (see Caira *et al.* 1999)			
Macrobothridium sp. (see Caira *et al.* 1999)			
PROTEOCEPHALIDEA			
Monticellia lenha Woodland 1933 (see Caira *et al.* 1999)			
Proteocephalus exiguus La Rue, 1911 (see Caira *et al.* 1999)			
Proteocephalus perplexus La Rue, 1911 ex. *Amia calva* Linnaeus	Hay Bay, Lake Ontario, Ontario, Canada	8 w.m.; 2 seg. c.s.; 2 sc. l.s.; 1 spcm. SEM	La Rue 1914
Proteocephalus pinguis La Rue 1911 ex. *Esox lucius* Linnaeus	Connecticut River and Haddam Lake, Connecticut, USA	3 w.m.; 6 seg. c.s.; 1 spcm. SEM	
PSEUDOPHYLLIDEA			
Diphyllobothrium cordatum (Leuckart 1863) ex. *Phoca vitulina* De Kay	Kvichak River, Lovelock, Alaska, USA	9 w.m.; 2 seg. c.s.; 4 seg. l.s.; 4 sc. c.s.; 1 spcm. SEM	
TETRABOTHRIIDEA			
Tetrabothrius cylindraceus Rudolphi 1819 ex. *Larus atricilla* Linnaeus	Chadwich Beach, Englewood, Florida, USA	1 w.m. (USNPC 89331); 3 w.m.; 2 seg. c.s.; 1 sc. l.s.; 1 spcm. SEM	
Tetrabothrius sp. ex. *Larus occidentalis* Audubon	San Diego, California, USA	1 w.m. (USNPC 89332); 4 w.m.; 1 seg. c.s.; 1 sc. l.s.; 1 spcm. SEM	
TRYPANORHYNCHA			
Floriceps minacanthus Campbell and Beveridge 1987 (see Caira *et al.* 1999)			
Grillotia similis (Linton 1908) Caira and Gavarrino 1990 (see Caira *et al.* 1999)		10 w.m.	
Gymnorhynchus isuri Robinson 1959 ex. *Isurus oxyrinchus* Rafinesque	Atlantic Ocean off Montauk, New York, USA	3 w.m.; 3 seg. c.s.; 3 sc. c.s.; 1 spcm. SEM	Caira and Bardos 1996; Robinson 1959
Mixodigma leptaleum Dailey and Vogelbein 1982 ex. *Megachasma pelagios* Taylor, Compagno and Struhsaker	Hakata Bay, Fukuoka, northern Kyushu, Japan	2 w.m.; 2 seg. w.m.; 1 seg. c.s.; 1 sc. c.s.; 1 sc. l.s.; 2 spcm. SEM	Dailey and Vogelbein 1982; Caira *et al.* 1997
Otobothrium n. sp. ex. *Negaprion acutidens* Rüppell	Buffalo Creek, Darwin, Northern Territory, Australia	1 w.m.; 1 seg. c.s.; 1 spcm SEM	
Tentacularia sp. ex. *Prionace glauca* (Linnaeus)	Atlantic Ocean off Montauk, New York, USA	8 w.m.; 4 seg. c.s.; 1 spcm. SEM	

* When specimens alone, or a combination of specimens and literature, were used for coding, the host and locality data listed above relate only to the specimens used. When coding was done solely from the literature, the host and locality data listed above are from that literature.

†Unless otherwise indicated, specimens examined were vouchers. Abbreviations for specimen sources are listed in the Materials and methods. Unless otherwise indicated, specimens examined are in the senior author's collection at the LRP.

Abbreviations: c.s., cross-section; l.s., longitudinal section; sc., scolex; sec., section; seg., segment; spcm., specimen; w.m., whole mount; w.w., whole worm. SEM, scanning electron microscopy; TEM, transmission electron microscopy.

Appendix 14.2 Character list.

Character 1:   Rhyncheal apparatus (RA): 0 = absent; 1 = present. (1)*

Character 2:   Construction of the tentacle hooks of the RA: 0 = hollow; 1 = solid; 9 = N/A (not applicable in taxa lacking bothria and in taxa lacking an RA).

Character 3:   Metabasal armature of the tentacles of the RA: 0 = heteroacanthous; 1 = homeoacanthous; 9 = N/A (not applicable in taxa lacking bothria and in taxa lacking an RA).

Character 4:   Chainettes on tentacles of the RA: 0 = absent; 1 = throughout length of tentacle; 2 = restricted to base of tentacle; 9 = N/A (not applicable in taxa lacking bothria and in taxa lacking an RA).

Character 5:   Point of attachment of retractor muscles of tentacles to the tentacle bulbs of the RA: 0 = anterior; 1 = posterior; 2 = middle; 9 = N/A (not applicable in taxa lacking bothria and in taxa lacking an RA).

Character 6:   Course of tentacle sheaths of the RA: 0 = straight; 1 = sinuous; 2 = coiled; 9 = N/A (not applicable in taxa lacking bothria and in taxa lacking an RA).

Character 7:   Posterior extent of bothria relative to tentacle bulbs of the RA: 0 = not reaching anterior margin of bulbs; 1 = extending beyond posterior margin of bulbs; 9 = N/A (not applicable in taxa lacking bothria and in taxa lacking an RA).

Character 8:   Extension of scolex in lateral plane: 0 = absent; 1 = present. (2)

Character 9:   Configuration of scolex: 0 = undivided; 1 = 2 pedicels; 2 = 4 stalks. (3)

Character 10:   Apical modification of scolex proper (in adult): 0 = absent; 1 = present. (4)

Character 11:   Invagination of apical modification of scolex proper: 0 = non-invaginable; 1 = invaginable; 9 = N/A. (5)

Character 12:   Aperture on apical modification of scolex proper: 0 = absent; 1 = present; 9 = N/A. (6)

Character 13:   Opening of aperture on apical modification of scolex proper: 0 = not changeable in diameter; 1 = changeable in diameter; 9 = N/A. (7)

Character 14:   Field of multiple papillae surrounding posterior margin of apical modification of scolex proper: 0 = absent; 1 = present; 9 = N/A. (10)

Character 15:   Apical organ on scolex: 0 = absent; 1 = present. (11)

Character 16:   Hooks on apical organ: 0 = absent; 1 = present; 9 = N/A (not applicable in taxa lacking an apical organ).

Character 17:   Apical organ hook arrangement: 0 = 1 or more continuous rings; 1 = distinct dorso-ventral groups; 9 = N/A (not applicable in taxa lacking an apical organ and in taxa lacking apical organ hooks).

Character 18:   Apical organ hook shape: 0 = similar in shape; 1 = dissimilar in shape; 9 = N/A (not applicable in taxa lacking an apical organ and in taxa lacking apical organ hooks).

Character 19:   Discrete muscle bundles associated with apical organ hook bases: 0 = absent; 1 = present; 9 = N/A (not applicable in taxa lacking an apical organ and in taxa lacking apical organ hooks).

Character 20:   Lateral hooklets on apical organ: 0 = absent; 1 = present; 9 = N/A (not applicable in taxa lacking an apical organ, in taxa lacking apical organ hooks, and in taxa lacking apical organ hooks that are arranged in distinct dorso-ventral groups).

Character 21:   Internal glandularity of apical organ: 0 = nonglandular; 1 = glandular; 9 = N/A. (12)†

Character 22:   Glandular papillae on surface of apical organ: 0 = absent; 1 = present; 9 = N/A (not applicable in taxa lacking an apical organ).

Character 23:   Muscularity of apical organ: 0 = diffuse musculature; 1 = conspicuous musculature; 9 = N/A. (13)†

Character 24:   Aperture on apical organ: 0 = absent; 1 = present; 9 = N/A. (14)

Character 25:   Retraction of apical organ: 0 = non-retractable; 1 = retractable; 9 = N/A. (15)

Character 26:   Invagination of apical organ: 0 = non-invaginable; 1 = invaginable; 9 = N/A. (16)

Character 27:   Apical organ tentacles: 0 = absent; 1 = present; 9 = N/A. (17)†

Character 28:   Position of attachment of scolex muscle bundles to apical organ: 0 = at periphery; 1 = at centre; 2 = throughout entire margin of apical organ; 3 = muscle attachment absent; 9 = N/A (not applicable in taxa lacking an apical organ).

Character 29:   Scolex form: 0 = without specialized attachment organs; 1 = with pseudosegments; 2 = with bothria; 3 = with acetabula. (19)†

Character 30:   Number of cruciform pseudosegments: 0 = 4; 1 = 5; 9 = N/A (not applicable in taxa lacking pseudosegments).

Character 31:   Hooks on pseudosegments: 0 = absent; 1 = hollow; 2 = solid; 9 = N/A (not applicable in taxa lacking pseudosegments).

Character 32:   Elaborate recurvature of pseudosegment: 0 = absent; 1 = present; 9 = N/A (not applicable in taxa lacking pseudosegments).

Character 33:   Number of bothria on scolex: 0 = 2; 1 = 4; 9 = N/A (not applicable in taxa lacking bothria).

Character 34:   Bothrial pit: 0 = absent; 1 = present; 9 = N/A (not applicable in taxa lacking bothria).

Character 35:   Bothrial notch: 0 = absent; 1 = present; 9 = N/A (not applicable in taxa lacking bothria).

Character 36:   Acetabular attachment: 0 = completely sessile; 1 = free posteriorly only; 2 = free anteriorly only; 3 = free both anteriorly and posteriorly; 9 = N/A. (20)

Character 37:   Anterior acetabular thickening forming auricle: 0 = absent; 1 = present; 9 = N/A (not applicable in taxa lacking acetabula).

Character 38:   Pointed extensions on margin of acetabulum: 0 = absent; 1 = present; 9 = N/A (not applicable in taxa lacking acetabula).

Character 39:   Rounded extensions on lateral margin of acetabulum: 0 = absent; 1 = present; 9 = N/A (not applicable in taxa lacking acetabula).

Character 40:   Configuration of posterior margin of acetabulum: 0 = complete; 1 = 1 notch; 2 = 2 notches; 9 = N/A. (21)†

Character 41:   Configuration of acetabulum: 0 = undivided; 1 = bifid; 9 = N/A. (21)†

Character 42:   Specialized anterior region of acetabulum: 0 = absent; 1 = in form of a sucker; 2 = in form of a loculus; 9 = N/A. (22)

Character 43:   Relative size of specialized anterior region compared to total size of acetabulum: 0 = less than 1/2; 1 = greater than 1/2; 9 = N/A (not applicable in taxa lacking acetabula, and in taxa lacking a specialized anterior region of the acetabulum).

Character 44:   Accessory sucker on specialized anterior region loculus: 0 = absent; 1 = 1; 2 = 3; 9 = N/A. (23)†

Character 45:   Thickening of anterior rim of accessory sucker on specialized anterior region loculus: 0 = absent; 1 = present; 9 = N/A (not applicable in taxa lacking acetabula, in taxa lacking a specialized anterior region of the acetabulum, and in taxa possessing a specialized anterior region of the acetabulum but lacking an accessory sucker on this region).

Character 46:   Posterior margin of distal surface of specialized anterior region loculus of acetabulum: 0 = fused; 1 = free; 9 = N/A (not applicable in taxa lacking acetabula, and in taxa lacking a specialized anterior region of the acetabulum).

Character 47:   Pointed muscular extensions on left and right posterolateral margins of distal surface of specialized anterior region loculus of acetabulum: 0 = absent; 1 = present; 9 = N/A. (24)†

Character 48:   Muscular extensions on outer posterolateral margin of proximal surface of specialized anterior region loculus of acetabulum: 0 = absent; 1 = present; 9 = N/A. (25)†

Character 49:   Number of columns of loculi on acetabulum: 0 = none; 1 = 1; 2 = 2; 3 = 3; 9 = N/A. (26)

Character 50:   Number of rows of loculi on acetabulum: 0 = none; 1 = 1; 2 = 2; 3 = 3; 4 = greater than 3; 9 = N/A. (27)

Character 51:   Anterior and posterior unpaired loculi on acetabulum: 0 = absent; 1 = anterior and posterior loculus present; 2 = only anterior loculus present; 9 = N/A. (28)†

Character 52:   External horizontal septa between loculi on acetabulum: 0 = absent; 1 = present; 9 = N/A. (29)

Character 53:   Longitudinal muscle bundle subdivisions on acetabulum: 0 = absent; 1 = present; 9 = N/A. (30)

Character 54:   Subloculi: 0 = absent; 1 = present; 9 = N/A. (31)

Character 55:   Marginal loculi on acetabulum: 0 = absent; 1 = incomplete; 2 = complete; 9 = N/A. (32)

Character 56:   Fusion of recurved posterior margin of acetabulum to midline of distal acetabular surface: 0 = absent; 1 = present; 9 = N/A. (33)†

Appendix 14.2 Continued.

Character 57:   Form of circular muscles at centre of distal acetabular surface: 0 = absent; 1 = complete ring; 2 = laterally interrupted ring; 3 = 2 incomplete rings; 9 = N/A. (34)†

Character 58:   Constriction narrowing the middle of acetabulum: 0 = absent; 1 = present; 9 = N/A (not applicable in taxa lacking acetabula).

Character 59:   Glandular structure at centre of acetabulum: 0 = absent; 1 = present; 9 = N/A. (35)

Character 60:   Gland cells throughout acetabular musculature: 0 = absent; 1 = present; 9 = N/A. (36)

Character 61:   Folding of acetabulum: 0 = unfolded; 1 = highly folded; 9 = N/A. (37)

Character 62:   Attachment of medial posterior margins of adjacent acetabula: 0 = separate; 1 = fused to one another; 9 = N/A. (38)

Character 63:   Velum between medial margins of adjacent acetabula: 0 = absent; 1 = present; 9 = N/A. (39)

Character 64:   Auxiliary tissue covering distal acetabular surface: 0 = absent; 1 = present; 9 = N/A. (40)

Character 65:   Cephalic peduncle: 0 = absent; 1 = with inconspicuous posterior boundary; 2 = with conspicuous posterior boundary. (41)†

Character 66:   Velum encircling cephalic peduncle at posterior margin: 0 = absent; 1 = present; 9 = N/A (not applicable in taxa lacking a cephalic peduncle).

Character 67:   Armature of cephalic peduncle: 0 = absent; 1 = multiple columns of triradiate spines; 9 = N/A. (42)

Character 68:   Extensively folded cephalic peduncle: 0 = absent; 1 = present; 9 = N/A (not applicable in taxa lacking a cephalic peduncle).

Character 69:   Modification of anterior segments: 0 = absent; 1 = laterally expanded; 2 = forming a suctorial trough. (43)†

Character 70:   Scolex with pseudobothridia: 0 = absent; 1 = present; 9 = N/A. (44)

Character 71:   Acetabular hooks: 0 = absent; 1 = present; 9 = N/A. (45)

Character 72:   Number of acetabular hook pairs: 0 = 1; 1 = 2; 9 = N/A. (46)

Character 73:   Relative sizes of medial and lateral hooks: 0 = similar in size; 1 = medial hook larger than lateral hook; 9 = N/A. (47)

Character 74:   Number of prongs per medial hook: 0 = 1; 1 = 2; 2 = 3; 9 = N/A. (48)

Character 75:   Number of prongs per lateral hook: 0 = 1; 1 = 2; 2 = 3; 9 = N/A. (49)

Character 76:   Relative lengths of medial and lateral hook bases: 0 = bases equal; 1 = medial base longer than lateral base; 2 = lateral base longer than medial base; 9 = N/A. (50)†

Character 77:   Dark matrix on bases of medial and lateral hooks: 0 = absent; 1 = present; 9 = N/A. (51)

Character 78:   Continuity of hook channels in multipronged hooks: 0 = 1 continuous channel within hook; 1 = two separate channels within hook; 9 = N/A. (52)

Character 79:   Hook orientation: 0 = directed posteriorly; 1 = directed anteriorly (then posteriorly); 9 = N/A. (53)

Character 80:   Hook prong colour: 0 = tawny; 1 = yellow; 2 = brown; 9 = N/A. (54)†

Character 81:   Accessory piece between bases of medial and lateral hooks: 0 = absent; 1 = present; 9 = N/A. (55)

Character 82:   Hook talon: 0 = absent; 1 = present; 9 = N/A. (56)

Character 83:   Axial hook prong tubercle: 0 = absent; 1 = present; 9 = N/A. (57)

Character 84:   Spiniform microtriches on scolex proper: 0 = absent; 1 = blade-like; 2 = serrate; 3 = maiziform; 4 = pectinate; 5 = tridentate; 6 = falcate. (58)†

Character 85:   Filiform microtriches on scolex proper: 0 = absent; 1 = short; 2 = long. (59)

Character 86:   Spiniform microtriches on modified apex of scolex proper: 0 = absent; 1 = blade-like; 2 = serrate; 3 = maiziform; 4 = pectinate; 5 = tridentate; 6 = falcate; 9 = N/A. (60)†

Character 87:   Filiform microtriches on modified apex of scolex proper: 0 = absent; 1 = short; 2 = long; 9 = N/A. (61)

Character 88:   Spiniform microtriches on apical organ: 0 = absent; 1 = blade-like; 2 = serrate; 3 = maiziform; 4 = pectinate; 5 = tridentate; 6 = falcate; 9 = N/A. (62)†

Character 89:   Filiform microtriches on apical organ: 0 = absent; 1 = short; 2 = long; 9 = N/A. (63)

Character 90:   Spiniform microtriches on distal acetabular surface: 0 = absent; 1 = blade-like; 2 = serrate; 3 = maiziform; 4 = pectinate; 5 = tridentate; 6 = falcate; 9 = N/A. (64)†

Character 91:   Filiform microtriches on distal acetabular surface: 0 = absent; 1 = short; 2 = long; 9 = N/A. (65)

Character 92:   Spiniform microtriches on proximal acetabular surface: 0 = absent; 1 = blade-like; 2 = serrate; 3 = maiziform; 4 = pectinate; 5 = tridentate; 6 = falcate; 9 = N/A. (66)†

Character 93:   Filiform microtriches on proximal acetabular surface: 0 = absent; 1 = short; 2 = long; 9 = N/A. (67)

Character 94:   Spiniform microtriches on distal bothrial surface: 0 = absent; 1 = blade-like; 2 = serrate; 3 = maiziform; 4 = pectinate; 5 = tridentate; 6 = falcate; 9 = N/A. (68)†

Character 95:   Filiform microtriches on distal bothrial surface: 0 = absent; 1 = short; 2 = long; 9 = N/A. (69)

Character 96:   Spiniform microtriches on proximal bothrial surface: 0 = absent; 1 = blade-like; 2 = serrate; 3 = maiziform; 4 = pectinate; 5 = tridentate; 6 = falcate; 9 = N/A. (70)†

Character 97:   Filiform microtriches on proximal bothrial surface: 0 = absent; 1 = short; 2 = long; 9 = N/A. (71)

Character 98:   Spines on proximal bothrial surface: 0 = absent; 1 = present; 9 = N/A. (72)

Character 99:   Spiniform microtriches on cephalic peduncle: 0 = absent; 1 = blade-like; 2 = serrate; 3 = maiziform; 4 = pectinate; 5 = tridentate; 6 = falcate; 9 = N/A. (73)†

Character 100:   Filiform microtriches on cephalic peduncle: 0 = absent; 1 = short; 2 = long; 9 = N/A. (74)

Character 101:   Spiniform microtriches on pedicels: 0 = absent; 1 = blade-like; 2 = serrate; 3 = maiziform; 4 = pectinate; 5 = tridentate; 6 = falcate; 9 = N/A. (75)†

Character 102:   Filiform microtriches on pedicels: 0 = absent; 1 = short; 2 = long; 9 = N/A. (76)

Character 103:   Spiniform microtriches on stalks: 0 = absent; 1 = blade-like; 2 = serrate; 3 = maiziform; 4 = pectinate; 5 = tridentate; 6 = falcate; 9 = N/A. (77)†

Character 104:   Filiform microtriches on stalks: 0 = absent; 1 = short; 2 = long; 9 = N/A. (78)

Character 105:   Spiniform microtriches on strobila: 0 = absent; 1 = blade-like; 2 = serrate; 3 = maiziform; 4 = pectinate; 5 = tridentate; 6 = falcate; 9 = N/A. (79)†

Character 106:   Filiform microtriches on strobila: 0 = absent; 1 = short; 2 = long; 9 = N/A. (80)

Character 107:   Distribution of spiniform microtriches on segment: 0 = over entire surface; 1 = restricted to anterior region; 2 = restricted to posterior margin; 3 = restricted to laciniations; 9 = N/A (not applicable in taxa lacking spiniform microtriches on strobila).

Character 108:   Ornamentation on surface of strobila: 0 = absent; 1 = scutes; 2 = leaf-like structures; 3 = scale-like structures; 4 = annulations. (81)†

Character 109:   Discrete circular muscles distinguishing cortex from medulla: 0 = absent; 1 = present.

Character 110:   Discrete longitudinal muscle bundles in segment cross-section: 0 = absent; 1 = present.

Character 111:   Segment sex: 0 = monoecious; 1 = dioecious. (82)

Character 112:   Segmental margins: 0 = acraspedote; 1 = craspedote only; 2 = laciniate. (83)

Character 113:   Segmental apolysis: 0 = anapolytic; 1 = apolytic; 2 = euapolytic; 3 = hyperapolytic. (84)

Character 114:   Shape of mature segment: 0 = longer than wide; 1 = wider than long. (85)

Character 115:   Anterior extent of primary field of testes: 0 = to anterior margin of segment; 1 = stopping short of anterior 1/3 of segment. (86)†

Character 116:   Posterior extent of primary field of testes: 0 = not reaching ovary; 1 = extending to, or overlapping with ovary; 9 = N/A. (87)

Character 117:   Postvaginal field of testes: 0 = absent; 1 = present; 9 = N/A. (88)

Character 118:   Postovarian field of testes: 0 = absent; 1 = present; 9 = N/A. (89)

Appendix 14.2 Continued.

Character 119: Number of columns of testes anterior to cirrus sac (or hermaphroditic sac) in dorso-ventral view: 0 = 1; 1 = 2; 2 = greater than 2. (90)

Character 120: Number of layers of testes in cross-section anterior to cirrus sac (or hermaphroditic sac): 0 = 1; 1 = greater than 1. (91)

Character 121: Position of bulk of vas deferens in relation to its exit from cirrus sac (or hermarphroditic sac): 0 = anterior; 1 = posterior; 2 = only lateral; 3 = anterior, lateral and posterior; 4 = only anterior and lateral; 5 = only lateral and posterior. (92)†

Character 122: Vas deferens size: 0 = minimal; 1 = extensive. (93)

Character 123: External seminal vesicle: 0 = absent; 1 = present. (94)

Character 124: Position of junction of proximal vas deferens with external seminal vesicle: 0 = near cirrus sac (or hermaphroditic sac); 1 = near ovarian bridge; 9 = N/A (not applicable in taxa lacking an external seminal vesicle).

Character 125: Internal seminal vesicle: 0 = absent; 1 = present. (95)

Character 126: Cirrus sac shape: 0 = straight; 1 = bent anteriorly; 2 = bent posteriorly. (96)

Character 127: Spiniform microtriches on cirrus: 0 = absent; 1 = present. (97)

Character 128: Position of genital pores: 0 = lateral; 1 = medial; 2 = sublateral. (98)

Character 129: Position of genital pores in segments along strobila: 0 = irregularly alternating; 1 = not alternating.

Character 130: Genital pore position relative to ovarian isthmus: 0 = anterior to isthmus; 1 = posterior to isthmus; 2 = at same level as isthmus; 9 = N/A. (99)†

Character 131: Degree of association of openings of cirrus and vagina: 0 = opening into common atrium; 1 = opening separately; 2 = vagina joining cirrus in hermaphroditic sac; 9 = N/A. (100)

Character 132: Position of opening of vagina relative to cirrus: 0 = anterior to cirrus; 1 = posterior to cirrus; 2 = at same level; 9 = N/A. (101)

Character 133: Continuous sheath surrounding cirri in adjacent segments: 0 = absent; 1 = present. (102)

Character 134: Vaginal sphincter: 0 = absent; 1 = present. (103)

Character 135: Vaginal course relative to genital pore: 0 = extending from genital pore directly to ovary; 1 = extending anterior to genital pore prior to extending posteriorly to ovary; 9 = N/A. (104)

Character 136: Vaginal course relative to cirrus sac (or hermaphroditic sac): 0 = not crossing cirrus sac (or hermaphroditic sac); 1 = crossing cirrus sac; 9 = N/A (not applicable in taxa with dioeceous segments).

Character 137: Position of vagina in segment: 0 = medial; 1 = lateral; 2 = perpendicular to long axis of segment. (105)†

Character 138: Course of vagina: 0 = straight; 1 = sinuous. (106)†

Character 139: Seminal receptacle: 0 = absent; 1 = present. (107)

Character 140: Ovary shape in dorso-ventral view: 0 = H; 1 = inverted A; 2 = V; 3 = inverted U; 4 = radiating lobes; 5 = sheet; 6 = irregularly lobed; 7 = single mass. (108)

Character 141: Ovary shape in cross-section: 0 = bilobed; 1 = tetralobed; 2 = radiate; 3 = sheet; 4 = trilobed; 5 = ring. (109)†

Character 142: Form of ovarian margins: 0 = smooth; 1 = follicular; 2 = lobulated; 3 = digitiform. (110)†

Character 143: Ovary left/right symmetry: 0 = symmetrical; 1 = asymmetrical. (111)†

Character 144: Vitelline condition: 0 = 2 vitelline masses; 1 = 2 vitelline chords; 2 = multiple, separate vitelline follicles; 3 = single mass. (112)†

Character 145: Vitelline position: 0 = cortical; 1 = medullary. (113)

Character 146: Vitelline element position relative to ring of longitudinal muscle bundles: 0 = only peripheral; 1 = only paramuscular; 2 = encroaching on muscle bundles from medial position; 3 = only medial; 4 = encroaching on muscle bundles from periphery; 9 = N/A (not applicable in taxa lacking longitudinal muscle bundles in mature segments).

Character 147: Position of single vitellarium: 0 = anterior to ovary; 1 = posterior to ovary; 9 = N/A (only applicable in taxa possessing a vitellarium in form of a single mass).

Character 148: Distribution of vitelline element(s): 0 = lateral; 1 = ventral; 2 = circumsegmental; 3 = associated with ovary; 4 = medial; 5 = lateral and dorsal. (114)†

Character 149: Number of lateral vitelline elements in cross-section: 0 = 2; 1 = 4; 2 = many; 9 = N/A (not applicable in taxa lacking laterally distributed vitelline elements).

Character 150: Anterior extent of vitelline elements: 0 = to anterior margin of segment; 1 = not reaching anterior margin of segment; 9 = N/A. (116)

Character 151: Posterior extent of vitelline elements: 0 = overlapping with ovary to any extent; 1 = extending posterior to ovary; 9 = N/A. (117)†

Character 152: Lateral interruption of vitelline elements by ovary: 0 = not interrupted; 1 = interrupted; 9 = N/A (not applicable in taxa possessing vitelline elements that are distributed ventrally or are associated with the ovary, in taxa possessing lateral vitelline elements that do not extend posterior to the ovary, and in taxa possessing a vitellarium in the form of a single mass).

Character 153: Dorso-ventral interruption of vitelline elements by ovary: 0 = absent; 1 = present; 9 = N/A (only applicable in taxa that possess vitelline elements that are distributed circumsegmentally, and taxa with ventrally or laterally and dorsally distributed vitelline elements that extend posterior to the ovary).

Character 154: Anterior extent of uterus in mature segment: 0 = extending approximately to anterior margin of segment; 1 = extending into anterior half of segment but not reaching anterior margin; 2 = not extending into anterior half of segment. (119)

Character 155: Uterine pore: 0 = absent; 1 = present.

Character 156: Uterus shape: 0 = vertically elongate sac; 1 = anterio-posteriorly bissaccate; 2 = laterally bisaccate; 3 = sinuous sac; 4 = horizontally elongate sac.

Character 157: Position of uterus relative to vagina in cross section: 0 = uterus ventral to vagina; 1 = uterus dorsal to vagina; 9 = N/A. (120)

---

* Numbers in parentheses following character descriptions indicate the character number used in Caira *et al.* (1999) for that particular character.

† Indicates characters that were adapted from Caira *et al.* (1999) with minor changes in wording or addition of character states.

Abbreviation: N/A, not applicable.

Appendix 14.3 Data matrix.

CHARACTER NUMBER*

```
                                                                                          1  1111111111111111111 1111111111111111111 1111111111111111111
                 11111111112 22222222233333333334 44444444455555555556 66666666677777777778 88888888899999999990 2222222223333333334 4444444455555555556 6666666666777777777
TAXON  12345678901234567890 12345678901234567890 12345678901234567890 12345678901234567890 12345678901234567890 12345678901234567890 12345678901234567890 12345678901234567

Acanma 09999990009999099999 99999999399999910000 02010100139100000000 00102000001001100000 0017?999901119999?1? 99990?90000020011020 ?0090110000000010100 12021990110991007
Acan1  09999990009999099999 99999999399999910000 02011100139100000000 00102000001001100000 00101999912179999912 99990290000120011020 01090010000000000100 01021990100990007
Acan2  09999990009999099999 99999999399999910000 02011100139100000000 00002000001001100000 0010?999901129999?1? 99991200000020011010 01090010000001000100 02121990110990007
Acan3  09999990009999099999 99999999399999910000 02010100139100000000 00102000001001100000 0010?999901129999912 99990?90000120011021 01090010000001000100 02121990100990007
Acanpa 09999990009999099999 99999999399999910000 02010100139100000000 00102000001001100000 00101999901219999912 99990290000020011021 01090010000000000000 00021990100990000
Acesth 09999990009999099999 99999999399999910000 02010000139100000000 00002000001011110002 00001999901119999911 99990?90010110011021 00090010000001000100 02021390200990000
Adelob 09999990010090109999 11100002399999910000 00999999009909000000 00900990000999999999 999021?00????99999999 99990290110120011021 10110000000100000100 03021290200990000?
Antero 09999990010110109999 17101007399999590000 00999999009909000000 00900990000999999999 99912120217179999999 99991210000120019000 00110010000200001000 10021990111191007
Anthco 09999990209999099999 99999999399999930000 00999990009909093000 09902000000999999999 999??99991?129999912 9912?270000220011020 00090110000000100000 02021990201091000
Anthla 09999990209999099999 99999999399999930000 00999990009901101000 09900990000999999999 999??999912129999999 99121200000220011020 20090110000000000100 02021990101091000
Anceal 09999990209999099999 99999999399999930000 01099999009909200000 09900990000999999999 99902999910129999999 99??00290000020000010 01090210000000000100 12021990201190000
Anceca 09999990209999099999 99999999399999930000 01099990009909200000 0990????00099999999 999??99991?129999?? 99?????00101200011021 00090210000000000100 12021390201190000
Ancedu 09999990209999099999 99999999399999930000 01099990009909200000 09901000000999999999 999??99990212999999?? 99??1?000?0110010020 40090010000001000100 11021?90201090000
Balano 09999990009999099999 99999999399999910000 02009000009909000000 00101000001001100100 0001299991???99999?2 99990290010110011020 00090010000000000000 01021390201091000
Biburs 09999990009999099999 99999999399999930000 01099999009909010000 09901000001001100000 999??99991111999999 999?????00000210011020 20090110000000100000 12021990200991000
Bilocu 09999990009999099999 99999999399999910000 02009000129100000000 00?02000001101110100 000??999912179999919 99991?100?00200110201 00090100000000010001 02021?90111092000
Callev 09999990010999099999 99999999399999910000 02009100139100000000 09902000001101120100 100019999012199999999 11990?90000020011010 00090010000001100010 00121990100991000
Callha 09999990009999099999 99999999399999910000 02010000139100000000 00102000001101120100 000019999112129999912 99990290000110011010 40090010000001100000 00721990010990000
Call1  09999990009999099999 99999999399999910000 02020100139100000000 00100999001101100100 000019999019129999999 999912300102100110201 01090010000000000110 02021390200991000
Call2  09999990009999099999 99999999399999910002 02020100139100000000 00100999001101100100 000029999012199999999 999912000002200110201 00090110000000000000 02021990101091000
Callri 09999990009999099999 99999999399999910000 02010100139100000000 00002000001101100100 100019999011199999912 99990290000110011010 01090010000000000010 00021990101991000?
Callvi 09999990009999099999 99999999399999910001 02010100139100000000 00102000001101100100 100019999011299999? 99990290000120011021 01090010000000000010 00021990010991000
Calypt 09999990009999099999 99999999399999910000 01199999009909000000 00001000000999999999 99912999902029999901 99991200000111011021 00090010000000000110 12021990200991000
Cardio 09999990009999099999 99999999399999910000 01099999220000200000 00002000000999999999 999??999911117999999? 99990290000020011021 00090110000000100000 12021990201091000
Carpob 09999990209999099999 99999999399999930000 00999990009909200000 09900990000999999999 999??99994211999999999 99010290000020000020 10090110000000010100 ?2021990201092000
Cathet 09999991010091099999 99999999099999999999 99999999999999999999 99991000009999999999 9990000099999999999900 99990090000100011020 00090110000001000100 00021992900001000
Caul1  09999990009999099999 99999999399999999999 00999991402999999999 00002000000999999999 99902999912029999999 99902090000020011120 00090010000001000100 12021990201091000
Caul2  09999990009999099999 99999999399999930000 00999992410000000000 00000990000999999999 9990299991202999999 99990290000020011010 31090110000000001100 02021990101091000
Ceph1  09999990010100109999 10100102399999950000 00999990009909000000 09009990000999999999 99912120212129999999 99990290010120011020 401100?0000100000000 0302129021099100?
Ceph2  09999990010100109999 11101000399999500000 00999990009909000000 09009990000999999999 99902100002029999999 99990290??0120011011 ??110010000100000000 ?2021?9021099?0??
Cerato 09999990009999099999 99999999399999950000 02009100009909000000 00102000000999999999 99902999912129999910 99990290000010011020 00090110000000010110 11021990201091000
Chimae 09999990009999099999 99999999399999910000 01099990009909200000 00000999000999999999 99901999901?99999999 99990200100201012021 10090010000000000005 30020992011001100
Clisca 09999990209999099999 99999999399999930000 01199999009909000000 09901000000999999999 99912999994217999999 99020290000010011020 00090110000000100000 02021990200991000
Clismo 09999990009999099999 99999999399999930000 01099990009909000000 10001000000999999999 99902999902127999999 99990290000120011021 00090110000000110000 02021990201091000
Corrug 09999990010100109999 00110002399999900000 00999999009909000000 00909099000999999999 999??????????99999 9999??00000220019001 20110010000100000006 0012199011119?0??
Crosso 09999990009999099999 99999999399999910000 01099990009909000000 00000999000999999999 99902999912129999999 99990290010211011020 00090010000000000000 11021290211091000
Dicran 09999990009999099999 99999999399999910000 02011100139100000000 00101000001001110000 0100?99990?17?9999?1? 99991?00000020011020 00090010000000000100 02021992901001000
Dinobo 09999990009999099999 99999999399999910000 02010110009909000000 01100999000999999999 999??99991210999999 9999??0110010011021 00090010000000000100 13021290200990000
Dioeco 09999990009999099999 99999993400000000000 00999993400000000000 00000990000999999999 99911999901019999999 99902001110010999221 20091210099910990110 00001993999990020
Diphyl 09999990009999099999 99999999299900099999 99999999999999999999 99909990009999999999 999??79999999?????099 9999??00101010019021 10100001100100000000 00020992000991137
Discob 09999990009999099999 99999993399999910000 00999990009900000000 00000999000999999999 99902999912129999999 99990290000220019010 00110010000100001000 12021990201091000?
Discul 09999990009999099999 99999998099999999999 99999991001099999999 99900999999999999999 9990099999999999999901 99990290000011011020 00090012000000000100 02021992900000000
Ditrac 09999990009999109999 00000001299900199999 99999999999999999999 99992100099999999999 9990299??99991202102 9999??0000010009020 100907110?100000102 00021992900992001
Duplmi 09999990109999099999 99999999399999910000 00999991490010000000 09901000000999999999 999??99997?99999992 ?299??4000120011021 11090012000000010104 23021995901011000
Dupl1  09999990109999099999 99999999399999910000 00999993400001000000 09009990000999999999 9997?9999017199999 01990290000120011021 30090010000000100174 23021995901011000
Dupl2  09999990109999099999 99999993399999910000 00999993400001000000 09902000000999999999 9997?99990101999999 0199??4000120011021 30090012000000000124 23021995901011000
Echei1 09999990211110109999 17001000399999950000 00999992410000000000 09009090000999999999 9990111??77119999999 9911029000120010021 10090010000000000100 12021390200990000
Echei2 09999990211110109999 ?10010007399999950000 00999992210000000000 09009090000999999999 99902120002129999999 99020290000120000010 31090010000000000100 00121990100990000
Echei3 09999990211110109999 ??????07399999950000 00999992410000000000 09002090000999999999 9990212?71212999991? 97120290000120010010 10090010000000000100 12021990100990?00?
Echimx 09999990009999111111 00100002999000599999 99999999999999999999 99992010099999999999 99910990299991X41001 99990290000010009010 01090011?00100000100 00021990101090001
Echisp 09999990009999111111 00100002999000599999 99999999999999999999 99992010099999999999 99941990X99990141001 99990290000010019000 11090011?00100100100 00021990101091000
Eoxxcu 09999990010100110000 17100002399999900000 00999990009909000000 00901000000999999999 99912127?17179999??? 99990?90000010111121 10090010000000000013 00031914999990000
Enioco 09999990010170109999 17?07?0239999990000 00999990009909000000 00900990200999999999 99910107?101099999 99991200000110011020 31110010000100000700 02021990010992000?
Flexib 09999990009999099999 99999993399999930000 01099999009909110000 00002000000999999999 999??999912119999?1? 99990290000020011020 00090110000001000100 12021990101092000
Floric 10012100009999099999 99999992299900199999 99999999999999999999 99992000009999999999 99902999999994242042 99990?93010120011121 20091000021000000100 0?021292901011007
Gastro 09999990009999099999 99999993399999910000 02010111009909000000 01101000000999999999 99902999902127999999 99990290010101010121 00090010000000100100 02021391910990000
Glypho 09999990009999099999 99999993399999900000 00999991490100000000 09001000000999999999 99902999902?79999902 99990290010120011121 00090010000000100000 13021392901001000?
Grillo 10011000009999099999 99999992299900199999 99999999999999999999 99992000009999999999 99901999999994141001 9999??0110010011121 31100000002100000100 12020192901001000
Gymnor 10011000009999099999 99999992299900199999 99999999999999999999 99902000009999999999 99902999999990202002 99990290110011011121 20100000002000000100 12020092901011001
Horn1  09999990010110109999 11011002399999990000 00999990009909000000 00000999100999999999 99912020012129999999 99990290000120011021 ?01100?0000200000000 02021992900001007
Horn2  09999990010110109999 1?0?1002399999910000 00999990009909000000 00000999100999999999 99912027?12129999999 99991220000120011020 00110000000200000100 02021990200990000
Horn3  09999990010170109999 1?0??07239999910000 00999990009909000000 00000999100999999999 99912127?70212999999 99991220000120019000 00110010000200000100 02021990210991007
Lecplt 09999990010110109999 1?10110239999900000 00999990009909000000 09909990000999999999 999???????12?79999999 99990290000120019010 00110110000101001100 02021990111191007
Lecsp  09999990010110109999 11101102399999900000 00999990009909000000 09909990000999999999 99902020072?79999999 99990290000120019010 40110010000100001100 02021990111191007
Litamp 09999990009999109999 00110003101199999999 99999999999999999999 99909990099999999999 99902990299999999999 99990290000120011020 20090010000200000103 02021990201101007
Litams 09999990009999109999 00110007110099999999 99999999999999999999 99909990099999999999 99902090299999999999 99991000000120011020 30090110000200000103 02021992901011007
Litdai 09999990009999109999 00110003112099999999 99999999999999999999 99909990099999999999 99902991299999999999 999912?0000120011010 11090010000200000103 02021992901011000
Macrob 09999990009999111111 00100000299000599999 99999999999999999999 99992000009999999999 9997?990299994242042 99990290010011019021 21090011100100000102 00021290200190001
Marsup 09999990209999099999 99999999399999930000 01099999009909000000 09900990009999999999 99902999912129999999 9912029000120011020 00090110000000000100 12021990201092000
Mixodi 11021100009999099999 99999992299900199999 99999999999999999999 99992000009999999999 999??99999999117?20? 99992??00000220011021 10090000000100000000 10021992901001000?
Monory 09999990009999099999 99999993399991?0000 02009000009909000000 00001000000999999999 99902999910109999910 9999??0000101011021 00090010000000100 12021992900001000
Montic 09999990009999099999 99999993399990?0000 00999990009909000000 00901000000999999999 9991?99991?1?99999? 9999??0000001011020 21090010000Y00000100 02021990200990000
Myzoce 09999990009999099999 99999993399999930000 02?09000??9?0?0000 ?000???700099999999 99912999971219999999 99990294010110011020 30090010000000000100 12021392901012000
Oryg1  09999990209999099999 99999993399999930000 01099990009909000010 19901000000999999999 999??99993112999999? 99??????1000120011021 20090010000000100000 12021990200991000
Oryg2  09999990209999099999 99999993399999930000 01099990009909000010 09901000000999999999 99902999993212999991 9912029101012001121 00090010000000100000 12021390201091000
Otobot 10000200009999099999 99999992299901099999 99999999999999999999 99992100099999999999 99901999999994141002 99990290000100110020 30091000021000000000 10021992211011000
Pachyb 09999990009999099999 99999993399999900000 00999990009909000000 00900990011000000900 0091299912999999000 99990290010020010020 00090110000000100000 12021290200992000
Paraba 09999990009999099999 99999993399999910000 01099990009909000000 00001000000999999999 9990?999932229999902 99990290?010110011020 00090010000000000100 12021790201191000
Paraex 09999990009999099999 99999999399999910000 01099990009909000000 00001000000999999999 99902999932229999902 99990290100010011010 00090010000000000100 12021990201191000
Parapr 09999990209999099999 99999999399999910000 01099990009909000000 00001000000999999999 99902999922229999902 99990290100120011010 00090110000000000100 12021990101192000
Pedigl 09999990009999099999 99999999399999910000 02009000009909000000 00900990010011100100 000129990129999999 99990290000010011020 00090010000000010100 11021990201092000?
Pedilo 09999990009999099999 99999999399999910000 02009000009909000000 00102000001001100100 0000?99991212999991? 99990?90000020011020 01090110000000010000 12021990101092000
Pedise 09999990009999099999 99999999399999910000 02009000009909000000 00009990010011000100 00019999111999999999 99990?90000020011020 00090010000000010100 12021990101092000?
Phorma 09999990009999099999 99999993399999910000 02010000129201000000 00101000001001100000 0100199990111999912 99991200000020011020 00090000000000100011 02021990010092000?
Phor1  09999990009999099999 99999993399999910000 02009000129201000000 00101000001002210000 0100199990111999912 99991200000020011020 10090010000000000100 02021990111091000
Phor2  09999990009999099999 99999993399991?0000 02009000129201000000 00101000001002210000 0107?999901129999912 99991200000020011020 00090170000000000?1 02021990110990200?
Phor3  09999990009999099999 99999993399999910000 02009000129101000000 00101000001002210000 0100299990102999912 99991200000020011020 00090070000000100001 02021990010099200?
Phylla 09999990009999099999 99999993399999910000 11099999009909000000 10000999000999999999 99900999902029999999 99990291000110011021 20090010000000100000 12021990200991000
Phylri 09999990009999099999 99999993399999910000 11099990009909000000 10000999000999999999 99900999912129999999 9999??7100011011021 20090010000000100000 12021990200991000
Pingui 09999990009999099999 99999993399999910000 00999999113999999999 00012000001001100? 00002999121?99999? 9999??0010001011021 41090010000000100100 02021390200990000
Platau 09999990009999099999 99999993399999910010 02011000139100000000 00010000001001110000 11001999902020000 99991200000020010020 00090010000000010000 02021990209200000
Plathy 09999990009999099999 99999993399999910C10 02011100139100000000 00010100000100110000 01001999901119999999 99991100000020011020 00090110000000100010 02021992900001000
```

```
                                                                          1  1111111111111111111  1111111111111111111  11111111111111111
            11111111112 22222222233333333334 44444444455555555556 66666666677777777778 88888888889999999990 00000000011111111112 22222222233333333334 44444444455555555555
TAXON*      12345678901234567890 12345678901234567890 12345678901234567890 12345678901234567890 12345678901234567890 12345678901234567890 12345678901234567890 12345678901234567

Platpa  09999990009999099999 99999999399999910010 02011100139100000000 00101000001001110000 11001999901119999912 99991200000020011020 00090110000000100000 02021992911001000
Platsp  09999990009999099999 99999999399999910010 02011100139100000000 00101000001001110000 11001999901119999912 99991200000020011020 00090110000000100010 02021992900001000
Poly1   09999990010110109999 11000110399999900000 00999999009909000000 00900999000999999999 99902520002529999999 99999029000012001900 00111010000100000110 10021990111191007
Poly2   09999990010110109999 11000112399999900000 00999999009909000000 00900999000999999999 99902520012129999999 99999029000010019000 00111010000100000100 10021990111109000?
Poly3   09999990010110109999 17000117399999900000 00999999009909000000 00900999000999999999 99902020002129999999 99999029000011019000 00111010000100100103 10021990111191017
Potamo  09999990009999099999 99999999399999910010 02009000009909000000 00101000001000000900 009??99991?1?99999912 99991200000020011020 01090010011000101001 00011990010991009
Prosob  09999990009999099999 99999999399999920000 00999900009909000001 00901000001000000900 99912990212??99999912 99991200000020011021 00090110000000100000 02021992901001000
Protex  09999990009999099999 00100000399999900000 00999999009909000000 00901000000999999999 99912990212129999902 99991000100000011020 40090010000Y00000003 02021990100990000
Protpe  09999990009999099999 99999999399999900000 00999999009909000000 00901000000999999999 99902999912129999991? 99991200100100011020 50090010000000000110 02021990200990000
Protpi  09999990009999919999 00110000?399999900000 00999999009909000000 00901000000999999999 999??99???????99999?? 99999??0100101011021 00090010000000000000 02021990200990000
Pseu1   09999990211110109999 10001007399999930000 00999999009909000000 00900999000999999999 99902120211129999999 99010190000120010020 30090010000000000100 10021990100990000
Pseu2   09999990211110109999 17001?00399999930000 00999999009909000000 00900999000999999999 9999??127?12119999999 99020290000110010021 50090010000000000100 10021990200990000
Pseush  09999990210090109999 ?010000239999930000 00999999009909000000 00900999000999999999 9999?1?1?1?129999999 99??0290000110010021 20090010000000000010 12021990200990000
Rhinco  09999990209999099999 99999999399999930000 00999999241000000100 09900200000999999999 99902999912129999912 99120290000120000010 31090210000000000100 11021990101091007
Rhinsp  09999990209999099999 99999999399999930000 00999999241000000100 09900200000999999999 99902999912129999912 99120290000120000010 00090010000000000100 10021990100990000
Rhinur  09999990209999099999 99999999399999930000 00999999241000000000 09900200000999999999 9999???99912129999991? 99??0290000120010020 01090210000000000100 12021990100990000
Rhbdes  09999990209999099999 99999999399999930000 00999999240000000000 09900999000999999999 99902999911129999999 99020290000010010020 20091210000000100111 12121990100990000
Rhodob  09999990209999099999 99999999399999930000 00999999009909100000 00901000000999999999 99902999902029999902 99020290000010010020 41090010000000000100 13021290200990000
Scypho  09999990009999099999 99999999399999930000 01099999009909000000 00901000000999999999 99902999997129999902 99999??0110101011020 20090010000000000110 12011390200991000
Spinil  09999990009999099999 99999999399999930000 02009000009909000000 09902000001000000911 009??99991119999912 1?99029001002?011020 00090010000000000100 12021390201092000
Tentac  11101010009999099999 99999992999109099999 99999999999909000000 999921000009999999999 9996?9999999626202?? 9999??0000011011121 21090000007?01000?70 12021992901010027
Tetbcy  09999990009999099999 99999999399999911000 00999999009909000000 00101000000999999999 99910999910109999910 99991000110101011121 20090002120200002100 03031904999991040
Tetbsp  09999990009999099999 99999999399999911000 00999999009909000000 00001000000999999999 9999???99?????9999999? 99999??0710170?1011121 00090002120200000120 ?3031904999997027
Tetrag  09999990010090109999 11100002399999900000 00999999009909000000 00900999000999999999 99902??007?029999999 99990290110010000021 30110110000100000107 52021390211190017
Thysan  09999990009999099999 99999999399999910000 12009010009909000000 10001000000999999999 99902999922229999999 99990292110111010121 00090110000000100000 12020490200990000?
Triloc  09999990009999099999 99999999399999910000 01099999210900000000 00900999000999999999 99902999902129999999 99991210??0030110020 00090011010200100002 ?002??90?1109?002
Tyloce  09999990010090109999 11100002399999900000 00999999009909000000 00900999000999999999 99902??00????99999999 99990290010120011021 11110000000100000100 03021292910001007
Uncibi  09999990009999099999 99999999399999910000 02011100129100000000 00102000001001110000 0010?999902029999912 99991200000020011021 00090110000000110101 02021990100992000?
Yorker  09999990109999099999 99999999399999910000 02009000009909000000 09902000001010010911 00901999912119999912 12990290000020000020 00090017000000001000 12021990111092000?
Zyxibo  09999990009999099999 99999999399999910000 00999992211900000000 00000099000999999999 9999?79999101099999999 99990290000010110020 10090010000000000100 00021990211091000
Ng1     09999990009999099999 99999999399999910000 00999999222000000000 00000099000999999999 99912999991?129999999 999902907?0110000021 50090010000000000113 02021?90201091000
Ng2     09999990009999099999 99999999399999910000 00999999009909000000 00000099000999999999 99902999912129999999 99990290000120000010 10090000000100000000 00021990101092000?
Ng3     09999990009999099999 99999999399999910000 00999999009909000000 00000099000999999999 99901999902129999999 99990290000120019000 401100?0000100000006 40121990?0119?0??
Ng4     09999990010100109999 17101007399999930100 00999999009909000000 00002000000999999999 99902027?02129999902 99990290000120019001 20110000000100000006 40021990101197077
Ng5     09999990010090109999 11100002399999900000 00999999009909000000 00900999000999999999 99902??0012129999999 99990290000120019000 00110100000101000100 10021990111191007
```

*Both character states are given for polymorphic taxa with the characters noted as:
X=1&2 for Echimx (95), Echisp (89) and Y=0&1 for Montic (132), Protex (132).
Unknown character states are represented by a "?".
Characters that do not apply to a taxon are coded with character state "9".

†Taxon abbreviations (listed in the order in which they appear):

Acanma, *Acanthobothrium marplatensis*
Acan1, *Acanthobothrium* n. sp. 1
Acan2, *Acanthobothrium* n. sp. 2
Acan3, *Acanthobothrium* n. sp. 3
Acanpa, *Acanthobothrium parviuncinatum*
Acesth, *Acanthobothroides thorsoni*
Adelob, *Adelobothrium* sp.
Antero, *Anteropora* n. sp.
Anthco, *Anthobothrium* c.f. *cornucopia*
Anthla, *Anthobothrium laciniatum*
Anceal, *Anthocephalum alicae*
Anceca, *Anthocephalum cairae*
Ancedu, *Anthocephalum duszynskii*
Balano, *Balanobothrium* n. sp.
Biburs, *Bibursibothrium gouldeni*
Bilocu, *Biloculuncus pritchardae*
Callev, *Calliobothrium evani*
Callha, *Calliobothrium hayhowi*
Call1, *Calliobothrium* n. sp. 1
Call2, *Calliobothrium* n. sp. 2
Callri, *Calliobothrium riseri*
Callvi, *Calliobothrium violae*
Calypt, *Calyptrobothrium* sp.
Cardio, *Cardiobothrium beveridgei*
Carpob, *Carpobothrium* n. sp.
Cathet, *Cathetocephalus* sp.
Caul1, *Caulobothrium* n. sp. 1
Caul2, *Caulobothrium* n. sp. 2
Ceph1, *Cephalobothrium* n. sp. 1
Ceph2, *Cephalobothrium* n. sp. 2
Cerato, *Ceratobothrium xanthocephalum*
Chimae, *Chimaerocestos* n. sp.
Clisca, *Clistobothrium carcharodoni*

Clismo, *Clistobothrium montaukensis*
Corrug, *Corrugatocephalum ouei*
Crosso, *Crossobothrium laciniatum*
Dicran, *Dicranobothrium spinulifera*
Dinobo, *Dinobothrium* sp.
Dioeco, *Dioecotaenia* sp.
Diphyl, *Diphyllobothrium cordatum*
Discob, "*Discobothrium*" n. sp.
Discul, *Disculiceps galapagoensis*
Ditrac, *Ditrachybothridium macrocephalum*
Duplmi, *Duplicibothrium minutum*
Dupl1, *Duplicibothrium* n. sp. 1
Dupl2, *Duplicibothrium* n. sp. 2
Echei1, *Echeneibothrium* sp. 1
Echei2, *Echeneibothrium* sp. 2
Echei3, *Echeneibothrium* sp. 3
Echimx, *Echinobothrium mexicanum*
Echisp, *Echinobothrium* n. sp.
Ecoccu, *Echinococcus granulosus*
Enioco, *Eniochobothirum* sp.
Flexib, *Flexibothrium ruhnkei*
Floric, *Floriceps minacanthus*
Gastro, *Gastrolecithus planus*
Glypho, *Glyphobothrium zwerneri*
Grillo, *Grillotia similis*
Gymnor, *Gymnorhynchus isuri*
Horn1, *Hornellobothrium* n. sp. 1
Horn2, *Hornellobothrium* n. sp. 2
Horn3, *Hornellobothrium* n. sp. 3
Lecplt, *Lecanicephalum peltatum*
Lecsp, *Lecanicephalum* sp.
Litamp, *Litobothrium amplifica*
Litams, *Litobothrium amsichensis*

Litdai, *Litobothrium daileyi*
Macrob, *Macrobothridium* sp.
Marsup, *Marsupiobothrium gobelinus*
Mixodi, *Mixodigma leptaleum*
Monory, *Monorygma* sp.
Montic, *Monticellia lenha*
Myzoce, *Myzocephalus* sp.
Oryg1, *Orygmatobothrium* sp. 1
Oryg2, *Orygmatobothrium* sp. 2
Otobot, *Otobothrium* n. sp.
Pachyb, *Pachybothrium hutsoni*
Paraba, *Paraorygmatobothrium barberi*
Paraex, *Paraorygmatobothrium exiguum*
Parapr, *Paraorygmatobothrium prionacis*
Pedig1, *Pedibothrium globicephalum*
Pedilo, *Pedibothrium longispine*
Pedise, *Pedibothrium servattorum*
Phorma, *Phoreiobothrium manirei*
Phor1, *Phoreiobothrium* n. sp. 1
Phor2, *Phoreiobothrium* n. sp. 2
Phor3, *Phoreiobothrium* n. sp. 3
Phylla, *Phyllobothrium lactuca*
Phylri, *Phyllobothrium riseri*
Pingui, *Pingucollum pingucollum*
Platau, *Platybothrium auriculatum*
Plathy, *Platybothrium hypoprioni*
Platpa, *Platybothrium parvum*
Platsp, *Platybothrium* sp.
Poly1, *Polypocephalus* n. sp. 1
Poly2, *Polypocephalus* n. sp. 2
Poly3, *Polypocephalus* n. sp. 3
Potamo, *Potamotrygonocestus magdalenensis*

Prosob, *Prosobothrium* sp.
Protex, *Proteocephalus exiguus*
Protpe, *Proteocephalus perplexus*
Protpi, *Proteocephalus pinguis*
Pseu1, *Pseudanthobothrium* n. sp. 1
Pseu2, *Pseudanthobothrium* n. sp. 2
Pseush, *Pseudanthobothrium shetlandicum*
Rhinco, *Rhinebothrium corymbum*
Rhinsp, *Rhinebothrium spinocephalum*
Rhinur, *Rhinebothrium urobatidium*
Rhbdes, *Rhinobothroides moralarai*
Rhodob, *Rhodobothrium* sp.
Scypho, *Scyphophyllidium* sp.
Spinil, *Spiniloculus* n. sp.
Tentac, *Tentacularia* sp.
Tetbcy, *Tetrabothrius cylindraceus*
Tetbsp, *Tetrabothrius* sp.
Tetrag, *Tetragonocephalum* sp.
Thysan, *Thysanocephalum* sp.
Triloc, *Trilocularia acanthiaevulgaris*
Tyloce, *Tylocephalum* n. sp.
Uncibi, *Uncibilocularis* sp.
Yorker, *Yorkeria* n. sp.
Zyxibo, *Zyxibothrium kamienae*
Ng1, new genus 1 n. sp.
Ng2, new genus 2 n. sp.
Ng3, new genus 3 n. sp.
Ng4, new genus 4 n. sp.
Ng5, new genus 5 n. sp.

Notes added in proof

i.   *Chimaerocestus* should be *Chimaerocestos* throughout
ii. *Duplicibothrium* n. sp. 1 (Dupl1) and *Duplicibothrium* n. sp. 2 (Dupl2) have recently been described as *Duplicibothrium paulum* and *Duplicibothrium cairae*, respectively, by Ruhnke *et al.* (2000).

Chapter 15

The Aspidogastrea: An archaic group of Platyhelminthes

Klaus Rohde

General characteristics of the Aspidogastrea

The Aspidogastrea is a small group of platyhelminths comprising four families, 12 genera and approximately 80 species. Among the genera, *Sychnocotyle* from Australian freshwater turtles was described recently (Ferguson *et al.* 1999), a second one, *Rohdella*, from freshwater teleosts in Southeast Asia, 15 years ago (Gibson and Chinabut 1984), and another, *Rugogaster*, from the rectal glands of chimaerid fishes in the northern Pacific, 26 years ago (Schell 1973), indicating that the taxonomy of the group is unlikely to be completely known (for a recent taxonomic discussion see Rohde in press).

The main characteristics of the larvae are illustrated in Figure 15.1. The larvae of all species that have been examined have a posterior sucker, a pharynx and a simple or bifurcate intestine. Some species have a false oral sucker (not lined by a capsule but separated from the parenchyma of the main body by a septum extending from the pharynx to the tegument), others lack such a sucker. Excretory pores are paired and located at the level of the anterior margin of the posterior sucker, connected to two excretory ducts each with three flame bulbs in those species that have been examined for this character (and lateral flames, described only in larval *Multicotyle purvisi*). Some species have a short posterior appendage, some have eye-spots, and some have ciliary patches; other species lack these characters (for a detailed discussion and references see Rohde 1972, 1994a).

Adults are characterized by a ventral attachment apparatus that consists of a row of transverse thickenings, i.e., rugae (in *Rugogaster*), suckers (in *Stichocotyle*), deep cavities separated by transverse septa (in *Multicalyx*), or an adhesive disc subdivided into three or four rows of alveoli (suckerlets, in species of the family Aspidogastridae) (Figure 15.2). Testes are multiple, paired or single, a cirrus sac is present or absent, the gonopore is always anterior, a Laurer's canal and 'marginal organs' (ampullae of marginal glands) are present or absent. In species of *Lophotaspis* the adhesive disc bears well-developed papillae, and in *Rohdella* the terminal parts of the male and female reproductive ducts are joined, forming an hermaphroditic duct.

Species of the families Rugogastridae, Stichocotylidae and Multicalycidae infect chondrichthyans (Holocephali and Elasmobranchii), whereas species of the Aspidogastridae infect teleosts and chelonians. The life cycles of all species that have been worked out include molluscs as intermediate hosts and, in some species, as facultative final hosts. Encapsulated juveniles of *Stichocotyle nephropis*, whose life cycle is not known, were found in crustaceans.

Phylogenetic relationships of the Aspidogastrea with other Platyhelminthes

Several characteristics suggest a sister-group relationship with the Digenea. These include presence of a Laurer's canal and a ventral/posterior sucker, as well as life cycles using mollusc and vertebrate hosts. DNA analyses have consistently supported such a sister-group relationship, although Blair (1993) has suggested that the Aspidogastrea may be the sister-group of all other Neodermata, a suggestion not supported by any of the other studies.

Phylogenetic relationships of aspidogastrean groups with each other

Previous analyses of the Aspidogastrea and their position within the Neodermata using phylogenetic systematics and based on morphological and host characteristics are by Brooks *et al.* (1985a,b), Gibson (1987), Brooks (1989a,b) and Pearson (1992). Gibson's (1987) phylogenetic tree is illustrated in Figure 15.3. The author stressed that some of the characters for his analysis are questionable, including the assumption that a molluscan host and a bifid intestine are plesiomorphic for the trematodes. Brooks *et al.* (1989) used 15 characters, assuming that a bifurcate gut is 'unambiguously a synapomorphy of the Trematoda and not just an autapomorphy of the Digenea'. They further assume that *Rugogaster* is the sister-group of the other aspidogastreans, that the amphistomatous condition is plesiomorphic for all trematodes, and that the mollusc plus vertebrate two-host life cycle is plesiomorphic for all trematodes. Pearson (1992) criticized the phylogeny of Brooks *et al.* and argued, on the basis of outgroup comparison, that the Aspidogastridae and not the Rugogastridae is the sister-group of the other aspidogastreans (following Gibson in this respect), that *Rugogaster* and *Stichocotyle* are derived multicalycids, and that a single caecum is plesiomorphic for digeneans and aspidogastreans.

To test these hypotheses, I performed maximum parsimony analyses using a heuristic search as implemented in PAUP (Swofford 1998), for which characteristics of the various genera were compiled in a data matrix (Table 15.1). It should be noted that some of these characters are not useful for a phylogenetic analysis and were therefore excluded from the analyses. Thus, ciliary patches and eye-spots (characters 1 and 2) are probably plesiomorphic characters that have been lost repeatedly, the bifurcate intestine of the larva (4) is not a character independent of the bifurcate intestine of the adult; the number of testes (11), loss of a cirrus-sac (12) and presence of anterior sucker-like structures of the adult (13) are probably convergent or plesiomorphic characters. In all analyses, Digenea were used as the outgroup. The results are illustrated in Figure 15.4, which shows a clade consisting of *Rugogaster*, *Stichocotyle* and *Multicalyx*, i.e., the genera infecting elasmobranchs, a clade consisting of the Cotylaspidinae (*Cotylogaster*, *Cotylaspis* and *Lissemysia*) and Rohdellinae (*Rohdella*), and a clade consisting of the Aspidogastrinae (*Multicotyle*, *Sychnocotyle*, *Lobatostoma*, *Aspidogaster* and *Lophotaspis*).

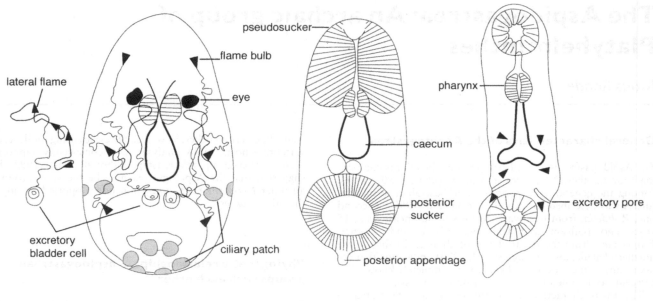

Multicotyle purvisi   Lobatostoma manteri   Rugogaster hydrolagi

Figure 15.1 Larvae of *Multicotyle purvisi, Lobatostoma manteri* and *Rugogaster hydrolagi*. Note the presence of ciliary patches and eyes in *Multicotyle* and absence in the other species. All species have a posterior sucker, and *Multicotyle* lacks an anterior pseudosucker. The caecum is bifurcate in *Rugogaster*, but simple in the other species. Excretory pores are at the anterior margin of the posterior sucker. The protonephridia have been best studied in *Multicotyle*; they have three pairs of flame bulbs, as well as lateral flames. Redrawn and modified from Rohde (1971b, 1972, 1973b) and Schell (1973).

There is more to phylogeny than cladistics

In this section, I discuss some characters of the aspidogastreans (and in one case of digenean trematodes) which are suggestive of phylogenetic relationships of platyhelminths with other phyla, although they are insufficient as firm evidence on their own.

Traditionally, discussions of the phylogenetic relationship of the Platyhelminthes with other invertebrate groups are based on free-living turbellarians, because parasitic forms are considered as more derived, with the loss of many or all archaic characters, and therefore less suitable for comparative studies at the level of phyla. However, studies based on DNA or combined morphological and DNA data have consistently shown that the Neodermata on their own (Littlewood *et al.* 1999a,b; Litvaitis and Rohde 1999, and further references therein), or Neodermata jointly with several taxa of parasitic turbellarians (Littlewood *et al.* 1999a) have branched off from free-living forms very early in their evolutionary history. Therefore, they may have retained archaic characters which may well provide clues as to the phylogenetic relationships of the phylum with other phyla.

A recent paper by Balavoine (1998) based on *Hox* genes and a re-evaluation of evidence from 18S rDNA sequences has again raised the possibility that platyhelminths (excluding the acoelans; see Ruiz-Trillo *et al.* 1999) are not near the base of the Metazoa, as postulated by the planuloid/acoeloid theory. It suggests that they are derived animals with a secondarily reduced complexity, which are closely related to the Annelida and other coelomate and segmented animals, as postulated by the archecoelomate theory (for a discussion and references see Balavoine 1998). Occurrence of a very special type of cleavage, i.e., quartet spiral cleavage, in the Annelida, Mollusca and one platyhelminth group, the Polycladida (versus a dual spiral or bilateral cleavage in the Acoela, see Boyer

and Henry 1998 and Boyer in Ruiz-Trillo *et al.* 1999), gives morphological support to this view, although it is also compatible with the planuloid theory. Spiral cleavage may not be as restricted within the Platyhelminthes as usually assumed. Figure 15.5 shows early cleavage in the aspidogastrean *Lobatostoma manteri*, originally described as 'irregular' by Rohde (1973b). Despite the great variability of cell arrangement at the six- and eight-cell stage, the pattern in some eggs shows similarities with spiral cleavage. There are mega- and microblasts rotated at an angle to each other. The possibility exists that the pattern is a rudimentary spiral one.

Another character which may suggest a relationship with 'higher' animals, is the well-developed 'pseudosegmentation' of aspidogastreans. During development from the larva to the adult, the posterior sucker becomes divided into a large number of suckerlets (Figure 15.6), rugae or deep cavities divided by transverse septa. By definition, segmentation is restricted to animals possessing a coelom, which is not found in the platyhelminths. However, 'pseudosegmentation' in the aspidogastreans is not just superficial, but involves the marginal glands and the nervous system (Figure 15.7). In each 'pseudosegment', a complex arrangement of nerves connects the various connectives. In the anterior part, several internal and external commissural rings are found; the brain is a very large internal commissure (Figure 15.8) suggesting that it has evolved very early in evolutionary history as a central organ for processing information and not as a peripheral clumping of nerves close to aggregations of sensory receptors. A high degree of complexity, not expected among 'primitive' acoelomates, is also indicated by sheath-like structures around parts of the posterior ventral connective (Figure 15.9), as well as by an extraordinary variety of sensory receptors in larvae which actively seek a host or the host's habitat and in adults of those species which live free in the intestine. Some of the receptors occur in very large numbers (Rohde 1994b, and further references

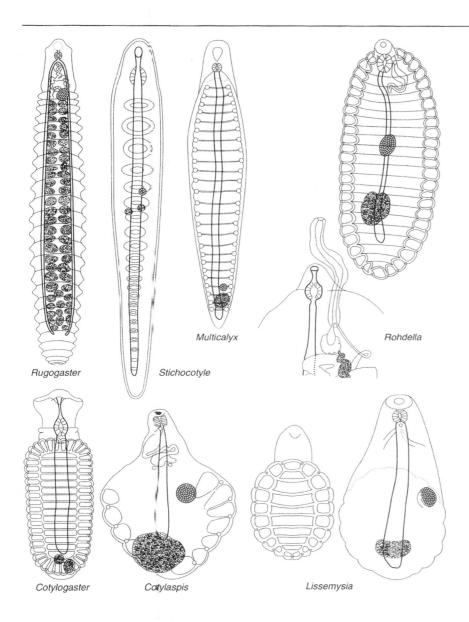

Rugogaster

Stichocotyle

Multicalyx

Rohdella

Cotylogaster

Cotylaspis

Lissemysia

Figure 15.2 Diagrams of aspidogastrean genera with the most important characters. Note rugae, small ventral sucker, bifurcate intestine and multiple testes in *Rugogaster*, a row of well-separated suckers, a simple caecum and two testes in *Stichocotyle*, a row of deep cavities separated by septa, a single caecum and testis in *Multicalyx*, and an adhesive disk subdivided into alveoli, a single caecum, and one or two testes in all other species. *Rohdella* possesses an hermaphroditic duct, and *Lophotaspis* has large papillae on the adhesive disc. Redrawn and modified from Rohde (in press) after various sources.

Continued overleaf

therein). Attention should also be drawn to the fact that a circulatory system, the so-called lymphatic system, is present in several digenean groups (but not in the aspidogastreans). For example, the amphistome digenean *Paramphistomum bathycotyle* has a pair of strongly branched lymphatic ducts (Figure 15.10) which contain lymphatic liquid (Figure 15.10A), 'free' nuclei, i.e., nuclei without cytoplasm visible under the light microscope (Figure 15.10B,C), and various other inclusions (Figure 15.10D–F). The left and right ducts do not communicate. In other species, two or three pairs of longitudinal ducts are present which pulsate.

These characters are not definitive evidence for the coelomate connection of platyhelminths, but they show an extraordinary degree of complexity and some suggestive similarities of the aspidogastreans (and digeneans) with coelomates.

The very complex life cycles of digeneans, involving a mollusc and vertebrate host and several 'larval' stages (miracidium, mother–sporocyst, daughter–sporocyst/redia, cercaria, metacercaria) has led to much speculation about their origin. Rohde (1971a) has suggested that they have evolved from the much simpler life cycle of aspidogastreans: maturation without

alternation of generations occurred in the original mollusc host; proto-digeneans were able to survive in vertebrates when ingested; they became more and more dependent on the vertebrate host and developed various adaptations to ensure infection of that host, such as multiplicative stages in the mollusc and free dispersal stages (cercariae). This hypothesis seems to contradict the suggestion of Littlewood *et al.* (1999b, and further references therein) that not molluscs but vertebrates were the original hosts. This view can be reconciled with Rohde's suggestion by assuming that the original host was indeed a vertebrate or proto-vertebrate and that molluscs harbouring non-multiplicative larval stages as found in extant aspidogastreans, were incorporated secondarily to assure transmission to the vertebrate host. Multiplicative and dispersal (cercarial) stages were incorporated later to make the infection of vertebrates more effective. Alternatively, the cladistic analysis, on which the conclusions of Littlewood *et al.* are based, may be incomplete because of our ignorance of the sister-group of the Neodermata. If this sister-group should be found among various turbellarians parasitic in vertebrates and invertebrates (fecampiids, *Urastoma*, *Ichthyophaga*,

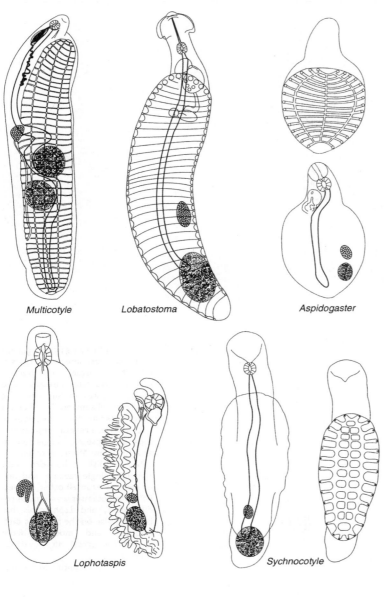

Multicotyle

Lobatostoma

Aspidogaster

Lophotaspis

Sychnocotyle

Figure 15.2 Continued.

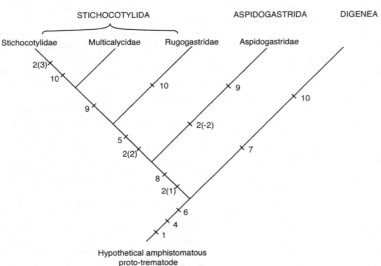

Hypothetical amphistomatous
proto-trematode

Figure 15.3 Phylogeny of the Aspidogastrea according to Gibson (1987). 1, amphistomatous juvenile with simple posterior sucker. 2(1), hypertrophy and linear subdivision of posterior sucker by transverse septa. 2(2), increased linear growth and subdivision with anterior subdivisions developing into small, fused suckers. 2(-2), lateral growth and vertical subdivision of posterior sucker by septa. 2(3), secondary suckers independent. 4, molluscan host. 5, selachian host (?preceded by crustacean host). 6, teleost host. 7, multiple generations. 8, atrophy of oral sucker. 9, reduction of bifid gut to monocaecal condition. 10, loss of marginal bodies.

	1 1234567890	111111111 123456789
Digenea	110101 000	111000101
Rugogaster	0011x1 100	0100x1110
Stichocotyle	xxx0x00100	01101xx10
Multicalyx	00x0000100	1x01x1x10
Rohdella	00x0000001	11011x101
Cotylogaster	1000000001	101111101
Cotylaspis	x0x0x0001	11011x101
Lissemysia	xxx0x0C001	100xxxx01
Multicotyle	1100000010	010111101
Lobatostoma	0010100010	110111101
Aspidogaster	0010100010	111111101
Lophotaspis	11x0x0C010	1001xx101
Sychnocotyle	1100000010	10011x101

Table 15.1 Data matrix used to reconstruct phylogenetic relationships. Source Rohde (in press, and various authors therein). Presence of character coded as 1, absence as 0, not known as x. 1, larva with ciliary patches; 2, larva with eye-spots; 3, larva with false oral sucker; 4, larva with bifurcate intestine; 5, larva with posterior appendage; 6, adult with ventral sucker; 7, adult with bifurcate intestine; 8, adult with single row of suckers, rugae or septa; 9, adult with adhesive disc subdivided into four rows of alveoli; 10, adult with adhesive disc subdivided into three rows of alveoli; 11, adult with single testis; 12, adult with cirrus sac; 13, adult with anterior sucker-like structure; 14, adult with marginal bodies; 15, adult with septate oviduct; 16, tegument with microtubercles; 17, adult with Laurer's canal; 18, hosts chondrichthyan fishes; 19, hosts teleost fishes and turtles. Note: larvae of Digenea are miracidia (characters 1–3) and cercariae (character 4).

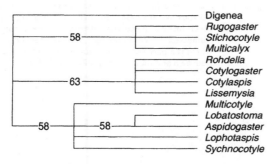

Figure 15.4 Cladogram based on morphological and host characteristics (Table 15.1) using PAUP. Digenea is the outgroup, characters 1, 2, 4, 11, 12, 13 are excluded; maximum parsimony, fast heuristic search, 1000 bootstrap replicates. Fifty per cent majority rule consensus tree. Numbers indicate branch support.

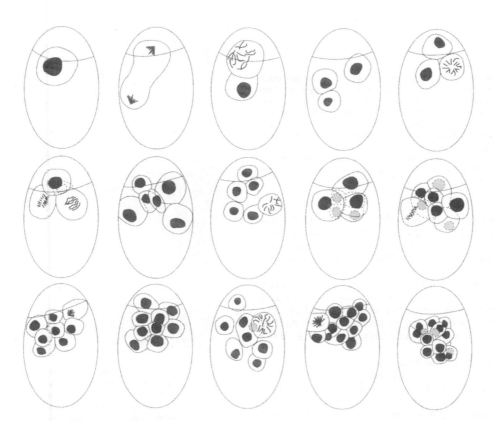

Figure 15.5 Early cleavage of *Lobatostoma manteri*. Note great variability but presence of mega- and microblasts at the six- and eight-cell stage in some eggs, the former rotated somewhat with regard to the latter. Redrawn and modified from Rohde (1973b).

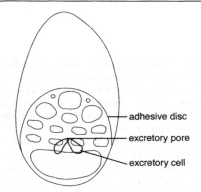

Figure 15.6 Larva of *Multicotyle purvisi* at an early stage of development of the adhesive disc. Only the first rows of alveoli have been formed. Redrawn and modified from Rohde (1971b).

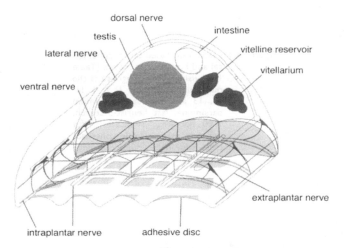

Figure 15.7 *Multicotyle purvisi*, adult. Reconstruction of the posterior nervous system based on serial sections impregnated with silver nitrate. Note the main posterior connectives connected by commissures in the dorsal part of the body, and the intra- (inside the adhesive disc) and extraplantar nerves (outside the adhesive disc) in the ventral part of the body. Note also the regular arrangement of the nerves corresponding to the rows of alveoli of the adhesive disc, suggesting 'pseudosegmentation'. Redrawn and modified from Rohde (1971c).

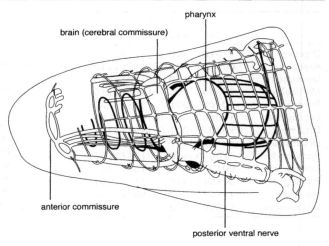

Figure 15.8 *Multicotyle purvisi*, adult. Reconstruction of the anterior nervous system based on serial sections impregnated with silver nitrate. Note the internal and external connectives connected by commissures, and the cerebral commissure which is a particularly well-developed internal commissure. Redrawn and modified from Rohde (1971c).

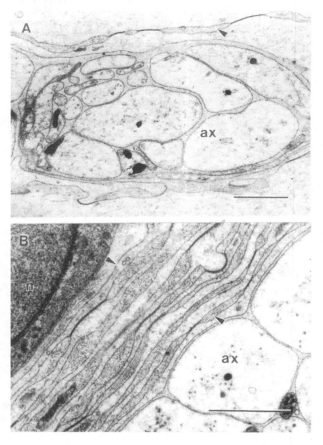

Figure 15.9 *Multicotyle purvisi*, adult. Transmission electron micrographs of cross-sections through (**A**) a small nerve and (**B**) the posterior ventral connective showing several concentric layers of lamellae surrounding the axons in a sheath-like fashion. Abbreviations: ax, axon; n, nucleus; arrowheads indicate lamellae near nerve in **A** and around part of posterior ventral nerve in **B**. Scale bars 1 μm (originals).

Notentera), it may well be that the proto-trematode was indeed a mollusc parasite (or parasite of many invertebrates including molluscs) and that the digenean life cycles have evolved as suggested by Rohde. This would imply a very early origin of parasitic platyhelminths, including trematodes, which finds some support in the recent study of Wang *et al.* (1999) which, based on 50 genes and although not including platyhelminths, estimated divergence time of chordates and arthropods at 993 ± 46 Ma, and of nematodes and the line leading to arthropods and chordates at 1177 ± 79 Ma (Figure 15.11A,B).

It should be noted that fecampiids, as well as *Urastoma*, *Ichthyophaga* and *Notentera* are parasites of crustaceans, molluscs, vertebrates and annelids, respectively, that DNA analyses have shown that they form one monophylum, and that ultrastructure of spermiogenesis and protonephridia (as well as some DNA analyses) suggest a sister-group relationship of these taxa (plus possibly some others) and the Neodermata

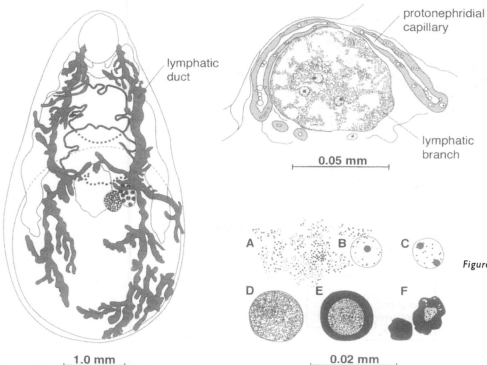

lymphatic duct

protonephridial capillary

lymphatic branch

0.05 mm

A   B   C

D   E   F

0.02 mm

1.0 mm

Figure 15.10 Lymphatic system of the amphistome trematode *Paramphistomum bathycotyle* reconstructed from serial sections (left), a cross-section through a lymphatic branch closely associated with an excretory capillary (top right), and various inclusions (bottom right). Redrawn and modified from Lowe (1966).

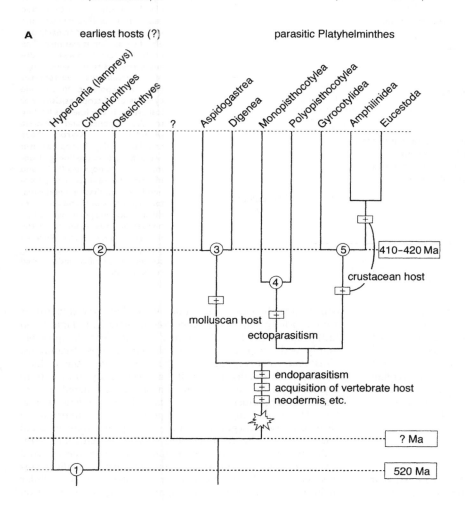

A   earliest hosts (?)   parasitic Platyhelminthes

Hyperoartia (lampreys)
Chondrichthyes
Osteichthyes
?
Aspidogastrea
Digenea
Monopisthocotylea
Polyopisthocotylea
Gyrocotylidea
Amphilinidea
Eucestoda

410–420 Ma

crustacean host

molluscan host

ectoparasitism

endoparasitism
acquisition of vertebrate host
neodermis, etc.

? Ma

520 Ma

Figure 15.11 **A**) Origin of parasitism in the Platyhelminthes acc. to Littlewood *et al.* (1999b). 1, No extant lamprey parasitized, therefore parasitism after divergence of Hyperoartia from other Gnathostomata (Chondrichthyes + Osteichthyes?); 2, Chondrichthyes + Osteichthyes lineages diverge c. 410 Ma; 3, Trematodes diverge with 2; 4, Both monogenean lineages diverge with 2, therefore must have diverged from one another before 2; 5, Cestodarians diverge with 2.

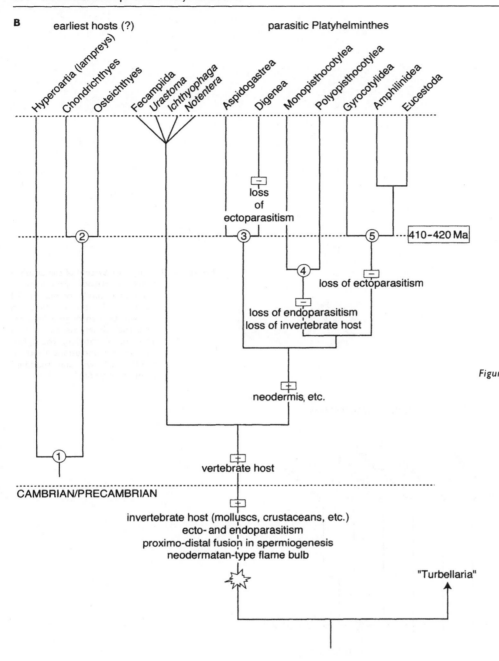

B

earliest hosts (?)   parasitic Platyhelminthes

loss
of
ectoparasitism

410–420 Ma

loss of ectoparasitism

loss of endoparasitism
loss of invertebrate host

neodermis, etc.

vertebrate host

CAMBRIAN/PRECAMBRIAN

invertebrate host (molluscs, crustaceans, etc.)
ecto- and endoparasitism
proximo-distal fusion in spermiogenesis
neodermatan-type flame bulb

"Turbellaria"

Figure 15.11 **B**) Origin of parasitism in the Platyhelminthes if a sister-group relationship of Neodermata and Fecampiida, etc. and an early origin of parasitism is assumed. The proto-neodermatan was an endo- and ectoparasite of a wide range of invertebrates; vertebrates were incorporated in the life cycle later. Ectoparasitism was lost twice, in the lineages leading to the Digenea and Cestoda (Amphilinidea, Gyrocotylidea and Eucestoda). Endoparasitism was lost once, in the lineage leading to the Monogenea (Poly- and Monopisthocotylea). The proto-neodermatan had a neodermatan-type flame bulb and proximo-distal fusion in spermiogenesis; a neodermis was acquired in the lineage leading to the Neodermata, but not in that leading to the fecampiids and related turbellarians.

(e.g., Watson and Rohde 1993a,b; 1995a; Joffe and Kornakova 1998; Kornakova and Joffe 1999; Littlewood *et al.* 1999a). It should also be noted that the Aspidogastrea have retained some archaic characters. They are ecto- as well as endoparasites of molluscs, most species have a wide host range (although the report of an aspidogastrean in an ascidian needs verification), and some species were shown to be able to survive for a long time in simple media outside a host (references in Rohde 1972, 1994a; Ferguson *et al.* 1999).

Conclusions

Combined DNA and morphological evidence has shown that the Aspidogastrea is the sister-group of the Digenea, both constituting the class Trematoda. Very likely, the trematodes are the sister-group of the other Neodermata (e.g., Figure 6 in Littlewood *et al.* 1999a). Because the Neodermata (possibly jointly with some parasitic turbellarian groups: fecampiids, *Urastoma*, *Ichthyophaga* and perhaps some others) have evolved from free-living turbellarians very early in evolutionary history, the Aspidogastrea must be considered a very ancient group, which may have retained some archaic characters. These characters include a possible rudimentary spiral cleavage, 'pseudosegmentation', which is not only external but extends to the glandular and nervous systems, an extraordinary variety of sensory receptors in some species, and a very complex nervous system. Furthermore, some digeneans possess a pulsating circulatory system with cellular inclusions. These characters support the suggestion based on comparisons of *Hox* genes, that platyhelminths are secondarily simplified coelomates. The life cycle is simple, including an intermediate mollusc and final vertebrate host, although some species can mature in the mollusc host. It is likely that the much more complex life cycles of digenean trematodes have evolved from that of the Aspidogastrea. Morphological and host characteristics on their own do not resolve phylogenetic relationships between aspidogastrean genera, although it is likely that the three families Rugogastridae, Stichocotylidae and Multicalycidae form one clade, which corresponds to the order Stichocotylida proposed by Gibson and Chinabut (1984). Combined DNA and morphological data may resolve phylogenetic relationships between genera.

ACKNOWLEDGEMENTS

Financial support was provided by the University of New England and the Australian Research Council. Sandy Hamdorf and Louise Streeting helped with re-drawing the figures, and Sandy Higgins typed the table. In particular, I wish to thank Tim Littlewood and Rod Bray for organizing the Symposium.

Chapter 16

The Digenea

Thomas H. Cribb, Rodney A. Bray, D. Timothy J. Littlewood, Sylvie P. Pichelin and Elisabeth A. Herniou

The Digenea is probably the largest group of internal metazoan parasites. According to the databases maintained at the Parasitic Worms Division of the Natural History Museum, London, there are, at present, about 150 recognized families containing nearly 2700 nominal genera and about 18 000 nominal species. The general description of a digenean, although exceptions abound, is of a platyhelminth with a complex life-cycle, with three hosts in the cycle, the first being a mollusc, the second a wide variety of invertebrates and vertebrates, and the final 'definitive' host being a vertebrate. The unique intra-molluscan stages are closely associated with the mollusc, often highly specific to a host-species or lower taxon, and are generally viewed as consisting of three 'generations', the latter two having been produced asexually by parthenogenesis or budding (the precise process remaining controversial). The final stage emerges from the mollusc as the motile form known as the 'cercaria', a stage unique to the Digenea. The cercaria develops into a metacercaria in its second intermediate host, which is ingested by the definitive host, in which the sexually reproductive stage occurs. Included in the group are major human (e.g., schistosomes) and domestic animal (e.g., paramphistomes) pathogens as well as numerous minor pathogens and worms whose pathogenic effects have not been established. The ubiquity of these parasites and the large numbers of host-species which may be affected by the life-cycle of a single digenean species, indicate that these worms are likely to be major players in many ecosystems.

Historical background

Fascinating studies of the history of our understanding (and misunderstanding) of helminths are to be found in Hoeppli (1959) and Grove (1990). Hippocrates (c. 460–374 BC) was aware of tapeworms, but it appears that flukes were not recognized until somewhat later, the first mention of the liver-fluke being by Jean de Brie in 1379 (de Brie 1379). The earliest mention of a fluke under a name now recognizable was probably that of *Hirudinella marina* Garcin 1730 (Garcin 1736) now recognized as *Hirudinella ventricosa* (Pallas 1774) (see Gibson and Bray 1977). Although Linnaeus (1758) listed two *Fasciola* species, only one – *F. hepatica* – has stood the test of time as a digenean. Cercariae were discovered by Müller (1773), who considered them independent adult worms. Dujardin (1845) recognized the four genera *Amphistoma*, *Monostoma* Zeder 1800, *Holostomum* Nitzsch 1819 and *Distoma* based on the presence/absence and position of suckers and the strigeid-type forebody. Monticelli (1888) recognized similar groups to those suggested by Dujardin (1845), giving them the 'family' names Amphistomeae, Diplostomeae, Distomeae and Monostomeae. In 1899, Looss divided up *Distoma* and erected 59 new genera. He recognized two major branchings of the Malacocotylea (Digenea), the Metastatica (the family Holostomidae) and the Digenea 's. strict.' (the families Distomidae, Rhopaliadae, Schistosomidae, Gasterostomidae, Didymozoonidae and Monostomidae; Looss 1899). The dawning realization of the complexity of the 'alternation of generations' (an idea developed by Steenstrup 1842; translated 1845) in digenean life-cycles is described by Grove (1990), culminating in the demonstration of the life-cycle of *F. hepatica* almost simultaneously by Leuckart (1881, 1882) and Thomas (1881, 1882).

The name Trematoda was introduced by Rudolphi (1808). The first major systematic, and (as we would see it) phylogenetic decision was by van Beneden (1858) who divided the Trematoda into two major groups, 'Monogénèses' and 'Digénèses', initiating the use of life-cycle characteristics in trematode systematics. The first use of the orthography Digenea was, apparently, by Carus (1863), who defined the group as having a life-cycle with alternation of generations, with larvae in molluscs and the sexual phase in vertebrates. La Rue (1957) summarized many of the early developments in trematode systematics, stressing the importance of life-cycle information, and pointing out the problems associated with early attempts based solely on adult comparative morphology.

Poche (1926) recognized two major grouping in the Digenea, i.e., the Gasterostomata (including only the Bucephalidae) and the Prosostomata (Figure 16.1). He apparently considered the Aspidogastridae derived from (or a sister-group of) the Haplosplanchnidae. Some sections of his 'pedigree' show similarities to more recent systems, e.g., the Hemiurida (including the Didymozoidae). La Rue (1957) produced the first modern system, based on the division of the class into Anepitheliocystidia and Epitheliocystidia (Figure 16.2), characterized by the lack or presence, respectively, of an epithelium-lined excretory vesicle and the details of its formation. At that time there had been no transmission electron microscope studies of the digenean excretory bladder, but in the early 1970s it became clear that these taxa could not be sustained based on this morphological criterion (Powell 1972, 1973, 1975). The Bucephalidae was included in the Brachylaemata by La Rue (1957), following several earlier authors, who dismissed the gasterostomate condition as a major distinguishing feature.

Relationships of the Digenea with other neodermatans

Sinitsin (1911) believed that classifying trematodes as Platodes [= Platyhelminthes] was based on a misunderstanding, and that they were derived from 'more highly organized coelomic invertebrates' such as Trochelminthes or Arthropoda. He also reckoned that the Monogenea were derived from the Digenea by simplification of the life-cycle. Janicki (1920) introduced his 'cercomer' theory, suggesting that the cercarial tail, the monogenean opisthaptor and the cestode cercomer were homologues, leading him to divide the Platyhelminthes into the parasitic Cercomeromorphae and the free-living Turbellaria, and to suggest the derivation of the cestodes from the digeneans. Until Baer (1931) and Bychowsky (1937), it was widely reckoned, or at least implied by systematic arrangements, that the Monogenea and Digenea + Aspidogastrea were sister-taxa within the class Trematoda. Bychowsky (1937, Figure 4) illustrated his 'evolutionary scheme' showing the Cestoda, Cestodaria, Gyrocotyloidea and Monogenea to be a monophyletic

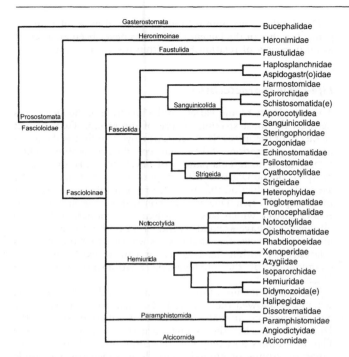

Figure 16.1 Poche's (1926) pedigree (Stammbaumes) of the Digenea in a dendrogram. Poche inserted families on internodes or leading directly to other families, these have been interpreted as sister-taxa to the distal taxa. Poche recognized two major groupings in the Digenea, i.e., the Gasterostomata (including only the Bucephalidae) and the Prosostomata. Poche's orthography is retained.

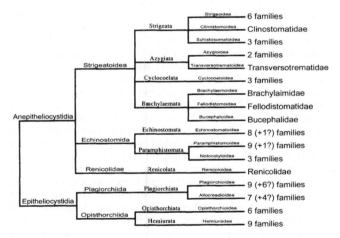

Figure 16.2 La Rue's (1957) classification of the Digenea. This is the first modern system of the group, and is based on the division of the class into Anepitheliocystidia and Epitheliocystidia (characterized by the lack or presence, respectively, of an epithelium-lined excretory vesicle and the details of its formation).

'superclass' Cercomeromorphae, diverging from the Rhabdocoela in the early Palaeozoic. The 'class Trematoda' is derived separately from the Rhabdocoela and slightly later in the Palaeozoic. A similar diphyletic system for the parasitic Platyhelminthes was proposed by Llewellyn (1965, Figure 8), but in this case the trematodes diverged earlier and the cestodes were considered to be derived from the Monogenea. Brooks

(1982) in introducing cladistic methods to platyhelminth phylogeny found, in contrast, that the major parasitic taxa, along with the Temnocephala, are monophyletic forming the 'subphylum' Cercomeria, which was defined on nine autapomorphies in a later paper (Brooks et al. 1985b). The latter authors retained Janicki's name Cercomeromorphae for the Monogenea + cestode groups, and considered the digenean ventral sucker rather than the cercarial tail as the homologue of the cercomer. Lebedev (1987) strongly disagreed on this latter point. Ehlers (1984) also considered the parasitic platyhelminths monophyletic, and gave the taxon the name Neodermata, expanding greatly on his finding in 1985. In Ehlers' system the Temnocephalida are members of the 'Dalyellioida', the sister-group to the Neodermata. Molecular evidence for the monophyly of the Neodermata was first introduced by Baverstock et al. (1991), and by the time Littlewood et al. (1999a) produced a detailed combined molecular and morphological phylogeny of the Platyhelminthes, they were able to comment that this monophyly 'could be considered beyond doubt'.

The internal topology of the neodermatan tree is also now well established, the major remaining controversy being around the monophyly or paraphyly of the Monogenea. The Digenea and Aspidogastrea form the Trematoda, a sister-taxon to the remaining neodermatans. Although some fairly recent authors have considered the possibility that the Aspidogastrea are a derived group within the Digenea (see Cable 1974), the monophyly of the Digenea now appears very well established and not disputed. The autapomorphies of the Digenea are considered to be: 1) internal parasitism of a mollusc; 2) the unique series of asexual generations in the mollusc, i.e., mother sporocyst and redia; 3) tiers of ectodermal cells on the miracidium; 4) the cercaria; 5) early stages (miracidium, mother sporocyst) without a digestive system; and 6) the complex life-cycle involving several hosts, including vertebrate definitive host (Ehlers 1985a; Brooks et al. 1985b; Gibson 1987; Pearson 1992; Ax 1996; Littlewood et al. 1999a). Controversial autapomorphies include: i) bifurcate gut (Brooks et al. 1985b); and ii) ventral sucker in cercaria and adult (Ehlers 1984).

Pearson (1972), in summarizing earlier scenarios for the evolution of the Digenea, pointed out several attempts to do so with 'step-by-step' evolution from free-living turbellarians, and lists eight authorities who suggested that digeneans were derived from dalyellioid rhabdocoels or, at least, a rhabdocoele-like ancestor. If monophyly of the Neodermata is accepted, then the basic point of these scenarios is contradicted. Littlewood et al. (1999b) put forward a view that the most parsimonious interpretation, bearing in mind the monophyly of the parasitic platyhelminth groups, is that early neodermatans were endoparasites of vertebrates, and that the molluscan host was acquired by the ancestor of the Trematoda, and has been lost in few cases (e.g., Aporocotyle).

Relationships within the Digenea: conflict and inconsistencies

Brooks et al. (1985a) were the first group to undertake a cladistic phylogenetic study of the Digenea. Many of the groupings discovered were uncontroversial, often ones previously suggested by such workers as La Rue (1957), and a large group of families was unresolved, with a 12-fold polytomy at one point. One major finding of Brooks et al. (1985a) did, however, cause controversy. They found that the North American turtle worm Heronimus was the sister to the remaining Digenea. The apparent lack of the redial 'generation' in

Heronimus suggested to Brooks *et al.* that presence of a redia is an autapomorphy of all of the rest of the Digenea. Gibson (1987) was the first to question the basic premise and to point out that Crandall (1960) described a thin sac around the cercarial embryos within the miracidium which, he thought, might represent an 'incipient' or, more likely, a 'retrogressive' redia. Pearson (1992), however, mounted the most sustained attack on the phylogeny of Brooks *et al.* (1985a) (some 85 pages long). Pearson analysed the study of Brooks *et al.* character by character and claimed that he had shown that 97 of the 212 characters were invalid as used, resulting in an increase in homoplasy by 18% to 37% with a concomitant decrease in resolution from 82% to 32%. For example, Brooks *et al.* (1985a) found three apomorphs for *Heronimus*, but Pearson reckoned there were 13. Unfortunately, neither Brooks *et al.* (1985a), Pearson (1992) nor Brooks and McLennan (1993a) presented a matrix.

Molecular approaches to inference of phylogeny within the Digenea

The use of molecular approaches to determine phylogenetic relationships of digeneans has grown very rapidly in the 1990s. These studies fall into two broad classes – studies of specific problems of relationship of closely related taxa within families and broader studies of the efficacy of different genes and the relationships between more distantly related taxa (see Table 16.1). There have been numerous studies exploring the relationships within the Schistosomatidae and isolated studies on Campulidae and Nasitrematidae, Didymozoidae, Echinostomatidae, Fellodistomidae, Lepocreadiidae and Paragonimidae. These studies used, variously, sequence information from large (28S-like, LSU) and small (18S-like, SSU) subunit rRNA, and ITS1 and ITS2.

A smaller number of studies has considered combinations of families whose relationships are unresolved. Barker *et al.* (1993b) used SSU rDNA to suggest that the key taxon *Heronimus* was neither the sister-group to the Digenea nor closely related to the Paramphistomidae. Barker *et al.* (1993a) explored the value of the D1 domain of LSU rRNA and concluded that whereas it was useful for phylogenetic inference within and between closely related families it had little value for inference of deep relationships. Blair and Barker (1993) used SSU rRNA to test hypotheses about the relationships of the Gyliauchenidae and concluded that they were most closely related to the Lepocreadiidae rather than the Paramphistomidae. Campos *et al.* (1998) used SSU ribosomal rRNA to conclude, contrary to Barker *et al.* (1993b), that the Heronimidae was the sister-group to the remainder of the Digenea. Schulenburg and Wägele (1998) analysed published SSU rDNA sequences and concluded that they were unsuitable for the inference of phylogeny within the Digenea. Blair *et al.* (1998) used the V4 region of SSU rRNA to examine relationships between ten families of putative hemiuroid trematodes. They found no evidence for the inclusion of the Azygiidae in this assemblage and evidence that the Lecithasteridae was poorly supported as a family. Schulenburg *et al.* (1999) explored the phylogenetic utility of the ITS1 of rDNA and concluded that it was unsuitable for the exploration of relationships between major lineages. In a forerunner to the study reported in this paper Littlewood *et al.* (1999a) used SSU rDNA to examine all major groups of platyhelminths. Their study incorporated only 11 digenean species but still suggested, among other findings, that the Schistosomatidae was the most basal of the taxa that they examined.

Despite this now quite substantial body of molecular phylogenetic studies of the Digenea, there has never been an attempt at an overview of relationships within the entire group.

Life-cycle characteristics

The complexity of and variety in the life-cycle of the Digenea is perhaps the most striking feature of the group. Two distinct elements of the life-cycle are noteworthy. The first is the alternation of generations and asexual reproduction within the first intermediate host. This pattern of development is an autapomorphy for the Digenea. It is tempting to suggest that it is the asexual reproduction that has been responsible for the success of the species-rich Digenea, in contrast to the species-poor Aspidogastrea which lack any asexual reproduction. Because all digeneans have asexual reproduction in the first intermediate host, however, this character is not one that impinges on relationships within the Digenea or can be explored by the examination of the phylogenetic relationships within the Digenea.

The second feature of the digenean life-cycle is its plasticity, which has allowed the infection of a vast array of vertebrate and invertebrate animals. In terms of numbers of species, the Digenea may be the single largest component of the Platyhelminthes and they have certainly infected the greatest range of groups of animals. The development of a credible phylogeny for the Digenea creates opportunities for the investigation of the evolution of life-cycle patterns and the mapping and analysis of life-cycle traits.

Knowledge of the digenean life-cycle

When Yamaguti (1975) published *A synoptical review of life histories of digenetic trematodes of vertebrates* there was already a mountain of information to review. The work summarized by Yamaguti commenced with the ground-breaking information published by Leuckart (1881, 1882) and Thomas (1881, 1882, 1883) on *Fasciola hepatica*, the first life-cycle known for a digenean. Yamaguti found over 1700 studies to review from the 90 years that followed. These dealt with representatives of 69 families and an enormous number of species. Despite this vast amount of information there remained about as many families (as recognized by the classification that Yamaguti had himself published slightly earlier; Yamaguti 1971) for which there was no, or very inadequate, information. Since Yamaguti's review, a number of the taxa for which information was missing have had life-cycles elucidated. For example, we have useful new information on the Angiodictyidae (Lotz and Corkum 1984), Didymozoidae (Anderson 1999) and Mesometridae (Jousson *et al.* 1998b). However, for none of these families can information yet be said to be complete. There has, however, been an apparent drop-off in interest in the elucidation of life-cycles. Although new information continues to appear, particularly on taxa that are already well-known, there seems little drive to pursue the more poorly known taxa. There have been few attempts (Gibson and Bray 1994) to synthesize the available data regarding the life-cycles of trematodes.

A complete summary of the state of knowledge of digeneans is presently impossible because of the combination of the size of the literature, the extent to which it is scattered, and the uncertainty about the relationships of the taxa involved. Despite these problems, we have attempted a review of the literature using the family as the key taxonomic level to explore knowledge. We have used knowledge of the cercaria

Table 16.1 Summary of previous phylogenetic studies of digeneans employing molecular data. Throughout, 28S and 18S refer to large subunit (LSU) and small subunit (SSU) ribosomal RNA (rRNA) or DNA (rDNA), respectively. Early studies on ribosomal subunits tended to work directly from the RNA, whereas recently it is more common to work on the gene (DNA) coding for the RNA; mt indicates mitochondrial. See original references for full details.

	Taxa analysed	Locus/Technique	Alignment length	Reference
Intra-family relationships				
Schistosomatidae	*Schistosoma mansoni*	28S rRNA D1, D2 & D3 domains	1992	Ali *et al.* (1991)
Schistosomatidae	*Schistosoma haematobium* group species	RAPD		Kaukas *et al.* (1994)
Schistosomatidae	6 spp. *Schistosoma* + *Schistosomatium douthitti* + 1 spirorchiid + 1 sanguinicolid	rDNA (ITS2); mt COI	398; 372	Bowles *et al.* (1995b)
Schistosomatidae	4 spp. *Schistosoma* + *Heterobilharzia americana*	28S rRNA D1, D2 & D3 domains	1345	Littlewood and Johnston (1995)
Schistosomatidae	7 spp. *Schistosoma* + *Schistosomatium douthitti*	28S rRNA D1 domain; SSU rRNA V4 region	177; 329	Barker and Blair (1996)
Schistosomatidae	*Schistosoma malayensis* relative to other Asian schistosomes	rRNA ITS1 & 2; mt COI	375; 358; 375	Blair *et al.* (1997)
Didymozoidae	11 spp. of didymozoids; lecithasterid = outgroup	ITS2 rDNA	342	Anderson and Barker (1998)
Fellodistomidae and Lepocreadiidae	Lepocreadiidae - 3 spp. *Lepidapedon*, 1 sp. *Opechona*; Fellodistomidae - 1 sp. *Fellodistomum* and 1 sp. *Steringophorus*	18S rRNA	1637	Lumb *et al.* (1993)
Fellodistomidae and Lepocreadiidae	3 spp. of Fellodistomidae; Lepocreadiidae, Opisthorchiidae and Schistosomatidae for outgroup comparison	partial 18S rRNA	470	Bray *et al.* (1994b)
Campulidae	6 genera of Campulidae and Nasitrematidae; *Fasciola* and *Dicrocoelium* = outgroups	mtDNA ND3; tRNAlys	363; 67	Fernández *et al.* (1998a)
Campulidae	3 genera of Campulidae and Nasitrematidae in comparison with Fasciolidae, Acanthocolpidae and Schistosomatidae	SSU rRNA	1952	Fernández *et al.* (1998b)
Echinostomatidae	7 spp. in 37-collar-spine group	rDNA ITS1, 5.8S, ITS2	447, 158, 437	Morgan and Blair (1995)
Echinostomatidae	Species of *Echinoparyphium*, *Pseudechinoparyphium*, *Neoacanthoparyphium* and *Hypoderaeum*	rDNA ITS1	376	Grabda-Kazubska *et al.* (1998)
Echinostomatidae	37-spined echinostomes	rDNA (ITS1, 5.8S, ITS2)	1006	Sorensen *et al.* (1998)
Paragonimidae	*Paragonimus miyazakii, P. macrorchis, P. ohirai* and *P. westermani* complex	rDNA ITS1	up to 748	van Herwerden *et al.* (1999)
Mesometridae	5 spp. of Mesometridae	rDNA (ITS1, 5.8S, ITS2)	1325	Jousson *et al.* (1998a)
Plagiorchiidae	7 spp. including 5 Leptophallinae spp.	28S rDNA	1265	Tkach *et al.* (1999)
Inter-family relationships				
	Echinostomatidae, Fasciolidae, Gyliauchenidae, Heronimidae, Opisthorchiidae, Paramphistomidae and Schistosomatidae	18S rDNA	1958	Barker *et al.* (1993b)
	Didymozoidae, Fasciolidae, Gyliauchenidae, Hemiuridae, Lepocreadiidae, Paramphistomidae and Schistosomatidae	28S rRNA D1 domain	228	Barker *et al.* (1993a)
	Echinostomatidae, Fasciolidae, Gyliauchenidae, Lepocreadiidae, Opisthorchiidae, Paramphistomidae, Schistosomatidae (*Lobatostoma* = outgroup)	18S rRNA	1970	Blair and Barker (1993)
	71 platyhelminth species including 21 spp. of trematodes; one new trematode sequence	18S rRNA	not stated	Campos *et al.* (1998)
	34 spp. of 11 families of putative Hemiuroidea (Hemiuridae, Lecithasteridae, Isoparorchiidae, Accacoeliidae, Derogenidae, Syncoeliidae, Hirudinellidae, Sclerodistomidae, Didymozoidae, Azygiidae, Ptychogonimidae) + Bivesiculidae	18S rRNA V4 region	372	Blair *et al.* (1998)
	Unidentified digenean from isopod in comparison with published sequences from Echinostomatidae, Fasciolidae, Fellodistomidae, Gyliauchenidae, Heronimidae, Lepocreadiidae, Opisthorchiidae, Paramphistomidae, Schistosomatidae, Telorchiidae and Aspidogastrea	18S rDNA	not stated	Schulenburg and Wägele (1998)
	Unidentified digenean from isopod in comparison with published sequences from Echinostomatidae, Lepocreadiidae, Schistosomatidae and Telorchiidae	rDNA ITS1	427	Schulenburg *et al.* (1999)

Table 16.1 Continued.

Taxa analysed	Locus/Technique	Alignment length	Reference
All major groups of platyhelminths incorporating 11 digeneans (Bucephalidae, Echinostomatidae, Fasciolidae, Gyliauchenidae, Heronimidae, Lepocreadiidae, Opisthorchiidae, Paramphistomidae and Schistosomatidae)	18S rDNA	1358 alignable positions	Littlewood *et al.* (1999a)
13 spp. of Fellodistomidae (*sensu lato*), Gorgoderidae, Bucephalidae, Gymnophallidae, Zoogonidae, Brachylaimidae, Bivesiculidae, Opecoelidae and Aspidogastrea	18S rDNA V4 region	322	Hall *et al.* (1999)

as the criterion of whether there is life-cycle information available. Knowledge of the cercaria usually brings with it at least the identity of the first intermediate host and of the final asexual generation. This criterion does, however, allow for lack of knowledge of the miracidium/mother sporocyst generation.

We find that there is life-cycle information for at least 70 families of digeneans – mainly those summarized by Yamaguti (1975). In addition, we identify about the same number for which we cannot find published life-cycles. In general, however, we conclude that the gaps do not reflect serious weaknesses in our knowledge of the digenean life-cycle *in toto*; although we may have life-cycles for only half the families, it is clear that most of the major patterns have been identified. Many of the unknown families are small and include only one to a handful of species. Many of these families may disappear once properly reconsidered. For example, the Atractotrematidae, for which there is no life-cycle information, are probably no more than specialized haploporids. Further, although many families may be unknown in their life-cycles, their taxonomic affinities based upon the morphology of the sexual adult makes the general pattern of the life-cycle highly predictable. Thus, although comprehensive life-cycle information is not available for the Accacoeliidae, Bathycotylidae, Dictysarcidae, Didymozoidae, Hirudinellidae, Sclerodistomidae, Sclerodistomoididae and Syncoeliidae, the morphology of these taxa makes it clear that all belong to the Hemiuroidea. Life-cycle studies of other hemiuroid families (Bunocotylidae, Derogenidae, Hemiuridae, Isoparorchiidae and Lecithasteridae) make it possible to predict confidently that all the unknown families will share the characteristic cystophorous cercaria. Similarly, although life-cycles may not be known for the Bolbocephalodidae, Brauninidae and Proterodiplostomidae, their clear diplostomoid affinities make their life-cycles broadly predictable.

Of the taxa for which nothing is known, we identify the gorgoderid subfamily Anaporrhutinae of elasmobranch fishes, the Campulidae of cetaceans, the Liolopidae of amphibians, reptiles and mammals, and the Enenteridae and Gyliauchenidae of herbivorous teleost fishes as those for which lack of information is most obvious and serious. Each includes enough species that they cannot be considered rare or insignificant, and for each, ideas about the nature of the life-cycle are at best informed guesses.

A few families have been very closely studied indeed. There are many life-cycles for such dominant families such as the Heterophyidae, Strigeidae, Echinostomatidae, Diplostomidae, Schistosomatidae and Notocotylidae. Dense sampling in these families reveals that, typically, the life-cycle pattern for a family is both highly characteristic and quite conservative. From this it follows that just a few well-known life-cycles can usually be taken to give a good idea of the nature of a life-cycle in a given family, assuming that we are confident of the monophyly of

that family. However, strikingly modified or abbreviated cycles such as that of the lepocreadiid *Lepidapedon elongatum* (Lebour 1908) elucidated by Køie (1985) in which the cercaria bears little resemblance to the typical lepocreadiid cercaria, demonstrate that caution is still required. Dense sampling of life-cycles is called for if intrafamilial relationships are to be explored. Within families there may be very considerable variation in the use of first intermediate hosts. For example, by using gastropods, bivalves and polychaetes, sanguinicolids demonstrate the danger in concluding much about the identity of hosts from only one or two life-cycles (see also examples in Gibson and Bray 1994). In contrast, sanguinicolid cercariae can be recognized relatively easily, regardless of the identity of the first intermediate host.

An important feature of the literature on digenean life-cycles is the extent to which the miradicium and mother sporocyst are overlooked. This generation (especially its larva, the miradicium) may be highly characteristic within digenean taxa. For example, notocotylids have highly modified miracidia which are injected into the body cavity of their gastropod hosts (Murrills *et al.* 1985), opisthorchioid eggs must be eaten to hatch and penetrate the gut of their hosts (e.g., Cribb 1986), hemiuroid miracidia have characteristic spines at their anterior ends (e.g., Kechemir 1978; Murugesh and Madhavi 1990; but see Schell 1975), and bucephalid miracidia have their cilia in characteristic tufts (e.g., Kniskern 1952). Despite the potential value and interest of these characters, the miradicium and mother sporocyst is very frequently omitted from otherwise 'complete' life-cycle accounts. Clearly this relates to the difficulty of the study of these stages. Miracidia are tiny and can usually only be studied after the collection and embryonation of the eggs (Semenov 1991). Those that hatch only after the ingestion of the egg are even more difficult to examine. The mother sporocyst generation is rarely encountered in nature because most infections in molluscs are identified by the observation of emerged cercariae. Usually, by the time cercarial emergence has begun the mother sporocyst generation has disappeared. Typically, the mother sporocyst is studied in experimental infections which are slow and difficult. It seems likely that there is much to be learned from the more careful study of this generation.

Aims

Many of the previous articles on the evolution of the Digenea were produced before the advent of cladistic techniques, and made many important and lasting advances in our understanding. The criticism of this approach is that it is susceptible to being anecdotal and that it develops theories based on assumptions, such as for example, that the molluscs were the primitive hosts of the group, or that the group was derived from rhabdocoel turbellarians. The approach adopted here is to attempt to base discussion of digenean evolution on a relatively objectively derived phylogeny. Thus here we present an explicit matrix and analysis that will allow future criticism and development and complement this with a molecular data set. After an attempt to reconstruct a matrix from the results and cladograms in Brooks *et al.* (1985a) and Brooks and McLennan (1993a), we decided to code a matrix ourselves from the primary literature and with direct reference to specimens.

The data sets

Our approach here is to use several types of information ('combined evidence'). This view is based on the growing theoretical support for this approach (de Queiroz *et al.* 1995; Huelsenbeck *et al.* 1996; see also critique by Miyamoto and Fitch 1995 and response by Kluge 1998) and the simple pragmatism that approaches using a restricted set of characters seem to have no prospect of resolving the relationships of the Digenea.

Morphological and life-cycle characters

Morphological and life-cycle data on the digeneans is widely scattered in the literature, although there are some useful compilations (e.g., Yamaguti 1971, 1975). Detailed systematic treatments by La Rue (1957) and Odening (1974) also supply useful data. Recently, Brooks *et al.* (1985a) and Pearson (1992) have also provided much information. Nevertheless, none of these useful sources, even the latter two, gives data in a format immediately useful for cladistic analysis. All have been used in compiling the matrix, but the matrix (as shown in Appendix 16.1) has been built from our own assessment of character homologies and coding.

We have used primary literature and examined specimens for coding the matrix where necessary, but have found general texts such as Yamaguti (1971, 1975), Schell (1985) and the series of volumes in Russian edited by Skrjabin very useful in checking the general validity of character states in families.

In the main, coding is at the family level, but there are exceptions. In the case of the Gorgoderidae, the subfamilies Gorgoderinae and Anaporrhutinae are coded separately due to the distinct differences in the alimentary system, female proximal genitalia and host-group. For the Zoogonidae, the subfamily Zoogoninae is coded because the female system differs so distinctly from the Lepicophyllinae. The superfamily Opisthorchioidea, rather than its constituent families, is coded, because the main families (Opisthorchiidae, Heterophyidae and Cryptogonimidae) are so similar.

There has been some dichotomy of opinion in the past between those who consider the life-cycle, and particularly the cercaria, as supplying the best data (e.g., La Rue 1957; Pearson 1972; Cable 1974; Bozhkov 1982) and those who stress the value of the adult (e.g., Gibson 1987). We have followed recent workers (e.g., Odening 1974; Brooks *et al.* 1985a; Galaktionov and Dobrovolskij 1998) who have emphasized the importance of all life-cycle stages.

Molecular characters

As a starting point in producing a complementary independent molecular data set, we have chosen to utilize nucleotide sequence data from the nuclear SSU rRNA gene. The ubiquity, range of nucleotide variability across closely related and disparate taxa, subsequent utility at a variety of phylogenetic levels and ease in determination using modern molecular techniques make it an ideal first choice molecule for a primary systematic survey for many large clades of platyhelminths (see Littlewood and Olson 2001, and other molecular-based studies presented in this volume). Ideally, more than one gene would be used to contribute to a molecular-based data set, so as to avoid being confronted with the problems in interpreting the gene-tree as a species-tree (e.g., Doyle 1997; Page and Charleston 1998).

A total of 75 digenean and three aspidogastrean species were sampled from as many disparate groups as could be found. These data were added to 20 digenean and two aspidogastrean sequences found on GenBank/EMBL (see Table 16.2 for full listing, origin of taxa and accession numbers). Our sampling represents about 36% of a total of about 150 recognized families.

Methodology

Taxon sampling, character definitions and coding are given below for each data set. All phylogenetic analyses were conducted using PAUP* (v. 40b2; Swofford 1998). Maximum parsimony (MP) analyses were conducted on all data sets employing ten random additions of taxa and the TBR branch swapping algorithm. Phylogenies from the SSU data were additionally estimated using minimum evolution (ME; Swofford *et al.* 1996) under a Logdet model (Lockhart *et al.* 1994).

Morphological data and analysis

The list of morphological character-states and argumentation relating to them are given in Appendix 16.1. The full nucleotide alignment is available via anonymous FTP from FTP.EBI.AC.UK under directory pub/databases/embl/align accession number DS42620. The matrix of morphological characters is given in Appendix 16.2.

Evidence from morphology, ultrastructure and molecules of the monophyly of the Trematoda and the sister-group relationship between Aspidogastrea and Digenea is very strong (e.g., Ehlers 1985a; Rohde 1990; Littlewood *et al.* 1999a) and we have had no hesitation in adopting the aspidogastreans as our outgroup. The morphology of aspidogastreans varies greatly, so we have coded three taxa for use as outgroups, namely the family Aspidogastridae Poche, 1907, the genus *Rugogaster* Schell, 1973, and the species *Stichocotyle nephropis* Cunningham, 1884.

Whereas we have no concern that the Aspidogastrea is the true sister-taxon to the Digenea, we have considerable concerns over its effectiveness. That is, it is entirely possible that the two groups have undergone so much parallel and convergent evolution that the Aspidogastrea has little value in polarizing the character-states found in the Digenea. Thus, for all the character-states relating to sporocysts, rediae and cercariae the Aspidogastrea have been coded in the matrix as '9', indicating inapplicability. In the analysis, however, we have changed all '9' codes to '?', indicating unknown. Neither solution is satisfactory, in that '9' creates a positive character-state for a condition that is simply not applicable. For example, to code the Aspidogastrea and the Bivesiculidae as '9' for metacercarial behaviour (encysts or does not encyst) would be misleading because the Aspidogastrea lacks a cercaria in the first place, whereas the Bivesiculidae only lacks a metacercarial stage. Our solution of analysing such conditions as unknowns has the disadvantage that it allows the parsimony algorithm to interpret them as any condition. However, we have concluded that this approach is less flawed than the first.

Our concerns over the effectiveness of the Aspidogastrea as an outgroup fall in two main areas. First, we have suspicions regarding the homology of the ventral suckers of the two taxa. Although the ventral suckers of digenean and aspidogastrean marita may be strikingly morphologically different, they do arise as very similar structures in the cercaria and cotylocidium respectively. However, two facts led us to be suspicious of their homology. First, a number of digenean taxa lack ventral suckers entirely (Bivesiculidae, Notocotylidae, Angiodictyidae) which creates the *a priori* possibility that the ventral sucker evolved independently within the

Table 16.2 Full listing, origin and accession numbers of taxa incorporated in molecular analysis. §, indicates SSU sequence is new.

Family	Species	Abbreviation	Accession	Host	Locality
Outgroup					
Aspidogastridae	Aspidogaster conchicola	ASP1	AJ287478§	Quadula pustulosa (Mo)	Tennessee River, Onile, USA
	Lobatostoma manteri	ASP2	L16911		
	Multicotyle purvisi	ASP4	AJ228785	Siebenrockiella crassicollis (R)	Malaya
Multicalycidae	Multicalyx sp.	ASP3	AJ287532§	Callorhynchus milii (H)	off Hobart, Tasmania, Australia
Rugogastridae	Rugogaster sp.	RUG	AJ287573§	Callorhynchus milii (H)	off Hobart, Tasmania, Australia
Ingroup					
Accacoeliidae	Accacoelium contortum	ACC	AJ287472§	Mola mola (T)	off Skegness, Lincolnshire, UK
Acanthocolpidae	Cableia pudica	ACA1	AJ287486§	Cantherines pardalis (T)	HI, Qld, Australia
	Stephanostomum baccatum	ACA2	AJ287577§	Eutrigla gurnardus (T)	North Sea, UK
Angiodictyidae	Neohexangitrema zebrasomatis	ANG1	AJ287544§	Zebrasoma scopas (T)	HI, Qld, Australia
	Hexangium sp.	ANG2	AJ287522§	Siganus fuscescens (T)	HI, Qld, Australia
Apocreadiidae	Homalometron synagris	APO1	AJ287523§	Scolopsis monogramma (T)	HI, Qld, Australia
	Neoapocreadium splendens	APO2	AJ287543§	Scolopsis monogramma (T)	HI, Qld, Australia
Atractotrematidae	Atractotrema sigani[2]	ATR	AJ287479§	Siganus lineatus (T)	LI, Qld, Australia
Azygiidae	Otodistomum cestoides	AZY	AJ287553§	Raja montagui (E)	North Sea, UK
Bivesiculidae	Bivesicula claviformis	BIV1	AJ287485§	Epinephelus quoyanus (T)	LI, Qld, Australia
	Paucivitellosus fragilis	BIV2	AJ287557§	Crenimugil crenilabis (T)	HI, Qld, Australia
Brachycoelidae	Mesocoelium sp.	BRA	AJ287536§	Bufo marinus (A)	Brisbane, Qld, Australia
Bucephalidae	Prosorhynchoides gracilescens	BUC	AJ228789		
Bunocotylidae	Opisthadena sp.	BUN	AJ287549§	Kyphosus cinerascens (T)	HI, Qld, Australia
Campulidae	Zalophotrema hepaticum	NAS	AJ224884		
Cephalogonimidae	Cephalogonimus retusus[3]	CEP	AJ287489§	Rana ridibunda (A)	Kokaljane, Bulgaria
Cryptogonimidae	Mitotrema anthostomatum[1]	ACN	AJ287542§	Cromileptes altivelis (T)	HI, Qld, Australia
Cyclocoelidae	Cyclocoelum mutabile	CYC	AJ287494§	Calidris canutus (B)	Fair Isle, UK
Derogenidae	Derogenes varicus	DER	AJ287511§	Hippoglossoides platessoides (T)	North Sea, UK
Dicrocoeliidae	Dicrocoelium dendriticum	DIC	Y11236		
Didymozoidae	Didymozoon scombri	DID	AJ287500§	Scomber scombrus (T)	North Sea, UK
Diplodiscidae	Diplodiscus subclavatus[4]	DIP	AJ287502§	Rana ridibunda (A)	Kokaljane, Bulgaria
Diplostomidae	Diplostomum phoxini[5]	DIT	AJ287503§	Phoxinus phoxinus (T)	Aberystwyth, UK
Echinostomatidae	Echinostoma caproni	ECH	L06567		
Enenteridae	Enenterid sp.1	ENE1	AJ287507§	Kyphosus vaigiensis (T)	HI, Qld, Australia
	Enenterid sp.2	ENE2	AJ287508§	Kyphosus cinerascens (T)	HI, Qld, Australia
Fasciolidae	Fasciola gigantica	FAS1	AJ011942		
	Fasciola hepatica	FAS2	AJ004969		
	Fasciolopsis buski	FAS3	L06668		
Faustulidae	Antorchis pomacanthi	FAU1	AJ287476§	Pomacanthus sexstriatus (T)	HI, Qld, Australia
	Bacciger lesteri	FAU2	AJ287482§	Selenotoca multifasciata (T)	Moreton Bay, Australia
	Trigonocryptus conus	FAU3	AJ287584§	Arothron nigropunctatus (T)	HI, Qld, Australia
Fellodistomidae	Fellodistomum fellis	FEL1	Z12601		
	Tergestia laticollis	FEL2	AJ287580§	Trachurus trachurus (T)	North Sea, UK
	Steringophorus margolisi	FEL3	AJ287578§	Spectrunculus grandis (T)	Porcupine Seabight, NE Atlantic
	Olssonium turneri	FEL4	AJ287548§	Alepocephalus agassizi (T)	Porcupine Seabight, NE Atlantic
Gorgoderidae	Degeneria halosauri	GOR1	AJ287497§	Halosauropsis macrochir (T)	Porcupine Seabight, NE Atlantic
	Gorgodera sp.	GOR2	AJ287518§	Rana ridibunda (A)	Kokaljane, Bulgaria
	Xystretrum sp.	GOR3	AJ287588§	Cantherhines pardalis (T)	HI, Qld, Australia
Gyliauchenidae	Robphildollfusium fractum	GYL1	AJ287571§	Sarpa salpa (T)	off Perpignan, France
	Gyliauchen sp.	GYL2	L06669		
Haploporidae	Pseudomegasolena ishigakiense	HAP	AJ287569§	Scarus rivulatus (T)	HI, Qld, Australia
Haplosplanchnidae	Hymenocotta mulli	HSP1	AJ287524§	Crenimugul crenilabis (T)	HI, Qld, Australia
	Schikhobalotrema sp.	HSP2	AJ287574§	Scarus rivulatus (T)	HI, Qld, Australia
Hemiuridae	Dinurus longisinus	HEM1	AJ287501§	Coryphaena hippurus (T)	off Kingston, Jamaica
	Lecithochirium caesionis	HEM2	AJ287528§	Caesio cuning (T)	HI, Qld, Australia
	Lecithocladium excisum	HEM3	AJ287529§	Scomber scombrus (T)	North Sea, UK
	Merlucciotrema praeclarum	HEM4	AJ287535§	Cataetyx laticeps (T)	Porcupine Seabight, NE Atlantic
	Plerurus digitatus	HEM5	AJ287562§	Scomberomorus commerson (T)	HI, Qld, Australia
Heronimidae	Heronimus mollis	HER	L14486		
Heterophyidae	Cryptocotyle lingua[1]	HET1	AJ287492§	Littorina littorea (Mo)	Isle of Sylt, Germany
	Haplorchoides sp.[1]	HET2	AJ287521§	Arius graeffei (T)	Lake Wivenhoe, Qld, Australia
Lecithasteridae	Lecithaster gibbosus	LEC	AJ287527§	Merlangius merlangus (T)	North Sea, UK
Lepocreadiidae	Austroholorchis sprenti	LEP1	AJ287481§	Sillago ciliata (T)	Moreton Bay, Qld, Australia
	Lepidapedon rachion	LEP2	Z12607	Melanogrammus aeglefinus (T)	North Sea, Uk
	Lepidapedon elongatum	LEP3	Z12600		
	Preptetos caballeroi	LEP4	AJ287563§	Naso vlamingii (T)	HI, Qld, Australia
	Tetracerasta blepta	LEP5	L06670		

Table 16.2 Continued.

Family	Species	Abbreviation	Accession	Host	Locality
Mesometridae	Mesometra sp.	MES	AJ287537§	Sarpa salpa (T)	off Perpignan, France
Microphallidae	Levinseniella minuta	MIC1	AJ287531§	Hydrobia ulvae (Mo)	Belfast, UK
	Maritrema oocysta	MIC2	AJ287534§	barnacle (C)	Belfast, UK
	Microphallus primas	MIC3	AJ287541§	Carcinus maenas (C)	Belfast, UK
	unidentified	MIC4	AJ001831		
Monorchiidae	Ancylocoelium typicum	MON1	AJ287474§	Trachurus trachurus (T)	North Sea, UK
	Provitellus turrum	MON2	AJ287566§	Pseudocaranx dentex (T)	HI, Qld, Australia
Nasitrematidae	Nasitrema globicephalae[6]	CAM	AJ004968		
Notocotylidae	Notocotylus sp.	NOT	AJ287547§	Lymnaea palustris (Mo)	Leckford Estate, Stockbridge, UK
Opecoelidae	Gaevskajatrema halosauropsi	OPE1	AJ287514§	Halosauropsis macrochir (T)	Porcupine Seabight, NE Atlantic
	Macvicaria macassarensis	OPE2	AJ287533§	Lethrinus miniatus (T)	HI, Qld, Australia
	Peracreadium idoneum	OPE3	AJ287558§	Anarhichas lupus (T)	North Sea, UK
Opisthorchiidae	Opisthorchis viverrini	OPI	X55357		
Opistholebetidae	Opistholebes amplicoelus	OPS	AJ287550§	Tetractenos hamiltoni (T)	off Stradbroke Is., Qld, Australia
Orchipedidae	Orchipedum tracheicola	ORC	AJ287551§	Cygnus olor (B)	Glasgow, UK
Pachypsolidae	Pachypsolus irroratus[3]	PAC	AJ287554§	Lepidochelys olivacea (R)	off Oaxaca, Mexico
Paramphistomidae	Calicophoron calicophorum	PAR	L06566		
Paragonimidae	Paragonimus westermani	PAG	AJ287556§	Canis familiaris (Ma)	Chunchon, S. Korea
Philophthalmidae	Philophthalmid sp.	PHI	AJ287560§	Pyrazus ebeninus (Mo)	Moreton Bay, Qld, Australia
Plagiorchiidae	Glypthelmins quieta	PLA1	AJ287517§	Rana catesbeiana (A)	Keith Co., Nebraska, USA
	Haematoloechus longiplexus	PLA2	AJ287520§	Rana catesbeiana (A)	Gage Co., Nebraska, USA
	Rubenstrema exasperatum	PLA3	AJ287572§	Crocidura leucodon (Ma)	Kokaljane, Bulgaria
	Skrjabinoeces similis	PLA4	AJ287575§	Rana ridibunda (A)	Kokaljane, Bulgaria
Sanguinicolidae	Aporocotyle spinosicanalis	SAN	AJ287477§	Merluccius merluccius (T)	North Sea, UK
Schistosomatidae	Schistosoma haematobium	SCH1	Z11976		
	Schistosoma japonicum	SCH2	Z11590		
	Schistosoma mansoni	SCH3	X53017		
	Schistosoma spindale	SCH4	Z11979		
Sclerodistomidae	Prosogonotrema bilabiatum	SCL	AJ287565§	Caesio cuning (T)	HI, Qld, Australia
Strigeidae	Ichthyocotylurus erraticus	STR	AJ287526§	Coregonus autumnalis (T)	Lough Neagh, UK
Syncoeliidae	Copiatestes filiferus	SYN	AJ287490§	Trachurus murphii (T)	off New Zealand
Tandanicolidae	Prosogonarium angelae	TAN	AJ287564§	Eurhisthmus lepturus (T)	Moreton Bay, Qld, Australia
Transversotrematidae	Crusziella formosa	TRA1	AJ287491§	Crenimugil crenilabis (T)	HI, Qld, Australia
	Transversotrema haasi	TRA2	AJ287583§	Caesio cuning (T)	HI, Qld, Australia
Zoogonidae	Deretrema nahaense	ZOO1	AJ287498§	Thalassoma lunare (T)	LI, Qld, Australia
	Lepidophyllum steenstrupi	ZOO2	AJ287530§	Anarhichas lupus (T)	North Sea, UK
	Zoogonoides viviparus	ZOO3	AJ287590§	Callionymus lyra (T)	North Sea, UK

Notes: in combined analysis morphology coded as: [1]Opisthorchiidae; [2]Haploporidae; [3]Plagiorchiidae; [4]Paramphistomidae, [5]Diplostomidae; [6]Campulidae.
Abbreviations: A, Amphibian; B, Bird; C, Crustacean; E, Elasmobranch; H, Holocephalan; HI, Heron Island, Great Barrier Reef (GBR); LI, Lizard Island, GBR; Ma, Mammal; Mo, Mollusc; R, Reptile; T, Teleost.

Digenea. The second is the striking similarity between cestode suckers and those of digeneans. There seems to be no suggestion that these structures are homologous so they establish the ease with which neodermatans may 'invent' suckers in response to the need to attach. The second concern relates to characters associated with the intramolluscan stages of the life-cycle. Because the Aspidogastrea lack most of these, we suspected that the resolving power of our analysis may have been diminished by having so many character-states effectively unpolarized.

To explore these issues we ran maximum parsimony (MP) analyses of four different codings of the data, depending on whether we used several selected characters ordered or unordered, and coded for homology or not of the digenean ventral sucker and the aspidogastrean ventral disc:

- *Run 1.* With the ventral sucker coded as homologous with the aspidogastrean arrangement – all characters unordered. We consider this the underlying data set, and combined it with the molecular data set for the combined evidence analysis.
- *Run 2.* As 1, but with three characters ordered (numbered as in Appendix 16.1) – (1) Adult. Ovary position; (2) Adult. Sperm reception; and (3) Life-cycle. Miracidium flame-cells.
- *Run 3.* With the ventral sucker coded as non-homologous with the aspidogastrean arrangement and all characters unordered.
- *Run 4.* As 3, but with three characters ordered – as in 2.

Neither bootstrap nor Bremer support were determined for the nodes as the data set suffered from insufficient phylogenetically informative characters for the number of taxa to make this computationally reasonable.

SSU data and analysis

All DNA preparation, amplification and sequencing methodologies were as those described in Littlewood *et al.* (2000) for the SSU gene. Both strands were sequenced fully, contig assembly was conducted using Sequencher™ (v.3.1.1, GeneCodes Corporation), sequences were aligned by eye using GDE (Smith, S.W. *et al.* 1994) and regions of ambiguity removed using previously published criteria (Littlewood *et al.* 1999a). Reference to the secondary structures presented by the Ribosomal Database Project II (Maidak *et al.* 1999) was made to aid alignment.

Combined evidence

The compatibility of the MP solutions was determined using constraint analysis followed by Templeton's test, whereby the number of changes required to map one data set onto the topology suggested by the other data set is tested against an unconstrained solution. In practice, only the molecular data could be constrained to mirror the topology offered by morphology and not vice versa, as the molecular-based solution suggested many paraphyletic taxa which in turn were coded as monophyletic clades in morphology.

Results

Morphological analyses

The four different analyses of the morphological data set produced surprisingly similar results. For many of the analyses the strict consensus tree gave very little resolution. We restrict our comments here, therefore, to the majority-rule trees. In every tree the Bivesiculidae + Transversotrematidae were identified as the sister-group to the remainder of the Digenea. In all analyses the Diplostomoidea + Schistosomatoidea were identified as the next most basal taxon. The lack of variation between these analyses serves to allay our concerns about the effectiveness of the Aspidogastrea as an outgroup. The different outgroups and interpretation produced no major rearrangement of key taxa. On this basis we conclude that it is sound to use the least modified data set (Run 1) for combination with the molecular data set in the analysis of combined evidence. The 50% majority-rule consensus tree is shown in Figure 16.3.

SSU analysis

The Logdet paralinear minimum evolution (ME) and a MP solution are shown in Figure 16.4. Two ME solutions were found with a heuristic search but differed only in their relative placement of two taxa (see Figure 16.4a). Both ME and MP solutions show remarkable similarity in terms of overall tree topology and resolution. In each case the earliest divergent digenean clade includes the Diplostomidae, Strigeidae, Sanguinicolidae and Schistosomatidae followed by the Bivesiculidae. The relatively short internal branches on the ME tree and the low Bremer support and bootstrap resampling values on the MP tree indicate rapid divergence of the remaining taxa, with some uncertainty as to whether the larger clades are stable or not. Certainly, the [Azygiidae + Accacoeliidae + Didymozoidae + Derogenidae + Sclerodistomidae + Bunocotylidae + Lecithasteridae + Hemiuridae] tend to fall into a clade which is the sister-group to a clade comprising the [Opecoelidae + Transversotrematidae]. Other robust clades include the [Zoogonidae + Faustulidae], the [Cyclocoelidae + Philophthalmidae + Echinostomatidae + Fasciolidae], the [Lepocreadiidae + Gyliauchenidae + Enenteridae], the [Campulidae + Nasitrematidae], and the [Plagiorchiidae + Brachycoeliidae + Cephalogonimidae]. Families appearing as non-monophyletic included the Lepocreadiidae, the Gorgoderidae, the Plagiorchiidae and finally the Acanthocolpidae, in which the problematical taxon *Cableia* appeared consistently as the sister-group to the Monorchiidae. Except in the case of *Cableia*, most of these 'failings' suggest an inadequacy of phylogenetic signal rather than robust contraindications of taxonomy.

Combined evidence analysis

Combining independent phylogenetic data sets can be approached from different philosophical positions. Whilst we advocate the 'combined evidence' approach we recognize the need for testing 'combinability' through the criteria of 'conditional combination' (*sensu* Huelsenbeck *et al.* 1996; Cunningham 1997). Our data set is too large to conduct some of the tests advocated for conditional combination and only a Templeton's Test (Larson 1994) was performed. The

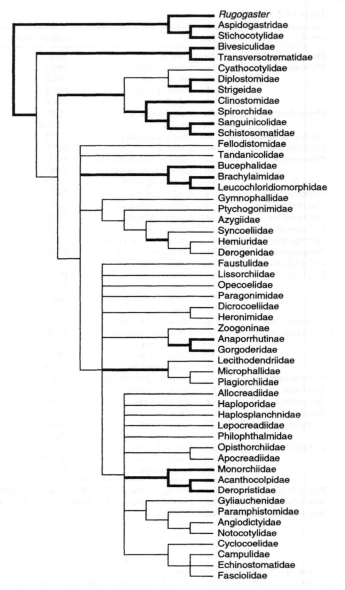

Figure 16.3 Phylogeny according to morphological data. A 50% majority-rule consensus trees derived from 147 677 equally most-parsimonious solutions in 10 heuristic searches; nodes and branches supported in the strict consensus solution are emboldened; length = 214, CI = 0.287 excluding uninformative sites, RI = 0.641, number of parsimony informative characters = 51, all characters unweighted and rooted against the Aspidogastrea (Run 1; see text for further details). Neither bootstrap support nor Bremer support could be determined with this data set.

data sets offered highly significantly different topologies (*P* < 0.0001) and would not pass the criteria for conditional combination. These differences in topology indicate clear differences in phylogenetic signal, but we do not know where the sources of homoplasy in either data set are. We were keen to see the resolution and topologies offered by combining both data sets, and thereby follow the rationale of Kluge (1989, 1998) wherein all data were combined unconditionally. The taxa used in the molecular analysis were coded for their morphology and a combined data set was analysed with

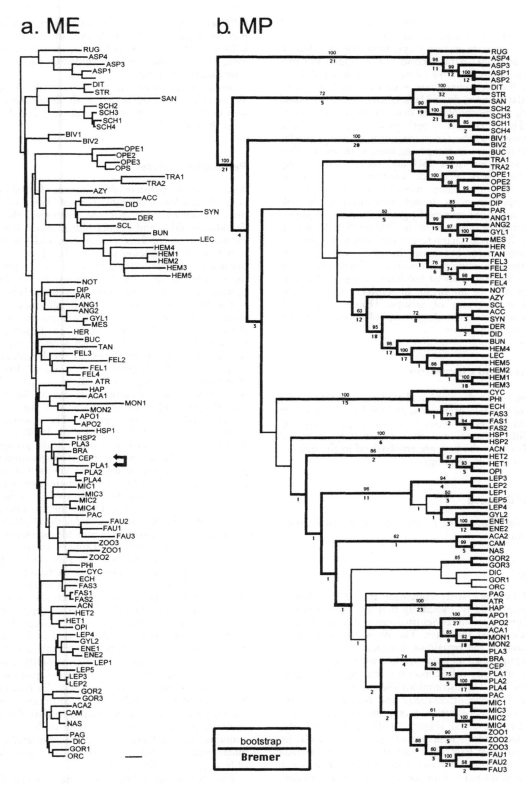

Figure 16.4 Phylogenetic solutions using SSU rDNA data alone: **a)** Logdet paralinear minimum evolution solution; two trees were found after 10 replicate heuristic searches but only one is shown; the two trees differed only in the relative position of *Cephalogonimus* and *Glypthelmins* which were interchangeable (marked by bold arrows), ME score = 2.223, scale bar represents 0.01 changes. **b)** Maximum parsimony solutions showing the 50% majority-rule consensus of 228 equally most-parsimonious trees with nodes and branches supported in the strict consensus emboldened; length = 4369, CI = 0.259 excluding uninformative sites RI = 0.564, number of variable positions = 814; number of parsimony informative sites = 577; > 50% bootstrap support (n = 1000) and Bremer support are presented above and below nodes, respectively. Tree rooted on functional outgroups represented by five aspidogastrean sequences held as a monophyletic clade. For full details of taxa, see Table 16.2.

MP; coding of families was taken directly from Appendix 16.2 except for those taxa indicated in Table 16.2.

We combined the data in two ways. First, we included all morphological and SSU data; second, we excluded the life-cycle characters. In this way we could qualitatively judge the influence of the life-cycle characters, even whilst influenced by the molecular data where, due to lack of characters, we could not do this with the morphological data set alone. The results are shown in Figure 16.5. It is perhaps advisable that these solutions are viewed as the strict consensus solutions (nodes and branches in bold) when comparing them with one another and with the solutions from SSU and morphology alone. Certainly the combined evidence solution (Figure 16.5a) shows greater resolution than morphology alone (Figure 16.3), but the topology seems strongly influenced by the molecular data and no doubt reflects the greater number of phylogenetically informative positions in the SSU set (full discussion below). Where life-cycle characters were removed, gains in resolution in some areas but losses in others were evident. Again, where the tree gains resolution it is clearly under the influence of the SSU set. Loss of resolution suggests that life-cycle characters may be important in resolving relationships among the most divergent of groups. However, we will leave a full assessment of homology, homoplasy and character mapping for a future analysis.

Interpretation and discussion

We have chosen to use the combined evidence solution as the basis for most of our interpretations, and whilst we appreciate the gene tree and species tree based on morphology are arguing for markedly different topologies, the conflicts, polytomies and incongruence between the two are emphasized in the light of our understanding of morphology.

Monophyly of Digenea

All trees find the Digenea to be monophyletic, with strong bootstrap support.

Basal Digenea

The combined evidence strict consensus tree (CES) (Figure 16.5a) has a basal tritomy, indicating that the Transversotrematidae and Diplostomoidea+Schistosomatoidea are distinct from the remainder of the class. The Bivesiculidae is more derived, but basal to the remaining Digenea. Thus there are three clades distinguished as basal or close to basal. The combined evidence majority-rule consensus tree (CEM) (Figure 16.5a) supports the Diplostomoidea+Schistosomatoidea as basal, with the Transversotrematidae and Bivesiculidae progressively less basal. The strict maximum parsimony SSU rDNA molecular tree (MS) (Figure 16.4b, bold) similarly, finds the Diplostomoidea+Schistosomatoidea basal, with the Bivesiculidae next most basal, but with the Transversotrematidae in a large unresolved derived clade. This position is not notably resolved in the maximum parsimony molecular majority-rule tree (MMR) (Figure 16.4). On the other hand, the morphological trees (all analyses) (Figure 16.3 for R1, others not illustrated), where resolved, indicate variously that the Bivesiculidae or Transversotrematidae is basal. Thus there are three strong candidates for most-basal digenean, and the final solution is still not resolved. There is no evidence for the basal status of Heronimidae as suggested by Brooks et al.

(1985a) from morphology and by Campos et al. (1998) from molecular analysis. None of the other groups discussed by Gibson (1987), i.e., Bucephalidae, Paramphistomidae, Azygiidae, Ptychogonimidae or Fellodistomidae, is identified as a potential basal taxon.

An implication of the position of the Bivesiculidae and Transversotrematidae is that both should be considered sole members of their own superfamilies, if not their own orders *sensu* Brooks.

Resolution of clades

Apart from the putative identification of candidate most-basal taxa, the resolution of higher taxa of Digenea is poor. In CEM there is, for example, a clade containing the Heronimidae, Fellodistomoidea and Hemiuroidea, which has less than 5% bootstrap support. This relationship, and a number of others, is not at all convincing so it is, therefore, not provident to discuss the higher classification of the Digenea from these results. On the other hand, some mid-level taxa (at about the current superfamily level) appear well resolved.

The Hemiuroidea is resolved in all studies, and the basal status of the Azygiidae in the Hemiuroidea is confirmed (see Blair et al. 1998). In contrast to Brooks, we find the Bivesiculidae has no relationship with the Hemiuroidea, despite the striking similarity of the bivesiculid cercaria to that of azygiids. Internally the clade is divided into two, along the same lines as in Blair et al. (1998), with the Sclerodistomidae, Accacoeliidae, Syncoeliidae, Derogenidae and Didymozoidae in one clade and the Hemiuridae and Lecithasteridae in the other. The monophyly of the Hemiuridae (*sensu* Gibson and Bray 1979) is not confirmed, as a *Lecithaster* species is included in the group. A key taxon missing from this analysis is the Ptychogonimidae. The combination of unspecialized follicular vitellarium and the unique use of scaphopods as first intermediate hosts lead us to predict a basal position within the Hemiuroidea for this family. Blair et al. (1998) found some evidence for this interpretation.

The Paramphistomoidea includes the Diplodiscidae, Paramphistomidae, Angiodictyidae and Mesometridae, as suggested earlier by La Rue (1957) and others. Inclusion of the Angiodictyidae and Mesometridae in this clade implies that the absence of the ventral sucker in those taxa is a secondary loss. The relatively derived position of this entire clade may suggest that the oral sucker has been lost secondarily in the clade. The enigmatic genus *Robphildollfusium* also appeared in this clade. This taxon shows considerable morphological similarity to the Gyliauchenidae, and indeed we listed it as a gyliauchenid in Table 16.2. This relationship is currently under study.

The Opisthorchioidea is strongly supported as including the Cryptogonimidae, Heterophyidae and Opisthorchiidae as suggested earlier by many workers, e.g., La Rue (1957). Known life-cycles of opisthorchioids are distinctive and remarkably unvarying across the superfamily, so that this relationship is of no surprise. Similarly the strong clade Echinostomatoidea contains the Cyclocoelidae, Echinostomatidae, Philophthalmidae and Fasciolidae – a group of families long suspected of having a close relationship.

The weak clade Fellodistomoidea contains Tandanicolidae and Fellodistomidae. Hall et al. (1999) found weak support for a sister-group relationship between the Tandanicolidae and the Gymnophallidae and that that clade was closest to the Fellodistomidae. The Gymnophallidae is not represented in our present database.

The Acanthocolpoidea contains the Acanthocolpidae, Campulidae and Nasitrematidae, as predicted by Cable (1974).

a. all data

b. life-cycle characters excluded

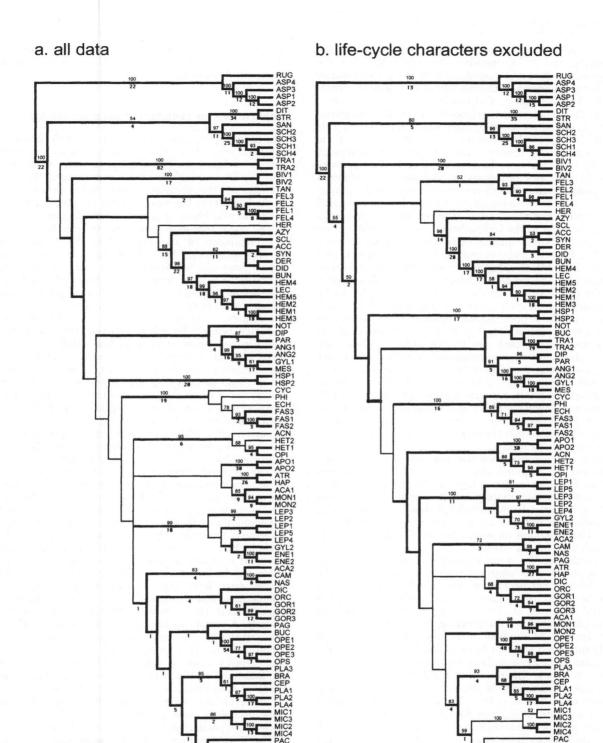

Figure 16.5 Combined evidence. Although the molecular and morphological solutions do not pass Templeton's test for conditional combination (*P* < 0.0001), a combined evidence solution of morphology and SSU rDNA data was estimated with parsimony using all morphological and SSU rDNA data in 10 heuristic searches. **a)** The 50% majority-rule consensus of 408 equally most-parsimonious trees are shown; length = 4609, CI = 0.259 excluding uninformative sites, RI = 0.586. **b)** The 50% majority-rule consensus of 120 equally most-parsimonious trees when life-cycle characters (3, 30–56 inclusive from Appendix 16.1) were removed; length = 4516, CI = 0.257 excluding uninformative sites, RI = 0.575. Strict consensus solutions for each tree are represented by emboldened branches and nodes; > 50% bootstrap support (*n* = 1000) and Bremer support are presented above and below nodes respectively. The trees are each rooted on functional outgroups represented by five aspidogastrean taxa held as a monophyletic clade. For full details of taxa, see Table 16.2.

Fernández *et al.* (1998a) analysed relationships within the Campulidae and Nasitrematidae and concluded that the Nasitrematidae should be considered part of the Campulidae.

The Lepocreadioidea, with strong support, is seen to encompass the Lepocreadiidae, Gyliauchenidae and Enenteridae. This result reaffirms the relationships of the enigmatic Gyliauchenidae as proposed by Blair and Barker (1993). This relationship implies that absence of the oral sucker in the Gyliauchenidae is the result of secondary loss. The loss appears to parallel that which may have occurred in the Paramphistomoidea. In both cases the loss may be associated with adoption of infection of predominantly herbivorous hosts and a change in the diet of the trematodes. Of the lepocreadiids incorporated in this study, *Preptetos*, often reported from herbivorous fishes, is the sister to the Gyliauchenidae + Enenteridae clade that is also reported almost exclusively from herbivorous fishes. The Lepocreadiidae is paraphyletic. This clade is also resolved in the MS tree, but not in the morphological trees. Strikingly, the Apocreadiidae, which has been considered a subfamily of the Lepocreadiidae or related to that family by morphology and life-cycle (see Cribb and Bray 1999), is not grouped with the Lepocreadioidea, even in CEM. In fact, in CEM, it groups with the Haploporoidea and Monorchioidea.

As would be expected, the Haploporoidea includes the Haploporidae and Atractotrematidae, families which are clearly closely related and have been considered synonymous (Nasir and Gómez 1976; Cribb *et al.* 1998). Because of the similarity in life-cycles between haploporids (Martin 1973; Shameem and Madhavi 1991) and haplosplanchnids (Cable 1954), it might have been expected that these two taxa would be closely related. However, our results find little support for such a relationship.

There is strong support for the sister relationship between *Cableia* and the Monorchiidae. *Cableia* has variously been placed in the Lepocreadiidae, Opecoelidae, Enenteridae and, most recently, the Acanthocolpidae (Bray *et al.* 1996), but has never been considered a monorchiid. It shows no indication of the modifications of the terminal genitalia found in the monorchiids. This is perhaps the most novel relationship of a single taxon indicated by this study.

The Plagiorchioidea is not supported strongly, but there is good support for a subgroup within it consisting of the Plagiorchiidae, Brachycoelidae and Cephalogonimidae. It should be noted that many authors (including Yamaguti 1971 and Tkach *et al.* 2001, this volume) would consider our four plagiorchiid species to belong in different, but closely related, families. Monophyly of the Microphallidae is not as strongly supported as might be expected. Pachypsolidae is the sister-group to the Zoogonidae + Faustulidae, but the relationship is not strong. On the other hand, the monophyly of Zoogonidae + Faustulidae is strongly supported, confirming the results of Hall *et al.* (1999). The Zoogonidae is paraphyletic, but with the zoogonines the sister to the monophyletic faustulids, not the lepidophyllines as indicated by Hall *et al.* (1999). Morphological consideration would probably support this latter relationship. Opecoelidae and Opistholebetidae are, not surprisingly, strongly related. See Tkach *et al.* (2001, this volume) for an analysis of the relationships of plagiorchioid families.

If the distribution of a ventral sucker is mapped on the phylogeny (Figure 16.6), then it is shown to have a very patchy appearance, completely absent in three areas – the Bivesiculidae, the Notocotylidae, Angiodictyidae and Mesometridae, and Bucephalidae. The relatively basal position of the Bivesiculidae raises the intriguing possibility that it is absent primitively. The apparently derived positions of the other families carries the implication that they have lost the ventral sucker. In the case of the Bucephalidae we speculate that loss of the ventral sucker is related to the movement of the mouth to a mid-ventral position.

Implications for evolution of the digenean life-cycle

The most important implication for the understanding of the digenean life-cycle is in the identification of putatively basal groups of digeneans. The simplest summary of the findings is that the Diplostomoidea + Schistosomatoidea, Transversotrematidae and Bivesiculidae all show indications of being close to the base of the Digenea. This has the potential to identify a plesiomorphic life-cycle pattern for the Digenea.

The taxon appearing most commonly as the sister-group to the remainder of the Digenea is the Diplostomoidea + Schistosomatoidea. This taxon includes three families of blood-flukes (the Schistosomatoidea) with two-host life-cycles in which the cercaria penetrates the definitive host directly, and the Diplostomoidea in which the cercariae penetrate and form metacercariae in second intermediate hosts which are eaten by the definitive host. If it is assumed that a two-host life-cycle appeared before a three-host life-cycle, then the schistosomatoid life-cycle becomes an immediate candidate for being the basal life-cycle pattern in the Digenea. Several observations argue against this, however. First, it requires that all the other digenean life-cycles are derivable from this life-cycle. Life-cycles in which the cercaria is eaten directly by the definitive host (e.g., Bivesiculidae, Azygiidae, some Fellodistomidae), attach directly under the scales of the definitive host (Transversotrematidae), or encyst in the open to be eaten by the definitive host (e.g., Fasciolidae, Paramphistomidae and Notocotylidae) are difficult to derive from such a life-cycle. Closer to hand, interpretation of the schistosomatoid life-cycle as primitive would also require that the three-host diplostomoid life-cycle has been derived by terminal addition to the two-host life-cycle. Rather, we think it likely that the schistosomatoid life-cycle represents an ancient abbreviation of a three-host life-cycle. This view has been proposed by, at least, La Rue (1951) and Pearson (1972). Analogous abbreviation that is clearly far less ancient is seen in the life-cycles of many families in which eggs may be produced in the second intermediate host (e.g., Allocreadiidae, Cryptogonimidae, Hemiuridae and Opecoelidae). In some of these cases it appears that the vertebrate definitive host has been lost entirely. A partial further test of this interpretation awaits in the resolution of the phylogenetic position of the Clinostomidae. This family, which infects piscivorous birds and reptiles, has a three-host life-cycle typical of the Diplostomoidea, but a cercaria barely distinguishable from those of sanguinicolids. Should the Clinostomidae prove to be more closely related to the Schistosomatoidea than the Diplostomoidea, then the idea that the schistosomatoid life-cycle is a secondary abbreviation would be supported.

The Transversotrematidae is a small family of about ten species found ectoparasitic beneath the scales of teleost fishes. This site is reached directly by the attachment of a highly specialized fork-tailed cercaria. Cribb (1988), Brooks *et al.* (1989) and Cribb *et al.* (1992) have discussed the possible position of this family. The isolated position occupied by this family near the base of the Digenea has the clear implication that this life-cycle pattern has not been derived recently from any other life-cycle incorporated in our present data set. If anything, we might conclude that the family shared a common ancestor with the Diplostomoidea + Schistosomatoidea. In contrast to that clade, however, there is no evidence of the key characteristic of that group: the active penetration of a host.

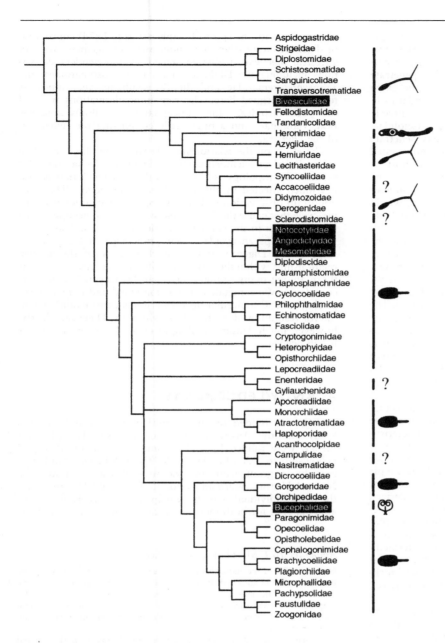

Figure 16.6 Combined-evidence tree for the Digenea showing distribution of complete absence of ventral sucker at all life-cycle stages (boxed taxa) and fork- and simple-tailed tailed cercariae. In addition to the marked families, the cercaria is unknown for the Atractotrematidae and the Opistholebetidae, but can be predicted to have a simple tail.

Penetration of a host is accomplished by highly specialized cercariae armed with penetration glands. In contrast, transversotrematids lack penetration glands (they never penetrate) and have instead processes at the base of the tail that allow them to recognize and attach to potential hosts when they contact them (Whitfield *et al.* 1975). It is, however, possible that the Transversotrematidae represent an abbreviated progenetic life-cycle.

The third apparently basal taxon to be considered is the Bivesiculidae. These trematodes occur in the intestines of marine fish. Their life-cycles are simple in that cystocercous cercariae are eaten directly by the definitive host, and their adult morphology is intriguing in that they lack oral and ventral suckers. Direct ingestion of the cercaria appears to have little in common with the route of transmission of members of the Schistosomatoidea + Diplostomoidea or the Transversotrematidae. For the bivesiculid life-cycle to be derived from either of those life-cycles would require complex patterns of losses.

'Szidat's Rule' suggests that primitive parasite taxa will tend to be found in primitive host taxa. Whereas the reliability of the 'rule' is certainly arguable, it is interesting to consider its application to the three putative taxa considered above. The Bivesiculidae and Transversotrematidae are both restricted to teleost fishes, but not to particularly primitive ones. Both families are probably most common in Perciformes. The Schistosomatoidea + Diplostomoidea is intriguing in being represented in all vertebrate classes except perhaps the Amphibia. They are most species-rich in birds and mammals, but sanguinicolids are found in both teleosts and elasmobranchs. It appears that no lineages of this clade occur as adults in the gastrointestinal tract of fishes. If the Schistosomatoidea + Diplostomoidea is indeed the most basal taxon within the Digenea, then its absence from the guts of the largest and oldest group of hosts is surprising to say the least!

Our analysis suggests that all the remaining digeneans in our data set form a clade derived at some point from ancestors in common with at least one of the taxa discussed above. This clade contains the great majority of digenean families and species. There is no one life-cycle pattern that characterizes this clade. It is possible, however, to identify several major patterns. None of these trematodes has cercariae that penetrate the definitive host directly. Instead, cercariae may be eaten directly (e.g., Azyglidae, some Fellodistomidae, Tandanicolidae), encyst as metacercariae in the open to be eaten directly by the definitive host (e.g., Angiodictyidae, Fasciolidae, Haploporidae, Paramphistomidae) or cercariae may penetrate second intermediate hosts to form encysted or unencysted metacercariae which are eaten with the intermediate host by the definitive host (many families). Two key points follow. The definitive host is always infected by ingestion, and in a number of groups there is a range of life-cycle strategies. Thus, within the Hemiuroidea, cystophorous cercariae may be eaten directly by the definitive host in the Azygiidae, whereas in perhaps all other members of that superfamily there is interpolated in the life-cycle a second intermediate host that eats the cercaria. Similarly, in the Echinostomatoidea some representatives encyst in the open (Fasciolidae), whereas others may encyst in or on second intermediate hosts (Echinostomatidae). These transitions are discussed in some detail by Pearson (1972). The relationships within this huge clade are too poorly resolved for definitive conclusions to be possible, but we conclude (cautiously) that the basal life-cycle pattern for this group was the direct ingestion of the cercaria as still seen in a number of taxa and that external encystment and three-host life-cycles developed independently a number of times and in a number of different ways.

A feature of most of the cercariae in the derived clade is that they are simple-tailed, whereas all the most basal taxa have fork-tails (Figure 16.6). Fork-tailed cercariae are characteristically the fastest swimming cercariae, a feature perhaps associated with the fact that they seek intermediate and definitive hosts actively. Cercariae that encyst in the open have simple tails and are typically much slower but perhaps more efficient swimmers. There are, however, numerous families of derived digeneans with simple tails that penetrate intermediate hosts. A hypothesis that explains this is that these are digeneans that evolved through a phase of encystment in the open (and consequent loss of the fork-tail) before adopting second intermediate hosts. The combination of the highly derived position that we have found for the Bucephalidae and its very unusual fork-tail suggests that the condition of its tail may have been arrived at secondarily.

A consequence of some of these ideas is the need for a reappraisal of how the life-cycle of the Diplostomoidea arose (if it is assumed that the life-cycle of the Schistosomatoidea is an abbreviation of the diplostomoid form). There is an apparent conflict in the fact that, whereas ingestion of a free-swimming cercaria or metacercaria that has encysted in the open appears to be the fundamental starting point for most digenean life-cycles, the Diplostomoidea (and the Schistosomatoidea) is uniformly characterized by the penetration of the intermediate host (definitive host in the case of Schistosomatoidea) by a highly specialized, fast-swimming, fork-tailed cercaria. There appears to be no evidence within the clade of how a two-host life-cycle might have been expanded by the interpolation of a second intermediate host as seen elsewhere (e.g., Hemiuroidea, Echinostomatoidea).

A key outcome of the analysis above is the conclusion that a search for an extant life-cycle pattern that is basal for the Digenea is likely to remain fruitless. We hypothesize that there may be three basal taxa that have life-cycles that are not derivable directly from each other and may be best considered as independent derivations from a proto-digenean life-cycle that no longer exists. Strikingly, this interpretation fits

very neatly with the spirit of Cable's (Cable 1974) conception of digeneans appearing in an unknown arrangement from an ancestral digenean. Cable envisaged six independent derivations including the Diplostomoidea + Schistosomatoidea, the Transversotrematidae and the Bivesiculidae, very much as found and argued here. The remaining derivations were of the Hemiuroidea, the Bucephalidae + Fellodistomidae + Brachylaimidae, and a huge clade comprising the remainder of the Digenea.

The effect on the relationships derived when life-cycle data are included or excluded is shown in Figure 16.5a and b. In essence, the topology is changed little, and resolution is also not much affected by the loss of these characters. The most important observation is that it is apparently characters of the life-cycle that cause the Transversotrematidae to occupy such a basal position in the combined evidence analysis. This observation perhaps encapsulates our knowledge of the relationships of the Digenea. It is clear that many of the deepest relationships within the group are still quite uncertain, whereas those among many terminal taxa can be accepted with some confidence. We conclude that the combined and parallel morphological and molecular approach holds considerable potential for resolution of relationships within the Digenea. Clearly, however, much denser sampling of taxa, characters and genes is still needed.

ACKNOWLEDGEMENTS

T.H.C. was funded by the Australian Biological Resources Study and the Australian Research Council. D.T.J.L. and E.A.H. were funded by a Wellcome Trust Senior Research Fellowship to D.T.J.L. (043965/Z/95/Z). S.P. was funded in part by a grant from the Zoology Research Fund, The Natural History Museum London. We would like to thank the many individuals who have helped to supply us with valuable material for both morphological and molecular analyses.

Appendix 16.1 Morphological characters.

Pearson (1992) at the end of his paper reviewing the phylogeny presented by Brooks *et al.* (1985a) stated 'It is surely too early to attempt a phylogenetic classification of the Digenea'. While presenting our attempt at just that in this chapter, we have some sympathy for Pearson's viewpoint, in that we have found the exercise to reveal many shortcomings in the understanding of the homologies of digenean features.

We have considered the many characters used by Brooks *et al.* (1985a) and criticized by Pearson (1992), and found many too difficult to use. Those we have selected for use do not lack problematical aspects, and we have discussed some of these below. It would not be an exaggeration to say that all of the characters present some problem or other. In many cases we have resorted to the majority or democratic method of coding (for a discussion of this method see Sluys 2001, this volume).

Evidence of the monophyly of the Trematoda and the sister-group relationship between Aspidogastrea and Digenea is very strong, and we have had no hesitation in adopting the aspidogastreans as our outgroup. The morphology of aspidogastreans varies greatly, so we have coded three taxa for use as outgroups, namely the family Aspidogastridae, the genus *Rugogaster* and the species *Stichocotyle nephropis*.

There is, similarly, strong evidence for the monophyly of the Neodermata and the sister-group relationship of the Cestoda/Monogenea and the Trematoda.

1. Adult. Ovary position: 0, ovary anterior to testes; 1, ovary between testes; 2, ovary posterior to testes.
   State 2 occurs in most hemiuroid families. State 1 includes forms with ovary between symmetrical testes (Notocotylidae), those with the ovary between tandem testes (Brachylaimidae, Clinostomidae, Bathycotylidae) and ones where the condition is variable (Cyclocoelidae).

2. Adult. Sperm reception: 0, uterine seminal receptacle and Laurer's canal present; 1, uterine seminal receptacle and Juel's organ formed from Laurer's canal; 2, blind seminal receptacle replaces Juel's organ and uterine seminal receptacle.

Gibson and Bray (1979) suggested this series, particularly in relation to the Hemiuroidea. They also pointed out the occasional occurrence of the derived conditions in other digenean and aspidogastrean groups. It appears that a Juel's organ is present in *Stichocotyle* (see character 4).

3. Life-cycle: Miracidium flame-cells: 0, 3–4 pairs; 1, 2 pairs; 2, 1 pair; 3, none seen.

Aspidogastrean larvae have 3–4 pairs of flame-cells (Schell 1978; Rohde 1994a). All digeneans appear to have fewer than this number. We have coded a few taxa as '3, none seen' where the tiny size of the miracidium correlates with a complete absence of reports of flame-cells in the miracidia.

4. Adult. Canalicular seminal receptacle: 0, absent; 1, present.

The canalicular seminal receptacle was defined by Gibson and Bray (1979) as 'a large proximal dilation of Laurer's canal'. It is a common arrangement in the Digenea and may have a variety of forms (see Pearson 1992, Figure 24) but is always closely associated with the proximal end of Laurer's canal. It is apparent that the female proximal system is somewhat plastic in both the Digenea and the Aspidogastrea. Most aspidogastreans have a uterine seminal receptacle and a normal Laurer's canal, but a few develop a Juel's organ like arrangement (see Gibson and Bray 1979). In *Stichocotyle* the seminal receptacle is illustrated as an oval organ, but Laurer's canal is not present, so it must be either a Juel's organ or a blind seminal receptacle: in fact, the texture in the illustrations suggest a Juel's organ-like arrangement (see character 2).

5. Adult. Cirrus-sac 0, present: 1, absent.

A cirrus-sac like organ is reported from many, but not all, aspidogastreans. The cirrus-sac is lost within many families, e.g., *Allassogonoporus* in Lecithodendriidae, *Pseudobacciger* in Faustulidae. In the opecoelid subfamily Opecoelinae the cirrus-sac is often reduced to a vestige.

6. Adult. Egg operculum: 0, operculate. 1, not operculate.
The aspidogastreans have operculate eggs.

7. Adult. Egg: 0, round in section: 1, triangular.
The triangular egg is, apparently, found only in campulids and nasitrematids.

8. Adult. Embryonation: 0, egg embryonated when laid; 1, egg unembryonated when laid. This feature varies in the aspidogastreans. We have seen embryonation clearly in uterine eggs in specimens of *Aspidogaster limacoides* Diesing, 1834 (see also Williams 1942), *A. enneatis* Eckmann, 1932, *Rohdella siamensis* Gibson and Chinabut 1984 (see also Gibson and Chinabut 1984) and *Cotylogaster basiri* Siddiqi and Cable, 1960 and it is known in other aspidogastrids (Rohde 1994), whereas uterine eggs are known to remain unembryonated in *Cotylaspis insignis* Leidy, 1857, *Multicotyle purvisi* Dawes, 1941 and *Sychnocotyle kholo* Ferguson, Cribb and Smales 1999 (see Kearn 1998; Ferguson *et al.* 1999). For this reason this character is placed as a query for the Aspidogastrea. The query for *Stichocotyle* is due to lack of information. The eggs of *Rugogaster* embryonate *in utero* (Schell 1973). Ferguson *et al.* (1999) pointed out that in the Aspidogastrinae the known cases of non-embryonation *in utero* are in species that inhabit freshwater turtles. It is likely that the stage of embryonation is affected also by such features as water temperature, salinity and oxygen content as has been shown for the pseudophyllidean cestode *Triaenophorus* Rudolphi, 1793 (see Kuperman 1973).

9. Adult. Excretory vesicle shape: 0, Y- or V-shaped; 1, I-shaped. Evidence from outgroups and ontogeny (La Rue 1957) indicate that a divided (Y- or V-shaped) excretory vesicle is the plesiomorphic condition. The excretory vesicle shape appears to be fairly reliable as a criterion for family membership in many cases, but in most families there are exceptional cases, such as divided vesicles in families where an I-shaped vesicle is almost universal (e.g., *Allolepidapedon* Yamaguti, 1940 and a few other genera in the Lepocreadiidae, *Koiea* Bray and Campbell, 1995 in the Zoogenidae). The V- and Y-shaped vesicle may co-occur within families (e.g., Fellodistomidae), so we have coded these two conditions together.

10. Adult. Genital atrium: 0, shallow; 1, long and narrow. The derived condition is characteristic of the families Acanthocolpidae, Deropristidae and Apocreadiidae.

11. Adult. Genital pore position: 0, in forebody or close to ventral sucker; 1, at or close to posterior extremity.

12. Adult. Opening of male and female pores: 0, male and female pores open separately, usually into genital atrium; 1, hermaphroditic duct present. The derived condition is found in Diplostomidae, Haploporidae and all Hemiuroidea except Hirudinellidae. There is some problem in differentiat-

ing this derived feature from the long, narrow genital atrium exhibited by such families as the Acanthocolpidae and Deropristidae (see above). In the case of these forms there is no sac-like organ associated solely with the male-duct (i.e., a cirrus-sac [apart from the prostatic sac in Azygiidae]) and the hermaphroditic duct is often within an hermaphroditic sac or sinus-sac.

13. Adult. Intestine: 0, caecum single; 1, bifurcate.
Both conditions occur in the outgroups, with the single caecum in the Aspidogastridae and *Stichocotyle*, and the bifurcate intestine in *Rugogaster*.

14. Adult. Oral sucker: 0, always absent; 1, present at any stage.
Pearson (1992) has shown convincingly that the oral sucker does not occur in the Aspidogastrea. He also discussed the evidence for the 'muscular organ at the entrance of the gut' being a pharynx in such families as the Bivesiculidae, Gyliauchenidae, Paramphistomidae and Notocotylidae.

15. Adult. Ovary: 0, rounded or oval; 1, elongate; 2, follicular.
State 1 is found in Didymozoidae and Isoparorchiidae, families not included in our main analysis. State 2 is an autapomorphy of the Syncoeliidae.

16. Adult. Pharynx: 0, present at some stage of ontogeny; 1, always absent.
The derived condition is that found in the Gorgoderinae and the blood-flukes, but the plesiomorphic condition is found in the gorgoderid *Degeneria*, which is one of the forms sequenced.

17. Adult. Seminal vesicle: 0, internal seminal vesicle (ISV) and external seminal vesicle (ESV) present; 1 ISV only present; 2 ESV only present. Both seminal vesicles may be present in some plagiorchiid genera (Tkach *et al.* 1999).

18. Adult. Sexuality: 0, monoecious; 1, dioecious.
We interpret that true dioecy occurs as an autapomorphy only in the Schistosomatidae. The only other candidates are certain didymozoids, but a more or less complete continuum from monoecious to dioecious occurs in this group.

19. Adult. Sinus-sac: 0, absent; 1, present.
The derived condition is that found in most hemiuroid families (Gibson and Bray 1979).

20. Adult. Spined cirrus 0, absent; 1, present.
The families with the derived character are Acanthocolpidae, Deropristidae, Fasciolidae and Monorchiidae. Distinct cirrus-spination occurs in some plagiorchiid species.

21. Adult. Tegumental spines: 0, absent; 1, present. Tegumental spines being consistently absent in the Aspidogastrea, the plesiomorphic character-state is clear. The state within families is, unfortunately, not invariant, e.g. the zoogonid genus *Deretrema* Linton 1910 has several species lacking tegumental spines in a family where tegumental spination is overwhelmingly common (Bray 1987). On the other hand, the systematic importance of tegumental spines has been shown by the recent work of Hall *et al.* (1999), in which on the basis of rDNA sequences, the subfamily Faustulidae (with spines) is removed from the Fellodistomidae (now lacking spines) and raised to family status as a sister-taxon to the Zoogonidae (almost always with spines, see above).

22. Adult. Testes shape: 0, round or ovoid; 1, tubular. Derived in Heronimidae and Didymozoidae.

23. Adult. Testes: 0, two; 1, numerous; 2, single.
Aspidogastreans may have one (*Aspidogaster* von Baer, 1826; *Cotylaspis* Leidy, 1857, *Lissemysia* Sinha, 1935, *Lobatostoma* Eckmann, 1932, *Lophotaspis* Looss, 1901, *Rohdella*, *Sychnocotyle*) or two (*Cotylogaster* Monticelli, 1892, *Multicotyle* Dawes, 1941) testes, and this feature is therefore treated as a query.

24. Adult. Tribocytic organ: 0, absent; 1, present. Derived condition in Cyathocotylidae, Diplostomidae, Strigeidae and Proterodiplostomidae.

25. Adult. Uterus: 0, mainly restricted to area anterior to gonads; 1, passing postovarian, usually to posterior end of body; 2, relatively short, close to posterior extremity with gonads. This character suffers from much within-group variation.

26. Adult. Ventral sucker position: 0, at posterior end of body; 1, in mid-body. The larval condition in those aspidogastreans we know is 0 (depending of course on the views relating to the homology of the posterior sucker in the aspidogastrean larva and the digenean ventral sucker), but the condition in the adult aspidogastrean is more difficult to interpret. Where the ventral disc is continuous it reaches from the anterior to the mid-body to the posterior extremity. The apparent ventral sucker in *Rugogaster* is, in fact, the anterior part of the ventral disc represented by a series of rugae (Rohde and Watson 1992c) and in *Stichocotyle* the ventral disc is represented by a series of suckers.

27. Adult. Ventral sucker: 0, absent; 1, present.
The primitive or derived condition of the ventral sucker in the Digenea will be judged by the views relating to the homology of this organ with the ventral disc of the aspidogastreans. Several studies have shown the development of the ventral disc in aspidogastrids from the posterior sucker-like organ of an amphistomatous larva (e.g., Williams 1942; Rohde 1975b; Tang and Tang 1980). This posterior larval sucker is strikingly similar to the digenean ventral sucker and has been invoked to indicate the primitiveness of digeneans with a similar amphistomatous structure (see Gibson 1987). In contrast, Ehlers (1984) considered the ventral (and oral) suckers in cercaria and adult as apomorphies for the Digenea and Pearson (1992: 106) put forward the notion that the ventral and oral suckers are primitively absent in the Digenea. Whilst Pearson morphological and histological evidence relating to the oral sucker is convincing, his arguments relating to the ventral sucker are predicated on a phylogenetic scenario rather than morphology or ultrastructure and postulates homoplasious acquisition of a ventral sucker. While it is true that suckers seem to be readily acquired in platyhelminths, this interpretation does not satisfy the criterion of parsimony. Pearson (1992: 121) does not follow up his suggestions in his phylogenetic reconstruction of the Aspidogastrea, making no mention of the ventral sucker in the Digenea, in contrast to Gibson (1987).

The above discussion relates to the Aspidogastridae, and it should be noted that the situation is different (at least superficially) in other aspidogastreans. In *Stichocotyle* there is no ventral disc, as such, but a serial row of separated sucker-like organs (up to 30) on the ventral surface (Odhner 1910). The early larval condition is not known. A similar serial repetition of suckers occasionally occurs in digeneans (e.g., *Jeancadenatia brumpti* Dollfus, 1947) (Dollfus 1947). In ventral view, *Rugogaster* appears to have a single separate ventral sucker in addition to a continuous series of (up to 25) rugae as described, e.g., by Schell (1973) and Amato and Pereira (1995). Rohde and Watson (1992c) studied the ultrastructure of *Rugogaster* and showed that the basal lamina ('capsule' of light microscope studies) encloses not only the 'ventral sucker', but also the series of rugae. Their observations caused them to state that the 'ventral sucker plus rugae of *Rugogaster* must be considered to be homologous with the ventral disc of Aspidogastridae'. The hatched larva of *Rugogaster* is amphistomatous, and more or less indistinguishable from that of the aspidogastrids (Schell 1973).

As there is some disagreement over this feature we have analysed our results with the ventral sucker as both plesiomorphic and derived in the Digenea.

28. Adult. Ventral sucker ontogeny if present: 0, always present; 1, lost in adult. Derived condition in Cyclocoelidae and Heronimidae.

29. Adult. Ventral sucker ontogeny: 0, always present; 1, rudimentary [or absent] in cercaria. Derived condition found in many taxa e.g., Cyathocotylidae and Opisthorchioidea. Both conditions are found in some families (e.g., Hemiuridae); in such cases we have coded the family as '0'.

30. Adult. Vitellarium: 0, lateral and follicular; 1, median, lobate or few masses; 2, filamentous tubules; 3, median, branched. Character-state 1 is that found in the hemiuroids, gorgoderids, zoogonines and gymnophallids. In all cases there is some evidence that these are homoplasiously reduced states relative to state 0. This is particularly evident in the Gymnophallidae where the vitellarium can be at one extreme, two small patches of follicles (e.g., *Pseudogymnophallus* Hoberg, 1981), and at the other, two or one compact masses (e.g., *Gymnophalloides* Fujita, 1925, *Parvatrema* Cable, 1953). Character-state 3 is that found in the Heronimidae.

31. Life-cycle. Cercarial penetration gland openings dorsal to oral sucker: 0, absent; 1, present (or, if worms do not encyst, indistinguishable from penetration glands).
We have coded the character of cercarial penetration glands twice (31 and 32) to account for three possibilities: cercariae that never penetrate for which penetration glands are thus inappropriate; cercariae that have penetration glands opening immediately dorsal to the oral sucker; and cercariae that have penetration glands either passing through or completely enclosed in the oral sucker. The two forms of penetration glands are polarized by the Aspidogastrea only in-as-much as the Aspidogastrea lacks a cercaria and so by default lacks both conditions. We have felt unable to put these characters into a transformation series, as we suspect that this will almost certainly over-simplify the relationship between the two forms. Pearson (1972) has discussed the ubiquity of frontal glands in sexually adult digeneans. It seems likely that penetration glands opening dorsal to the oral sucker are directly related to these glands and, thus, may be relatively uninformative as a character. In contrast, penetration glands that pass

through or are contained in the oral sucker are seen not only in most of the Diplostomoidea + Schistosomatoidea but also in the Bucephalidae. Strangely, the Cyathocotylidae is seemingly distinct from the remainder of the Diplostomoidea in having penetration glands outside the oral sucker.

32. Life-cycle. Cercarial penetration glands penetrate oral sucker: 0, absent; 1, present. See above.

33. Life-cycle. Asexual generations: 0, absent; 1 present.
This character serves to define the Digenea.

34. Life-cycle. Cercarial behaviour: 0, escapes from first intermediate host; 1, eaten in first intermediate host.
We argue that this character cannot be polarized by reference to the Aspidogastrea because that taxon lacks a cercaria. As a result, the Aspidogastrea is coded here as '9' (inapplicable) but run as '?' (unknown). The effect of this is that the polarity of this character is not forced by the analysis. We concede that another position is arguable. Because the vertebrate definitive host in the aspidogastrean life-cycle is usually infected by ingestion of a mollusc, the Aspidogastrea could be coded as '1'.

35. Life-cycle. Cercaria: dorsal body fin-fold: 0, absent; 1, present.
This character is coded as present only for the Clinostomidae, Sanguinicolidae and Spirorchiidae. The character is compelling in that it seems likely that these three families are related. It must be admitted, however, that not all members of these families possess unequivocal dorsal fin-folds. Thus, in the Spirorchiidae *Vasotrema robustum* (Stunkard, 1928) appear to lack a fin-fold (Wall 1951), whereas *Spirorchis parvus* (Stunkard 1923) does have one (Wall 1941).

36. Life-cycle. Cercaria: cystogenous glands: 0, absent; 1, tegument; 2, excretory vesicle. This character relates to the origin of the material that forms the metacercarial cyst.
Opisthorchioids and gorgoderids are known to produce this material from the wall of excretory vesicle, whereas all other families for which we have found information have cystogenous glands in the body of the cercaria. Many taxa remain uncoded for this character, and we think it likely that both general types of cystogenous glands may include non-homologous conditions.

37. Life-cycle. Cercaria: excretory duct ciliation: 0, present; 1, absent.

38. Life-cycle. Cercaria: eye-spots: 0, present; 1, absent.

39. Life-cycle. Cercaria: flame-cells in tail: 0, present; 1, absent.

40. Life-cycle. Cercaria: fork-tail body enclosure: 0, absent; 1, cystocercous; 2, cystophorus. The cystocerous condition is shared by the Bivesiculidae and Azygiidae (Hemiuroidea), whereas the cystophorous condition is apparently shared by all the other hemiuroid families except for the Ptychogonimidae in which the cercaria is apparently reduced secondarily.

41. Life-cycle. Cercaria: alimentary system: 0, single caecum; 1, bifurcate gut; 2, not developed. This character has problems associated with it in that the distinction between the three characters may be indistinct. Thus, guts of species that will ultimately be bifurcate may appear as a simple sac or one with feint bilobation. The distinction might be phylogenetically important if the ontogenetic change reflects the phylogenetic history of the structure. Coding of the gut as 'not developed' has the potential to reflect a likely evolutionary shift towards small cercariae in which the gut develops only after the formation of the metacercaria. There is, however, a considerable grey area of guts that are partly developed.

42. Life-cycle. Cercaria: final excretory pore: 0, secondary; 1, tertiary.
All cercariae appear to have a secondary terminal excretory pore on the body formed by the fusion of independent excretory systems. In a few taxa (Heronimidae, Paramphistomidae) a tertiary pore is formed *de novo*.

43. Life-cycle. Cercaria: simple-tail enclosure: 0, absent; 1, cystocercous.
This character appears to characterize only the gorgoderid subfamily Gorgoderinae. Within this subfamily not all species and genera have the character-state, although absence in such taxa as *Pseudophyllodistomum johnstoni* Cribb, 1987 (Cribb 1987) may well be a secondary reversal.

44. Life-cycle. Cercaria: 0, absent; 1 present.
This character serves to define the Digenea.

45. Life-cycle. Cercaria: primary excretory pore: 0, terminal on tail; 1, mid-tail or anterior.

46. Life-cycle. Cercarial tail: 0, forked; 1, simple.

47. Life-cycle. Cercarial tail: 0, present; 1, absent.

48. Life-cycle. Cercaria stylet: 0, absent; 1, present.

49. Life-cycle. Miracidium: 0, hatches outside 1st intermediate host; 1, emerges in 1st intermediate host.

50. Life-cycle. Metacercarial behaviour: 0, does not encyst; 1, encysts.
Encystment here is restricted to the process by which the metacercaria forms a cyst around itself, rather than being encapsulated by its host.

51. Life-cycle. Miracidium: eye-spots: 0, present; 1, absent.
52. Life-cycle. Mother-sporocyst form: 0, sac-like; 1, deeply branched.
   The sac-like form is apparently the norm for digeneans.
53. Life-cycle. Redia: collar: 0, absent; 1, present.
54. Life-cycle. Redia: form: 0, forked; 1, not forked.
   The forked condition is so far restricted to the Bivesiculidae.
55. Life-cycle. Redial appendages: 0, absent; 1, present.
56. Life-cycle. Ultimate parthenita: 0, redia; 1, daughter-sporocyst.
   This is an interesting and difficult character. Its use carries the implication

that there has been an evolutionary change from one to the other. The most likely direction is from redia to sporocyst because this requires losses rather than invention of a gut *de novo*, and because there is actual ontogenetic evidence for such loss as summarized by Pearson (1992). The difficulty associated with the character is that the ontogenetic loss summarized by Pearson makes coding difficult for some families. Nonetheless, the character is usually very clear and likely to contain phylogenetic information. We have coded for the dominant condition for each family.

Appendix 16.2 Matrix of morphological characters: several taxa were used only in the combined evidence analysis. For characters, see Appendix 16.1. Characters shown as '9' (inapplicable) were converted to '?' (unknown) prior to analysis.

	1	2	3	4	5	6	7	8	9	10	11	12	13	14	15	16	17	18	19	20	21	22	23	24	25	26	27	28	29	30	31	32	33	34	35	36	37	38	39	40	41	42	43	44	45	46	47	48	49	50	51	52	53	54	55	56
Aspidogastridae	0	0	0	0	0	0	0	?	0	0	0	0	0	0	0	?	0	0	9	0	0	?	0	0	0	1	0	0	0	?	?	0	9	9	9	0	0	0	9	9	9	9	9	0	9	9	9	9	0	9	0	9	9	9	9	9
Stichocotyle	0	1	?	0	0	0	0	?	0	0	0	0	0	0	0	1	0	0	0	0	0	0	0	1	?	1	0	0	9	9	0	9	9	9	0	9	9	9	9	9	9	9	9	0	9	9	9	0	9	0	9	?	?	9	9	9
Rugogaster	0	0	0	0	0	0	0	?	0	0	0	1	0	0	0	1	0	0	0	0	0	0	1	0	1	0	1	0	0	9	9	0	9	9	0	9	9	9	9	9	9	9	0	9	9	9	9	0	9	0	9	9	9	9	9
Acanthocolpidae	0	0	2	0	0	0	0	0	1	1	0	0	1	1	0	0	1	0	0	1	0	0	0	1	1	0	0	0	1	0	1	0	0	?	0	0	1	9	1	0	0	1	1	1	0	1	?	1	1	0	0	1	0	0
Accacoeliidae	2	0	?	0	1	0	0	?	0	0	0	1	1	1	0	0	?	0	1	?	0	0	0	0	1	1	0	0	2	?	?	1	?	?	?	?	?	?	?	?	1	?	?	?	?	0	?	?	?	?	?
Allocreadiidae	0	9	2	1	0	0	0	1	1	0	0	0	1	1	0	0	0	0	0	0	0	0	0	1	1	0	0	0	1	0	1	0	0	?	1	0	1	9	1	0	0	1	1	1	0	1	0	0	0	1	0	0
Anaporrhutinae	0	9	?	1	1	0	0	0	1	0	0	0	1	1	0	0	2	0	0	9	0	0	0	0	1	1	0	?	1	?	1	?	?	?	?	?	?	?	1	?	?	?	0	?	?	?	?	?	?
Angiodictyidae	2	0	?	0	1	0	0	1	0	0	0	1	0	0	0	1	0	0	0	9	0	0	0	0	0	9	0	9	9	0	0	0	1	0	0	?	?	?	?	9	1	?	0	1	?	1	0	0	0	1	?	?	?	?	?
Apocreadiidae	0	9	2	1	1	0	0	1	1	0	0	1	0	0	1	0	0	2	0	0	9	1	0	0	0	0	1	0	1	0	0	0	?	?	0	1	9	1	0	0	0	1	1	0	0	0	1	0	0
Azygiidae	0	0	2	0	1	0	0	0	0	0	0	1	1	1	0	0	2	0	0	9	0	0	0	0	0	1	1	0	0	0	1	0	0	0	0	1	1	0	1	0	1	0	0	0	0	1	1	0	0	0
Bivesiculidae	0	0	2	0	0	0	0	1	0	0	0	1	0	0	0	0	0	0	0	0	1	0	2	0	1	9	0	9	9	0	0	0	0	0	0	0	0	0	1	1	0	9	1	0	0	0	0	0	9	0	?	0	0	1	0
Brachycoeliidae	0	?	?	?	0	0	0	0	0	?	0	0	1	1	0	0	?	0	0	?	1	0	0	0	1	1	1	0	0	0	?	?	1	?	?	0	?	1	?	9	1	?	0	1	?	1	0	1	1	?	?	?	?	?
Brachylaimidae	1	?	3	0	0	0	0	0	1	0	1	1	0	0	1	0	0	1	0	0	0	0	1	0	0	0	1	0	1	0	1	9	1	0	0	1	1	9	1	0	0	1	0	1	0	9	9	9	1
Bucephalidae	0	0	3	0	0	0	0	1	0	1	0	0	1	0	0	1	0	0	0	1	0	0	0	1	0	0	0	1	9	0	9	9	0	0	1	1	1	0	0	9	1	0	0	0	0	1	1	1	9	9	9	1
Bunocotylidae	2	1	?	0	?	0	0	1	0	0	0	1	1	0	0	?	0	1	?	0	0	0	1	1	0	0	1	?	?	1	0	?	?	?	2	2	?	?	9	1	?	0	0	0	?	?	?	?	0	1	0	0
Campulidae	0	0	?	0	0	0	1	?	1	0	0	0	1	1	0	0	1	0	0	0	1	0	0	0	1	0	0	0	1	1	0	0	?	?	1	?	?	?	?	?	?	?	?	1	?	?	?	?	?	?	?	?	?	?	?
Clinostomidae	1	9	1	1	0	0	0	1	0	0	1	0	1	1	0	0	1	0	0	0	0	0	0	0	0	0	1	0	1	0	1	0	1	0	?	0	0	0	0	0	9	1	0	0	0	0	0	?	0	0	1	0	0
Cyathocotylidae	0	0	1	0	0	0	0	1	0	0	1	0	1	1	0	0	1	0	0	0	0	0	0	0	1	1	1	0	1	0	1	0	1	0	0	0	?	0	1	0	0	1	0	0	0	0	1	0	0	9	9	9	1
Cyclocoelidae	1	9	2	1	0	0	0	1	0	0	0	1	0	0	0	1	0	0	0	0	0	0	0	9	1	1	0	0	1	0	1	0	1	0	1	1	0	7	1	9	9	1	0	1	0	1	0	0	1	0	1	1	0
Derogenidae	2	?	2	2	0	1	0	0	0	0	0	1	1	1	0	0	2	0	0	9	0	0	0	0	1	0	0	0	1	0	0	1	1	0	2	1	0	9	1	0	0	0	0	1	0	0	1	0	0
Deropristidae	0	9	?	1	0	0	0	1	1	0	0	1	1	0	0	1	0	0	0	0	1	0	0	0	1	1	0	0	0	0	1	0	1	0	0	?	0	0	1	9	1	0	1	0	1	0	1	0	1	0	1	0	0
Dicrocoeliidae	2	9	2	1	0	0	0	0	1	1	0	0	1	1	0	0	1	0	0	1	0	0	0	0	0	1	0	0	1	0	1	0	?	0	0	1	1	9	2	2	0	1	1	0	1	1	0	0	9	9	9	1
Didymozoidae	2	0	?	0	1	0	0	0	0	0	0	1	1	1	1	0	2	0	0	9	0	0	1	0	0	1	1	0	?	2	?	?	?	?	?	?	?	?	1	?	?	?	?	?	0	?	?	?	?	?	?
Diplostomidae	0	0	1	0	1	0	0	1	0	0	1	1	1	1	0	0	2	0	0	9	0	0	0	0	1	1	1	1	0	0	0	1	1	0	0	?	?	0	0	1	9	1	0	0	0	0	1	0	0	9	9	9	1
Echinostomatidae	0	0	2	0	0	0	0	1	0	0	0	0	1	1	0	0	1	0	0	0	0	0	0	0	1	1	0	0	0	0	1	0	1	0	0	1	0	9	1	9	0	0	1	1	0	0	0	1	0	0	0	1	0
Enenteridae	0	0	?	?	0	0	0	?	1	0	0	0	1	1	0	0	?	0	0	?	1	0	0	?	1	0	0	0	0	0	?	0	?	1	?	?	?	?	?	?	?	?	?	?	?	?	?	?	?
Fasciolidae	0	0	2	0	0	0	0	1	1	0	0	0	1	1	0	0	1	0	0	0	1	0	0	0	0	1	1	0	0	0	0	0	1	0	0	0	0	1	9	1	?	0	1	?	1	0	0	0	0	1	0	0	1
Faustulidae	0	9	?	1	0	0	0	?	0	0	0	1	1	0	0	1	0	0	1	0	0	0	0	0	1	1	0	0	0	1	0	1	0	0	0	0	1	?	0	1	?	0	1	0	0	0	1	0	1	0	9	9	9	1
Fellodistomidae	0	0	?	0	0	0	0	?	0	0	0	1	1	0	0	1	0	0	0	9	0	0	0	0	0	1	1	0	0	0	0	0	0	1	1	0	1	?	0	0	1	?	0	0	0	7	0	0	0	7	0	9	9	9	1
Gorgoderinae	0	0	2	0	1	1	0	0	1	0	0	0	1	1	0	1	2	0	0	9	0	0	0	0	1	1	0	0	1	1	0	0	2	?	1	1	9	0	?	1	1	?	0	1	0	1	0	1	1	0	9	9	9	1
Gyliauchenidae	0	9	?	1	0	0	0	1	1	0	0	0	1	1	0	0	2	0	0	0	0	0	0	0	0	1	0	0	1	0	?	0	1	0	?	?	?	?	?	?	?	?	1	?	?	?	?	?	?	?	?
Gymnophallidae	0	0	?	0	1	0	0	0	0	0	0	1	1	0	0	0	2	0	0	9	0	0	1	0	0	0	1	1	0	0	1	0	1	0	0	0	0	1	0	0	0	1	0	0	0	1	0	?	0	9	9	9	1
Haploporidae	0	9	?	1	?	0	0	0	1	0	0	1	1	0	0	0	0	0	9	1	0	0	0	1	0	0	0	1	0	0	0	1	0	1	0	1	9	1	0	1	1	0	0	0	0	1	0	7	0	0	0	0
Haplosplanchnidae	0	9	?	1	1	0	0	1	1	0	0	0	1	0	0	2	0	0	9	0	0	2	0	0	1	1	0	0	0	0	1	0	1	9	0	0	0	1	0	1	0	1	?	9	9	9	1
Hemiuridae	2	1	2	0	1	0	0	0	0	0	0	1	1	1	0	0	2	0	0	9	0	0	0	0	1	1	1	0	0	0	1	0	0	0	0	1	1	0	2	1	0	9	1	0	0	0	0	1	0	1	0	0
Heronimidae	0	9	2	1	0	0	0	1	0	0	0	0	1	1	0	0	1	0	0	0	0	0	0	1	0	0	1	9	1	0	1	0	2	0	0	1	0	0	9	1	0	1	0	1	0	0	9	9	9	9	1
Lecithasteridae	2	2	2	0	1	0	0	0	0	0	0	1	1	1	0	0	2	0	1	9	0	0	0	0	1	1	0	1	1	0	1	1	0	0	0	0	1	1	0	2	1	0	9	1	0	0	0	0	1	0	0	1	0	0
Lecithodendriidae	0	9	3	1	0	0	0	0	0	0	0	1	1	0	0	1	0	0	0	0	1	0	0	0	0	1	1	0	0	0	1	0	0	?	?	1	9	1	0	1	?	1	0	1	1	1	0	0	9	9	9	1
Lepocreadiidae	0	9	2	1	0	0	0	1	0	0	0	1	1	0	0	1	0	0	0	0	0	1	0	0	0	0	1	0	1	0	1	0	0	?	0	0	1	9	1	0	0	0	1	0	0	0	1	0	0	0
Leucochloridiomorphidae	0	0	?	?	0	?	0	?	1	0	1	0	1	1	0	0	2	0	0	0	0	0	0	0	0	0	1	0	0	0	0	1	0	0	0	0	0	0	1	9	1	1	0	0	?	?
Lissorchiidae	0	9	?	0	0	0	0	?	1	0	0	1	1	0	0	1	0	0	0	0	0	0	0	1	0	0	0	1	0	1	0	0	?	?	1	?	9	1	?	0	1	0	1	0	1	?	1	1	?	9	9	9	1
Mesometridae	2	0	?	0	1	0	0	0	0	0	0	1	0	0	0	0	2	0	0	9	1	0	0	0	0	0	9	0	9	9	0	7	?	1	0	0	1	0	?	0	1	9	1	0	0	0	1	0	0	1	0	0
Microphallidae	0	9	?	1	0	0	0	1	0	0	0	1	0	0	0	1	0	0	1	0	0	0	1	0	0	0	1	0	0	0	?	?	1	1	9	1	0	0	0	1	0	0	1	1	1	0	0	9	9	9	1
Monorchiidae	0	0	?	0	0	0	0	1	1	0	0	0	1	1	0	0	1	0	0	1	0	0	1	1	0	0	0	1	1	0	1	0	0	?	0	0	1	9	1	0	0	0	1	1	0	0	1	0	0	1	0	0
Notocotylidae	1	0	3	0	0	0	0	1	0	0	0	1	1	0	0	0	0	0	0	0	0	0	0	0	0	0	0	9	9	0	0	0	1	0	0	1	0	0	1	9	1	0	0	1	1	0	0	0	0
Opecoelidae	0	0	2	0	0	0	0	1	0	0	0	0	1	1	0	0	1	0	0	0	0	0	0	0	1	0	0	0	1	0	1	0	0	?	0	1	9	1	0	0	0	0	1	0	0	1	1	0	0	0
Opistholebetidae	0	9	?	1	0	0	0	?	0	0	0	0	1	1	0	0	1	0	0	0	0	0	0	0	0	0	0	1	1	0	?	?	1	?	?	1	?	?	?	?	?	?	?	1	?	?	?	?	?
Opisthorchioidea	0	9	2	1	0	0	0	0	0	0	1	0	1	1	0	0	2	0	0	0	0	0	1	0	0	1	1	0	1	0	1	0	1	0	0	2	1	0	1	9	0	0	1	1	1	0	0	1	0	0	0
Orchipedidae	0	9	?	1	1	0	0	?	0	0	0	1	1	0	0	2	0	0	9	0	0	0	1	0	0	1	0	0	1	0	1	0	?	?	1	?	?	2	?	?	?	?	1	?	?	?	0	?	?	?	?	?
Paragonimidae	0	9	2	1	0	0	0	1	0	0	0	0	1	1	0	0	1	0	0	9	1	0	0	0	0	0	1	1	0	0	0	0	1	0	0	?	1	1	1	9	1	0	0	1	?	1	0	0	1	0	1	0	0
Paramphistomidae	0	0	2	0	1	0	0	1	1	0	0	0	1	0	0	0	1	0	0	0	0	0	0	9	0	0	0	0	0	1	0	0	1	0	0	0	1	0	1	9	1	0	1	0	1	0	0	0	1	0	0	0
Philophthalmidae	0	0	2	0	1	0	0	1	0	0	0	0	1	1	0	0	1	0	0	0	0	0	0	0	1	1	0	0	1	0	1	0	1	0	0	?	0	1	1	9	1	0	0	1	0	1	0	0	1	0	0	0
Plagiorchiidae	0	9	2	1	0	0	0	1	0	0	0	0	1	1	0	0	1	0	0	0	0	0	0	0	1	1	0	0	0	0	1	0	0	?	0	1	9	1	0	1	0	1	1	1	0	9	9	9	1
Ptychogonimidae	0	0	?	0	1	0	0	?	0	0	0	1	1	1	0	0	2	0	1	9	0	0	0	0	1	1	0	0	0	1	?	?	0	?	?	?	?	1	?	?	?	?	0	0	?	9	9	9	1
Sanguinicolidae	2	0	2	0	0	1	0	1	0	0	1	0	1	1	0	0	1	0	0	1	0	0	1	0	9	0	9	9	0	0	0	1	1	0	0	1	0	7	1	9	1	0	0	0	0	0	9	1	0	9	9	9	1
Schistosomatidae	9	9	1	1	0	1	0	1	1	0	0	9	1	1	0	1	?	1	0	9	1	0	0	1	0	0	1	0	7	0	0	1	1	0	0	0	0	0	0	0	9	1	0	9	9	9	1
Sclerodistomidae	2	0	?	0	1	0	0	0	0	0	0	1	1	1	0	0	2	0	0	9	0	0	0	0	1	1	0	0	1	0	0	?	1	?	?	1	?	?	?	?	?	?	?	?	0	0	?	?	?
Spirorchiidae	2	0	1	0	0	0	0	0	0	0	1	0	1	0	1	0	1	2	0	0	0	0	0	0	1	0	0	1	1	0	0	1	9	1	0	0	0	0	0	9	0	0	9	9	9	1
Strigeidae	0	0	1	0	1	0	0	1	0	0	1	0	1	1	0	0	2	0	0	9	0	0	0	0	1	1	1	0	1	0	1	0	0	1	0	9	1	0	0	0	0	0	0	9	9	9	1
Syncoeliidae	2	0	?	0	1	0	0	0	0	0	1	1	1	1	2	0	2	0	1	9	0	0	0	1	0	1	1	0	?	?	1	?	1	?	0	?	0	1	?	?	?	?	1	?	?	?	0	7	?	?	?
Tandanicolidae	0	0	?	0	0	0	0	0	1	0	0	0	1	0	0	1	0	0	1	0	0	0	9	1	0	0	0	1	1	0	0	?	?	1	0	0	?	?	?	?	?	?	0	9	9	9	9	1
Transversotrematidae	0	0	2	0	1	0	0	1	1	0	0	0	1	0	0	0	2	0	0	9	1	0	0	0	0	9	1	9	0	0	0	0	1	0	9	1	0	0	0	0	0	9	0	0	1	1	0
Zoogoninae	0	?	?	2	1	0	0	0	0	0	0	1	0	0	1	0	0	0	0	0	0	1	1	0	0	1	1	0	0	1	0	1	0	1	0	7	1	?	1	1	1	0	1	1	?	?	?	1
Cableia	0	0	?	0	0	0	0	?	1	0	0	1	0	0	1	0	0	1	1	0	0	0	0	1	1	0	?	0	7	0	?	?	?	?	?	?	?	1	?	1	1	1	0	?	?	?	?
Robphildollfusium	0	?	?	1	0	0	0	1	1	0	0	0	1	0	0	0	2	0	0	0	0	0	0	1	1	0	?	0	7	0	?	?	?	1	?	?	?	?	?	?	?	?	?	1	?	?	?	?
Lepidophyllinae	0	?	?	1	0	0	0	?	1	0	0	0	1	1	0	0	1	0	0	0	1	0	0	0	1	1	0	?	0	7	0	?	1	?	?	?	?	?	?	?	?	?	?	?	1	?	1	?	?	?	?

Molecular phylogeny of the suborder Plagiorchiata and its position in the system of Digenea

Vasyl V. Tkach, Jan Pawlowski, Jean Mariaux and Zdzislaw Swiderski

The suborder Plagiorchiata is a large group of digenean trematodes which includes a number of family-level taxa (Skrjabin and, Antipin 1958; Yamaguti 1958, 1971; Brooks *et al.* 1985a, 1989; Sharpilo and Iskova 1989). Despite the extensive amount of published information, the limits of the suborder are not clear, and have always been a subject of controversy.

Although several systematic schemes of the group based on morphological characters of adult and/or larval stages have been proposed (La Rue 1957; Odening 1964, 1971, 1974; Cable 1974), the phylogeny of the suborder has been explored formally only recently (Brooks *et al.* 1985a, 1989; Brooks and McLennan 1993a). These studies designated the plagiorchiates as the most derived group of digeneans. Several consecutive systematic rearrangements made by Brooks *et al.* (1985a, 1989) and Brooks and McLennan (1993a) resulted in a division of Plagiorchiata into two infrasuborders, Plagiorchiatea and Opecoelatea. The former clade included four superfamilies, Plagiorchioidea, Dicrocoelioidea, Gorgoderoidea and Telorchioidea, and the latter was comprised of Opecoeloidea and Microphalloidea. However, for several reasons (for discussion see Tkach *et al.* 2000) the phylogeny of Plagiorchiata was relatively poorly resolved in these cladistic works. Thus, the taxonomic content of the suborder and phylogenetic relationships among different families and genera belonging to it, remained unclear.

During the past several years, nucleotide sequences of several nuclear and mitochondrial genes have been used as an alternative/additional data source for phylogenetic studies in different groups of parasitic Platyhelminthes, particularly digeneans (e.g., Barker *et al.* 1993; Blair and Barker 1993; Rohde *et al.* 1993a, 1995; Littlewood and Johnston 1995; Barker and Blair 1996; Mollaret *et al.* 1997; Rollinson *et al.* 1997; Blair *et al.* 1998; Fernández *et al.* 1998; Jousson *et al.* 1998; Mariaux 1998; Littlewood *et al.* 1999a; Litvaitis and Rohde 1999) and have proven to be highly useful in phylogenetic reconstruction. Therefore, in our previous paper (Tkach *et al.* 2000) we attempted to explore the phylogenetic relationships of different taxa within Plagiorchiata on the basis of partial 28S nuclear rDNA sequences. In that work, we found that representatives of the superfamilies Opecoeloidea, Dicrocoelioidea and Gorgoderoidea appeared outside the clade of Plagiorchiata, and the plagiorchiates were divided into two large clusters representing superfamilies Microphalloidea and Plagiorchioidea. However, we did not consider the position of Plagiorchiata in the Digenea, because only few external taxa were examined. Also, only one representative of Troglotrematata was used to test the sister-group relationships of the two suborders and representatives of suborder Renicolata, another potential sister-group of plagiorchiates, were not examined.

The present study significantly updates the analysis of Tkach *et al.* (2000), first of all regarding the diversity of digenean taxa represented. Partial 28S rDNA sequences from representatives of more plagiorchiatan taxa as well as 14 digenean families outside Plagiorchiata were obtained in order to assess their phylogenetic relationships with this group. Additionally, representatives of Nanophyetidae (Troglotrematata), and Renicolidae (Renicolata) were used to test their sister-group relationships with Plagiorchiata.

The main aims of this study were to: 1) analyse with the enlarged set of taxa the taxonomic content of Plagiorchiata and phylogenetic relationships of key family-level taxa in this suborder; 2) explore the phylogenetic position of Plagiorchiata in relation to some other large digenean groups; and 3) test the existing hypotheses on the probable sister-groups of this suborder.

Materials and methods

Sample collection

Specimens belonging to 51 digenean species collected from different vertebrate hosts, were used in the study (Table 17.1). Sequence of the aspidogastrean *Multicotyle purvisi* (Littlewood *et al.* 1998a; GenBank no. AJ005796), was used to determine the basal digenean taxon in our data set.

The studied taxa set included representatives of all superfamilies recognized in Plagiorchiata by Brooks *et al.* (1985a, 1989). In many cases, species belonging to the type genera of corresponding families (for instance, *Plagiorchis*, *Lecithodendrium*, *Pleurogenes*, *Telorchis*, *Haematoloechus*, *Dicrocoelium*, *Brachycoelium*, *Gorgodera*, etc.), were used. Voucher specimens of most species are deposited in the collection of the Department of Parasitology, Institute of Zoology, Kiev.

DNA extraction, amplification and sequencing

DNA has been extracted, amplified and sequenced according to the protocols described by Tkach and Pawlowski (1999) and Tkach *et al.* (2000). Only single specimens were used for DNA extraction. The DNA fragment sequenced in our study localized at the 5′ end of the 28S rDNA. Digenean-specific forward primer digl2 (5′-AAG CAT ATC ACT AAG CGG-3′), and universal reverse primer L0 (5′-GCT ATC CTG AG(AG) GAA ACT TCG-3′) have been used for both amplification and sequencing.

The new sequences obtained in this study were deposited in the GenBank under the accession numbers AF151943, AF184248-AF184265 (Table 17.1).

Sequence analysis

Forward and reverse sequences were assembled using the Perkin Elmer Autoassembler II software, and then manually aligned using the Genetic Data Environment software, version 2.2 for Solaris OS (Larsen *et al.* 1993). Cladograms were constructed by parsimony analysis (MP) using PAUP* 4.0b2 (Swofford 1998). Heuristic searches with 10 replicates with random stepwise addition were used. Branch support was estimated with 1000 bootstrap replicates.

Results and discussion

The DNA fragment sequenced in our study included variable domains D1–D3 (Hassouna *et al.* 1984), as well as moderately variable and highly conserved regions. The whole alignment included 1306 sites. The sequences could not be aligned in

Table 17.1 Digenean species used in this study, their hosts, geographical origin of material and GenBank accession numbers for corresponding sequences. Specimens were collected in the Ukraine unless otherwise is stated.

Digenean taxa	Host species	Geographic origin	GenBank	Authors
Allassogonoporidae				
Allassogonoporus amphoraeformis	*Pipistrellus kuhli* (M)	Kherson region	AF151924	Tkach *et al.* 2000
Brachycoeliidae				
Brachycoelium salamandrae	*Salamandra salamandra* (A)	Zakarpatska region	AF151935	Tkach *et al.* 2000
Brachylaemidae				
Brachylaima thompsoni	*Blarina brevicauda* (M)	Wisconsin, USA	AF184262	This study
Dicrocoeliidae				
Dicrocoelium dendriticum	*Marmota bobak* (M)	Kharkiv region	AF151939	Tkach *et al.* 2000
Lyperosomum transcarpathicum	*Sorex minutus* (M)	Zakarpatska region	AF151943	This study
Diplostomidae				
Alaria alata	*Nyctereutes procyonoides* (M)	Kherson region	AF184263	This study
Echinostomatidae				
Euparyphium melis	*Nyctereutes procyonoides* (M)	Kherson region	AF151941	Tkach *et al.* 2000
Echinoparyphium cinctum	*Anas platyrhynchos*	Kherson region	AF184260	This study
Encyclometridae				
Encyclometra colubrimurorum	*Natrix natrix* (R)	Kiev region	AF184254	This study
Eucotylidae				
Tamerlania zarudnyi	*Corvus monedula* (B)	Chernigiv region	AF184248	This study
Gorgoderidae				
Gorgodera cygnoides	*Rana lessonae* (A)	Geneva, Switzerland	AF151938	Tkach *et al.* 2000
Haematoloechidae				
Haematoloechus variegatus	*Rana arvalis* (A)	Ivano-Frankivsk region	AF151916	Tkach *et al.* 2000
Haematoloechus asper	*Rana arvalis* (A)	Ivano-Frankivsk region	AF151934	Tkach *et al.* 2000
Haematoloechus abbreviatus	*Bombina variegata* (A)	Zakarpatska region	AF184251	This study
Lecithodendriidae				
Lecithodendrium linstowi	*Nyctalus noctula* (M)	Sumy region	AF151919	Tkach *et al.* 2000
Prosthodendrium chilostomum	*Nyctalus noctula* (M)	Sumy region	AF151920	Tkach *et al.* 2000
Prosthodendrium longiforme	*Myotis daubentoni* (M)	Kiev region	AF151921	Tkach *et al.* 2000
Prosthodendrium hurkovaae	*Myotis daubentoni* (M)	Kiev region	AF151922	Tkach *et al.* 2000
Pycnoporus heteroporus	*Pipistrellus kuhli* (M)	Kherson region	AF151918	Tkach *et al.* 2000
Pycnoporus megacotyle	*Pipistrellus kuhli* (M)	Kherson region	AF151917	Tkach *et al.* 2000
Leucochloridiidae				
Leucochloridium perturbatum	terrestrial snail	Poland	AF184261	This study
Macroderidae				
Macrodera longicollis	*Natrix natrix* (R)	Kiev region	AF151913	Tkach *et al.* 2000
Microphallidae				
Maritrema subdolum	*Tringa erythropus* (B)	Kherson region	AF151926	Tkach *et al.* 2000
Maritrema neomi	*Neomys anomalus* (M)	Zakarpatska region	AF151927	Tkach *et al.* 2000
Monorchiidae				
Monorchis monorchis	*Diplodus vulgaris* (F)	near Corsica	AF184257	This study
Nanophyetidae				
Skrjabinophyetus neomydis	*Neomys anomalus* (M)	Zakarpatska region	AF184252	This study
Notocotylidae				
Notocotylus attenuatus	*Aythya ferina* (B)	Kherson region	AF184259	This study
Paramonostomum anatis	*Tringa erythropus* (B)	Kherson region	AF184258	This study
Opecoelidae				
Opecoeloides furcatus	*Mullus surmuletus* (F)	near Corsica	AF151937	Tkach *et al.* 2000
Macvicaria mormyri	?fish	near Corsica	AF184256	This study
Gaevskajatrema perezi	?fish	near Corsica	AF184255	This study
Plagiorchiidae				
Plagiorchis vespertilionis	*Myotis daubentoni* (M)	Kiev region	AF151931	Tkach *et al.* 2000
Plagiorchis elegans	*Lanius collurio* (B)	Kiev	AF151911	Tkach *et al.* 1999
Plagiorchis koreanus	*Nyctalus noctula* (M)	Sumy region	AF151930	Tkach *et al.* 2000
Plagiorchis muelleri	*Eptesicus serotinus* (M)	Chernigiv region	AF184250	This study
Astiotrema monticelli	*Natrix natrix* (R)	Kiev region	AF184253	This study
Paralepoderma cloacicola	*Natrix natrix* (R)	Kiev region	AF151910	Tkach *et al.* 2000
Metaleptophallus gracillimus	*Natrix natrix* (R)	Kiev region	AF151912	Tkach *et al.* 2000
Leptophallus nigrovenosus	*Natrix natrix* (R)	Kiev region	AF151914	Tkach *et al.* 2000
Haplometra cylindracea	*Rana arvalis* (A)	Zakarpatska region	AF151933	Tkach *et al.* 2000
Lecithopyge rastellus	*Bombina variegata* (A)	Ivano-Frankivsk region	AF151932	Tkach *et al.* 2000

Continued

Table 17.1 Continued.

Pleurogenidae				
Pleurogenes claviger	*Rana temporaria* (A)	Kiev region	AF151925	Tkach *et al.* 2000
Parabascus semisquamosus	*Pipistrellus kuhli* (M)	Kherson region	AF151923	Tkach *et al.* 2000
Prosthogonimidae				
Prosthogonimus ovatus	*Pica pica* (B)	Chernigiv region	AF151928	Tkach *et al.* 2000
Psilostomidae				
Psilochasmus oxyurus	*Anas platyrhynchos* (B)	Kherson region	AF151940	Tkach *et al.* 2000
Renicolidae				
Renicolidae sp.	*Cerithium vulgatum* (snail)	near Corsica	AF184249	This study
Schistosomatidae				
Bilharziella polonica	*Anas platyrhynchos* (B)	Kherson region	AF184265	This study
Heterobilharzia americana		Louisiana, USA	Z46506	Littlewood and Johnston 1995
Schistosoma haematobium		Strain 'Mali'	Z46521	Littlewood and Johnston 1995
Schistosoma spindale		Sri Lanka	Z46505	Littlewood and Johnston 1995
Strigeidae				
Apharyngostrigea cornu	*Ardea cinerea* (B)	Kherson region	AF184264	This study
Telorchiidae				
Telorchis assula	*Natrix natrix* (R)	Kiev region	AF151915	Tkach *et al.* 2000
Opisthioglyphe ranae	*Rana arvalis* (A)	Ivano-Frankivsk region	AF151929	Tkach *et al.* 2000
Troglotrematidae				
Nephrotrema truncatum	*Neomys anomalus* (M)	Zakarpatska region	AF151936	Tkach *et al.* 2000

hypervariable regions. These regions, as well as a very few sites containing undetermined bases were deleted from the analysis and 991 unambiguously aligned sites were finally considered, of which 371 were parsimony informative. The alignment is available from FTP.EBI.AC.UK in directory /pub/databases/embl/align under number DS39899.

In order to analyse the phylogenetic relationships among digeneans, it was necessary to define first the basal group in our taxa set. Because the sister-group relationships of Aspidogastrea and Digenea have been examined and confirmed previously in several morphological and molecular phylogenetic studies (e.g., Ehlers 1985a; Brooks 1989a; Rohde *et al.* 1993b; Littlewood *et al.* 1999a; Litvaitis and Rohde 1999), prior to the main analysis, we aligned the digenean sequences with a sequence of the aspidogastrean *Multicotyle purvisi*. Part of the sequence was unalignable because of great differences, therefore a reduced alignment containing 619 sites was used for this analysis. The trees resulting from the maximum parsimony (MP) analysis with 10 replicates in the random stepwise addition (not shown), revealed the basal position of the Schistosomatidae in relation to other digenean taxa represented in our data matrix. Although the bootstrap support was less than 50%, the position of this branch did not change under varying search conditions. This topology was also in agreement with data of Littlewood *et al.* (1999a); thus, in further analyses of the relationships within the Digenea we used four species of Schistosomatidae as outgroup (Figure 17.1).

Analysis of phylogeny of the Plagiorchiata and closely related groups was mainly based on a reduced set of taxa to achieve better branch support (Figure 17.2). In this analysis, nine most basal families were excluded, and representatives of Opecoelidae were used as an outgroup. The alignment length in this case was the same as with the larger set of taxa, and 286 characters were parsimony-informative. In this analysis, the family Opecoelidae has been chosen as an outgroup because its basal position in relation to rest of the taxa has been well supported in the analysis of Tkach *et al.* (2000) and did not change with the current data set (Figure 17.1). It should be noted that the branch topology did not alter when large or reduced data sets were used.

The analysis using the aspidogastrean *Multicotyle purvisi* as an outgroup has placed schistosomatids in the basal position among the digenean taxa included in our study. It should be noted that, despite the much smaller data set, our result is in agreement with the conclusions of studies based of complete 18S rDNA sequences and morphological characters (Littlewood *et al.* 1999a). The inclusion in the further studies of representatives of some other potential candidates for most primitive digeneans (e.g., see Gibson 1987) may add new information in the discussion concerning the basal digenean group.

The parsimony analysis of the whole data set in PAUP* with gaps treated as missing characters, produced two shortest trees (L = 1848, CI = 0.38, RI = 0.62), the consensus of which is shown on Figure 17.1. The only difference between the two trees was the relative position of *Heterobilharzia americana* and *Bilharziella polonica*, in the outgroup.

The most basal ingroup clade, including two 100% supported branches, consists of members of the Diplostomidae + Strigeidae and Leucochloridiidae + Brachylaimidae, respectively. Next to this cluster is a strongly supported clade formed by the Echinostomatidae and Psilostomidae (Figure 17.1). The sister-group (Figure 17.1) relationship of these taxa is in agreement with the existing systematic views, although members of the Psilostomidae lack such prominent morphological feature as the collar of spines characteristic for the Echinostomatidae.

The echinostomatid/psilostomid branch is followed by the Notocotylidae, including representatives of *Notocotylus* and *Paramonostomum* which demonstrated a high sequence similarity and formed a 100% supported clade. The two genera are similar in their morphology despite the great differences in size and absence of the ventral tegumental papillae in *Paramonostomum*. It should be mentioned that in the phylogeny presented by Brooks *et al.* (1985a) Notocotylidae appeared as one of the most basal digenean taxa.

Relationships between several other taxa present in this part of the tree are not clearly established, as the bootstrap support for their nodes was low. It should be noted, however, that the branch topology among more derived taxa did not change when the reduced taxa set was used. Despite the addition of many new taxa, a refined alignment and some changes in site selection, the current study supports the results obtained by

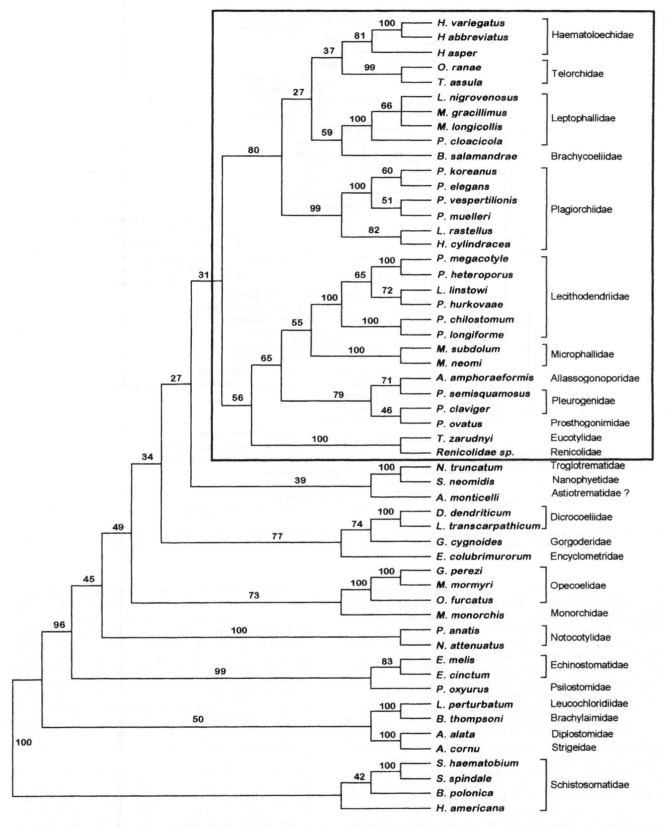

Figure 17.1 Strict consensus of two shortest trees (L = 1848, CI = 0.38, RI = 0.62) obtained in maximum parsimony analysis with the stepwise addition option and 10 replicates. Bootstrap values are indicated above branches (1000 replicates). A rectangle indicates the taxa belonging to Plagiorchiata, as defined in this study.

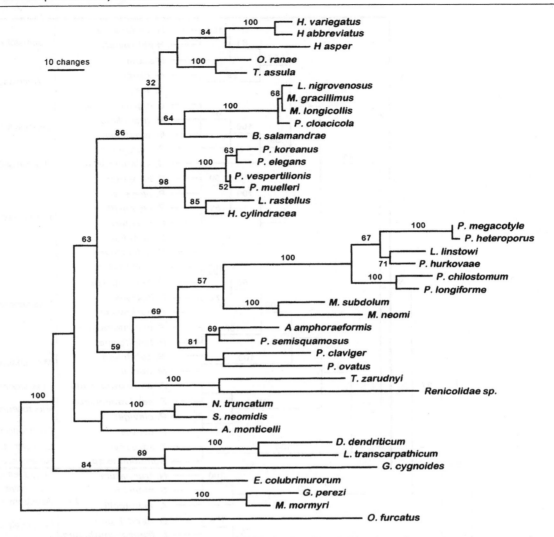

Figure 17.2 Strict consensus of two most-parsimonious trees (L = 1111, CI = 0.4851, RI = 0.6521) obtained in maximum parsimony analysis with the step-wise addition option and 10 replicates, with a reduced set of taxa. Numbers above branches are bootstrap values (> 50%).

Tkach *et al.* (2000). Some loss of resolution at the suprafamilial level can be explained by addition of some long-branch taxa (two more representatives of Opecoelidae, species of Encyclometridae, Renicolidae, Eucotylidae, etc.) into the analysis.

Similar to the results of Tkach *et al.* (2000), the Opecoelidae appear in the tree basal to representatives of two families (Troglotrematidae and Nanophyetidae) belonging to suborder Troglotrematata as well as to all other taxa included by Brooks *et al.* (1985a, 1989) in Plagiorchiata.

The families Monorchiidae and Opecoelidae appeared in a 73% supported clade; *Monorchis* basal to the three species of Opecoelidae (Figure 17.1). This position of Monorchiidae, although relatively well supported in our tree, should be evaluated with care. It cannot be excluded that inserting some other related taxa of fish digeneans may change this topology. It should be mentioned that in the study of Brooks *et al.* (1985a), Monorchiidae were listed among Digenea *incertae sedis* and in the phylogeny by Brooks *et al.* (1989) this family was placed far from opecoeliids in a different suborder, Hemiuriformes.

Three superfamilies, namely the Opecoeloidea, Gorgoderoidea and Dicrocoelioidea, out of six to seven recognized

by Brooks *et al.* (1985a, 1989) in the Plagiorchiata, appeared in our trees outside the main cluster of plagiorchiates (Figures 17.1 and 17.2). As has been stated by Tkach *et al.* (2000), upon the exclusion of Opecoeloidea and Gorgoderoidea from Plagiorchiata the suborder includes almost only digeneans parasitic as adults in terrestrial vertebrates, except for a very few species of Microphallidae and several genera of Macroderoididae whose phylogenetic relationships and systematic position are as yet uncertain and have to be studied. However, the families Prosthogonimidae, Lecithodendriidae and Microphallidae which Brooks *et al.* (1989) also included in Opecoelatea, appear within one of the subclades of Plagiorchiata in the trees obtained from DNA sequences (Figures 17.1 and 17.2).

It is particularly interesting that *Encyclometra colubrimurorum*, a representative of Encyclometridae, is situated in a strongly supported (84%) clade with Gorgoderidae and Dicrocoeliidae (Figure 17.2). The phylogenetic relationships of Encyclometridae have always been unclear. Odening (1960) provisionally placed Encyclometridae in the suborder Plagiorchiata, but in his later work (Odening 1968) changed his opinion suggesting that they belong to Opisthorchiata on the basis of the protonephridial system organization. Yamaguti (1971) considered them as a subfamily within the Plagiorchiidae, while Sharpilo and Iskova (1989) did not include the Encyclometridae in the Plagiorchiata. This latter viewpoint is now supported by the molecular data.

The troglotrematates, which are represented in our data by *Nephrotrema truncatum* (Troglotrematidae) and *Skrjabinophyetus neomidis* (Nanophyetidae), differ from plagiorchiates primarily by having a redial generation in their life cycles. The viewpoint of Yamaguti (1971) that the Nanophyetidae are more closely related to the Plagiorchioidea than to the Troglotrematoidea, is thus not supported by our data.

The position of *Astiotrema monticelli* outside the Plagiorchiidae and Plagiorchiata is of special interest. A majority of the authors (for instance, Yamaguti 1971; Sharpilo and Iskova 1989) included *Astiotrema* in the Plagiorchiidae as a type-genus of the subfamily Astiotrematinae. The morphology of *A. monticelli* differs from other representatives of the genus occurring in Europe (e.g., *A. emydis*, *A. trituri*) by short caeca reaching to only about the middle of the body, short vitelline fields and some other characters. There is some evidence from the morphology of adults and cercariae suggesting that *A. trituri*, the only species known from amphibian hosts, is not closely related to other representatives of the genus. Regarding this, Prudhoe and Bray (1982: 117) wrote: '... in the concept presented by Yamaguti (1971) *Astiotrema* would appear to represent a composite group of genera, not yet recognised, and is in need of a very careful revision based on an understanding of the systematics of the family Plagiorchiidae. Even a superficial comparison of the descriptions of *Astiotrema reniferum*, *A. monticelli* and *A. trituri* will throw some doubt on their being congeneric'. Thus, our results suggest two possiblities: 1) *Astiotrema* does not belong to the Plagiorchiidae; and 2) this genus is polyphyletic and some of its members may belong not only to different genera, but even to different families. Inclusion of *Astiotrema trituri* and some *Astiotrema* from turtles in future studies would provide necessary information to clarify this systematic problem.

The close phylogenetic relationships between *Astiotrema* and Troglotrematata (Figures 17.1 and 17.2) seem surprising because of differences in their morphology and life cycles. The topology of these trees indicates, rather, the necessity to look for other closely related taxa in order to clarify the phylogenetic affinities of *A. monticelli*.

The remaining taxa form a large clade, rather weakly supported in the large tree (Figure 17.1) and with 63% bootstrap support in a tree obtained upon exclusion of basal digenean taxa (Figure 17.2). The bootstrap support for this clade in the present analysis is lower than in the analysis of Tkach *et al.* (2000). As mentioned above, this may be a result of the inclusion of several new long-branch taxa (Figure 17.2). This clade may correspond, in our opinion, to the suborder Plagiorchiata, being separated from other taxa by the clade of the Troglotrematata (Troglotrematidae + Nanophyetidae). The troglotrematates differ from plagiorchiates in such important biological features as the possession of a redial generation in their life cycles (daughter sporocysts in all plagiorchiates) and miracidia hatching in the external environment (miracidia of plagiorchiates hatch in the intestine of a first intermediate host). Although support for the Troglotrematata being a closest basal taxon to Plagiorchiata, is weak (Figures 17.1 and 17.2), it should be noted that this position was stable in different analyses. The taxa comprising the next closest branch (Dicrocoeliidae, Gorgoderidae and Encyclometridae), also differ from Plagiorchiata by certain morphological/life-cycle features. Dicrocoeliidae, unlike all plagiorchiates, are characterized by the ovary being situated posterior to the testes and by the absence

of the gut in cercariae. Gorgoderidae are different in many characters such as absence of the pharynx and cirrus-sac (only external seminal vesicle present) in adults, eggs hatching in the external environment, cystocercous cercariae, etc. Encyclometridae, as demonstrated by Odening (1968), possess a protonephridial system, very different from that in Plagiorchiata. In addition, a vast majority of plagiorchiates have a Y- or V-shaped excretory vesicle, while in the Troglotrematata, Dicrocoeliidae and Gorgoderidae it is I-shaped. This latter character, however, may be considered rather as a trend because the estimation of the excretory vesicle shape depends on the life-cycle stage, physiological state and age of digeneans and is not consistent in the literature.

In our analyses of the present set of digenean taxa the Plagiorchiata always appeared as a highly derived group which is generally in agreement with the results of Brooks *et al.* (1985a, 1989) and Brooks and McLennan (1993a) who used morphological/life-cycle characters. However, it is possible that this situation may change with the addition of other digenean groups. The relatively weak support for several branches (Figure 17.1) also emphasizes the uncertainty of some of the relationships, especially at higher taxonomical levels.

The subtree of Plagiorchiata consists of two large clusters (Figure 17.2). One of them is strongly supported (bootstrap values 86%) and includes Plagiorchiidae, Leptophallidae, Haematoloechidae, Telorchiidae, and Brachycoeliidae, while the other has only 59% support and includes Renicolidae, Eucotylidae, Lecithodendriidae, Microphallidae, Prosthogonimidae, Pleurogenidae and Allassogonoporidae. The first cluster corresponds to the superfamily Plagiorchioidea, and the second includes taxa of the Microphalloidea plus the strongly supported long branch of Renicolidae/Eucotylidae. It is probable that the latter branch represents an independent superfamily-level lineage of Plagiorchiata. More taxa should be studied to better resolve these relationships. It should be noted, however, that Bayssade-Dufour *et al.* (1993) on the basis of their comparative studies of cercarial chaetotaxy suggested close phylogenetic relationships of Lecithodendriidae, Microphallidae and Renicolidae and assumed that these families have a common origin and may be grouped together in the superfamily Microphalloidea. On the other hand, these authors (see also Bayssade-Dufour 1979) demonstrated distinct differences in cercarial chaetotaxy between Microphalloidea and Plagiorchioidea. These views are well supported by our molecular data.

Microphalloidea + Renicolidae/Eucotylidae

The most important point revealed by the present analysis in comparison with the results of Tkach *et al.* (2000) is the position of the family Renicolidae which has been considered by Brooks *et al.* (1989) as a sister-group of Plagiorchiata. According to our data, Renicolidae belongs to the Plagiorchiata being related to the members of the Microphalloidea.

The sister-group relationship of the Renicolidae with Eucotylidae deserves more detailed consideration, because the phylogenetic affinities and systematic position of Eucotylidae have always been uncertain; for instance, Brooks *et al.* (1985a) listed this family among Digenea *incertae sedis*. Its position as a sister-group of the Renicolidae is somewhat surprising because these taxa have not been considered as closely related in the literature. However, this relationship is strongly supported by bootstrap values and could not be changed by varying analysis conditions or choice of taxa. It should be mentioned that both renicolids and eucotylids are parasites of bird kidneys and share many morphological features; thus, their divergence from a common ancestor is conceivable. According to the diagnoses of Yamaguti (1971), the Tanaisiinae (represented by *Tamerlania zarudnyi* in our data) have even more morphological similarities with the Renicolidae than with the Eucotylinae; these include absence of the cirrus-sac (present in Eucotylinae) and presence of seminal receptacle (absent in the Eucotylinae). Therefore, it would be interesting to add the representatives of the Eucotylinae into analyses to test the monophyly of the Eucotylidae and clarify their relationships with the Renicolidae.

Microphalloidea

The rest of the families in this clade have been considered as closely related by most authors, and are grouped within the superfamilies Prosthogonimoidea and Microphalloidea in the system of Odening (1964) or Lecithodendrioidea and Microphalloidea in the system of Brooks *et al.* (1985a, 1989). However, Sharpilo and Iskova (1989) did not include Microphallidae in the Plagiorchiata.

The distinct separation of the Lecithodendriidae and Pleurogenidae by molecular data supports the systematic arrangement proposed by Odening (1959), who withdrew the subfamily Pleurogeninae from the Lecithodendriidae and raised its status to the family level. Odening's viewpoint has been accepted by some authors (e.g., Sharpilo and Iskova 1989), but rejected by others (Yamaguti 1971; Prudhoe and Bray 1982). In our trees (Figures 17.1 and 17.2) three genera of lecithodendriids always formed a strongly supported (100% bootstrap values) monophyletic group, showing the close phylogenetic relationships of *Lecithodendrium* and *Pycnoporus*. *Prosthodendrium hurkovaae* which has been previously allocated to either *Prosthodendrium* or *Pycnoporus*, appears in the molecular tree together with *Lecithodendrium* which may indicate a need for a re-evaluation of the morphological characters used for generic differentiation in this group.

The position of *Allassogonoporus* as a sister-group of *Parabascus* is somewhat unexpected from the viewpoint of some traditional systematic schemes. This species was included by Skarbilovich (1948) in the subfamily Allassogonoporinae, family Lecithodendriidae. Later, Odening (1964) raised the status of this subfamily to the family level because of obvious morphological differences between the two groups. Yamaguti (1971) placed Allassogonoporinae back in Lecithodendriidae, but in the monograph of Sharpilo and Iskova (1989), the Allassogonoporidae was again considered as a separate family. Meanwhile, representatives of the pleurogenid subfamily Parabascinae, especially *Parabascus*, demonstrate the highest morphological similarity to *Allassogonoporus*; they differ only by presence of the cirrus-sac in *Parabascus* instead of a seminal vesicle lying free in the parenchyma in Allassogonoporidae. Based on the morphology of the adult digeneans and results of molecular study, *Allassogonoporus* and other members of Allassogonoporidae should be allocated to the family Pleurogenidae. Interestingly, Yamaguti (1958) already placed *Allassogonoporus amphoraeformis* (as *Moedlingeria amphoraeformis*) into the subfamily Parabascinae, but later (Yamaguti 1971) changed his opinion and relocated this species into *Allassogonoporus*, Allassogonoporidae.

One more genus from bats, *Ophiosacculus*, is also characterized by the presence of a long seminal vesicle lying freely in the parenchyma and, due to this feature, has been included by Sharpilo and Iskova (1989) in the Allassogonoporidae. However, the difference in the position of the genital pore (median in *Ophiosacculus*, marginal in representatives of Allassogonoporidae) suggests that this systematic arrangement should be considered with caution.

Although the trees obtained by the current analysis suggest that *Parabascus* and *Allassogonoporus* may belong to a group different from *Pleurogenes*, the type-genus of Pleurogenidae, which appears in a common clade with *Prosthogonimus* (Figures 17.1 and 17.2), this topology is rather weakly supported and may be a result of long-branch attraction. When representatives of Renicolidae and Eucotylidae are withdrawn from the analysis (also see Tkach *et al.* 2000) *Allassogonoporus* + *Parabascus* appeared as a subclade of Pleurogenidae. Moreover, after addition of other representatives of Pleurogenidae (genera *Prosotocus, Pleurogenoides*) into the analysis, the four pleurogenids and *Allassogonoporus* always formed a monophyletic clade with *Prosthogonimus* situated basal to them. Further studies with more pleurogenid taxa should better resolve these relationships.

Plagiorchioidea

Five strongly supported clades are found in this group with branch topology identical in all trees obtained from larger or smaller taxa sets and different analyses. These clades were also present in the trees obtained by Tkach *et al.* (2000), and correspond to the families Brachycoeliidae, Telorchiidae, Plagiorchiidae, Haematoloechidae and Leptophallidae (the former subfamily Leptophallinae with rank raised to family). Content and relationships of these groups was discussed by Tkach *et al.* (2000), so we will only briefly mention here the main points of interest.

The molecular analysis strongly supports the status of Haematoloechidae as a separate family (Figures 17.1 and 17.2). In the trees, *H. asper* is basal to the closely related species, *H. variegatus* and *H. abbreviatus*.

The species *Lecithopyge rastellus* has been considered by Prudhoe and Bray (1982) and some other authors as a member of *Dolichosaccus*, subfamily Opisthioglyphinae (belonging to Telorchiidae according to Prudhoe and Bray (1982) and to the Plagiorchiidae according to a majority of other authors). However, by molecular analysis *L. rastellus* is not closely related to *Opisthioglyphe* (Figure 17.2; for more detailed discussion see Tkach *et al.* 2000). The relationships of *L. rastellus* with the typical *Dolichosaccus* from Australian anuran amphibians are still obscure and need to be studied.

The position of *Opisthioglyphe ranae* in the 100% supported common clade with *Telorchis* is of great systematic interest. Different authors included the genus *Opisthioglyphe* in the Omphalometridae (Yamaguti 1971) or in Plagiorchiidae (Sharpilo and Iskova 1989; Bayssade-Dufour and Grabda-Kazubska 1993) as a separate subfamily Opisthioglyphinae. However, Prudhoe and Bray (1982) recognized the close affinities between *Telorchis* and *Opisthioglyphe* by placing them in the subfamilies Telorchiinae and Opisthioglyphinae of the Telorchiidae. Although not accepted by some later authors (Sharpilo and Iskova 1989), this systematic scheme is also supported by the close similarity of cercarial chaetotaxy and morphology in *Telorchis* and *Opisthioglyphe* (Dimitrov *et al.* 1989; Grabda-Kazubska and Moczon 1990; Grabda-Kazubska and Lis 1993). The result of the molecular analysis is consistent with the viewpoint of Prudhoe and Bray (1982) on this matter.

The systematic position of Brachycoeliidae has long been questioned. Odening (1964) placed this family in Plagiorchiata, but his views were not accepted by several other authors (Yamaguti 1971; Brooks *et al.* 1985a, 1989; Sharpilo and Iskova 1989). In our analysis, however, the position of *Brachycoelium salamandrae* within the Plagiorchioidea is well supported (Figures 17.1 and 17.2).

The relationships of the genera *Macrodera*, *Leptophallus*, *Metaleptophallus* and *Paralepoderma* which appear in the molecular tree as a very strongly supported clade, have been discussed in detail by Tkach *et al.* (1999). Convincing evidence coming from the adult and cercarial morphology has been found (Tkach *et al.* 1999) to support this initially unexpected result of molecular study.

Conclusions

Firstly, according to the results of our study, the utilized fragment at the 5′ end of 28S rDNA is a good target for the studies of the digenean phylogenetic relationships at levels from genus to superfamily, but gives generally lower resolution at higher taxonomic levels.

Secondly, the possible systematic implications of our molecular study can be summarized, as follows:

- At higher taxonomic level, the most basal group among the digenean taxa present in our data set, is family Schistosomatidae. Lineages of Diplostomidae/Strigeidae and Brachylaimidae/Leucochloridiidae are most basal of the remaining taxa. Plagiorchiata appear as the most derived large group in the current data set. A relatively weak support for some branches at higher level should be, however, noted and changes of some topologies may be expected with the expansion of the data set.

- At the subordinal and family levels, the suborder Plagiorchiata as considered in previous phylogenies (Brooks *et al.* 1985a, 1989) is polyphyletic according to our results. Molecular data also suggest placement of the families Gorgoderidae, Dicrocoeliidae and Encyclometridae outside the Plagiorchiata. The representatives of the Renicolidae, considered to be one of the sister-groups of the Plagiorchiata, were clustered within one of the two main clades of Plagiorchiata, together with the superfamily Microphalloidea.

- At familial and generic levels this study indicated a variety of noteworthy issues. With the present set of examined taxa, Plagiorchiata consists of two to three main lineages each containing several families. The addition of several important taxa in future studies may increase the number of suprafamilial groupings. The family Eucotylidae has proved to be closely related to Renicolidae, both including parasites of bird kidneys. The representative of *Astiotrema*, traditionally included in Plagiorchiidae, appeared outside this family and outside the whole clade of Plagiorchiata. Molecular data placed the Brachycoeliidae into the Plagiorchioidea and demonstrated close phylogenetic affin-

ities between *Macrodera* and representatives of the Leptophallinae. *Opisthioglyphe* and *Telorchis* form a separate, strongly supported clade within the Plagiorchioidea. The Pleurogenidae and Allassogonoporidae do not belong to the Lecithodendriidae, as was suggested by Yamaguti (1971) and some other authors.

Thirdly, the present molecular phylogeny, despite its preliminary nature, clearly indicates a need for a systematic review of several families of Plagiorchiata such as Plagiorchiidae, Lecithodendriidae, Telorchiidae, Pleurogenidae and others. In some of the above-mentioned cases, when the molecular results seemed to be contradictory with current systematic views, strong evidence from the morphology and/or life-cycle was found to support the molecular phylogeny.

However, the aim of our work was to reveal phylogenetic relationships using molecular characters rather than to establish taxonomic status or rank of the taxa considered. The general outline and phylogeny of Plagiorchiata presented in our study, may be considered as a basis for future more detailed phylogenetic analyses of families and genera in this taxonomically diverse group. Along with some strongly supported clades in the current molecular phylogeny, there are also some poorly resolved relationships. A majority of these questionable points can probably be resolved by future studies including additional taxa. Ideally, these studies will combine molecular data with the morphological/life-cycle information to achieve a better understanding of this group of digeneans.

ACKNOWLEDGEMENTS

The authors thank José Fahrni (University of Geneva, Switzerland) for his valuable assistance in running the automated sequencer, Olivier Jousson (University of Geneva), Oksana Grebin', Vadim Kornyushin, Yuri Kuzmin, Alexander Fedorchenko (all at the Institute of Zoology, Kiev, Ukraine), Teresa Pojmanska (Institute of Parasitology, Warsaw, Poland) and Scott Snyder (University of Wisconsin, Oshkosh, USA) for providing specimens for our study and/or help in field collecting of material. Nedezhda Iskova (Institute of Zoology, Kiev) kindly helped to determine some digeneans from birds. We would like to thank Colomban de Vargas, Juan Montoya, José Fahrni (all at the University of Geneva), Timothy Littlewood and Peter Olson (The Natural History Museum, London), for their methodological assistance and useful discussions of the subjects considered in this study. Rodney Bray (The Natural History Museum, London) made useful remarks and kindly helped to improve the English in the manuscript. The study was supported financially by the Swiss National Science Foundation Programme of Cooperation in Science and Research in Central and Eastern Europe Countries grant 7IP051825, and by the Polish State Research Committee grant 6 P04C009 17. The completion of the analysis and preparation of the manuscript for publication was supported by a Royal Society fellowship at The Natural History Museum, London for V.V.T.

Chapter 18

The Schistosomatidae: Advances in phylogenetics and genomics

Scott D. Snyder, Eric S. Loker, David A. Johnston and David Rollinson

From an evolutionary viewpoint, the Schistosomatidae is a fascinating group within the Digenea due to its two-host life-cycle, the presence of separate sexes, and the colonization of the vasculature of birds, mammals and crocodiles. Together with the hermaphrodite families the Sanguinicolidae (parasites of fishes) and the Spirorchiidae (parasites of turtles), they collectively represent the 'blood-flukes', and gain entry to their definitive hosts by direct penetration of the body surface. The family contains the genus *Schistosoma*, the most intensively studied of trematode parasites, well known for the widespread diseases they cause in man and domestic animals. Thirteen genera are currently recognized in the family (Table 18.1), comprising around 85 species (Basch 1991). However, the relationships between *Schistosoma* and the 12 remaining genera of schistosomatids have received scant attention. The genera differ markedly in the degree of sexual dimorphism shown (Carmichael 1984; Basch 1991; Morand and Müller-Graf 2000). Most noticeably, some genera possess males which lack a gynaecophoric canal (e.g., *Dendritobilharzia*), whereas in the majority of other genera the canal is well developed, with the larger, muscular male grasping the more slender female. Striking differences in the number of testes in the adult male worms are also apparent; for example, in the case of *Trichobilharzia* there can be more than 500 testes, whereas two to seven occur in *Schistosoma*. Recently, Morand and Müller-Graff (2000) conducted a comparative analysis of genera within the family. They suggested that there is a trade-off between investment in muscle (in relation to sexual dimorphism) and number of testes (other variables being controlled for).

This chapter considers existing morphological and current molecular phylogenies, and examines possible mechanisms involved in the speciation and evolution of the Schistosomatidae. There is no doubt that biogeography and host capture have had profound influences on evolution within this family. Examination of evolutionary patterns and processes may help to reveal the origin and radiation of the medically important genus *Schistosoma*. Within this genus, molecular studies continue to elucidate the relationships between species and the significance of inter-'specific' reproductive pairing (Rollinson *et al*. 1997). Considerable information concerning the nuclear and mitochondrial genomes of *Schistosoma* sp. is now being generated by international efforts, and here we summarize recent data acquisition and highlight some of the opportunities genome research will provide for future comparative studies within the Platyhelminthes.

Relationships of genera within the family

The substantial social and economic impact of schistosomes has created a great deal of interest in the biology and control of these parasites, but until recently little attention has been given to schistosome evolution. Most studies have focused on members of the genus *Schistosoma* (Rollinson *et al*. 1997), leaving the relationship of these parasites to the other 12 genera

in the family open to speculation. In 1984, Carmichael used 24 morphological characters to divide the family into three clades (Figure 18.1). This PhD thesis provided an intriguing look into schistosome relationships, but suffered a little from a lack of close examination of characters in rare specimens and difficulties in polarizing ambiguous and multi-state characters. Recently, Morand and Müller-Graf (2000) have incorporated Carmichael's data set into a more detailed morphological analysis of the family, although their phylogeny has not been included in the present synopsis.

The most comprehensive molecular phylogeny of the Schistosomatidae to date includes 10 of the 13 genera within the family (Snyder and Loker 2000). In this study, analysis of approximately 1100 bases of the large subunit ribosomal RNA gene revealed two clades, one comprising the genera *Schistosoma* and *Orientobilharzia*, with the other containing the eight remaining genera (Figure 18.2).

Although substantial differences exist between this molecular phylogeny and the morphological phylogeny of Carmichael (1984), several similarities in tree topology are worthy of mention. The clade formed by the genera *Bilharziella*, *Trichobilharzia*, *Gigantobilharzia* and *Dendritobilharzia* is identical in both the analyses (Figures 18.1 and 18.2). All four taxa parasitize birds and pulmonate snails, are distributed globally, and display reduced sexual dimorphism when compared to other members of the family. Species in the genus *Bilharziella* have a reduced gynaecophoric canal, whereas the gynaecophoric canals of *Trichobilharzia* and *Gigantobilharzia* are either reduced or absent. Members of these three genera do display some size difference between males and females, whilst there is no sex-specific size difference in *Dendritobilharzia*. Similarities among these four genera have been long recognized, and they have been considered as the sole representatives of the subfamily Bilharziellinae (Price 1929).

In addition to a lack of sexual dimorphism, the genus *Dendritobilharzia* is highly unusual in that it inhabits the arterial system of its definitive host, a habitat known to be exploited by only one other schistosome, the mammalian blood-fluke *Schistosoma hippopotami* (McCully *et al*. 1965). Platt and Brooks (1997) presented a hypothesis in which *Dendritobilharzia* was basal to all other members of the Schistosomatidae, their arterial habitat being considered as an evolutionary remnant of close affiliation to the arterial-dwelling turtle blood-flukes of the family Spirorchiidae. According to this hypothesis, *Dendritobilharzia* species remained arterial parasites as the lineage ancestral to all other members of the Schistosomatidae invaded the venous circulation and diversified. The aforementioned phylogenetic analyses refute this hypothesis and indicate that the genus *Dendritobilharzia* is among the most derived schistosomes, the lineage colonizing the arterial system independently of other blood-flukes (Figures 18.1 and 18.2).

Orientobilharzia and *Schistosoma* are recognized as closely related taxa in both the morphological and molecular phylogenetic analyses (Figures 18.1 and 18.2). Carmichael (1984) considered only generic-level relationships among taxa,

Table 18.1 Geographical ranges of major taxa comprising the Schistosomatidae.

Taxon	Geographical range
Austrobilharzia	Cosmopolitan
Bilharziella	Cosmopolitan
Dendritobilharzia	Cosmopolitan
Gigantobilharzia	Cosmopolitan
Macrobilharzia	Cosmopolitan
Ornithobilharzia	Cosmopolitan
Trichobilharzia	Cosmopolitan
Bivitellobilharzia	Africa and Asia
Griphobilharzia	Australia
Heterobilharzia	South-eastern United States
Orientobilharzia	Asia
Schistosoma haematobium	Africa
Schistosoma mansoni	Africa
Schistosoma japonicum	Asia
Schistosomatium	Northern North America

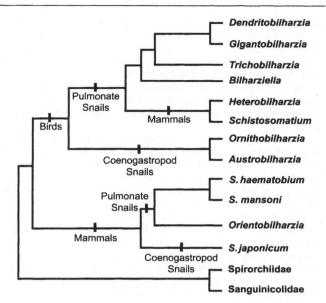

Figure 18.2 Phylogeny of the Schistosomatidae based on molecular characters (after Snyder and Loker 2000). Patterns of intermediate and definitive host use are mapped on the tree, and provide equivocal evidence for the ancestral molluscan and vertebrate hosts of the Schistosomatidae. Patterns of host utilization demonstrate that host switching has played an important role in the evolution of schistosomes.

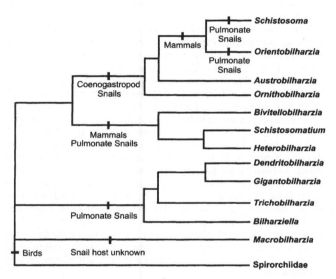

Figure 18.1 Phylogeny of the Schistosomatidae based on morphological characters (after Carmichael 1984). Patterns of intermediate and definitive host use are mapped on the tree, and argue for an origin of the Schistosomatidae in birds. Evidence for the ancestral snail host of schistosomes is equivocal.

whereas Snyder and Loker (2000) incorporated three species of the genus *Schistosoma* in their analysis. By including representatives of the three major *Schistosoma* lineages (the African terminal-spined and lateral-spined groups and the Asian *Schistosoma japonicum* complex), the latter authors revealed *Schistosoma* to be paraphyletic, with *Orientobilharzia* more closely related to the African *Schistosoma* species than is *S. japonicum*. This finding suggests that both *Orientobilharzia* and *Schistosoma* should undergo closer taxonomic examination.

Two members of the genus *Orientobilharzia* were originally described as *S. bomfordi* (by Montgomery 1906) and *Schistosoma turkestanicum* (by Skrjabin 1913). Price (1929) used the large number of testes and the condition of the ovaries to remove these parasites from *Schistosoma* and include them in the genus *Ornithobilharzia*, a genus that previously included only avian schistosomes. Mammalian schistosomes were later removed from the genus *Ornithobilharzia* and placed in the

new genus *Orientobilharzia* by Dutt and Srivastava (1955). These parasites were not replaced into the genus *Schistosoma* due to the larger number of testes in members of *Orientobilharzia*, but were considered to be most closely related to *Schistosoma* (Dutt and Srivastava 1961). The recent molecular phylogeny (Snyder and Loker 2000) supports the break up of *Ornithobilharzia* into genera comprised of avian and mammalian parasites, but suggests that the make-up of the genus *Schistosoma* should be altered to include taxa now in the genus *Orientobilharzia*. Alternatively *Schistosoma* might well be split into multiple genera to avoid the apparent paraphyly of the genus. LeRoux (1958) suggested such a split, proposing at least six genera to encompass the species now included in *Schistosoma* and *Orientobilharzia*.

The position of *S. japonicum* and *Orientobilharzia turkestanicum* as basal to the African *Schistosoma* has important implications for our understanding of the evolution of the genus *Schistosoma* (including *Orientobilharzia*). Davis (1980, 1992) made the case that *Schistosoma* originated in Gondwanaland and drifted on the Indian Plate to reach and colonize Asia within the past 50 million years. If this were so, the extant African lineages should be expected to reflect their Gondwanan origin by occupying a basal position in the *Schistosoma* clade, a view not supported by the molecular data (Snyder and Loker 2000). Both the *Schistosoma japonicum* species complex and all members of the genus *Orientobilharzia* are Asian in distribution. Thus, an Asian origin of the genus *Schistosoma*, with a subsequent colonization of Africa, becomes the most plausible interpretation of the molecular phylogenetic data.

Both morphological and molecular phylogenies concur that the mammalian flukes *Schistosomatium douthitti* and *Heterobilharzia americana* are sister-taxa (Figures 18.1 and 18.2). Both are monotypic genera parasitic in lymnaeid snails and are restricted in distribution to North America. The relationship of these two taxa to other members of the family differs widely between the two studies. Carmichael (1984) considered the clade comprised of *Schistosomatium* and *Hetero-bilharzia* to be the sister-group to *Bivitellobilharzia*, a genus encompassing two species that parasitize African and Asian elephants. Snyder and Loker (2000) were unable to obtain specimens of *Bivitellobilharzia* for their analysis, but considered *Schistosomatium* and *Heterobilharzia* to be members of a larger clade otherwise comprised of avian parasites.

Numerical measures of support for the placement of *H. americana* and *S. douthitti* in the otherwise avian clade were not strong (Snyder and Loker 2000). However, the authors noted the presence of eye-spots in the cercariae of these two species. All other members of the avian clade are known or thought to have eye-spots. Cercariae of *Ornithobilharzia* have not yet been described in detail. Penner and Wagner (1956) indicate that *Ornithobilharzia* cercariae are more closely related to members of the genus *Austrobilharzia* (where cercariae possess eye-spots), although this source makes no specific reference to the presence/absence of eye-spots. In contrast, members of the genera *Orientobilharzia* and *Schistosoma* do not have cercarial eye-spots. This morphological character was thought to lend independent support to the inclusion of *Heterobilharzia* and *Schistosomatium* in the avian clade.

Both cercarial and adult morphology have been used to argue for the close relationship among *Bivitellobilharzia* (not yet available for molecular analysis), *Orientobilharzia* and *Schistosoma* (Dutt and Srivastava 1961). The historical restriction in the distribution of these three taxa to Africa and Asia could also support this view. Inclusion of *Heterobilharzia* and *Schistosomatium* in a clade distinct from the other mammalian schistosomes indicates that the parasitism of mammals has arisen at least twice in this family. The restriction of these two genera to North America, and the apparent lack of similar parasites in the intensively surveyed mammals of other continents, indicates that this lineage arose after the break-up of Laurasia. In this view, a schistosome of birds captured North American mammalian hosts and gave rise to the evolutionary lineage that is today represented by *Heterobilharzia* and *Schistosomatium* (Snyder and Loker 2000).

Carmichael (1984) found the clade comprised of the avian parasites *Bilharziella*, *Trichobilharzia*, *Gigantobilharzia* and *Dendritobilharzia*, along with the avian schistosome *Macrobilharzia*, to be basal to the entire family. This topology would indicate that birds were the ancestral hosts of schistosomes and that mammals have been colonized more recently in evolution. Available molecular data neither support, nor contradict, this conclusion; providing a tree topology that gives no indication of the ancestral vertebrate host of schistosomes. In this view, the aforementioned clade of four bird schistosomes (exclusive of *Macrobilharzia*, which was not available for molecular analysis) is highly derived within the Schistosomatidae.

The molecular analysis of schistosome phylogeny (Snyder and Loker 2000) also provides equivocal information concerning the ancestral schistosome molluscan host. *Austrobilharzia* and *Ornithobilharzia* comprise avian parasites that develop in marine coenogastropod (prosobranch) snails. The clade giving rise to these taxa occupies a basal position in the primarily avian clade (Figure 18.2). Similarly, *S. japonicum* occupies a basal position in the mammalian clade and is restricted to development in amphibious fresh water coenogastropod snails. However, if all intermediate host information is optimized onto this tree, it is equally plausible that ancestral schistosomes were originally parasites of pulmonate as they were of coenogastropod snails. The work of Carmichael (1984) argues for an ancestral pulmonate host for the Schistosomatidae.

Future phylogenetic studies within the Schistosomatidae

Lack of clarity from the molecular study concerning the ancestral hosts of the Schistosomatidae, and the need to assess characters based on partial specimens and literature descriptions in the morphological work, indicates that considerable detail of the phylogeny and evolution of the Schistosomatidae remains to be elucidated. Of primary importance in these future investigations will be the inclusion of the only reported schistosome from a cold-blooded definitive host, the crocodilian parasite *Griphobilharzia amoena* (Platt *et al.* 1991), which is of pivotal importance in determining the early evolutionary history of the Schistosomatidae, Spirorchiidae and Sanguini-colidae. Recent molecular studies provide two competing hypotheses about the relationships of crocodiles, birds and turtles to one another. In the first, crocodiles and birds form a clade that is the sister-taxon to the turtles (Zardoya and Meyer 1998). In the second, crocodiles and turtles form a clade that is the sister to birds (Hedges and Poling 1999). In both studies the crocodilian, chelonian and avian clade is the sister-taxon to the mammals.

The close evolutionary affinity of turtles, crocodiles and birds is intriguing, especially when coupled with evidence that schistosomes and the turtle-dwelling Spirorchiids are sister-taxa in the Digenea (Brooks *et al.* 1985; Cribb *et al.* 2001, this volume). This leads to an expectation that mammalian schistosomes would be more derived/distant than crocodilian and avian schistosomes. Inclusion of a crocodilian schistosome in a comprehensive phylogeny would provide considerable information to test the hypothesis erected by Platt *et al.* (1991) that schistosomes arose in the cold-blooded crocodilians before acquiring homeothermic bird and mammal hosts. Viewing host–parasite relationships in a coevolutionary context, evidence might also provide support to one of the competing hypotheses of crocodile, turtle and bird relationships. Moreover, circumstantial evidence links *G. amoena* to a life-cycle involving pulmonate snails (Platt *et al.* 1991), making it conceivable that a greater knowledge of the life-cycle and phylogenetic position of this unusual parasite might shed light on the evolution of intermediate host utilization among schistosomes.

Considerable information about the evolution of mammalian blood-flukes might be gained by the inclusion of the elephant schistosome, *Bivitellobilharzia*, in future analyses. If, as proposed above, *Bivitellobilharzia* is more closely related to *Orientobilharzia* and *Schistosoma* than to *Heterobilharzia* and *Schistosomatium*, the phylogenetic position of this genus may help to indicate the ancestral molluscan hosts of African and Asian mammalian schistosomes. *Bivitellobilharzia* has been reported to develop in pulmonate snails (Vogel and Minning 1940). A position of this genus basal to *Schistosoma* and *Orientobilharzia* would indicate a pulmonate host origin for this group with subsequent host capture of coenogastropod snails by members of the *S. japonicum* complex. The relative phylogenetic position of the two species of *Bivitellobilharzia*, which occur respectively in Asia and Africa, would also be useful in testing the hypothesis that the genus *Schistosoma* arose in Asia and later colonized Africa (Snyder and Loker 2000).

Schistosoma species

Of the 19 species of *Schistosoma* currently recognized, five are known to infect man and are responsible for much human suffering. *Schistosoma* is the only genus of mammalian schistosomes to infect man. It is also the genus that shows the greatest diversity in terms of geographical distribution, hosts parasitized and species recognized. The species can be split into four distinct groups (*S. mansoni*, *S. haematobium*, *S. japonicum* and *S. indicum* species groups) based on the compatibility of the species with particular host genera, geographical distribution, as well as common egg morphologies. For the most part, molecular phylogenies based on ribosomal RNA genes, mitochondrial DNA and randomly amplified polymorphic DNA (RAPDs) agree with observations based on morphological or life-history characteristics (see Rollinson *et al.* 1997 for synopsis). It is clear that the parasites of man do not form a monophyletic group, and that close relationships exist between parasites within the species groups – especially in the *S. haematobium* group of species. In all analyses the *S. japonicum* group appears to be distantly related to the other species groups (Johnston *et al.* 1993; Littlewood and Johnston 1995; Barker and Blair 1996).

There are still species that have not been placed on the molecular phylogeny including three of the four currently placed in the *S. indicum* group. The one that has been studied, *S. spindale*, clusters closely with the *S. haematobium* group (Johnston *et al.* 1993; Littlewood and Johnston 1995). Després *et al.* (1995) obtained the ITS2 nucleotide sequence from a single worm of *S. hippopotami* and suggested that this species diverged very early, before the separation of the *Schistosoma* species with terminal spined eggs (*S. haematobium* group) from those species with lateral spined eggs (*S. mansoni* group). Further molecular analysis of this species is warranted. The other intriguing example that requires molecular analysis is *S. sinensium*, a species which has clear affinities with the *S. japonicum* group, but which has a prominent lateral spine to the egg (a characteristic of the *S. mansoni* species group).

Schistosome genome analysis: a new resource for studies on platyhelminth biology and phylogenetics

Given the wealth of data now being accumulated by the *Schistosoma* Genome Project (SGP), it is pertinent to consider how this may impact on future phylogenetic (and other) analyses within both the Schistosomatidae and the wider platyhelminth fauna.

It will be abundantly clear from the above discussion, and from other chapters in this volume, that sequence data (primarily from the ribosomal RNA gene complex) have made a major contribution to our understanding of the evolutionary radiation of the Platyhelminthes. Advances in molecular biology, especially in PCR and automated sequencing, mean that the generation of molecular sequence data has become a routine process and it is now possible to generate data sets covering numerous taxa, either to complement morphology/life history data through 'total evidence' approaches, to test hypotheses generated from the analysis of such data, or to supply character sets for taxa where more 'traditional' approaches cannot be used (through incomplete knowledge of life history, difficulties in collecting specimens from elusive, rare or protected hosts, etc.).

Since the first platyhelminth DNA sequence was deposited in the public databases in 1987 (Lanar *et al.* 1986), sequences have been generated from a very broad spectrum of platyhelminth taxa. In January 2000, the public databases contained over 16 500 platyhelminth sequences representing approxi-

mately 150 families, 375 genera and 650 species (http://www.ncbi.nlm.nih.gov/Taxonomy/taxonomyhome.html). However, as Table 18.2 clearly demonstrates, such simple, summary statistics give a misleading impression of the true breadth of our knowledge. If we exclude: a) sequences from the major seven taxa represented (of which *Schistosoma* is, by some 50-fold, the largest contributor); and b) sequences from approximately 15 genes that are commonly used for phylogenetic and population based studies, we are left with under 20 sequences representing other genes. Thus, whilst we can, with confidence, place numerous taxa onto molecular phylogenetic trees, our understanding of the molecular mechanisms underlying the biochemical, metabolic, behavioural, developmental and morphological adaptations responsible for this radiation is virtually non-existent. It is in this context that the SGP can provide valuable reference data for studies on the wider platyhelminth fauna. The SGP (see Williams and Johnston 1999 for synopsis) is one of five parasite genome projects initiated by the World Health Organization (WHO) in 1993 in recognition of the fact that the best prospects for identifying new drug targets and candidate vaccine antigens came from rational design strategies that were based on a detailed understanding of the molecular mechanisms underlying parasite biology, pathology, diversity, development and drug resistance (Johnston *et al.* 1999). Despite the obvious biomedical remit of the SGP, it should not be viewed simply in these terms, but rather as a set of biological and informatics resources with great potential for exploitation by the wider platyhelminth research community.

With a haploid genome size of 270 MB (approximately one-tenth the size of the human genome) containing an estimated 40–60% repetitive sequences and 20 000 expressed genes (Simpson *et al.* 1982), comprehensive genome analysis and full genome sequencing would be a massive undertaking, far beyond the financial resources currently available. Consequently, analysis has focused on goals that are achievable within the current funding framework; gene discovery, physical and chromosomal mapping and informatics.

Table 18.2 Categories of platyhelminth DNA sequences in the public databases (January 2000).

Category	Number of sequences
Total number of platyhelminth sequences	16 672
Sequences from:	
Schistosoma	14 631
Echinostoma	265
Echinococcus	210
Dugesia (= *Girardia*)	108
Fasciola	106
Paragonimus	93
Taenia	78
All other platyhelminth taxa,	1181
of which:	
rRNA genes and spacer regions	962
homeotic genes	57
cytochrome oxidase genes	53
elongation factor genes	22
transposable elements	17
actin/myosin/tubulin	12
spliced leader sequences	12
NADH dehydrogenase	12
cysteine protease genes	10
satellite sequences	6
all other sequences	18

Gene discovery

Before the start of the SGP, six years of molecular analysis had accumulated some 200 *Schistosoma* sequences in the public databases. Thus, vaccine and drug development programmes were being undertaken with knowledge of about 1% of the parasite's gene repertoire. A rapid increase in the size of the known gene catalogue was clearly the priority, and generation of expressed sequence tags (ESTs) provided the means to achieve this.

ESTs are short (100–600 nucleotide) sequences generated by single pass, automated DNA sequencing of the 5′ or 3′ end of randomly selected cDNA clones. Comparison of these sequences to the public databases reveals whether the clone represents an already identified *Schistosoma* gene, or a new addition to the catalogue. If a new EST has homology with another sequence on the database, a tentative identity or function can be assigned to the gene/clone from which it was derived. The annotated EST sequences are deposited in the databases and serve as tags to the source clones from which they were produced (which are available on request to anyone wishing to characterize them further). These clones and sequences represent a potentially valuable resource for the whole platyhelminth community, since if there is sufficient evolutionary conservation to allow the EST to be identified, there is likely to be enough to allow the *Schistosoma* sequence to be used as a probe, or as a source of PCR primers, to isolate the homologue from other helminths.

Production of ESTs from cDNA libraries representing different stages of the life-cycle allows the identification of different, stage-specific or stage-limited, transcripts (Santos *et al*. 1999). Where database homology allows the putative identification of ESTs, this can provide fascinating insights into parasite biology. For example, as described above, the presence or absence of cercarial eye-spots is an important character in schistosome systematics. *Schistosoma* cercariae do not possess eye-spots, but are known to respond to both shadow and increasing illumination (Saladin 1982), and ciliary bodies near the anterior end of the cercaria have been proposed as putative photoreceptors (Short and Gange 1975). Analysis of ESTs from cercarial cDNA libraries has revealed two sequences with homology to proteins involved in mammalian light perception mechanisms (Santos *et al*. 1999). In-situ hybridization of these sequences to cercariae would identify the anatomical location of expression, and detailed characterization could help elucidate the nature of helminth light-response mechanisms.

ESTs represent a very rapid and cost-effective means of discovering genes; as of January 2000, there were 12 500 *S. mansoni* and 1400 *S. japonicum* sequences, from diverse developmental stages, in DBEST, the EST database (http://www.ncbi.nlm.nih.gov/dbEST/dbEST_summary.html) and 14 600 *Schistosoma* sequences overall in the public domain. However, 14 600 sequences does not equate to 14 600 different genes; since clone selection for EST production is random, and highly expressed genes are represented many times within a cDNA library, clones representing the same gene may be picked again and again. Cluster analysis is used to determine the number of unique sequences present. At the most recent analysis (December 1999), 10 193 *S. mansoni* sequences formed some 5750 clusters, suggesting that at least 25% of *Schistosoma* genes had already been tagged. Redundancy in the data sets does not necessarily represent wasted effort. For example: 1) it allows patterns of gene expression to be mapped across the life-cycle and the identification of potentially stage-specific or stage-limited genes; 2) it reveals the most commonly generated sequences (amongst these are several which possess no database homology/known function; these are clearly of major importance to the parasite and so worthy of early attention; 3) it identifies families of closely related genes, permitting analysis of their evolutionary history; and 4) it locates polymorphic sites within sequences, which may suggest new genes (even those of unknown function), with value in population or wider phylogenetic studies.

Physical and chromosomal mapping

The physical map of a genome is an assembly of cloned genomic DNA fragments which together cover the entire genome with minimal overlap, and whose relative positions and orientation within the assembly are known. It represents a basic resource for the positional cloning of genes of interest, for the analysis of upstream and downstream regulatory elements and for the investigation of genome organization. Work is currently underway to construct a physical map of the *Schistosoma* genome. A large insert, genomic DNA library containing some 21 000 clones and eight-fold genome coverage, has been constructed for *S. mansoni*, using a bacterial artificial chromosome (BAC) vector (LePaslier *et al*. 2000). The map is generated by successive rounds of hybridization of BAC-derived probes onto high-density filters of the library, to identify overlapping clones and thus walk along the genome. Because the schistosome genome is so large, the mapping effort is initially focused on chromosome 3, to provide basic information and prove the feasibility of the approach. Although a large genome provides huge logistical problems from sequencing and mapping perspectives, it does also have advantages. The *Schistosoma* genome is sufficiently large that the chromosomes are clearly visible by light microscopy. Thus, the karyotype has been unambiguously determined (Short and Menzel 1979); *Schistosoma* possesses seven pairs of autosomes and one pair of sex chromosomes, the female being heterogametic (ZW), the male homogametic (ZZ). Size, shape, position of the centromere and C banding can differentiate the individual chromosomes, and there are subtle differences in chromosome shape and banding between strains and species (Grossman *et al*. 1981; Short *et al*. 1989).

Techniques are available to localize cloned DNA fragments onto the chromosomes by fluorescent in-situ hybridization (FISH) (Hirai and LoVerde 1995), and some 100 markers have been analysed to provide a first-generation chromosome map. These markers will provide essential landmarks to position and orientate the physical map. Additional markers are being mapped onto the karyotype, both to increase overall coverage and to verify the correct assembly of the physical map.

The utility of the physical and chromosome maps extends far beyond *Schistosoma* itself. First, karyotypes have been determined for numerous members of the Schistosomatidae and for some other Strigeata, and hypotheses have been erected to explain both the evolution of schistosome sex chromosomes from the autosomes of hermaphrodite ancestors, and the origins and pathways of karyotype variation within the family (see Short 1983 for synopsis). Once the physical map of the *S. mansoni* genome is completed, it will provide an essential blueprint for comparative analysis, allowing defined regions of the *Schistosoma* genome to be mapped to specific locations within the genomes of other species, thus testing these hypotheses. Second, schistosomes are virtually unique within the Platyhelminthes in possessing separate, morphologically distinct sexes, with sex known to be determined at the chromosome level. Other dioecious Platyhelminthes include specific genera within three subfamilies of the Didymozoidae (Digenea) (Anderson and Barker 1998), two small, and unrelated cestode families (the Dioecotaeniidae and Dioecocestidae) (Khalil *et al.* 1994), the symbiotic rhabdocoel *Kronborgia* (Christensen and Kanneworff 1964) and the tricladids *Sabussowia dioaca* and *Cercyra teissieri* (closely related but each with hermaphrodite congeners) (Sluys 1989b). In these other dioecious Platyhelminthes, the degree of sexual dimorphism varies greatly, and relatively little (if anything) is known about karyotype or mechanisms of sex determination. There is some evidence that the sex of dioecious didymozoids and cestodes may be determined by the presence of other individuals; the first worm to colonize a host becoming female, while later colonizers have their 'femininity inhibited' and thus become males (Ishii 1935; Schmidt and Roberts 1977). Vestigial traces of organs of the opposite sex can sometimes be found in didymozids, as indeed they can in *Schistosoma*, where the influence of one sex on the reproductive maturation of the other is very well documented (see Basch 1991 for review). The fact that separate sexes appear to have evolved, independently, at least eight times within the phylum argues for a relatively simple underlying mechanism. Since the Platyhelminthes are considered to be amongst the most ancestral bilateral metazoans (Salvini-Plawen 1978), with schistosomes amongst the earliest diverging lines (Littlewood *et al.* 1999a), a detailed analysis of the composition of schistosome sex chromosomes, and identification of the genes and regulatory elements on them, may provide fundamental insights into the evolutionary origins and mechanisms of metazoan sex determination. Third, comparative genomics may also provide valuable short cuts to the isolation of otherwise elusive genes from other platyhelminth taxa. If the *Schistosoma* homologue of the gene is known to be positioned close to a housekeeping gene that is conserved across the taxa and gene order (synteny) is also preserved, then the housekeeping gene can be used to screen large fragment DNA libraries to retrieve the gene from the other taxon. If the gene is flanked by housekeeping genes, long-range PCR can also be used.

Genomic sequencing

The *Schistosoma* genome is too large currently to contemplate full genomic sequencing. However, more focused approaches are practical. DNA sequences have recently been obtained for the mitochondrial coding regions of *S. mansoni*, *S. japonicum* and *S. mekongi*, and partial sequences have been generated for *S. haematobium*, *S. bovis*, *S. curassoni*, *S. intercalatum*, *S. mattheei*, *S. rodhaini* and *S. margrebowiei*. Complete, or near-complete data are also available for some other helminths (*Paragonimus westermani*, *Fasciola hepatica* and *Taenia crassiceps*). One of the most striking things to emerge from these data is that there are significant differences in gene order between the African and Asian schistosomes, with *S. japonicum* and *S. mekongi* sharing the same gene order as *Paragonimus*, *Fasciola* and *Taenia* (Blair *et al.* 1999; Le *et al.* 2000). These results provide further evidence of the deep phylogenetic divide between the African and Asian schistosomes, and corroborate the molecular data presented earlier in this chapter which suggest that *Schistosoma* evolved in Asia and subsequently colonized Africa, rather than the other way round as had previously been proposed (Davis 1980, 1992). Examination of mitochondrial gene order within the *S. indicum* species group will be particularly interesting given its apparent phylogenetic position between the 'African' and 'Asian' *Schistosoma*

species (Johnston *et al.* 1993; Littlewood and Johnston 1995) and its geographic distribution. Conflicting hypotheses exist concerning the date of colonization of 'India/Asia' by the ancestors of this group; Davis (1980) proposes an early colonization, during the spread from 'Africa' to 'Asia', whereas Barker and Blair (1996) interpret molecular phylogenies as suggesting a much later colonization during the dispersal of modern man from Africa.

Informatics

As with any genome initiative, the data generated by the SGP is only as useful as the ease with which it is accessible to the wider community. In addition to all EST and other sequence data being deposited in the NCBI and EMBL databases, the SGP maintains its own, public domain informatics resources. A WWW server (http://www.nhm.ac.uk/hosted_sites/schisto/) holds information on current projects, cluster analysis results, resources, protocols, etc. together with Network membership lists, policy documents and meeting reports. A *Schistosoma* genome database, SchistoDB, is also curated (http://www.nhm. ac.uk/hosted_sites/schisto/informatics/SchistoDB_info.html). This contains all *Schistosoma* DNA and protein sequences (with updated BLAST homology search results), cluster analysis results, all *Schistosoma* references in MEDLINE and other genome-related information. As the physical map grows, the database will also incorporate mapping data. SchistoDB uses the ACeDB database engine, which was devised for the *Caenorhabditis elegans* Genome Project and which allows the display, querying and analysis of molecular, genetic, genomic and text data through intuitive graphical displays and 'hypertext' links to associated data in the database (Durbin and Thierry Mieg 1991–). ACeDB has become a *de-facto* standard for the curation and distribution of genomic data and is available for Windows, Linux and Unix systems.

Conclusions

The dioecious nature of schistosome blood-flukes remains an evolutionary enigma around which a wealth of literature has been generated (e.g., Combes 1991; Després and Maurice 1995). Platt and Brooks (1997) outlined an ambitious research effort that would use phylogenetic methods to test some of these hypotheses on the origin of separate sexes. In addition to this effort, the speciose genera *Trichobilharzia* and *Gigantobilharzia* have fascinating life-history variations that make them prime candidates for evolutionary analysis. Work on these and other schistosome taxa has long suffered from fragmented specimens and incomplete descriptions (Farley 1971). Within the genus *Schistosoma* itself, competing hypotheses of geographical origin now need to be addressed. A renewed collection effort, accompanied by molecular analysis and morphological study that includes electron microscopy, will dramatically improve our understanding of these fascinating worms. It is anticipated that the wealth of data being generated by the *Schistosoma* Genome Initiative will provide a blueprint to guide these studies, and will also provide valuable data and resources to facilitate a wide range of analyses within the wider platyhelminth fauna.

Section IV

Characters and techniques

Chapter 19

Protonephridia as phylogenetic characters

Klaus Rohde

It is essential for a cladistic analysis that the only characters to be used are those likely to be homologous (e.g., Rohde 1990, 1996, further references therein). Rieger and Tyler (1985) summarized homology criteria as follows: 1) homologous structures have a similar position relative to other structures, as well as similarities in the position of substructures; 2) even dissimilar structures may be homologous if they are connected by a sequence of intermediate forms; 3) structures can be interpreted as homologous if their distribution in a group of organisms coincides with other similarities; the first and most important criterion should always be used first. A homology analysis should be supplemented by a functional analysis using the following criteria (Rieger and Tyler 1985): 4) convergence of similar structures becomes more likely with increasing correlation between the similarity of structures and certain environmental conditions; 5) convergence becomes less likely if there are many 'solutions' to a certain problem; 6) different ontogenetic origins of structures suggest convergence; and 7) common selection pressure suggests convergence.

In the following, I examine the usefulness of protonephridial ultrastructure for establishing phylogenetic relationships in the platyhelminths using the above homology criteria. Previous discussions are by Ehlers (1985a), Ehlers and Sopott-Ehlers (1986), Rohde (1990, 1991, a brief account also in 1988), and (Rohde *et al.* 1995), but the vastly improved database now available makes a new evaluation necessary. Ruppert and Smith (1988) discussed the functional organization of filtration nephridia in various animal groups, and recent discussions of function and evolution of protonephridia including those of platyhelminths are by Bartolomaeus and Ax (1992) and Xylander and Bartolomaeus (1995). A brief, somewhat out-of-date account was provided by Hertel (1993).

A brief glossary of terms is meant to make the descriptions of protonephridia easier to understand for those not familiar with details of protonephridial ultrastructure, and is presented in Appendix 19.1.

Functional differences

The flame bulbs of most species have a filtration apparatus constructed as a 'weir', i.e., a structure consisting of ribs (rods) which are outgrowths of the terminal, or the terminal as well as the proximal canal cells; the ribs are connected by a 'membrane' of extracellular substance through which filtration occurs. In some species, the weir consists of 'membrane'-covered slits in the cytoplasm lining the lumen of the cell. In the haplopharyngid *Haplopharynx rostratus*, careful search for a weir was negative (Rohde and Watson 1998). There are neither ribs nor 'membranes', but large numbers of exocytotic vesicles are aggregated around the capillaries of the three terminal cells which form a terminal complex. Many of these vesicles were seen to be connected to the surface membrane, i.e., in open communication with the capillary lumen. Hence, excretion and/or osmoregulation in this species is apparently by exocytosis, a process which may also play a role in some species with a weir (Rohde and Watson 1998). The macrostomid *Paromalostomum proceracauda* has ribs which, how-ever, are not connected by a 'membrane' and – as in *H. rostratus* – there are many vesicles (Brüggemann 1986a). Among some other macrostomids, weirs of various types are common, and thus absence of a weir in *Haplopharynx* and *Paromalostomum* appears to be due to secondary reduction. In *Urastoma cyprinae*, ribs of the terminal and proximal canal cells are tightly coiled around each other and a 'membrane' could not be seen, but may well be present. Conspicuous in this species is an extensive system of tubules (54–66 nm in diameter), some appearing 'empty' and some filled with a granular material, which may play a role in excretion/osmoregulation. Microtubules (of different diameter) are present in (some) other species, but such an extensive system was never seen in other species. Hence, this character appears to be an autapomorphy of *Urastoma*.

Differences in the arrangement of the terminal and proximal canal cells, and in the structure of the weir

Acoelomorpha

The acoelans and nemertodermatids lack protonephridia, which may be the primitive condition or due to secondary reduction. The 'pulsatile' epidermal cells described by various authors from acoelans and nemertodermatids are degenerating epidermal cells, and not excretory ones (Ehlers 1992a,c). However, Ehlers (1992d) described 'dermonephridia', i.e., specialized epidermal cells with a 'probable excretory function' from the acoelan *Paratomella rubra*. Even if experimental evidence for such a function should be found, the cells are certainly not homologous with the protonephridial cells of catenulids and other 'turbellarians'.

Catenulida (Figure 19.1, type I; Figures 19.2–19.4)

In *Catenula, Stenostomum* and *Suomina* (see Rohde and Watson 1993a, 1994a; also Silveira 1993) two lateral cytoplasmic cords of the terminal cell have transverse outgrowths that interdigitate with each other, connected by a filtration 'membrane' and supported by more internally located longitudinal leptotriches (outgrowths of the basal cytoplasm of the terminal cell). The flame is composed of two cilia closely adjoined to each other, with one short vertical and one long lateral rootlet, the latter extending along the cytoplasmic cords of the weir. These cords also contain some microtubules. Terminal cells are joined to capillaries lined by several canal cells which contain loosely scattered cilia (lateral flames). These cilia have two keels opposite each other (Rohde and Watson 1993a). Keels were also seen along parts of the cilia in flame bulbs of *Stenostomum* (Rohde and Watson 1993a). In *Retronectes*, the weir, according to Ehlers (1985a), is formed by longitudinal ribs connected by a filtration 'membrane', and there also are two cytoplasmic cords containing lateral rootlets of the cilia and microtubules. My own examination of a *Retronectes* sp. showed two (and sometimes three) cilia of the terminal cell, but I have never seen a weir consisting of

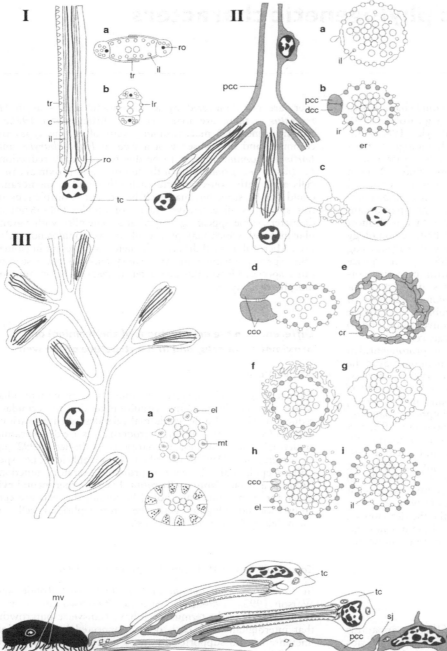

Figure 19.1 Diagrams of representative longitudinal and transverse sections of main types of flame bulbs. I. Type I. Catenulida. a, *Catenula*, *Suomina* and *Stenostomum*; b, *Retronectes*. II. Type II. Most Rhabditophora except Lecithoepitheliata and Rhabdocoela. a, *Microstomum*; b, *Macrostomum*; c, Götte's larva (Polycladida); d, Proseriata (*Monocelis*); e, *Urastoma*; f, Prolecithophora (*Archimonotresis*); g, Tricladida; h, Monogenea and Trematoda; i, Cestoda. III. Type III. a, Rhabdocoela; b, Lecithoepitheliata. Abbreviations: c, cilia; cco, cytoplasmic cords; cr, coiled rods; dcc, distal canal cell; el, external leptotrich; er, external rib; il, internal leptotrich; ir, internal rib; lr, longitudinal rib; mt, microtubule; pcc, proximal canal cell; ro, rootlet of cilium; tc, terminal cell; tr, transverse rib. Proximal canal cells are shaded, terminal cells are white. (Redrawn and modified from Rohde 1991.)

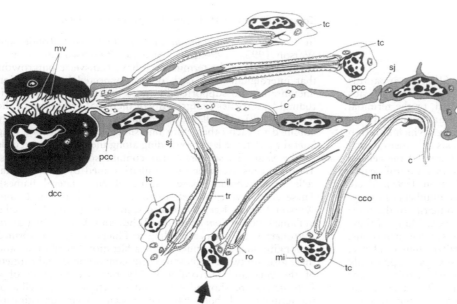

Figure 19.2 Flame bulbs and capillaries of regenerating *Stenostomum* sp. Abbreviations: mi, mitochondrion; mv, microvilli; sj, septate junction; other abbreviations as in Figure 19.1. Terminal cells white, proximal canal cells shaded, distal canal cell black. Arrow indicates free flame bulb in parenchyma. (Redrawn and modified from Rohde and Watson 1993a.)

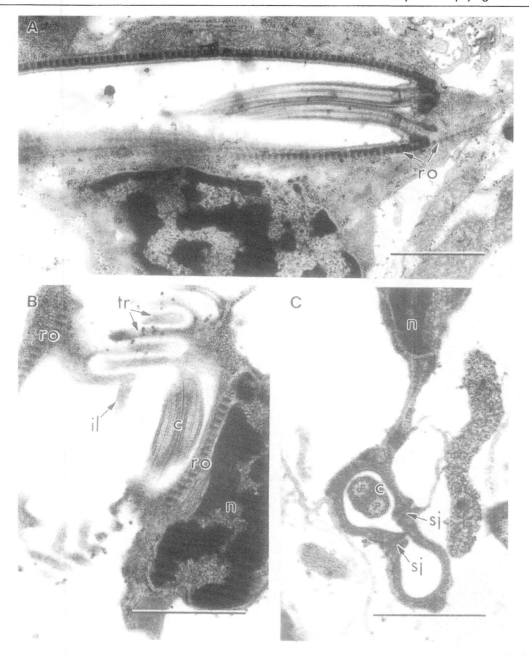

Figure 19.3 Transmission electron micrographs of flame bulb of *Suomina* sp. **A)** Longitudinal section. Note the long ciliary rootlets in the wall of the flame bulb, and shorter ones basal to the basal bodies. **B)** Oblique-longitudinal section. Note the interdigitating transverse ribs and internal leptotriches. **C)** Cross-section through middle of flame bulb. Note the two septate junctions in wall of flame bulb and tightly adjoined cilia. Abbreviations: n, nucleus; sj, septate junction; other abbreviations as in Figure 19.1. Scale bars 1 μm. (Originals.)

ribs and a filtration 'membrane', and rootlets in the cytoplasmic cords are very weakly developed.

Macrostomida (Figure 19.1, type IIa,b; Figures 19.5 and 19.6)

The macrostomids have a great variety of flame bulbs. In *Microstomum* sp. the weir consists of longitudinal outgrowths of the terminal cell and two (or perhaps more) terminal cells are connected to one proximal canal cell which grows for a short distance into the cytoplasmic cylinders of the terminal

cells (Rohde 1991; Rohde and Watson 1991a). In *Macrostomum tuba*, the weir consists of interdigitating outgrowths of the terminal and proximal canal cells, and a further, distal canal cell extends a tongue of cytoplasm internal to the tube of the proximal canal cell, partly surrounding the flames (Watson *et al.* 1991). The flame bulb of *Macrostomum spirale* resembles that of *M. tuba*, but the second canal cell extends a thick cytoplasmic cord along the terminal complex, and the perikaryon of the terminal cell is more distant from the flame in its distal regions (Kunert 1988); *Paromalostomum proceracauda* has a terminal complex that consists of three

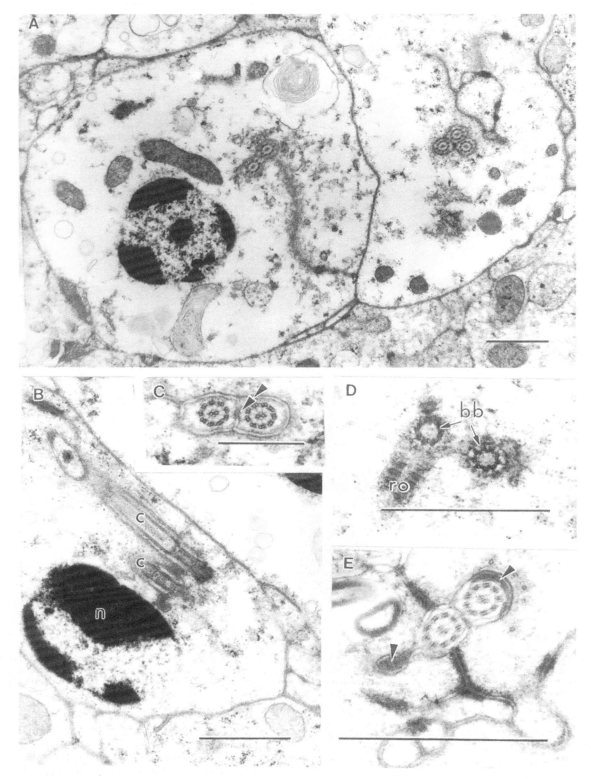

Figure 19.4 Transmission electron micrographs of protonephridium of *Retronectes* sp. **A)** Transverse section through two flame bulbs. Note three cilia on the right, and two cilia on the left. **B)** Longitudinal section through flame bulb. **C)** Cross-section through flame bulb. Note two cilia tightly adjoined by a structure resembling a septate junction (double arrowhead). **D)** Cross-section through basal bodies of the two cilia forming the flame of a flame bulb. Note ciliary rootlet. **E)** Cross-section through flame bulb. Note dense material (arrowheads) (ciliary rootlets?). Abbreviations: bb, basal body; n, nucleus; ro, rootlets of cilia; c, cilia. Scale bars 1 μm (A,B,D,E) and 0.5 μm(C). (Originals.)

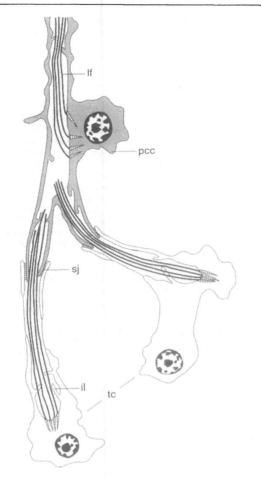

Figure 19.5 Terminal part of protonephridium of *Microstomum* sp. Terminal cells white, proximal canal cells shaded. Abbreviations: lf, lateral flame; sj, septate junction; other abbreviations as in Figure 19.1. (Redrawn and modified from Rohde and Watson 1991a.)

terminal cells surrounded by a filamentous sheath. Each cell has a weir consisting of ribs which are not connected by a filtration 'membrane'. The flames of the three terminal cells reach into a joint basket lumen, and there are many vesicles in their cytoplasm (Brüggemann 1986a).

Haplopharyngida (Figure 19.7)

Only one species has been examined, i.e., *Haplopharynx rostratus* (see Rohde and Watson 1998). This species has a terminal protonephridial complex that consists of three terminal cells. A weir consisting of ribs connected by a filtration 'membrane' does not exist, but there are some cytoplasmic outgrowths of the terminal cells into the lumen. The flames of two of the terminal cells protrude into the lumen of the third, centrally located, one. The complex is surrounded by a sheath containing many filaments. It resembles that of the macrostomid *Paromalostomum proceracauda* supporting the view that macrostomids and haplopharyngids are closely related.

Polycladida (Figure 19.1, type IIc; Figure 19.8)

Götte's larvae of *Stylochus mediterraneus* was examined by Watson *et al.* (1992a) and Wenzel *et al.* (1992). The two protonephridia, one on each side of the body, are formed by a terminal and a canal cell. The weir is formed by the terminal cell: cytoplasmic outgrowths of the terminal cell interdigitate with each other forming convoluted slits that constitute the weir. Cilia of the terminal and lateral flames arise along the terminal and canal cells.

Prolecithophora (Figure 19.1, type IIf; Figure 19.9)

Watson and Rohde (1997) examined three species of prolecithophorans (families Cylindrostomidae and Pseudostomidae). All species have a weir consisting of scattered short filtration slits in the wall of the terminal cell, a nucleus close to

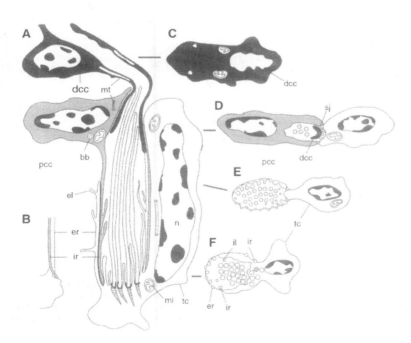

Figure 19.6 Terminal part of protonephridium of *Macrostomum tuba*. Terminal cell white, proximal canal cell shaded, distal canal cell black. Abbreviations: bb, basal body; mi, mitochondrion; n, nucleus; sj, septate junction; other abbreviations as in Figure 19.1. (Redrawn and modified from Watson *et al.* 1991.)

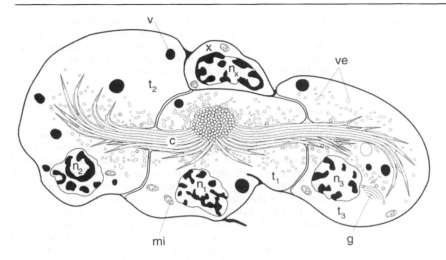

Figure 19.7 Terminal complex of *Haplopharynx rostratus*. Abbreviations: g, Golgi complex; mi, mitochondrion; n₁–n₃, nuclei of terminal cells 1–3; nₓ-nucleus of cell associated with terminal complex; t₁–t₃, terminal cells 1–3; x, cell associated with terminal complex; v, vacuole; ve, vesicle; other abbreviations as in Figure 19.1. (Redrawn and modified from Rohde and Watson 1998.)

the weir, cilia arising at various levels along the terminal cell, and a wall strengthened by long, cross-striated ciliary rootlets but lacking bundles of microtubules. Similar flame bulbs with slits, nucleus close to the weir, and supporting rootlets in the wall were also demonstrated by Ehlers and Sopott-Ehlers (1997a) in another three species of Prolecithophora (families Plagiostomidae, Cylindrostomidae and Pseudostomidae); only one of these species had a few longitudinal microtubules in the wall of the terminal cell (but no bundles of microtubules). In *Archimonotresis limophila* (fam. Protomonotresidae), Ehlers (1989) found a fundamentally different type of flame bulb. It consists of longitudinal ribs connected by a filtration 'mem-

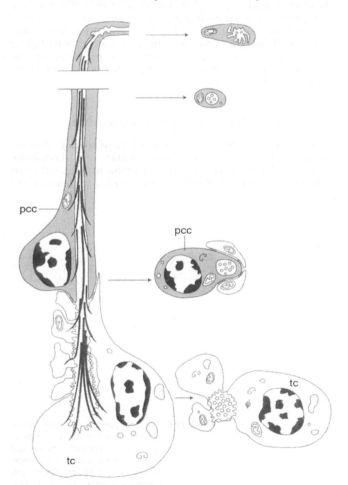

Figure 19.8 Protonephridium of Götte's larva of *Stylochus* sp. (Polycladida). Abbreviations as in Figure 19.1. (Redrawn and modified from Watson *et al.* 1992a.)

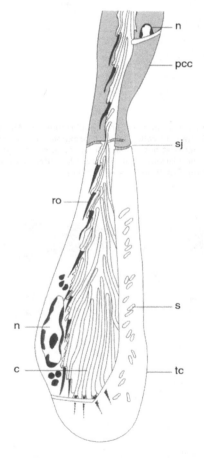

Figure 19.9 Terminal part of protonephridium of Prolecithophora (most species). Terminal cells white, proximal canal cells shaded. Abbreviations: n, nucleus; s, slits; sj, septate junction; other abbreviations as in Figure 19.1. (Redrawn and modified from Watson and Rohde 1997.)

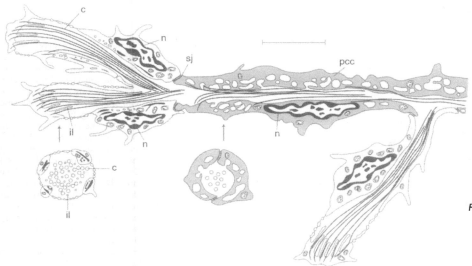

Figure 19.10 Terminal part of protonephridium of *Artioposthia* sp. Abbreviations: n, nucleus; sj, septate junction; other abbreviations as in Figure 19.1. (Redrawn and modified from Rohde and Watson 1992a.)

brane', supported by longitudinal bundles of microtubules and with a 'network' of segregated cytoplasmic areas external to the weir. It resembles the flame bulbs of many rhabdocoels.

Tricladida (Figure 19.1, type IIg; Figure 19.10)

In *Artioposthia* sp., Rohde and Watson (1992a) found terminal complexes consisting of one or two terminal cells, each with a nucleus located in the lateral wall of the flame bulb, and joined to probably two proximal canal cells that line the proximal canal. The weir is restricted to the proximal part of the terminal cells and consists of convoluted slits separated by thick columns of cytoplasm. Cross-striated rootlets running in various directions in the wall of the flame bulbs and, to a lesser degree, that of the capillary, probably have a supportive function. There are two septate junctions in the wall of the proximal canal. Slits forming the weir were also reported for *Dugesia tigrina* by McKanna (1968), for *Romankenkius libidinosus* by Rohde *et al.* (1990b), and for *Bdellocephala brunnea* by Ishii (1980). In the last species they are formed by interdigitating processes of a stellate cell. Development of slits from such interdigitations may well be a characteristic of all triclads, although they were not seen in other species.

Lecithoepitheliata (Figure 19.1, type IIIb)

Like many rhabdocoels, the Lecithoepitheliata have clusters of non-nucleate flame bulbs with a weir consisting of a single row of longitudinal ribs connected by a weir 'membrane'. Nuclei are present in the canals into which the flame bulbs open. In contrast to kalyptorhynchs, typhloplanids, dalyelliids and temnocephalids, the ribs of lecithoepitheliates lack microtubules but have filaments instead. In cross-section, the ribs are triangular. The proximal canal lacks a junction (Timoshkin *et al.* 1989; Rohde and Watson 1991b).

Proseriata (Figure 19.1, type IId; Figure 19.11)

The flame bulbs of the proseriate *Monocelis* sp. resemble those of the trematodes and monogeneans in having a weir formed by interdigitations of the terminal and proximal canal cells and connected by a 'membrane', internal leptotriches arising from the terminal cell, and two very thick cytoplasmic cords extending from the canal cell along the weir. The proximal protonephidial capillaries possess a septate junction. The inner surface area of the capillaries is enlarged by lamellae and a reticulum of vacuoles. However, whereas in the Neodermata the internal ribs are outgrowths of the terminal cells, and the external ribs are outgrowths of the proximal canal cell, this is reversed in the Proseriata. A maximum of two supporting cells were found along the weir (Rohde *et al.* 1988a). Ehlers and Sopott-Ehlers (1987) described the protonephridium of *Invenusta paracnida* with a similar structure.

Dalyelliida (Figure 19.1, type IIIa; Figure 19.12)

Typically, clusters of non-nucleated flame bulbs open into nucleate canals with lateral flames. The weir is formed by a single row of longitudinal ribs connected by a 'membrane'. The ribs contain microtubules and are more or less round in cross-section. In the umagillid *Syndisyrinx punicea* and the pterastericolid *Pterastericola pellucida*, many flame bulbs are located in the cytoplasm of a single perikaryon. As in other rhabdocoels, internal leptotriches are absent, and bundles of microtubules extend along the flame bulbs (e.g., Brüggemann 1986a, 1987, 1989; Rohde *et al.* 1990b, 1993a; Lumbsch *et al.* 1995). In *Luriculus australiensis*, Rohde *et al.* (1993a) did not find clusters of flame bulbs along protonephridial canals, although they did not exclude the possibility that one perikaryon is connected to more than one flame bulb. Brüggemann (1989) suggested that the terminal cell of *Vejdovskya pellucida* may have a nucleus located in a long cytoplasmic process of the cell, but found no clear evidence for it (see also Rohde *et al.* 1988b: *Gieysztoria* sp.; Rohde *et al.* 1992a: one species each of Umagillidae and Pterastericolidae). N. Watson (personal communication) found bundles of microtubules in the flame bulbs of three species of Graffillidae and two species of Provorticidae; in three of these species (one graffillid and two provorticids), a nucleus was seen near the tip of the flame bulb.

Temnocephalida (Figure 19.1, type IIIa; Figure 19.13)

Temnocephala sp., *Didymorchis* sp., *Craspedella* sp. and *Actinodactylella blanchardi* were examined by Rohde

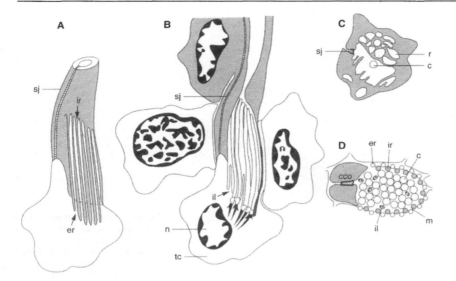

Figure 19.11 Terminal part of protonephridium of *Monocelis* sp. (Proseriata). Terminal cells white, proximal canal cell shaded. Note two supporting cells along flame bulb. **A)** Surface view of flame bulb showing filtration apparatus. **B)** Longitudinal section through flame bulb showing terminal cell, two supporting cells and proximal canal cell. **C)** Cross-section through proximal canal. **D)** Cross-section through filtration apparatus. Abbreviations: m, 'membrane'; r, reticulum of vacuoles; n, nucleus; sj, septate junction; other abbreviations as in Figure 19.1. (Redrawn and modified from Rohde *et al.* 1988a.)

(1986a, 1987a) and Rohde *et al.* (1988b). The flame bulb of temnocephalids is identical with that of most other rhabdocoels in having a weir formed by a single row of longitudinal ribs supported by bundles of microtubules, the lack of internal leptotriches and of a junction in the capillaries. (See also the study by Williams (1981b) on *Temnocephala novaezealandiae*.)

Typhloplanida (Figure 19.1, type IIIa)

Rohde *et al.* (1987b) examined *Mesostoma* sp., and Watson (1998a) examined four species of Typhloplanida (Trigonostomidae). All have the typical rhabdocoel flame cell consisting of a single row of ribs supported by aggregations of microtubules, but in three of the species there is a 'reduced definition of their filtration ribs'. N. Watson (personal communication) examined one species of Byrsophlebidae, two species of Promesostomidae, one species of Solenopharyngidae and four species of Typhloplanidae. All of these species have a single row of ribs in the weir supported by bundles of microtubules, and none has a nucleus near the flame bulb. External leptotriches were seen in one typhloplanid.

Kalyptorhynchia (Figure 19.1, type IIIa)

Early studies of kalyptorhynchs are by Rohde *et al.* (1987a, 1988b). The flame bulb resembles those of dalyelliids and

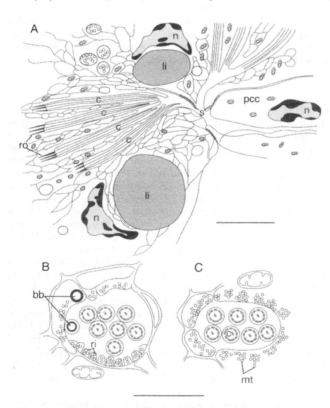

Figure 19.12 Terminal part of protonephridium of *Pterastericola pellucida* (Dalyelliida). **A)** Longitudinal section through bundles of flame bulbs. **B)** Cross-section through proximal part of flame. **C)** Cross-section through distal part of flame. Abbreviations: li, lipid granule; n, nucleus; ri, rib; other abbreviations as in Figure 19.1. (Redrawn and modified from Rohde *et al.* 1992a.)

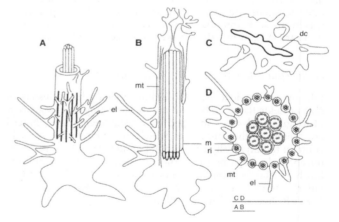

Figure 19.13 Flame bulb of *Temnocephala* sp. (Temnocephalida). **A)** Surface view of flame bulb showing filtration apparatus. **B)** Longitudinal section through flame. **C)** Cross-section through proximal canal. **D)** Cross-section through filtration apparatus. Abbreviations: dc, dense cytoplasm; m, 'membrane'; ri, rib; other abbreviations as in Figure 19.1. (Redrawn and modified from Rohde 1990.)

temnocephalids. In *Baltoplana magna* (Schizorhynchia, Karkinorhynchidae), the proximal canal cell (which lacks a septate junction) overlaps the distal part of the terminal cell. The terminal cell has a single row of longitudinal ribs containing bundles of microtubules, and internal leptotriches are absent. The flame bulb differs from those of the other rhabdocoels previously examined in consisting of a single nucleated terminal cell; they are not arranged in bundles connected to nucleated proximal canals (Rohde and Watson 1994b). Watson and Schockaert (1997) made an ultrastructural study of protonephridia of several species of kalyptorhynchs, and found three types of flame bulb. Species of Schizorhynchia and two species of Cicerinidae (Eukalyptorhynchia) have a perikaryon of the terminal cell connected to the proximal canal cell by a septate junction. In two other species of Cicerinidae, there is a junction between the flame bulb and proximal canal cell, but a nucleus of the terminal cell could not be seen. Species of two other families of Eukalyptorhynchia lack a terminal perikaryon and the flame bulb is continuous with the proximal canal cell and without a junction, as in most other rhabdocoels, i.e., the Typhloplanida, Dalyelliida and Temnocephalida. In all species, the weir is formed by longitudinal ribs supported by bundles of microtubules. The authors consider the first type as plesiomorphic for any rhabdocoel, and the last type as most derived – a suggestion further supported by the demonstration of apparently degenerating nuclei in the flame bulbs of two kalyptorhynch species (Watson 1999a).

Fecampiida (Figure 19.14)

Watson *et al.* (1992b) examined the protonephridial system of larval *Kronborgia isopodicola*. Cytoplasmic processes of the

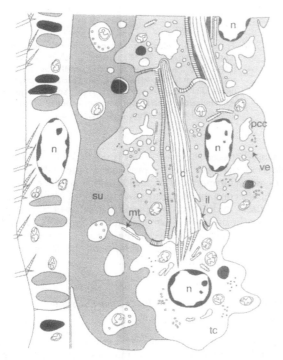

Figure 19.14 Protonephridium of *Kronborgia isopodicola*. Epidermis on left of diagram. Abbreviations: n, nucleus; su, subepidermal space; ve, vesicle; other abbreviations as in Figure 19.1. (Redrawn and modified from Watson *et al.* 1992b.)

terminal and proximal canal cells interdigitate, connected by septate junctions. The lumen of the flame bulb is connected to the subepidermal cavity by the intercellular space between terminal and canal cells crossed by a septate junction.

Urastoma Figure 19.1, type IIe; Figure 19.15)

Rohde *et al.* (1990a) described the flame bulb of *U. cyprinae*. The terminal cell has a nucleus located basal to the flame, and an extraordinarily large number of tubules in the cytoplasm. Outgrowths of the terminal and proximal canal cells lined by thick plasma membranes, are coiled around each other and a 'membrane' between these outgrowths may well be present, but could not be seen, perhaps because of the tightly packed coils. The proximal canal possesses a septate junction. The flame bulb does not resemble those of prolecithophorans, in which *Urastoma* is usually included, nor those of any rhabdocoels.

Ichthyophaga (Figure 19.16)

Unpublished observations by K. Rohde indicate that the flame bulb of *Ichthyophaga* sp. has internal and external ribs, as well as internal and external leptotriches. It strongly resembles those of the Neodermata, although serial sections were insufficient to locate the perikarya of the cells involved in the formation of the weir. Protonephridial capillaries contain two septate junctions.

Neodermata

All species examined have a weir formed by the terminal and proximal canal cells. The internal ribs are continuous with the terminal, and the external ribs are continuous with the proximal canal cell, although internal and external ribs sometimes cannot be clearly distinguished (Rohde 1982).

Trematoda (Aspidogastrea and Digenea) and Monogenea (Poly- and Monopisthocotylea) (Figure 19.1, type IIh; Figure 19.17)

The proximal protonephridial capillaries contain a septate junction, which extends along the weir between two thick cytoplasmic cords, and larger ducts contain several junctions. The surface of protonephridial capillaries lining the lumen is enlarged by lamellae and/or a reticulum of interconnected vacuoles. The following Aspidogastrea have been studied: *Multicotyle purvisi* (Rohde 1971a,b, 1972), *Lobatostoma manteri* (Rohde 1989a), *Rugogaster hydrolagi* (Watson and Rohde 1992). For some studies of Digenea, see Jeong *et al.* 1980; Pan 1980; Orido 1987; Ferrer 1986; Soboleva *et al.* 1988; Mattison *et al.* 1992; Niewiadomska and Czubaj 1996. For studies of Monogenea, see Rohde (1993; for two species of Dactylogyridae see Rohde *et al.* 1989c, for *Gyrodactylus* see Rohde 1989b, for a species of Loimoidae and a species of Calceostomatidae see Rohde *et al.* 1989a; for species of Polyopisthocotylea see Rohde 1973a, 1975a, 1980. Rohde (1993) evaluated evidence from ten genera of Digenea belonging to nine families, of three genera of Aspidogastrea belonging to two families, of two genera of Monogenea Polyopisthocotylea belonging to two families, and of seven genera of Monogenea Monopisthocotylea belonging to seven families. In all of them, a septate junction in the proximal canal and in the flame bulb is present. Such junctions are, however, absent in *Udonella caligorum* and *Anoplodiscus cirrusspiralis*, both species of

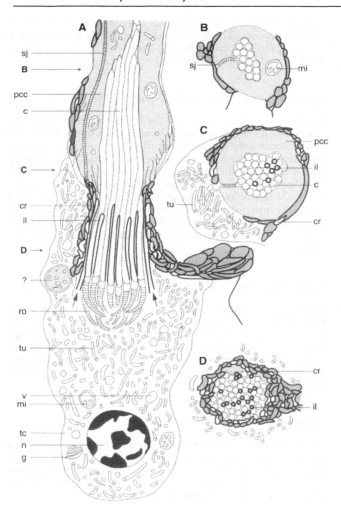

Figure 19.15 Terminal part of protonephridium of *Urastoma*. **A)** Diagrammatic longitudinal section. **B–D)** Cross-sections at various levels. Abbreviations: g, Golgi complex; mi, mitochondrion; n, nucleus; sj, septate junction; tu, tubule; v, vacuole; other abbreviations as in Figure 19.1. (Redrawn and modified from Rohde *et al.* 1990a.)

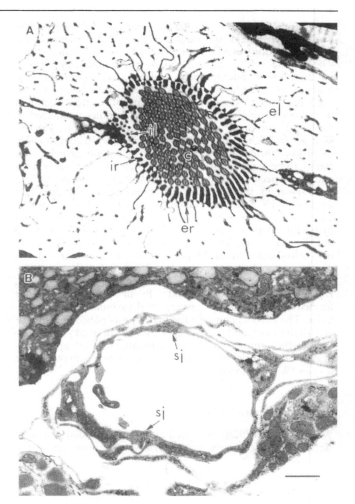

Figure 19.16 Transmission electron micrographs of flame bulb of *Ichthyophaga* sp. **A)** Cross-section through middle of flame bulb. Note external and internal ribs and external and internal leptotriches. **B)** Cross-section through protonephridial capillary. Note two septate junctions. Abbreviation: sj, septate junction; other abbreviations as in Figure 19.1. Scale bars 1 μm. (Originals.)

Monopisthocotylea (Rohde *et al.* 1992b). The presence of an apparently rudimentary junction in the latter species indicates that its loss is secondary. The capillaries of these two species also lack lamellae, which have been demonstrated in species of nine genera of Digenea (eight families), three genera of Aspidogastrea (two families), three genera of Polyopisthocotylea (three families), and three genera of Monopisthocotylea (three families). In species of one genus (as well as *Anoplodiscus* and *Udonella*) lamellae are absent (Rohde *et al.* 1989a; Rohde 1993). Trematodes and monogeneans also have well-developed external leptotriches which are absent in *Udonella* and the cestodes (*ibid.*).

Oncomiracidia of the polyopisthocotylean monogenean *Zeuxapta seriolae* and of the monopisthocotylean *Encotyllabe chironemi* were examined by Rohde (1997) and Rohde (1998), respectively. In both species, the weir has all the characteristics of adult monogeneans, except for the small number (or absence) of external leptotriches, and the much smaller number of cilia and the smaller size of the flame bulb than in adults. Also, the second species lacks lamellae in the capillaries. An important

difference between the two species is the lack of a well-defined bladder in the first and its presence in the second species. Rohde (in Whittington *et al.* 2000) examined the oncomiracidium of a second monopisthocotylean, i.e., of *Neoheterocotyle rhinobatidis*, and found the same structure as in *E. chironemi*, except for the presence of lamellae in the capillaries. He did not observe a connection between the bladder and the outside, suggesting that the protonephridium of the larva does not open until after infection of a host.

Cestoda (Gyrocotylidea, Amphilinidea and Eucestoda)
(Figure 19.1, type IIi; Figure 19.18)

The protonephridial capillaries do not contain a septate junction, and the surface of the capillaries lining the lumen is enlarged by small microvilli, which are very regularly spaced in the Eucestoda, less regularly in the Amphilinidea, and of irregular shape and spacing in the Gyrocotylidea. Rohde (1993; see also Xylander 1987a,b) evaluated evidence from

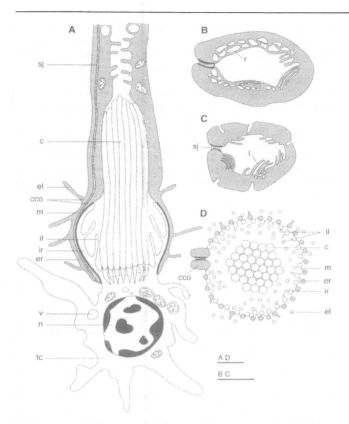

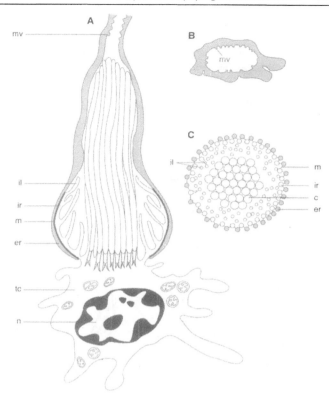

Figure 19.17 Terminal part of protonephridium of Trematoda/Monogenea. Terminal cell white, proximal canal cell shaded. **A)** Longitudinal section through flame bulb. **B**, **C)** Cross-sections through proximal canal. **D)** Cross-section through filtration apparatus. Abbreviations: l, lamellae; m, 'membrane'; n, nucleus; r, reticulum of vacuoles; sj, septate junction; v, vacuole; other abbreviations as in Figure 19.1. (Redrawn and modified from Rohde 1990.)

Figure 19.18 Terminal part of protonephridium of Cestoda. Terminal cell white, proximal canal cell shaded. **A)** Longitudinal section through flame bulb. **B)** Cross-section through proximal canal. **C)** Cross-section through filtration apparatus. Abbreviations: m, 'membrane'; mv, microvilli; n, nucleus; other abbreviations as in Figure 19.1. (Redrawn and modified from Rohde 1990.)

species of 16 genera of eucestodes belonging to 13 families, of one genus of Gyrocotylidea, and of two genera of Amphilinidea, for all of which microvilli in the protonephridial capillaries had been demonstrated. Postlarval *Gyrocotyle urna* also has lamellate protonephridial ducts (Xylander 1992b). Lack of a septate junction had been demonstrated in nine genera of Eucestoda belonging to eight families, and in two amphilinids and one gyrocotylid. Hence, these characters can be considered as characteristic of these taxa. (For some studies of Eucestoda see Šlais 1973; Swiderski *et al* 1975; Lumsden and Hildreth 1983; McCullough and Fairweather 1991; Pospekhova *et al.* 1993; Korneva *et al.* 1998; for studies of adult and larval Amphilinidea see Rohde and Georgi 1983; Xylander 1986a; Rohde and Watson 1987b; for studies of Gyrocotylidea see Xylander 1986a, 1987a; for a review see Coil 1991).

Development

The development of the filtration apparatus is known in few species (Figure 19.19). In the amphilinid *Austramphilina elongata*, at an early stage of development, a hollow cytoplasmic cylinder of the proximal canal cell establishes contact with the terminal cell, held to it by a septate junction (Figure

19.19B,D,F). Inner outgrowths (internal leptotriches) of the terminal cell grow into its lumen. Those leptotriches in contact with the cytoplasmic cylinder of the proximal canal cell become the internal ribs; along the lines of contact, the cytoplasm of the canal cell breaks up into external ribs, connected to the internal ones by a 'membrane' of extracellular matrix (Figure 19.19E,G) (Rohde and Watson 1988a). In the developing cercaria of the digenean trematode *Philophthalmus* sp., Rohde and Watson (1992b) observed that the perikarya of terminal and proximal canal cells are close together, and that sheet-like outgrowths of the terminal cell are externally surrounded by cytoplasm of the proximal canal cell. Internal outgrowths and external cytoplasm are connected by many 'membranes', i.e., desmosome-like structures. Internal sheets break up into internal ribs, and the external cytoplasm breaks up into external ribs; external and internal ribs are connected by a filtration 'membrane'. Palmberg (1990a,b) used autoradiography and immunochemistry to study the development of various cell types from stem cells in *Microstomum lineare* and *Stenostomum leucops*. She observed developing flame cells in both species. In the first species, there were bundles of cilia surrounded by cytoplasmic rods, but details of developmental stages were not given.

Relevant here are observations on the protonephridia of the regenerating catenulid *Stenostomum*. Serial ultrathin sections showed that free terminal cells with all the characteristics of fully developed terminal cells attached to canal cells, float freely in the intercellular spaces. Apparently, they attach

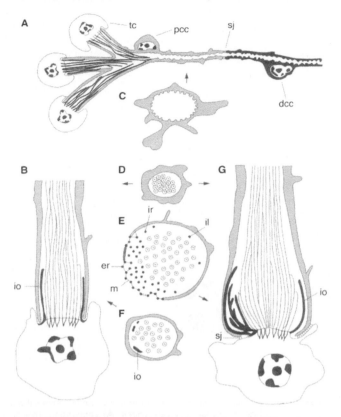

Figure 19.19 Developing protonephridium of *Austramphilina elongata*. (Cestoda, Amphilinidea). **A)** Fully developed bundle of flame bulbs formed by three terminal cells and one proximal canal cell. **C)** Cross-section through proximal canal cell. **B, D–G)** Developing stages of flame bulb. Abbreviations: io, inner outgrowth; m, 'membrane'; sj, septate junction; other abbreviations as in Figure 19.1. (Redrawn and modified from Rohde and Watson 1988a.)

themselves to canal cells at a later stage of regeneration (Rohde and Watson 1993a).

Korneva *et al.* (1998) made an ultrastructural study of the 'secondary excretory system' in different stages of the procercoid of the pseudophyllidean cestode *Triaenophorus nodulosus*, but there are no data on the development of the terminal and proximal canal cells relevant to this review.

Application of the homology and functional criteria to platyhelminth protonephridia

Criterion 1

Since flame bulbs are scattered throughout the body, the first part of this criterion (similar position relative to other structure) cannot be applied (it can be applied to similarities in the position of excretory pores, for example in the monopisthocotylean and polyopistocotylean Monogenea). On the other hand, because of their very great complexity, similar position of substructures in many cases indicates that certain types of flame bulbs are synapomorphies for certain taxa. This is likely for the following taxa: the catenulids *Catenula*, *Stenostomum* and *Suomina*, and perhaps *Retronectes* although

the last genus lacks transverse ribs; the macrostomid *Paromalostomum proceracauda* and the haplopharyngid *Haplopharynx rostratus*; most Prolecithophora; the Tricladida Terricola and Paludicola; the Kalyptorhynchia, Typhloplanida, Dalyelliida, and Temnocephalida; the Aspidogastrea and Digenea; the Monopisthocotylea and Polyopisthocotylea; the Gyrocotylidea, Amphilinidea and Eucestoda. It is also possible for the Prolecithophora and Tricladida; the Kalyptorhynchia, Typhloplanida, Dalyelliida, Temnocephalida and Lecithoepitheliata; the Dalyelliida, Temnocephalida and the prolecithophoran *Archimonotresis limophila*; the Fecampiida, *Urastoma, Ichthyophaga* and the Neodermata; the Proseriata, Neodermata, Fecampiida, *Urastoma* and *Ichthyophaga*.

Criterion 2

There are no fossils which represent a series of intermediate forms, and the few studies of the development of protonephridia are insufficient for phylogenetic conclusions. However, the demonstration of three different types of flame bulb in kalyptorhynchs, one with nucleated terminal cells, one with non-nucleated flame bulbs separated from the proximal canal cells by a junction, and a third type with a flame bulb continuous with the proximal canal cell, suggests that non-nucleated flame bulbs have developed from nucleated terminal cells at least in the kalyptorhynchs, and possibly in other rhabdocoels as well (Watson and Schockaert 1997). This suggestion is supported by the demonstration of apparently degenerating nuclei in the flame bulbs of two kalyptorhynch species (Watson 1999a). Also, the variety of flame bulbs with weirs in the macrostomids and lack of a weir in one macrostomid and a haplopharyngid suggests that lack of a weir (and excretion/osmoregulation by exocytosis instead) is a secondary development.

Criterion 3

Almost all taxa listed under Criterion 1 as having flame bulbs likely to be synapomorphic for them, also have other characters suggesting monophyly, such as similar spermatozoa and spermiogenesis, sensory receptors, tegument and epidermis, etc. (for details, see the data matrix in Littlewood *et al.* 1999a).

Criterion 4

It is unlikely that convergence is responsible for the similarity of flame bulbs in many taxa, because there is no (or little) correlation between certain environmental conditions and a similar structure of flame bulbs. For example, many taxa living in the same, marine environment have dissimilar flame bulbs (e.g., Catenulida, Macrostomida, Haplopharyngida, Polycladida, Prolecithophora, Proseriata, Typhloplanida, Kalyptorhynchia, Dalyelliida), parasitic dalyelliids (*Provortex*) and free-living ones have the same type of flame bulb, symbiotic Temnocephalida and free-living Dalyelliida share the same type of flame bulb, Tricladida Terricola and Paludicola share the same type of flame bulb, larval trematodes from molluscs have the same type of flame bulb as adults from vertebrates, marine and freshwater Monogenea have identical flame bulbs, and eucestodes and gyrocotylids from the intestines of various vertebrates have the same type of flame bulb as amphilinids from the body cavity of vertebrates and their free-swimming larva.

Criterion 5

The discussion above has shown that there is a great number of 'solutions' to the problem of excreting waste and/or regulating the liquid content in the body, which considerably enhances the value of flame bulbs as phylogenetic tools.

Criterion 6

As pointed out above, little is known about the ontogeny of flame bulbs. (See Criterion 2.)

Criterion 7

This criterion is similar to Criterion 4. Common selection pressure (for example, for better adaptation to a parasitic way of life, or for life in the marine benthos) is unlikely to be responsible for similar flame bulbs, as pointed out under Criterion 4.

PAUP analyses of data

Data matrices based on the discussion above are given in Tables 19.1–19.3. Different approaches are used to define characters; the approaches differ in that the characters used vary in the degree of likelihood of being homologous. The most conservative approach is used in Table 19.1: characters are defined as particular types of flame bulbs (composed of a number of substructures) found in certain taxa which are almost certain to be homologous, because they are identical in all their substructures (1st homology criterion). In Table 19.2, the assumption is made that similar flame bulbs in some taxa are homologous, i.e., there is an intermediate degree of likelihood of homology, because flame bulbs are identical in most but not all substructures. Table 19.3 uses the least conservative approach: various substructures of flame bulbs are considered to be independent characters, although homology of these structures is not always certain and in some cases is even unlikely (e.g., internal leptotriches and formation of filtration apparatus by interdigitating cytoplasmic processes in several taxa). The fact that three different approaches can be used, demonstrates that characters can by no means always be clearly defined and that there is always an element of subjectivity in the choice of characters. Choice of the most suitable

Table 19.1 Data matrix used to reconstruct phylogenetic relationships. Presence of character coded as 1, absence coded as 0. Characters (types of flame bulb) 1. *Catenula*-type (Figure 19.2); 2. *Retronectes*-type (Figures 19.1, 19.4); 3. *Microstomum*-type (Figure 19.5); 4. *Macrostomum*-type (Figure 19.6); 5. *Paromalostomum–Haplopharynx*-type (Figure 19.7); 6. Polycladida–Tricladida-type (Figures 19.8, 19.10); 7. Prolecithophora-type 1 (Figure 19.9); 8. Polecithophora-type II (*Archimonotresis*-type) (Figure 19.1); 9. Lecithoepitheliata-type (Figure 19.1); 10. Proseriata-type (Figure 19.11); 11. Dalyelliida-type (Figures 19.12, 19.13); 12. Fecampiida-type (Figure 19.14); 13. *Urastoma*-type (Figure 19.15); 14. *Ichthyophaga*-type (Figure 19.16); 15. Trematoda/Monogenea-type (Figure 19.17); 16. Cestoda-type (Figure 19.18).

Characters	*1* *1234567890*	*111111* *123456*
Catenulida	1000000000	000000
Retronectes	0100000000	000000
Acoelomorpha	0000000000	000000
Macrostornida	0011100000	000000
Haplopharyngida	0000100000	000000
Polycladida	0000010000	000000
Prolecithophora	0000001100	000000
Tricladida	0000010000	000000
Lecithoepitheliata	0000000010	000000
Proseriata	0000000001	000000
Dalyelliida	0000000000	100000
Temnocephalida	0000000000	100000
Typhioplanida	0000000000	100000
Kalyptorhynchia	0000000000	100000
Fecampiida	0000000000	010000
Urastoma	0000000000	001000
Ichthyophaga	0000000000	000100
Aspidogastrea	0000000000	000010
Digenea	0000000000	000010
Polyopisthocotylea	0000000000	000010
Monopisthocotylea	0000000000	000010
Gyrocotylidea	0000000000	000001
Amphilinidea	0000000000	000001
Eucestoda	0000000000	000001

Table 19.2 Data matrix used to reconstruct phylogenetic relationships. Presence of character coded as 1, absence coded as 0. Characters (types of flame bulb) 1. Catenulida/*Retronectes*-type (two cilia of flame); 2. *Microstomum*/Lecithoepitheliata/*Archimonotresis*/Rhabdocoela-type (weir formed by single row of longitudinal ribs); 3. *Macrostomum*/Proseriata/Fecampiida/*Urastoma*/*Ichthyophaga*/Neodermata-type (one row of longitudinal ribs each from, terminal and proximal canal cells); 4. Polycladida/Tricladida-type (short slits formed by interdigitating ribs); 5. *Paromalostomum–Haplopharynx*-type (joint lumen of several terminal cells); 6. Prolecithophora-type (short slits of terminal cells).

Characters	*123456*
Catenulida	100000
Retronectes	100000
Acoelomorpha	000000
Macrostomida	011010
Haplopharyngida	000010
Polycladida	000100
Prolecithophora	010001
Tricladida	000100
Lecithoepitheliata	010000
Proseriata	001000
Dalyelliida	010000
Temnocephalida	010000
Typhloplanida	010000
Kalyptorhynchia	010000
Fecampiida	001000
Urastoma	001000
Ichthyophaga	001000
Aspidogastrea	001000
Digenea	001000
Polyopisthocotylea	001000
Monopisthocotylea	001000
Gyrocotylidea	001000
Amphilinidea	001000
Eucestoda	001000

Table 19.3 Data matrix used to reconstruct phylogenetic relationships. Presence of character coded as 1, absence coded as 0. ? - not known. Characters (types of flame bulbs). 1. Flames consisting of two cilia; 2. Flames consisting of many cilia; 3. Weir formed by interdigitating ribs; 4. Weir formed by terminal cells only; 5. Weir formed by terminal and proximal canal cells; 6. Weir consists of short slits; 7. Proximal canal with septate junction; 8. Flame bulbs non-nucleated and arranged in bundles; 9. Nucleated flame bulbs; 10. Ribs of weir supported by bundles of microtubules; 11. Weir with filtration membrane; 12. Excretion/osmoregulation entirely by exocytosis; 13. Proximal canal with many more or less regularly arranged short microvilli; 14. Proximal canal with lamellae; 15. Flame bulbs forming a terminal complex with joint distal lumen; 16. Flame bulb with external leptotriches; 17. Flame bulb with internal leptotriches.

	1 *1234567890*	*1111111* *1234567*
Characters		
Catenulida	1011000010	1000001
Retronectes	10?1000010	1000000
Acoelomorpha	0000000000	0000000
Macrostomida	0111100010	1100101
Haplopharyngida	0100000010	0100101
Polycladida	0111000010	1000001
Prolecithophora	0101010011	1000011
Tricladida	0111010010	1000001
Lecithoepitheliata	0101000100	1000000
Proseriata	0110101010	1001001
Dalyelliida	0101000111	1000010
Temnocephalida	0101000101	1000010
Typhloplanida	0101000101	1000010
Kalyptorhynchia	0101000111	1000010
Fecampiida	0110101010	1000000
Urastoma	0110101010	1000001
Ichthyophaga	0110101010	1000011
Aspidogastrea	0110101010	1001011
Digenea	0110101010	1001011
Polyopisthocotylea	0110101010	1001011
Monopisthocotylea	0110101010	1001011
Gyrocotylidea	0110100010	1011001
Amphilinidea	0110100010	1010011
Eucestoda	0110100010	1010001

method will depend on how many characters (e.g., in addition to protonephridial ones) are available for a phylogenetic analysis. The following analyses show that the most conservative approach (Table 19.1) is more successful in resolving relationships between taxa than the other two approaches, although the other approaches may conceivably be more successful when a wider range of characters (including non-protonephridial ones) is used. The reason is that they permit resolution over a wider range of taxa.

Data were analysed using PAUP 4.0.0d65 (Swofford 1998). Acoelomorpha were excluded from all analyses, and bootstrap replicates were 1000 throughout. Bootstrapping with heuristic search of the data in Table 19.1 showed the following monophyla: Macrostomida/Haplopharyngida (branch support 65), Polycladida/Tricladida (64), Dalyelliida/Temnocephalida/

Typhloplanida/Kalyptorhynchia (66), Aspidogastrea/ Digenea/ Polyopisthocotylea/Monopisthocotylea (67), Gyrocotylidea/ Amphilinidea, Eucestoda (62). Catenulida including *Retronectes* appeared as the sister-group of all other taxa.

Bootstrapping with heuristic search of the data in Table 19.2 showed the catenulids as the sister-group of the rest (66), and a single monophylum Polycladida/Tricladida (66).

Bootstrapping with heuristic search of the data in Table 19.3 showed catenulids as the sister-group of the rest (69), and a single monophylum consisting of the Lecithoepitheliata plus 'dalyellioids' (56).

Conclusions

The above discussion has shown that the ultrastructure of protonephridial flame bulbs is indeed a useful phylogenetic tool in the Platyhelminthes, but that flame bulbs on their own are insufficient to resolve many of the phylogenetic relationships between taxa.

ACKNOWLEDGEMENTS

Financial support was provided by the University of New England and the Australian Research Council. Louise Streeting and Sandy Hamdorf helped with re-drawing of the figures. Reinhard Rieger supplied the specimen of *Retronectes*, Nikki Watson made her unpublished data on rhabdocoels available to me, and carried out preparative work for Figures 19.3, 19.4 and 19.6. I also wish to thank Tim Littlewood and Rod Bray for organizing the symposium.

Appendix 19.1 Glossary of terms associated with protonephridial ultrastructure.

Cytoplasmic cord: cytoplasmic extension along the filtration apparatus
Distal canal cell: cell forming the protonephridial capillary beyond the proximal canal cell
External leptotrich: cytoplasmic outgrowth of the terminal or proximal canal cell into the tissue spaces around the flame bulb
External rib (rod): longitudinal cytoplasmic cord forming part of the filtration apparatus and located externally to the internal ribs
Filtration apparatus: part of the flame bulb which contains a 'membrane' of extracellular matrix through which filtration of the excretory fluid occurs (in some species a 'membrane' is lacking and excretion is entirely by exocytosis)
Flame: bundles of cilia in the flame bulbs
Flame bulb: part of the protonephridium containing the flame and the filtration apparatus
Internal leptotrich: cytoplasmic outgrowth of the terminal cell into the lumen of the flame bulb
Internal rib (rod): longitudinal cytoplasmic cord forming part of the filtration apparatus and located internally to the external ribs
Lateral flame: cilia arising from the proximal or distal canal cells and extending into the protonephridial capillaries
Proximal canal cell: cell between terminal and distal canal cells and usually contributing to the formation of the flame bulb
Terminal cell: cell at the proximal end of the protonephridial duct which forms (part of) the flame bulb
Weir: see 'filtration apparatus'

Chapter 20

Insights from comparative spermatology in the 'turbellarian' Rhabdocoela

Nikki A. Watson

The Rhabdocoela is a large taxon within the Platyhelminthes, and its members demonstrate a wide variety of forms, characteristics and life styles. Hyman (1951) included the Catenulida and Macrostomida within it (along with the Typhloplanoida and Temnocephalida), while Ehlers (1985a) excluded Catenulida and Macrostomida, but did include the parasitic groups (Neodermata), based on the presumed synapomorphy of a bulbous pharynx. I use Rhabdocoela to include only those species traditionally classified into Kalyptorhynchia, Typhloplanida, Dalyelliida and Temnocephalida, exploring evidence from sperm studies that may be informative about relationships between the principal and subordinate taxa.

Ehlers' phylogenetic scheme included 'Typhloplanoida' and 'Dalyellioida' within Rhabdocoela, and neither was thought to be monophyletic. 'Typhloplanoida' (with prominent anterior rhabdoid tracts or a proboscis) included Typhloplanida (rhabdoid tracts) and Kalyptorhynchia (anterior proboscis and axonemes fully incorporated in sperm body). 'Dalyellioida', lacking such tracts or proboscis, included Dalyelliida (doliiform pharynx) and Temnocephalida (anterior tentacles and/or posterior suckers). Subsequent studies have failed to find evidence in support of the two higher taxa, and in fact only two orders, Temnocephalida and Kalyptorhynchia, are reasonably well defined. Within the other two orders, it may in fact be safest to consider species only as members of a particular family, when gathering data for phylogenetic purposes. Moreover, even some 'families' are of doubtful solidarity, since they are based largely on gross anatomical characters, especially the gonads, which often exhibit very flexible arrangements. This emphasizes a real dilemma in the establishment of robust phylogenies in the Platyhelminthes, i.e., the lack of extensive comparative data on morphological characters of established homology.

Sperm and spermiogenesis data contribute to phylogenetic analyses

Data on sperm characteristics, particularly those determined at the electron microscopical (EM) level, have been used for many years to infer relationships within various phyla. In 1976, Bacetti and Afzelius declared 'comparative studies have shown that interspecific variations in sperm structure are enormous and without parallel in other cell types'. More than 20 years later, EM studies of sperm continue to expand the known ranges of this variation.

The male gametes of members of the Platyhelminthes are particularly rich in complexity and diversity, compared with those of many other phyla, and their sperm and sperm development (spermiogenesis) provide many characters for phylogenetic analysis. The parasitic groups, and in particular the Monogenea, were the focus of many of the early EM studies, but more recent years have seen several research groups also targeting free-living taxa (see reviews by Justine 1995; Watson and Rohde 1995b; Watson 1999b and earlier references in each). Many useful data have now been gathered, and some clear sperm synapomorphies have emerged. The number of species examined and the families into which they are classified does, however, still represent a small percentage of the phylum and there remains much to be done. Nevertheless, it is probably true to say that some data about sperm characteristics are known from a larger number of platyhelminth taxa than are data about any other single ultrastructural characteristic. This fact alone makes these data useful components of a morphological database for cladistic analysis.

Sperm are undoubtedly homologous structures (across the phylum and most likely throughout the Animal Kingdom), as are some of the components and developmental events that are shared between taxa. On the other hand, it is also likely that some characters are convergent, having arisen through common selection pressure for sperm to be efficient and effective in their ultimate goal of fertilization of the ovum. The possibility of convergence does not preclude the use of sperm characteristics, and indeed the concomitant study of spermiogenesis helps to detect developmental occurrences. It does, however, dictate that the homology of any character be clearly established, and that its distribution be mapped widely, so that instances of parallel evolution may be detected.

General description of sperm and spermiogenesis in rhabdocoel species

Sperm

In common with mature sperm from some other platyhelminth groups, sperm of most rhabdocoel species have the following general construction:

> 'The sperm shaft is filiform and lined with cortical longitudinal microtubules. There is no acrosome, nor recognizable head, middle piece or tail region, but two free flagella are attached at or near the anterior end. An elongated nucleus extends for most of the length of the shaft. Numerous mitochondria, or one or more fused mitochondrial rods or derivatives, and various kinds of dense granules and/or dense bodies are distributed in various arrangements within the cytoplasmic region. Sperm of some species have no flagella, and some have flagellar axonemes that are fused or incorporated with the sperm body.'

While the elements listed above are common to most species, their appearances and arrangements vary greatly across taxonomic divisions, providing characters that can be used in a morphological database. In rhabdocoels, sperm that have neither flagella nor incorporated axonemes are the simplest, while many other sperm have complex structures associated with the attachment regions of their flagella. Sperm of some species also have up to five distinctly different kinds of dense bodies, in various arrangements in the cytoplasm, and some have complex, spiral elaborations of the anterior end. Many of these characters represent synapomorphies, providing phylogenetic resolution at various levels.

Spermiogenesis

In rhabdocoel species, this follows a sequence common also to proseriates, triclads and lecithoepheliates – other turbellarian orders that are also placed in the taxon Trepaxonemata (Ehlers 1985a). Members of Trepaxonemata are clearly defined by the presence of a unique complex central element in the sperm axonemes (see Figures 20.2, 20.3, 20.5, etc.). The parasitic groups, constituting the Neodermata, also belong to the Trepaxonemata, but are distinguished from the turbellarian members by a different mode of sperm development. In neo-dermatans, development is termed proximo-distal (Justine 1991a), while in most turbellarians it is distal-proximal as outlined below.

The general pattern of sperm development (spermiogenesis) in rhabdocoels (and other turbellarian members of Trepaxonemata) is therefore as follows:

'Following meiosis of spermatocytes, early spermatids remain joined together in isogenic clusters. The nucleus of each spermatid moves to the apical region, and a zone of differentiation (ZD) develops between the nuclear envelope and the apical cell membrane. In this ZD, two flagella grow out from basal bodies that are normally oriented at 180° to one another. The basal bodies flank a multipartite structure known as an intercentriolar body (ICB), composed of a series of disc-shaped parallel plates of varying electron density. Striated rootlet structures extend from the sides of the basal bodies towards or alongside the nucleus. The ZD is then carried distally, as the spermatid shaft elongates, while nucleus and other components move out from the residual cytoplasm to their final positions along the shaft. A row of longitudinal microtubules, originating in the ZD, extends along most of the shaft, immediately beneath the plasma membrane.'

In some taxa, several distinctive events or structures are superimposed on this common developmental sequence. These constitute synapomorphies providing phylogenetic distinction at various levels.

Synapomorphic characters based on sperm and spermiogenesis

As yet, we do not know of any sperm or spermiogenesis characters that unite the four orders of the Rhabdocoela. The archetypal rhabdocoel sperm probably closely resembled that of proseriates, and fitted the basic description given above. In the other neoophoran turbellarian orders, Lecithoepitheliata, Prolecithophora and Tricladida, sperm are specialized to vary-

ing degrees (see reviews by Watson and Rohde 1995b; Watson 1999b), and it therefore seems unlikely that the ancestral rhabdocoel sperm was derived from any of those types.

Within the Rhabdocoela, however, several synapomorphies relating to sperm or spermiogenesis have emerged. Some unite taxa from three of the four orders, some unite members of a suborder, some unite genera of a single family, and some unite only species of a genus. Thus they provide data for various levels of phylogenetic resolution. The following section describes these apomorphic characters, briefly outlines their distribution, and mentions the level of resolution that they afford, based on the limited number of species that have been examined to date. The known distributions of some of these characters within examined rhabdocoel species are listed in Table 20.1. Gaps in the table indicate that a particular character was not mentioned, either as absent or present, nor can its presence be determined from published micrographs.

The greatest impediment to the use of these data at the present time is the relatively very small number of species in which sperm and spermiogenesis have been examined. Furthermore, unfortunately, most of the earlier investigations did not report on many of the characters now known to be informative, or did not document them in sufficient detail. There are even significant numbers of families from which no member has yet been studied. Characters that are presently known only from a single genus may well be shown in future to be present in related genera, and those now known to be common to a particular family may prove also to be present in related families.

1. Small granules (approximately 30–50 μm diameter) arranged in a single layer of longitudinal rows along most of the sperm shaft, directly beneath the cortical microtubules (see Figures 20.1, 20.2 and 20.13) (character 1; Table 20.1). These have been documented in many (but not all) species of Typhloplanida, Dalyelliida and Temnocephalida, in both flagellate and aflagellate sperm. They do not occur in the sperm of any members of Kalyptorhynchia. In part of the sperm of many species they have been shown to occupy a semicircular distribution, with a much larger dense body occupying the space between the ends of the semicircle (Figure 20.2). Their appearance and location are highly consistent across taxa, and they represent a strong synapomorphy. They are absent, however, from *Brinkmaniella* sp., *Provortex* spp., as well as all members of Trigonostomidae, Typhloplanidae, Umagillidae, Pterastericolidae and

---

Figure 20.1 *Pogaina* sp. (Dalyelliida). Longitudinal section of sperm. Note nucleus (N), row of separate mitochondria (M), and longitudinal rows of small granules (arrows) beneath the cortical microtubules. Scale bar, 500 nm.

Figure 20.2 *Pogaina* sp. (Dalyelliida). Transverse section near anterior of sperm. Note free flagella (F), mitochondrion (M), nucleus (N) with connection to plasma membrane (arrow), granules (arrowhead) beneath cortical microtubules, and dense body (D) within the circle of granules. Scale bar, 200 nm.

Figure 20.3 *Adenorhynchus balticus* (Typhloplanida). Longitudinal section near anterior end of sperm. Note axonemal spur where flagellum (F) is embedded in sperm shaft, and small dense bodies (D). Scale bar, 250 nm.

Figure 20.4 *Adenorhynchus balticus* (Typhloplanida). Flagellar attachment region of spermatid before flagellar rotation. Note dense heel (arrow) on basal body of flagellum (F). Arrowhead points to nucleus of another spermatid – note attachment to plasma membrane. I, intercentriolar body. Scale bar, 200 nm.

Figure 20.5 *Adenorhynchus balticus* (Typhloplanida). Flagellar region of spermatid after rotation. Note axonemal spurs forming in place of dense heels. Arrowhead points to central group of microtubules, compressed during rotation from their original cortical location between the flagella (F). Scale bar, 200 nm.

Figure 20.6–20.10 *Adenorhynchus balticus* (Typhloplanida). Semi-serial micrographs from near anterior of a single spermatozoon (6 = most anterior). Note dense bodies (D), axonemal spurs, basal bodies (arrowheads) and central microtubules (arrows) derived from compressed cortical group. Scale bar, 200 nm.

Figure 20.11 *Thylacorhynchus ambronensis* (Schizorhynchia). Near tip of spermatid. Note dense heel (arrowhead) at end of basal body of spermatid. F, flagellum. Scale bar, 200 nm.

Figure 20.12 *Baltoplana magna* (Schizorhynchia). Transverse section through spermatid near anterior tip after flagellar rotation. Note dense heels (arrowheads) at ends of both basal bodies, but only one developed flagellum (F). Scale bar, 200 nm.

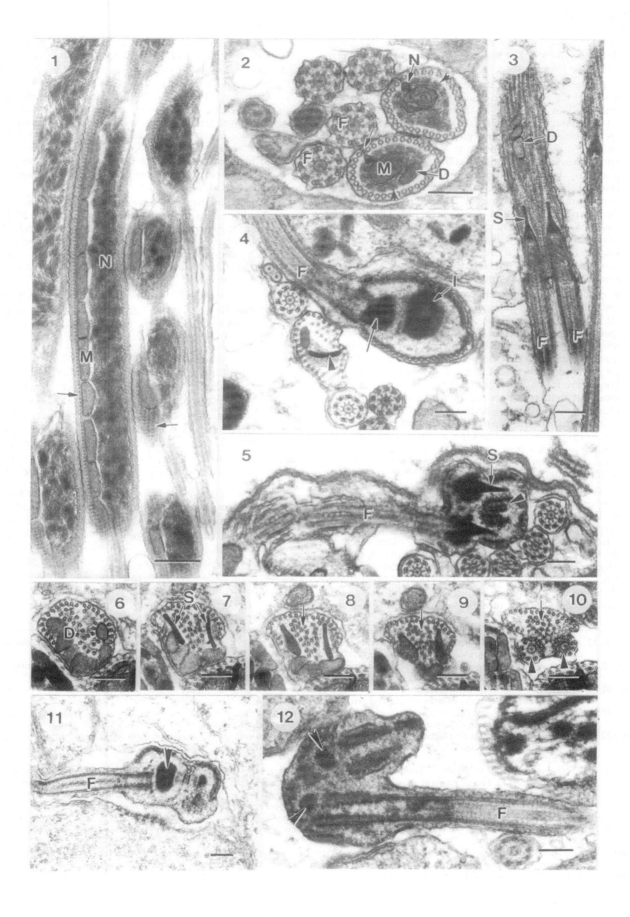

Luridae, which may be due to the loss of the feature in those groups, or they may be not part of the clade in which the feature arose.

2. A pointed, spur-shaped structure (axonemal spur) at the base of the sperm axoneme where the axoneme is embedded in the sperm shaft (Figures 20.3 and 20.7–20.9) (character 2; Table 20.1). This originates during spermiogenesis as a dense globular heel (Figures 20.4 and 20.5) that first appears at the base of each basal body when the spermatid is still rounded or pyriform, prior to shaft elongation (but see also character 4 below). The globular heels gradually become attenuated to pointed structures (Figure 20.5), the pointed end facing the anterior or distal end of the spermatid. The structure has been documented in most examined flagellated sperm of typhloplanids, dalyelliids and temnocephalids, with the known exceptions of the symbiotic dalyelliid species, *Pterastericola* spp. (Pterastericolidae) and *Syndisyrinx punicea* (Umagillidae), and possibly other umagillids.

3. A group of longitudinal microtubules lying between the embedded ends of the sperm axonemes (Figures 20.6–20.10) (character 3; Table 20.1). They originate, during spermiogenesis, from that part of the ring of cortical microtubules that is initially situated between the emergent flagella, and that becomes compressed, as flagella rotate around the shaft, into a distinctive, collapsed horse-shoe formation (Figure 20.5). This feature has the same distribution as does the axonemal spur described in character 2 above, and neither feature has been found independently of the other. There is, however, no obvious functional reason for the linkage, so it is possible that they are independent features, inherited from a common ancestor, and that species do or did exist in which only one feature is/was present.

4. Dense heel on the basal bodies during spermiogenesis, that may or may not develop into an axonemal spur (character 4; Table 20.1). The transient form, disappearing late in spermiogenesis, has been documented during spermiogenesis in the schizorhynchs *Baltoplana magna* (Figure 20.12), *Karkinorhynchus bruneti*, and three species of *Thylacorhynchus* (Figure 20.11), and in some proseriates (see e.g., Sopott-Ehlers 1990, 1994a). The feature may therefore be plesiomorphic in the Rhabdocoela, and have become further modified into the axonemal spur

structure (character 2), in a branch containing some typhloplanids, dalyelliids and temnocephalids.

5. A membranous connection between the nuclear membrane and the sperm plasma membrane, often fine and attenuated (Figures 20.2, 20.13 and 20.34) (character 5; Table 20.1). This feature has been documented in some flagellated typhloplanids, dalyelliids and temnocephalids, but not in any aflagellated species, nor in any kalyptorhynchs. Detection of this character requires good fixation of material, and careful observation. Most authors have not referred to the feature, but their micrographs sometimes reveal it.

6. Flagellar axonemes. (Numerals used in this section refer to flagellar character distribution listed in Table 20.1.) The ground pattern of platyhelminth sperm undoubtedly included two free flagella, and this also appears to be the plesiomorphic condition inherited by the Rhabdocoela. Within Rhabdocoela, a variety of modifications exist:

- Two free flagella (1). All temnocephalids as well as some members of Typhloplanida and Dalyelliida retain this primitive feature (e.g., Figures 20.2, 20.4, 20.13 and 20.19).

- Two completely fused axonemes (2a). Sperm of all kalyptorhynchs have axonemes that fuse with the shaft during spermiogenesis in such a manner that they lie completely surrounded by the cortical microtubules. All examined species of Eukalyptorhynchia thus have sperm with two deeply fused axonemes (Figures 20.14, 20.23 and 20.25). Sperm of the typhloplanid *Strongylostoma radiatum* also (independently?) have deeply fused axonemes (Figure 20.15).

- Two superficially fused axonemes (2b). The axonemes are surrounded by the sperm cytoplasm and plasma membrane, but not by the cortical microtubules, which have merely been pushed to a more internal position. This feature is known in one species of Typhloplanidae (Figures 20.16 and 20.17), one Graffillidae (Watson and Jondelius 1995) and several species of *Pterastericola* (Jondelius 1992; Watson *et al.* 1993). It is also known to exist in at least two other

Figure 20.13 *Adenorhynchus balticus* (Typhloplanida). Transverse section of sperm in sperm duct. Note nuclei (arrowheads) connected to plasma membrane, free flagella (F), granules (G) and dense bodies (D). Scale bar, 500 nm.

Figure 20.14 *Duplacrorhynchus heyleni* (Eukalyptorhynchia). Transverse section of sperm. Note two fully incorporated axonemes (A), single rod-like mitochondrion (M), small nuclear profile (N), and dense bodies (D). Scale bar, 200 nm.

Figure 20.15 *Strongylostoma radiatum* (Typhloplanida). Transverse section of sperm near anterior end. Note two fully incorporated axonemes (A), some remnant internal microtubules (arrowheads) and dense body (D). Scale bar, 200 nm.

Figure 20.16 *Phaenocora anomalocoela* (Typhloplanida). Transverse section of sperm near anterior end. Note dense bodies (D), small nuclear profile (N), and two superficially fused axonemes (A) not enclosed within the ring of cortical microtubules. Scale bar, 300 nm.

Figure 20.17 *Phaenocora anomalocoela* (Typhloplanida). Transverse section, more proximal region of sperm after one axoneme has terminated. Note nucleus (N), mitochondrion (M), dense body (D) and superficially fused axoneme (A) not enclosed within the ring of cortical microtubules. Scale bar, 250 nm.

Figure 20.18 *Baltoplana magna* (Schizorhynchia). Transverse section of sperm. Two profiles on left are near anterior, containing only the single incorporated axoneme (A). Profile on right shows single axoneme (A), mitochondrial rod (M) and nucleus with several dense chromatin rods (arrow). Scale bar, 200 nm.

Figure 20.19 *Actinodactylella blanchardi* (Temnocephalida). Transverse section of sperm near flagellar attachment zone. Note free flagella (F), and spiral configuration of cortical microtubules (arrowheads). Scale bar, 200 nm.

Figure 20.20 *Actinodactylella blanchardi* (Temnocephalida). Longitudinal section of posterior of sperm with split ends (arrowhead). Note nucleus termination (N). Scale bar, 1 μm.

Figure 20.21 *Actinodactylella blanchardi* (Temnocephalida). Transverse section of posterior of sperm, split into several strands (arrowheads) each containing one or more of the cortical microtubules. Note electron-dense nucleus of adjacent sperm. Scale bar, 200 nm.

Figure 20.22 *Temnocephala dendyi*. Note transverse section of narrow, folded strip of the posterior end of one sperm (arrowhead), containing cortical microtubules. Scale bar, 500 nm.

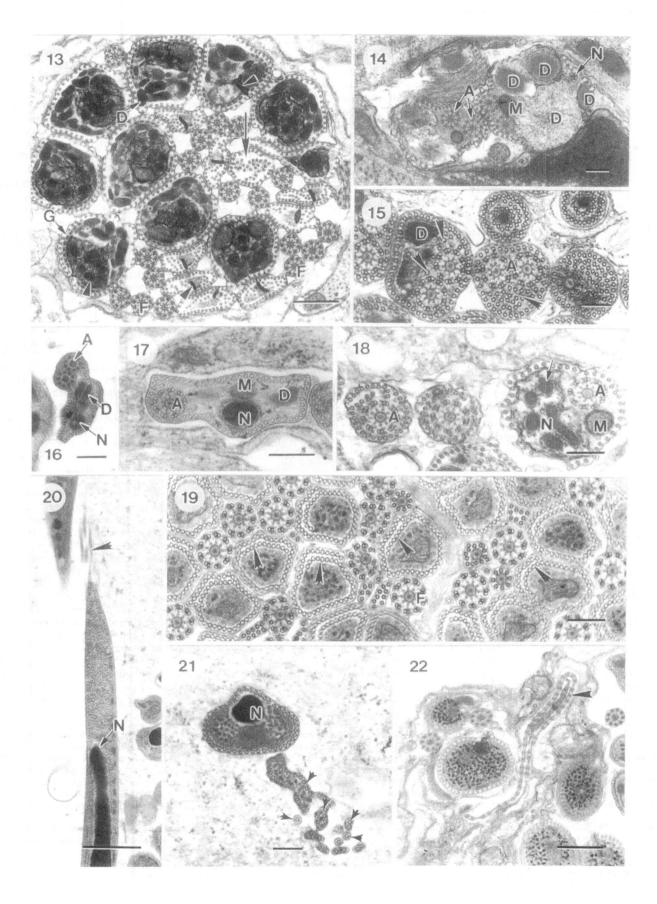

unidentified typhloplanids (personal observation). It is likely that such superficial fusion has taken place independently of the deep fusion in kalyptorhynch sperm. A particular kind of fusion appears to represent a synapomorphy for the species of *Pterastericola*, and the same mechanism or an independent one may also be a synapomorphy for another group of taxa.

- One fused axoneme (3). In all examined species of Schizorhynchia, one of the two initially free flagella does not develop beyond a small bud (Figure 20.12), and subsequently degenerates leaving no trace in mature sperm (Figure 20.18). In one species classified as a cicerinid eukalyptorhynch (*Nannorhynchides herdlaensis*), one of the two axonemes degenerates in the same way, while in a closely related species (*Toia calceformis*) both basal bodies are abortive (see Watson 1998b). This strongly implies a close relationship of these two species with the Schizorhynchia.

- No sperm axonemes (4). Sperm of some typhloplanids, some dalyelliids, and one kalyptorhynch lack any axonemal structures, either as free flagella or as incorporated axonemes. While this includes all examined species of Trigonostomidae (Typhloplanida) from five genera, and the one examined species of Luridae, it only applies to some of the species classified in the dalyelliid families Graffillidae and Provorticidae. In the trigonostomids and in the three *Provortex* spp., centrioles and a vestigial ICB are present during early spermiogenesis, but in three *Paravortex* spp. and in *Luriculus australiensis* not even these rudimentary structures appear to be present. The loss of axonemes has probably occurred independently several times, and/or the current classification may not reflect phylogeny. Nonetheless, the distribution of non-axonemal sperm may still be informative at a lower taxonomic level.

7. Spiralling of cortical microtubules (character 7; Table 20.1). In all examined species of Temnocephalida, with the exception of the scutariellid *Troglocaridicola* sp., there is a region of the sperm shaft just posterior to the flagellar insertion point in which the longitudinal cortical microtubules form an overlapping row, best seen in transverse sections (Figure 20.19). This spiralling arrangement is caused by the very regular manner in which the compressed, central group of microtubules (described in character 3 above) rejoins the outer, non-compressed group of microtubules. In the other rhabdocoel taxa exhibiting the

central compressed group of microtubules, the manner of recovery of the full cortical arrangement is not regular, and these microtubules are simply scattered in the interior shaft in this region (Figures 20.7–20.10, 20.13 and 20.15). The spiral feature represents a clear synapomorphy for the Temnocephaloidea, including Didymorchiidae, Diceratocephalidae, Actinodactylellidae and Temnocephalidae. Its apparent absence in the Scutariellidae reinforces other morphological data that suggest the Scutariellidae is the sister-group to the other Temnocephalida (see Joffe and Cannon 1998a, and other references cited therein).

8. Splayed or split posterior end (character 8; Table 20.1). In all members of Temnocephalida in which the posterior end of the sperm has been identified, the cytoplasm either spreads laterally, still lined with microtubules (Figure 20.22), or it splits into a variable number of 'threads', each containing some microtubules (Figures 20.20 and 20.21). The two alternative arrangements appear to be expressions of a single tendency, since both are found within one genus, *Temnocephala*, and the feature constitutes a synapomorphy for the Temnocephalida.

9. Spiral anterior projection (character 9; Table 20.1). In the few instances where sperm penetration of an ovum has been observed by electron microscopy, it is the end that contains the flagellar basal bodies or their remnants that penetrates first (see Figure 20.27), and is therefore designated as anterior. A cork-screw-like arrangement of dense material on the outer surface of this end has been documented in *Thylacorhynchus ambronensis* (Schizorhynchia) (Watson and Schockaert 1996), three species of *Pogaina* (Dalyelliida) (Watson 1999c), and *Troglocaridicola* sp. (Temnocephalida) (Iomini et al. 1994). Such a reinforced anterior structure probably aids rigidity during locomotion, and penetration of the ovum. It may also allow the sperm to anchor in the walls of female sperm storage organs, and avoid displacement by sperm from subsequent matings. The feature has probably arisen several times independently, and also occurs in some monogeneans and probably all eucestodes (Bâ and Marchand 1995) (also probably not homologous). However, its distribution in rhabdocoels is, nevertheless, informative at lower levels of phylogenetic resolution. For example, two other species of *Thylacorhynchus* are now known to have a similar structure to that found in *T. ambronensis*, although the pitch of the spiral ridges varies between species, being 400 nm in *T. conglobatuus* (Figure 20.30), 600 nm in *T. ambronensis*, and 800 nm in *T. pyriferus* (personal observation). In all three species, the nucleus and mitochondria also spiral tightly around the single axoneme along most of the length of the shaft (Figure 20.31). A similar projection occurs on mature sperm of *Proschizorhynchus pectinatus* (personal observation),

---

Figure 20.23 Rogneda patula (Eukalyptorhynchia). Sperm. Note two fully incorporated axonemes (A), row of dense bodies, each with an electron-dense core and floccular exterior (arrowhead), mitochondrion (M), and three other kinds of dense bodies (D). Scale bar, 300 nm.

Figure 20.24 Ancistrorhynchus ischnurus (Eukalyptorhynchia). Longitudinal section of autosperm. Note conspicuous row of electron-lucent bodies (D), each with dense spot. These bodies were absent from allosperm found beneath the epidermis of the same individual. Scale bar, 500 nm.

Figure 20.25 Limipolycystis sp. (Eukalyptorhynchia). Sperm transverse and longitudinal sections. Note two incorporated axonemes (A), separate mitochondria (M), very large dense body (D) and at least two other smaller kinds of dense bodies (arrowheads). Scale bar, 300 nm.

Figure 20.26 Duplacrorhynchus heyleni (Eukalyptorhynchia). Sperm longitudinal section. Two rows of dense bodies (arrowheads), each with a denser core and floccular outer region. Scale bar, 200 nm.

Figure 20.27 Rogneda patula (Eukalyptorhynchia). Sperm just penetrating an ovum. Note ovum nucleus with nuclear pores (P), anterior end of sperm (S) with two axonemes, and clear (lysed?) area (arrowhead) surrounding penetrating sperm. The remainder of this particular sperm was outside the ovum; the ovum contained no other parts of this or any other sperm. Scale bar, 2 μm.

Figure 20.28 Pogaina sp. (Dalyelliida). Longitudinal section anterior tip of sperm. Note spiral ridge (arrowhead). Scale bar, 200 nm.

Figure 20.29 Pogaina natans. (Dalyelliida). Longitudinal section of anterior region of sperm, showing spiralling, backward-facing ridge (arrowhead). Scale bar, 1 μm.

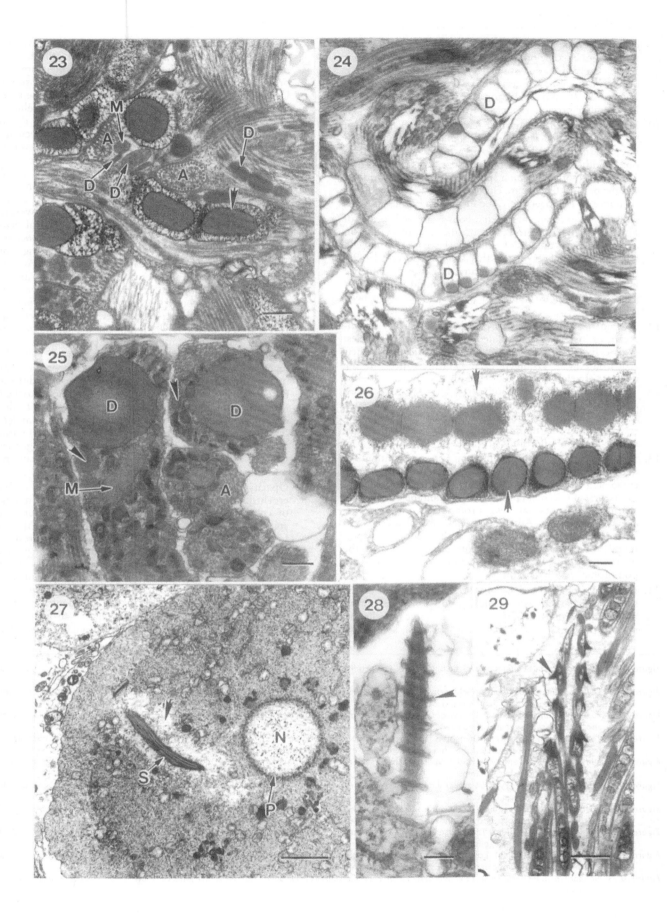

classified in the same family (Schizorhynchidae). A common origin at some point in Schizorhynchidae is therefore highly likely. Of four studied species of *Pogaina*, three develop a persistent anterior spiral in mature sperm (see Figures 20.28 and 20.29 for two species), although all have similar precursor ridges during spermiogenesis.

10. Flagella joined to one another (character 10; Table 20.1). Of four examined species of *Pogaina* (Dalyelliida), two have sperm in which the two free flagella are joined to one another by double-bridge structures (see Figure 20.32). Two other *Pogaina* spp. do not have such bridges, but the pairs of flagella are closely adjoined. This feature is therefore of phylogenetic value at the level of genus.

11. Archway over emergent flagella. Four examined species of *Pogaina* all have a projection or archway of cytoplasm extending over the distal area of the bases of flagella where they emerge from the shaft (Watson 1999c) (Figures 20.36–20.38). The feature differs in prominence between the species, and has not been found in any other genus, although no supposed close relatives have been examined.

12. Marginalization of heterochromatin. During early spermiogenesis in two closely related eukalyptorhynch species, *Nannorhynchides herdlaensis* and *Toia calceformis*, heterochromatin is arranged in a band just beneath the nuclear envelope (Watson 1998b) (see Figure 20.39). The formation is clearly homologous in the two species, and reflects their origin from a common ancestor. Examination of other species in the same family may reveal a wider distribution of the character, and be informative about relationships at this level.

13. Network of microvilli around elongating spermatids. All five examined species of Trigonostomidae from five different genera (see Table 20.1 for species and references) have a network of microvilli partly surrounding the outgrowing spermatids (somewhat less pronounced in *Beklemischeviella contorta*). All these species also produce basal bodies in a diplosomal formation in the ZD of early spermatids, but no axonemal elements develop. Clearly, the microvilli and the lack of sperm axonemes represent synapomorphies for this taxon.

14. Distinctive dense bodies. So-called dense bodies in the sperm are characteristic of most groups of turbellarians, and completely absent in neodermatans. Some turbellarian taxa also appear to lack them (e.g., triclads), some have only one or two different kinds of dense bodies, while in others up to five distinct kinds can be identified. Their arrangements within the sperm shaft are often precisely ordered both in cross-sectional and longitudinal positions. Dense bodies appear to be of Golgi origin, as is the material in the acrosome of other animal groups. Platyhelminth sperm lack an acrosome (with the exception of the Nemertodermatida), and Hendelberg (1983) speculated that dense bodies might play a role during penetration of the ovum. Sperm that had penetrated the oocytes of the proseriate *Notocaryoplanella glandulosa* indeed lacked the dense bodies that were conspicuous in autosperm of this species, and Sopott-Ehlers (1986) therefore concurred with the suggestion of Hendelberg. However, I have also found that the most conspicuous kind of dense body in the autosperm of a kalyptorhynch *Ancistrorhynchus ischnurus* (see Figure 20.24) was missing from allosperm located just beneath the epidermis (Watson 1999b). Moreover, allosperm in the bursa or around oocytes of three other kalyptorhychs also lacked at least one kind of dense body that was present in autosperm in the same individual (personal observation on *Alcha* sp., *Rogneda patula* and *Phonorhynchoides* sp.). Different types of dense bodies may therefore be utilized at several stages of the sperm journey from testis to ovum. The phylogenetic potential of these bodies has probably been underestimated, since at this stage they are only differentiated visually by electron microscopy, and no histochemical or immunological methods have been applied to their discrimination. Nonetheless, the distribution of at least one distinctive kind (see Figures 20.23 and 20.26) within the eukalyptorhynch family, Polycystidae, is being documented, as part of taxonomic revision (personal communication, T.J. Artois and others, 1998).

15. Opacity of the nucleus, appearance/arrangement of heterochromatin. In mature sperm, the degree and manner of condensation of heterochromatin varies (often conspicuously) between species. In sperm of many kalyptorhynchs and some typhloplanids, distinct heterochromatin strands, first appearing early in spermiogenesis, lie parallel with the longitudinal axis (e.g., Figure 20.18). In an extreme case, dense nuclear rods are individually surrounded by double membranes (*Toia calceformis*; see Watson 1998b, and Figure 20.33). By contrast, in the sperm nucleus of most dalyelliids, heterochromatin is of uniform opacity, or appears finely granular or reticulate.

Figure 20.30 *Thylacorhynchus conglobatus* (Schizorhynchia). Longitudinal section of sperm, showing spiral ridges (arrowheads) on anterior projection. Scale bar, 1 μm.

Figure 20.31 *Thylacorhynchus conglobatus* (Schizorhynchia). Longitudinal section of sperm shaft. Note that nucleus (N) and mitochondria (M) spiral around the single incorporated axoneme (A). Scale bar, 500 nm.

Figure 20.32 *Pogaina natans* (Dalyelliida). Transverse section part of group of four sperm. Note granules (arrowheads), and two axonemes joined to one another by bridging material (arrow). Scale bar, 200 nm.

Figure 20.33 *Toia calceformis* (Eukalyptorhynchia). Transverse section of sperm. Note widely separated cortical microtubules, and nuclear rods (arrowheads) each separately enclosed within a double nuclear envelope. Scale bar, 500 nm.

Figure 20.34 *Didymorchis* sp. (Temnocephalida). Sperm transverse section. Note uniformly opaque nucleus (N) with connection to plasma membrane (arrowheads). Scale bar, 500 nm.

Figure 20.35 *Proxenetes fasciger* (Typhloplanida). Longitudinal section of sperm. Note strongly twisted nucleus (N), and numerous lucent dense bodies (arrowhead). Scale bar, 1 μm.

Figure 20.36 *Pogaina kinnei* (Dalyelliida). Transverse section through spermatid zone of differentiation. Note early dense formations on basal bodies (arrow), intercentriolar body (I), and lateral cytoplasmic extensions (arrowheads). Scale bar, 1 μm.

Figure 20.37 *Pogaina* sp. (Dalyelliida). Transverse section through outgrowing flagellum (F) of spermatid. Note nucleus (N) and cytoplasmic archway (arrowhead) over the flagellum. Scale bar, 500 nm.

Figure 20.38 *Pogaina* sp. (Dalyelliida). Longitudinal section through zone of differentiation of early spermatid. Note outgrowing flagella (F), and distal, overhanging cytoplasmic extension (arrowhead). Scale bar, 500 nm.

Figure 20.39 *Nannorhynchides herdlaensis* (Eukalyptorhynchia). Part of group of early spermatids. Note marginal position of heterochromatin (arrowhead) in the nucleus. Scale bar, 1 μm.

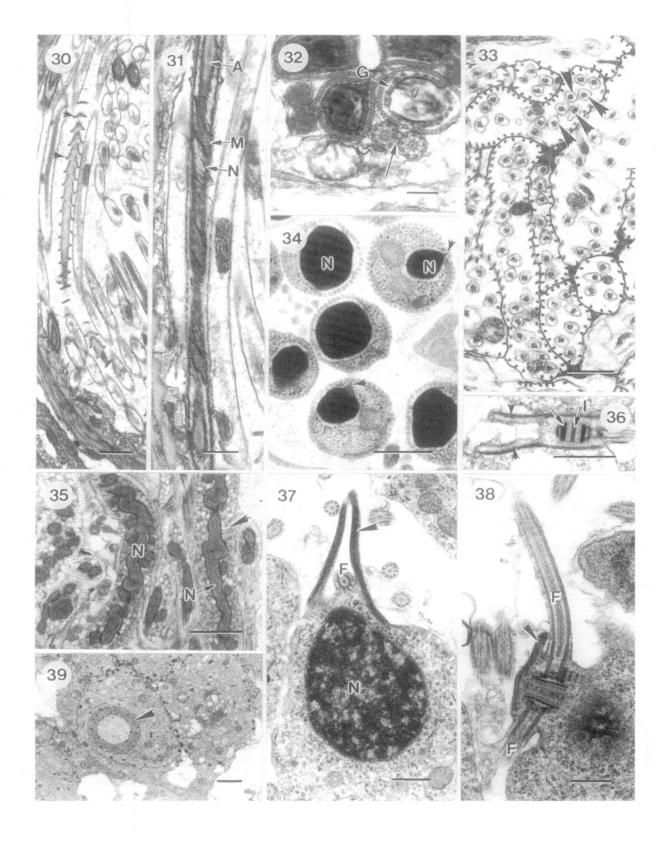

Related taxa display similar nuclear profiles, and these may well reflect specific arrangements of genetic material (e.g., Joffe *et al.* 1998a), and therefore be of phylogenetic value. At least three species of trigonostomids (*Trigonostomum setigerum*, *Proxenetes fasciger* and *Ceratopera* sp.) share another nuclear feature, viz. pronounced twisting along parts of the shaft, especially in bursal sperm (see Figure 20.35).

16. Length of flagella/axonemes. This character is known to vary, even within a genus (e.g., *Pogaina* spp.; see Watson 1999c); free flagella range in length from as short as one-eighth the length of the shaft to longer than the shaft. The lengths of incorporated axonemes also vary to a similar degree, and the two incorporated axonemes of each sperm are frequently of different lengths. The character is constant within a species, and undoubtedly reflects ancestry, but it is not an easy character to assess or quantify using transmission electron microscopy. Tubulin immunocytochemistry has been used successfully to determine the position and length of axonemes (see e.g., Iomini *et al.* 1995), but is best suited to species that can be cultured or found naturally in abundance, to provide living material and large numbers of sperm for investigation.

17. Numbers and arrangements of mitochondria. Mitochondrial material in the sperm of rhabdocoel species may be present as many small discrete mitochondria, arranged apparently randomly or in a very orderly longitudinal row or rows, or it may be fused into one or a small number of mitochondrial rods that may be relatively unmodified, or modified considerably so as to be only recognizable as of mitochondrial origin by the presence of the double membrane envelope. The arrangement of mitochondrial elements is similar in related species, and therefore of phylogenetic value, so we should continue to document it. For example, in most examined schizorhynchs there is a single long mitochondrial rod, while some eukalyptorhynch species have sperm with multiple mitochondria, and others have a single rod. However, some degree of fusion appears to have occurred several times, possibly independently across taxa. Furthermore,

while it is quite feasible to show that individual mitochondria exist, it is more difficult to be sure that an apparently single rod is not made up of several long, closely adjoined units.

Phylogenetic conclusions

Cladistic analyses have not yet been carried out on rhabdocoel taxa using only sperm and spermiogenesis data. To date, insufficient numbers of species have been studied in the required detail, to determine the distribution of informative characters for analysis by computerized means. Nevertheless, it is still useful to look critically and qualitatively at the data, 'by hand', and to come to some realistic conclusions.

1. Temnocephalida

(i) Sperm data corroborate the monophyly of all examined Temnocephalida: they share the synapomorphy of a split or splayed proximal end of sperm.

(ii) The synapomorphy, spiral region of microtubules, is shared by all except *Troglocaridicola* sp., implying that Scutariellidae is the sister-taxon to the remaining temnocephalids (Temnocephaloidea of Joffe and Cannon 1998a).

(iii) Temnocephalida is part of a larger taxon that includes many dalyelliids and typhloplanids (see point 3 below).

2. Kalyptorhynchia

(i) Kalyptorhynchia is characterized by the sperm synapomorphy of flagella completely fused with the sperm shaft, i.e., fully incorporated axonemes.

(ii) All species of Schizorhynchia examined constitute another clade, sharing the synapomorphy of limited growth and eventual loss of one axoneme during spermiogenesis.

Table 20.1 Distribution of some selected characters in the sperm/spermiogenesis of rhabdocoel species.

Higher taxon	Family	Species	Characters present*										Source of data
			1	2	3	4	5	6	7	8	9	10	
	Cicerinidae	*Cicerina remanei*	0	0	0		0	2a	0	0	0	N/A	Watson 1999b
		Nannorhynchides herdlaensis	0	0	0	0	0	3	0	0	0	N/A	Watson 1998b
		Ptyalorhynchus coecus	0				0	2a	0	0	0	N/A	Watson 1999b
		Toia calceformis	0	0	0	0	0	4	0	0	0	N/A	Watson 1998b
	Gnathorhynchidae	*Ancistrorhynchus ischnurus*	0	0	0	0	0	2a	0	0	0	N/A	Watson 1999b
		Odontorhynchus sp.	0				0	2a	0			N/A	Rohde et al. 1987a
		Prognathorhynchus typhlus						2a	0			N/A	Hendelberg 1986
	Koinocystidae	*Brunetia camarguensis*	0	0	0	0	0	2a	0	0	1	N/A	Watson 1999b (referred to as *Utsurus camarguensis*)
Eukalyptorhynchia	Polycystidae	*Alcha* sp.	0	0	0		0	2a	0	1	0	N/A	pers. obs. and LUC Zoology Research Group
		Austrorhynchus sp.	0	0	0	0	0	2a	0	0	0	N/A	pers. obs. and LUC Zoology Research Group
		Cincturorhynchus sp.	0	0	0		0	2a	0	0	0	N/A	pers. obs. and LUC Zoology Research Group
		Duplacrorhynchus heyleni	0	0	0	0	0	2a	0	0	0	N/A	pers. obs.
		Gyratrix hermaphroditus, Gyratrix sp.	0	0	0	0	0	2a	0	0	0	N/A	Hendelberg 1986; Rohde et al. 1987a; Sopott-Ehlers and Ehlers 1998a; Watson 1999b
		Lacertorhynchus devochti	0	0	0	0	0	2a	0	0	0	N/A	pers. obs. and LUC Zoology Research Group
		Limipolycystis sp.	0	0	0		0	2a	0	0		N/A	pers. obs.
		Phonorhynchus helgolandicus						2a	0			N/A	Hendelberg 1986
		Polycystis ali	0	0	0		0	2a	0	0	1	N/A	pers. obs.
		Paulodora sp.	0	0	0	0	0	2a	0	0	1	N/A	Watson 1999b (referred to as *Polycystella* sp.)
		Polycystis naeglii	0				0	2a	0			N/A	L'Hardy 1988
		Rogneda patula	0	0	0		0	2a	0	0		N/A	Watson 1999b
		Typhlopolycystis sp.	0	0	0		0	2a	0	0	1	N/A	pers. obs.
		Undescribed polycystid	0	0	0		0	2a	0	0		N/A	pers. obs.

Continued

Table 20.1 Continued.

Higher taxon	Family	Species	Characters present*										Source of data
			1	2	3	4	5	6	7	8	9	10	
Schizorhynchia	Diascorhynchidae	Diascorhynchus rubrus	0	0	0	0	0	3	0	0	1	N/A	Watson 1999b
	Karkinorhynchidae	Baltoplana magna	0	0	0	1	0	3	0	0	0	N/A	Watson and L'Hardy 1995
		Cheliplanilla caudata	0					3	0			N/A	Watson 1999b
		Cheliplana remanei	0	0	0	0	0	3	0			N/A	Watson 1999b
		Karkinorhynchus bruneti	0			1	0	3	0			N/A	pers. obs.
	Schizorhynchidae	Carcharodorhynchus sp.1	0				0	3	0			N/A	Watson 1999b
		Carcharodorhynchus sp.2	0				0	3	0			N/A	Watson 1999b
		Limirhynchus danicus	0			0	0	3	0			N/A	Watson 1999b
		Proschizorhynchus pectinatus	0				0	3	0		1	N/A	pers. obs.
		Schizochilus marcusi	0	0	0	0	0	3	0			N/A	Watson 1999b
		Thylacorhynchus ambronensis	0	0	0	1	0	3	0	0	1	N/A	Watson and Schockaert 1996
		Thylacorhynchus conglobatus	0	0	0	1	0	3	0	0	1	N/A	Watson 1999b
		Thylacorhynchus pyriferus	0	0	0	1	0	3	0	0	1	N/A	Watson 1999b
Typhloplanida	Byrsophlebidae	Maehrenthalia sp., M. agilis	1	1	1	1	1	1	0	0	0	0	Watson and Jondelius 1995; Watson 1999a
	Promesostomidae	Adenorhynchus balticus	1	1	1	1	1	1	0	0	0	0	Watson 1999b
		Brinkmaniella sp.	0	1	1		0	1	0			0	Watson 1999b
	Solenopharyngidae	Anthopharynx sacculipenis	1		1	1		1					Sopott-Ehlers 1994b
		Aulopharynx sp.	1			1	1	1		1	0	0	Watson 1999b
	Trigonostomidae	Beklemischeviella contorta	0	N/A	N/A	N/A	0	4	N/A			N/A	Watson 1999b
		Ceratopera sp.	0	N/A	N/A	N/A	0	4	N/A	0	0	N/A	Watson 1999b
		Proxenetes fasciger	0	N/A	N/A	N/A	0	4	N/A			N/A	Watson 1999b
		Ptychopera westbladi	0	N/A	N/A	N/A	0	4	N/A	0	0	N/A	Sopott-Ehlers and Ehlers 1997
		Trigonostomum setigerum	0	N/A	N/A	N/A	0	4	N/A			N/A	Watson 1999b
	Typhloplanidae	Bothromesostomum personatum	0	1	1	1	0	1					Cifrian et al. 1988a
		Castrada natans	0	1			0	1	0			0	Watson 1999b
		Phaenocora anomalocoela	0	1	1	1	0	2b	0		0	N/A	Watson and Rohde 1994a
		Strongylostoma radiatum	0	1	1	1	0	2a	0	1	0	N/A	Watson 1999b
Dalyelliida	Dalyelliidae	Gieysztoria sp.		1	1	1	1	1	0	0	0	0	Watson and Rohde 1995a
		Halammovortex nigrifrons	1			1	1	1					Sopott-Ehlers 1998
		Jenseaia angulata	1	1	1	1		1					Sopott-Ehlers 1997b
	Graffilliidae	Bresslauilla relicta	1			0	1	2b	0	0	0	1	Sopott-Ehlers and Ehlers 1986; Watson and Jondelius 1995; Sopott-Ehlers and Ehlers 1995a
		Graffilla buccinicola	1										pers. obs.
		Paravortex cardii	1	N/A	N/A	N/A		4	N/A			N/A	Cifrian et al. 1988b
		Parvortex sp. 1, P. karlingi, P. cardii	1	N/A	N/A	N/A		4	N/A			N/A	Noury-Sraïri et al. 1989a
	Luridae	Luriculus australiensis	0	N/A	N/A	N/A		4	0			N/A	Rohde and Watson 1993b
	Provorticidae	Pogaina sp.	1	1	1	1	1	1	0	0	1	0/1	Watson 1999b; Watson 1999c
		Pogaina kinnei	1	1	1	1	1	1	0	0	1	1	Watson 1999b; Watson 1999c
		Pogaina natans	1	1	1	1	1	1/2b	0	0	1	1	Watson 1999b; Watson 1999c
		Pogaina suecica	1	1	1	1	1	1	0	0	1/0	0/1	Watson 1999b; Watson 1999c
		Provortex pallidus	0	N/A	N/A	N/A		4	N/A			N/A	Watson 1999b
		Provortex psammophilus	0	N/A	N/A	N/A	0	4	N/A			N/A	Sopott-Ehlers 1995
		Provortex tubiferus	0	N/A	N/A	N/A	0	4	N/A			N/A	Sopott-Ehlers and Ehlers 1986; Sopott-Ehlers and Ehlers1995b
	Pterastericolidae	Pterastericola spp.	0	0	0	0	0	2b	0			0	Jondelius 1992: Watson et al. 1993
	Umagillidae	Anoplodium stichopi											Sopott-Ehlers and Ehlers 1986
		Cleistogamia longicirrus	0			0		1				0	Rohde et al. 1988c
		Seritia stichopi	0			"1"		1				0	Rohde et al. 1988c; Watson 1999b
		Syndesyrinx punicea	0	0	0	0		1				0	Rohde and Watson 1988b; Li et al. 1992
Temnocephalida	Actinodactylellidae	Actinodactylella blanchardi		1	1	1	0	1	1	1			Watson and Rohde 1995a
	Diceratocephalidae	Diceratocephala boschmai	1	1	1	1	0	1	1	1	0	0	Watson et al. 1995a
	Didymorchiidae	Didymorchis sp.	1	1	1	1	1	1	1	1		0	Watson and Rohde 1995a; Rohde 1987c
	Scutariellidae	Troglocaridicola sp.	1	1	1	1		1	0	1	1	0	Iomini et al. 1994
	Temnocephalidae	Temnocephala spp.	1	1	1	1	1	1	1	1	0	0	Williams 1983, 1984, 1994; Justine et al. 1987; Watson et al. 1995a
		Craspedella spp.	1	1	1	1		1	1	1	0	0	Watson et al. 1995a; Rohde 1987c
		Decadidymus gulosus	1	1	1	1	0	1	1	1	0	0	Watson et al. 1995a
	??	Undescribed symbiotic turbellarian from brittlestar, Amphipholus squamata	1	N/A	N/A	N/A		4	N/A			N/A	Watson 1999b

* Characters identified as follows:
1. Small granules in longitudinal rows beneath the cortical microtubules of mature sperm.
2. Axonemal spur–a pointed, spur-shaped structure at the base of the sperm axonemes, originating as a dense globular heel during spermiogenesis.
3. A group of longitudinal microtubules between the embedded ends of sperm axonemes, derived from part of the cortical row by compression during flagellar rotation.
4. A dense heel developing on the ends of basal bodies during spermiogenesis.
5. A fine connection between the nuclear and plasma membranes along much of the sperm length.
6. Flagellar axonemes: (1) two free flagella; (2a) two axonemes completely fused with the sperm shaft; (2b) two superficially fused axonemes; (3) one fused axoneme; (4) no sperm axonemes.
7. Spiral arrangement of cortical microtubules just posterior to the flagellar insertion region.
8. Splayed or split posterior end of sperm.
9. Spiral anterior projection.
10. Paired flagella joined to one another.

(iii) Two species classified in the Eukalyptorhynchia (viz. *Toia calceformis* and *Nannorhynchides herdlaensis*) show similar axonemal reduction, and may therefore constitute a sister-taxon to the Schizorhynchia.

(iv) All species of Kalyptorhynchia lack all four distinctive characters found in many other rhabdocoels (see characters 1, 2, 3, and 5 in Table 20.1). If non-sperm data are assumed to validate the Rhabdocoela as a whole (as concluded by Littlewood *et al.* 1999a), then Kalyptorhynchia represents the sister-taxon to the remaining rhabdocoels.

3. Dalyelliida and Typhloplanida

Sperm data do not support the existence of these two higher taxa, but neither do they unequivocally indicate an alternative division. Rather they reveal:

(i) somewhat distant positions for Luridae, Trigonostomidae, Umagillidae, Pterastericolidae and *Provortex* species;

(ii) somewhat closer relationships between some 'typhloplanid' and some 'dalyelliid' families than within the traditional bounds of Typhloplanida and Dalyelliida. The five above-mentioned taxa (in 3(i)) lack all of the characters (1, 2, 3, 5) found in various combinations in other rhabdocoels classified in Typhloplanida, Dalyelliida, and Temnocephalida. The trigonostomids, *Luriculus* and the *Provortex* spp. may have lost characters 2 and 3 along with the loss of flagella, but alternatively they may never have possessed these characters;

(iii) existence of a large taxon (T1) comprised of (at least) all those investigated rhabdocoels that share features 2 and 3 (spur on embedded ends of axonemes that originates as a dense heel during spermiogenesis, and central group of microtubules near the flagellar insertion zone, originating from compression of a subset of the cortical microtubules). Apart from in the aflagellate species where they may have been lost, these features are otherwise known to be lacking only in pterastericolids and umagillids (as well as all kalyptorhynchs);

(iv) within T1, existence of a lower taxon, T2, the members of which share two synapomorphies: rows of granules beneath the cortical microtubules (character 1) and the connection between the nuclear and plasma membranes (character 5, subsequently lost or not detected in some temnocephalids). T2 excludes the Typhloplanidae.

4. Apomorphies

As mentioned earlier, several distinctive apomorphies each unite various smaller groups of species or genera. These include:

(i) the spiral anterior projection in several schizorhynch species, and, independently, in *Pogaina* species;

(ii) increasing tendency to fusion of flagella with one another in *Pogaina* species;

(iii) archway over outgrowing flagella during spermiogenesis in *Pogaina* species;

(iv) marginalization of chromatin in spermatid nuclei of two cicerinids (*Toia calceformis* and *Nannorhynchides herdlaensis*);

(v) microvillus network around spermatids, and non-axonemal sperm in Trigonostomidae; and

(vi) a distinctive kind of dense body found in several species of Koinocystidae and Polycystidae (Eukalyptorhynchia); it has a more dense core, surrounded, especially at the ends or corners, by a more expanded and distinctly floccular region.

Consideration of molecular data from rhabdocoels

The most recent, reasonably comprehensive phylogenetic analysis of the Platyhelminthes (Littlewood *et al.* 1999a) included molecular data from ten rhabdocoels (not including *Kronborgia isopodicola*, which is undoubtedly not a rhabdocoel, based on all morphological and molecular grounds). This is the largest number of rhabdocoel taxa for which molecular data are available at this time, and although the number is generally considered to provide quite a good coverage of a taxon in molecular studies, it is, nonetheless, less than 12% of the taxa now covered by sperm data. Furthermore, the study included species from only nine of 34 families in Rhabdocoela. Moreover, the selection of taxa was, of necessity, based on expediency – those that were available to the authors – rather than on perceived 'representativeness' or 'basal' status of the sequenced species within a taxon. In fact, several of the species used would be considered by specialists in those taxa to be rather atypical. For example, many species of Graffillidae (including the one analysed), and all members of Pterastericolidae are symbiotic rhabdocoels, with many specialized morphological characteristics, not typical of dalyelliids. Within the Typhloplanida, trigonostomids are remarkable for the uniformity of their sperm (as well as other characters), but they, as well as the pterastericolids and the selected graffillid, *Mariplanella frisia*, are conspicuous in the sperm data matrix for the lack of some characters shared by many other rhabdocoels. Indeed, in their analyses *Mariplanella frisia* consistently appeared 'outside' a clade containing the other two typhloplanids plus the temnocephalid and the dalyelliid, and the two symbiotic 'dalyelliids' appeared as a sister-taxon to the other dalyelliids plus typhloplanids. *Gyratrix hermaphroditus* is certainly a common kalyptorhynch species, readily available and of cosmopolitan distribution, but it is also unusual in that is actually a species complex. I am even aware of significant differences in the morphology of sperm between different strains (personal observation). There is, therefore, some justification for interpreting with caution phylogenies that use such species to represent a higher taxon. The morphological database used in that study also suffers from a similar limitation. For example, the characters listed as present for the various taxa are not always shared by all members of that taxon. The decision to designate a character as present is therefore based on an *a priori* decision about the status of the character within the taxon. If it is in fact derived within the taxon, as well as within other taxa, then taxa may be wrongly considered as closely related on the basis of a character that has evolved in parallel. This issue is especially important when so few species have been studied or sequenced.

The morphological database of Littlewood *et al.* (1999a) did incorporate two of the sperm characters most widely distributed amongst rhabdocoels, viz. the small cortical granules (character 1 in Table 20.1 here; character 36 in Littlewood *et al.*), and a combination of two characters – the dense heel that becomes an axonemal spur, and the internal microtubules that result from a particular manner of flagellar rotation (characters 2 and 3 in Table 20.1; character 35 in Littlewood *et al.*). These sperm characters were, in fact, considered by those authors to be the only unambiguous morphological features uniting the Typhloplanida, Dalyelliida and Temnocephalida, and that clade was strongly supported by morphological, molecular and 'total evidence' data. However, as pointed out in the section above (phylogenetic conclusions), the two characters used by Littlewood *et al.* (characters 35 and 36, but actually three characters since there is no justification for combining characters 2 and 3 of Table 20.1 here) are by no means present in or synapomorphic for all dalyelliids, typhloplanids and temnocephalids as is, however, implied by their morphological data matrix. In fact, all three characters are known to be absent from the Trigonostomidae and the Pterastericolidae, only one of the characters is present in the Graffillidae, and one is missing from Typhloplanidae. These are all families in which the sequenced species are placed.

Other Issues

Rate of evolution

Sperm data illustrate very clearly that the rate of evolution of some characters may be highly variable. For example, the five trigonostomids examined from five different genera have very similar sperm, barely distinguishable between genera, while sperm of four examined species of *Pogaina* are quite distinct, and trends in several characters are discernible. Such divergence in variability may be due to different processes of speciation. Two of the *Pogaina* species are currently sympatric, and reproductive isolation events may have initiated speciation in this genus, favouring sperm diversity, while the sperm uniformity in the trigonostomids may imply geographic isolation as the initiator of speciation. Such differences in the rate of evolution also have implications for the interpretation of molecular data.

Functional considerations

The interpretation of variation in sperm characters in the flatworms is seriously hampered by an almost total absence of knowledge about sperm and fertilization biology. The works of Vreys, Michiels and coworkers on some readily cultured triclads are notable exceptions (e.g., Vreys and Michiels 1997; Vreys *et al.* 1997a,b,c; Michiels and Newman 1998). The small size of most rhabdocoels, and the lack of established culture techniques, present enormous challenges in this area. All have internal fertilization, but genital tracts are notoriously complex and often highly variable between taxa. The route of sperm entry can vary from hypodermic impregnation to deep deposition into a storage bursa. The digestion of allosperm is also refined to various degrees.

Selection pressures on sperm are enormous: they must be competitive amongst their fellow ejaculates in the race to reach the oocyte, by moving rapidly through tissues or along variously muscular female ducts; they must resist being digested in the female ducts, or dislodged, either by fellow sperm or sperm of subsequent inseminations from different individuals, or by female choice mechanisms. Sperm competition is therefore likely to be a principal engine driving diversity. Indeed, sperm and female reproductive tract morphology are among the most rapidly evolving characters known in insects (Presgraves *et al.* 1999), and this is probably also true in the Platyhelminthes.

With very few exceptions, platyhelminths are hermaphrodites (often simultaneously so), although varying degrees of protandry are common. This situation presents conflicts of interest between the female and male roles and may, therefore, also be an important factor driving evolution and diversity of sperm. For example, aflagellate sperm may be incapable of directed movement and therefore reliant on muscular contractions of the female tract to reach their goal. Rather than as a result of positive selection for successful sperm traits, such sperm may therefore originate from selection of the female function for increased female choice and greater control over digestion of unwanted sperm.

The presence of enormous variation in different types of dense bodies in turbellarian sperm is also a complex and puzzling issue. Dense bodies are present in sperm of acoelans, macrostomids, prolecithophorans, lecithoepitheliates and proseriates, and are therefore most likely plesiomorphic in the Rhabdocoela. However, they are totally lacking from all neodermatans, all triclads, and some rhabdocoel species, e.g., three examined species of Karkinorhynchidae (Schizorhynchia), and three species of *Thylacorhynchus* (Schizorhynchidae), while some species have sperm that are packed with up to five different kinds of dense bodies. The total number of different kinds of dense bodies in rhabdocoels alone may exceed 20. Obviously, sperm can achieve their goals without dense bodies, but their presence and proliferation in the majority of turbellarians implies that they play important roles in the success of sperm. They may carry nourishment for the sperm, the means to resist digestion in the female tract, and/or enzymes for penetration of the recipient epidermis, intercellular matrix, and egg membranes. If some types of dense bodies are directly associated with better mobility or resistance to digestion, then their absence in other sperm may result from the opposing selection pressures for greater female choice. Whatever their functions, dense bodies offer great potential as characters in data matrices for phylogenetic estimations, but until we can categorize them more precisely, using more than just standard EM visibility, this potential will not be realised.

Presence of symbionts and the possibility for horizontal transfer

Bacteria, viruses and kinetoplastid flagellates have been reported from within and amongst male gametes in testes, sperm ducts and female receptacles (see Watson 1999b and references therein). They may, therefore, be sexually transmitted, and be potential agents of horizontal transfer of genetic information, through interaction with the host genome. This raises the possibility of disjunct character distribution amongst species, and could make the phylogenetic interpretation of morphological or molecular data difficult.

Conclusions

Compared with molecular data, sperm data from electron microscopy studies are much more difficult and time-consuming to gather. In order to document the range of informative characters, one needs to section and examine extensively, in testis tissues, seminal vesicles and female storage organs. So, while at this stage more rhabdocoels have been studied for sperm traits than for molecular analysis, the situation may well be very quickly reversed, especially since the number of laboratories gathering molecular data is increasing, while those studying turbellarian sperm are few and diminishing.

Nevertheless, sperm data make an important contribution to a wider morphological database. They may also be plotted along cladograms derived from other data, as an independent measure of their likelihood. At present, the very sparse coverage of taxa with either homologous morphological data or molecular data dictate extreme caution in the interpretation of phylogenies based upon them. Consensus amongst published trees is small, although most do indicate monophyly of the Rhabdocoela (without the Neodermata, which is itself an unambiguous clade). Undoubtedly, many more taxa must be examined before we arrive at an acceptable tree for the phylum.

In summary, within Rhabdocoela, sperm data provide clear synapomorphies for Kalyptorhynchia, Schizorhynchia, Temnocephalida and Temnocephaloidea. While also defining several other smaller taxa, they indicate that the traditional division between Typhloplanida and Dalyelliida may not be valid.

ACKNOWLEDGEMENTS

The majority of the turbellarian studies carried out at The University of New England (UNE) Armidale, Australia, over the past 10 years or so have been funded by grants from the Australian Research Council. They have also relied heavily on the generosity and taxonomic expertise of a number of valued colleagues, including Lester Cannon, Ulf Jondelius, Jean-Pierre L'Hardy, Ernest Schockaert and Alain De Vocht, who provided much of the material for our studies. The undescribed turbellarian from the brittlestar *Amphipholus squamata* was collected by Dimitri Deheyn. Ulf Jondelius sponsored a collecting trip to the Kristineberg Marine Station, Sweden in 1995. Ernest Schockaert generously hosted my visits of four months in 1995 and five weeks in 1998 to Limburgs Universitair Centrum (LUC), Belgium, where we concentrated on investigations of kalyptorhynchs. Micrographs were printed by Zoltan Enoch (UNE).

Chapter 21

Spermatozoa as phylogenetic characters for the Platyhelminthes

Jean-Lou Justine

Spermatozoa were used as phylogenetic characters for the Platyhelminthes in early studies such as those of Hendelberg (1969). The use of cladistic methods gave a new direction to these studies. Sperm characters were included in the definition of several major groups by Ehlers (1984, 1985a,b, 1986) and Brooks (Brooks 1989a; Brooks and McLennan 1993b). A critical assessment of sperm characters in the systems of Ehlers and Brooks was given later (Justine 1991a), and sperm characters were used for the construction of a phylogeny of the monogeneans, mainly the monopisthocotyleans (Justine 1991b, 1993). Sperm characters of the monogeneans have been incorporated, along with other morphological characters, into wider analyses of monogeneans (Boeger and Kritsky 1993, 1997, 2001, this volume). Sperm characters of the cestodes (Bâ and Marchand 1995) were also given a cladistic interpretation (Justine 1998b), and were incorporated into a phylogenetic analysis of the group (Hoberg et al. 1997c). Although sperm characters have been used in a variety of phyla for phylogenetic reconstruction (Jamieson et al. 1995), the wide use of sperm characters in modern phylogenies is certainly a distinctive feature of the phylogenetics and taxonomy of Platyhelminthes, to an extent not seen in other invertebrate groups. Absence of fossils, lack of reliable, homologous morphological characters and extensive variation of sperm are probably responsible for this exception (Justine 1998c).

Recent reviews list hundreds of species for which information is available about sperm ultrastructure (Justine 1995, 1998b; Watson and Rohde 1995b; Watson 1999b, 2001, this volume). This information will not be repeated here; emphasis will be placed on new interpretation and data published subsequent to the recent reviews.

This chapter includes several distinct parts:

- comments on synapomorphies proposed from sperm ultrastructure, at various ranks within the Platyhelminthes;

- an update of analyses of sperm ultrastructure in the Monogenea (Justine 1991b, 1993), using a new matrix and cladistic analysis from this matrix;

- an update of an analysis of sperm ultrastructure in the Cestoda (Justine 1998b), with comments about the relationships of the Cyclophyllidea and closely related taxa; and

- a summary of recent results obtained with the use of new methods in addition to ultrastructure.

Comments about sperm-based synapomorphies for various taxa

Stacking of membranes and monophyly of the Prolecithophora

Ehlers (1988) proposed a synapomorphy for the Prolecithophora, the presence of conspicuous stacks of membranes in the spermatozoon. He pointed out that the absence of axonemes and of granules also characterized the Prolecithophora, but he wanted to emphasize a positive character rather than negative characteristics. Ehlers mentioned that the synapomorphy was not present in *Urastoma* and, therefore, denied its position within the Prolecithophora; this has been confirmed by further ultrastructural studies (Noury-Sraïri et al. 1989b; Watson 1997a) and molecular systematics (Littlewood et al. 1999a).

The synapomorphy, proposed from data available on *Acanthiella, Archimonotresis, Cylindrostoma, Hydrolimax, Plagiostomum* and *Pseudostomum* (Ehlers 1988), was later confirmed by studies on *Cylindrostoma* (Watson and Jondelius 1997), *Multipeniata* (Schmidt-Rhaesa 1993), *Allostoma* (Lanfranchi 1998), *Ulianinia* and *Reisingeria* (Jondelius et al. 2001, this volume).

The 9 + '1' axoneme and monophyly of the Trepaxonemata

The name Trepaxonemata was proposed from the structure of the 9+'1' axoneme (Figure 21.1A) found in non-acoelan Platyhelminthes. The Trepaxonemata includes the Polycladida, Seriata, Prolecithophora, 'Typhloplanoida', 'Dalyellioida' and Neodermata (Ehlers 1984, 1985a,b, 1986). This is an historical example of a major taxon based on an ultrastructural character of spermatozoa. The name means 'axoneme in spiral', and refers to the structure of the central core of the 9+'1' axoneme. In contrast to most axonemes of eukaryotes, in which the 9+2 pattern (Figure 21.1A) prevails, the trepaxonematan axoneme shows a central core that appears spiralled in longitudinal sections. In the 9+2 pattern, the figure 2 refers to the presence of two singlet microtubules in the centre, and the figure 9 refers to nine peripheral doublets. In the 9+'1' pattern, the figure 1 refers to the presence of a single structure, and the apostrophes indicate that the single structure found is not a microtubule (Justine and Mattei 1981) (a real 9+1 pattern, with nine peripheral doublets and a single central microtubule exists in certain rare cases, but not in the Trepaxonemata). The ultrastructure of the trepaxonematan central core has been substantiated by early studies (Shapiro et al. 1961; Henley et al. 1969; Silveira 1969, 1975). Recent electron microscopic immunocytochemical studies (Iomini and Justine 1997; Iomini et al. 1998), using antibodies against tubulin and several of its post-translational modifications, demonstrated that the central core does not contain tubulin, therefore confirming the distinctive character of the trepaxonematan 9+'1' pattern; although these studies dealt only with a digenean, the strong homogeneity in the ultrastructural appearance of this feature ensures that these conclusions are valid for all trepaxonematans.

The trepaxonematan 9 + '1' structure is found in spermatozoa of all members of the Trepaxonemata. The rare exceptions can be classified in two categories: variations from 9+'1' axoneme, and aflagellate spermatozoa.

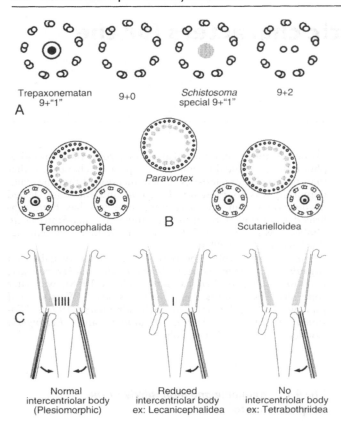

Figure 21.1 Diagrams of selected spermatozoal synapomorphies. **A)** Axonemal structures, transverse sections. The 9+'1' trepaxonematan structure is found in all Trepaxonemata, with the exception of the didymozoids (9+0 pattern) and *Schistosoma* (special 9+'1' pattern). The 9+2 pattern is found in spermatozoa of nemertodermatids and certain species of Acoela, and in somatic axonemes of all Platyhelminthes. The central core of the trepaxonematan 9+'1' axoneme does not contain tubulin, in contrast to the two microtubules in the centre of the 9+2 axoneme. **B)** Sperm structures in groups having small granules beneath the microtubule row. Note that small granules are found both in biflagellate (Temnocephaloidea and Scutari-elloidea) spermatozoa and aflagellate (*Paravortex*) spermatozoa. Note also presence of spiral of microtubules in the Temnocephalida, but not in other groups. **C)** Structure of zone of differentiation in the Lecanicephalidea and other groups of Eucestoda. Note that the intercentriolar body can be complete (several plates), reduced to one plate, or absent. (See also Table 21.4.)

Variations from the 9+'1' pattern have been found only in the digeneans: certain species of didymozoids, but not all, have sperm axonemes with nine doublets around an empty centre, i.e., a 9+0 pattern (Justine and Mattei 1983b, 1984), and schistosomes (Justine and Mattei 1981; Justine *et al.* 1993) have nine doublets around a fuzzy dense core, i.e., a 9+'1' pattern different from the trepaxonematan 9+'1', sometimes termed 9+0. In the latter case, it is noteworthy that only the genus *Schistosoma*, but not the closely related genus *Trichobilharzia* (unpublished observations on *T. regenti* by J.-L. Justine and P. Horak) nor the other blood-flukes (Sanguinicolidae and Spirorchidae; see Justine 1995), has an axonemal structure different from the 9+'1' pattern.

Aflagellate spermatozoa apparently arose independently in various groups such as the trigonostomids (Typhloplanida)

(Sopott-Ehlers and Ehlers 1997; Watson 1999b), certain Dalyelliida such as lurids (Rohde and Watson 1993b), graffillids (Noury-Sraïri *et al.* 1989a) and provorticids (Sopott-Ehlers and Ehlers 1995b; Watson 1999b) (see Watson 2001, in this volume for more details), and, within the Neodermata, the monogenean *Diplozoon* (Justine *et al.* 1985b). Incidentally, secondary reduction to aflagellate spermatozoa, within the Trepaxonemata, has been proposed as a synapomorphy for the Provorticidae + Graffillidae (Sopott-Ehlers and Ehlers 1995b).

In Ehlers' (1984, 1985a) system, the Macrostomida (the members of which have aflagellate spermatozoa) is the sister-group of the Trepaxonemata. In a recent paper, Sopott-Ehlers and Ehlers (1999) confirmed sister-group relationships between the Macrostomorpha (including Macrostomida and *Haplopharynx*) and Trepaxonemata, both composing the Rhabditophora, but indicated that the 9+'1' pattern could be either a synapomorphy of the Trepaxonemata or of the Rhabditophora. In a study combining molecular and morphological data (Littlewood *et al.* 1999a), the Macrostomida was found, with the Haplopharyngida, to be close to the Polycladida (an unambiguous trepaxonematan taxon), and validity of the taxon Trepaxonemata was thus rejected. In the present author's opinion, if the phylogeny proposed by Littlewood *et al.* (1999a) is correct, it would be more logical to hypothesize secondary loss of axonemes in the Macrostomida within the Trepaxonemata, defined by the clear basal autapomorphy of the 9+'1' axoneme, than to reject the Trepaxonemata.

Spiral pattern, cork-screw structure, monophyly of the Temnocephalida, and relationships between scutariellids and temnocephalids

The Temnocephalida is one group of Platyhelminthes for which sperm ultrastructure was used relatively early for defining a synapomorphy (Justine *et al.* 1987; Justine 1991a): the spiral pattern of the peripheral microtubule row (Figure 21.1B). This character has been found in all Temno-cephalida studied so far (references in Watson and Rohde 1995a; Watson *et al.* 1995a; Batista *et al.* 1999), with the exception of the scutariellid *Troglocaridicola* sp. (Iomini *et al.* 1994) (Figure 21.1B). This species, in contrast, has a terminal cork-screw structure not found in other species. The Scutarielloidea are now considered the sister-group of all other Temnocephalida, grouped as the Temnocephaloidea (Joffe *et al.* 1998b). The spiral arrangement of the microtubules is one of three apomorphies proposed for the Temnocephaloidea, and the cork-screw structure is one of three apomorphies proposed for the Scutarielloidea (Joffe *et al.* 1998b).

Small granules beneath the microtubular row as a synapomorphy for a monophylum comprising the Temnocephalida, Dalyelliida and Typhloplanida

Small granules (c. 25 nm in diameter, and not bound by a membrane) beneath the microtubular row (Figure 21.1B) have been described in a variety of 'turbellarian' species. Importantly, these granules are found both in spermatozoa with axonemes and in aflagellate sperm, making them a character independent from the general structure of the spermatozoon. According to Watson (1999b, and 2001, this volume), these granules are found in most, but not all, of the Typhloplanida, Dalyelliida and Temnocephaloidea (including

Scutariellidae). The presence of these granules has been proposed as a synapomorphy for a group including these three taxa, and this group is independently supported by DNA analysis (Littlewood *et al.* 1999a). Absence of this character in certain taxa may be explained in certain cases by poor conditions of fixation (e.g., they were not visible in studies of temnocephalids subjected to long periods of fixation; Justine *et al.* 1987). Reversal may be hypothesized in other cases.

Proximo-distal fusion, the Neodermata, and their sister-groups

The Neodermata includes the main groups of parasitic Platyhelminthes, i.e., the Monogenea, Digenea, Aspidogastrea, Gyrocotylidea, Amphilinidea and Eucestoda. An abundant literature has been devoted to the search for the sister-group of the Neodermata in recent years, and sperm structure has taken an important part in this discussion.

In the system of Brooks (Brooks 1989a; Brooks and McLennan 1993b), the Udonellidea was regarded as the sister-group to the main parasitic groups (known as Cercomeridea, and including the Monogenea, Digenea, Aspidogastrea, Gyrocotylidea, Amphilinidea and Eucestoda); Cercomeridea + Udonellidea together formed the Neodermata, and Temnocephaloidea was the sister-group to the Neodermata. These relationships have since been rejected. In a critical study of synapomorphies based on sperm ultrastructure, Justine (1991a) claimed that proximodistal fusion was a synapomorphy of the Cercomeridea, but not of the Udonellidea, because a typical zone of differentiation was not found in spermatids of *Udonella*. Studies of ultrastructure (Rohde *et al.* 1989b; Rohde and Watson 1993c) and a molecular analysis (Littlewood *et al.* 1998) have demonstrated that *Udonella* is, in fact, a monopisthocotylean monogenean, a group in which typical proximodistal fusion is not found. Furthermore, there is no sperm synapomorphy shared by the Temnocephaloidea and the Neodermata, and a sister-group relationship between them has been rejected (Rohde 1990; Ehlers and Sopott-Ehlers 1993). Sister-group relationships in the system of Brooks are therefore falsified. In the rest of this chapter, the Neodermata designates the main groups of parasitic Platyhelminthes, listed above, sharing the synapomorphy of proximo-distal fusion during spermiogenesis.

Several species have been proposed as sister-groups to the Neodermata on the basis of sperm structure. These are *Kronborgia isopodicola* (Watson and Rohde 1993b), *Urastoma cyprinae* (Watson 1997a) and *Notentera ivanovi* (Joffe *et al.* 1997; Joffe and Kornakova 1998; Kornakova and Joffe 1999). As stated by Watson (1999b), 'all the three species exhibit forms of proximo-distal development that may indicate close links to the Neodermata'. Jondelius (1992) proposed that the Pterastericolidae and the Fecampiidae together form a taxon that is the sister-group to the Neodermata, but a later study of spermiogenesis in *Pterastericola astropectinis* demonstrated that fusion of axonemes was not proximo-distal (Watson *et al.* 1993). Kornakova and Joffe (1999) recently proposed that Urastomidae was the sister-group to the Neodermata, both united in the taxon Mediofusata, the name indicating the presence of a proximo-distal fusion. They furthermore proposed that the Fecampiida (including *Kronborgia* and *Notentera*) constituted the sister-group to the Mediofusata, both united in the Revertospermata. However, recent DNA studies did not support these relationships (Littlewood *et al.* 1999a,b), and the Fecampiida and *Urastoma* (no DNA data exist for *Notentera*) appeared close to the Tricladida, in a large group comprising the Proseriata,

Kalyptorhynchia, Dalyelliida, Temnocephalida, Typhloplanida and Lecithoepitheliata, which was the sister-group to the Neodermata. If true, these relationships imply either that proximo-distal fusion has evolved at least twice, or that the process of proximo-distal fusion in *Kronborgia/Urastoma/Notentera* is not homologous with that of the Neodermata.

An update of the spermatozoal matrix for the Monopisthocotylea and Polyopisthocotylea

As expressed in previous studies, there is no spermatozoal synapomorphy uniting the Monopisthocotylea and the Polyopisthocotylea (Justine 1991b, 1993, 1995, 1998a). It has been possible, however, to recognize synapomorphies based on sperm structure, respectively for the Monopisthocotylea and for the Polyopisthocotylea (Justine 1991a,b, 1993).

Construction of the spermatozoal matrix

The matrix of Justine (1993) was used as a basis for constructing a new matrix. The same characters and character-states were used (Table 21.1), and drawings of character-states can be found in Justine (1991b). A few modifications were made in taxa already included in the matrix. The matrix line of *Calicotyle*, previously based on a study of *Calicotyle kroyeri*

Table 21.1 Spermatozoa of monogeneans: list of characters and character-states (from Justine 1993).

1. Lateral microtubules in the spermatozoon's principal region (0: absent; 1: present)
2. Dorsoventral microtubules in the spermatozoon's principal region (0: present; 1: absent)
3. Intercentriolar body (0: present; 1: absent)
4. Striated roots (0: present; 1: absent)
5. Number of axonemes during spermiogenesis (0: 2 axonemes; 1: 1 axoneme + 1 altered axoneme; 2: 1 axoneme + 1 disappearing axoneme; 3: 1 axoneme from beginning)
6. Distal region containing only the nucleus in mature spermatozoon (0: absent; 1: present)
7. Cytoplasmic middle process and flagella (0: separate, then fused; 1: fused from the start)
8. External ornamentation of the cell membrane (0: present; 1: absent)
9. Number of centrioles in the spermatozoon (0: 2 centrioles; 1: 1 centriole)
10. Centriole adjunct (0: absent; 1: present)
11. Axoneme structure in mature sperm (0: circular; 1: non-circular)
12. Axonemal b microtubules during spermiogenesis (0: complete; 1: incomplete)
13. Lateral crest on mature sperm (0: absent; 1: present)
14. External microtubules associated with the spermatid (0: absent; 1: present)
15. Anterior region of the nucleus (0: not coiled; 1: coiled)
16. A bead-like giant mitochondrion (0: absent; 1: present)
17. Microtubules in the spermatid's zone of differentiation (0: present; 1: absent)
18. Ontogeny of microtubules in the zone of differentiation (0: persisting; 1: disappearing)
19. Centriole made up of diverging singlets (0: absent; 1: present)
20. Axoneme central core (0: normal 9+'1'; 1: hollow)
21. External elements associated with the zone of differentiation (0: absent; 1: present)
22. One single peripheral microtubule during spermiogenesis (0: absent; 1: present)

(Tappenden and Kearn 1991), was updated from the study on *Calicotyle australiensis* (Watson and Rohde 1994b). The matrix line of *Heterocotyle*, previously based on the study of *Heterocotyle* sp. (Justine and Mattei 1983a) was updated with the study on *Heterocotyle capricornensis* (Watson and Chisholm 1998), and finally split into two lines. Information about *Euzetrema knoepffleri* was updated according to Fournier and Justine (1994), and about *Pseudodactylogyrus* sp. according to Mollaret and Justine (1997). Taxa added were *Isancistrum subulatae* (Malmberg and Lilliemark 1993), *Monocotyle helicophallus* (Watson and Rohde 1994b), *Troglocephalus rhinobatidis*, *Neoheterocotyle rhinobatidis* and *Merizocotyle australiensis* (Watson 1997b), *Udonella* sp. (Rohde and Watson 1993c), *Pseudodactylogyroides marmoratae* and *Sundanonchus micropeltis* (Mollaret *et al.* 1998).

The matrix mainly deals with monopisthocotylean monogeneans. Recent studies on polyopisthocotyleans, including *Metamicrocotyla macracantha* (Baptista-Farias *et al.* 1995), *Pseudodiplorchis americanus* (Cable and Tinsley 1993), *Octomacrum lanceatum* (Hathaway *et al.* 1995), *Polystoma* sp. (Li *et al.* 1998), *Gonoplasius* sp. (Rohde and Watson 1994c), *Polylabroides australis* (Rohde and Watson 1994d), *Atriaster heterodus* (Santos *et al.* 1997), *Neopolystoma* sp. (Watson and Rohde 1995c), *Concinnocotyla australensis* and *Pricea multae* (Watson *et al.* 1995b) did not necessitate changes to the matrix line of the polyopisthocotyleans.

The matrix (Table 21.2) has 36 taxa and 22 characters. Redundant taxa and autapomorphies, that do not contribute to the present analysis, are nevertheless listed in the matrix because they could be useful for the construction of another matrix including other characters, and for detection of synapomorphies when other species are studied. Redundant taxa were then grouped using MacClade (Maddison and Maddison 1992) 'Utilities' Menu, 'Filtering redundant taxa'. Twelve taxa were considered redundant. The new matrix obtained (Table 21.3) has 24 taxa and 22 characters. *Tetraonchoides*, *Isancistrum* and *Pseudodactylogyroides* were removed from the analysis because they have too many missing data (spermiogenesis not observed). The analysis thus dealt with 21 taxa.

Parsimony analysis

The use of a small set of characters, here restricted to sperm ultrastructure only, is not recommended for resolving the phylogeny of a wide group. However, cladistic studies based only on sperm provided the first cladistic analysis of the monogeneans (Justine 1991b, 1993), and those characters were later used in analyses combining morphological and spermatological characters (Boeger and Kritsky 1993, 1997). Therefore, an analysis was performed on the present set of data, keeping in mind that it could be more useful in a wider set of characters. A parsimony analysis computed with PAUP* 4.0 (Swofford 1998) produced 110 equally parsimonious trees. Following 10 000 replicates of bootstrapping, the 50% majority-rule consensus tree (Figure 21.2) had a length of 32 steps, CI of 0.75, CI excluding uninformative characters of 0.68, and showed a high degree of polytomy.

Comments on monopisthocotylean relationships

Interrelationships of monopisthocotylean monogeneans inferred from sperm ultrastructure alone can be summarized as follows (Figure 21.2):

Table 21.2 Matrix of spermatozoal data of monopisthocotylean monogeneans, including all taxa (alphabetical order). Characters: see Table 21.1.

	1 1234567890	1111111112 1234567890	22 12
Acanthocotyle	0111001100	0000001?00	01
Amphibdella	0111301111	1000101?00	00
Amphibdelloides	0111301111	1000101?00	00
Anoplodiscus	0111301110	0000001?01	00
Caballerocotyla	0111001100	0000010100	01
Calceostoma	0111311110	1111001?00	00
Calicotyle	0111010000	0000000000	10
Cichlidogyrus	01??3?????	0?00????00	00
Cleithrarticus	0111301010	0100001?00	00
Digenea	0000000000	0000000000	00
Dionchus	0111001100	0000010100	01
Diplectanum	0111311111	0000001?00	00
Encotyllabe	0111001100	0000010100	01
Euzetrema	01??00??0?	0?000??00	00
Furnestinia	0111311111	0000001?00	00
Gyrodactylus	01??00??0?	0?000??00	00
Heterocotyle capricornensis	0111010000	0000000000	00
Heterocotyle sp.	0112100000	0000000000	00
Isancistrum	01??0??0?	0???????0	??
Lamellodiscus	01??31???1	0?000??00	00
Loimosina	0111110000	0000000000	00
Macrogyrodactylus	0111001100	0000001?00	00
Megalocotyle	0111001100	0000010100	01
Merizocotyle	0111000100	0000001?00	00
Monocotyle	0111000100	0000000000	10
Myxinidocotyle	0111001100	000000 1?100	1
Neoheterocotyle	0111000000	0000000000	00
Polyopisthocotylea	1000000000	0000000000	00
Pseudodactylogyroides	01??0??11	0?????00	??
Pseudodactylogyrus	0111301011	0100001?00	00
Sundanonchus	0111301?11	0?00?0??0	0?
Tetraonchoides	01??30?11?	0?0000??00	0?
Tetraonchus	01??3?????	0?00????00	0?
Trochopus	01??00??0?	0?000??00	0?
Troglocephalus	0111000100	0000000000	00
Udonella	0111001100	0000000100	00

• The Monocotylidae (and Loimoidae) form a paraphyletic group, basal to all other taxa. Among them, a few species form a monophylum. It is noteworthy that among seven monocotylids, none was redundant, thus showing a high heterogeneity in this family. In a molecular analysis of the monogeneans, paraphyly of the Monocotylidae was also suggested (Mollaret *et al.* 1997).

• A group composed of the Capsalidae, Dionchidae, Acanthocotylidae, Gyrodactylidae, Myxinidocotylidae and Udonellidae is polytomous, and basal to the branch that includes monogeneans with uniflagellate spermatozoa. From the description of spermatozoa of *Lagarocotyle* (Hathaway *et al.* 1993; Kritsky *et al.* 1993), the Lagarocotylidae would also be included in this group. On the basis of sperm ultrastructure, but without data on spermiogenesis, *Udonella* had been considered close to the capsalids (Justine *et al.* 1985a). Ultrastructural studies of various organs and of spermiogenesis (Rohde *et al.* 1989b; Rohde and Watson 1993c), and recent molecular analyses (Littlewood *et al.* 1998), clearly confirm the

Table 21.3 Matrix of spermatozoal data of monopisthocotylean monogeneans, showing redundant taxa grouped together. Alphabetical order. *Tetraonchoides, Isancistrum* and *Pseudodactylogyroides* were excluded because of missing data (spermiogenesis not observed).

	1 1234567890	1111111112 1234567890	22 12
Acanthocotyle and *Trochopus*	0111001100	0000001?00	01
Amphibdella and *Amphibdelloides*	0111301111	1000101?00	00
Anoplodiscus	0111301110	0000001?01	00
Caballerocotyla and *Dionchus* and *Encotyllabe* and *Megalocotyle*	0111001100	0000010100	01
Calceostoma	0111311110	1111001?00	00
Calicotyle	0111010000	0000000000	10
Cichlidogyrus and *Cleithrarticus* and *Tetraonchus*	0111301010	0100001?00	00
Digenea	0000000000	0000000000	00
Diplectanum and *Furnestinia* and *Lamellodiscus*	0111311111	0000001?00	00
Euzetrema and *Gyrodactylus* and *Macrogyrodactylus*	0111001100	0000001?00	00
Heterocotyle capricornensis	0111010000	0000000000	00
Heterocotyle sp.	0111210000	0000000000	00
Loimosina	0111110000	0000000000	00
Merizocotyle	0111000100	0000001?00	00
Monocotyle	0111000100	0000000000	10
Myxinidocotyle	0111001100	0000001?10	01
Neoheterocotyle	0111000000	0000000000	00
Polyopisthocotylea	1000000000	0000000000	00
Pseudodactylogyrus and *Sundanonchus*	0111301011	0100001?00	00
Troglocephalus	0111000100	0000000000	00
Udonella	0111001100	0000000100	00

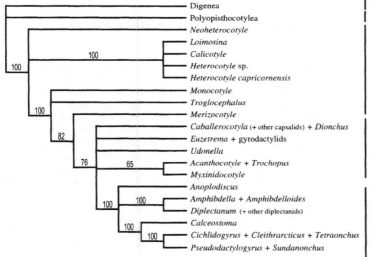

Figure 21.2 Majority-rule consensus tree of monopisthocotylean relationships obtained by a parsimony analysis of the matrix in Table 21.3. Bootstrap values are indicated above branches.

position of *Udonella* within the monopisthocotylean monogeneans.

- All families showing uniflagellate spermatozoa (Anoplodiscidae, Amphibdellatidae, Diplectanidae, Calceostomatidae, Ancyrocephalidae, Pseudodactylogyridae, Sundanonchidae, Tetraonchidae) are grouped in a monophylum. Justine (1991b) proposed the name Monoaxonematidea for this grouping, which also includes the Tetraonchoididae, excluded from the present analysis because of unsufficient data.

An update of the spermatozoal matrix of Eucestoda

The matrix

A data matrix of characters of spermatozoa of the Eucestoda was presented in a recent review (Justine 1998b), and characters of spermatozoa have been used, in addition to morphological characters, for the construction of a phylogeny of the eucestodes (Hoberg *et al.* 1997c).

A new matrix is now proposed, and includes the following modifications. The Lecanicephalidea was the single order in which sperm structure was unknown. We examined spermiogenesis and sperm ultrastructure in the lecanicephalid *Tetragonocephalum* sp. from *Himantura* sp. (unpublished observations by J.-L. Justine, O. Raikova and L. Euzet, and Figure 21.1C). New data are available for the Tetraphyllidea *Phyllobothrium lactuca* (Sène *et al.* 1998, 1999), and the Cyclophyllidea *Vampirolepis microstoma* (Bâ and Marchand 1998), *Paranoplocephala omphalodes* (Miquel and Marchand 1998b), *Dipylidium caninum* (Miquel *et al.* 1998b), *Anoplocephaloides dentata* (Miquel and Marchand 1998a; Miquel *et al.* 1998a), *Catenotaenia pusilla* (Hidalgo *et al.* 1999), and *Taenia* spp. (Tian *et al.* 1998). Data on spermiogenesis in *Mesocestoides litteratus* (Miquel *et al.* 1999) advocate creation of a new line in the matrix for this taxon (see below).

Recent publications reported the presence of striated roots in several cyclophyllidean species, *Dipylidium caninum* (Miquel *et al.* 1998b) and *Anoplocephaloides dentata* (Miquel and Marchand 1998a; Miquel *et al.* 1998a). The structures labelled as striated roots by these authors do not show the large diameter and the regular transverse striated pattern of the striated roots found in most cestodes. The character 'absence of striated roots' proposed by Justine (1998b) is now re-coded as 'absence of *typical* striated roots'. The presence of the thin striated structures found in certain cyclophyllids is probably of phylogenetic interest at the level of intra-ordinal relationships within the cyclophyllideans.

A typical intercentriolar body (Figure 21.1C) is a characteristic of the zone of differentiation found in the Neodermata, and is certainly plesiomorphic within the Eucestoda. It is a prominent structure, made up of several plates, found in nearly all polyopisthocotylean monogeneans and digeneans (notable exceptions are didymozoids and schistosomes). This structure is also found in most Trepaxonemata (e.g., Proseriata; Sopott-Ehlers 1990). In the Eucestoda, the intercentriolar body was previously coded (Justine 1998b) as present (plesiomorphic) or absent (apomorphic). In some Eucestoda, a simpler structure, made up of a single plate, is found in certain groups. In the Proteocephalidea, *Nomimoscolex* has a single plate ('electron-dense rod' of Sène *et al.* 1997), but *Proteocephalus* has a normal intercentriolar body. In the Tetraphyllidea Onchobothriidae, *Acanthobothrium filicolle filicolle* has a single plate ('structure discoïde' of Mokhtar-Maamouri 1982), but the structure is normal in *Acanthobothrium filicolle benedenii*. In the Tetraphyllidea Phyllobothriidae, the intercentriolar body is present (and bent) in *Phyllobothrium gracile*. A single plate is found in the Lecanicephalidea and *Mesocestoides*. The intercentriolar body is absent in the Tetrabothriidea and the Cyclophyllidea (see Justine 1998b for all references not otherwise specified for this paragraph). The intercentriolar body (Figure 21.1C) is tentatively coded as a multi-state character in Table 21.4. It is clear that a progressive reduction of the intercentriolar body occurs in the higher cestodes, although polymorphism of certain taxa hampers easy proposal of synapomorphies.

In addition, a mistake was detected in the matrix printed in the review by Justine (1998b), for the line concerning *Tetrabothrius* (Tetrabothriidea), based on the paper by Stoitsova *et al.* (1995). Although the text correctly states (p. 399) that the two centrioles each have a striated root, the matrix (Table VIII) erroneously indicates that striated roots are absent (derived character) in the Tetrabothriidea.

The new matrix for characters of cestode spermatozoa is presented in Table 21.4.

Interrelationships of the Cyclophyllidea, Mesocestoididae and Tetrabothriidea suggested by sperm structure

Recent studies on *Mesocestoides* suggest relationships that are different from the position generally accepted for this family, i.e., within the Cyclophyllidea (Khalil *et al.* 1994). Provided that the Cyclophyllidea, Tetrabothriidea and Mesocestoididae are terminal groups within the Eucestoda (Hoberg *et al.* 1997c), the following relationships (see Figure 21.3) are suggested from an analysis of spermatozoal characters only.

The Cyclophyllidea is the terminal taxon and is characterized by three synapomorphies: the absence of typical striated roots; absence of flagellar rotation; and presence of a single axoneme in the zone of differentiation. The Cyclophyllidea and Tetrabothriidea are united by two synapomorphies: the presence of twisted microtubules, and the presence of an axonemal sheath. This latter character, however, shows reversal in certain cyclophyllids. *Mesocestoides*, which has none of the apomorphies cited above, is considered basal to the Tetrabothriidea + Cyclophyllidea. Interestingly, this is also the conclusion of a molecular analysis of the Cestoda based on partial 18S rDNA sequences, in which *Mesocestoides* was found basal to the Tetrabothriidea + Cyclophyllidea; it was thus suggested that the Mesocestoididae should be removed from the Cyclophyllidea, as the Mesocestoidata (Mariaux 1998). Although sperm structure alone strongly endorses this separation, an analysis using other characters (see Hoberg *et al.* 2001, this volume) is needed to reach a definite conclusion for the position of the Mesocestoididae.

New techniques for the study of sperm structure

Most spermatozoal characters useful for phylogeny in the Platyhelminthes have been defined with the use of transmission electron microscopy. However, the reconstruction of a long, filiform spermatozoon using electron microscopy alone is not reliable without the use of extensive serial sectioning, which is extremely time-consuming. Visualization of the axonemes by immunocytochemical labelling of tubulin and labelling of the nucleus by DNA-specific fluorescent probes readily shows the general structure of the spermatozoon. Use of these techniques has been reviewed elsewhere (Justine 1999), and is herein only briefly summarized. Features demonstrated include relative lengths of axonemes and microtubules, shifts of axonemes, position of the nucleus (Iomini *et al.* 1995; 1997, 1998; Iomini and Justine 1997; Mollaret and Justine 1997; Raikova *et al.* 1997, 1998b; Justine *et al.* 1998; Raikova and Justine 1999), and the number of spermatids in isogenic groups (Mollaret and Justine 1997), a character of possible use for phylogeny (Justine 1995). These techniques should now be used for species in which electron microscopy has not clarified the position, number, possible shift or disappearance of axonemes, such as certain monocotylid monogeneans and the diphyllid cestodes.

In addition to simply showing the axonemes and other microtubular structure, antibodies against tubulin can be used for discriminating between microtubular structures on the basis of their composition of the numerous variants of tubulin, including its post-translational modifications (see also Raikova *et al.* 2001, this volume). A major phylogenetic finding was that cortical microtubules of the Neodermata are not acetylated, whereas cortical microtubules of certain species

Table 21.4 Possible synapomorphies of spermatozoa and spermiogenesis at the ordinal level in the Eucestodes. The order of groups follows a tree based on morphological and spermatological characters, with the hypothesis of the Tetraphyllidea paraphyletic (as Onchobothriidae and Phyllobothriidae) (Hoberg *et al.* 1997c). Modified from Justine (1998b), with new data concerning the Lecanicephalidea (unpublished observations), the Mesocestoididae (Miquel *et al.* 1999) and corrections concerning the striated roots. Two proposals are given for the coding of the intercentriolar body, binary or multistate. Abbreviations: Gy, Gyrocotylidea; Am, Amphilinidea; Ca, Caryophyllidea; Spa, Spathebothriidea; Pse, Pseudophyllidea; Hap, Haplobothriidea; Dip, Diphyllidea; Try, Trypanorhyncha; Onc, Tetraphyllidea Onchobothriidae; Phy, Tetraphyllidea Phyllobothriidae; Lec, Lecanicephalidea; Pro, Proteocephalidea; Nip, Nippotaeniidea; Tet, Tetrabothriidea; Mes, Mesocestoididae; Cyc, Cyclophyllidea.

Character	Gy	Am	Ca	Spa	Pse	Hap	Dip	Try	Onc	Phy	Lec	Pro	Nip	Tet	Mes	Cyc	Remarks
Mitochondrion 0, present; 1, absent	0	0	1	1	1	1	1	1	1	1	1	1	1	1	1	1	Synapomorphy for the Eucestoda
Crested body 0, absent; 1, present	0	0	?	?	1	?	1	?	1	1	1	1	?	1	1	1	Synapomorphy for the Eucestoda (or for a part of the Eucestoda only: basal to the Caryophyllidea, the Spathebothriidea, or the Pseudophyllidea?)
Intercentriolar body, binary coding 0, present; 1, absent	0	0	0	?	0	0	0	0	0	0	0	0-1	?	1	0	1	Also coded as multistate character below
Intercentriolar body, coded as a multistate character 0, present; 1, absent	0	0	0	?	0	0	0	0	0-1	0	1	0-1	?	2	1	2	Synapomorphy for higher groups, but polymorphism in several taxa
Typical striated root 0, present; 1, absent	0	0	0	?	0	0	0	0	0	0	0	0	?	0	0	1	Synapomorphy for the Cyclophyllidea
Peripheral microtubules 0, parallel; 1, twisted	0	0	0	0	0	0	0	0	0	0	0	0-1?	0?	1	0	1	Synapomorphy for the Tetrabothriidea + Cyclophyllidea
Periaxonemal sheath 0, absent; 1, present	0	0	0	?	0	0	0	0	0	0	0	0	?	1	0	0-1	Synapomorphy for the Tetrabothriidea + Cyclophyllidea, with reversal in certain Cyclophyllidea?
Flagellar rotation 0, present; 1, absent	0	0	0	?	0	0	0	0	0	0	0	0	?	0	0	1	Synapomorphy for the Cyclophyllidea
Number of axonemes in zone of differentiation 0, 2 axonemes; 1, 1 axoneme	0	0	0	?	0	0	0	0	0	0	0	0	?	0	0	1	Synapomorphy for the Cyclophyllidea
Number of axonemes in mature spermatozoon 0, 2 axonemes; 1, 1 axoneme	0	0	1	0	0	0	0-1	0	0	1	1	0-1?	1	1	1	1	A homoplastic character, with convergence in several unrelated groups.
Two types of cortical microtubules 0, 1 type; 1, two types	0	0	0	?	0	0	0	0	1	1	0	0	?	0	0	0	A character found only in the Onchobothriidae and Phyllobothriidae. Possible synapomorphy for the Tetraphyllidea?
Proximo-distal fusion 0, present; 1, absent	0	0	0	?	0	0	0	0	0	0	0	0	?	0	0	0-1	Synapomorphy for certain Cyclophyllidea?

of Acoela show acetylation, therefore suggesting non-homology between the two structures (Justine *et al.* 1998).

Conclusion: spermatozoal versus other characters

In addition to well-established spermatozoal synapomorphies (such as the absence of mitochondria in sperm of the Eucestoda, which clearly separates them from the Gyrocotylidea and Amphilinidea), recent years have seen multiple cases of 'reciprocal illumination' between spermatozoal characters and other features. This includes: the relationships of the Tetrabothriidea and Cyclophyllidea (sperm and 18S DNA); inclusion of *Udonella* within the Monopisthocotylea (sperm, protonephridia, 18S and 28S DNA); recognition of the Temnocephaloidea + Dalyelliida + Typhloplanida as a monophylum (sperm and 18S DNA); low support for monophyly of the Monogenea (sperm and 18S and 28S DNA); and position of the Mesocestoididae vs. the Cyclophyllidea (sperm and 18S DNA). Other cases are expected in the future.

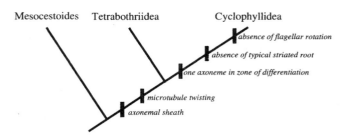

Figure 21.3 Relationships between *Mesocestoides*, the Tetrabothriidea and the Cyclophyllidea (Eucestoda) suggested by sperm characters. Axonemal sheath shows reversal in certain Cyclophyllidea. (See also Table 21.4.)

Summary

This chapter reviews cases in which ultrastructural characters of spermatozoa have proved useful for definition of synapomorphies, and for phylogenetic discussion, concerning: monophyly of the Prolecithophora; monophyly of the Trepaxonemata; definition of a monophylum including the

Temnocephalida, Dalyelliida and Typhloplanida; monophyly of the Temnocephalida; relationships between scutariellids and temnocephalids; monophyly of the Neodermata; the possible sister-group(s) of the Neodermata; internal relationships of the Monopisthocotylea; phylogenetic affinities of *Udonella*; internal relationships of the Eucestoda; and monophyly of the Cyclophyllidea. A matrix of spermatozoal characters is constructed for the monopisthocotylean monogeneans. A parsimony analysis, from sperm characters alone, reveals three main taxa: a paraphyletic group including the Monocotylidae + Loimoidae; a paraphyletic group including the Capsalidae, Dionchidae, Acanthocotylidae, Myxinidocotylidae, Gyrocotylidae and *Udonella*; and a monophyletic group including all families in which the sperm has a single axoneme (Anoplodiscidae, Amphibdellatidae, Diplectanidae, Calceostomatidae, Ancyrocephalidae, Pseudodactylogyridae, Sundanonchidae and Tetraonchidae). A matrix of spermatozoal characters of the Cestoda is constructed, with new observations on the Lecanicephalidea. If sperm characters alone are considered, the Mesocestoididae appear basal to the Tetrabothriidea and Cyclophyllidea. In addition to ultrastructure, new techniques, such as immunocytochemistry and fluorescent labelling of the nucleus, have recently been used to provide a better understanding of sperm ultrastructure in long, filiform sperm cells. These techniques have shown that, unlike the axonemal microtubules, the central structure of the trepaxonematan $9 + '1'$ axoneme does not contain tubulin, thus confirming its distinctive character.

ACKNOWLEDGEMENTS

Barrie Jamieson kindly edited the English. Jan Hendelberg and Nikki Watson provided useful comments.

Chapter 22

Comparative neurobiology of Platyhelminthes

Maria Reuter and David W. Halton

According to a reconstructed phylogenetic system for flatworms, the stem form of the Bilateria gave rise to the Platyhelminthes and the Eubilateria (Figure 22.1) (Ax 1984). While the monophyletic or polyphyletic status of the Platyhelminthes and the relationship to the protostome and deuterostome lines is under discussion (see Smith *et al.* 1986; Koopowitz *et al.* 1995; Haszprunar 1996; Tyler 2001, this volume), recent general phylogenetic schemes advocate a flatworm-like common ancestor for all metazoan phyla (Barnes *et al.* 1988; Willmer 1990).

The taxonomic value of the morphological characters of flatworms has long been a subject for evaluation. In such discussions, the phylogenetic significance of the neuronal cytoarchitecture, the potential homology of neurons and the organization of the flatworm nervous system have all been debated (Hyman 1951; Bullock and Horridge 1965; Reisinger 1972, 1976; Ax 1985; Ehlers 1985; Joffe and Kotikova 1991; Elvin and Koopowitz 1994; Koopowitz *et al.* 1995; Reuter and Gustafsson 1995; Haszprunar 1996; Joffe and Cannon 1998a). Flatworms are the most primitive extant animals in which it is possible to find a central nervous system and cephalic ganglion or *archaic brain* and discrete muscles. Analysis of the neuromuscular organization of flatworms may help to better understand the phylogenetic status of the platyhelminths and the relationships between flatworm taxa of different orders.

Extant platyhelminths have been evolving in their own right for at least 600 million years, and have become distinctly specialized (Moore and Willmer 1997). They comprise a diverse phylum of free-living turbellarians, largely ectoparasitic monogeneans, and endoparasitic digeneans and cestodes. Their life patterns are extremely varied, and the demand for neuromuscular adaptations is obvious. The absence of an endocrine system means that integration of bodily functions and behaviours in flatworms is accommodated by the nervous system. Basically, the nervous system in all flatworm taxa, except the Acoelomorpha (Raikova *et al.* 2001, this volume), is differentiated into a central nervous system (CNS) comprising a main ganglion in the form of the bilobed brain or cephalic ganglion, from which extend paired longitudinal nerve cords intersected by transverse commissures, and a peripheral nervous system (PNS) of several nerve plexuses, notably submuscular, subepidermal and infraepithelial, and the more-or-less autonomic pharyngeal–stomatogastric and genital plexuses. However, brain morphology, cytoarchitecture, organization plans and total design often show great variation. Thus, primitive features and organization plans occur alongside more advanced features, and there are derived plans in lower as well as higher flatworm taxa. This raises the fundamental question: do the variations reflect an independent evolution of similar structural features and organization plans in distantly related groups, i.e., *parallelism*? or, since a high degree of morphological diversity is typical for archaic groups of organisms, does the diversity reflect the versatility that shaped the basis for evolution, i.e., *primitiveness* (Mamkaev 1986).

In comparisons between taxa and the evaluation of similarities of individual morphological characters, certain precautions are necessary to avoid false homologies, i.e., similarities that have arisen independently due to parallelism. Resemblance between animal taxa is not always due to close common

ancestry. As stated by Moore and Willmer (1997), similarities between two species may indeed be due to limited divergence from a common ancestor. Alternatively, similarities may arise through convergence or parallel evolution, defined by Mayr (1969: 243) as: 'the development of similar characters separately in two or more lineages of common ancestry'. In other words, descendants have inherited the potential to express a character, and two or more lineages change in similar ways when faced with similar problems (Moore and Willmer 1997).

According to current studies on the cytoarchitecture, the nervous systems of flatworms share features with protostome as well as deuterostome lines (Keenan *et al.* 1981; Koopowitz 1989, Elvin and Koopowitz 1994; Koopowitz *et al.* 1995). Thus, Koopowitz *et al.* (1995) questions whether the 'Turbellaria' should be sited at the base of the Bilateria at a position similar to that of the flatworm-like common ancestor (Barnes *et al.* 1988).

In discussing the organization of the flatworm nervous system, it has been stressed that morphologically similar types of nervous systems may have arisen independently several times, thus demonstrating striking parallelism (Figure 22.2). This may be due to certain general trends in the evolution of the nervous system and the influence of specialized forms of the body (Kotikova 1986, 1991; Joffe 1990).

Phylogenetic and functional aspects of flatworm nervous systems have recently been reviewed by Reuter and Gustafsson (1995) and Halton and Gustafsson (1996). The morphological variations of the nervous system in the 'turbellarian' taxa have been reviewed and discussed by Ehlers (1985) and Rieger *et al.* (1991). In this present review, the following aspects of the nervous system will be considered with respect to phylogenetic implications:

- neuronal homologies and data on neuronal cell types that suggest advanced features at this low phylogenetic level;

- the organization of the nervous system inclusive of the stomatogastric nervous system in different flatworm taxa;

- ultrastructural features of the nervous system; and

- neuroactive substances.

Neurons

Homology of neurons

Single neurons

The introduction of fluorescence methods in conjunction with confocal scanning laser microscopy (CSLM) has revolutionized the ability to image neuronal circuitry. Constancy in the position of neurons in closely related species has been observed and homologies suggested (for review, see Joffe and Kotikova 1991; Elvin and Koopowitz 1994). The most reliable results have been obtained following the injection of dyes into individual cells (for review, see Elvin and Koopowitz 1994; Koopowitz *et al.* 1995). This method has revealed corresponding morphologically distinct neuron types in the polyclads,

BILATERIA

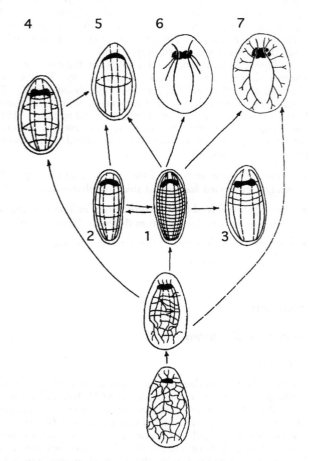

Figure 22.1 Phylogenetic relationship of Bilateria. (From Willmer 1990, after Ax 1984.)

Figure 22.2 The types of nervous system in Platyhelminthes — plexus, cordal, orthogonal: 1, orthogonal; 2, regular; 3, uniform; 4, irregular; 5, concentrated; 6, fan-like; 7, network. (For details, see Kotikova 1991.)

Notoplana acticola and *Alloeoplana* and in the rhabdocoel, *Mesostoma ehrenbergii*. Heteropolar bi- and multipolar, isopolar bi- and multipolar as well as unipolar cells have been found in the cerebral ganglion of these worms. In addition, bilaterally symmetrical, dye-coupled circuits have been demonstrated (Figure 22.3A,B). In the triclad, *Bdelloura candida* heteropolar multipolar cells were reported by Hanström (1926). The features shared by Polycladida, Tricladida and Rhabdocoela may constitute synapomorphies, and perhaps reflect properties of ancestral flatworm nervous systems (Elvin and Koopowitz 1994).

Furthermore, the heteropolar multipolar cells are unlike the classical isopolar multipolar cells of cnidarian nerve nets, and seem more akin to the heteropolar multipolar cells of vertebrates (Koopowitz 1989; Elvin and Koopowitz 1994). Other features reminiscent of vertebrate neurons are dendritic spines. The authors address the question as to whether the similarity of the heteropolar multipolar cell type with cells found in chordates may tie the flatworms closer to the deuterostome line. In most invertebrate phyla, cells in the ganglion rind are monopolar. Compared to higher invertebrates, the polyclads have more bi- and multipolar cells. Bi- and multipolar cells are frequently reported in fluorescence and CSLM studies of the proserate and triclad brains (Joffe and Reuter 1993; Reuter *et al.* 1995a,b,c, 1996a), and heteropolar multipolar cells are common in parasitic flatworms (Gustafsson and Wikgren 1981). However, multi- or bipolar heteropolar cells have not so far been observed in CSLM studies of the brains in worms from the lower flatworm taxa, Catenulida and Macrostomida (Wikgren and Reuter 1985; Reuter *et al.* 1996a). The presence of specialized cell types, such as heteropolar multipolar cells, in only advanced flatworms undoubtedly raises the question as to whether the 'stem-bilaterian' had these cell types, or whether they evolved independently in flatworms and higher animals.

Most observations of homologous cells concern closely related species and confirm postulated relationships. Homology has been suggested for the pairs of bipolar neuropeptide F (NPF)- and RFamide-immunoreactive cells positioned anterior to the pharynx and for the NPF-immunoreactive cells of the pharyngeal nerve ring that occupy the same positions in the proserate taxa, Otoplanidae and Monocelididae (Joffe and Reuter 1993; Reuter *et al.* 1994). A pair of large peptidergic cells observed caudodorsally in the brain of all investigated proserates may be an additional homologous pair (Joffe and Reuter 1993; Reuter *et al.* 1995b).

Joffe and Kotikova (1991) have suggested homology for the four pairs of large identifiable catecholamine-positive cells in the brain of Dendrocoelidae and Planariidae. In Dugesiidae, only two pairs have been found; a fact regarded as a sign of a

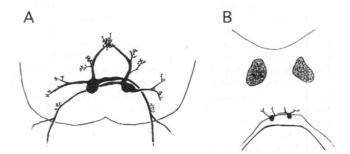

Figure 22.3 **A)** Dye-coupled multipolar heteropolar neurons in the posterior part of the brain of *Notoplana acticola* (Polycladida). **B)** Dye-coupled multipolar heteropolar neurons in the posterior part of the brain of *Mesostoma ehrenbergii* (Rhabdocoela) (Koopowitz *et al.* 1995).

more primitive status of this taxon. In the large genus, *Promesostoma* (Rhabdocoela), neurons in the so-called anterior complex (AnDo) have been demonstrated in all species investigated thus far. This characteristic differentiates this genus from other Typhloplanoidea (Joffe and Kotikova 1991). Furthermore, homologies of neurons in closely related species of Seriata and of Rhabdocoela have been suggested (for reviews, see Joffe and Kotikova 1991; Kotikova 1994).

Evidence for homologous catecholaminergic neurons in Temnocephalida and a representative of the putative sister-group, Dallyelliida has recently been obtained (Joffe and Cannon 1998a). This study provides the first evidence of the persistence of the homologous neurons through significant evolutionary changes of the organs they innervate. Promising results for identification of homologous sensory neurons have been obtained using staining with DiO (3,3′-dioctadecyloxacarbocyanine perchlorate) (Joffe and Cannon 1998c; Joffe *et al.* 1998; see also Cannon and Joffe 2001, this volume).

Groups of neurons

Another type of homology of neurons is that of cell groups in proseriates has been suggested for the catecholaminergic ventromedial (VVM) neurons in *Monocelis* and *Archilopsis* (Joffe and Kotikova 1991), and the ventromedial NPF- and RFamide-immunoreactive neurons adjoining the transverse commissures in the monocelidids and the otoplanid, *B. balticus* (Joffe and Reuter 1993). However, great caution is needed in the interpretation, since the varying position of these cells in the monocelidid and otoplanid species may be related to the variability of the submuscular plexus of which they form a part.

Marker neurons

The use of marker neurons for the nerve cords has been discussed in a study of the orthogon in the proseriate, *Bothriomolus balticus* (Joffe and Reuter 1993). Catecholaminergic and serotoninergic neurons are of special interest in this connection. The majority of the platyhelminths have a group of catecholamine-positive neurons (L-group) situated in postcerebral lateral zones of the body (Joffe and Kotikova 1991; Reuter and Eriksson 1991). According to Joffe (1991), these neurons constitute a group that is homologous in all members of Rhabditophora. For the Rhabditophora and Catenulida, the possession of the catecholaminergic L-group neurons, associated to the main nerve cords (see Organization, below, for explanation) is probably a synapomorphy (Joffe 1991).

In addition, postcerebral serotoninergic neurons seem to mark the main nerve cords. These cells either send their processes only to the main cords, so that they reach other cords only via the main one, or they direct one process (or processes) to the main cord and the other (or others) to another cord (Joffe and Reuter 1993). In the Catenulida, Macrostomida, Proseriata and Tricladida, serotoninergic cells are associated to the MCs (see Reuter and Gustafsson 1995; Reuter *et al.* 1998a). Studies on the distribution of serotonin-immunoreactive cells in the Rhabdocoela are in progress (E. Kotikova and M. Reuter, unpublished observations).

Organization

The demands of the 'lifestyle' of a particular organism influence the development of the body musculature. Therefore, the neuromuscular organization of the organism reflects more the functional morphology than the phylogeny. The evolutionary possibilities are restricted by the inherited conditions seen as certain trends in the evolution of the nervous system. Among modern flatworms, we can find none corresponding to an ancestral form. In the simplest, but perhaps not the most archaic, flatworms, i.e., in the Acoelans (see Raikova *et al.* 2001, this volume), conduction is mainly carried out by nerve nets, i.e., plexuses. According to Ax (1963), an internal nervous system with a brain, several pairs of longitudinal cords, and transverse commissures is more probable for the turbellarian archetype than that of a superficial nerve plexus. In addition to the nerve nets, the presence of thin radial-symmetric longitudinal nerves characterizes many simple flatworms. Reisinger (1925) introduced the term orthogon for the simple geometric, ladder-like baseplan of the flatworm nervous system, with a brain and a set of longitudinal nerve cords crossed by transverse commissures (Figure 22.4). The orthogon is regarded as the basic plan for the protostome nervous system developed from an insunk 'Hautnervenplexus'. Thus, the ancestral form 'the pro-plathelminth' had a 'Hautnervenplexus' which, in concert with the evolution of bilateral symmetry, gave rise to the orthogon, i.e., the basic condition in flatworms (Reisinger 1976). The orthogonal pattern occurs in members of different flatworm taxa. This orthogonal plan, however, shows variation in the number of cords and in the total design (Bullock and Horridge 1965). As to the evolution of the orthogon, the following three trends have been suggested. First, a reduction in the number of longitudinal cords; second, a strengthening of the remaining ones (Reisinger 1970, 1972); and, third, a concentration of plexal

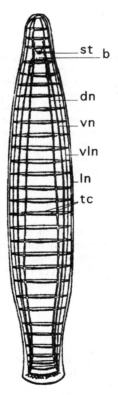

Figure 22.4 Orthogonal nervous system of *Minona trigonopora* Ax. Abbreviations: b, brain; dn, dorsal longitudinal nerve cord; ln, lateral longitudinal nerve cord; st, statocyst; tc, transversal commissure; vln, ventrolateral longitudinal nerve cord; vn, ventral longitudinal nerve cord. (After Reisinger 1972.)

fibres to longitudinal cords (Joffe and Reuter 1993). As a consequence of these trends, a common pattern can be discerned in all flatworms, apart from the Acoela — CNS composed of a brain and one pair of longitudinal nerve cords (MCs) (see Main cords, below) and a PNS composed of minor cords and plexuses (for discussion, see Reuter et al. 1998a).

Brain

The structure of the brain varies from a few cells positioned laterally to a fibrillar neuropile to a distinct bilobed ganglion connected by commissures. Features of interest for phylogenetic discussion have been found in Acoela (see Raikova et al. 2001, this volume), Catenulida, Polycladida and Rhabdocoela.

Catenulida

Here the brain differs from the most common construction of a flatworm brain. Frontally, lateral lobes emerge from the central brain lobes connected by neuropilar fibres (see Figure 22.5A; opposite p. 244). To determine if these lobes are predominantly sensory or represent genuine parts of the brain needs further research (for review see Rieger et al. 1991). In *Stenostomum leucops*, the sensory frontal lobes do not form part of the brain *sensu stricto* (Wikgren and Reuter 1985; Reuter et al. 1993; Reuter 1994). They innervate the sensory pits, which are connected to the body by strong muscles (M. Reuter and O. Raikova, unpublished observations). According to Ehlers (1985), the construction of the catenulid brain may be an autapomorphy for the taxon Catenulida.

Polycladida

The most complex flatworm brains are found in the polyclads. Their encapsulated brain appears to be constructed of an outer rind of somata and a more or less well-separated inner core of neuropile similar to that of other protostomes (for review, see Ehlers 1985; Rieger et al. 1991). Specialized groups of cells – globuli cells – occur outside the capsule. The vertebrate features, such as dendritic spines and the abundance of heteropolar multipolar neurons in the polyclads, *Alloeoplana* and *Notoplana* (see Koopowitz et al. 1995), were dealt with in the text detailing the homology of single neurons (p. 239).

Rhabdocoela

Architectural features in the brain of *M. ehrenbergii* similar to those of the polyclad brain may represent homologies between brains of the two taxa (Koopowitz et al. 1995). The authors suggest that the shared features can be regarded as synapomorphies, and represent primitive and general features of flatworm nervous systems.

Nerve cords

As already mentioned, the basic orthogonal plan shows variation in the number of cords and in the total design. The variability in the organization of the nervous system in closely related flatworm species makes phylogenetic evaluation a task that demands great caution by the investigator.

One pair of main cords and a varying number of minor cords characterize most flatworms. The terminology of the longitudinal nerve cords has been confusing (e.g., they have

been named dorsal, ventral, lateral, ventrolateral and marginal). To avoid confusion, two concepts were introduced, namely, the main nerve cords and the minor cords (Joffe and Reuter 1993; Reuter and Gustafsson 1995). The main nerve cords have been defined as the pair of longitudinal nerve cords that, independent of lateral or ventral position: i) start with strong rootlets in the cerebral ganglion; ii) consist of wide fibre bundles; and iii) are associated with more neurons (particularly aminergic marker neurons) than the other cords. Cords in other positions are thinner and have less pronounced contact with the brain; they have collectively been named minor cords.

Using the above-mentioned criteria for the main nerve cords, a working hypothesis was presented (Reuter et al. 1995a, 1998), namely, that the main nerve cords in flatworms are homologous; in other words, they correspond to each other and they have a common phylogenetic origin. Parallelism in the shape of the orthogon may, nevertheless, exist.

Main cords (MCs)

What importance can be attributed to the MCs? Do they form a stable characteristic or not? Support of the special status of the MCs has been obtained from studies of the neuroanatomy of Catenulida, Macrostomida, Lecithoepitheliata, Proseriata, Tricladida, and of parasitic flatworms.

Catenulida and Macrostomida. In *Stenostomum leucops* (Catenulida) and *Microstomum lineare* (Macrostomida) the lateral cords have the status of MCs, i.e., they are the only cords with strong roots from the brain (Figure 22.5A,B; opposite p. 244); they are wide and adjoined by 5-hydroxytryptamine(HT)-immunoreactive and catecholaminergic neurons (Wikgren and Reuter 1985; Reuter et al. 1986; Joffe and Kotikova 1991).

Seriata, Proseriata. In a study of the taxon Seriata, a surprising variety in the position of the different nerve cords was observed. Striking variations in dominance and position were noted in two families of Proseriata, with four pairs of cords being demonstrated in both families. However, the use of the above-mentioned criteria for MCs showed that, independent of their lateral or ventral position, the MCs of Monocelididae and Otoplanidae correspond to each other. They start from the neuropile, with broad multifibre roots extending posteriorly, and are associated with catecholaminergic and 5-HT-immunoreactive cells (Figure 22.5C,D). No other cords have such a strong connection to the brain (Joffe and Reuter 1993; Reuter et al. 1995a). Furthermore, the MCs of Monocelididae and Otoplanidae have the same type of contact with the pharyngeal nervous system. In both groups, a pair of bipolar cells located anteriorly to the pharynx connects the MCs with the pharyngeal nervous system.

Seriata, Tricladida. Support for a correspondence of the MCs in flatworms has been obtained even from studies of several species of the taxon Tricladida. Recent immunocytochemical (ICC) studies of the neuroanatomy of planarians have shown that their ventral longitudinal nerve cords fulfil the stipulated criteria for MCs (see Figure 22.5E) (see Reuter et al. 1998).

When the position of the main cords in flatworm taxa is compared, it seems that the lateral position characterizes Catenulida, Macrostomida and the proseriate taxon Monocelididae, whilst a ventral position is typical for flatworms of higher taxa. The evidence, for a correspondence of the MCs in species of the taxons Catenulida, Macrostomida,

Lecithoepitheliata and Seriata, based on the criteria for MCs, supports the suggested common origin of MCs in flatworms.

Rhabdocoela. With regard to the Rhabdocoela, the evidence is incomplete. The organization of the nervous system of the taxon Rhabdocoela has been studied mainly by Kotikova and Joffe (Kotikova 1986, 1991; Joffe 1990; Joffe and Kotikova 1991). In most species, ventral and lateral cords begin with common roots, and seem to be composed of about the same number of nerve fibres (see Figure 22.5). As mentioned in the text about marker neurons (p. 241), the most prominent nerve cords, i.e., the MCs, contain catecholaminergic nerve fibres and receive processes from the postcerebral catecholaminergic-positive L-group neurons. Data about the localization of 5-HT-immunoreactive cells in Rhabdocoela have hitherto been published only for *Gyratrix hermaphroditus*, where they adjoin the beginning of the MCs (Reuter *et al.* 1988). The difficulties in defining the most prominent cords in rhabdocoels shows that the distribution of serotoninergic marker neurons in lateral and ventral cords needs to be clarified. Furthermore, future research will clarify if the differences in numbers of sensory fibres in ventral and lateral cords of similar diameter can be used as a marker for nerve cords (Joffe and Cannon 1998b).

Neodermata. Studies of the general neuroanatomy of parasitic flatworms (for references, see Gustafsson 1992; Halton *et al.* 1992; Reuter and Gustafsson 1995; Gustafsson *et al.* 1994, 1995) provide support for the special status of main cords. In the main part of adult and larval stages of parasitic Digenea, Monogenea and Cestoda, main cords can easily be distinguished from minor cords (Maule *et al.* 1990; for review, see Gustafsson and Reuter 1992). In procercoids of *Diphyllobothrium dendriticum*, the nervous system consists of only the two 5-HT-immunoreactive longitudinal MCs connected by a brain commissure (Figure 22.6A; opposite p. 244) (Wikgren 1986). In cercariae of *Schistosoma mansoni* and in adult cestodes, a pair of central main cords projects from the bilobed brain, while peripheral nerves occur in association with the peripheral musculature (Skuce *et al.* 1990; Gustafsson and Reuter 1992). During the life-cycle of the eye-fluke, *Diplostomum pseudospathaceum*, the transverse ring commissures increase in number during the change from the cercarial to the metacercarial stage, but the relationship between the MCs and the minor cords remains unchanged (Figure 22.6B) (Grabda-Kazubska *et al.* 1991).

Joffe and Reuter (1993) discussed the assumption of 'total homology' of the main cords in the Plathelminthes, but found the assumption slightly oversimplified the situation. In the evolution of cords two processes have to be taken in account: i) a shift of the main cords accompanied by formation of new cords from the plexus and fusion of other cords; and ii) a redistribution of nerve processes and perikarya between the cords. The correspondence of cords is determined by the homology of the neurons associated with these cords (for discussion, see Joffe and Reuter 1993). Therefore, the homology of the MCs may be partial because a redistribution of neurons has to be taken in account, especially when Rhabdocoela are concerned, in which two similarly wide pairs of cords begin with common roots.

MCs in asexual multiplication. The role of MCs in asexual multiplication stresses their unique character. In *Stenostomum leucops* (Catenulida) and *Microstomum lineare* (Macrostomida), in which chains of individuals are formed, the first indication of the development of a new individual (zooid) is always the appearance of 5-HT-immunoreactive cells associated with the parental MCs (see Figure 22.7A;

opposite p. 244). Thus, the MCs form a centre for the interactions, which determine the position of the new zooids (Reuter and Palmberg 1989). Nerve fibres growing from the MCs towards the centre then form a new brain commissure. Also, a recent ICC study of *Girardia (Dugesia) tigrina* points to the importance of MCs (Reuter *et al.* 1996b). The development of the new brain after fission involves two processes. First, an outgrowth of the original MCs sends processes transversally towards the centre, indicating the position of the new brain commissure. Thereafter, new nerve cells in front of the commissure differentiate from neoblasts and fasciculate with the fibres from the old MCs (Figure 22.7B) (Reuter *et al.* 1996b).

Minor cords

The number and position of the minor cords show great variability, indicative of their more dynamic nature and close relationship to the submuscular nerve net.

Correspondence between lateral nerve cords and between dorsal nerve cords has been suggested for the closely related taxa within Proseriata. In the proseriate families Monocelididae and Otoplanidae, the organization of the lateral cords (LCs) and dorsal cords (DCs) are similar, but not identical (Joffe and Reuter 1993; Reuter *et al.* 1995b). The LCs start from the brain, run parallel to the roots of the MCs, and diverge not far from the brain in an anterior and a posterior branch (see Figure 22.5C,D). However, the position is more dorsal in the Otoplanidae than in the Monocelididae (Figure 22.7C).

The DCs run near the body surface and join in the frontal end of the body. They lack roots from the brain, the same features and positions seeming to characterize the DCs in all the orthogonal types described for rhabditophorans (Kotikova 1991). Thus, the topographical architecture of DCs may represent a synapomorphic feature of the Rhabditophora.

Transverse commissures

In Catenulida, orthogonal circumferential or transverse commissures are observed in *Xenostenostomum* (Reisinger 1976) but not in *Stenostomum* (Wikgren and Reuter 1985). In Macrostomida, regularly spaced thin circumferential commissures connect the nerve cords, and a strong transverse commissure, 'the postpharyngeal commissure' occurs behind the mouth and pharynx. It may, according to Ehlers (1985), be an autapomorphy for the taxon Macrostomida. Present studies, however, show both catecholaminergic-positive and 5-HT-immunoreactive fibres connecting the longitudinal cords behind the pharynx also in several typhloplanid rhabdocoels (Figure 22.8). These postpharyngeal commissures connecting ganglionic cell clusters at the cords are regarded as an example of parallel evolution (E. Kotikova and M. Reuter, unpublished observations).

The most regular construction of commissures forming the orthogonal structure is found in the taxon Seriata. Regular circumferential commissures characterize the Proseriata. In the triclad Maricola, circumferential routes are observed (Rieger *et al.* 1991), whilst in other planarians, transverse commissures connect the MCs. In addition, lateral branches from the MCs run towards the body periphery (Reuter *et al.* 1995a,b,c, 1996a; Mäntylä *et al.* 1998). The transverse commissures and the lateral branches originate from ganglion-like condensations along the MCs, whose position indicates an origin from crossing points of MCs and circumferential commissures

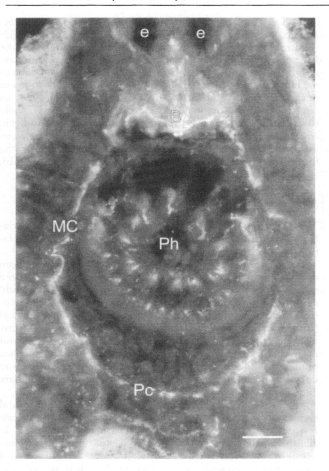

Figure 22.8 *Bothromesostoma essenii* 5-HT-IR in main nerve cords (MC) connected by postpharyngeal commissure (Pc). Abbreviations: Ph, pharynx; B, fibres in brain; e, cerebral eyes. Scale bar, 20 μm.

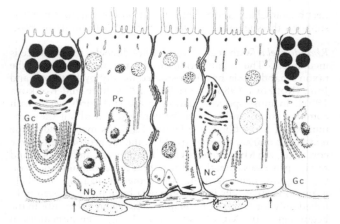

Figure 22.9 *Microstomum lineare*. Schematic drawing of the gastroderm showing neuronal cells (Nc) in the gastroderm, above the basal lamina (arrow). Abbreviations: Gc, granular club cells; Nb, neoblast; Pc, phagocytic cells; M, subepithelial muscle (Reuter and Palmberg 1987).

(Figure 22.5E) (Reuter *et al.* 1998). In dalyellioid rhabdocoels, reduction of commissures has taken place. The phenomenon has been reviewed and discussed in detail by Kotikova and Joffe (1988).

Plexuses

Three differently located peripheral nerve plexuses are found in flatworms: the submuscular; the subepidermal or subepithelial; and the infraepithelial plexus (for definitions, see Ehlers 1985; Rieger *et al.* 1991). The first two plexuses are observed in most taxa, whilst an infraepithelial plexus is observed in *S. leucops* (Catenulida) (M. Reuter, unpublished observations), Acoelomorpha (Ehlers 1985), and Polycladida (Koopowitz and Chien 1974, 1975). An additional inner or parenchymal plexus associated with the inner muscles occurs in larger flatworms such as Tricladida, Polycladida and Neodermata (Kotikova and Kuperman 1977; Ehlers 1985). The presence of sensory nerves and terminals penetrating the epithelium is common to all flatworms. The nerve nets within and immediately beneath the epidermis may combine sensory and motor functions, whilst the submuscular nerve plexus is believed to have a purely motor role. Confocal microscopy studies of lecithoepitheliates, proseriates and planarians have revealed that a dynamic interrelationship occurs between the

submuscular plexus and the orthogonal commissures; in particular, transverse orthogonal nerve fibres contact or constitute part of the submuscular plexus (for review, see Reuter *et al.* 1998).

Plexuses specialized for different adaptive purposes are observed in many flatworm taxa, notably polyclads, planarians, monogeneans, digeneans and cestodes (Reisinger 1972). These differences reflect a variety of functional morphologies.

The pharyngeal and stomatogastric nervous system

The stomatogastric nervous system of flatworms is of interest from an evolutionary point of view since it is regarded as plesiomorphic (Ehlers 1985a), and may therefore show preserved archaic features.

Flatworms are the most primitive animals in which there is an alimentary tract, with mouth, pharynx and intestine, but without an anus. As a stomatogastric infra- and/or subepidermal nervous system occurs in lower as well as higher flatworms, a stomatogastric nervous system is regarded as archaic and may thus show preserved features which are of interest from an evolutionary point of view. The infraepithelial position occupied by nerve cells and fibres in the intestine of *M. lineare* (Macrostomida) (Figure 22.9) appears to be a plesiomorphic feature (Reuter and Palmberg 1987). In more advanced flatworms, a subepidermal intestinal nerve plexus of irregular and reticular structure without differentiated tracks is observed (Reisinger 1976; Baguñà and Ballester 1978; Reuter and Palmberg 1987; Reuter *et al.* 1995b). The absence of an intestinal nervous system in many typhloplanides lacking intestinal muscles has been interpreted as retrograde evolution, depending on their type of nourishment (Reisinger 1976). The same holds true for cestodes where all traces of an intestine are lacking. According to Reisinger (1976), the presence of an autonomous nervous system consisting of a pharyngeal and an intestinal part in most flatworms proves the relationship between Platyhelminthes and Spiralia.

Immunocytochemical studies show that only peptidergic nerve fibres and cells occur in the gastrodermis of *M. lineare* (Macrostomida) (Figures 22.9 and 22.10; opposite p. 245) and that the intestinal plexus in all investigated flatworms is mainly innervated by peptidergic neurons (Reuter *et al.* 1995a,b,c).

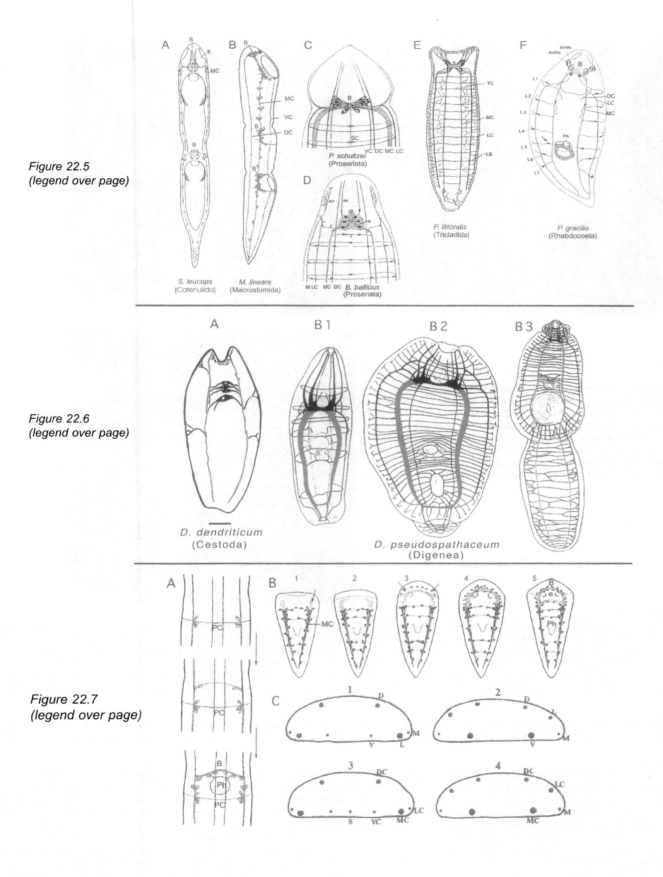

Figure 22.5
(legend over page)

Figure 22.6
(legend over page)

Figure 22.7
(legend over page)

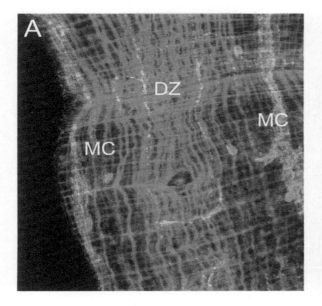

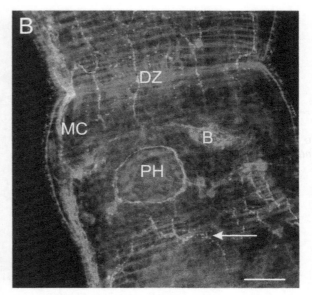

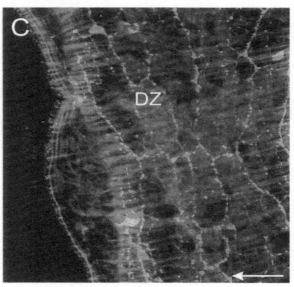

Figure 22.10

Peptidergic intestinal nerve net in zooid of *Microstomum lineare*. Double staining; nerves in green stained with anti-FMRFamide; muscles in red stained with TRITC-labelled phalloidin. Optical sections visualizing: A) level of subepidermal muscles close to surface; B) level of pharynx (PH). Note the connection to the intestinal nerve net (arrow). C) Level of peptidergic intestinal nerve net (arrow). Abbreviations: B, brain; MC, main nerve cord; DZ, division zone between zooids. Scale bar, 10 µm.

Figure 22.5
Nerve main cords (MC) marked in red in schematic drawings of the organization of the orthogon in: A) *Stenostomum leucops*, Catenulida (Wikgren and Reuter 1985);B) *Microstomum lineare*, Macrostomida (Reuter et al. 1986); C) *Promonotus schultzei*, Proseriata, Monocelididae (Reuter et al. 1995a);D) *Bothriomolus balticus*, Proseriata, Otoplanidae (Joffe and Reuter 1993); E) *Procerodes littoralis*, Tricladida (Reuter et al. 1995b); F) *Promesostoma paragracilis*, Rhabdocoela. (After Joffe and Kotikova 1991.) Abbreviations: An, anterior nerves; AnDo, anterior dorsal neurons; AnVe, anterior ventral neurons; B, brain; fl, frontal lobes; ICP, innervation of ciliary pits; DC, dorsal nerve cord; L1–L7, L-group neurons; LB, lateral branch; LC, lateral nerve cord; M, marginal nerve; MC, main nerve cord; Ph, pharynx; SC, single nerve cord; TC, transverse commissure.

Figure 22.6
A) *Diphyllobothrium dendriticum* procercoid. Anti-serotonin positive cells and fibres, main nerve cords marked in red. Scale bar, 20 µm. (After Wikgren 1987.) B 1–3) The orthogon during the life-cycle. Main nerve cords marked in red. (After Grabda-Kuzubska et al. 1991.)

Figure 22.7
A) Pattern of serotonin immunoreactivity during development (arrows) of a new brain in zooids of Microstomum lineare. The formation of a new brain is first indicated by accumulation of serotoninergic (grey) neurons at the parental main nerve cords connected by a postpharyngeal commissure (PC). Abbreviations: B, brain; Ph, pharyngeal nerve ring (Reuter and Palmberg 1989). B) Schematic drawing showing the two processes involved in regeneration of the brain ganglion of Girardia (Dugesia) tigrina. Days 1–2, first processs. 1, 5-HT-immunoreactive nerve fibres sprouting from old main nerve cords (MC) (thin arrows); 2, fibres growing transversely toward the centre close to blastemal border. Days 3–5, second process. 3, Neoblasts showing 5-HT-immunoreactive in front of developing anterior brain commisure (broad arrows). 4–5, 5-HT-immunoreactive cells fasciculating with the anterior brain commisure. Ph, pharynx (Reuter et al. 1996b).C) Topographic position and naming of the longitudinal nerve cords in transverse sections of (1) Monocelididae and (2) Otoplanidae according to Joffe and Reuter (1993), and renaming of the cords: (3) Monocelididae and (4) Otoplanidae (Reuter et al. 1995a). Abbreviations: D and DC, dorsal cords; L and LC, lateral cords; M, marginal cords; MC, main cords; S, single cord; V and VC, ventral cords.

The observation indicates that peptidergic substances might have had an important role in the ancestral stock that gave rise to flatworms.

Ultrastructure

Neurons

The most striking ultrastructural feature of flatworm neurons is their highly secretory nature (see Reuter and Gustafsson 1995). A variety of vesicles are present in the cells and fibres. They range in shape from spherical to compressed, and in appearance from clear to dense. A single type of vesicle often dominates in the perikarya. More variations in vesicle types occur in the fibres, where small clear vesicles are often seen alongside the dominating vesicle type. Vesicle types have been used as markers for different cell types in the brain (small clear vesicles (sv) as cholinergic, dense-cored vesicles (dcv) as aminergic, and large dense vesicles or elementary granules (ldv) as peptidergic or neurosecretory in nature). However, the results from electron microscopic immunocytochemistry show that a classification of neurons with respect to neuronal vesicle types and function, on the basis of ultrastructure only, is at best problematic (Reuter and Gustafsson 1995).

Neurons containing dcv (Figure 22.11A) dominate the ultrastructural picture of the flatworm nervous system. The core varies in size and density. Dcv-containing neurons occur in the brain, in the intestine and in presumed sensory receptors. According to Reuter and Gustafsson (1989), the dcv-containing neuron may be considered an archaic neuropara-neuronal cell and a likely progenitor cell type for conventional neurons.

As to the ultrastuctural differentiation, the neurons of the catenulids and the macrostomids seem to be structurally least differentiated (Moraczewski 1981; Reuter 1990), and resemble the neuroblasts of higher animals (Moraczewski 1977, 1981). In Acoela, more cellular types are present than in Catenulida and Macrostomida and, as in advanced flatworms, the population of vesicles is more heterogeneous. Ganglion cells containing solely small clear vesicles have been found in Acoela, Tricladida and Rhabdocoela (Morita and Best 1966; Lentz 1967; Bedini and Lanfranchi 1991, 1998; Reuter and Gustafsson 1995).

'Glial' cells

A second category of cells found in the nervous system of some flatworms are the neuroglia. The observations are scattered among different taxa: planarians (Figure 22.11B), Lecithoepitheliata, Acoela and Typhloplanidae (for review, see Bedini and Lanfranchi 1991, 1998; Reuter and Gustafsson 1995; Mäntylä et al. 1998). These cells are believed to perform supporting, isolating and metabolic functions. Multilayered lamellae (Figure 22.11C) are found around neurons and/or nerve fibres of proseriates, planarians, temnocephalids and in the polyclad, Notoplana acticola (for review, see Reuter and Gustafsson 1995; Mäntylä et al. 1998).

In the lower flatworm taxa Catenulida and Macrostomida, neither glial cells nor any form of wrapping tissue has been observed. The scattered observations of 'glial' cells and lamellae in members of several flatworm taxa indicate that neuroglia may have evolved independently in flatworms and represent an example of convergence.

Synapses and non-synaptic release sites

A conspicuous variation in the morphology of the vesicle release sites is observed (Figure 22.12). It is difficult to draw a boundary between synaptocrine and paracrine release in flatworms (Reuter and Gustafsson 1989). The variability of synaptic structures in lower flatworms indicates that a rich structural assortment was available in the stem form. The following trends are observed, namely increases in: i) the density of synaptic membrane structures; ii) the number of multiple and serial synapses; and iii) the specialization of paramembraneous densities (Reuter and Gustafsson 1995).

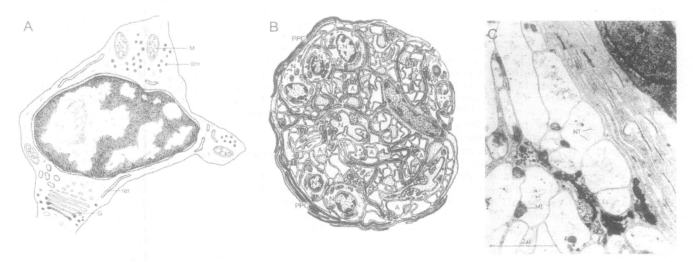

Figure 22.11 Ultrastructural features of flatworm nervous system. **A)** Drawing illustrating secretory character of flatworm neuron. Abbreviations: dcv, dense-cored vesicles; G, Golgi complex; M, mitochondrion; rer, granular endoplasmic reticulum. **B)** Diagram of the fine structure inclusive neuroglia in the ventral cord in a planarian. Abbreviations: A, axon; G, glial cell; N, nerve cell; PPC, processes of parenchyma cells (Golubev 1988). **C)** Section through nerve of Multicotyle purvisi. Abbreviations: LA, lamellae of nerve sheath; MI, mitochondrion; N, nucleus; NT, neurotubule (Rohde 1971c).

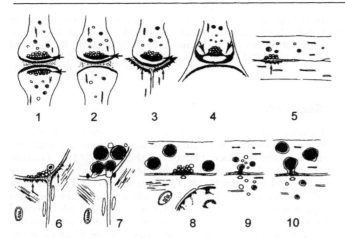

Figure 22.12 Diagram showing different types of synapses and non-synaptic release sites in flatworms. **1.** Two-way synapse with paramembraneous densities (arrows). **2.** Single synapse with paramembraneous density (arrow). **3.** Shared synapse with paramembraneous density (large arrow) and postsynaptic densities (small arrows). **4.** Synapse with paramembraneous density connected by strings to presynaptic density. **5.** En passant synapse and electrotonic junction. **6.** Shared synapse-like structure towards muscle fibres, postsynaptic densities (arrows). **7.** Release via protrusions (arrows) through axolemma towards muscles. **8.** Synapse-like structure at cell body (empty arrow). **9.** Release via omega figure of small dense vesicles (empty arrow). **10.** Release via omega figure from large dense vesicle (empty arrow).

Synaptocrine release

Several types of true chemical synapses have been observed in the worms. Chemical synapses are either polarized or unpolarized, i.e., the pre- and postsynaptic membranes are structurally similar or different, with unpolarized 'two-way' synapses having been observed in microturbellarians and acoelans (Reuter and Palmberg 1989; Bedini and Lanfranchi 1991). The polarized synapses can be divided into the following categories: single, shared, en passant, complex serial synapses, and neuromuscular junctions.

The types of presynaptic vesicles vary in size and density, but sv and dcv are the most common types. In the presynaptic terminal, a specialized structure – the 'paramembraneous density' – is often seen proximally from the presynaptic thickening of the membrane. In acoels, a paramembraneous density resembling the presynaptic T-bar of arthropod synapses has been observed (Bedini and Lanfranchi 1991). A paramembraneous density connected by strings of dense material to several thickenings of the presynaptic membrane has been observed in geocentrophorans (Lecithoepitheliata), and also in *Procerodes littoralis* (Tricladida) (Böckerman *et al.* 1994; Mäntylä *et al.* 1998). In the neuromuscular junctions, the cytoplasmic extensions of the muscle make contact with the axon (Chien and Koopowitz 1972; Moraczewski 1981), or they occur at nonfilamentous areas of the muscles (Reuter 1990).

Somato-axonic synapses and axo-dendritic contacts with 'spine apparatus', comparable to those described in vertebrates, have been identified in the rhabdocoel *Stronglyostoma simplex* (Bedini and Lanfranchi 1998). Greater electron density of the synaptic structures as well as serial synapses occur more frequently in advanced flatworms, indicating a greater degree of specialization of their nervous system (Reuter and Lindroos 1979; Gustafsson 1984).

Paracrine or non-synaptic release

Omega profiles and synapse-like structures lacking the morphological characteristics of conventional synapses have been observed in lower turbellarians (Catenulida, Macrostomida) and in Cestoda (for review, see Reuter 1990; Reuter and Gustafsson 1995).

The types of release sites observed indicate that chemical signalling in flatworms is transmitted as neurotransmitters or neuromodulators, either directly to their targets at synaptic and non-synaptic release sites, or as local chemical mediators via intercellular spaces and extracellular matrix (ECM). Signalling by hormones in a classic sense has to be excluded in these worms, which lack a blood vascular system. According to Fujita *et al.* (1988), the dominance of neuronal transmission in acoelomates is a feature that links them to the Spiralia, forming the protostomian series.

Neuroactive substances

The application during the past 10 years of ICC studies using well-characterized antisera has revealed a broad anatomical distribution of several candidate transmitter or modulatory substances in the nervous systems of flatworms (Table 22.1). These include classical transmitters such as acetylcholine, 5-HT, noradrenaline and dopamine, together with histamine and the amino acids, glutamate and γ-aminobutyric acid (GABA), and a growing number of larger regulatory molecules, the neuropeptides. Of the known gaseous messengers, only nitric oxide (NO) has been demonstrated, albeit indirectly, in the nervous systems of both free-living and parasitic flatworms.

Acetylcholine

Where examined, a physiological role for acetylcholine (ACh) as an inhibitory neurotransmitter in flatworms is indicated by the fact that their muscular activity is reduced by cholinomimetics and by cholinesterase (ChE) inhibitors, ultimately producing flaccid paralysis. There are at least two exceptions to this response, namely the monogenean, *Diclidophora merlangi*, which in over 100 motility trials failed to produce a consistent inhibitory or excitatory response to ACh (Maule *et al.* 1989), and the demonstration of the excitatory actions of ACh on the isolated muscle fibres of the marine turbellarian, *Bdelloura candida* (Blair and Anderson 1996). There is also no evidence to suggest that the flatworm cholinergic receptor resembles the vertebrate nicotinic or muscarinic subtypes (Pax and Bennett 1991).

5-Hydroxytryptamine

5-Hydroxytryptamine appears to be the dominant biogenic amine in all of the flatworm taxa examined, and there is good experimental evidence that it serves a variety of functions, most notably that of excitatory neurotransmission. Thus, the application of exogenous 5-HT has been shown to induce motility in muscle preparations of turbellarians (*Procerodes*), monogeneans (*Diclidophora merlangi*), trematodes (*Fasciola hepatica, Schistosoma mansoni*) and cestodes (*Hymenolepis diminuta*). In addition, data exist which indicate the presence of a functional 5-HT receptor on the muscle membranes, but one with a pharmacological profile unlike that of any mammalian 5-HT receptor subtype (Mansour 1984; Pax

Table 22.1 Neuroactive substances demonstrated in selected flatworms. (After Halton and Gustafsson 1996.)

	ChE	NADr	DA	5-HT	HA	Glu	GABA	NO	NPs (n)
Turbellaria									
Dugesia	*+	+	+	+	+	.	+	+	+ (6)
Microstomum	.	.	+	+	+	.	.	.	+ (10)
Stenostomum	.	+	.	+	.	.	.	.	+ (7)
Monogenea									
Gyrodactylus	.	.	.	+	.	.	.	.	+ (5)
Microcotyle	.	+	+	.	.	.	.	.	
Diclidophora	+	.	.	+	.	.	.	.	+ (12)
Trematoda									
Fasciola	+	+	+	+	.	+	+	.	+ (8)
Haplometra	+	.	.	+	+	.	.	.	+ (14)
Schistosoma	+	+	+	+	.	+	.	.	+ (24)
Cestoda									
Diphyllobothrium	+	+	+	+	+	.	.	+	+ (16)
Hymenolepis	+	+	+	+	+	+	.	+	+ (8)
Moniezia	+	.	.	+	.	.	+	.	+ (8)

Key: ChE, cholinesterase; NAdr, noradrenaline; DA, dopamine; 5-HT, serotonin; HA, histamine; Glu, glutamate; NO, nitric oxide; NPs, neuropeptides (number identified).
*+, present; ·, information not available.

et al. 1984; Maule et al. 1989; Thompson and Mettrick 1989). In many instances, the distribution pattern of putative 5-HT staining in flatworms is distinct from that of the cholinergic and peptidergic neuronal pathways, but parallels (where it has been recorded) the staining for catecholamines (Shishov 1991).

Catecholamines

The catecholamines, noradrenaline (NA) and dopamine (DA) have been detected in flatworms. Age-dependent changes occur in the numbers of DA-reactive cells in the brain of the polyclad, Notoplana acticola, but whether this is a result of neurogenesis or induced catecholamine synthesis in pre-existing cells is unclear (Hauser and Koopowitz 1987). Small amounts of NA and DA have been shown to be present biochemically and histochemically in Schistosoma mansoni, Fasciola hepatica and Hymenolepis diminuta (Chou et al. 1972; Bennett and Gianusos 1977; Gianusos and Bennett 1977). In the plerocercoids of Diphyllobothrium dendriticum, the distribution pattern for catecholaminergic neurons overlaps that of serotoninergic neurons and, to a lesser degree, peptidergic components (Gustafsson and Eriksson 1991); in Diplostomum pseudospathaceum, the time of appearance of catecholamine-reactivity in developing cercariae has been related to the onset of their motility (Niewiadomska et al. 1996). Catecholamines have been shown to be myoactive in flatworms, inhibiting or exciting motility in Schistosoma mansoni, Fasciola hepatica and Diclidophora merlangi (Holmes and Fairweather 1984; Pax et al. 1984; Maule et al. 1989), but as yet their precise functions remain to be determined.

Histamine

Whilst there is indirect evidence to indicate that histamine may serve as a neurotransmitter in arthropods and molluscs, evidence of its presence and/or its synthesis in flatworms is either poor or non-existent. However, there are two notable exceptions, Mesocoelium monodi and Haplometra cylindracea (see Mettrick and Telford 1963; Eriksson et al. 1996). Small numbers of histaminergic fibres have been demonstrated immunocytochemically in the turbellarians, Microstomum lineare and Polycelis nigra, and in the cestode, Diphyllobothrium dendriticum, where they are scattered mainly in the longitudinal nerve cords (Wikgren et al. 1990). Histamine has also been demonstrated immunocytochemically in the turbellarian, Dugesia tigrina, but not in the CNS; instead it was found only in the photoreceptor cells in the cerebral eye, with preliminary evidence that its occurrence may exhibit some circadian variation (Panula et al. 1995).

Glutamate

Glutamate is a candidate neurotransmitter in at least two cestodes, eliciting strong excitatory responses where examined, i.e., Gyrocotyle fimbriata and Hymenolepis diminuta (Keenan and Koopowitz 1982; Thompson and Mettrick 1989). Intense glutamate-like immunoreactivity (Glu-IR) has been demonstrated in cell bodies and fibres in and around the longitudinal nerve cords of adult Hymenolepis diminuta, with somewhat weaker staining in the ring commissures (Webb and Eklove 1989). In Mesocestoides corti tetrathyridia and adult Fasciola hepatica, a much wider distribution of Glu-IR has been recorded in nerve cells and fibres throughout both the central and peripheral nervous systems (Brownlee and Fairweather 1996). There are also reports of glutamate-like immunoreactivity in the trematodes, Trichobilharzia ocellata and Schistosoma mansoni (Solis-Soto and de Jong Brink 1994).

γ-Amino butyric acid (GABA)

Although GABA is acknowledged as an important inhibitory neurotransmitter of widespread occurrence in both vertebrates and invertebrates, there are only three reports of its presence in flatworms. Using two independent immunocytochemical methods, involving a monoclonal antibody specific for

GABA, combined with different fixation protocols, Eriksson and Panula (1994), Eriksson *et al.* (1995a) and Eriksson *et al.* (1995b) have demonstrated extensive GABA-immunoreactivity in fibres in the central nervous system of the turbellarians, *Polycelis nigra* and *Dugesia tigrina*, in the longitudinal nerve cords and nerve nets of *Moniezia expansa*, and in the longitudinal cords and lateral nerves in the posterior portion of *Fasciola hepatica*. High-performance liquid chromatography (HPLC) analysis confirmed the presence of GABA in the two parasitic worms, but revealed a much higher concentration of the amine in the cestode. The apparent widespread occurrence of GABA in the CNS of all flatworms thus far examined, and the comparatively high concentration and capacity for its synthesis in these animals, suggest a functional role, perhaps as an inhibitory neurotransmitter. In the marine polyclad turbellarian, *Notoplana acticola* treatment with GABA at concentrations as low as 1 μM causes a decrease in electrical activity in the ventral nerve cords, an effect which is reversible with picrotoxin, bicuculline or strychnine, indicating the presence of GABA receptors in flatworms (Keenan *et al.* 1979).

Nitric oxide (NO)

NO is generated from arginine via NO synthase (NOS), and since NADPH-diaphorase is believed to be identical to NOS in the mammalian nervous system, its histochemical demonstration is used widely to identify NO-producing neurons. Using this approach, NADPH-diaphorase-positive neurons have been demonstrated in *Dugesia tigrina* and in adult *Hymenolepis diminuta*, with reactive fibres occurring in the MCs and brain commissure and in nerve terminals innervating the acetabula and rostellar sac (Eriksson *et al.* 1996; Gustafsson *et al.* 1998; Lindholm *et al.* 1998).

Neuropeptides

A wide range of neuropeptide immunoreactivities has been described in the nervous systems of flatworms by virtue of their cross-reactivity with antisera raised largely against vertebrate neuropeptides. Homologues to some 26 mammalian and six invertebrate peptides have so far been identified (see Halton *et al.* 1994 for details), suggesting there are numerous bioactive peptides in platyhelminths. However, in most cases, the antigens responsible for the recorded immunostaining are unknown. The exceptions are the six native flatworm neuropeptides that have been isolated and their amino acid sequences determined (Table 22.2), thus enabling the generation of highly specific autologous antisera. The peptides comprise two families of invertebrate neuropeptides: i) neuropeptide F (NPF), namely NPF

(*M. expansa*), a 39-amino-acid residue peptide first isolated from the cestode, *Moniezia expansa* (Maule *et al.* 1991) and NPF (*A. triangulata*), an analogous peptide of 36-residue length recovered and sequenced from the terrestrial planarian, *Artioposthia triangulata* (Curry *et al.* 1992); and ii) FMRFamide-related peptides (FaRPs), whose primary structures are YIRFamide (from *Bdelloura candida*), GYIRFamide (*Dugesia tigrina, Bdelloura candida*), RYIRFamide (*Artioposthia triangulata*) and GNFFRFamide (*Moniezia expansa*) (Maule *et al.* 1993, 1994; Johnston *et al.* 1995, 1996). Using antisera to these authentic flatworm peptides has revealed a widespread distribution of peptide-immunoreactivity across flatworm species, with intense staining in cells and fibres throughout both the CNS and PNS.

The physiological functions of NPF-related peptides and of FaRPs in flatworms remains unclear, though their occurrence throughout the flatworm nervous system implies a fundamental role in nerve–muscle physiology. NPF in the molluscan sea hare, *Aplysia californica* induces prolonged presynaptic neuroinhibition in the abdominal ganglia (Rajpara *et al.* 1992), and all of the FaRPs so far isolated and sequenced from flatworms have been shown to be myoexcitatory in a concentration-dependent and reversible manner when applied exogenously to isolated muscle cells or muscle-strips from free-living and parasitic flatworms (Day *et al.* 1994, 1997; Johnston *et al.* 1996; Marks *et al.* 1996; Moneypenny *et al.* 1997). However, the effects recorded showed species differences with respect to the potencies of the four peptides (Table 22.3). Thus, the turbellarian peptides, YIRFamide, GYIRFamide and RYIRFamide were more potent in eliciting an excitatory response on turbellarian, monogenean and digenean muscle fibres than the cestode FaRP, GNFFRFamide. For example, using isolated schistosome muscle fibres, Day *et al.* (1997) showed that turbellarian-derived peptides possessing the YIRFamide motif were 100-fold more potent in inducing contractions than the cestode GNFFRFamide. In contrast, of the four native flatworm peptides examined, only GNFFRFamide significantly induced contraction of cestode muscle (A. Maule, unpublished results). The structural differences between the cestode hexapeptide amide, GNFFRFamide and the turbellarian YIRFamide-containing FaRPs point to significant functional differences with respect to excitation of flatworm muscle (Day *et al.* 1997).

The high degree of potency of turbellarian FaRPs inducing contractions in monogenean and digenean muscle suggests that these peptides closely resemble the endogenous ligand for the monogenean and digenean receptor, but not that of the cestode muscle receptor. This may imply a closer phylogenetic relationship between turbellarians and the monogeneans and digeneans than between and turbellarians and cestodes. However, as yet no monogenean or digenean neuropeptide has been

---

Table 22.2 Primary structures of all known flatworm neuropeptides.

A. FaRPs (FMRFamide-related peptides — Phe-Met-Arg-Phe-amide)

1. *Dugesia tigrina* (freshwater turbellarian)	GYIRFamide
2. *Bdelloura candida* (marine turbellarian)	GYIRFamide; YIRFamide
3. *Artioposthia triangulata* (land turbellarian)	RYIRFamide
4. *Moniezia expansa* (cyclophyllidean tapeworm)	GNFFRFamide

B. Neuropeptide F (NPFs) (C-terminus shown)

1. *Artioposthia triangulata*	-YQIYLRNVSKYIQLRGRPRFamide
2. *Moniezia expansa*	-LRDYLRQINEYFAIIGRPRFamide

Underscoring in **A** and **B** indicates homologous residues among the FaRPs and NPFs, respectively.

Table 22.3 Order of potencies of FaRPs on muscle of flatworm taxa.

GYIRFa* > RYIRFa > YIRFa (*Bdelloura*** - turbellarian)[1]
GYIRFa > RYIRFa > GNFFRFa (*Procerodes*** - turbellarian)[2]
YIRFa > GYIRFa = RYIRFa > GNFFRFa (*Diclidophora* - monogenean)[3]
GYIRFa > YIRFa > RYIRFa > GNFFRFa (*Schistosoma*** - digenean)[4]
RYIRFa > GYIRFa = YIRFa > GNFFRFa (*Fasciola* - digenean)[5]
GNFFRFa > GYIRFa = RYIRFa (*Grillotia* - cestode)[6]

*XYIRFa is considered to be the ancestral flatworm FaRP structure, where X is variable.
**Isolated muscle fibres employed; muscle strips in all others cases.
[1]Johnston *et al.* (1996); [2]Moneypenny (1999); [3]Moneypenny *et al.* (1997); [4]Day *et al.* (1997); [5]Marks *et al.* (1996); [6]A. Maule (unpublished results).

isolated and structurally characterized. Nonetheless, it is clear that FaRP receptors are quite well conserved across flatworms and are able to discriminate between neuropeptides on the basis of their N-termini. Moreover, in the light of these data, it would seem that the ancestral bilaterian flatworm FaRP structure is XYIRFamide, where X is a variable residue. In contrast to other invertebrate groups – most notably molluscs, arthropods and nematodes, where there is a broad spectrum of FaRP isoforms – the complement of these neuropeptides in flatworms appears to be restricted to very few molecules. As such, their value in flatworm phylogenetics would seem to be very limited.

Conclusions

In conclusion, it is clear that the phylogenetic significance of neurobiological data of flatworms is restricted. Some features may be due to parallelism, whilst others confirm postulated relationships of closely related species. However, as a consequence of distinct trends in the neuronal evolution, a common pattern in the organization of the nervous system and its chemistry has been observed in all flatworms, apart from Acoela; a CNS composed of a brain and a pair of longitudinal nerve cords, and a PNS composed of minor cords and plexuses, all of which express a complement of classical and peptidic messenger molecules. Features possibly due to parallelism are: 1) morphological similarities superimposed on the basic orthogonal plan; 2) advanced features in neuronal cytoarchitecture shared by polyclads, rhabdocoels and vertebrates; 3) 'glial' cells and lamellae scattered in members of several flatworm taxa; and 4) presynaptic paramembraneous structures reminiscent of those in arthropods and vertebrates. Nevertheless, the fundamental question posed in the introduction remains to be answered. Do these variations in the flatworm nervous systems reflect independent evolution or primitiveness?

Chapter 23

The use of life-cycle characters in studies of the evolution of cestodes

Ian Beveridge

There has been a long history of the use of life-cycle characters in attempts to elucidate the phylogeny both of the cestodes and related groups of platyhelminth parasites. Perhaps the most notable historical example is the so-called 'cercomer theory' of Janicki (1920), used primarily for evidence of a link between the cestodes and the trematodes. The first comprehensive modern attempt to compare cestode life-cycles was undertaken by Joyeux and Baer (1961). Detailed studies during the 1950s and 1960s of cestode eggs and larval cestode morphogenesis resulted in significant reviews by Rybicka (1966) on the structure of cestode eggs and Voge (1967) and Šlais (1973) on cestode larvae. Subsequently, two major but not entirely concordant attempts to utilize these data in analyses of the phylogeny of the cestodes were made by Freeman (1973) and Jarecka (1975). A summary was provided by Ubelaker (1983a,b) in reviews of egg and metacestode development, and was subsequently updated by Burt (1987).

O'Grady (1985) was the first worker to attempt to codify the features of platyhelminth life-cycles for cladistic analyses, though Rohde (1990) criticized some of the data used in this analysis. Subsequently, Brooks *et al.* (1991) and Hoberg *et al.* (1997c, 1999a) have incorporated life-cycle characters into cladistic analyses of the orders of cestodes and of families within the cestode order Cyclophyllidea. In the most recent analyses of cestode orders (Hoberg *et al.* 1997c), 25% of the 49 characters used were based in life-cycle characters, while in the analysis of taxa within the order Cyclophyllidea (Hoberg *et al.* 1999a), 24% of the 42 characters used were based on life-cycles. Therefore, features of cestode life-cycles currently make an important contribution to phylogenetic analyses. Due to the format of the publications cited, detailed discussion of the characters used has been limited, data being based primarily on the reviews of Freeman (1973) and Jarecka (1975) with the addition of newer information from more recent publications.

This chapter takes the opportunity to review the development of the study of cestode life-cycles (with particular reference to characters which are of phylogenetic significance), to look at the coverage of cestode life-cycles, the robustness of individual characters, to pose the more general question of how significant features of cestode life-cycles are in reconstructing the phylogeny of the class, and finally to attempt to define critical areas of deficiency in our knowledge.

The classification of cestode orders utilized is that of Khalil *et al.* (1994). This classification provides a number of advantages, one of which is that the traditional cyclophyllidean family Dilepididae (e.g., Yamaguti 1959; Schmidt 1986) is here subdivided into several families of which the Dilepididae, Paruterinidae, Metadilepididae and Dipylidiidae have representatives whose life-cycles have been elucidated. The polyphyletic nature of the former Dilepididae was widely recognized (e.g., Freeman 1973) and earlier extrapolations of life-cycle characteristics from individual species to the general pattern exhibited within the family need therefore to be treated with caution.

The groups examined within the order Cyclophyllidea are those utilized by Hoberg *et al.* (1999a), differing slightly from

the usage of Khalil *et al.* (1994). Specific differences include the use of the subfamilies Anoplocephalinae, Linstowiinae, Inermicapsiferinae and Thysanosomatinae. They were placed within the Anoplocephalidae by Beveridge (1994) purely for convenience, in spite of the fact that the subfamilies for which life-cycles are known, exhibit major differences in the development of their metacestodes (Stunkard 1961; Voge 1967). Similarly, the Davaineidae was split into the subfamilies Davaineinae, Ophryocotylinae and Idiogeninae based on the presence or absence of a paruterine organ and egg capsules. In the analysis of cyclophyllidean taxa by Hoberg *et al.* (1999a) the family Gryporhynchidae was also utilized.

The subfamily Gryporhynchinae was erected by Spasskii and Spasskaya (1973) for the genera *Bancroftiella, Clelandia, Cyclorchida, Cyclustera, Dendrouterina, Parvitaenia* and *Proorchida. Gryporhynchus* was considered under the category 'larval and collective names' by Khalil *et al.* (1994: 676), but represents a group of taxa within the Dilepididae as defined by Bona (1994) which are distinctive morphologically and can be utilized constructively as a separate entity when considering features of life-cycles.

Before any examination of the available data is undertaken, it needs to be pointed out that a major obstacle in any analysis of features of the life-cycles of cestodes is the highly scattered nature of the literature. The life-cycles of more than 200 species have been elucidated to the point of providing some information on the morphogenesis of the metacestode, thereby potentially providing a significant resource of information for phylogenetic analysis (Table 23.1). In addition, for cestode taxa in which complete life-cycles are not known or data on morphogenesis is lacking, the morphology of the fully developed metacestode is informative. Groups in which complete life-cycles are lacking but metacestodes are known are also indicated in Table 23.1. However, there is no synoptical treatment of this literature, as exists for the digeneans (Yamaguti 1975) or the parasitic nematodes (Anderson 1992), and the brief summary of Freeman (1973) is useful but by no means comprehensive. It is therefore laborious not only to compile characteristics for analysis but also to decode and interpret character matrices in publications unless the characters are explained in considerable detail and numerous references are cited in support of the analyses presented.

The lack of or fragmentary nature of data on cestode life-cycles is an obstacle bemoaned by all previous reviewers of larval cestode development (Voge 1967; Freeman 1973; Šlais 1973; Jarecka 1975), and continues to be a significant obstacle. Although many life-cycles have been at least partially elucidated (Table 23.1), the concentration of research effort on certain groups is striking. Thus, approximately 25% of all life-cycles elucidated are those of species belonging to a single cyclophyllidean family, the Hymenolepididae, within which development of the metacestode is relatively uniform. The cyclophyllidean subfamilies or families Anoplocephalinae, Taeniidae and Dilepididae again are relatively well studied such that cyclophyllidean life-cycles constitute approximately 68% of those elucidated to date. Of the non-cyclophyllidean

Table 23.1 Numbers of species of cestode in each order in which life-cycles have been elucidated to include data on morphogenesis of the metacestode in the intermediate host.

Cestode taxon	No. of cestode species
Gyrocotylidea	0*
Amphilinidea	2
Caryophyllidea	20
Spathebothriidea	0
Pseudophyllidea	19
Haplobothriidea	1
Diphyllidea	0*
Trypanorhyncha	3
Tetraphyllidea	0*
Lecanicephalidea	0*
Proteocephalidea	21
Nippotaeniidea	1
Tetrabothriidea	0*
Cyclophyllidea	
Mesocestoididae	0*
Catenotaeniidae	1
Nematotaeniidae	0
Paruterinidae	6
Metadilepididae	0
Dilepididae	13
Gryporhynchidae	3
Progynotaeniidae	1
Acoleidae	0
Dioecocestidae	0
Hymenolepididae	49
Taeniidae	16
Amabiliidae	3
Dipylidiidae	1
Davaineinae	2
Raillietiniinae	9
Idiogeninae	0
Ophryocotylinae	0*
Anoplocephalinae	21
Linstowiinae	6
Inermicapsiferinae	0
Thysanosomatinae	7

* Partial elucidation of life-cycles.

taxa, the Proteocephalidea, Pseudophyllidea and Caryophyllidea are relatively well understood from the point of view of life-cycles, while knowledge of other orders is scanty. Of some significance is the fact that of the 21 life-cycles of proteocephalideans elucidated, 16 are species of *Proteocephalus* or the related genus *Ophiotaenia*, whilst the remainder are of species of *Gangesia*, *Corallobothrium* or *Megathylacoides*. Thus, within the order there is a significant concentration of life-cycle studies, compared with a complete lack of knowledge for the remaining 40 or so genera recognized (Rego 1994).

Also of importance are the orders for which no life-cycle data are currently available, the most significant (in terms of number of genera and species) of which is the Tetraphyllidea, with 52 genera recognized by Euzet (1994). Within the Trypanorhyncha (46 genera, in Campbell and Beveridge 1994) only three life-cycles have been completed experimentally, those of *Grillotia erinaceus*, *Lacistorhynchus tenuis* and *L. dollfusi* (see Ruszkowski 1934; Sakanari and Moser 1989). In these life-cycles, tanned, operculate eggs release a coracidium which is ingested by a copepod. Other trypanorhynch life-cycles which have been partially elucidated and which resemble them are *Callitetrarhynchus nipponicus* (see

Nakajima and Egusa 1969, 1973) and *Poecilancistrum caryophyllum* (see Mattis 1986). However, in the eutetrarhynchids *Parachristianella monomegacantha* (see Mudry and Dailey 1971) and *Prochristianella hispida* (see Mattis 1986), the egg is not operculate and ingestion by a copepod is followed by hatching in its gut. Early analyses of cestode relationships (e.g., Freeman 1973) utilized the life-cycle of *G. erinaceus* as 'typical' for the Trypanorhyncha, a situation which would now be considered highly misleading. How much greater diversity exists within the life-cycles of this order remains to be determined.

The brief overview presented here therefore highlights not only the fragmentary nature of the information available on cestode life-cycles but also the potential dangers inherent in assuming that a small number of life-cycles which might have been elucidated in any one order is typical of the remainder of the order. Hence, the fragmentary data represents a major problem for analyses dealing with the higher groups of cestodes.

Outgroups

Current analyses of cestode relationships are invariably based on outgroup comparisons (Hoberg *et al.* 1997c, 1999a). A potential criticism of the earlier reviews of Freeman (1973) and Jarecka (1975) could be the lack of such explicit outgroup comparisons. However, the characters used in their analyses are compared with the condition in a group of cestodes which are acknowledged universally as being 'primitive', providing a clear polarity to character use. In both recent morphological (Hoberg *et al.* 1997c, 1999a) and molecular (Mariaux 1998) studies of the cestodes, comparisons have been made using the Gyrocotylidea and Amphilinidea as the outgroups. While relatively well known from a morphological perspective, their life-cycles are poorly understood with no complete gyrocotylidean cycles known and with the two amphilinidean cycles which have been completely elucidated differing. The anoperculate egg of *Amphilina foliacea* is apparently ingested by a gammarid (Dubinina 1974, 1982), whilst the anoperculate egg of *Austramphilina elongata* hatches immediately upon release into water to release a lycophore which actively penetrates the body of freshwater shrimps and crayfish (Rohde and Georgi 1983). Parts of the life-cycle of *Nesolecithus africanus* are also known (Gibson *et al.* 1987). The problems associated with a limited knowledge of these life-cycles have not inhibited the coding of the small numbers of characters used to date (Hoberg *et al.* 1997c), but it clearly represents a significant obstacle to more detailed studies. It also suggests that additional characters may become available from such studies. Active penetration of the intermediate host by the larva occurs in many digeneans and in a modified form in monogeneans, but not in eucestodes. The character of ingestion of egg or larva could therefore represent a synapomorphy for all eucestodes and some amphilinidians (see also Brooks *et al.* 1985b: 11).

Until more information is available on the life-cycles of amphilinideans and gyrocotylideans, it may be useful or necessary to utilize digeneans and/or monogeneans as outgroups, as has been done in part by Hoberg *et al.* (1997c).

Characters and their 'robustness'

Apart from the patchy nature of the data available, the 'robustness' of the available characters also warrants consideration. Rohde (1996) has argued forcefully that the homology of morphological characters needs to be well established prior to use in cladistic analyses, and it is pertinent to apply the criterion of homology as well as coverage to the characters used to date.

Of the 22 characters based on cestode life-cycles used by Hoberg et al. (1997c, 1999a) in recent analyses, 13 relate to the structure of the egg. Many of the characters utilized are based on the reviews by Rybicka (1966) and Fairweather and Threadgold (1981a) together with more recent studies by Conn (1985a, 1988), Conn et al. (1984), Jones (1988), Swiderski (1972, 1994), Swiderski and Tkach (1997), Tkach and Swiderski (1997, 1998) and Brunanska (1999). The ontogeny of the various egg layers has been relatively well established on an ultrastructural basis, the homology of the layers has been determined (Rybicka 1966), and most of the characters are relatively non-controversial. Rohde's (1996) criticisms are therefore unlikely to apply to these characters. Embryonation of eggs in utero, egg tanning and whether or not eggs are operculate continue to represent problems, the latter occurring primarily with pseudophyllideans (see Bray et al. 1999), amphilinideans (see Rohde and Georgi 1983) and spathebothrideans (see Sandeman and Burt 1972). Swiderski (1994) has highlighted problems associated with the terminology of 'ciliated embryophore' for pseudophyllideans and 'embryophore' for the entire envelope. It appears therefore that some problems need to be resolved with respect to these characters.

Tanning of egg shells appears to be a characteristic which can be coded quite simply. However, Coil (1987) has demonstrated that the histochemistry of shell formation in Austramphilina elongata differs from both eucestodes and digeneans and warrants more detailed examination. Use of amphilinids as an outgroup in this instance may therefore need particular care. Similarly, codings need to be re-examined continuously to keep pace with the availability of new data.

Amongst the cyclophyllideans, there is not only great diversity in egg structure but also a series of ultrastructural studies from which additional characters might be derived (see for example Fairweather and Threadgold 1981a, Figure 2). The potential uses of egg structures have also been emphasized by Bona (1994) and Hoberg et al. (1999a). A cursory examination suggests that the electron-dense 'capsule shell' of hymenolepidids and dioecocestids may represent a synapomorphy, that the embryophore of Catenotaenia may in fact be a pyriform apparatus and hence provide a synapomorphy with the anoplocephalines and that rewording the character 'embryophore striated' to 'embryophore composed of numerous blocks' would provide a synapomorphy for the taeniids with Anoplotaenia and possibly Dasyurotaenia, two cyclophyllidean genera of uncertain taxonomic position (Khalil et al. 1994). Clearly, much more detailed and extensive investigations of egg structure are required. In addition, it would have to be demonstrated that the features observed are not homoplasies functioning to facilitate infection of a particular group of intermediate hosts, be it mammal, insect, arachnid or crustacean (see discussion in Fairweather and Threadgold 1981a). An example might be the embryophoric 'blocks' of taeniids and Anoplotaenia, the interstitium between which dissolves under the action of the enzymes pepsin or trypsin (Laws 1968; Beveridge et al. 1975). In spite of minor reservations, therefore, egg ultrastructure – at least within the Cyclophyllidea – appears to provide a series of characters which, to date, has not been extensively exploited in phylogenetic analyses.

The remaining phylogenetic characters derived from cestode life-cycles relate to the development of the metacestode and are: i) the presence of a cercomer; ii) the development of a primary lacuna; iii) the retraction/invagination of the scolex;

iv) the presence of an 'apical organ' in the metacestode; and v) histological/ultrastructural features of the metacestode.

Cercomer

The presence or absence of a cercomer has been an important (albeit controversial) character in analyses of cestode evolution, and its use has been reviewed by various authors (Freeman 1973; Jarecka 1975; Jarecka et al. 1981). The presence of a cercomer is universally recognized as a characteristic feature of all basal eucestodes, though its loss during ontogenesis in some species can present technical difficulties – a problem occurring particularly within the Proteocephalidea (see Freeman 1973). In the Trypanorhyncha, Ruszkowski (1934) and Mattis (1986) described a rather short, stout cercomer in the species they studied, and this may account for the discrepancy with the observations of Mudry and Dailey (1971) on the lack of a cercomer in Lacistorhynchus dollfusi (reported as L. tenuis).

Greatest diversity and controversy exists within cyclophyllidean life-cycles, in many of which a cercomer is lacking (Freeman 1973). In some davaineids and dilepidids the cercomer may be highly reduced or lost during development providing difficulties for the coding of the character (Hoberg et al. 1999a). In a number of dilepidids (seven genera, 13 species), a prominent, elongate cercomer disintegrates as the scolex is retracted, forming numerous globules which persist in the mature cysticercoid (Gabrion and Helluy 1982), while in several davaineids, Davainea, Skrjabinia (formerly a subgenus of Raillietina), the cercomer is short and rounded, or is absent in the mature cysticercoid (Wetzel 1932, 1934; Jones and Horsfall 1936; Voge 1960). Similarly, within the Paruterinidae, species exist with (Metroliasthes lucida; see Jones 1936) and without (Cladotaenia, Paruterina; see Freeman 1957, 1959) cercomers (and primary lacunae). Jarecka (1975) and Jarecka et al. (1981) have argued that in certain cyclophyllidean cestodes (e.g., taeniids), the cercomer is in fact the entire outer wall of the metacestode. Its differentiation as cercomer is based on ultrastructural evidence, namely that it is covered with microvilli. This interpretation is at variance with traditional concepts of a tail-like appendage to the metacestode and has yet to achieve general acceptance. Hence, considerably more information is required before the character of presence of a cercomer can be used with confidence, particularly in the Cyclophyllidea. Furthermore, Hoberg et al. (1999a) have suggested that the loss of the cercomer within the Cyclophyllidea may be homoplasious and may have occurred independently in several groups, such that even if its characteristics are fully elucidated, it may prove not to be a useful phylogenetic character.

Primary lacuna

The development of a primary lacuna occurs in some cyclophyllideans during early morphogenesis of the metacestode, and its demonstration is equally beset with technical problems. In the Linstowiinae, for example, a primary lacuna was reported in Oochoristica vacuolata by Hickman (1963), O. deserti by Milleman (1955) and Atriotaenia procyonis by Gallati (1959) but was 'indistinct' in O. osheroffi (see Widmer and Olsen 1967) and not evident in O. anolis (see Conn 1985b). Conn (1985b) suggested that the lack of serial histological sections may have resulted in the apparent absence of the structure in O. anolis. In Dipylidium caninum, the primary lacuna is a transitory structure (Venard 1938; Marshall 1967) and is subsequently filled with parenchymal cells. Šlais (1973) observed that in the studies by Freeman (1959) on *Cladotaenia* spp., metacestodes were first examined eight days after infection, during which time a temporary lacuna could have formed and disappeared.

In the Dilepididae, a primary lacuna has been described in many species (e.g., *Anomotaenia constricta*; see Gabrion 1975), yet in *Dilepis undula* there is apparently no such primary lacuna (Gulyaev 1997), and Jourdane (1972, 1977) did not describe a primary lacuna in *Choanotaenia crassiscolex* or *C. estavarensis*, though the structure is described as a 'hollow ball stage' in the development of their congener *C. infundibulum* by Horsfall and Jones (1937). Inconsistencies thus exist in a number of cyclophyllidean families, and these need to be clarified before the character can be utilized with confidence in phylogenetic analyses.

Scolex

In basal cestodes, the scolex invariably develops externally, whilst in higher cestodes the scolex is either inverted or withdrawn during the process of development. Few characteristics have been as controversial as the significance of the difference between withdrawal and inversion of the scolex, which were used as key features by Voge (1967) in her review of cestode larval development. Freeman (1973) summarized observations indicating that with certain proteocephalideans (e.g., *Ophiotaenia perspicua*), the scolex may begin to evert and, in doing so, passes through a transitory stage during which it appears to be 'withdrawn'. Herde (1938: 288) stated that 'older specimens show a tendency to keep the scolex continuously invaginated', suggesting that the variation is related primarily to developing metacestodes. Whilst Freeman's (1973) caution concerning the care with which observations need to be made is laudable, it appears in the literature that the term 'invaginated' has often been used to include metacestodes in which the scolex is actually withdrawn and a potentially useful character may have been lost or at least obscured. The character is of greatest potential use in the cyclophyllidean family Dilepididae, an analysis of the distribution of the character having been made by Gabrion and Helluy (1982). For the purposes of analysis, Hoberg *et al.* (1999a) divided the family into the Dilepididae and Gryporhynchidae. The dilepidid genera *Amoebotaenia, Choanotaenia, Dilepis, Eurycestus, Lateriporus, Paricterotaenia, Rauschitaenia, Trichocephaloidis, Sacciuterina* and *Anomotaenia* exhibit a scolex which is 'withdrawn' (Joyeux and Baer 1961; Shapkin and Gulyaev 1973; Gabrion and Gabrion 1976; Jourdane 1977; Krasnoshchekov and Tomilovskaya 1978; Gabrion and Helluy 1982), whilst those of the gryporhynchid genera *Neogryporhynchus, Valipora* and *Paradilepis* have scoleces which are invaginated (Jarecka 1970; Gabrion and Helluy 1982). Gabrion and Helluy (1982) designated these two different forms of metacestode as 'monocercus' and 'cercoscolex' respectively, following Jarecka (1970) and Gabrion (1981).

Further clarification of this character may assist in phylogenetic analyses within the Cyclophyllidea, since, if the proteocephalideans are used as an outgroup, the withdrawn scolex would represent a synapomorphy linking the hymenolepidids, anoplocephalines, davaineids and dilepidids. Clarification may involve re-examination of certain life-cycles since in some instances (e.g., Jones 1936) it is difficult to ascertain from the published drawings whether the scolex is withdrawn or inverted. The phenomenon of scolex retraction in hymenolepidids has been studied in considerable detail at the histological and ultrastructural level, indicating that a number of morphological changes occur in the wall of the cysticercoid during or following scolex retraction (e.g., Caley 1974). Comparable studies of cestodes with inverted scoleces are few in number.

The 'apical organ'

An apical organ develops in the metacestodes of tetraphyllideans (see for example Mudry and Dailey 1971), trypanorhynchans (Nakajima and Egusa 1969; Mudry and Dailey 1971), tetrabothriideans (Hoberg 1987) and in the cyclophyllidean genus *Catenotaenia* (see Joyeux and Baer 1945), yet regresses to disappear in the adult stage. In most cyclophyllideans, the apical organ becomes the rostellum. In proteocephalideans, the structure may or may not persist (Freeman 1973). The character has been used by Hoberg *et al.* (1997c, 1999a) in phylogenetic analyses. However, there are few histological (Wood 1965; Fischer and Freeman 1969) and no ultrastructural studies of this organ. Therefore it may be argued that the homology of apical organs present in various orders is not yet firmly established, and that further clarification of their structure and function is required before they can be used reliably in phylogenetic analyses.

Histology and ultrastructure

Histological and ultrastructural features of the metacestode have been relatively little utilized, and offer considerable scope for future analyses. The 'fibrous' layer present in the wall of the cysticercoids of hymenolepidids, anoplocephalines, davaineids and certain dilepidids has been used as a synapomorphy by Hoberg *et al.* (1999a). Studies of the histology of metacestodes have however been largely restricted to the classical 'cysticercoids' and 'cysticerci' limiting the possibilities of making comparisons across all taxa within the Cyclophyllidea, let alone the wider array of cestode orders.

Similarly, ultrastructural studies of tegumental structures offer promise but the data currently available are too restricted to permit a general phylogenetic analysis. The presence of microvilli, knob-like microtriches and microtriches similar to those found on adult cestodes has been described by a number of authors (e.g., Collin 1970; Baron 1971; Rees 1973a,b; Crowe *et al.* 1974; Caley 1976; Gabrion and Gabrion 1976; Richards and Arme 1984a,b; Rogan and Richards 1987), including the occurrence of different types of projections on different regions of the metacestode tegument. Generally, microtriches develop on those parts of the metacestode which persist in the definitive host, whilst the components of the metacestode which will be shed upon infection of the definitive host bear microvilli (Rogan and Richards 1987). In some cestodes it has been shown that microvilli are shed during development and microtriches develop either *de novo* from the tegument or from knob-like microtriches. The distributions of microvilli and microtriches on the developing cysticercoid have been utilized in comparisons of hymenolepidid and dilepidid cysticercoids and in arguments concerning the nomenclature of various regions of the developing metacestode (Jarecka *et al.* 1981). Ultrastructural studies on the cysticercoid of *Anomotaenia constricta* by Gabrion and Gabrion (1976) revealed a number of characteristics in common with that of *Tatria octacantha* described by Rees (1973a), in contrast to similarities between *Tatria* and taeniids suggested by Freeman (1973). These studies unfortunately, have not been pursued even though they clearly provide evidence which might help resolve disputed relationships. Jarecka *et al.* (1981, 1984) have argued that hymenolepidid metacestodes are withdrawn into two envelopes, as are some dilepidid cysticercoids

(Gabrion and Gabrion 1976), whereas other dilepidids are withdrawn into a single envelope (Jarecka 1970). Similar studies are required on a wider range of cestodes to determine whether or not these characteristics are valid, but the data assembled to date appear to be strong.

Ultrastructural and histological studies of metacestodes therefore appear to hold considerable promise of novel characters for phylogenetic analysis, but until more comprehensive studies become available, the current data are not amenable to analysis.

Usefulness in analyses

The recent attempts to utilize features of cestode life-cycles in cladistic analyses do allow a preliminary assessment (even if indirect) on the value of these characters in analyses. The obvious caveats of course are that the characters have been correctly polarized and coded, and that the deductions made with respect to utility are from a robust tree.

In the analysis of the cestode orders by Hoberg *et al.* (1997c), four life-cycle characters supported the clade containing the Eucestoda, namely the presence of a hexacanth embryo, the ciliated embryophore, the presence of an apical organ during development of the metacestode and the structure of the metacestode (Figure 23.1). The nature of the egg membranes, egg tanning and the oligolecithal nature of eggs supported the grouping of the Tetraphyllidea + Lecanicephalidea + Proteocephalidea + Tetrabothriidea + Cyclophyllidea, whilst the absence of protonephridia in the egg links these taxa with the Diphyllidea and Trypanorhyncha. The structure of the metacestode was utilized at several places within the tree (Figure 23.1). By contrast, the presence of an operculum, delayed embryogenesis, presence of a primary lacuna and scolex retraction did not appear as significant synapomorphies. Life-cycle characters in this instance therefore make a contribution to phylogenetic analysis of the cestode orders. This conclusion is supported by an analysis of the orders of cestodes based solely on life-cycle characters (Figure 23.2; Tables 23.2 and 23.4). The basal groups in this analysis, the Pseudophyllidea and Haplobothriidea, are those cestodes with a free-swimming coracidium (the Trypanorhyncha are polymorphic for this character) and a plerocercoid. The 'higher

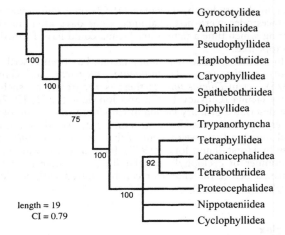

Figure 23.2 Cladogram of cestode orders derived from life-cycle characters, based on those utilized by Hoberg et al. (1997c) (Table 23.2). Differences from their data matrix involve adding polymorphisms for the Trypanorhyncha based on the studies of Mudry and Dailey (1971) and Mattis (1986) (characters 5–7, 9), for the Amphilinidea for character 7 based on the studies of Rohde and Georgi (1983), and for the Pseudophyllidea for character 2 based on Bray et al. (1999). The polarity of character 7 was reversed based on the argument that delayed embryogenesis is the plesiomorphic characteristic in the Digenea and Monogenea. Character 13 (presence of a cercomer) is new. Features for characters 9 and 12 for the Diphyllidea have been added based on the work of Ruszkowski (1928). The 70% majority-rule tree is shown, based on the consensus of 864 trees; trees 19 steps in length; consistency index (CI) 0.79; rescaled consistency index 0.70.

Table 23.2 Characters based on features of life-cycles used in recent analyses of the relationships of cestode orders.

Characters	Character no. (Hoberg et al. 1997c)
Egg characters	
1. Egg polylecithal (0) or oligolecithal (1)	19
2. Egg with operculum (0) or without (1)	20
3. Egg tanned (0) or not (1)	21
4. Egg with single membrane (0) or with inner and outer membranes (1)	22
5. Embryophore ciliated (0), non-rigid (1), rigid (2)	24
6. Hexacanth absent (0), present (1)	26
7. Hexacanth with protonephridia (0) or without (1)	23
8. Embryogenesis not delayed (0) or delayed (1)	25
Metacestode characters	
9. Scolex of final metacestode stage not invaginated or retracted (0), or invaginated or retracted (1)	35
10. Primary lacuna absent (0) or present (1)	36
11. Apical organ present (0) or absent (1)	37
12. 'Metacercoid' absent (0), bothriate (1), acetabulate (2), caudate (3), neoplerocercus (4), acetabulo or glanduloplerocercoid (5)	38
13. Cercomer absent (0) or present (1)	–

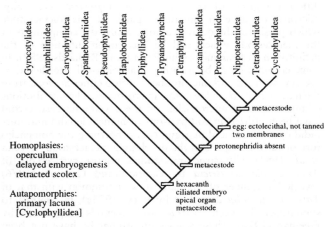

Figure 23.1 Phylogenetic analysis of the orders of the Cestoda from Hoberg et al. (1997c) based on characters of adult morphology and life-cycles with the synapomorphic characters derived from life-cycle features (Table 23.2) mapped on to the tree.

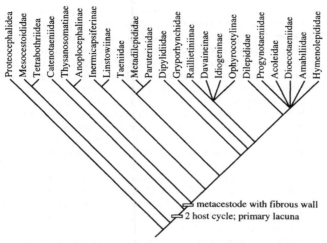

Figure 23.3 Phylogenetic analysis of the taxa within the Cyclophyllidea from Hoberg et al. (1999a) based on adult morphology and adult life-cycles with synapomorphic characters derived from life-cycle features (Table 23.3) mapped on to the tree.

Table 23.3 Characters based on features of life-cycles used in recent analyses of the cestode order Cyclophyllidea.

Characters	Character no. (Hoberg et al. 1999a)
Egg characters	
1. Embryophore non-striated (0) or striated (1)	18
2. Pyriform apparatus absent (0) or present (1)	19
3. Polar processes of embryophore absent (0) or present (1)	20
4. Egg capsules thin-walled, uterine in origin, absent (0) or present (1)	21
5. Egg capsules not fibrous (0) or fibrous (1)	22
Metacestode characters	
6. Three-host life-cycle (0) or two-host life cycle (1)	31
7. Primary lacuna absent (0) or present (1)	32
8. Cercomer present (0) or absent (1)	33
9. Apical structure in metacestode only (0) or absent (1)	34
10. Metacestode without fibrous layer in wall (0) or with fibrous layer (1)	35
11. Scolex invaginated (0) or withdrawn (1)	–

Table 23.4 Characters used in analysis of relationships of cestode orders based on features of life-cycles. Data derived primarily from Hoberg et al. (1997c), modified according to caption for Figure 23.2. Numbers for characters are those from Table 23.2.

	1	2	3	4	5	6	7	8	9	10	11	12	13
Gyrocotylidea	0	0	0	0	0	0	0	0	?	?	0	?	?
Amphilinidea	0	1	0	0	0	0	0	P	0	0	0	0	P
Spathebothriidea	0	0	0	0	1	1	0	0	0	0	1	3	1
Caryophyllidea	0	0	0	0	P	1	0	0	0	0	1	3	1
Pseudophyllidea	0	P	0	0	P	1	0	P	0	0	1	1	1
Diphyllidea	0	1	?	?	1	1	?	0	1	?	?	4	?
Trypanorhyncha	0	P	0	P	P	1	1	P	P	0	1	4	1
Tetraphyllidea	1	1	1	1	1	1	1	1	0	0	P	5	0
Lecanicephalidea	1	1	1	1	1	1	1	1	0	?	1	5	0
Haplobothriidea	0	0	0	0	1	1	0	1	0	0	1	1	1
Nippotaeniidea	1	1	1	1	1	1	1	1	0	?	1	2	?
Proteocephalidea	1	1	1	1	1	1	1	1	P	0	P	2	1
Tetrabothriidea	1	1	1	1	1	1	1	1	0	?	1	5	?
Cyclophyllidea	1	1	1	1	1	1	1	1	1	P	P	2	1

P = 0, 1.

Table 23.5 Characters used in analysis of relationships of cestode taxa belonging to the Cyclophyllidea based on features of life-cycles. Numerals for characters are those from Table 23.3, omitting character 6.

	1	2	3	4	5	7	8	9	10	11
Proteocephalidea	0	0	0	0	0	0	0	P	0	0
Mesocestoididae	0	0	0	0	0	0	0	0	0	0
Catenotaeniidae	0	1	0	0	0	0	1	1	0	0
Paruterinidae	0	0	0	0	0	0	P	1	0	0
Dilepididae	0	0	P	P	0	P	0	P	0	1
Dipylidiidae	0	0	0	1	0	1	0	1	0	P
Gryporhynchidae	0	0	0	0	0	1	0	1	0	0
Progynotaeniidae	0	0	0	0	0	1	0	1	1	1
Hymenolepididae	0	0	P	0	0	1	0	P	1	1
Taeniidae	1	0	0	0	0	1	1	1	0	0
Amabiliidae	0	0	P	0	0	1	1	1	1	1
Davaineinae	0	0	0	0	0	1	0	1	1	1
Idiogeninae	0	0	0	0	0	?	?	1	?	1
Ophryocotylinae	0	0	0	0	0	?	0	1	1	1
Raillietiniinae	0	0	0	0	1	1	0	0	1	1
Anoplocephalinae	0	1	0	0	0	1	0	0	1	1
Linstowiinae	0	0	0	1	0	1	1	0	0	0
Inermicapsiferinae	0	0	0	0	1	?	?	0	?	?
Thysanosomatinae	0	0	0	0	0	1	0	0	?	?

P = 0,1.

cestodes' are united by non-tanned eggs, a non-ciliated embryophore, lack of protonephridia in the hexacanth and oligolecithal eggs, that is, a series of characters related to the structure of the egg.

In the case of cyclophyllidean families (Hoberg et al. 1999a) however, none of the five characters of the egg appeared as a major synapomorphy. The number of hosts in the life-cycle and the presence of a primary lacuna united all cyclophyllideans to the exception of the mesocestoidids and nematotaeniids, whilst the structure of the metacestode united all except these two plus the catenotaeniids (Figure 23.3; Tables 23.3 and 23.5). The presence of a cercomer and an apical organ were treated as homoplasies.

When the data relating to life-cycles were extracted and analysed separately (Figure 23.4) there was poor resolution of taxa. The strict consensus tree (not shown) provided no resolution at all, while the majority-rule tree had only three clades which appeared in more than 80% of 1000 trees. The number of hosts in the life-cycle was excluded, even though it proved a useful character in the analysis of Hoberg et al. (1999a). Using the Proteocephalidea as the outgroup, they designated a three-host life-cycle as plesiomorphic. However, both two- and three-host life-cycles occur within the Proteocephalidea (see Biserkov and Genov 1988; Scholz 1999) such that this character is actually polymorphic in the outgroup. It is unlikely that repolarizing the character would

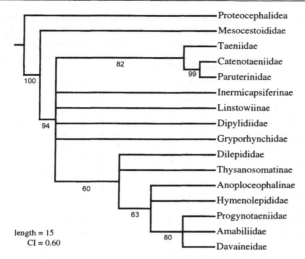

Proteocephalidea
Mesocestoididae
Taeniidae
Catenotaeniidae
Paruterinidae
Inermicapsiferinae
Linstowiinae
Dipylidiidae
Gryporhynchidae
Dilepididae
Thysanosomatinae
Anoploceophalinae
Hymenolepididae
Progynotaeniidae
Amabiliidae
Davaineidae

length = 15
CI = 0.60

Figure 23.4 Cladogram of cyclophyllidean taxa derived from life-cycle characteristics utilized by Hoberg *et al.* (1999a) (Table 23.3). Subfamilies of the Davaineidae were not separated; the families Nematotaeniidae, Metadilepididae, Acoleidae and Dioecocestidae were excluded due to lack of information. The number of hosts in the life-cycle (character number 31 of Hoberg *et al.*, 1999a) was excluded; an additional character based on whether the scolex is retracted or invaginated has been added. The majority-rule tree based on the first 1000 equally parsimonious trees is shown. Tree length 15; consistency index (CI) 0.60; rescaled consistency index 0.43.

influence the outcome of the analysis. As well as excluding one potentially useful character, the subfamilies of the Davaineidae were not separated in the current analysis, and four families were excluded due to lack of information. In spite of these differences, the separate analysis of life-cycle characteristics supports the conclusions drawn from mapping life-cycle characters on to the cladogram derived from all characters (Hoberg *et al.* 1999a).

Life-cycle characters therefore appear to have contributed little to the analysis of taxa within the Cyclophyllidea. This at first sight is surprising given the diversity of characters available for analysis within the order. However, a number of egg characters as currently defined represent autapomorphies and, as suggested earlier, significant aspects of the ultrastructure of eggs have not been investigated as potentially useful characters. Similarly, in the case of the presence of a primary lacuna and cercomer, conflicting interpretations of the cercomer, its loss during development and the transitory nature of the primary lacuna represent significant difficulties which need to be overcome before these characters can be utilized confidently. A negative judgement of the utility of developmental characteristics in analyses of relationships of taxa within the Cyclophyllidea would, in spite of the evidence, seem premature.

Conclusions

The conclusions which one might draw from an examination of the utility of life-cycles in reconstructing the phylogeny of the cestodes are, at first sight, obvious. The lack of a synopsis of life-cycle data is a clear impediment to further analysis but is one which can be remedied with relative ease. The fragmentary nature of life-cycle data is another obvious deficiency which is not easily remedied. Clearly, elucidation of the life-cycles of marine cestodes of the orders Tetraphyllidea, Lecanicephalidea, Diphyllidea and Trypanorhyncha, as well as the order

Spathebothriidea, are critical, but the lack of current data is not due to the lack of effort by workers on these groups. It may therefore be imprudent to advocate a massive research assault on the elucidation of the life-cycles of sometimes small, and often inconspicuous, cestode taxa. Even so, it appears that the elucidation of amphilinidean and gyrocotylidean life-cycles as well as those of the 'marine' cestodes will remain a very high priority.

By contrast, the ultrastructure of eggs and metacestodes, as indicated above, offers considerable opportunities and appears to be a field which has, to date, not been exploited in a systematic fashion. The database of ultrastructural data has proved to be extremely important in phylogenetic analyses of the phylum (Ehlers 1985a), but has not received the same level of attention within the Cestoda (Brooks 1989a). The recent review by Justine (1998b) has illustrated how ultrastructural data on spermiogenesis can be utilized to provide important additional data on the evolution of cestodes. The same may also be true of eggs and metacestodes.

Finally, it is important to note two points in the development of our understanding of the importance of cestode life-cycles. The first is the change in attitude from the view expressed implicitly (Stunkard 1975) that the unravelling of the features of cestode life-cycles would, *ipso facto*, reveal the evolutionary plan of the cestodes. The inability of such data alone to determine unequivocally the phylogenetic relationships of cestodes is evident in the analyses presented above, but could not better be exemplified than by the cestode *Anoplotaenia dasyuri*, a tapeworm of the Tasmanian devil, whose entire life-cycle has been elucidated, including morphogenesis of the metacestode together with ultrastructural studies on the egg and metacestode (Beveridge *et al.* 1975) yet which cannot (Khalil *et al.* 1994) be assigned with confidence to a cyclophyllidean family. Rather, developmental data provide a series of useful characteristics which can be utilized along with features of adult morphology, spermiogenesis and molecular data in providing insights into cestode evolution. Second, it is important to acknowledge the continuing changes in evolutionary hypotheses based on adult morphology. The sequential dismemberment of the traditional 'Dilepididae', perhaps facilitated by Freeman's (1973) observations of the multiplicity of life-cycle types, into the Dipylidiidae, Paruterinidae, Metadilepididae, Gryporhynchidae and Dilepididae, greatly facilitates and influences studies on cestode life-cycles and their evolution. A similar splitting of the traditional Anoplocephalidae into Anoplocephalinae, Thysanosomatinae, Linstowiinae and Inermicapsiferinae based on life-cycle characteristics (Stunkard 1961) has not yet achieved a satisfactory conclusion, but has at least initiated further analyses. Thus life-cycle studies and studies of adult morphology are interdependent and need to proceed hand in hand.

Thus, there is an abundance of available data which, when suitably aggregated, could be mined more extensively for characters. In spite of obvious deficiencies in the range of cestodes for which life-cycles are known, the most pragmatic approach to increase the utility of this area of analysis would appear to be to examine in greater detail the ultrastructural features of the eggs and metacestodes, particularly those of cyclophyllideans.

ACKNOWLEDGEMENTS

I wish to thank Drs R.A. Bray and D.T.J. Littlewood for the invitation to prepare this review, and Dr T. Mattis for kindly providing a copy of his Ph.D. thesis which proved invaluable. Dr T. Scholz very kindly provided reprints and access to a review prior to publication, while Drs V. Tkach and B. Georgiev provided much help in obtaining the papers cited.

Chapter 24

Embryology and developmental genes as clues to flatworm relationships

Maximilian J. Telford

The position of the flatworms within the animal kingdom has long seemed fundamental to our understanding of the evolution of the Bilateria due to the widely held idea that they are the most basally branching triploblastic animals, and hence progenitors of all other animal body plans. The standard theory of the emergence of the triploblasts postulates the appearance of increasingly complex animals, starting with the evolution of a bilaterally symmetrical acoelomate worm from a radially symmetrical gastrula-like planuloid ancestor (represented by the cnidarian planula). From these acoelomate worms the pseudocoelomates were thought to have evolved and subsequently to have given rise to coelomate protostomes and deuterostomes. Although schemes differ in their particulars, the position of the flatworms as the most basal triploblasts is broadly accepted such that their apparent simplicity and lack of anus and body cavity is considered to be primitive. Any flatworm characters shared by other triploblasts were also considered primitive characters such as the spiral cleavage seen in the spiralian taxa. An alternative to this acoeloid scheme, championed by European workers (e.g., Siewing 1980a), reversed the logic and emphasized the apparently derived nature of flatworm embryogenesis, in particular the spiral mode of cleavage, and suggested that flatworms are related to the other spiralian taxa (molluscs, annelids, etc.) and hence have secondarily lost the coelom and anus. This distinction has enormous consequences for our understanding of the evolution of the Metazoa.

Recent results from analyses of molecular sequence data seemed convincingly to disprove the acoeloid theory – removing the flatworms from their pivotal position as the most primitive triploblasts from which body plan all other animals are derived – and support their simplified nature (presumably secondarily derived) within a spiralian clade (Aguinaldo *et al.* 1997; Carranza *et al.* 1997; Littlewood *et al.* 1998b). This state of affairs was not destined to last long, however, and further molecular studies have reinstated a single clade of flatworms, the acoels, which had not been included in the previous analyses, as the most basal triploblasts unrelated to the other flatworms. This result means that once again we have a representative of the starting point of triploblast evolution with which to compare all other animals (Ruiz-Trillo *et al.* 1999). It is these two results that I wish to consider in this chapter. First, I will consider recent literature and show that there are detailed similarities, both morphological and molecular, providing support for the installation of the flatworms (the Rhabditophora at least) within the spiralian clade. Second, I will consider the evidence for the placement of the acoels at the base of the triploblastic metazoans.

Embryology and the spiralian characteristics of the flatworms

Recent molecular analyses divide the triploblasts into three principal lineages: the deuterostomes, the ecdysozoans (moulting animals); and the lophotrochozoans (lophophorates and spiralians – or eutrochozoans) (Aguinaldo *et al.* 1997). The spiralian clade contains a number of phyla (molluscs, annelids, sipunculids, echiurans and nemerteans) with a larval stage called a trochophore that develops in a stereotypical way. Here, I discuss evidence to suggest that flatworm embryogenesis is so similar to that of the trochozoan taxa that there is independent support for the idea that they should be considered members of this clade.

Almost all of the experimental evidence from flatworms is derived from studies of archoophoran turbellarians, principally of the polyclad *Hoploplana inquilina* (e.g., Boyer 1989; Henry *et al.* 1995; Reiter *et al.* 1996). Polyclad early development has been intensively studied because eggs and spermatozoa can be harvested, *in-vitro* fertilization is possible, and because their entolecithal holoblastically cleaving eggs are easy to study. There is good reason to believe that the situation seen in this clade is the ancestral condition of the phylum and that the neoophoran condition with yolk-cell-filled ectolecithal eggs and blastomere anarchy is derived (Ellis and Fausto-Sterling 1997). Several neoophorans still show clear traces of the spiral cleavage pattern, and molecular evidence demonstrating the monophyly of the archoophorans plus neoophorans confirms the derivation of one developmental pattern from the other, suggesting that the observations in *Hoploplana* may be generalized to the rest of the phylum.

Early divisions and fate maps

From the earliest events of embryogenesis there are striking and complex similarities between polyclads and other spiralian phyla such as molluscs and annelids (Figure 24.1). Cell lineage studies using a fluorescent lineage tracer, DiI, have shown conservation of the earliest blastomere fates between higher spiralians and flatworms. Work on *Hoploplana* shows that the initial cleavage plane is not left–right symmetrical or equatorial, but it divides the embryo roughly into left-ventral (AB) and right-dorsal (CD) halves, with the second division dividing the left-ventral blastomere approximately into a left (A) and a ventral (B) blastomere and the right-ventral blastomere into a right (C) and a ventral (D) blastomere (Henry *et al.* 1995; Boyer *et al.* 1998). This situation is exactly that seen in, for example, the molluscan trochophore (Figure 24.1A).

Spiral cleavage pattern

At the third division, when the four macromeres formed as described above divide to give rise to a quartet of micromeres (1a–1d) which come to lie above them, the spirality of the cleavage pattern becomes apparent. The following cleavages will alternate between a clockwise and anticlockwise direction when producing subsequent quartets of micromeres (2a–2d, 3a–3d and 4a–4d) (Figure 24.1A). The result of this lateral displacement of division products is that the micromeres do not lie directly above the macromeres that produce them, but

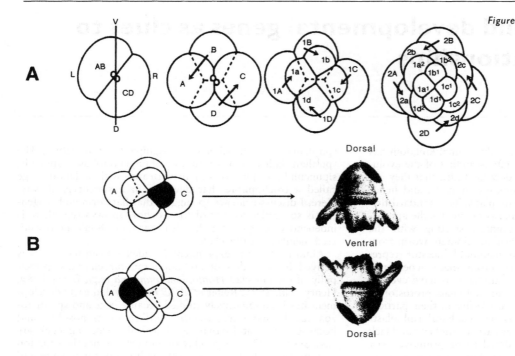

Dorsal

Ventral

Dorsal

Figure 24.1 Similarities between flatworm and typical spiralian early development. **A)** Cleavage stages from two cells to 16 cells, as seen in spiralia and flatworms. The presumptive larval axes do not follow the early planes of division such that the AB blastomere is left-ventral and the CD blastomere is right-dorsal. The alternating clockwise and anticlockwise directions of cleavage are indicated by arrows in the four-, eight- and 16-cell stages. **B)** Results from first quartet micromere deletion experiments are identical in molluscs and flatworms. Deletion of lateral (a and c quadrant) micromeres causes the anticlockwise adjacent quadrant to become dorsal such that the quartet with a deleted micromere is usually left. Left-hand structure (left eyes) are seen to be missing in the developed larvae, as represented by dark patches on the larvae. This inherent chirality of effect is seen in both molluscs and flatworms. Micrographs reproduced with permission from C. Nielsen, *Animal evolution: interrelationships of the living phyla.* © Oxford University Press, Oxford, 1995.

above and to the left or right (depending on a clockwise or anticlockwise division) such that they lie in a furrow between the macromeres. In radial cleavage, the divisions are not skewed to the left or right, so the micromeres lie directly above their parent macromere. This is the most evident characteristic of a spirally cleaving organism, and is seen in the flatworms (both the archoophoran polyclads and in certain basal neoophoran worms such as the proseriate *Monocelis* and the triclad *Dendrocoelum*; Ellis and Fausto-Sterling 1997).

Importantly, it seems clear that the *radial* cleavage pattern is ancestral. Radial cleavage is seen in the diploblastic cnidarians, in the deuterostomes and in at least some of the ecdysozoan phyla; chaetognaths, priapulids, nematomorphs (modified radial; Bresciani 1991) and some arthropods where there is little yolk (e.g., Hertzler and Clark 1992); attempts to show that arthropod cleavage is spiral (e.g., Anderson 1973), in order to fit in with the discredited Articulata hypothesis which derives arthropods from annelids, are unconvincing. The cleavage pattern is unknown in the ecdysozoan Loricifera and Kinorhyncha and is idiosyncratic in the nematodes (Schierenberg 1997). This distribution of radial cleavage suggests that spiral cleavage is a derived character, and gives cladistic support to the relationship between the flatworms and the annelids, molluscs, etc. (Valentine 1997).

Origin of mesentoderm

The derivation of the mesentoderm (a source of the majority of both the mesoderm and the gut) from the 4d micromere is the other great constant in spiralian development, and the similarity in origin of mesentoderm in flatworms further strengthens the link between flatworms and the spiralians. There are changes in the timing of this event with a tendency for earlier determination during evolution in both the molluscs and the

annelids, though the ancestral timing seems to be at the 64-cell stage in these taxa (van den Biggelar *et al.* 1997). In flatworms the mesentoblast is appropriately derived from the fourth quartet of micromeres, albeit one cell division later than in typical spiralians (Ellis and Fausto-Sterling 1997). The 3D macromere divides to give the macromere 4D and the micromere 4d that produces mesentoderm in spiralians (Boyer *et al.* 1996a). In flatworms, it has been suggested that 4d divides once more, horizontally, and only the upper of these cells, $4d^1$, produces mesoderm. This seems likely to be an insignificant difference as the timing of determination of mesoderm is seen to be subject to change, for example, within the molluscs (see above).

Mesectodermal as well as mesendodermal origin of mesoderm

There is also a secondary source of mesoderm found in the flatworms from a cell from the second quartet of micromeres, 2b. This lineage produces some mesoderm and also ectoderm in contrast with the mesentoderm produced by the 4d; this dual origin of mesoderm is also seen in molluscs and, along with a generally similar fate map, has been cited as a link between the two clades (Boyer *et al.* 1996a).

Induction of the mesentoblast and determination of the body axes

Results from blastomere ablation experiments more firmly link the polyclads to spiralians such as molluscs. Cell ablation experiments in equal-cleaving molluscs have demonstrated the involvement of the animal pole micromeres in inducing the 3D mesentoderm mother cell to produce the 4d mesentoblast. This interaction also determines the future dorsoventral axis of the embryo and therefore makes the embryo bilaterally symmetrical. In experiments in the gastropod molluscs *Lymnaea* and *Patella*, deletion of a single *median* cell of the first quartet inevitably results in the quadrant opposite that deleted producing mesoderm and determining the dorsal side of the larva. This shows the importance of these micromeres in establishing the dorsoventral axis and in making the animal bilateral (Arnolds *et al.* 1983). Strongly resembling this is the observation in *Hoploplana* that deletion of multiple animal micromeres results in a radially symmetrical embryo – there has been no determination of the dorsoventral axis or bilaterality – suggesting that homologous mechanisms are involved (Boyer 1989).

More striking still is that deletion of *lateral* first quartet cells in the mollusc almost always results in loss of left-lateral structures (the anticlockwise neighbouring micromere almost always designating the dorsal quadrant), presumably due to some inherent chirality in the embryo (Arnolds *et al.* 1983) (Figure 24.1B). Deletion of lateral animal pole micromeres in *Hoploplana* similarly has a large bias towards the loss of left-lateral structures (Boyer 1989).

Larval types and life-cycles

In most turbellarian groups, development is direct but three types of larvae do exist: the Müller's and Götte's larva of the polyclads, and the Luther's larva of the catenulid genus *Rhynchoscolex*. The only obvious difference between Müller's and Götte's larvae is in the number of ciliated lobes – eight in the Müller's and four in the Götte's. Suggestions that the four-lobed larva develops later into the eight-lobed have been disproved by following four-armed larvae through to metamorphosis. There seems no prospect of comparing larval forms within the flatworms in order to understand their intraphyletic relationships as the Müller's and Götte's larva co-occur in the polyclads and nowhere else. The Luther's larva strongly resembles the adults, but has more ciliation; this led Nielsen (1995) to suggest that it should be regarded as a specialized juvenile rather than a true larva. The idea that polyclad larvae represent the ancestral mode of development within the phylum, and can be compared with other animal larvae in order to understand interphylum relationships, might be slightly more fruitful. Although not necessarily constituting synapomorphies, there are basic similarities between the mature polyclad larva and trochophore larvae. These include the apical ciliary tuft which is incorporated into the brain of both spiralians and flatworms (unlike that of deuterostome larvae; Nielsen 1995), a mouth formed from the blastopore (although inevitably so in flatworms as they have no anus), and a band of motile cilia. This last character might conceivably be demonstrated to be homologous to the prototroch of trochophore larvae if it could be shown by cell-labelling experiments to be derived from specific micromeres (1a–1d and 2a–2c), as is found in both the annelids and molluscs (discussed in Henry and Martindale 1998). Perhaps a more convincing larval synapomorphy comes from the suboral rejectory cell seen in polyclads (the Müller's larva of *Pseudoceros canadensis* in this case) with detailed similarities to various trochophore larva of polychaete annelids and the pilidium of nemerteans (Lacalli 1987).

In establishing homology of any attribute of living things, it is important to demonstrate such complex similarities that the attributes in question can most parsimoniously be interpreted as similar due to homology rather than convergence. The complexity of the similarities I have described between early events in the development of polyclad flatworms and

typical spiralians certainly seem to suggest similarity due to common ancestry, and the fact that they differ in these respects from other metazoans supports the molecular phylogenetic placement of the flatworms.

Hox genes

The detailed similarities described above between coelomate spiralians and a model flatworm embryo seem convincing, and are congruent with the recent molecular phylogenetic studies (Aguinaldo *et al.* 1997; Carranza *et al.* 1997; Littlewood *et al.* 1998b; Ruiz-Trillo *et al.* 1999). Further support for a derived position of the flatworms on the spiralian branch has come from comparisons of their complements of *Hox* genes (de Rosa *et al.* 1999). *Hox* genes are a family of transcription factors involved in patterning the anteroposterior axis of animal embryos. They all contain a highly conserved 60-amino acid DNA-binding motif known as the homeobox, and seem to have arisen by tandem duplication in a basal metazoan, their presence being confined to the Metazoa. It also appears that additional duplications of some members of the *Hox* gene family have occurred in specific animal lineages such that certain *Hox* genes are confined to certain clades and have been used as characters diagnostic of those clades. It has also been suggested that even ubiquitous *Hox* genes can be used as diagnostic characters if they have differences specific to a certain lineage (Balavoine 1997, 1998).

The *Lox5* gene of spiralians is a likely orthologue of the *fushi tarazu* (*ftz*) gene of ecdysozoa and possibly *Hox6* of deuterostomes (unpublished observation) but has, downstream of the homeodomain, a conserved peptide not seen in the other main metazoan clades. This '*Lox5*' peptide appears to be a reliable diagnostic character for the spiralians. It might be hoped that any taxa suspected of being spiralian sister-groups – perhaps the rotifers – would have this peptide, so confirming the suspected relationship (de Rosa *et al.* 1999).

The problem with using this kind of character as diagnostic of a clade is that it is, as yet, not possible for it to be polarized. The *ftz/Hox6/Lox5* gene does not exist in any outgroup of the triploblasts such as the cnidaria and, as the three major branches of triploblasts all apparently have a different state for this character all seem equally likely, *a priori*, to represent the ancestral state. It does not seem possible to say that the peptide seen in spiralians is not an ancestral character, however unlikely this may seem.

Perhaps more promising are the lineage-specific duplications of *Hox* genes whereby certain *Hox* genes exist only in specific taxa and are hence convincingly 'typical' of the clades in which they exist. For example, there appear to have been independent duplications of *Ultrabithorax/Abdominal* A (*Ubx/Abd*-A) type genes in the ecdysozoan and spiralian lineages such that certain *Hox* paralogs exist only in one or other clade. These lineage-specific gene duplications would make convincing synapomorphies. The same is true of the *Abdominal B* (*Abd* B) type genes which are expressed in the posterior regions of metazoan embryos. Independent duplications of *AbdB*-like genes appear to have occurred in the ecdysozoans, the spiralians and the deuterostomes. Although it is still possible to propose scenarios in which, for example, the two *AbdB*-like genes of spiralians (*Post-1* and *Post-2*) are plesiomorphic characters lost in other clades and hence their presence in a putative sister taxon of the spiralia is uninformative, finding several of these spiralian 'signatures' seems most parsimoniously explained by common ancestry.

The importance of this preamble should become clear when it is known that flatworms (rhabditophorans at least) have a

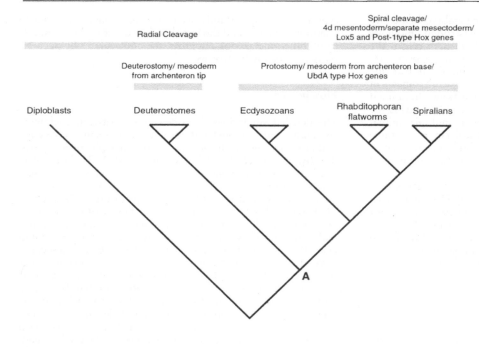

Figure 24.2 Tree showing likely relationships of the flatworms relative to the diploblasts, ecdysozoans, spiralians and deuterostomes and common features of different clades. Only the cleavage pattern is rooted by outgroup comparison using the cnidarians, and as such spiral cleavage is the only clear synapomorphy on this tree. The triploblast common ancestor at the branching point marked 'A' could have been protostomous or deuterostomous, could have formed mesoderm from archenteron tip or archenteron base (discussed in Telford 1993), and could have had spiralian or ecdysozoan or deuterostomian type *Hox* genes. In principle, the basal acoels can now be used as an outgroup to polarize all of these characters; however, the result from 18S rDNA studies might be incorrect and any similarities to the spiralians may be due to close relationship rather than common possession of ancestral characters. Attempts to determine the position of the acoels by looking for signature *Hox* genes, for example, should consider this difficulty of interpretation.

Lox5 gene, a spiralian-like *UbdA* peptide-containing gene (de Rosa *et al*. 1999) and a spiralian *AbdB*-type gene (personal communication, Taisaku Nogi, Himeji Institute of Technology, Japan). Although it is still theoretically possible that all of these characters are plesiomorphic, when considered together with the molecular phylogenetic evidence and the detailed similarities of early development, there seems to be overwhelming evidence that the flatworms do not represent the basal taxa, as has long been assumed (Figure 24.2).

Expression of Hox genes

The expression patterns of several *Hox* genes, as well as those of several other developmentally important genes, have been reported in whole and regenerating planarians. Two conclusions can be drawn from these comparisons. First, the *Hox* genes are involved in axis respecification during regeneration as well as, presumably, in normal embryonic development. Experiments also show that planarians express *Hox* genes continuously as fully developed adults, suggesting that their totipotent neoblasts are provided with continuous information about their axial position. In addition, the *Dugesia* homologue of the *Drosophila melanogaster* gene *vasa* – involved in *Drosophila* in specifying the totipotent germline – is expressed not only in the planarian germline (both ovary and testis) but is also expressed in somatic cells in the mesenchymal space. These *vasa*-expressing mesenchymal cells disappeared when regenerative capacity was eliminated by X-rays, and increased in the blastema of regenerating planarians. This strongly suggests that flatworm *vasa* may be involved in maintaining the totipotency of the neoblasts as well as that of the germline (Shibata *et al*. 1999).

The second point of note is that the regions specified by these developmental genes show fundamental similarities between flatworms and other metazoans. Studies of expression

of *Hox* genes show gradations of expression along the length of the intact adult planarians. The relative positions of anterior expression boundaries of planarian *Hox* genes are the same as those of their orthologues in *Drosophila* and mice (Bayascas *et al*. 1998; Orii *et al*. 1999).

Other genes with a conserved function in the metazoans are expressed with the expected pattern in the planaria. These include a dorsal expression of a *BMP4/dpp* (*Decapentaplegic*) homologue (expressed dorsally in other non-chordates but ventrally in the chordates, thus supporting the hypothesis of dorsoventral inversion in this lineage) (Orii *et al*. 1998). *Pax-6* is expressed in the eyes of planarians as elsewhere (Callaerts *et al*. 1999) and, as in vertebrates and arthropods, one sees anterior nervous expression of two *Orthodenticle* homologues and an *orthopedia* homologue (Umesono *et al*. 1999). As with the *Hox* genes, all of these genes appear to be expressed in intact adult planarians, as well as being expressed in the expected regions of regenerating animals. This suggests that, in order to permit the fantastic ability for regeneration as well as growth and degrowth, expression of developmentally important transcription factors must be continuously maintained.

After the previous discussions suggesting that the flatworms are not a primitive basal clade but derived bilaterians, the similarity of expression to that of other Bilateria might not be so surprising. However, results of expression studies of flatworm developmental genes are often prefaced with the statement that flatworms are believed to be one of the most primitive living bilaterians, and so expression in this clade is supposed to represent the ancestral condition (Callaerts *et al*. 1999; Umesono *et al*. 1999). With respect to flies and mice, the flatworms are no more ancestral than molluscs or annelids.

The embryological comparisons and data from developmental genes support the 18S rRNA molecular phylogenies in suggesting a close relationship between the flatworms and spiralian phyla. Further data will be required to elucidate more exactly their position within this clade. 18S data suggest that flatworms, along with other minor phyla (e.g., rotifers and gastrotrichs) are basal within the lophotrochozoans. 18S rRNA studies, however, place lophophorates (bryozoans, phoronids and brachiopods) within a spiralian crown group which excludes the flatworm. The lophophorates do not have trochophore larvae or the derived characters of spiral cleavage or 4d mesentoderm shared by the spiralian taxa and the flatworms, suggesting that either the 18S rRNA data are giving misleading results, or that the lack of these features in lophophorates is in need of explanation.

The position of the acoels

The real impact of this reassessment of the position of the flatworms is the loss of a morphological outgroup of the pseudocoelomate and coelomate animal phyla. The many studies using flatworms as a representative starting point from which to understand evolution within the animals can no longer be justified unless the acoels can fill this gap.

The phylogenetic analysis of 18S rDNA that proposes that the acoels are in fact not flatworms but are the most basal bilaterian animals does seem to bring us back to the presumptions we had previously concerning the origins of the higher Bilateria (Ruiz-Trillo et al. 1999). We return to an acoeloid ancestral state, bilateral but without body cavity or separate mouth and anus, and we have a living representative of the ancestral state of the Bilateria with which to compare all other animals in order to understand how evolution has proceeded.

Consistent with the idea that acoels are distant from the rhabditophoran flatworms and primitive relative to the Bilateria in general, they appear morphologically simple with either a simple pharynx or none at all, and lack a permanent intestinal lumen, protonephridia and well-delimited gonads (Balavoine 1998; Ruiz-Trillo et al. 1999). The acoel cleavage pattern is also strikingly different, although still apparently spirally cleaving (Baguñà and Boyer 1990). Duets rather than quartets of micromeres are formed and, in addition, the development seems to be regulative (Boyer 1971), unlike typical spiralians (and polyclads) although regulative development is also seen in nermertean development (Henry and Martindale 1997). In contrast to typical spiralians where the initial cleavage divides the embryo into left-ventral and right-dorsal halves (see above), in the acoels the first cleavage plane divides the embryo into left-dorsal and right-ventral halves (Boyer et al. 1996b). The atypical counterclockwise cleavage that gives rise to the first duet of micromeres could account for this and indeed, when the first micromeres arise from a counterclockwise division in otherwise normal quartet spiralians the same left-dorsal, right-ventral division is seen, suggesting that this difference may not be of such fundamental importance as it might first appear (Boyer and Henry 1998). If, as seems likely, spiral cleavage is a derived character with radial cleavage being plesiomorphic, experiments are needed to confirm whether the pattern really is an unusual form of spiral cleavage, in which case the acoels are spiralians and not the most basal triploblasts

The likely sister-group of the acoels is not basal according to 18S rDNA

According to Ruiz-Trillo et al. (1999), the acoels are unrelated to their long supposed sister-group, the nemertodermatids (the two taxa have been linked in the Acoelomorpha based on a common unusual ciliary structure and lack of intestine (e.g.,

Littlewood et al. 1999a)). In fact, reports of work (as yet unpublished) showing that the nemertodermatids have duet cleavage would provide a very strong indication that the acoels and nemertodermatids are sister-groups. The importance of this is that if the separation of the acoels from the nemertodermatids is artefactual – perhaps due to the long branched acoels being pushed towards the base of the 18S rDNA tree – then the 18S rDNA result is called into question. If the nemertodermatids are in fact the sister-group of the acoels and yet are grouped by 18S rDNA with the flatworms rather than adjacent to the acoels at the base of the Bilateria, then the phylogenetic positions of the two acoelomorpha taxa are in direct conflict with each other. (Incidentally, differences in mitochondrial genetic code whereby the rhabditophoran ATA codon codes for isoleucine instead of methionine and AAA codes for asparagine instead of lysine support the distinction between the Rhabditophora on the one hand and the acoels, nemertodermatids and catenulids on the other (Telford et al. 2000)).

Acoel Hox genes

The other attractive route to test the idea of a basal position of the acoels would be to look for the 'signature' Hox genes typical of the spiralians discussed above. The problem I foresee with these experiments is if spiralian signatures were found in acoels, as some workers certainly anticipate. This would bring the results from 18S rDNA into apparent conflict with results from signature Hox genes. As I have emphasized, the spiralian signature Hox genes are consistent with a derived spiralian clade, but there is no absolute proof that they are derived characters. Finding these signatures in acoels might suggest that 18S rDNA analyses are incorrect and that acoels are the sister-group of the flatworms. On the other hand, if one accepts the 18S rDNA data (which agreed up to this point with the signature Hox data) one would accept the basal position of the acoels and be forced to conclude that the spiralian 'signatures' are in fact plesiomorphic. The question of the position of the acoels is still not convincingly resolved.

Conclusions

From comparisons of early development, results from molecular phylogenetic studies of the 18S rDNA and by consideration of the complements of signature Hox genes, it appears that the rhabditophoran flatworms at least are not basal Bilateria, as was required by the planuloid/acoeloid theories of metazoan evolution. Rather than being primitively acoelomate and lacking an anus, they have secondarily simplified body plans. Expression of a number of developmental genes in planarians demonstrates their adherence to the zootype, as expected of a member of the spiralia/lophotrochozoa. The question of the position of the acoels seems likely to provide scope for argument if based on unpolarized signature Hox genes, and equally if the acoels are shown to be the sister-group of the apparently derived nemertodermatids.

ACKNOWLEDGEMENTS

Many thanks to Tim Littlewood, Pete Olson and Guillaume Balavoine and three reviewers for helpful comments on the manuscript, although any remaining errors are my own.

Chapter 25

Small subunit rDNA and the Platyhelminthes: Signal, noise, conflict and compromise

D. Timothy J. Littlewood and Peter D. Olson

The strategies of gene sequencing and gene characterization in phylogenetic studies are frequently determined by a balance between cost and benefit, where benefit is measured in terms of the amount of phylogenetic signal resolved for a given problem at a specific taxonomic level. Generally, cost is far easier to predict than benefit. Building upon existing databases is a cost-effective means by which molecular data may rapidly contribute to addressing systematic problems. As technology advances and gene sequencing becomes more affordable and accessible to many researchers, it may be surprising that certain genes and gene products remain favoured targets for systematic and phylogenetic studies. In particular, ribosomal DNA (rDNA), and the various RNA products transcribed from it continue to find utility in wide-ranging groups of organisms. The small (SSU) and large subunit (LSU) rDNA fragments especially lend themselves to study as they provide an attractive mix of constant sites that enable multiple alignments between homologues, and variable sites that provide phylogenetic signal (Hillis and Dixon 1991; Dixon and Hillis 1993). Ribosomal RNA (rRNA) is also the commonest nucleic acid in any cell, and thus was the prime target for sequencing in both eukaryotes and prokaryotes during the early history of SSU nucleotide based molecular systematics (Olsen and Woese 1993). In particular, the SSU gene (rDNA) and gene product (SSU rRNA[1]) have become such established sources of taxonomic and systematic markers among some taxa that databanks dedicated to the topic have been developed and maintained with international and governmental funding (e.g., *The Ribosomal Database Project II*, Maidak *et al.* 1999; the *rRNA WWW Server*, van de Peer *et al.* 2000). One or more species from all metazoan phyla, except the Loricifera, have had their SSU genes characterized in part or fully, and if SSU rDNA appears to be a suitable target at the outset of a phylogenetic study then these databases (including EMBL and GenBank) often provide a head start. In addition to raw sequences, many of the databases include sequences aligned against models of secondary structure and/or against other sequences in the database. Addresses of the WWW pages for all these databases are given below[2]. In addition, the development of our knowledge of molecular evolution as it relates to phylogeny reconstruction has been influenced greatly by the genes chosen, and SSU rRNA has certainly played its part such that features of the gene have been characterized for many phyla (e.g., Abouheif *et al.* 1998; Zrzavý *et al.* 1998).

The first flatworm species to have had its SSU sequence partly determined, in three fragments, was *Dugesia tigrina* (Field *et al.* 1988). The first fully sequenced SSU gene was, perhaps not surprisingly considering its medical importance, from *Schistosoma mansoni* (Ali *et al.* 1991). Importantly, not all the early partial fragments found their way onto public databanks (GenBank/EMBL), but many partial sequences are now being replaced by full or nearly complete sequences.

In this chapter we begin with a brief review of some of the important features of SSU molecules and how phylogenetic studies utilizing the gene have affected our understanding of the phylogeny of the platyhelminths. We then take all the existing complete, or near-complete, SSU data available, and add to these new sequences that were determined previously for ordinal, or at least subphylum level, phylogenies. Our aims are to reconstruct phylogenies based on the available SSU data and to reveal the recurring signal and underlying noise in such reconstructions, predominantly at higher taxonomic levels. At the outset we do not advocate single gene phylogenies, gene trees interpreted without reference to other phylogenetic data, the preferred use of SSU rDNA, or indeed gene sequencing as the primary means by which molecular data can add to platyhelminth phylogenetics. However, the most diverse and widely sampled gene of flatworms serves as the most suitable starting point to which new molecular data may be added to the database and from which salutary lessons may be learned. We begin with a review of some of the important features and the rationale that guided us in assembling and utilizing the data set.

The monophyly of the Platyhelminthes contradicted by SSU rDNA

One of the first data sets that sampled the SSU rDNA broadly suggested that the Acoela were basal platyhelminths (Katayama *et al.* 1993), although the paper included only acoels, triclads and polyclads rooted against a species of yeast, *Saccharomyces cerevisae*, and an ascomycete fungus, *Neurospora crassa*. A more densely sampled analysis including other free-living and parasitic exemplars (Katayama *et al.* 1996) supported this finding, although the ingroup of platyhelminths was rooted against *Saccharomyces* and a collection of diploblasts. Carranza *et al.* (1997) broadened the sampling further still, largely in an attempt to test a tenet from early zoological studies, that platyhelminths are basal metazoans forming the likely sister-group of the other bilaterian phyla. Disturbingly, in these analyses of complete SSU rDNA involving various deuterostome and protostome triploblast taxa rooted against three diploblasts and a protozoan, the flatworms appeared as either paraphyletic or polyphyletic due to

---

[1] *Note*: Many people are familiar with the synonymy between 18S rRNA and SSU rRNA. However, throughout the chapter we avoid referring to the sedimentation coefficient (Svedberg, or S-unit) of the SSU molecules, e.g., 18S, as these are generally estimated or assumed and infrequently determined empirically. Although the SSU molecules are homologous with one another (or indeed paralogous in the case of some triclads; Carranza *et al.* 1996, 1999) many are so large that they are unlikely to have such low S values. In this study for instance, complete SSU sequences ranged in length from 1739 to 2906 bps.

[2] EMBL/EBI: http://www.ebi.ac.uk
GenBank: http://www3.ncbi.nlm.nih.gov/
Ribosomal Database Project: http://www.cme.msu.edu/RDP/html/index.html
rRNA WWW Server: http://rrna.uia.be/index.html

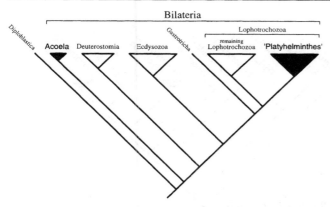

Figure 25.1 Simplified diagrammatic representation of an SSU rDNA-based maximum likelihood tree of 61 metazoan species published by Ruiz-Trillo et al. (1999). Acoels were consistently placed at the base of the Bilateria and never grouped with other platyhelminths, including a single nemertodermatid acoelomorph.

the placement of the catenulid and acoel species. The catenulid taxon appeared at the base of the Bilateria, and the authors drew attention to the long branches exhibited by the acoels. Metazoan wide sampling, using the same data, also suggested a paraphyletic assemblage (Zrzavý et al. 1998). Subsequently it was felt the position of the Acoela could not be confirmed without further sampling. Denser sampling of the Platyhelminthes, including the Acoela, retained the catenulids as sister-group to all other flatworms, but maintained the non-monophyly of the group with the acoels as a long-branching sister-group to all other bilaterians (Littlewood et al. 1999a). Most recently, a thorough analysis of metazoan taxa, including 18 species of acoels, allowed Ruiz-Trillo et al. (1999) to identify an acoel, Paratomella rubra, that was demonstrated to have evolved at a sufficiently slow rate such that long-branch effects in phylogenetic analysis (Felsenstein 1978; Siddall 1998) may be avoided. Nevertheless, analyses of SSU rDNA continued to keep the acoels apart from the other platyhelminth taxa; the catenulids, the one species of nemertodermatid and the Rhabditophora were retained as a monophyletic clade (Figure 25.1). Although denser sampling returned the catenulids as members of the Platyhelminthes, whether from a phylum-wide or Kingdom-wide perspective, the acoelomorph flatworms cannot be considered members of the Platyhelminthes sensu stricto based on SSU rDNA.

SSU rDNA and the position of flatworms among the Metazoa

Tyler (2001, this volume) discusses the affinities of flatworms with other phyla from broader perspectives, but here we briefly note the 'contribution' made by SSU rDNA. Historically, flatworms have been considered to represent basal triploblasts and yet, notwithstanding the contentious basal position of the Acoela in metazoan-wide analyses of SSU rDNA (Ruiz-Trillo, et al. 1999; Figure 25.1), the gene fails to support such a basal placement of the whole phylum. Members of the Rhabditophora, often used as representative platyhelminths, appear firmly ensconced within the Lophotrochozoa (Aguinaldo et al. 1997; Balavoine 1997; Adoutte et al. 1999; Ruiz-Trillo et al. 1999), or at best, when viewed conservatively, as unresolved members of the Protostomia (Abouheif

et al. 1998). Most recently, Giribet et al. (2000) argue for a 'Platyzoa' clade which is sister-group to the Trochozoa. Within the 'Platyzoa' a monophyletic assemblage of flatworms (with the notable exclusion of the nemertodermatids) is sister-group to the Gastrotricha, and together forms a clade with a monophyletic group of Gnathostomulida, Cycliophora, Monogononta, Acanthocephala and Bdelloida. Such results and the instability of these phylogenies dependent upon parameter settings, once again demonstrate that our understanding of metazoan interrelationships has a long way to go, requires new molecular evidence and a broader insight into the morphological, evolutionary and biological consequences of single gene-dominated schemes. In addition, as our understanding of molecular evolution, and our ability to resolve evolutionary history from it improves, so too will our estimates of phylogeny.

History of SSU and the interrelationships of flatworms

The quest to resolve the sister-group of the Neodermata certainly gave impetus to early molecular-based studies on platyhelminth systematics, and continues to do so to this date. Establishing the sister-group allows us to discuss the origins and evolution of parasitism among the obligate parasitic groups more objectively, or at least more rigorously within a cladistic framework. Whilst the monophyly of the Neodermata is well established on both morphological and molecular grounds, differences of opinion concerning character homology has resulted in a number of candidate sister-groups. Littlewood et al. (1999b) reviewed some of the more popular and compelling suggestions from morphological, SSU and LSU data, and concluded that SSU and LSU rejected some scenarios whilst suggesting novel ones as well. Nevertheless, identifying the sister-group to the Neodermata remains a challenging task.

The first study to employ SSU rRNA to examine the interrelationships of the Platyhelminthes was that of Baverstock et al. (1991) (Figure 25.2a). In their study of ten partial sequences rooted against man and Artemia (Crustacea), the small data set resulted in a reasonable degree of resolution; neodermatans were monophyletic, whilst monogeneans were shown to be more closely related to the cestodes, and for the first time it was demonstrated that the monogeneans were not monophyletic. The data broadly supported those topologies suggested by morphologists, but accommodated conflicting topologies since the data were generally too labile to strongly contradict one or another hypothesis. Blair (1993) used the database to place the aspidogastrean Lobatostoma manteri, but could not resolve the monophyly of the Trematoda. The apparent paraphyly of the Monogenea was noted again and in contrast to Baverstock et al. (1991), Blair provided strong support for the monophyly of the Cestoda. Additional taxa sampled from other platyhelminth groups allowed new questions on the phylogeny of the group to be addressed (Rohde et al. 1993b). These authors were cautious in their interpretation of the new data, particularly as competing hypotheses were again almost as likely to be accepted as the most parsimonious solutions offered by the data. However, these were the first indications that some key hypotheses on the interrelationships of platyhelminths founded on morphology were to be challenged by molecular data; e.g., the identity of the sister-group to the Neodermata and the apparent similarity in flame bulb and protonephridial ultrastructure between the Rhabdocoela and the Lecithoepitheliata (see Rohde et al. 1993b for full discussion). Additionally, the value of the growing SSU data set was

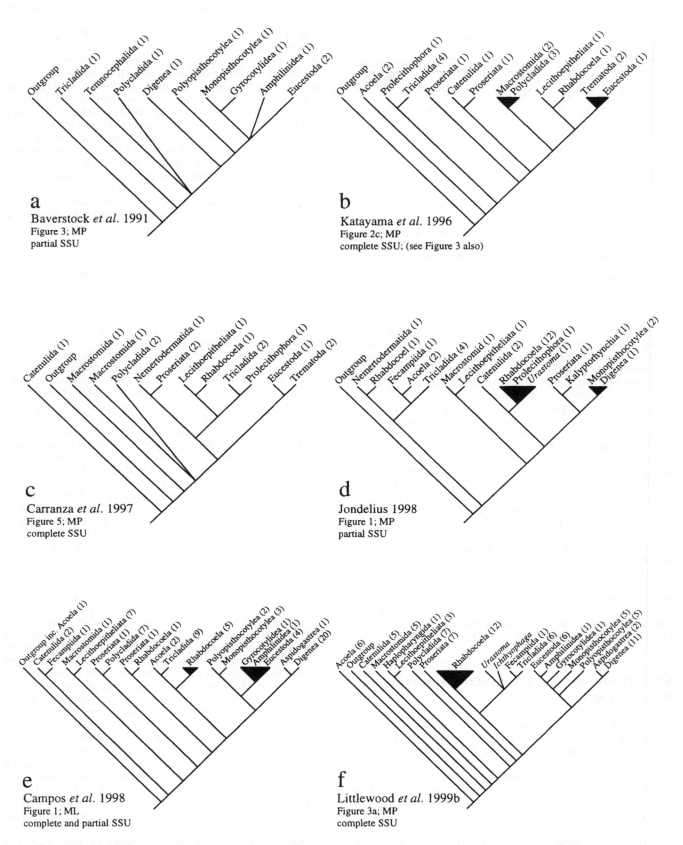

Figure 25.2 Recent hypotheses on the interrelationships of the Platyhelminthes based on SSU rDNA sequence data with the number of taxa representing each higher group indicated parenthetically. Filled triangles represent non-monophyletic groupings of constituent taxa. Abbreviations: ML, maximum likelihood; MP, maximum parsimony.

clear to those wishing to establish the phylogenetic positions of problematic taxa among the platyhelminths with data independent from morphology.

Riutort et al. (1992a) started to generate what is now a large SSU database on triclads (Baguñà et al. 2001, this volume) by considering the monophyly of selected subgenera, genera and families. Barker et al. (1993b) used SSU rDNA to estimate the position of the sole member of the Heronimidae within the Digenea, a topic that had been widely debated by morphologists. The debate continues even with the addition of more digenean SSU data (Cribb et al. 2001, this volume). The systematics and phylogenetics of other digenean taxa have also been reviewed with the addition of partial or complete SSU data (e.g., Lumb et al. 1993 on fellodistomids and lepocreadiids, Blair et al. 1998 on the Hemiuroidea). The efforts of Rohde (see Rohde 2001, this volume), Watson (see Watson 2001, this volume), Ehlers and Sopott-Ehlers (e.g., Ehlers 1995 Sopott-Ehlers 1998) among others, on the ultrastructure of various features such as protonephridia and spermatozoa, suggested new priorities for SSU sequencing in order to test controversial morphological synapomorphies. For example, the phylogenetic affinities of Kronborgia isopodicola was one such question, and the SSU data supported separation of the group (Fecampiida) rather than unification with any existing taxon (Rohde et al. 1994).

Acoels, bearing in mind their current controversial position among the Metazoa using SSU (Ruiz-Trillo et al. 1999), were still confidently viewed as platyhelminths when first introduced into the SSU data set (Katayama et al. 1993) and lived up to their expectation as basal flatworm taxa. In a search to detect the earliest divergent species of the genus Geocentrophora (Lecithoepitheliata) in Lake Baikal, Kuznedelov and Timoshkin (1993) pushed the SSU beyond its limits of resolution, failing to find sufficient differences between the partial SSU sequences. However, their limited results were consistent with existing taxonomic schemes. A subsequent analysis of these and additional data by the same authors (Kuznedelov and Timoshkin 1995) allowed one of the first 'turbellarian'-based assessments to be made with SSU. Monophyly of the Seriata (Tricladida and Proseriata) was challenged, and a monophyletic clade including the Kalyptorhynchia, Proseriata and Lecithoepitheliata (rather than the strictly bifurcating topology of Ehlers (1985a)) was first suggested. Katayama et al. (1996) continued the focus on 'turbellarian' orders, and provided the first comprehensive molecular-based analysis of their interrelationships. Although we show here only one of the solutions provided by Katayama and her co-workers (see Figure 25.2b; original Figure 2c), it was considered that the sequences of some taxa, two proseriates and a prolecithophoran, added more noise than signal. Whether the loss of resolution was due to poor sampling, analytical or sequencing error does not appear to have been suggested. Nevertheless, even without these apparently aberrant sequences, anomalies included paraphyly of the macrostomids and trematodes, acoels were most basal, and triclads were the sister-group to all other flatworms.

Carranza et al. (1997) utilized SSU data to question the monophyly of the Platyhelminthes and its position among major metazoan clades, as well as to infer interrelationships. Their conclusion that the phylum is not monophyletic depends largely on the anomalous position of the catenulid Stenostomum leucops, rather than the acoels. Macrostomids remained paraphyletic but did appear at the base of the Rhabditophora. This study showed the likelihood that the sister-group of a monophyletic Neodermata was a large clade comprised of 'turbellarian' taxa, although another potentially rogue sequence – this time from the acoelomorph Nemertinoides elongatus

(Nemertodermatida) – added some confusion to the otherwise 'equitable' phylogeny (Figure 25.2c). The same sequence continued to plague other studies (e.g., Littlewood et al. 1999a) until a second nemertodermatid SSU was determined. Jondelius had been working at the same time with partial SSU sequences and with the nemertodermatid Meara stichopi. However, the poor signal from partial sequences apparently added more confusion to the SSU trees, with his solution bearing even less resemblance to accepted or previously hypothesized schemes (Jondelius 1998) (Figure 25.2d). Also, during the latter part of the 1990s Campos et al. (1998) (Figure 25.2e) had gathered full and partial sequences from the literature and provided a more comprehensive treatment of groups. Once again, some groupings were unique, notably the grouping of Catenulida with Fecampiida and the Acoela with the Tricladida, whereas the interrelationships of the Neodermata made eminent sense in the light of morphology (e.g., Ehlers 1984). SSU alone has clearly been capable of enthralling and frustrating flatworm systematists.

The first phylum-wide study to incorporate morphological evidence and combine it with SSU for a cladistic treatment was that of Littlewood et al. (1999a). SSU data alone, involving 82 sequences, reflected many patterns seen with less densely sampled analyses, but was at least allowing fewer options; e.g., a completed sequence of Meara stichopi provided by Ulf Jondelius tempered the saltatory behaviour of the Nemertodermatida. The treatment also highlighted the conflict between a morphological and molecular analysis, with the two data sets arguing for statistically different phylogenetic solutions. Finally, prior to the present analysis, Littlewood et al. (1999b) added a few more taxa, adopted a refined morphological matrix from their previous study, and found the SSU data to be compatible with the results based on morphology (at least in terms of passing Templeton's test) where it had failed previously (Littlewood et al. 1999a). The SSU data set alone provided conflicting topologies depending on the analysis performed (maximum parsimony result is shown in Figure 25.2f), but many major clades were supported as monophyletic and, combined with morphology, the data provided a working model based explicitly on much of the available evidence. The same study reviewed the influence of a molecular and combined-evidence approach in establishing the elusive sister-group to the Neodermata.

Rooting the SSU rDNA tree

Controversy regarding the position of the acoels, particularly in their distance from the acoelomorph nemertodermatids, and the placement of the latter group, are just two reasons why we have chosen not to include acoels, or indeed nemertodermatids in our present analysis. SSU rDNA sequences from acoelomorphs are notoriously difficult to align with other flatworm taxa, and result in the exclusion of many more regions to maintain an ambiguity-free alignment than if they are excluded altogether. Thus, to determine the underlying phylogenetic patterns supported by SSU rDNA for the greatest number of taxa with the highest resolution, we have rooted our tree against the catenulids. Our hypotheses therefore reflect the interrelationships of the Rhabditophora (Ehlers 1984), the monophyly of which is more broadly accepted.

Why have we not chosen representatives from another phylum to root a tree of Platyhelminthes, or at least Rhabditophora + Catenulida? The sister-group to the Platyhelminthes is not certain from either morphological or molecular studies, and just as there are problems with the SSU sequences of basal platyhelminth taxa, there appear to be problems with sister-group candidates. For example, both xenoturbellids (Ehlers and Sopott-Ehlers 1997b; Lundin 1998) and gnathostomulids (Haszprunar 1996) have been considered basal Bilateria and/or sister-groups to the Platyhelminthes. However, SSU places xenoturbellids closer to the Mollusca (Norén and Jondelius 1997) and gnathostomulid SSU sequences have long branches and are placed variously among the Ecdysozoa (Littlewood et al. 1998b) or not (Zrzavý et al. 1998; Giribet et al. 2000). The number of outgroups we have chosen may not be ideal for any phylogenetic reconstruction but, following the criteria of Smith (1994), we know from previous analyses that the catenulids are suitable candidates to root the Rhabditophora as they are monophyletic within the ingroup in larger studies of SSU (e.g., Carranza et al. 1997; Littlewood et al. 1999a,b; Ruiz-Trillo et al. 1999) and are the likely sister-group to the rhabditophoran flatworms.

The data set and sampling

Many partial SSU rDNA sequences are available, but to attain the highest number of variable and phylogenetically informative sites we have restricted our analysis to complete or near-complete sequences. Furthermore, we have excluded certain complete sequences, despite their availability on GenBank at the time of analysis, for one or more of the following reasons: 1) SSU sequence appears more than once on GenBank for the same taxon; 2) alignment in highly conserved regions was difficult and suggested high probability of sequencing error; and 3) previous phylogenetic analyses indicated sufficient error in the sequence to compromise its utility.

Whilst we will not discuss the interrelationships of the constituent major clades of flatworms sampled, it is important to highlight the diversity of taxa that underlies them. Appendix 25.1 gives a complete listing of the 270 taxa used in this study and indicates the families from which the species have been classified for each major clade. As with the majority of sequencing studies, which require access to properly fixed or fresh material that has been identified by an expert prior to fixation or molecular analysis, opportunistic collecting tends to dominate the strategy. Furthermore, in this study, our sample reflects efforts, largely by us and in collaboration with others, to sample widely for studies concentrating on smaller clades of flatworms. Readers wishing to add SSU sequences from aco-elomorphs to this data set should see Ruiz-Trillo et al. (1999) for a listing of available sequences. An overview of the diversity of exemplar taxa follows:

- *Macrostomida and Haplopharyngida* – perhaps more accurately grouped as Macrostomorpha (Rieger 2001, this volume); this is a small group but poorly sampled in our analysis.

- *Lecithoepitheliata* – only three species within the same genus are represented. Campos et al. (1998) utilized more members of the same genus, but these were the partial sequences of Kuznedelov and Timoshkin (1993).

- *Polycladida* – although a highly diverse group we include just six sequences representing four families.

- *Rhabdocoela* – here, we include a variety of families (nine) from a variety of higher taxa that arguably should or could be treated separately (e.g., as in Littlewood et al. 1999a,b). However, many constituent taxa (e.g., Temnocephalida) are very poorly sampled, and taking the Rhabdocoela as

the group of interest allows us to argue for relatively diverse sampling.

- *Prolecithophora* – our data come largely from the studies dedicated to prolecithophoran interrelationships (Norén and Jondelius 1999; Jondelius et al. 2001, this volume), but include three new sequences that in total represent five families.

- *Tricladida* – the majority of taxa come from dense samplings of triclads by Carranza et al. (1998a,b), including nine families and recently reviewed by Baguñà et al. (2001, this volume). We add one new species.

- *Proseriata* – although most phylum-wide studies have included at least some proseriate sequences, here we provide the densest and most diverse sample that was used for a treatment on the interrelationships of the group (Littlewood et al. 2000; see also Curini-Galletti 2001, this volume).

- *Fecampiida + Urastomidae* – these genera represent a clade that has yet to be given a formal name. *Ichthyophaga* and *Urastoma* were each originally classified as Prolecithophora; Watson (1997a) and Noury-Sraïri et al. (1989b) demonstrated differences in sperm ultrastructure in *Urastoma*, and Littlewood et al. (1999a) showed, using SSU data, that *Ichthyophaga* fell outside the Prolecithophora. The fecampiid, *Kronborgia*, was shown to group with *Urastoma* and *Ichthyophaga* in Littlewood et al. (1999a,b). The fecampiid *Notentera ivanovi* was sequenced for this study and for another rather different perspective on flatworm phylogenetics (see Joffe and Kornakova 2001, this volume).

- *Monopisthocotylea* – nine families including five new sequences represent the densest sampling of SSU data for this group of monogeneans to date.

- *Polyopisthocotylea* – 13 families including nine new sequences represent the densest sampling of SSU data for this group to date; the majority of published monogenean sequences are from Littlewood et al. (1998a).

- *Amphilinidea* – the two families of amphilinideans, each represented by a single sequence are now supplemented with an additional amphilinid.

- *Gyrocotylidea* – two members of the single constituent family are included.

- *Eucestoda* – 27 families representing the 12 currently recognized orders (Khalil et al. 1994), as well as the nominal orders Diphyllobothriidea and Litobothriidea are included from a study on cestode interrelationships (Olson et al. in press).

- *Aspidogastrea* – three of the four families are represented, and we include three new sequences.

- *Digenea* – 55 families are sampled, and include 75 new sequences generated for this study and another concentrating on digenean interrelationships (Cribb et al. 2001, this volume).

New sequences presented herein were determined using techniques outlined in Littlewood et al. (1999a) or Olson and Caira (1999). Appendix 25.2 lists primers used by the authors for PCR

Table 25.1 Statistical summary of SSU rDNA data sets analysed.

Dataset	Number of positions*					%G	%A	%T	%C	Averages†		
	Included	Constant	Uninformative	Gapped	Informative					CI	RI	RC
Complete	1,215	409	208	363	598	26.6	27.6	25.5	20.3	0.57	0.57	0.24
Stems	727	237	128	213	362	29.8	22.7	25.8	21.6	0.57	0.58	0.27
Loops	488	172	80	150	236	21.7	34.8	25.1	18.4	0.52	0.57	0.20

* Total number of positions in alignment = 3587. Numbers of uninformative and informative positions based on parsimony.
† Values represent means of the character consistency index (CI), retention index (RI) and rescaled consistency index (RC) for all positions included in the data set analysed.

amplification of the complete SSU rDNA gene of platyhelminths, as well as primers for sequencing the PCR products.

Alignment

Variability of sequence lengths was extremely high, ranging from 1739 bps in the triclad, *Girardia tigrina*, to 2906 bps in the amphilinid tapeworm, *Gigantolina magna*. It is interesting that the neodermatan taxa possessed SSU sequences of greater length than those of the 'turbellarian' taxa without exception. In general, 'turbellarian' SSU sequences were ~1800 bps, digenean and monogenean sequences ~1950 bps, and cestode sequences ~2100 bps in length. Primarily these differences reflect modifications to variable domains of the gene (Figure 25.3), whilst the conserved core of the secondary structure model (e.g., Neefs *et al.* 1993) was alignable across the broad spectrum of taxa examined. To date only two species of flatworms, *Schistosoma mansoni* and *Spirometra erinaceieuropaei*, have had their secondary structure at least partially predicted (see Ali *et al.* 1991 and Liu *et al.* 1997, respectively), and it may be worth examining the model for other taxa. In particular, large insertions, notably among amphilinidean cestodes (Olson and Caira 1999), suggest that the mature SSU ribosomal RNA may take a wide range of forms among the flatworms.

It is well known that even small changes in alignment can have major effects on phylogeny reconstruction (e.g., Winnepenninckx and Backeljau 1996), and we have aimed to be highly conservative in our determination of positional homology. Furthermore, because the effects of missing data can have an undesirable influence on resulting trees (Barriel 1994; Wilkinson 1995), we have discarded most positions that required gaps to be inserted in the alignment for a large number of taxa. The result was that a majority (66%) of the 3587 positions in the full alignment[3] was discarded either for lack of positional homology or for the presence of insertion/deletions unique to small numbers of taxa. Ultimately, 1215 positions were included in the analyses of 270 taxa. This provided 806 variable positions, of which 598 were phylogenetically informative under the criterion of parsimony (see Table 25.1). Figure 25.3a provides a diagrammatic representation of the full alignment, indicating the variable domains as defined by Neefs *et al.* (1993), and the distribution of phylogenetically informative positions. Figure 25.3b shows in greater detail three regions of the alignment (*i, ii, iii*; as indicated by horizontal bars in Figure 25.3a) that together encompass all positions included the analysis. Using a 5 bp sliding window method of averaging, these three histograms depict the rescaled consistency indices (RC) of the characters (based on a maximum parsimony

consensus tree) as distributed across the alignment. From this there is no clear pattern to suggest that some regions of the molecule contain more reliable, or less homoplasious, sites than do others, with the obvious exception of the variable domains in which most sites had to be discarded altogether. Instead, sites showing high RC values are scattered across the more conserved regions of the gene alignable among the 270 taxa. An effective sequencing strategy therefore requires information from the entire gene to maximize the number of such positions.

Analysis

Large data sets are not amenable to all methods of analysis. In particular, maximum likelihood analysis is not possible unless restricted, for example, to 4-taxon statements (e.g., the quartet puzzling methods of Strimmer and von Haeseler 1996; Wilson 1999). Here we restrict ourselves to minimum evolution (ME) and maximum parsimony (MP) approaches and concentrate only on the interrelationships of major clades of flatworms. We have purposefully avoided providing details of lower level interrelationships, for example within triclads, prolecithophorans, digeneans, cestodes, etc., as these are dealt with elsewhere in this volume. Furthermore, the scope of the alignment across the Rhabditophora cannot accurately reflect the SSU signal, as many positions potentially informative within subsets of the taxa will have been excluded from the global alignment.

Numerous discussions on the philosophical merits of phylogenetic reconstruction methods exist in the literature (for example, see the journals *Systematic Biology, Molecular Biology and Evolution, Molecular Phylogenetics and Evolution,* and *Cladistics*). Here, we take two very different approaches commonly used to estimate phylogenetic patterns from nucleotide data. Maximum parsimony is a character-based approach that seeks the topological solution that incurs the fewest number of character-state changes. Minimum evolution is a distance-based algorithm that builds a topology based on pairwise distances estimated by a model of nucleotide substitution, that in turn attempts to compensate for the biases inherent to the sequence data (e.g., substitution rate variation and base-compositional bias). Considerable detail on the computational aspects of both methods can be found in Swofford *et al.* (1996). All phylogenetic analyses were conducted using PAUP* ver. 4.0 (Swofford 1998).

Treatment of gaps

Alignments of homologous genes invariably generate the need for gaps, or indels, as insertions and deletions are inferred from

---

[3] The full alignment may be obtained by anonymous FTP from FTP.EBI.AC.UK under directory pub/databases/embl/align, accession number DS42209.

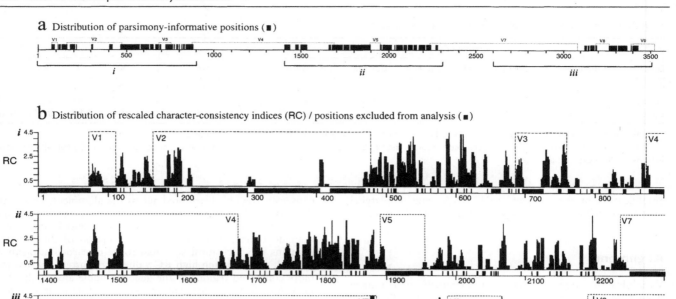

a Distribution of parsimony-informative positions (■)

b Distribution of rescaled character-consistency indices (RC) / positions excluded from analysis (■)

Figure 25.3 Graphical representations of the sequence alignment consisting of 270 platyhelminth SSU sequences. a) Complete alignment indicating the distribution of parsimony-informative positions (black columns), variable domains as defined by Neefs et al. (1993; dotted boxes) and the three alignment regions (*i, ii, iii*) shown in b (horizontal bars). b) Rescaled consistency-index for each character included in the analysis averaged over a 5 bp sliding window. Variable domains indicated by dashed boxes. Black columns below the x-axis indicate positions excluded from the analyses; note, however, that the method of averaging employed yields values even for excluded positions so long as they are within 4 bps of an adjacent position with a value >0.

multiple pairwise comparisons of sequences. The inclusion of gaps as fifth state characters, available in MP analysis only, has been demonstrated to provide additional valuable statements on homology (e.g., Giribet and Wheeler 1999) and some data sets utilizing SSU data rely on indels for finer phylogenetic resolution (e.g., echinoids; Littlewood and Smith 1995). In our alignment, treating gaps as fifth character states, or as missing data, had neither any effect on the number of phylogenetically informative positions, nor on the topology of the MP tree. Consequently, we have restricted our analyses to working with gaps treated as missing data for both MP and ME solutions.

Minimum evolution (Figure 25.4)

The log determinant model (Lake 1991; Lockhart *et al.* 1994) of nucleotide substitution was used to estimate genetic distances that were then analysed by the method of minimum evolution. Tree bisection-reconnection (TBR) branch-swapping was aborted after 18 hours, and $> 1.5 \times 10^6$ topological arrangements had been evaluated.

Maximum parsimony (Figure 25.4)

Characters were run unordered and taxa were added via random addition. Not a single heuristic search using TBR branch-swapping ever reached completion before the computer ran out of memory storing trees. Thus, we show the strict consensus of this same number of equally parsimonious trees (42 100) found within 13 hours of searching.

Rate categorization of sites

Although it is a controversial subject, there are logical reasons to justify selectively excluding positions from the analysis. One obvious reason is to reduce noise (random signal) by removing sites that are highly homoplasious based on either an a priori or a posteriori criterion. We chose to use the rescaled consistency index of the characters (based on the topology of the ME tree) as a measure by which to separate the characters into ten categories. We then examined the effects on tree topology and resolution of removing characters with low RC values (and thus high rates) through successive maximum parsimony analyses. The result (not shown) was that the structure of the parsimony-based tree in Figure 25.4 was largely supported, but with less resolution and with the occasional spurious arrangement as more and more characters were removed from the analysis. Similar to the *a posteriori* successive approximations approach (Farris 1969), we could have *differentially* down-weighted characters in high-rate categories (rather than down-weighting them to nought by removal). However, the subjectiveness of such a weighting scheme and the lack of any striking differences in the results after the complete removal of high-rate sites suggested to us that we were unlikely to enhance the signal through further analysis.

Analysis of consensus sequences (Figure 25.5)

Because our concerns herein are focused on the interrelationships among major clades, and not interrelationships within these clades, we considered the effects of reducing the terminal

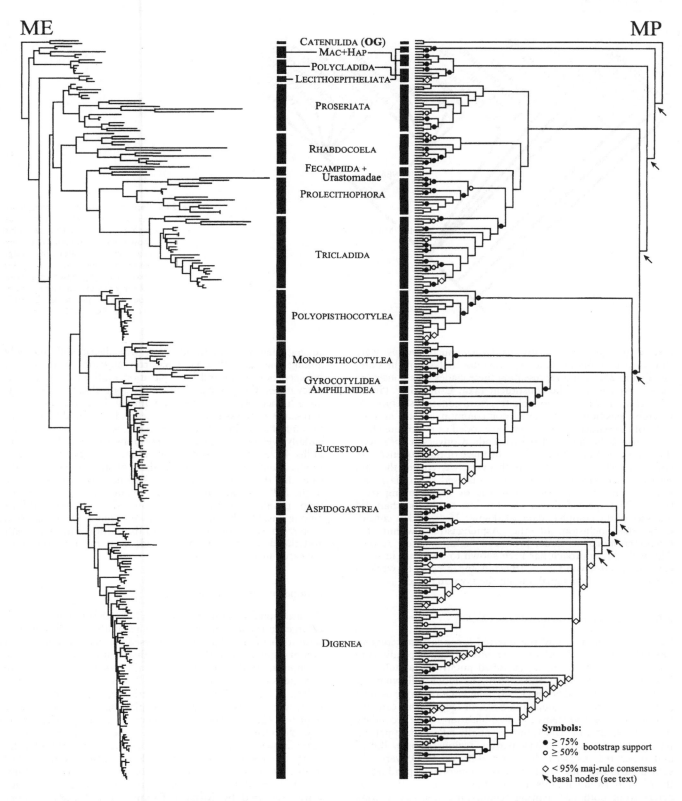

ME

MP

CATENULIDA (**OG**)
MAC+HAP
POLYCLADIDA
LECITHOEPITHELIATA

PROSERIATA

RHABDOCOELA

FECAMPIIDA +
Urastomadae

PROLECITHOPHORA

TRICLADIDA

POLYOPISTHOCOTYLEA

MONOPISTHOCOTYLEA

GYROCOTYLIDEA
AMPHILINIDEA

EUCESTODA

ASPIDOGASTREA

DIGENEA

Symbols:
● ≥ 75% bootstrap support
○ ≥ 50%
◇ < 95% maj-rule consensus
↖ basal nodes (see text)

Figure 25.4 Results of minimum evolution (ME) and maximum parsimony (MP) analyses (1215 characters). Catenulids designated as outgroup (OG) taxa. Left topology depicts the minimal tree based on a distance matrix (LogDet model of nucleotide substitution); right topology depicts the majority-rule (maj-rule) consensus of 42 100 equally parsimonious trees (EPTs; 5185 steps, CI = 0.26, RI = 0.77, RC = 0.2); heuristic search aborted after examining >2 × 10^9 topological arrangements. A vast majority of nodes were common amongst all EPTs; those found in less than 95% of the EPTs are indicated with open diamonds. Bootstrap support based on 26 973 replicates using a fast heuristic search algorithm. Arrows indicate the basal nodes used for examining the potential saturation of character state substitutions as shown in Figure 25.6B (see text).

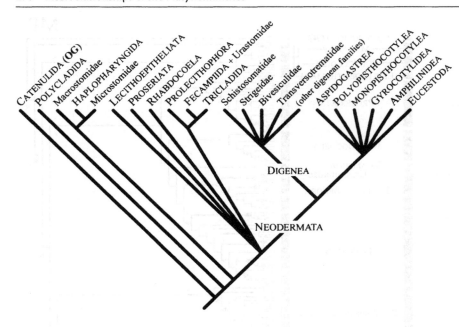

Figure 25.5 Results of maximum parsimony analysis of 75% consensus sequences representing 22 clades shown to be monophyletic by prior analysis. Catenulida designated as the outgroup (OG) taxon. Topology based on a strict consensus of 612 equally parsimonious trees (466 steps, CI = 0.7, RI = 0.66, RC = 0.46).

taxa into representative groups (where such groups were shown to be monophyletic via previous analysis of the data, i.e., Figure 25.4), and representing these clades by a consensus of the sequences of the constituent taxa. Using GDE (Smith S.W. *et al.* 1994), it is possible to create consensus sequences whereby positions with states not common among at least 75% of the sequences considered are coded as multistate characters using the standard IUPAC code. Conversely, positions that show the same state in 75% or more of the sequences are coded as such for the group. In this way, we reduced the data set to 22 consensus sequences and analysed them via maximum parsimony. A strict consensus of the resulting trees (Figure 25.5) provided considerably less resolution than did analyses of the complete data set (Figure 25.4), and in considering the inclusion of the Aspidogastrea within the Cercomeromorphae clade, produced highly unlikely results. Although this approach has been shown to be useful in some cases (e.g., Littlewood *et al.* 1997), it appeared to be weak in reducing conflicting signal among the taxa analysed.

Effects of secondary structure (Figure 25.6)

Using our alignment and the inferred secondary structure of one SSU sequence (*Pseudomurraytrema* sp.; see Appendix A in Olson and Caira 1999), we classified putatively homologous base positions as being either stems or loops following the rationale of Soltis and Soltis (1998), wherein loops were defined as being four or more unpaired bases in length. Our categorization of loops and stems is a simplification of the secondary structure features of rDNA, but essentially reflects base-pairing regions (stems) and non-base-pairing regions (loops). Bulges and 'other' regions (*sensu* Vawter and Brown 1993) were subsumed variously into 'stem' or 'loop' categories depending on their length. Even with this simplification, it was clear that characters from both base-pairing regions and non-base-pairing regions contain phylogenetic information; although of all phylogenetically informative positions, 60.5% appeared in stems and 39.5% in loops. Table 25.1 provides a statistical summary of the different data partitions. Stem regions were slightly G–T rich, whereas loop regions were comparatively A-rich. Chi-square analysis as implemented in PAUP*, however, did not

suggest that this nucleotide bias was distributed unevenly among the taxa (P = 1). Character statistics were similar between both stem and loop partitions, although the consistency index (CI) was slightly lower for loop characters suggesting a higher degree of saturation among these positions.

Because of the size of our data set and the inability to reach the end of a 'standard' heuristic search, we chose not to differentially weight stem and loop positions (see e.g., Dixon and Hillis 1993), but instead analysed the data partitions separately. Maximum parsimony analysis of stem bases only indicates that these regions contribute significantly to the structure of the MP topology, and provides greater resolution than the loop positions alone. Furthermore, dubious relationships among basal 'turbellarian' groups were found when analysing loop regions. Of course, because of the nature of the alignment and the diversity of taxa sampled, most included positions appeared in stem regions (60%).

Mutational saturation (Figure 25.7)

We attempted to examine the possibility that the antiquity of divergence events within the Platyhelminthes has resulted in saturation of the characters analysed. In such a situation, a plot of sequence divergence versus divergence time will become asymptotic at the time in which all sites free to vary have become saturated (see Figure 5.19 in Page and Holmes 1998). In Figure 25.7 we show a series of graphs that approximate the comparison above by plotting observed pairwise sequence substitutions (=divergence) against estimated pairwise distances (~divergence time). Because transitions occur more frequently than do transversions, we looked at both categories of substitutions separately. Pairwise substitution ratios (transitions or transversions as a proportion of the total number of observed differences) were calculated using *Seq_db* software (authored by Richard Thomas, The Natural History Museum). We also employed two different substitution models to estimate genetic distances: Log-determinant and general time-reversible (GTR; Swofford *et al.* 1996; Waddell and Steel 1997) including estimates of invariant sites and among-site rate heterogeneity (as estimated from the ME-based topology). Differences in the assumptions of the two models

stems

loops

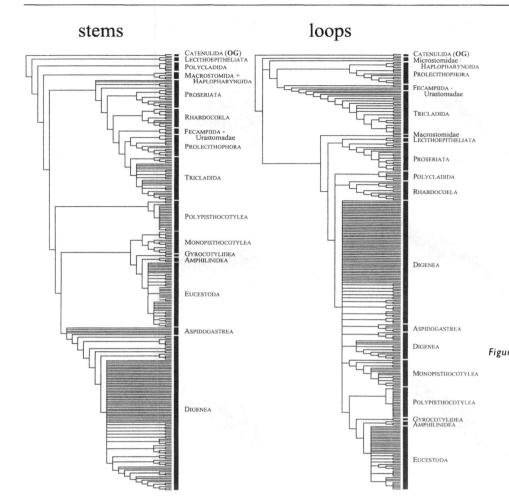

Figure 25.6 Results of separate maximum parsimony analyses of 'stem' (717 characters) and 'loop' (488 characters) positions as defined in the text. Catenulids designated as outgroup (OG) taxa. Left topology depicts the strict consensus of 32 200 equally parsimonious trees (EPTs) (2763 steps, CI = 0.3, RI = 0.79, RC = 0.24), and the right topology depicts the strict consensus of 32 000 EPTs (2336 steps, CI = 0.24, RI = 0.75, RC = 0.18).

generally result in quite different estimates of genetic distance (with the GTR model typically resulting in estimates of distance more disparate from observed values), and could thus lead to different conclusions regarding mutational saturation. The left column of plots in Figure 25.7 show comparisons based on all included positions, whereas plots in the right column show comparisons only from positions that change on the basal nodes (denoted by arrows in Figure 25.4) of the maximum parsimony consensus topology, where the potential for saturation would be expected to be highest due to the greater age of such early divergences. Results derived from all included positions show a tightly clumped, linear increase of divergence with distance, suggesting that neither transitions nor transversions are saturated. However, a different pattern is seen when only those positions observed to change on the basal nodes of the tree are considered in isolation. These plots show considerably greater scatter in all cases, and transitions appear to asymptote when genetic distances reach a value of ~0.1 using either model of nucleotide substitution, whereas transversions show a more linear rate of increase. Saturation of transitional substitutions along basal nodes may account in part for the instability and lack of support of these partitions in the tree.

Figure 25.7b illustrates further patterns of nucleotide substitution in the SSU data. A transition/transversion plot for pairwise comparisons of taxa demonstrates any potential bias towards one or other substitution type as well as presenting

further visualization of any saturation of substitutions as a function of time since divergence. Generally, transitions occur more frequently than transversions; in our data set the overall estimated transition:transversion ratio was 1.3 : 1. As in Figure 25.7a, plots that deviate from a linear relationship indicate saturation effects from multiple substitutions, erasing the record of previous changes.

Signal

SSU provides a phylogeny of the Platyhelminthes with certain relationships robust to the vagaries of reconstruction method and steadfast under the scrutiny of bootstrap analysis. It certainly seems to be the case that a denser sampling of taxa has yielded more robust phylogenies of the platyhelminths with more groups retaining monophyly than other studies to date, although some relationships have been identified even with a minimum number of taxa (e.g., see Figure 25.2). Taking the present study as the basis for discussing SSU and the platyhelminths, as it represents the most densely sampled data set and therefore the best molecular-based estimate to date (Hillis 1996, 1998; Graybeal 1998), monophyly of the following groups is found to be strongly supported: Neodermata, Trematoda, Digenea, Cestoda, Amphilinidea, Gyrocotylidea, Monopisthocotylea, and Polyopisthocotylea. The Gyrocotylidea is the

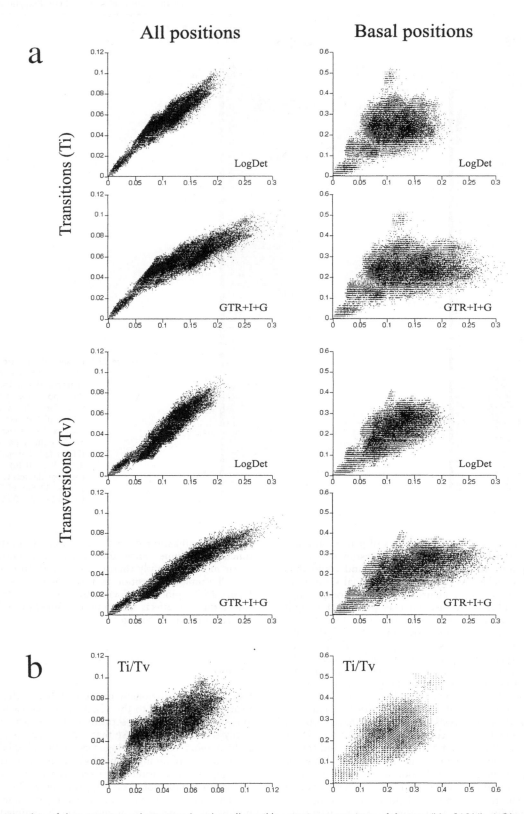

Figure 25.7 Scatter plots of character-state substitutions based on all possible pairwise comparisons of the taxa (N = 36 316). a) Observed transitions or transversions versus genetic distance as estimated by either of two nucleotide substitution models: log determinant (LogDet) or general-time reversible, including estimates of invariant sites and among-site rate variation (GTR + I + G). b) Scatter plots of observed transitions versus transversions. Plots in the left column are calculated from all included positions; those of the right column are calculated from only those characters observed to change along the basal nodes of the consensus tree (see arrows in Figure 25.4). For each pairwise comparison, the substitution value (transition or transversion) is relative to the total number of changes observed between the two taxa.

sister-group to a clade comprising the eucestodes and amphilinideans. Likewise, SSU confirms the monophyly of the Tricladida, Prolecithophora, Polycladida, Lecithoepitheliata, Macrostomida + Haplopharyngida, a clade comprising the parasitic 'turbellarian' genera, *Ichthyophaga, Kronborgia, Notentera* and *Urastoma*, and the non-neodermatan rhabdocoels; namely a clade comprised of the Dalyelliida, Kalyptorhynchia, Temnocephalida and Typhloplanida. Within this latter clade Kalyptorhynchia and Dalyelliida are also monophyletic (details of relationships within the Rhabdocoela are shown in Littlewood *et al.* 1999b, Figure 3). Figure 25.6 illustrates further the strength of the signal, regarding the monophyly of the major clades in stem and loop regions of the SSU data. The signal in the solution based on stem regions is largely consistent with that in the full analyses, whereas the solution provided by loop regions offered less resolution and less congruence.

In the full analyses, the rhabdocoels and proseriates are also each monophyletic, but with low bootstrap support that suggests that there is less signal in the SSU for the confident placement of these taxa. Indeed, although many clades appear to be monophyletic, few deeper branching nodes are well supported.

SSU data suggest that the sister-group to the Neodermata is a large clade comprised of all the 'turbellarian' taxa to the exclusion of the more basal macrostomids, haplopharyngids, lecithoepitheliates and polyclads. Interestingly, only with relatively dense sampling were the Proseriata both monophyletic and members of this larger 'turbellarian' sister-group clade (contrast Littlewood *et al.* 1999a,b).

Noise and conflict

Relationships amongst the earliest divergent platyhelminth taxa are not well resolved with available SSU data. The two methods of analysis provide contradictory solutions with respect to the Macrostomida + Haplopharyngida and Lecithoepitheliata vying for the position of the most basal rhabditophoran. Consequently, we cannot place the Polycladida firmly either. Within the remaining 'turbellarian' groups sampled, only the interrelationships of the rhabdocoels (Kalyptorhynchia, Temnocephalida, Typhloplanida and Dalyelliida) are in conflict due to the non-monophyly of the typhloplanids sampled. The remaining conflict involves the interrelationships of the Neodermata. As the Monogenea remain to be confirmed as truly monophyletic, we are no further in resolving the interrelationships of the cestodes, monogeneans and trematodes than we are when including morphology or LSU rDNA data (Littlewood *et al.* 1999b). This conflict is known from other ribosomal gene data (Mollaret *et al.* 1997; Justine 1998a) and in our study, in spite of denser sampling, there is conflict between the ME and MP results. Although the Monogenea are not monophyletic in either case, ME provides a more traditional scheme of neodermatan interrelationships with monogenean and cestode groups forming a clade ('Cercomeromorphae') that is the sister-group to the Trematoda (Aspidogastrea + Digenea). In contrast, MP analysis suggests that the polyopisthocotylean monogeneans are the sister-group to all other neodermatans. Considering the density of sampling so far, the problem of monogenean monophyly, which is contradicted only by ribosomal evidence, is not likely to be readily solved with the addition of more SSU data.

Influence of SSU data on combined evidence analyses

As a result of the breadth of sampling of the SSU gene among flatworms, this gene locus has been used to determine combined evidence phylogenetic solutions, usually in combination with morphologically based matrices, but occasionally in combination with other gene sequences. Combining data in phylogenetic analyses is a controversial topic (de Queiroz *et al.* 1995; Huelsenbeck *et al.* 1996), and such studies involving flatworms are few in number. A review indicates the relatively great influence SSU has on combined-evidence tree topologies. We are aware of few studies of platyhelminths where SSU has been combined with other systematic evidence and analysed using cladistics (e.g., Blair *et al.* 1998 on hemiuroid digeneans; Olson and Caira 1999 on cestodes; Littlewood *et al.* 1999b on the phylum). In cases where morphology alone has provided highly unresolved trees, the influence of SSU data is clearly overriding. One such example comes from the Digenea (Cribb *et al.* 2001, this volume), where an extensive morphological matrix coding many characters for numerous taxa results in a poorly resolved morphological tree when compared to that offered by SSU alone. The combined-evidence solution is largely similar to the tree derived from the SSU analysis. Such scenarios not only call for more morphological characters if possible, but more characters independent of the SSU gene in order to assess the possibility of interpreting a gene phylogeny. A similar example comes from the combined-evidence treatment of the Platyhelminthes (Littlewood *et al.* 1999b), where quite different scenarios are suggested each by morphology and SSU. The combined-evidence solution, legitimized by the compatibility of the two data sets, appears to be more similar to the SSU tree than the morphology tree. However, some morphologically based synapomorphies (e.g., those uniting the Monogenea) persist and highlight potentially homoplastic signal in the SSU data. These studies should not be used to fuel a debate on molecules versus morphology (Hillis 1987; Patterson *et al.* 1993). Character conflict demonstrates the need for additional data and/or an understanding of where the homoplasy lies in one or more data sets (Larson 1994; Hillis 1998).

Other studies have shown that SSU data can be as much in conflict with other genes as it can with morphology. In the Proseriata both SSU and LSU rDNA suggest alternative phylogenetic solutions for the group. In the absence of sufficient morphological signal the debate turns from molecules and morphology to gene versus gene. Such results highlight deficiencies in sample size, as only more taxa or more characters are likely to lead to congruence or a better estimate of phylogeny (Graybeal 1998; but see Naylor and Brown 1997).

Compromise

With a plethora of phylogenetic schemes available from SSU data in the literature, it is incumbent upon us to provide a solution that we feel reflects both the signal and the noise in the molecule. On the premise that the most densely sampled data set is best, but without advocating one phylogenetic reconstruction method over another, we have combined the tree solutions offered by our analyses into a strict consensus, shown in Figure 25.8, where conflict between the most parsimonious trees and the topology estimated by ME are reflected as polytomies. Although a conservative estimate, a considerable amount of structure remains nonetheless.

Amongst the Rhabditophora, the most basal clade is presently unresolved with SSU data alone, leaving us with a polytomy of macrostomorphs, lecithoepitheliates and polyclads. None of these groups is particularly well sampled in comparison to the other major groups, and if one were to rely solely on SSU data, further sequences may help resolve the polytomy.

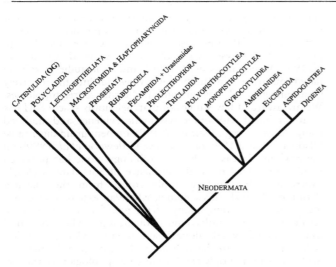

Figure 25.8 Compromise – phylogenetic relationships among the major clades of platyhelminths based on a strict consensus of the results of minimum evolution and maximum parsimony analyses shown in Figure 25.4.

Two other major clades of platyhelminths are resolved, namely the Neodermata and a clade comprising the proseriates, rhabdocoels, Fecampiida + Urastomidae, prolecithophorans and triclads. The latter clade has not been previously reported in any study of platyhelminth interrelationships. Less densely sampled analyses have generally resulted in the exclusion of the proseriates from such a clade (e.g., Littlewood *et al.* 1999b; Baguñà *et al.* 2001, this volume). The addition of Prolecithophora and more proseriates appears to have strengthened the case for this clade, although it is poorly supported by bootstrap resampling procedures. Interestingly, the parasitic 'turbellaria' nestle within the clade and therefore refute the monophyly of the Revertospermata (but see Kornakova and Joffe 1999; Joffe and Kornakova 2001, this volume). That the triclads and prolecithophorans are sister taxa has been found by those concentrating on the interrelationships of triclads using SSU data (Baguñà *et al.* 2001, this volume), but all other relationships within this clade appear to be new hypotheses.

Whilst SSU provides excellent resolution of and within the Neodermata, the monophyly of the Monogenea remains uncertain. Campos *et al.* (1998) suggested monophyly with SSU, but greater sampling leaves the group paraphyletic (Littlewood *et al.* 1999a,b, this study). The compromise solution does, however, support traditional relationships among the gyrocotylideans, amphilinideans and eucestodes.

Are these relationships correct? We can only hope to have demonstrated the signal that SSU data provides. As with any phylogeny, we can impose subjective decisions as to the value of particular nodes, or hopefully, judge the signal against additional apomorphies from independent data sets (for example, the reader may look elsewhere in this volume). It is also important to note that our compromise topology is probably the best estimate of what the SSU data presently provides and, amongst the resolved nodes including those with low bootstrap support, it may truly reflect the evolution of the gene whilst not necessarily reflecting the evolutionary history of the species. Conflict between species trees and gene trees is well known (e.g., Page and Charleston 1997; Slowinski and Page 1999), and it is important to bear this in mind when evaluating or utilizing single-gene phylogenies.

Clearly, there is need for greater SSU sampling of many of the 'turbellarian' groups, in particular the Macrostomida, Lecithoepitheliata, Temnocephalida, Kalyptorhynchia, Dalyelliida and Typhloplanida, not only for the placement of these taxa, but in order to evaluate the interrelationships within these groups. Whilst we have not covered the utility of SSU within some of the major clades in this chapter, the gene has clearly demonstrated great utility among constituent platyhelminth taxa; e.g., the triclads (Baguñà *et al.* 2001, this volume), the prolecithophorans (Jondelius *et al.* 2001, this volume), the monogeneans (Littlewood *et al.* 1998a), the cestodes (Mariaux and Olson 2001, this volume), and the digeneans (Cribb *et al.* 2001, this volume). As regards the broader relationships, the SSU data set is now generally well sampled, and attention spent on other genes and molecular markers will probably be more profitable.

ACKNOWLEDGEMENTS

The majority of sequences presented in this chapter were generated by generous funding from the Wellcome Trust (043965/Z/95/Z) to D.T.J.L. P.D.O. was funded by a Marshall-Sherfield Fellowship (Marshall Aid Commemoration Commission, UK) and through the Fellowship held by D.T.J.L. Both funding sources are gratefully acknowledged. We would like to thank Karen Clough, Elisabeth Herniou and Julia Bartley who provided technical assistance in generating novel sequence data for this contribution. We are grateful to David Blair and Robert Hirt for critical comments on the manuscript. Very many colleagues and friends kindly and unselfishly provided raw material for sequencing; their expertise comes from morphologically and taxonomically based backgrounds without which the molecular data would be uninterpretable, and this study would not have been possible.

Appendix 25.1 Taxonomic listing of platyhelminth species analysed and the GenBank/EMBL accession numbers of their complete SSU rDNA sequences. Accession numbers followed by '§' are new to this study.

CATENULIDA			LECITHOEPITHELIATA		
Stenostomidae	*Stenostomum leucops aquariorum*	AJ012519	Prorhynchidae	*Geocentrophora sphyrocephala*	D85089
				Geocentrophora sp.	U70080
	Suomina sp.	AJ012532		*Geocentrophora wagini*	AJ012509
MACROSTOMIDA			POLYCLADIDA		
Macrostomidae	*Macrostomum tuba*	U70082	Leptoplanidae	*Notoplana koreana*	D85097
	Macrostomum tuba	D85092		*Notoplana australis*	AJ228786
Microstomidae	*Microstomum lineare*	U70081	Planoceridae	*Planocera multitentaculata*	D83383/D17562
	Microstomum lineare	D85091	Discocelidae	*Discocelis tigrina*	U70079
HAPLOPHARYNGIDA			Pseudocerotidae	*Thysanozoon brocchii*	D85096
Haplopharyngidae	*Haplopharynx rostratus*	AJ012511		*Pseudoceros tritriatus*	AJ228794

Continued

Appendix 25.1 Continued.

RHABDOCOELA		
Dalyelliidae	*Microdalyellia rossi*	AJ012515
Graffillidae	*Graffilla buccinicola*	AJ012521
Pterastericolidae	*Pterastericola australis*	AJ012518
Temnocephalidae	*Temnocephala* sp.	AJ012520
Trigonostomidae	*Mariplanella frisia*	AJ012514
Typhloplanidae	*Bothromesostoma* sp.	D85098
	Mesocastrada foremani	U70082
	Mesostoma lingua	AJ243682
Polycystidae	*Gyratrix hermaphroditus*	AJ012510
	Arrawaria sp.	AJ243677
Diascorhynchidae	*Diascorhynchus rubrus*	AJ012508
Karkinorhynchidae	*Cheliplana* cf. *orthocirra*	AJ012507
PROLECITHOPHORA		
Baicalarctiidae	*Baicalarctica gulo*	AJ287483§
	Friedmaniella karlingi	AJ287513§
	Friedmaniella sp. *(rufula?)*	AJ287512§
Pseudostomidae	*Pseudostomum gracilis*	AF065426
	Pseudostomum klostermanni	AF065424
	Pseudostomum quadrioculatum	AF065425
	Reisingeria hexaoculata	AF065426
Cylindrostomidae	*Cylindrostoma fingalianum*	AF051330
	Cylindrostoma gracilis	AF065416
Plagiostomidae	*Plagiostomum cinctum*	AF065418
	Plagiostomum vittatum	AF051331
	Plicastoma cuticulata	AF065422
	Vorticeros ijimai	D85094
Ulianinidae	*Ulianinia mollissima*	AF065427
TRICLADIDA		
Procerodidae	*Ectoplana limuli*	D85088
	Procerodes littoralis	Z99950
Bdellouridae	*Bdelloura candida*	Z99947
Uteriporidae	*Uteriporus* sp.	AF013148
Geoplanidae	*Artioposthia triangulata*	AF033038
	Cenoplana caerulea	AF033040
	Australoplana sanguinea	AF033041
Bipaliidae	*Bipalium kewense*	AF033039
Rhynchodemidae	*Microplana nana*	AF033042
Planariidae	*Crenobia alpina*	M58345
	Polycelis nigra	AF013151
	Polycelis tenuis	Z99949
	Phagocata ullala	AF013149
	Phagocata sp.	AF013150
	Phagocata sibirica	AJ287559§
Dendrocoelidae	*Dendrocoelum lacteum*	M58346
	Dendrocoelopsis lactea	D85087
	Baikalobia guttata	Z99946
Dugesiidae	*Schmidtea mediterranea*	U31084
	Schmidtea polychroa	AF013152
	Romankenkius lidinosus	Z99951
	Cura pinguis	AF033043
	Dugesia subtentaculata	M58343
	Dugesia japonica	AF013153
	Girardia tigrina	AF013157
PROSERIATA		
Archimonocelididae	Archimonocelidinae n.gen.sp.1	AJ270150
	Archimonocelis crucifera	AJ270151
	Archimonocelis staresoi	AJ270152
	Calviria solaris	AJ270153
Coelogynoporidae	*Cirrifera dumosa*	AJ270154
	Coelogynopora gynocotyla	AJ243679
	Vannuccia sp.	AJ270162
Monocelididae	*Archiloa rivularis*	U70077
	Monocelis lineata	U45961
Monotoplanidae	*Monotoplana* cf. *diorchis*	AJ270159
Otoplanidae	*Archotoplana holotricha*	AJ243676
	Monostichoplana filum	AJ270158
	Otoplana sp.	D85090
	Parotoplana renatae	AJ012517
	Xenotoplana acus	AJ270155
Unguiphora	*Nematoplana coelogynoporoides*	AJ012516
	Nematoplana sp.	AJ270160

	Polystyliphora novaehollandiae	AJ270161
FECAMPIIDA		
Fecampiidae	*Kronborgia isopodicola*	AJ012513
'Fecampiid'	*Notentera ivanovi*	AJ287546§
'TURBELLARIA' INCERTAE SEDIS		
Urastomidae	*Urastoma cyprinae*	U70086
Genostomatidae	*Ichthyophaga* sp.	AJ012512
MONOGENEA – MONOPISTHOCOTYLEA		
Monocotylidae	*Calicotyle affinis*	AJ228777
	Dictyocotyle coeliaca	AJ228778
	Troglocephalus rhinobatidis	AJ228795
Capsalidae	*Encotyllabe chironemi*	AJ228780
	Benedenia sp.	AJ228774
	Capsala martinieri	AJ276423§
Gyrodactylidae	*Gyrodactylus salaris*	Z26942
Anoplodiscidae	*Anoplodiscus cirrusspiralis*	AJ287475§
Udonellidae	*Udonella caligorum*	AJ228796
Dactylogyridae	*Pseudohaliotrema sphincteroporus*	AJ287568§
	Pseudodactylogyrus sp.	AJ287567§
Sundanonchidae	*Sundanonchus micropeltis*	AJ287579§
Pseudomurraytrematidae	*Pseudomurraytrema* sp.	AJ228793
Microbothriidae	*Leptocotyle minor*	AJ228784
MONOGENEA – POLYOPISTHOCOTYLEA		
Polystomatidae	*Neopolystoma spratti*	AJ228788
	Polystomoides malayi	AJ228792
Diclybothriidae	*Pseudohexabothrium taeniurae*	AJ228791
Plectanocotylidae	*Plectanocotyle gurnardi*	AJ228790
Mazocraeidae	*Kuhnia scombri*	AJ228783
Allodiscocotylidae	*Metacamopia oligoplites*	AJ287538§
Neothoracocotylidae	*Paradawesia* sp.	AJ287555§
	Mexicotyle sp.	AJ287539§
Gotocotylidae	*Gotocotyla bivaginalis*	AJ276424§
	Gotocotyla secunda	AJ276425§
Diclidophoridae	*Diclidophora denticulata*	AJ228779
Discocotylidae	*Discocotyle sagittata*	AJ287504§
Diplozoidae	*Eudiplozoon nipponicum*	AJ287510§
Microcotylidae	*Bivagina pagrosomi*	AJ228775
	Cynoscionicola branquias	AJ287495§
	Microcotyle sebastis	AJ287540§
	Neomicrocotyle pacifica	AJ228787
Axinidae	*Zeuxapta seriolae*	AJ228797
Heteraxinidae	*Probursata brasiliensis*	AJ276426
CESTODA – AMPHILINIDEA		
Amphilinidae	*Austramphilina elongata*	AJ287480§
	Gigantolina magna	AJ243681
Schizochoeridae	*Schizochoerus liguloideus*	AF124454
CESTODA – GYROCOTYLIDEA		
Gyrocotylidae	*Gyrocotyle urna*	AJ228782
	Gyrocotyle rugosa	AF124455
CESTODA – EUCESTODA		
Caryophyllaeidae	*Caryophyllaeus laticeps*	AJ287488§
	Hunterella nodulosa	AF124457
Hymenolepididae	*Hymenolepis diminuta*	AF124475
	Hymenolepis microstoma	AJ287525§
	Wardoides nyrocae	AJ287587§
Echinobothriidae	*Echinobothrium fautleyi*	AF124464
Macrobothridiidae	*Macrobothridium* sp.	AF124463
Diphyllobothriidae	*Diphyllobothrium stemmacephalum*	AF124459
	Schistocephalus solidus	AF124460
Haplobothriidae	*Haplobothrium globuliforme*	AF124458
Lecanicephalidae	*Cephalobothrium* cf *aetobatidis*	AF124466
	Eniochobothrium gracile	AF124465
Tetragonocephalidae	*Tylocephalum* sp.	AJ287586§
Litobothriidae	*Litobothrium* sp.	AF124468§
	Litobothrium amplifica	AF124467
Nippotaeniidae	*Amurotaenia decidua*	AF124474§
	Nippotaenia mogurndae	AJ287545§
Monticellidae	*Gangesia parasiluri*	AJ287515§
Proteocephalidae	*Proteocephalus perplexus*	AF124472

Continued

Appendix 25.1 Continued.

Bothriocephalidae	*Bothriocephalus scorpii*	AJ228776			*Tergestia laticollis*	AJ287580§
Triaenophoridae	*Abothrium gadi*	AJ228773			*Steringophorus margolisi*	AJ287578§
	Anchistrocephalus	AJ287473§			*Olssonium turneri*	AJ287548§
	microcephalus			Gorgoderidae	*Degeneria halosauri*	AJ287497§
	Eubothrium crassum	AJ287509§			*Gorgodera* sp.	AJ287518§
Acrobothriidae	*Cyathocephalus truncatus*	AJ287493§			*Xystretrum* sp.	AJ287588§
Spathebothriidae	*Spathebothrium simplex*	AF124456		Gyliauchenidae	*Robphildollfusium fractum*	AJ287571§
Tetrabothriidae	*Tetrabothrius erostris*	AJ287581§			*Gyliauchen* sp.	L06669
	Tetrabothrius forsteri	AF124473		Haploporidae	*Pseudomegasolena ishigakiense*	AJ287569§
	Tetrabothrius sp.	AJ287582§		Haplosplanchnidae	*Hymenocotta mulli*	AJ287524§
Onchobothriidae	*Calliobothrium cf verticillatum*	AF124469			*Schickhobalotrema* sp.	AJ287574§
	Platybothrium auriculatum	AF124470		Hemiuridae	*Dinurus longisinus*	AJ287501§
Dasyrhynchidae	*Dasyrhynchus pillersi*	AJ287496§			*Lecithochirium caesionis*	AJ287528§
Gilquiniidae	*Gilquinia squali*	AJ287516§			*Lecithocladium excisum*	AJ287529§
Grillotiidae	*Grillotia erinaceus*	AJ228781			*Merlucciotrema praeclarum*	AJ287535§
	Grillotia heronensis	AJ287519§			*Plerurus digitatus*	AJ287562§
Hepatoxylidae	*Hepatoxylon* sp.	AF124462		Heronimidae	*Heronimus mollis*	L14486
Lacistorhynchidae	*Callitetrarhynchus gracilis*	AJ287487§		Heterophyidae	*Cryptocotyle lingua*	AJ287492§
Otobothriidae	*Otobothrium dipsacum*	AJ287552§			*Haplorchoides* sp.	AJ287521§
Pterobothriidae	*Pterobothrium lintoni*	AJ287570§		Lecithasteridae	*Lecithaster gibbosus*	AJ287527§
Sphyriocephalidae	*Sphyriocephalus* sp.	AJ287576§		Lepocreadiidae	*Austroholorchis sprenti*	AJ287481§
Tentaculariidae	*Tentacularia* sp.	AF124461			*Lepidapedon rachion*	Z12607
ASPIDOGASTREA					*Lepidapedon elongatum*	Z12600
Aspidogastridae	*Aspidogaster conchicola*	AJ287478§			*Preptetos caballeroi*	AJ287563§
	Lobatostoma manteri	L16911			*Tetracerasta blepta*	L06670
Multicalycidae	*Multicalyx* sp.	AJ287532§		Mesometridae	*Mesometra* sp.	AJ287537§
Multicotylidae	*Multicotyle purvisi*	AJ228785		Microphallidae	*Levenseniella minuta*	AJ287531§
Rugogastridae	*Rugogaster* sp.	AJ287573§			*Maritrema oocysta*	AJ287534§
DIGENEA					*Microphallus primas*	AJ287541§
Accacoeliidae	*Accacoelium contortum*	AJ287472§			unidentified	AJ001831
Acanthocolpidae	*Cableia pudica*	AJ287486§		Monorchiidae	*Ancylocoelium typicum*	AJ287474§
	Stephanostomum baccatum	AJ287577§			*Provitellus turrum*	AJ287566§
Angiodictyidae	*Neohexangitrema zebrasomatis*	AJ287544§		Nasitrematidae	*Nasitrema globicephalae*	AJ004968
	Hexangium sp.	AJ287522§		Notocotylidae	*Notocotylus* sp.	AJ287547§
Apocreadiidae	*Homalometron synagris*	AJ287523§		Opecoelidae	*Gaevskajatrema halosauropsi*	AJ287514§
	Neoapocreadium splendens	AJ287543§			*Macvicaria macassarensis*	AJ287533§
Atractotrematidae	*Atractotrema sigani*	AJ287479§			*Peracreadium idoneum*	AJ287558§
Azygiidae	*Otodistomum cestoides*	AJ287553§		Opisthorchiidae	*Opisthorchis viverrini*	X55357
Bivesiculidae	*Bivesicula claviformis*	AJ287485§		Opistholebetidae	*Opistholebes amplicoelus*	AJ287550§
	Paucivitellosus fragilis	AJ287557§		Orchipedidae	*Orchipedum tracheicola*	AJ287551§
Brachycoeliidae	*Mesocoelium* sp.	AJ287536§		Pachypsolidae	*Pachypsolus irroratus*	AJ287554§
Bucephalidae	*Prosorhynchoides gracilescens*	AJ228789		Paramphistomidae	*Calicophoron calicophorum*	L06566
Bunocotylidae	*Opisthadena* sp.	AJ287549§		Paragonimidae	*Paragonimus westermani*	AJ287556§
Campulidae	*Zalophotrema hepaticum*	AJ224884		Philophthalmidae	*Philophthalmid* sp.	AJ287560§
Cephalogonimidae	*Cephalogonimus retusus*	AJ287489§		Plagiorchiidae	*Glypthelmins quieta*	AJ287517§
Cryptogonimidae	*Mitotrema anthostomatum*	AJ287542§			*Haematoloechus longiplexus*	AJ287520§
Cyclocoelidae	*Cyclocoelum mutabile*	AJ287494§			*Rubenstrema exasperatum*	AJ287572§
Derogenidae	*Derogenes varicus*	AJ287511§			*Skrjabinoeces similis*	AJ287575§
Dicrocoeliidae	*Dicrocoelium dendriticum*	Y11236		Sanguinicolidae	*Aporocotyle spinosicanalis*	AJ287477§
Didymozoidae	*Didymozoon scombri*	AJ287500§		Schistosomatidae	*Schistosoma haematobium*	Z11976
Diplodiscidae	*Diplodiscus subclavatus*	AJ287502§			*Schistosoma japonicum*	Z11590
Diplostomidae	*Diplostomum phoxini*	AJ287503§			*Schistosoma mansoni*	X53017
Echinostomatidae	*Echinostoma caproni*	L06567			*Schistosoma spindale*	Z11979
Enenteridae	Enenterid sp.1	AJ287507§		Sclerodistomidae	*Prosogonotrema bilabiatum*	AJ287565§
	Enenterid sp.	2AJ287508§		Strigeidae	*Ichthyocotylurus erraticus*	AJ287526§
Fasciolidae	*Fasciola gigantica*	AJ011942		Syncoelidae	*Copiatestes filiferus*	AJ287490§
	Fasciola hepatica	AJ004969		Tandanicolidae	*Prosogonarium angelae*	AJ287564§
	Fasciolopsis buski	L06668		Transversotrematidae	*Crusziella formosa*	AJ287491§
Faustulidae	*Antorchis pomacanthi*	AJ287476§			*Transversotrema haasi*	AJ287583§
	Bacciger lesteri	AJ287482§		Zoogonidae	*Deretrema nahaense*	AJ287498
	Trigonocryptus conus	AJ287584§			*Lepidophyllum steenstrupi*	AJ287530§
Fellodistomidae	*Fellodistomum fellis*	Z12601			*Zoogonoides viviparus*	AJ287590§

Appendix 25.2a

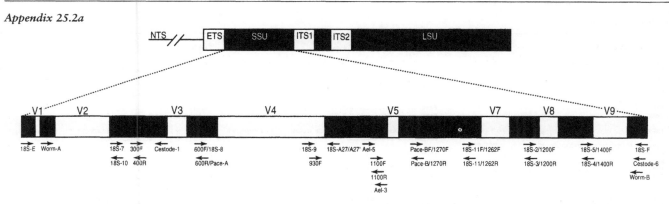

Ribosomal array showing relative positions of primers for the SSU gene locus. Abbreviations: ETS, external transcribed spacer; ITS1, ITS2, internal transcribed spacers; LSU, large subunit; NTS, non-transcribed spacer; SSU, small subunit; V1–V9, variable domains.

Appendix 25.2b SSU rDNA Primers. Conserved PCR/sequencing primers for the SSU rDNA gene used by the authors are listed below (format follows Simon *et al.* 1994) showing discrepancies observed among 22 platyhelminth exemplar sequences (listed below). Primer names and aliases are given followed by the direction of priming (→, 5′-3′; ← 3′-5′). The following line shows the size, definition, and annealing location of the primer based on the complete 1932 bp SSU sequence of *Pseudomurraytrema* sp. [GenBank No. AJ228793].

ID: Classification (exemplar taxon):

'Turbellaria'
Cate Catenulida (*Stenostomum leucops*)
Macr Macrostomida (*Macrostomum tuba*)
Leci Lecithoepitheliata (*Geocentrophora wagini*)
Poly Polycladida (*Discocelis tigrina*)
Pros Proseriata (*Polystyliphora novaehollandiae*)
Kaly Rhabdocoela: Kalyptorhynchia (*Cheliplana orthocirra*)
Daly Rhabdocoela: Dalyelliida (*Graffila buccinicola*)
Typh Rhabdocoela: Typhloplanida (*Mesocastrada foremani*)
Note Fecampiida (*Notentera ivanovi*)
Prol Prolecithophora (*Cylindrostoma gracilis*)
Tric Tricladida (*Phagocata ullala*)

Neodermata
Monp Monogenea: Polypisthocotylea (*Neomicrocotyle pacifica*)
Monm Monogenea: Monopisthocotylea (*Dictyocotyle coeliaca*)
Gyro Cestoda: Gyrocotylidea (*Gyrocotyle urna*)
Amph Cestoda: Amphilinidea (*Gigantolina magna*)
Spat Eucestoda: Spathebothriidea (*Spathebothrium simplex*)
Tetr Eucestoda: Tetraphyllidea (*Calliobothrium* cf. *verticillatum*)
Cycl Eucestoda: Cyclophyllidea (*Hymenolepis diminuta*)
Aspi Aspidobothrea (*Aspidogaster conchicola*)
Schi Digenea: Schistosomatidae (*Schistosoma mansoni*)
Fasc Digenea: Fasciolidae (*Fasciolopsis buski*)
Hemi Digenea: Hemiuridae (*Merlucciotrema praeclarum*)

18S-E (alias 18S-A) (→)

```
(35mer) 5′ CCGAATTCGTCGACAACCTGGTTGATCCTGCCAGT 3′
```

Comments: It was not informative to check this 'universal' 5′-end primer as it was itself used to amplify the SSU gene in a majority of the taxa above (and was thus incorporated into the PCR products sequenced).

WormA (→)

```
(21mer) 5′ GCGAATGGCTCATTAAATCAG 3′      [ 67-87]
Leci      .....A.........G.....
Daly      ..AT...........T.A...
Tric      ...G...........T.A...
Monp      A....................
```

18S-7 (→)

```
(22mer) 5′ GCCCTATCAACTGTCGATGGTA 3′      [ 295-316]
Cate      .A...........A.......
Macr      .A...........A.......
Leci      .A...................
Poly      ............TA..T.....
Pros      ............TA.......
Kaly      .AA.........A.G..A....
Daly      ............--A......G
Typh      .A......T............G
Note      .........CA.T........
Prol      .A......G..A....C....
Tric      .A..........T........
Monp      ............TA.......
Gyro      ............T........
Amph      ............A........
Spat      ............T........
Tetr      ............T........
Cycl      ............T........
Aspi      ............TA.......
Schi      .........T.T--.T.....
Fasc      .........T.T.........
Hemi      ............T.........
```

18S-10 (←)

```
(22mer) 5′ TACCATCGACAGTTGATAGGGC 3′      [ 316-295]
```

Comments: Reverse complement of 18S-7 above.

300F (→)

```
(17mer) 5′ AGGGTTCGATTCCGGAG 3′      [ 358-374]
Cate      .........T.......
Macr      ........C........
Typh      ..T..............
```

400R (alias 300R) (←)

```
(18mer) 5′ TCAGGCTCCCTCTCCGGA 3′      [ 385-368]
Cate      .......-..........
Poly      ...T..............
Kaly      .A................
Daly      .A................
Typh      .A................
Note      .......AA.........
```

Comments: 3′ end partially overlaps with 300F.

Cestode-1 (←)

```
(20mer) 5′ TTTTTCG-TCACTACCTCCCC 3′      [ 463-444]
Cate      .......-............T.
Macr      .......-............T.
Poly      .......T.............
Kaly      .......-............T.
Daly      ..C.CT.T.GG...TT..A.T
Typh      .......-............T.
Prol      .....T.-.............
Tric      ...A.T.-............A.
Monm      ...G..T-.T.........T.
Amph      .......-...C.........
Spat      .......-...C.........
Tetr      .......-...C.........
Cycl      .......-...C.........
Hemi      .......-............A..
```

600F /
18S-8 (→)

```
(18mer) 5′ GGTGCCAGCMGCCGCGGT        3′ [ 549-566]
(20mer) 5′       GCAGCCGCGGTAATTCCAGC 3′ [ 556-575]
Leci      ....................-...........
Poly      ....................-...........
Kaly      ....................C...........
Monp      ....................C...........
Gyro      ....................C...........
Amph      ....................C...........
Spat      ....................-...........
Tetr      ....................C...........
Cycl      ....................-..-.........
Aspi      ....................C...........
Schi      ....................C...........
Fasc      ....................C...........
Hemi      ....................C...........
```

600R (←)

```
(18mer) 5′ ACCGCGGCKGCTGGCACC 3′      [ 566-549]
```

Comments: Reverse complement of 600F.

Pace-A (←)

```
(18mer) 5′ GTGTTACCGCGGCTGCTG 3′      [ 571-554]
Cate      .AA...............
Macr      .AA...............
Leci      .AA..-............
Poly      .AA...............
Pros      .AA...............
Kaly      .A................
Daly      .AA...............
Typh      .AA...............
Note      .AA...............
Prol      .AA...............
Tric      .AA...............
Monp      .A................
Monm      .A................
Gyro      .A................
Amph      .A................
```

Continued

Appendix 25.2b Continued.

Pace-A (*continued from previous page*)

```
(18mer) 5' GTGTTACCGCGGCTGCTG 3'    [ 571-554]
Spat    .A................
Tetr    .A................
Cycl    .A....-...........
Aspi    .A................
Schi    .A................
Fasc    .A................
Hemi    .A................
```

18S-9 (→)

```
(18mer) 5' TTTGAGTGCTCAAAGCAG 3'    [ 863-880]
Cate    ..A...............
Macr    ..A...............
Leci    ..A.......T......A
Poly    ..G...........T...
Pros    ..A...............
Kaly    ..A...............
Daly    ..A...............
Typh    ..AA..............
Note    ..A.......T.......
Prol    ..G...............
Tric    ..A.......T.......
Gyro    .............C...
Amph    .............A...
Spat    .............C...
Tetr    .............C...
Cycl    .............T...
```

930F (→)

```
(20mer) 5' GCATGGAATAATGGAATAGG 3'    [ 904-923]
Leci    .............A....
Poly    ...............A..
Daly    ...............C..
Note    ...............A...
Prol    ...............A....A.
Tric    ...........A.......
Monp    ...........A.......
Amph    ...........A.......
Schi    ...............A...
Hemi    ...............A...
```

18S-A27 (←)

```
(21mer) 5' CCATACAAATGCCCCCGTCTG 3'    [ 997-977]
Cate    AA....G...........A.C.
Macr    AA....G...........A.C.
Leci    .A....G...........C...
Poly    .A....G.............
Kaly    ......G...........
Daly    ......TTT.......G.A.
Typh    ..................G.A.
Note    G......-...-........
Prol    GT.....G..........A.
Tric    G......T.........G.A.
Gyro    ........C.T........
Amph    ........C.T.....C.C.
Spat    ........C.T.....C...
Tetr    ........C.T.....C...
Cycl    ........C.T.....C...
Hemi    ..................T..

(A27')   ........C.T.....C...
```

Comments: A27' (Olson and Caira 1999) was a modification to match eucestodes.

Ael-5 (→)

```
(20mer) 5' TGTTTTCATTGACCATGAGC 3'    [ 1063-1082]
Cate    C..C.C....A.T..A..A.
Macr    C..C.C....A.T..A..A.
Leci    ......-.G..A.T..A..A.
Poly    .....C...A.T..A..A.
Pros    ...........A.T..A..A.
Kaly    ..........A.T..A..A.
Daly    ..C.C.....A.T..A..A.
Typh    ...C......A.T..A..A.
Note    ...C......A.T..A..A.
Prol    ..........A.T..A..A.
Tric    ..........A.T..A..A.
Monp    .............A.A....
```

Ael-5 (*continued from previous column*)

```
(20mer) 5' TGTTTTCATTGACCATGAGC 3'    [ 1063-1082]
Monm    ..........T....G....
Tetr    ...........G........
Cycl    ...........G........
Aspi    ...........T...G....
Schi    ............T.G....
Fasc    ............T.T....
Hemi    ............T.TG....
```

Comments: Design (DTJL) based on the sequence of *Austramphilina elongata* (Cestoda: Amphilinidea).

1100F (→)

```
(19mer) 5' CAGAGATTCGAAGACGATC 3'    [ 1089-1107]
Cate    ......G.............
Macr    ......G.............
Leci    ......G....NN....T
Poly    ......G.............
Pros    ......G.............
Kaly    ......G........G....
Daly    T.....G.............
Typh    ......G.............
Note    ......G.............
Prol    ......G.............
Tric    ......GA............
Monp    ......G.............
Monm    ......G.............
Gyro    G...GC.............
Amph    ......GC............
Spat    ......GC............
Tetr    ......GC............
Cycl    ......GC............
Aspi    ......G....T.....
Schi    ......T.............
Fasc    ......G.............
Hemi    ......GA............
```

1100R (←)

```
(18mer) 5' GATCGTCTTCGAACCTCTG 3'    [ 1107-1089]
```

Comments: Reverse complement of 1100F.

Ael-3 (←)

```
(20mer) 5' GTATCTGATCGTCTTCGAGC 3'    [ 1113-1094]
Cate    ..G...............A.
Macr    ..G...............A.
Leci    ......A...NN....A.
Poly    ..................A.
Pros    ..................A.
Kaly    ..........C.....A.
Daly    ..................A.
Typh    ..................A.
Note    ..................A.
Prol    ..................A.
Tric    ..................T.
Monp    ..................A.
Monm    ..................A.
Aspi    ..........A...A.
Schi    ..................AA
Fasc    ..................A.
Hemi    ..................T.
```

Comments: Design (DTJL) based on the sequence of *Austramphilina elongata* (Cestoda: Amphilinidea).

Pace-B / 1270R (←)

```
(20mer) 5' CCGTCAATTCCTTTAAGTTT 3'    [ 1260-1241]
(18mer) 5' CCGTCAATTCCTTTAAGT 3'    [ 1260-1243]
Leci    ...C................
```

Comments: Highly conserved reverse primer.

Pace-BF / 1270F (→)

```
(20mer) 5' AAACTTAAAGGAATTGACGG 3'    [ 1241-1260]
(18mer) 5' ACTTAAAGGAATTGACGG 3'    [ 1243-1260]
```

Comments: Reverse complements of Pace-B/1270R.

18S-11 / 1262R (alias 1055R) (←)

```
(21mer) 5' AACGGCCATGCACCACCACCC 3'    [ 1393-1373]
(15mer) 5' CGGCCATGCACCACC 3'    [ 1391-1377]
Cate    ...................T..
Macr    ...................T..
Daly    ...................TT.
Note    .....A...............
Monp    ...................A..
Monm    ..................T.A..
Gyro    ...................A..
Amph    ...................A..
Spat    ...................A..
Tetr    ...................A..
Cycl    ...................A..
Aspi    ...................A..
Fasc    ...................A..
Hemi    ...................T..
```

18S-11F / 1262F (alias 1055F) (→)

```
(21mer) 5' GGGTGGTGGTGCATGGCCGTT 3'    [ 1373-1393]
(15mer) 5' GGTGGTGCATGGCCG 3'    [ 1377-1391]
```

Comments: Reverse complements of 18S-11/1262R.

18S-2 / 1200F (→)

```
(25mer) 5' ATAACAGGTCTGTGATGCCCTTAGA 3' [1579-1603]
(16mer) 5' CAGGTCTGTGATGCCC 3'[ 1583-1598]
Tric    .......................A.
Hemi    ....................C...
```

18S-3 / 1200R (←)

```
(25mer) 5' TCTAAGGGCATCACAGACCTGTTAT 3'[ 1603-1579]
(16mer) 5' GGGCATCACAGACCTG 3'[ 1598-1583]
```

Comments: Reverse complements of 18S-2 / 1200F.

18S-5 / 1400F (→)

```
(25mer) 5' CCCTTTGTACACACCGCCCGTCGCT 3' [1779-1807]
(17mer) 5' TGYACACACCGCCCGTC 3'[ 1788-1804]
```

18S-4 / 1400R (←)

```
(19mer) 5' AGCGACGGGCGGTGTGTAC 3'    [ 1807-1789]
(15mer) 5' ACGGCGGTGTGTAC 3'    [ 1803-1789]
```

Comments: Truncated reverse complements of 18S-5 / 1400F.

Cestode-6 / WormB (←)

```
(20mer) 5' ACGGGAAACCTTGTTACGACT 3'    [ 1932-1913]
(21mer) 5' CTTGTTACGACTTTTTACTTCC 3'    [ 1924-1904]
```

Comments: 3' end primers designed to avoid misannealing of 18S-F in platyhelminth taxa.

18S-F (alias 18S-B) (←)

```
(30mer) 5' CCAGCTTGATCCTTCTGCAGGTTCACCTAC 3'
```

Comments: 'Universal' 3'-end primer. Mis-annealing in cestode taxa results in a ~400 bp PCR product when used together with 18S-E.

Chapter 26

Flatworm phylogeneticist: Between molecular hammer and morphological anvil

Boris I. Joffe and Elena E. Kornakova

Before the 1970s, the phylogenetics of the Platyhelminthes was based on light microscopic data. However, it was during the early 1970s, that the era of transmission electron microscopy began, and what may now be called 'the traditional scheme of flatworm phylogeny' (see Ehlers 1985a; Westheide and Rieger 1996) is to a great degree, based on ultrastructural data. The change from light to electron microscopy has not, however, caused changes in the methods of phylogenetic inference. Ultrastructural characters can be assessed using the same homology criteria that were the theoretical basis of phylogenetics from the very beginning of phylogenetic studies (Rieger and Tyler 1985). The development of effective methods for DNA sequencing caused a great increase in the number of studies using molecular data on phylogenetic questions. Although qualitative approaches to molecular phylogenetics do exist, the dominating methods of phylogenetic inference from molecular data are statistical in nature. Phylogenies that result from these studies mostly do not agree with traditional views in many important aspects, but no morphologist can now disregard molecular data when studying phylogeny. How one should deal with these controversies is an important question that will be discussed in this chapter.

One area where the answer to this question is to be sought is the method used for phylogenetic inference from molecular data. Several factors can cause errors in topologies of phylogenetic trees inferred from molecular data, some of them irrespective of the length of sequences analysed. These problems were recognized very early on, but 'have often been underestimated probably due to an over-optimistic attitude' (Philippe and Adoutte 1998).

Practical effects of factors which may cause errors in the topology of trees computed from molecular data depend upon the taxonomic range studied and the sequence used, and therefore should be discussed in each individual case. In this chapter, the matter will be discussed with regard to the controversy between morphological and 18S rDNA molecular data relating to the sister-group of the Neodermata (the parasitic Platyhelminthes), and relationships between the Proseriata and Tricladida.

Morphology: phylogenetic relationships of the Neodermata with 'turbellarians'

This question has been studied intensely using morphological methods, and was also the focus of several investigations using molecular techniques. It is one of the most impressive examples of the conflict between morphological and molecular data.

Monophyly of the parasitic Platyhelminthes (the Neodermata) has been demonstrated by morphological methods very convincingly. Several authors (Hendelberg 1962, 1969; L'Hardy 1988; Justine 1991a, 1995) noticed that spermiogenesis of the neodermatans has a polarity which is the reverse of that of free-living turbellarians (Figure 26.1) Two other synapomorphies support monophyly of the Neodermata: 1) the neodermis, that is, the syncytial epidermis with multiple connections between each insunk epidermal perikaryon and the surface layer; and 2) the presence of sensory receptors with electron-dense collars (see Ehlers 1985a; Rohde 1994b; also Tyler and Tyler 1997 for critical discussion of the epidermis). Fecampiidae, a peculiar family of parasitic turbellarians, was first tentatively suggested as possible sister-group of the Neodermata on the basis of the organization of the mature spermatozoa (Ehlers 1985a; Xylander 1989b). Watson and Rohde (1993a,b) found in the fecampiid *Kronborgia* the same polarity of spermiogenesis as in the Neodermata, and this pointed strongly to the Fecampiidae being the sister-group of the Neodermata. However, this opinion was soon rejected solely on the basis of a study of partial 18S rDNA sequences (Rohde *et al.* 1994). Watson (1997a) found in *Urastoma* a spermiogenesis that was even more similar to that of the Neodermata because of the presence of the so-called median process (see below), and tentatively raised the question of a possible sister-group relationship with the Neodermata. Monophyly of the turbellarians with Neodermata-like spermiogenesis and the Neodermata was vigorously supported in our study of another turbellarian with neodermatan-type spermiogenesis (Joffe and Kornakova 1998; Kornakova and Joffe 1999).

It was suggested (Kornakova and Joffe 1999) that all platyhelminths with neodermatan-type spermiogenesis should be united in the taxon Revertospermata (Figure 26.2) characterized by two synapomorphies: 1) distal orientation of the axonemes; and 2) the reverting migration by which the nucleus moves to a position distal to the basal bodies (Figure 26.1A,B). The plesiomorphic states are observed in turbellarian-type spermiogenesis in that first, the axonemes are directed proximally (which is especially clear when they are incorporated; L'Hardy 1988; Justine 1991a) and second, the initial relative positions of the nucleus and basal bodies (nucleus proximal to basal bodies) does not change during spermiogenesis (Figure 26.1C). Plesiomorphy of these features for the Neophora is implied by the occurrence of turbellarian-type spermiogenesis in the Polycladida. Such spermiogenesis is also well documented in all other phylogenetic branches of the Neophora with flagellated spermatozoa: Rhabdocoela, Proseriata, Tricladida (for reviews; see Watson and Rohde 1995b; Watson 1999b). The position of the nucleus proximal to the basal bodies during all of spermiogenesis is a feature that is also characteristic of the typical uniflagellate spermatozoa in which the axonemes have the primary distal orientation (e.g., in the Nemertodermatida; see Hendelberg 1977; Lundin and Hendelberg 1998).

Within the Revertospermata, Kornakova and Joffe (1999) distinguished two sister groups: Fecampiida and Mediofusata, the latter including the Neodermata and the Urastomidae + Genostomatidae (see below). The Revertospermata is supported by the following synapomorphy: the basal bodies remain at the proximal end of the spermatid when the shaft (i.e., the median cytoplasmic process) elongates. The plesiomorphic state is observed both in the Fecampiida and in

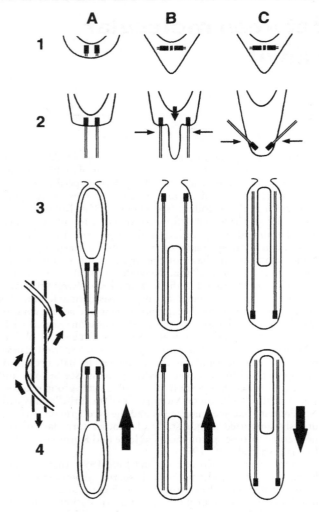

Figure 26.1 Relative positions of basal bodies, axonemes, and nucleus in spermiogenesis of *Notentera ivanovi* (variant with free flagella in early spermatids) (**A**), in the typical Neodermata and *Urastoma* (**B**), and in the Platyhelminthes with the turbellarian-type spermiogenesis (**C**; to simplify comparisons, the variant with incorporation of the axonemes is shown). 1, Early zone of differentiation with the basal bodies. 2, Later stage with developed axonemes. 3, Spermatid before detachment from the cytophore. 4, Mature spermatozoa (the figure on the left shows axonemes wound around the nucleus during the reverting migration in *N. ivanovi*). Arrows show movements of organelles which are not due to the growth of the spermatid; thin arrows indicate incorporation of free flagella; large arrows in the bottom row show the antero-posterior polarity of spermatozoa. (Modified from Kornakova and Joffe 1999.)

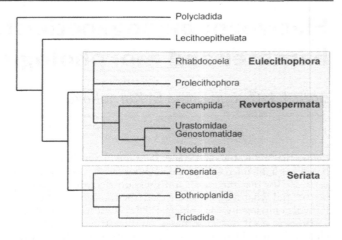

Figure 26.2 One of the possible morphological hypotheses about phylogeny of the Neoophora. This tree is based on traditional scheme of phylogenetic interrelationships of the neoophoran taxa modified by adding Eulecithophora (see Sopott-Ehlers 1997b) and Revertospermata.

those platyhelminths with the turbellarian-type spermiogenesis: the basal bodies move from the proximal pole as the shaft elongates. In some representatives of the Mediofusata the axonemes grow within the median process and the proximo-distal fusion does not take place (see a brief discussion in Justine 1998b; Kornakova and Joffe 1999), but the proximal position of the basal bodies is always retained.

Spermiogenesis of the Genostomatidae – *Genostoma* and *Ichthyophaga* – has not been studied in full, and this family is placed into the Revertospermata on the basis of clear morphological similarity to *Urastoma*. In particular, they share a

plicate pharynx and a common oro-genital pore at the posterior pole of the body. On the dorsal side of the body, *G. kozloffi* has a syncytial epidermis very similar to the typical neodermis of the Neodermata. Spermatozoa have a single incorporated axoneme, and receptors with electron-dense collars have not been observed (Hyra 1994; Tyler and Tyler 1997). *Ichthyophaga* was described and placed in the Genostomatidae by Syromjatnikova (1949); this worm also has an epidermis with insunk nuclei (Menizki 1963).

At present the hypothesis postulating monophyly of platyhelminths with neodermatan-type spermiogenesis, i.e., close relationships of the Neodermata with urastomid-fecampiid turbellarians appears the most well-based from the morphological point of view (see Littlewood *et al.* 1999a,b). Though mainly based on spermiogenesis, it also has some support from the absence of dense bodies in mature spermatozoa and organization of protonephridia (see Watson 1997a). Furthermore, all representatives of the fecampiid-urastomid clade always form a single cluster in trees computed from rDNA data (Littlewood *et al.* 1999a,b). If one assumes that they are not related to the Neodermata, one must assume that not only reverted polarity of spermiogenesis, but even such a specific feature as the median process has arisen in evolution twice. Independent evolutionary shift of the basal bodies to the opposite pole of the spermatid also does not seem likely because the position of the basal bodies is correlated with the polarity of ordered spatial arrangement of chromosomes found in the spermatozoa of platyhelminths (planarians) and some other animals (reviewed in Joffe *et al.* 1998a). Of course, non-spermiological characters remain necessary for further confirmation of the conclusions discussed above and for understanding the position of the Prolecithophora that has aflagellate spermatozoa.

Morphological data about more distant relations of the Neodermata are limited. Based on the ultrastructure of female gametes, Sopott-Ehlers (1997b) suggested that Neodermata, Rhabdocoela and Prolecithophora constitute the monophylum Eulecithophora (see also Falleni and Lucchesi 1992). Rohde *et al.* (1999) remarked that eyes with mitochondrial lenses have been noticed only in representatives of Neodermata, Rhabdocoela, and in *Urastoma*, though they showed reservations on the phylogenetic significance of this character.

Morphology: relationships between the Proseriata and Tricladida

The Proseriata has always been considered the sister-group for the Tricladida (Figure 26.2), and even a supposedly intermediate form, *Bothrioplana semperi*, is known. This point of view was finalized by Sopott-Ehlers (1985). Suggested synapomorphies of the joint taxon Seriata are: 1) serial arrangement of vitellaria and testes (both vitellaria and testes are follicular); and 2) tubiform plicate pharynx oriented horizontally and directed backward. Both features find some analogies in representatives of the Prolecithophora, due to which some authors considered the synapomorphies uniting the Seriata not sufficiently convincing. Trees computed from 18S rDNA data never support monophyly of the Seriata (Littlewood *et al.* 1999a,b). *B. semperi* clusters in such trees with the Proseriata, far away from the Tricladida (Baguñà *et al.* 2001, this volume).

Both planarians and proseriates have been intensely studied in recent years and a number of characters that distinguish these two taxa are known. By contrast, *B. semperi* is relatively rare and only a few morphological studies of this species have been carried out recently (Reisinger and Kelbetz 1964; Reisinger 1970; Reisinger *et al.* 1974; Kornakova and Joffe 1996). This allows comparison of *B. semperi* with proseriates and triclads across a number of selected characters. Most importantly in this context, this set of characters (based on preceding studies) had actually been determined before the monophyly of the Seriata was questioned on the basis of molecular data.

We have collected several specimens of *B. semperi* and studied several aspects of its organization, namely, the ultrastructure of cyrtocytes, position of muscle layers in the walls of the pharynx, and the structure of the GAIF-positive component of the nervous system. We then compared new and published data on the organization of *B. semperi* with data on Proseriata, Tricladida and other neoophoran orders (Table 26.1). Our study has not indicated new synapomorphies for the Seriata; rather, all we could learn about the organization of *Bothrioplana semperi* corresponds well with the hypothesis that it is an early offspring from a tricladid clade still retaining many characters of a proseriate-like ancestor.

Sources of artefacts in topology of optimal trees and their possible effects on positions of taxa in 18S rDNA platyhelminth trees

The main current method of molecular phylogenetics is computation of phylogenetic trees based on comparison of sequences of the same gene from different species using maximum parsimony (MP), maximum likelihood (ML), or minimum evolution (ME) algorithms. All these methods search for the tree(s) which infer minimal evolutionary changes, namely, with minimal number of nucleotide substitutions

Table 26.1 Phylogenetically informative characters in Proseriata, *Bothrioplana* and Tricladida. Apomorphic states are in capitals and bold; **; data are available for Monocelididae and Otoplanidae.

Character	Proseriata	Bothrioplana	Tricladida	Rhabdocoela	Prolecithophora
Synapomorphies of the Bothrioplanida + Tricladida					
1. Intestine bifurcates at the level of the pharynx which is directed backward and situated between the intestinal branches	no	**YES**	**YES**	not bifurcated, or relations with the pharynx are different	no
2. Collar receptors lost	no	**YES**	**YES**	no	no
3. Multiplied GAIF-positive cells in the lateral zones of the body	no**	**YES**	**YES**	no	? (no)
Synapomorphies of the Tricladida					
1. Cranial position of germaria	no	no	**YES**	no	no
2. Multiple serially arranged excretory pores	no	no	**YES**	no	no
3. Complex innervation of the pharynx including perikarya situated in the pharynx and several 'plexus-like' layers of nerve processes (plesiomorphic state: perikarya are mostly situated outside pharynx and send processes to a subdistal nerve ring)	no**	no	**YES**	pharynx is not of plicate type	? (no)
4. Embryonal pharynx	no	no	**YES**	no	no
Synapomorphies of the Proseriata					
1. Lamellated rhabdites are lacking	**YES**	no	no	no	? Yes
2. Ventro-medial serial GAIF-positive bipolar perikarya with transversal direction of the processes	**YES****	no	no	no	? (no)
Other characters					
Musculature of the internal wall of the pharynx (counting from the pharynx lumen)	longitudinal, circular	circular, longitudinal	circular, longitudinal	circular, longitudinal	circular, longitudinal (Combinata) or longitudinal, circular (Separata)
Cyrtocytes with perforated wall (supposedly plesiomorphic state: weir with two rows of ribs)	no	no	**YES**	no (anucleated cyrtocytes with one row of ribs)	**YES**

(MP), or with maximal probability of the necessary substitutions (ML), or with the minimal sum of distances between all observed and inferred stages of evolution (=sum of branch lengths of the tree, ME). The resulting trees may be collectively referred to as optimal trees. All these methods are based on certain assumptions about the evolution of DNA sequences, e.g., that sequences change due to independent substitution of single nucleotides with a relatively low rate and that this rate is equal in all species. If these assumptions are strongly violated, the topologies of optimal trees become a strongly misleading estimate of phylogeny (see Felsenstein 1978). It has been suggested many times that the discrepancy between the assumptions underlying the optimal algorithms and the real characteristics of sequence evolution is the main cause of the striking controversies between morphological and molecular data. We shall discuss those aspects of this problem which are more important in connection with platyhelminths.

Difference in rates of evolution (nucleotide change accumulation)

This is the best known source of errors in topology of trees computed by optimal algorithms first demonstrated by Felsenstein (1978). Species with higher rates ('fast clocks') tend to cluster together (though not necessarily all gather to a single cluster). Topology of the resulting tree shows further distortion because it is optimized with regard to the wrong positions of the faster clocks. The character of these changes depends upon the species set. Either fast clocks branch off earlier than they should, or they attract other branches which go off in a paraphyletic manner (one after another) from the 'stem' which ends with fast clocks (cf. Mooers *et al.* 1995). The tree shown in Figure 26.3 in general has the structure of the second type, while the neodermatan subtree shows the topology of the first type. With regard to unrooted trees with a single cluster of fast clocks, these two variants actually mean the same. In trees with a high number

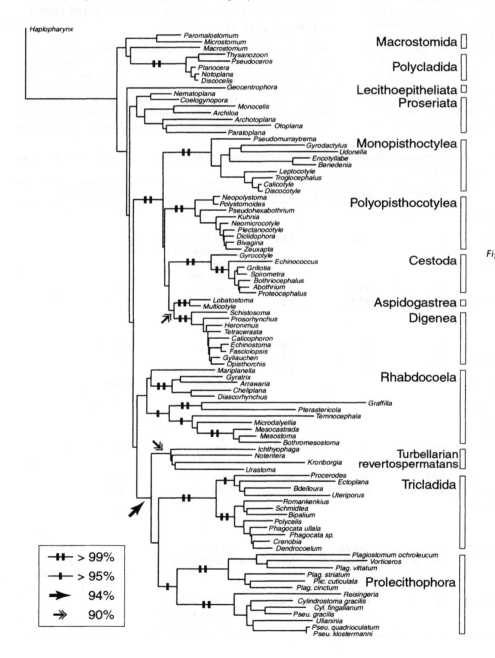

Figure 26.3 NJ tree of Rhabditophora computed from 18S rDNA sequences (distance: LogDet). Topology of the NJ tree is retained in the 50%-majority rule consensus bootstrap tree. Bootstrap values are shown for the major phylogenetic branches of the Neoophora which have high bootstrap support. High bootstrap support within these branches is marked only where necessary for comparison of the tree topology with the qualitative molecular data. The MP tree computed from the same data set (50%-majority rule consensus bootstrap tree) is similar to the NJ tree. The branches which have high bootstrap support in the NJ tree, also appear with high support in the MP tree. The differences between the two trees are limited to the regions with low bootstrap support: the branches comprising (i) Tricladida, Prolecithophora and turbellarian revertospermatans, (ii) Neodermata, (iii) Rhabdocoela and (iv) Proseriata (two branches) go off from the same level in MP tree (i.e., MP consensus tree is not resolved in this part); Lecithoepitheliata fail to cluster with other Neoophora.

of species, more than one fast clock cluster may be formed and such trees include several 'simple' subtrees joined to one another in various manners.

This effect, known as 'long-branch attraction', is often understood as major changes caused by a few fast-clock taxa. Actually, smaller differences may cause less dramatic (more local) changes which are even more difficult to notice. Provided that the average rate does not change along the sequence, the longer the sequences used, the more consistent becomes the effect of differences in the rate of evolution between taxa. Due to this, erroneous elements of tree topology may have high bootstrap and decay support. There are indications that rates of evolution vary greatly between genes from one genome, even when quite long sequences are considered (Philippe and Adoutte 1998). This means that to average the rate within a genome, long sequences from different genes are necessary, and it does not seem likely that rates will then become more similar between species. Serious differences in rates of rDNA evolution between platyhelminth taxa were apparent already even in the first studies of molecular phylogeny of flatworms (Riutort et al. 1992b), and their potential effect was noted (Joffe et al. 1995c).

We made a test data set including 100 18S rDNA sequences (see Appendix 26.1 for description of the data set and methods used to compute trees) and used it to compute phylogenetic trees for 94 rhabditophoran species included in this set. The topology of the trees computed using this data set (Figure 26.3) correspond to those published earlier (Littlewood et al. 1999a,b). A characteristic feature is the branch including turbellarians with neodermatan-type spermiogenesis, Tricladida and Prolecithophora. It is the only branch above the level of turbellarian orders and the Neodermata, which has a high bootstrap support in maximum parsimony and distance trees (see below). As a matter of fact, this branch appears in many trees discussed by different authors in this volume. To some degree (but not fully) such a clade may be supported by the organization of protonephridial flame bulbs (see Rohde 2001, this volume), but it does not agree with gonad evolution (see Rieger et al. 1991) and data on spermatozoa (see above). We will discuss here the technical aspect of this problem.

Higher taxa of Platyhelminthes show clearly different characteristic rates of evolution, and taxa with higher and lower rates are not distributed randomly over the tree (Figure 26.3). To tackle this problem, we estimated rates of evolution of neoophoran taxa in the test data set for the period starting from an early point of neoophoran evolution using the method (see Figure 26.4) suggested by Wilson et al. (1977). Within the Neoophora the rates vary by more than three-fold (Table 26.2), with differences between groups with higher and lower rates of evolution being statistically significant (Table 26.3)

Data on the rates of evolution confirmed correlation between the topology of the platyhelminth 18S rDNA tree and characteristic rates of evolution of taxa (Figure 26.5). Tricladida and Prolecithophora are the fastest clocks in this data set, and their association with a high probability might be due exclusively to long-branch attraction. Turbellarian revertospermatans also have a high rate, while the majority of the Neodermata and Proseriata have a low rate (Tables 26.2 and 26.3). These data provide serious reasons to conclude that the distant position of Neodermata and Proseriata from turbellarian revertospermatans and Tricladida, respectively, might depend upon differences in the rate of evolution. The fact that *Bothrioplana* clusters with the Proseriata (see Baguñà et al. 2001, this volume) might be related to a low rate of evolution of this species (0.038, cf. Table 26.2 for the Proseriata and Tricladida). It is noteworthy that Polyopisthocotylea have the same very low rate as Cestoda and

Table 26.2 Intragroup distances and rates of evolution of some platyhelminth taxa.

Group	Intragroup distances		Rate	
	Average	Maximum	Minimum	Maximum
Macrostomida	0.07961	0.10606	–	–
Polycladida	0.02348	0.03467	–	–
Proseriata	0.08502	0.13492	0.02737	0.09539
Tricladida	0.07201	0.12351	0.09252	0.12831
Prolecithophora	0.12664	0.20588	0.10304	0.15392
Separata	0.10248	0.14824	0.11264	0.15392
Combinata	0.07052	0.10901	0.10304	0.13858
Rhabdocoela	0.11523	0.19152	0.05354	0.15070
Kalyptorhynchia	0.05926	0.08015	0.05354	0.07093
FW-GP*	0.11876	0.19116	0.07423	0.15070
TRev**	0.10844	0.12631	0.07783	0.12054
Monopisthocotylea	0.07118	0.11123	0.07527	0.13638
Polyopisthocotylea	0.02441	0.03811	0.05384	0.06701
Cestoda	0.02248	0.03988	0.06713	0.08791
Digenea	0.03116	0.05423	0.04719	0.06507
Total***	–	–	0.02737	0.15392

*, clade formed by freshwater rhabdocoel groups, Graffillidae and Pterastericolidae.
**, clade formed by *Notentera, Kronborgia, Urastoma,* and *Ichthyophaga* (turbellarian revertospermatans).
***, Lecithoepitheliata are omitted because only one species was included in the species set; the rate is c. 0.05.

Table 26.3 Statistical significance of differences in the rates of sequence evolution between major taxa of neoophoran Platyhelminthes. Taxa are listed in an order corresponding to their characteristic rates of evolution of 18S DNA sequences (c.f. Table 26.3). FW-GP: the clade formed by freshwater rhabdocoel groups, Graffillidae, and Pterastericolidae. Numbers in parentheses indicate the number of species within each group. Statistical significance of differences was estimated using runs test; * and **, significant differences ($P > 0.95$ and $P > 0.99$, respectively).

N	Taxa	1	2	3	4	5	6	7	8	9	10
1	Proseriata (7)										
2	Kalyptorhynchia (4) (Rhabdocoela)	–									
3	Trematoda (11)	–	–								
4	Polyopisthocotylea (9)	–	–	–							
5	Cestoda * Gyrocotylida (7)	–	–	**	**						
6	Rhabdocoela (12)	–	–	–	–	–					
7	FW-GP (Rhabdocoela) (7)	*	**	**	**	–	–				
8	turbellarian Revertospermata (4)	–	*	**	**	–	–	–			
9	Monopisthocotylea (9)	**	**	**	**	–	–	–	–		
10	Tricladida (12)	**	**	**	**	**	**	**	–	–	
11	Prolecithophora (13)	**	**	**	**	**	–	–	–	–	–

Trematoda, while Monopisthocotylea demonstrate a notably higher rate (Tables 26.2 and 26.3; note also low intragroup variation in the Neodermata). This fact cannot be ignored in considerations of the paraphyletic status of the Monogenea (cf. Littlewood et al. 1999a,b).

The differences in rates of evolution suggest the performance of a simple experiment: to select from the data a small set of

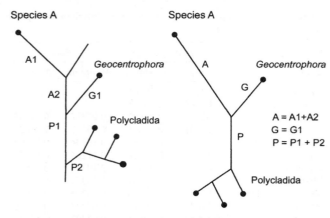

Figure 26.4 Scheme explaining comparison of rates of evolution based on the method suggested by Wilson *et al.* (1977). Rates may be compared based on lengths of branches to each species from the same point in the tree. Practically, this may be done by selecting two landmark species (or species groups) and constructing a series of trees which differ only in presence of one species (A in the figure). It is important that all species compared — but not the first landmark (Polycladida in this case) — belong to the sister-group of the taxon represented by the second landmark (*Geocentrophora*). In a tree wholly free from homoplasy, the distances will be as shown on the figure. In real trees, G1 and P1 + P2 vary depending on the measured species (A); therefore, it is important that the two landmarks are as closely related as possible.

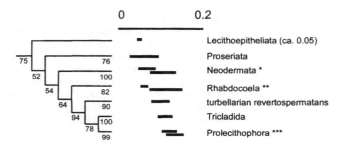

* Monopisthocotylea (right) and all others (left)

** Kalyptorhynchia (left) and Typhloplanidae, Dalyelliidae, Temnocephalida, Graffillidae, Pterastericolidae (right)

*** Combinata (left) and Separata (right)

Figure 26.5 Correlation of the topology of the tree and rates of evolution. The tree shown on the left is the same as in Figure 26.3, only branching within the taxa listed on the right is not shown. The range of variation of the rate of evolution for each taxon is shown in the middle (cf. Table 26.3). Bootstrap support is shown by numbers below branches of the tree. Note that support of morphologically recognized taxa is rather high; branches showing relations between taxa have low support, with exception for branches uniting the fastest clocks.

species with equal rates, e.g., c. 0.135. Unfortunately, trees based on data sets with a small number of species depend strongly upon individual species representing taxa (see next paragraph). Furthermore, to make such a set, one must combine the fastest proseriates with slowest triclads, etc., which greatly increases the danger of using aberrant taxa. Yet, even this way, the idea cannot be realized fully because all the sequenced Lecithoepitheliata are too slow, while the Prolecithophora are too fast. This is important because algorithms computing

optimal trees are sensitive to factors that can cause even the smallest differences in the length of the resulting tree. Nevertheless, it should be mentioned that this experiment failed to provide direct proof of a long-branch effect. In several experiments, the Neodermata tended to branch off earlier than the Proseriata, but the branch including representatives of the three fast-clock groups persisted (data not shown). Whether or not topology of these trees was still affected by distorting factors mentioned above, remains unknown.

Composition of species set

Phylogenies inferred by trees computed from molecular data greatly depend upon the species used. The presence or absence of even a single species, especially in smaller species sets, may dramatically change the topology of the tree computed from this set with one and the same method. Samples which differ in species representing higher taxa can give high bootstrap support for contradictory topologies, irrespective of the amount of information used (Lecointre *et al.* 1993; Philippe and Douzery 1994), an effect which occurs in trees based on complete 18S rDNA sequences. In platyhelminth trees where the Nemertodermatida are represented only by *Nemertinoides* this species invariably clusters with the Proseriata (Carranza *et al.* 1997; Littlewood *et al.* 1999a), including data sets in which only species with a low rate of evolution have been selected (Ruiz-Trillo *et al.* 1999). Another representative of the Nemertodermatida and a faster clock, *Meara*, in the absence of *Nemertinoides* tends to branch off early, mostly earlier than the Catenulida. If both *Nemertinoides* and *Meara* are present in the species set, they either form a separate branch situated near the catenulid branch (Littlewood *et al.* 1999b) or *Nemertinoides* clusters with Proseriata. Irrespective of the nemertodermatid problem, i.e., when *Nemertinoides* is not included in the test species set, the five proseriates used by Littlewood *et al.* (1999a) failed to collect in one branch, and seven proseriate species used in our test set also often failed to form one branch with 70% or even 50% bootstrap support (Littlewood *et al.* 1999b; see also our MP trees).

The strong effect of the composition of the species set on tree topology means that small, high-ranked taxa may cause a serious problem. In particular, extinction can cause unrecoverable damage to the molecular record of phylogeny or, at least, necessitate much longer sequences than that of 18S rDNA for reproducing the correct topology. Turbellarians with neodermatan-type spermiogenesis are represented by a few genera parasitizing hosts from different groups and showing quite different relations with the host. Bothrioplanida is represented by a single species. These are exactly the cases in which an insufficient set of species available for study from a relic group might cause a wrong position of this group in the tree. Indeed, if only *Kronborgia* is left in the data set, it takes a position of the sister-group to the Tricladida (rather than to Tricladida + Prolecithophora, as in Figure 26.3) with a relatively high bootstrap support of 72% (NJ tree, data not shown). The question remains, what positions would the turbellarian revertospermatans and the Bothrioplanida occupy in molecular trees if we had more representatives, and how would this affect the general topology of the tree?

Explosive radiation

Palaeontological data strongly suggest that the history of animals may include episodes of explosive radiation, i.e., relatively very quick formation of clades (Gould 1989). Trees computed from molecular data also often show very rapid divergence of major phylogenetic lineages in comparison to long branches supporting these lineages and/or terminal nodes. This is a matter of great practical importance, first because relationships between such lineages (i.e., between reliably recognized major subtaxa) are the most important ones for understanding the phylogeny of the studied taxon, and second because in case of rapid radiation followed by evolution with unequal rates between branches, the optimal algorithms tend to resolve tree topology only on the basis of differences in the rates of evolution (Philippe and Adoutte 1998; Stiller and Hall 1999). Third, even disregarding this effect, the length of sequence necessary for reliable resolution of the explosive part of the tree may be much bigger than might seem at first glance. Indeed, it is likely that the proportion of positions informative about the explosive part of the tree, is too small, such that the absolute number of informative positions remains insufficient even with considerably greater sequencing effort.

This is exactly the case with the 18S rDNA trees of the Platyhelminthes, which all show a very rapid divergence of the major clades of the Neoophora. In the tree inferred from 18S rDNA, the major clades of the Neoophora branch successively in the order corresponding to their characteristic rates of evolution (Figure 26.5). The fast-clock rhabdocoels and monopisthocotylean monogeneans are the only clades which occupy places corresponding to their systematic positions, despite differences in the rates of evolution from the other representatives of the taxa they belong to (the Rhabdocoela and the Neodermata, respectively). Bootstrap support also grows along the tree (Figure 26.5), which suggests that it is also affected by long-branch effects (the higher rates of evolution, the more consistently fast-clock clusters are reproduced). Trees inferred from 18S rDNA data also show rapid cladogenesis for the major taxa of the Neodermata, and the topology in this region is also problematic: in particular, the monophyly of the Monogenea (i.e., of the fast-clock Monopisthocotylea and the slow-clock Polyopisthocotylea) is never supported. Actually, the evolution at the morphological level reflects the evolution at the molecular level. Therefore, fast divergence (by and large) makes the same parts of the tree difficult for both molecular and morphological phylogenetics. The recently published platyhelminth trees, as well as the protist trees, distinguish most of the generally accepted monophyletic taxa (e.g., Prolecithophora with resolution of the Separata and Combinata, Tricladida, Rhabdocoela, Neodermata, Cestoda, Trematoda, Monopisthocotylea and Polyopisthocotylea, etc.), but fail to establish reliable interrelationships among them.

The gene studied

Only comparisons of 18S and 28S rDNA data may be carried out for flatworms, and only to a very limited extent (Littlewood *et al.* 1998a,b). Moreover, evolution of these sequences clearly cannot be considered as independent. Data available for some other groups are more diverse, and trees based on different genes are mostly rather different. According to Philippe and Adoutte (1998), these differences arise because evolution of most of the genes is independent and with regard to different genes, different systematic groups may represent 'fast clock' or 'slow clock' taxa. Only a study of several genes has allowed elucidation of the hypotheses which correspond reasonably well to data on a number of genes (Embley and Hirt 1998; Hirt *et al.* 1999), and these were hypotheses about certain branches rather than the tree topology as a whole. Independent evolution of genes is a universal phenomenon which does

not need to be proved for each particular taxon. Thus, the above discussion may be extrapolated to flatworms.

Conclusions

The discussion above leads to the following conclusions:

1. Lineage-dependent differences in the rate of evolution are one of the major factors determining tree topologies.
2. Because of this, the composition of the species-set strongly affects the tree topology (which is especially important in connection with groups with a small number of extant representatives).
3. In the case of explosive cladogenesis, differences in rates of evolution between clades may become the main factor determining the tree topology, and such a situation seems particularly probable with regard to some most important phylogenetic questions.
4. Trees based on different sequences show major incongruences, most probably because genes evolve with different rates in different groups. Accordingly, phylogenies inferred from optimal trees based on a single gene may be wrong.

The great practical importance of these factors may be best illustrated by the example of some protists (Microsporidia, *Giardia* and *Trichomonas*). Their basal phylogenetic position in 18S rDNA trees seemed to be supported by the amitochondrial condition in these taxa, and therefore appeared to be well proven. However, the idyllic concordance was ruined: secondary loss of mitochondria was convincingly inferred for these taxa by the presence of mitochondrial genes in the nuclear genome; the position of these taxa in 18S rDNA trees was strongly questioned by high rates of sequence evolution in all of them; evidence from other genes suggested that these taxa evolved independently in two different branches of eukaryotes and relatively late (for review, see Embley and Hirt 1998; Hirt *et al.* 1999; Stiller and Hall 1999).

Philippe and Adoutte (1998) resume their discussion of 'the pitfalls of molecular systematics' with the following conclusion: 'As described above, traditional molecular phylogenetic methods would require an unrealistic number of nucleotides to resolve this radiation [of protists] confidently. We think that *in fact these refinements will emerge from a few well chosen anatomical or biochemical similarities which, when analysed at the molecular level, will turn out to be true homologies'*. In other words, the following strategy is suggested: at the first step, a convincing putative synapomorphy for some phylogenetic branch should be identified based on some morphological, or physiological, or any other phenotypic character; at the second step, the significance of this synapomorphy should be finally confirmed by homology at the level of genes determining this character (as, for example, the homology of ribosomes in all living organisms is confirmed by the fundamental similarity of the nucleotide sequences in genes coding ribosomal RNAs).

This very important conclusion actually declares that the most promising approach in molecular phylogenetics is a combination of molecular data with the classical method of analysis based on nested synapomorphies (i.e., specific homologies). Molecular data open a rich source of homologies (between DNA sequences) which can be proved with much higher reliability than is possible using any non-molecular methods. Logical methods of analysis (Boolean for those who are keen to use mathematical methods) are free from the shortcomings that cause so many problems with optimal trees. In particular, the reliabililty of qualitative methods is determined by the density of systematic sampling (provided that homologies used are themselves reliable). Adding new data does not affect the results obtained earlier, whilst in the case of optimal trees adding new data theoretically (and often, practically) affects topology of the whole tree.

The discussion above shows that the problems arising in the studies of the phylogeny of protists and platyhelminths are very similar. Nevertheless, the conclusions made based on the analysis of the phylogeny of protists cannot be mechanically extrapolated to the phylogeny of the Platyhelminthes: these problems differ in time and diversity ranges under analysis. Yet, the main problem persists: even high bootstrap support does not guarantee that a given element of tree topology is correct and, vice versa, no indications of long-branch attraction can prove that a given element of tree topology is false. Therefore, qualitative markers (molecular synapomorphies) remain very important.

Qualitative markers in molecular phylogeny of the Platyhelminthes

Hitherto, only a few publications dealing with Platyhelminthes have paid special attention to qualitative molecular data. Telomere repeat, which had already been shown as a very constant marker of several phylogenetic lineages, has been studied in flatworms and some other lower metazoans (Joffe et al. 1996a, review and data on flatworms). A satellite repeat specific for some species from the Dugesia gonocephala complex has been found by Batistoni et al. (1998). The presence of two types of 18S rDNA genes in Dugesiidae and Terricola, but not in other Tricladids, has been proved (Carranza et al. 1998a), and a comparative study of the composition of the Hox gene family has been started (Balavoine 1997, 1998).

Most of the molecular data on flatworms now available is 18S rDNA sequences. Such data contain qualitative information of two types: i) related to secondary structure (cf. Hori and Osawa 1986); and ii) based on insertions and deletions. The latter is easier to deal with (albeit less promising) and will be used below. Studies in this direction were greatly encouraged by the existence of large insertions in segments E23 and 43 in 18S rDNA in all representatives of the Neodermata (van de Peer et al. 1998; Littlewood et al. 1999a).

It is well known that insertions and deletions in rDNA are gradually compensated to normalize the secondary structure of the molecule (van de Peer et al. 1998 and references therein). We have designed a computer program that assists in the search for putative compensated insertions and deletions (indels). This program moves a window along each sequence in the data set and compares the sequence within the window with the consensus sequence. If more than a threshold number of nucleotides are different from the consensus, the nucleotide corresponding to the current position of the window is marked. Groups of marked nucleotides may then be easily found in the alignment by eye. We used the window width of seven nucleotides and the threshold was three (e.g., windows with four or more nucleotides different from the consensus were marked). Data sets used for this purpose were limited to a single taxon (as Proseriata or Monopisthocotylea), but then the results were compared with our alignment of all rhabditophorans. We interpret as putative compensated indels sequences satisfying two conditions: 1) the supposed indel differs from the consensus by at least 50% of nucleotides for shorter indels (up to 10 nucleotides) or about 50% for longer indels; and 2) the region is not variable, i.e., other sequences show no or a few single nucleotide changes (at least, within the

taxon in question). Quite a proportion of putative indels found also included explicit one to two nucleotide-long insertions or deletions (Figure 26.6). This confirms that they are partially compensated insertions or deletions, rather than occasional clusters of single nucleotide changes. A good example is the putative indel which starts in Maricola from position 647/8 and affects 12–13 nucleotides (Figure 26.6A). Its nature is confirmed by: i) an explicit one-nucleotide insertion in Ectoplana; and ii) by the palindromic sequence 'gctcg'. It starts from position 648 in maricolans, but from position 653 in other triclads; in other Platyhelminthes the same or similar, mostly palindromic sequences also start from position 653. Explicit and putative compensated indels found by us are listed in Tables 26.4 and 26.5.

We showed in Table 26.4 only the more convincing indels, so they are only 'the tip of the iceberg'. Some species contain several indels, and always have the highest or nearly the highest rate of evolution in the corresponding taxon (data not shown). Importantly, this high rate estimate is not determined by partially compensated indels themselves: when the regions containing major putative compensated indels in Pseudograffilla and Pterastericola were taken out of the alignment, the rate estimates remained high (see Stiller and Hall 1999 for similar observations on 18S rDNA of protists). As mentioned above, 18S rDNA of neodermatans contains homologous long insertions, especially large in cestodes. Surprisingly, the same is observed in the region of 28S rDNA studied by Litvaitis and Rohde (1999)! The Neodermata is the only major taxon of the Neoophora which is based on a relatively long branch (Figure 26.3): it is exactly the branch to which insertions common for the Neodermata belong. Therefore, the possible role of indels in the evolution of sequences deserves more attention, especially in connection to fast-clock groups. If compensated indels are included in the data used for computation of optimal trees, they obviously increase estimated rate of nucleotide change, and make a disproportionately large contribution to monophyly of the clade which has this indel, but only increase noise with regard to the position of the group in optimal trees which are based on the assumption of random single nucleotide changes.

Indels shared by more than one species may have phylogenetic interest. Longer indels are reliable markers; short ones (one to two nucleotides) may be more easily compensated or appear independently and are therefore less reliable. Four 'co-localized' partially compensated insertions (one of them includes a five to six nucleotide-long explicit insertion; Figure 26.6B,C) are present in Graffilla and Pterastericola, and strongly testify to the monophyly of these two groups (Table 26.5). A characteristic indel also supports the monophyly of the studied maricolan planarians (Table 26.4). A consistent two-nucleotide insertion is shared by all Neodermata, as well as two large insertions that were mentioned above (Table 26.4). A less convincing indel (because of its location in a relatively variable region) is shared by monopisthocotylean monogeneans. It is also noteworthy that there are short indels supporting the monophyly of the Dalyelliidae (and Temnocephalida) + Typhloplanidae (in addition to the topology of trees computed from 18S rDNA data). Both groups are predominantly from freshwater or brackish water and both have a single ovary. A two to three nucleotide-long deletion supports monophyly of the above-mentioned clade with Pterastericolidae + Graffillidae (Table 26.4). Unfortunately, the number of rhabdocoels with sequenced 18S rDNA outside these putative clades (four kalyptorhynchs and the marine representative of the Typhloplanoida, Mariplanella) remains too small.

The last and very important question we have to discuss is: how much do the qualitative markers agree with the topologies of optimal trees and with morphological data? At present, no

A

```
                     640       650       660       670
cMacrostomida      gcgcaaattacccact-cccg-gcacg-g-ggaggtagtg
cPolycladida       gcgcaaattacccact-cccg-gcacg-g-ggaggtagtg
cProseriata        gcgcaaattacccact-cccg-gcacg-g-ggaggtagtg
Procerodes         gcgcaacatacccatt-GcTC-gAtGA-g-cgaggttgtg
Bdelloura          gcgtaacatgcccact-GcTC-gAtGA-g-cgaggctgtg
Uteriporus         gcgtaacatacccaat-GcTC-gAAGA-g-cgaggttgtg
Ectoplana          gcctaacatgcccaat-TGcTT-gAAAA-g-cgaggctgtg
Crenobia           gcgtaaattacccaat-accg-gctcg-g-tgaggtagtg
Dendrocoelum       gcgcaaattacccaat-accg-gctcg-g-tgaggtagtg
Polycelis          gcgcaaattacccaat-aaca-gctcg-t-tgaggtagtg
Phagocata sp.      Gcgcaaattacccaat-accg-g?tcg-g-tgaggtagtg
Phag. ullala       gcgtaaattacccaat-accg-gctcg-g-tgaggtagtg
Romankenkius       gcgtaaattacccaat-accg-gctcg-g-tgaggtagtg
Schmidtea type 1   gcgtaaattacccaat-accg-gttcg-g-tgaggtagtg
Bipalium type 1    gcgcaaattacccaat-accg-gctcg-g-tgaggtagtg
cCombinata         gcgtaaattacccact-ccgg-cacgg-g---gaggtagtg
cSeparata          gcgcaaattacccact-ctca-gtacg-a-ggaggtagtg
cRhabdocoela       gcgcaaattacccact-ctca-gcgag-a-ggaggtagtg
cMonopisthocot.    Gcgcaaattacccact-ctca-gaatg-a-ggaggtagta
cPolyopisthocot.   Gcgcaaattacccact-cccg-gcacg-g-ggaggtagtg
cCestoda           gcgcaaattacccact-ccca-gcacg-g-ggaggtggtg
cDigenea           acgcaaattacccaat-cccg-gcacg-g-ggaggtagtg
```

B

```
                     640       650       660       670       680
cMacrostomida      aaattacccact-cccg-gcacg-g-ggaggtagtg-acgaaaaat-aac
cPolycladida       aaattacccact-cccg-gcacg-g-ggaggtagtg-acgaaaaat-aac
cProseriata        aaattacccact-cccg-gcacg-g-ggaggtagtg-acgaaaaat-aac
cTricladida        aaattacccaat-accg-gctcg-g-tgaggtagtg-acaataaat-aac
cCombinata         aaattacccact-ccgg-cacgg-g---gaggtagtg-acaaaaaat-aac
cSeparata          aaattacccact-ctca-gtacg-a-ggaggtagtg-acgaaaaat-aac
Mariplanella       aaattacccact-tcca-gcgcg-g-agaggtagtg-acgaaaaat-aac
Gyratrix           aaattacccact-ctca-gctcg-a-ggaggtagtg-acgaaaaat-aac
Cheliplana         aaattacccact-ctcg-gctag-a-ggaggtagtg-acgaaaaat-aac
Diascorhynchus     aaattacccact-ctca-gctcg-a-ggaggtagtg-acgaaaaat-aac
Arrawaria          aaattacccact-ctca-gctag-a-ggaggtagtg-acgaaaaat-aac
Mesostoma          aaattacccact-ctca-gcgag-a-ggaggtagtg-acgaaaaat-aac
Bothromesostoma    aaattacccact-ctca-gagag-a-ggaggtagtg-acgaaaaat-atc
Mesocastrada       aaattacccact-ctca-gcgag-a-ggaggtagtg-acgaaaaat-aac
Microdalyellia     aaattacccact-ctca-gttcg-a-ggaggtagtg-acgaaaaat-aac
Temnocephala       aaattgcccact-ctca-gtgag-a-ggaggcagtg-acgataaat-aac
Graffilla          aacttaTTcactGct-a-gtgTT-aGTgaAAtagCCAacAGaGaat-acc
Pterastericola     aacttacccactGct-a-CtgTC-aGCgaggtagTgAGcgaTaaat-aac
cMonopisthocot.    Aaattacccact-ctca-gaatg-a-ggaggtagta-aagacaaat-atc
cPolyopisthocot.   Aaattacccact-cccg-gcacg-g-ggaggtagtg-acgaaaaat-aac
cCestoda           aaattacccact-ccca-gcacg-g-ggaggtggtg-acgaaaaat-acc
cDigenea           aaattacccaat-cccg-gcacg-g-ggaggtagtg-acgaaaaat-acg
```

C

```
                   1450......1460......1470......1480
cMacrostomida      ggtttt-cggaa-----aaagaggtaatgattaa
cPolycladida       ggttt-t-cggaa------catgaagtaatgattaa
cProseriata        ggttt-t-cggaa------ctgaagtaatgattaa
cTricladida        ggttt-t-cggaa------ctgaagtaatgattaa
cCombinata         ggttt-t-agaga------ccgaagtaatgattaa
cSeparataa         ggttt-t-aggaa------cagaagtaatgattaa
Mariplanella       ggttt-t-cggta------ccgaagtaatgattaa
Gyratrix           ggtttt-cggaa------tcgaagtaatgattaa
Cheliplana         ggtttt-cggat------ccgaagtaatgattaa
Diascorhynchus     ggtttt-cggtt------acgaagtaatgattaa
Arrawaria          ggtttt-cggaa------tagaagtaatgattaa
Mesostoma          ggtttt-cggaa------tcgaagtaatgattaa
Bothromesostoma    ggtttt-cggac------ccgaagtaatgattaa
Mesocastrada       ggtttt-aggac------ccgaagtaatgattaa
Microdalyellia     ggtttt-cggtt------ccgaagtaatgattaa
Temnocephala       ggtttt-cggta------tcaaagtaatgattaa
Graffilla          gatttt-cTAaGCATTATGTAgagtaatgattaa
Pterastericola     ggtttt-cAgtaAAACA-GcAgagcaatgattaa
cMonopisthocot.    Ggtttgatagac------ccgaagtaatgataaa
cPolyopisthocot.   Ggtttt-cggat------ctgaagtaatggttaa
cCestoda           ggtttt-cggat------ccgaagtaatgatcaa
cDigenea           ggtttt-cggat------ccgaagtaatggttaa
```

Figure 26.6 Three examples of putative compensated indels: in maricolan triclads (**A**) and in *Graffilla* and *Pterastericola* (**B, C**). In the taxon of interest (triclads and rhabdocoels, correspondingly) all species are shown, other taxa are represented by majority consensuses. Putative indels are shadowed. Nucleotides different from the group consensus are shown in capitals. Underlining shows that the palindromic sequence gctcg is shifted in maricolan species from the position it occupies in other triclads; in other platyhelminths this or a similar sequence (e.g., gcacg) has the same position.

Table 26.4 Explicit and putative compensated insertions and deletions in 18S rDNA of Platyhelminthes.

	Start position	End position	Number of positions**	Modified	Inserted /deleted (+/−)	Total modified	Not modified	Comment
Polycladida								
*all	1334	1334	I	−	+1	I	−	C
Proseriata								
*Otoplanidae	434	436	3	−	+3	3	−	RCA
Tricladida								
*all Tricladida	480	480	I	+1				C
*all Tricladida	c. 696	c. 714	c. 15	3 to 5	−3	6 to 8	9 or less	
*Maricola, all	648	457	9–10	6 to 8	I or 0	6 to 8	2–3	Palindromic GCTCG starts in the pos. 648 in Maricola but 653 in other triclads, the same or homologous mostly palindromic sequence in other flatworms also from 653; explicit one-nucleotide insertion in *Ectoplana* in pos. 647
Uteriporus	299	312	I	6	+2	8	3	probably shared by *Procerodes* and *Bdelloura* which share insertion in pos. 311: R, and have deletions in pos. 305–306
Uteriporus	696	717	14	4–5	−4, +3	10–11	3–4	overlaps an indel shared by all triclads
Uteriporus	1950	1968	18	9	+1	0	8	
Phagocata sp.	451	474	19	4	−3, +3	10	9	
Combinata								
*all Combinata	648	648	I	−	−I	I	−	
*all Combinata	2403	2403	I		−I	I		
*all Combinata	2409	2409	I		+I	I	−	Deletion in pos. 2403 and insertion in pos. 2409 (A) seemingly compensate one other
Reisingeria	2337	2348	12	7	−	7	5	
Reisingeria	2405	2413	9	5	+1	6	3	
Separata								
*All Separata	718	718	I	−	−I	I	−	
*All Separata	722	722	I	−	+I	I	−	Deletion in pos. 718 and insertion in pos. 722 (T) seemingly compensate one other
Rhabdocoela								
*Typhloplanidae,	1322	1322	I	−	+I	I	−	R
Dalyelliidae, Temnocephalida *?Typhloplanidae,	1688	1690	3	−	+3	3	−	Predominantly TTT; the difference from other studied rhabdocoels is clear, but in Platyhelminthes in general, the region is variable.
*Typhloplanidae, Dalyelliidae, Temnocephalida, Pterastericolidae, Graffillidae	1744	1746	3(2)	−	−3(2)	3–2	0(1)	Two-nucleotide deletion in *Microdalyellia* and *Temnocephala*
Neodermata								
* all	c. 900–940	c. 1290	up to c. +350	−	all	all	−	A long insertion in the E23 region. In the region 868–1339 flanked by invariant blocks, the actual number of nucleotides is 100–117 in the turbellarians and 196–442 in the neodematans. The size of the insertion is especially big and variable in the Cestoda, shorter and relatively constant in the Digenea.

Continued

Table 26.4 Continued.

	Start position	End position	Number of positions**	Modified	Inserted /deleted (+/−)	Total modified	Not modified	Comment
Neodermata								
*all	c. 2030	c. 2290	up to c. +250	−	all	all	−	A long insertion in the helix 43 region. In the region 2010–2211 flanked by invariant blocks, the actual number of nucleotides is 24–45 in the turbellarians and 70–198 in the neodermatans. The size of the insertion is especially big and variable in the Cestoda, more short and relatively constant in the Monopisthocotylea
*all	446	447	2	−	+2	2	−	with a few exceptions, GT
Digenea								
*Schistosoma	2237	2245	7	5	+2	7	−	All three species mentioned by Littlewood et al. (1999a)
Monopisthocotylea								
*all	302	303	2	−	+2	2	−	TY
*all	311	311	1	−	−1			? Compensates insertion in the pos. 302
*all	2252	2256	5	−	+5	5	−	Predominantly AACCA; possibly, a part of an initially longer insertion, starting in the Capsalidae from the pos. 2245
*Capsalidae	2245	2250	6	2	+2	4	2	flanks the insertion starting from the pos. 2252 in all Monopisthocotylea
Polyopisthocotylea								
*Polystomatidae	504	504	1	−	+1	1	−	In both studied species, T
Kuhnia	145	156	10	5	+2	7	3	
Kuhnia	421–422	431	9	1	+5–6, −1	7	2	
Pseudohexabothrium	1380	1387	8	3	+4	7	1	Possibly, duplication of . . . AGCA starting from 1373 in the vast majority of Platyhelminthes
Cestoda								
*Eucestoda	1390	1390	1	−	+1	1	−	T, not in Gyrocotyle!
Echinococcus	239	267	29	4	+25	29	−	

*, indels of potential taxonomic value.
**, positions included in the alignment because of insertions in single or a few species from other groups were excluded from this count.

Table 26.5 Explicit and putative compensated indels in 18S rDNA of Graffilla and Pterastericola.

	Start position	End position	Number of positions in other rhabdocoel turbellarians	Modified	Inserted/deleted	Total modified	Not modified
Graffilla	78	92	10	3	−3	6	4
(Pterastericola	78	92	10	5		5	5)
Pterastericola	78	127	39	17	17	22	
Graffilla	641	677	31	12	+3, −1	16	20
Pterastericola	647	676	25	7	+3, −1	11	17
Graffilla	1429	1434	6	4	4	2	
(Graffilla	1429	1437	9	4	−1	5	4)
Pterastericola	1429	1434	6	4	4	2	
Graffilla	1455	1466	6	5	+6	11	1
Pterastericola	1455	1466	6	3	+5	8	3

Table 26.6 List of species used for the data set; * (numbers in parenthesis show the number of species from the corresponding taxon included in the data set).

Species name and systematic position*	Accession	Species name and systematic position*	Accession
Acoela (1)		*Gyratrix hermaphroditus*	AJ012510
Paratomella rubra	AF102892	*Cheliplana* cf. *Orthocirra*	AJ012507
Nemertodermatida (2)		*Diascorhynchus rubrus*	AJ012508
Meara sp.	AF051328	*Arrawaria* sp.	AJ243677
Nemertinoides elongatus	U70084	**'Typhloplanoida'** (4)	
Catenulida (3)		*Mariplanella frisia*	AJ012514
Stenostomum leucops aquariorum	AJ012519	*Mesostoma lingua*	AJ243682
Stenostomum leucops	U95947	*Bothromesostoma* sp.	D85098
Suomina sp.	AJ012532	*Mesocastrada foremani*	U70082
Haplopharyngida (1)		**'Dalyellioida'** (3)	
Haplopharynx rostratus	AJ012511	*Microdalyellia rossi*	AJ912515
Macrostomida (3)		*Graffilla buccinicola*	AJ012521
Macrostomum tuba	D85092	*Pterastericola australis*	AJ012518
Paramalostomum fuscum	AJ012531	**Temnocephalida** (1)	
Microstomum lineare	D85091	*Temnocephala* sp.	AJ012520
Polycladida (5)		**'Turbellarian revertospermatans'** (4)	
Notoplana australis	AJ228786	*Notentera ivanovi*	AJ287546
Planocera multitentaculata	D83383	*Kronborgia isopodicola*	AJ012513
Discocoelis tigrina	U70079	*Urastoma cyprinae*	U70086
Thysanozoon brocchii	D85096	*Ichthyophaga* sp.	AJ012517
Pseudoceros tritriatus	AJ228794	**NEODERMATA** (36)	
Proseriata (7)		**Monogenea** (18)	
Monocelis lineata	U45961	**Monopisthocotylea** (9)	
Archiloa rivularis	U70077	*Gyrodactylus salaris*	Z26942
Nematoplana coeloginoporoides	AJ012516	*Pseudomurraytrema* sp.	AJ228793
Coelogynopora gynocotyle	AJ243679	*Troglocephalus rhynobatidis*	AJ228795
Archotoplana holotricha	AJ243676	*Calycotyle affinis*	AJ228777
Otoplana sp.	D85090	*Dictyocotyle coeliacea*	AJ228778
Parotoplana renatae	AJ12517	*Leptocotyle minor*	AJ228784
Tricladida (12)		*Encotyllabe hironemi*	AJ228780
Procerodes littoralis	Z99950	*Benedenia* sp.	AJ228774
Bdelloura candida	Z99947	*Udonella caligorum*	AJ228796
Uteriporus sp.	AF013148	**Polyopisthocotylea** (9)	
Ectoplana limuli	D85088	*Neopolystoma spratti*	AJ228788
Crenobia alpina	M58345	*Polystomoides malayi*	AJ228792
Dendrocoelum lacterum	M58346	*Pseudohexabothrium taeniurae*	AJ228791
Polycelis nigra	AF013151	*Diclidophora denticulata*	AJ228779
Phagocata ullala	AF013149	*Kuhnia scombri*	AJ228783
Phagocata	AF013150	*Plectanocotyle gurnardi*	AJ228790
Romankenkius lidinosus	Z99951	*Neomicrocotyle pacifica*	AJ228787
Schmidtea polychroa, Type 1 gene	AF013152	*Bivagina pagrosomi*	AJ228775
Bipalium kewense, Type 1 gene	AF033039	*Zeuxapta seriolae*	AJ228797
Lecithoepitheliata (1)		**Gyrocotylida** (1)	
Geocentrophora wagini (1)	AJ012509	*Gyrocotyle urna*	AJ228782
Prolecithophora (13)		**Cestoda** (6)	
Separata (6)		*Grillotia erinaceus*	AJ228781
Vorticeros ijimai	D85094	*Abothrium gadi*	AJ228773
Plicastoma cuticulata	AF065422	*Bothriocephalus scorpii*	AJ228776
Plagiostomum cinctum	AF065418	*Echinococcus granulosus*	U27015
Plagiostomum vittatum	AF065421	*Proteocephalus exiguus*	X99976
Plagiostomum striatum	AF065420	*Spirometra erinacei*	D64072
Plagiostomum ochroleucum	AF065419	**Trematoda** (11)	
Combinata (7)		*Echinostoma caproni*	L06567
Cylindrostoma gracilis	AF065416	*Fasciolopsis buski*	L06668
Cylindrostoma fingalianum	AF065415	*Heronimus mollis*	L14486
Ulianinia mollissima	AF065427	*Calicophoron calicophorum*	L06566
Reisingeria hexaoculata	AF065426	*Gyliauchen* sp.	L06669
Pseudostomum quadrioculatum	AF065425	*Tetracerasta blepta*	L06670
Pseudostomum klostermanni	AF065424	*Opisthorchis viverrini*	X55357
Pseudostomum gracilis	AF065423	*Schistosoma haematobium*	Z46521
Rhabdocoela (11)		*Prosorhynchoides gracilescens*	AJ228789
Kalyptorhynchia (4)		*Lobatostoma manteri*	L16911
		Multicotyle purvisi	AJ228785

meaningful answer to this question can be suggested. The number of known qualitative markers is small. A substantial proportion of evolutionary changes (of any kind) belong to long branches of trees which support taxa resolved well by both morphological methods and optimal trees. Not surprisingly, the few known molecular synapomorphies support phylogenetic branches that have a high bootstrap support in computed trees, but also agree with reliable morphological conclusions (or at least, do not contradict them). It is only in the long run that one may find out that some branches of optimal trees remain unsupported by molecular synapomorphies and/or that there are molecular synapomorphies refuting these branches of optimal trees.

Concluding remarks

Molecular data have already brought new results and interesting new ideas into flatworm phylogenetics. In some cases, tree topology is soundly supported by qualitative molecular data and morphological evidence either corresponds with them or at least does not contradict them strongly. Such questions may be considered finally solved. We would classify to this category: 1) monophyly of Rhabditophora supported by modified codon usage in mitochondrial genome (Telford 2001, this volume); 2) monophyly of the Dugesiidae + Terricola confirmed by the presence of a second type of rDNA gene (Carranza et al. 1998a); 3) monophyly of the Neodermata supported by two long and one short insertions; 4) the position of Udonella within the Monopisthocotylea (Monogenea) supported by partially independent quantitative 18S and 28S data (Littlewood et al. 1998a) and an indel; and 5) monophyly of the Graffillidae + Pterastericolidae supported by four co-localized, partially compensated indels.

In some cases qualitative support does exist (albeit insufficient), whilst morphological data at least do not strongly contradict the molecular data. The chances of finding morphological and molecular evidence for further support of such clades seem good, and we would exemplify this with Rhabdocoela.

Finally, in some cases phylogenetic conclusions are made based on details of topology of optimal trees that may have high support from bootstrap or quartet likelihood analysis, but are not supported by qualitative evidence and strongly contradict sound morphological data. That is really the situation between the hammer and anvil for a phylogeneticist, especially if morphological data also are not wholly unanimous. We hope that the evidence discussed above convincingly shows that results based solely on the topology of optimal trees are hardly sufficient for ruling out well-based morphological conclusions. There is another important reason to think so. In the macrophylogenetics of the Platyhelminthes, several problems exist where such molecular data disagree with morphological theories: non-monophyly of the Acoelomorpha (see also Tyler et al. 2001, this volume), suggested monophyly of the Polycladida and Macrostomida, sister-group of the Neodermata, monophyly of the Seriata, and paraphyly of the Monogenea. In each of these cases the morphological data may be misleading, though it appears barely possible that they are misleading in all cases at once.

ACKNOWLEDGEMENTS

We are grateful to the organizers of the symposium, who made it possible for us to take part in the Symposium Inter-

relationships of the Platyhelminthes, and to Dr T. Littlewood, who supplied us with the sequences used in this study (while corresponding publications were still in press) and also helped us to solve many other problems. Dr M. Riutort kindly sent us the sequence of B. semperi. We thank Dr P. Olson, Dr M. Telford and Prof. S. Tyler for their constructive criticism which helped us to present our results more convincingly.

Appendix 26.1

Molecular data set

The 100 species used in this data set, species name abbreviations, and GenBank accession numbers are listed in Table 26.6. We started from the alignment used by Littlewood et al. (1999a); new species were aligned by eye, and alignment was changed where necessary. Final alignment has been deposited with EMBL under accession ds42836. Alignment comprised 1469 unambiguosly aligned positions. All positions occupied by insertions in single species (that is, where majority consensus sequence had '–') were deleted, so that 1324 positions were used for computation of trees and calculation of rates.

Trees

Trees were computed using PAUP 4.0b1 (Swofford 1998). Algorithms used were the neighbour-joining method and maximum parsimony. Distances for the NJ tree were computed using the LogDet model option. ML and HKY models were also tried, but the resulting trees showed no difference worth mentioning from that in Figure 26.3. Other parameters had default values. Data were bootstrap resampled (1000 replicates). The NJ method is a quick analogue of the ME algorithm, actually no less effective than ME itself (Nei et al. 1998). In our case it indeed produced trees very similar to those obtained using more complex methods. Most importantly in the context of this article, it reproduced very well all the details of topology of the published molecular trees (Littlewood et al. 1999a,b; Littlewood and Olson 2001, this volume) which contradict the morphological data and are suspected as artifacts. Using more time-consuming algorithms, therefore, proved unnecessary. The MP tree was calculated by a heuristic search (10 addition replicates) using the branch and bound option. Data were bootstrap resampled (100 replicates). Correspondence between branches showing high bootstrap support in NJ, and MP trees confirmed once more that the methods used for the computation of trees were fully sufficient.

Rates of nucleotide substitution

Rates of nucleotide substitution were estimated for comparison using the method first suggested by Wilson et al. (1977) as the branch length in a corresponding tree (see Figure 26.4). Each such tree comprised all five species of Polycladida, Geocentrophora and the measured species from the other neoophoran taxa. Polycladida were chosen because they are generally considered the sister-group of the Neoophora (or a part of such sister-group). Geocentrophora was chosen because in computed trees it always branches off earlier than other neoophorans. For each set of seven species a maximum likelihood tree was computed using PUZZLE (Strimmer and von Haeseler 1996), by enumeration of trees using the HKY85 model of evolution and uniform rate distribution. Rates were calculated for all neoophoran species in our data set. The topology of all trees was identical (see Figure 26.4). Variation of the distance from central node to the shared node of the polyclad species and to Geocentrophora (0.037–0.058 and 0.041–0.062, respectively) was small in comparison to variation of the distance to the neoophoran species (0.023–0.154). This confirms that the estimation of distance was carried out correctly. Intragroup distances for Table 26.2 were calculated using the same program (PUZZLE) and options.

To compare rates of nucleotide substitution between taxa, we used the runs test (see, e.g., Wilrich and Hennig 1987). This non-parametric test is particularly appropriate for comparison of taxa because it does not use any assumptions about the distribution of the analysed character in the sets compared. To carry out numerous comparisons, we designed a computer program that counted R values for this test for all pairs of taxa based on the series of rates for individual species in each taxon.

Chapter 27

Towards a phylogenetic supertree of Platyhelminthes?

Mark Wilkinson, Joseph L. Thorley, D. Timothy J. Littlewood and Rodney A. Bray

With the steady expansion of phylogenetic data and hypotheses, more attention is now being paid to the synthesis of results from different phylogenetic studies into more comprehensive phylogenetic trees. Such synthesis will be needed to weld the results of disparate phylogenetic studies into a comprehensive Tree of Life, and thereby fulfil the ultimate goal of phylogenetics. On the road to this ultimate goal, ever larger trees will provide frameworks for ever broader comparative studies of evolutionary patterns and processes.

One approach to producing large trees is to produce and analyse large data matrices, for example by combining data from different studies. While this may have some appeal, as a strategy for assembling the Tree of Life, it has some serious practical limitations. Sanderson *et al.* (1998) listed as drawbacks the potential for combined data to include many missing entries when only a few taxa are common to the original data sets, the expense of gathering the data to fill the gaps, that some data types are not readily combinable, and that homology assessment becomes more difficult as matrices grow. In addition, the available software packages place limits on the sizes of trees that can be constructed. The rapid increase in tree space (the set of possible trees) with tree size ensures that only heuristic searches of tree space are possible for large-scale analyses, so that the efficiency of heuristic searches is expected to decrease with the size of the trees. Thus any increase in the sizes of trees that can be handled by phylogenetic inference software is likely to be accompanied by decreased ability to perform thorough analyses. Consequently, phylogenetic analysis will remain compartmentalized to a high degree, and if more comprehensive trees are to be had then a methodology for merging disparate phylogenetic trees into larger trees is needed. Possible methods, have been developed and applied under the banner of supertree construction.

Our original aim for this chapter was to attempt a synthesis of the diverse platyhelminth phylogenies included within the book, and thereby provide a summary of the interrelationships of the Platyhelminthes. However, our principal findings have been that currently available methods of supertree construction are either of too limited applicability, or that they have undesirable properties that necessitate a healthy degree of scepticism regarding their specific results and their general utility. As a result, our aims have shifted towards the methodological. In this chapter we use both hypothetical and platyhelminth examples to illustrate the potential and the limitations of current supertree methods, and we discuss the challenges this presents for constructing a platyhelminth supertree.

Materials, methods and terminology

The analyses reported in this chapter are of hypothetical trees or trees derived from the literature on Platyhelminthes phylogeny. NEXUS (Maddison *et al.* 1997) tree files were prepared manually, or interactively using MacClade (Maddison and Maddison 1992), or automatically by PAUP (Swofford 1993). Tree manipulations, construction of consensus trees, determinations of tree measures (size, information content) and the production of matrix representations of trees were performed with

RadCon (Thorley and Page 2000). Parsimony analysis of matrix representations of trees were performed with PAUP using exact methods (branch and bound) for smaller data sets (< 16 leaves) and heuristics (closest addition sequence, TBR branch swapping) for larger data sets.

For the purposes of this chapter we have adopted the following terminology. Throughout, we follow Sanderson *et al.* (1998) in referring to the trees that we seek to combine as *source* trees, and we use *leaves* as our preferred synonym for the tips of trees (base points, OTUs, terminal taxa, etc.). The *leaf set* of a tree is the set of leaves included in that tree. Leaves that are common to all of a set of source trees are the *common leaves*, and other leaves are *unique*. A *subtree* of a tree is produced by pruning leaves and the branches that subtend them from the tree. Our focus is upon *cladistic* relationships, as given by the branching patterns of rooted trees, and we do not consider branch lengths or unrooted trees. Rooted trees are sometimes treated as sets of *components* (clusters or clades, etc.) or as collections of resolved triplets (three-taxon statements). Components and resolved triplets are respectively the most and the least inclusive members of a class of cladistic information termed *n-taxon statements* (Wilkinson 1994). N-taxon statements are simple cladistic hypotheses of the general form 'some specific leaves are more closely related to each other than any of them are to some other specific leaves'. For example, the assertion that A and B are more closely related to each other than either of them is to C, D or E is a five-taxon statement. An n-taxon statement can be thought of as a tree with a single internal branch. These concepts are illustrated in Figure 27.1. Consensus terminology follows that of Wilkinson (1994), so that the method of Sokal and Rohlf (1981), commonly referred to as the strict consensus, is here described as a strict component consensus (SCC). SCC trees are strict in that they include only relationships that are true of all the trees the consensus represents, and the type of relationships represented are components. The strict reduced cladistic consensus (RCC) method (Wilkinson 1994) produces a profile of one or more consensus trees, each of which represents only n-taxon statements that are true of all the source trees. The profile jointly represents all such n-taxon statements (Wilkinson 1994). Cladistic information is conveyed by any restriction upon the possible cladistic relationships of a set of leaves, and we use the *cladistic information content* (CIC) measure of Thorley *et al.* (1998) to quantify the information in trees.

What is a supertree?

'Supertree' is a quite recent addition to the lexicon of systematics. It was first introduced in the literature on the mathematics and algorithmics of classification problems (Gordon 1986) and was defined as 'a dendrogram from which each of the original [source] trees can be regarded as samples'. Gordon drew a crucial distinction between sampling that either does or does not introduce distortion. For example, a sample of source trees that is derived from a supertree simply by pruning leaves from the supertree introduces no distortion. Consequently, the information in each source tree is contained within the supertree, each source tree is *consistent* with the supertree, and the supertree may be said to *extend* each of the source trees. In this case, the set of source trees are *compatible* (do not conflict), and the supertree includes all the cladistic information in the set of source trees. In contrast, if the sampling process introduces some distortion then the source trees may be *incompatible* (imply contradictory information about the supertree), in which case there is no supertree that extends all the source trees.

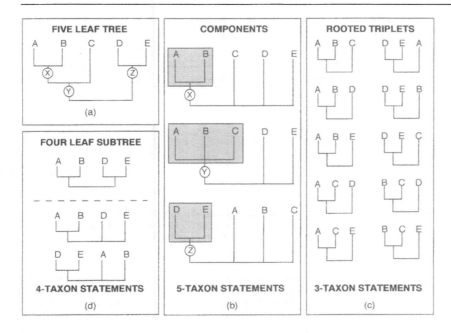

Figure 27.1 Types of trees and relationships. **a)** A tree showing cladistic relationships among the five leaves (A–E) with three internal branches labelled X–Z. **b)** Set of three components, AB, ABC and DE (shown in shaded boxes) that are present in (a). The leaves in a component are more closely related to each other than any of them are to any other leaves in the tree. Thus, each component corresponds to a single internal branch corresponding to an internal branch in the tree and each is equivalent to a five-taxon statement in which the non-members of the component are explicit, e.g. (AB)CDE for the first component. **c)** The set of ten rooted triplets (three-taxon statements) that are present in (a). **d)** A four-leaf subtree of (a) formed by pruning C and the two four-taxon statements that it includes and that are also included in (a).

In the context of phylogenetics, source trees would be expected to be compatible if they are correct. More typically, we are confronted with incompatible source trees. This entails either some inaccuracy in one or more of the source trees or mismatch between gene trees or gene and species trees due to, for example, lineage sorting or gene duplication (Slowinski and Page 1999). As we shall see, the distinction between consistent and inconsistent source trees is extremely important both for the applicability in practice and the desirability in principle of current approaches to supertree construction.

In a recent review, Sanderson *et al.* (1998) defined a supertree in its broadest sense as 'Any tree containing all the taxa found among the source trees' and a *strict* supertree to be a supertree that extends all the (compatible) source trees. We prefer to call the latter a *consistent* supertree. Under these definitions supertrees need bear no relation to the source trees other than including all their leaves. In contrast, a consistent supertree has the desirable property of including all the relationships in all the source trees.

One of Gordon's (1986) important results is that there may be more than one consistent supertree for a given set of compatible source trees. We follow Bryant and Steel (1995) in referring to the set of consistent supertrees as the *span* of the source trees. Gordon (1986) suggested using consensus methods to summarize multiple supertrees in the form of *consensus* supertrees. He discussed only strict consensus methods in this context, so that consensus supertrees would only represent what is true of all the consistent supertrees and therefore entailed by the set of source trees.

How are supertrees produced?

Supertree methodology is concerned with the construction of one or more supertrees, or their consensus, from a set of source trees. Typically the source trees have overlapping but non-identical sets of leaves. Two distinct approaches to supertree construction have been taken, and the two schools have produced methods with important differences. For the purposes of this chapter we distinguish these as *direct* and *indirect*

methods. In the following sections we provide brief (if somewhat superficial) summaries of the work that has been done in each of these schools. In the section on direct methods we also discuss an intuitive approach to combining trees at different taxonomic levels.

Direct methods

The first school of supertree research was founded by Gordon's (1986) seminal paper. Its approach has been to characterize and prove algorithms for combining source trees, to characterize the properties of the supertrees thus produced, and to prove theorems relating to the complexity and the generality of the supertree problem. Gordon (1986) described an algorithm for merging two compatible binary source trees that yields the strict component consensus of the span of consistent supertrees (SCCSCS). We refer to this as a direct method because it does not require enumeration of the span of consistent supertrees and the subsequent construction of their consensus, but proceeds more directly from source trees to consensus supertree. Gordon's method is quite simple and can readily be performed by hand. Two compatible binary source trees share the same subtree for their common leaves, and the SCCSCS is constructed by grafting the unique leaves to this subtree according to three simple rules. Figure 27.2 illustrates the method with Gordon's original example.

Gordon (1986) rightly emphasized two limitations of his method. The first, its restriction to only pairs of source trees, has been addressed by Steel (1992), who proved the existence of completely different, efficient (polynomial time) algorithms for determining if a set of source trees are compatible and for constructing their SCCSCS (see also Constantinescu and Sankoff 1995). Steel's methods are less easily applied manually than Gordon's and, unfortunately, no software implementing the method is currently available. The second limitation concerns the information content of the SCCSCS. Gordon (1986: 343) noted that there are seven possible consistent supertrees for the two source trees in his example (Figure 27.2, Trees 1 and 2), and that the SCCSCS 'conceals some information about the structure present in the seven possible supertrees

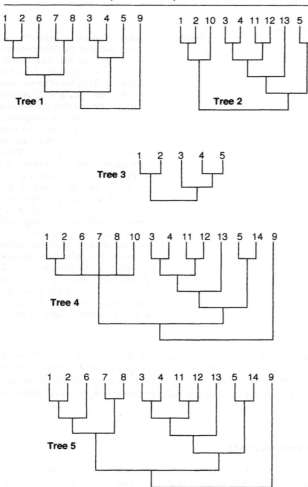

Figure 27.2 Two source trees (Tree 1 and Tree 2) their common subtree (Tree 3) and their SCCSCS (Tree 4) from Gordon (1986) and a reduced supertree (Tree 5) from Gordon. To construct the SCCSCS from the source trees, the common subtree of the source trees is used as a framework to which the unique leaves are grafted. Each branch on the common subtree is compared to the corresponding branch or branches on the source trees. The grafting of leaves is governed by the following three rules: 1) if both source trees have unique leaves originating from the branch, then insert a new node on the branch in the common subtree and graft all the unique leaves to the new node; 2) if only one source tree has unique leaves on the branch, graft all the unique leaves to this branch of the subtrees so as to preserve the relationships in the source tree; 3) if neither source tree has unique leaves arising from the branch, graft no leaves to the subtree.

– for example, that objects 7 and 8 are perceived as more similar to one another than they are to object 6 – and it may be that an alternative representation would be more informative'.

Wilkinson and Thorley (in preparation) have developed an algorithm for directly combining a pair of consistent source trees that produces the strict reduced cladistic consensus profile of the span of consistent supertrees. This 'reduced supertree profile' of Wilkinson and Thorley (1998) includes all the n-taxon statements common to the set of consistent supertrees. Applied to Gordon's example this method yields a profile of two consensus supertrees. The first is the SCCSCS (Figure 27.2; Tree 4)

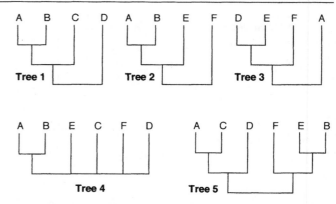

Figure 27.3 Three source trees (Trees 1–3) and two SCCSCSs (Trees 4–5). Tree 4 is the SCCSCS for Trees 1 and 2 and Tree 5 is the SCCSCS for Trees 1 and 3 (see text for explanation).

which has a CIC of 36.1318 bits. Despite excluding leaf 10, the second (Figure 27.2, Tree 5) has a higher information content (CIC = 38.202 bits) because it captures the additional information that Gordon recognized was concealed in the SCCSCS. It is not yet clear how this method can be extended to collections of more than two source trees.

A minimum requirement for the effective merging of two compatible source trees is that there are at least two common leaves. In addition, the relationships of the common leaves and unique leaves play a big part on the degree to which merging produces a more or less well-resolved SCCSCS and thus its information content. Trees 1, 2 and 3 in Figure 27.3 each contain 3.9069 bits of cladistic information. The SCCSCS of Trees 1 and 2 (Tree 4, Figure 27.3) is poorly resolved and actually contains less information than either source tree (CIC = 3.1699 bits). In contrast, the SCCSCS of Trees 1 and 3 (Tree 5, Figure 27.3) contains more information (CIC = 9.8842 bits) than the sum of the two source trees. In the first example it is not possible to combine information regarding the unique leaves of both source trees. The common leaves are close together and effectively act like a single leaf. In the second example, combination is possible and profitable as information entailed by the source trees but not included in either one of them is included in the consensus supertree. Here, the leaves are more distantly separated and the unique leaves do not occur on the same branch. It is difficult to recommend a general strategy of taxon sampling to promote well-resolved SCCSCSs. Denser sampling of common leaves should improve the chance of producing a well-resolved consensus supertree, but denser sampling of unique leaves can have the opposite effect if unique leaves occur in corresponding parts of more than one source tree.

Gordon (1986) also discussed the problem of constructing supertrees from incompatible source trees. He suggested using a consensus tree to represent agreement on the relationships of the common leaves and then grafting the unique leaves according to the rules used in SCCSCS construction. He envisaged that alternative consensus methods used in the first step would correspond to alternative supertree methods, and emphasized the need for studies of the properties of these methods. Unfortunately, the properties of the methods remain poorly understood, and it is doubtful that the methods Gordon outlined could be readily extended to merging more than two source trees. As Gordon (1986: 347) commented 'it may be that a completely different algorithm for merging two inconsistent sample [incompatible source] trees is markedly superior'.

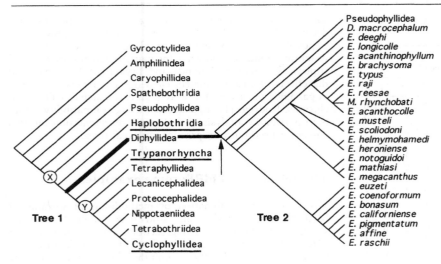

Tree 1

Gyrocotylidea
Amphilinidea
Caryophillidea
Spathebothridia
Pseudophyllidea
Haplobothridia
Diphyllidea
Trypanorhyncha
Tetraphyllidea
Lecanicephalidea
Proteocephalidea
Nippotaeniidea
Tetrabothriidea
Cyclophyllidea

Tree 2

Pseudophyllidea
D. macrocephalum
E. deeghi
E. longicolle
E. acanthinophyllum
E. brachysoma
E. typus
E. raji
E. reesae
M. rhynchobati
E. acanthocolle
E. musteli
E. scoliodoni
E. helmymohamedi
E. heroniense
E. notoguidoi
E. mathiasi
E. megacanthus
E. euzeti
E. coenoformum
E. bonasum
E. californiense
E. pigmentatum
E. affine
E. raschii

Figure 27.4 The taxon substitution approach to supertree construction. Tree 1 and Tree 2 are source trees of higher taxa of Eucestoda from Hoberg *et al.* (1997c) and of species of Diphyllidea from Ivanov and Hoberg (1999). The thickened line indicates the position at which the ingroup portion of Tree 2 (marked by an arrow) can be united with Tree 1 under the assumption that the Diphyllidea is monophyletic. Leaves underlined are a combination of outgroups whose inclusion in Tree 2 would allow the same taxon substitution without assuming monophyly of Diphyllidea.

In the face of conflicting source trees we can either attempt somehow to resolve the conflict, or to remove the conflict. Gordon's advocacy of strict consensus methods in this context takes the latter option. Conflict is not resolved by consensus methods, but is removed by collapsing conflicting relationships to polytomies and/or by pruning leaves. This is a conservative approach in that all relationships in the consensus supertree thus formed are not contradicted by any of the source trees.

In many phylogenetic analyses single leaves represent entire clades. Although not discussed explicitly in the supertree literature, this leads to the possibility of supertree construction by a simple process of *taxon substitution*, in which the leaf representing the clade is replaced by a tree for the clade. In Figure 27.4, Tree 1 is a hypothesis of the interrelationships of various higher taxa of eucestodes from Hoberg *et al.* (1997c) and Tree 2 is a hypothesis of interrelationships of the species within one of these higher taxa, the Diphyllidea, from Ivanov and Hoberg (1999). Using taxon substitution a supertree can be formed by replacing the Diphyllidea in Tree 1 with the ingroup part of Tree 2. However, this procedure is not without assumptions, and these depend in turn on the nature or the interpretation of the leaf representing the higher taxon. If the leaf is interpreted as a groundplan (reconstruction of the ancestor of the group), then we are effectively assuming that the group is monophyletic. If the leaf is a particular exemplar, then the same assumption of monophyly also justifies taxon substitution.

However, if we are not willing to assume monophyly, then simple taxon substitution may not be legitimate given the rules for merging compatible source trees. Rather, the form of the supertree will depend upon the sampling of outgroups. For example, Tree 2 has only a single outgroup, the Pseudophyllidea. It thus asserts that all the diphyllideans are more closely related to each other than they are to the Pseudophyllidea. This information is consistent with any placement of the diphyllidean species anywhere on Tree 1 above node X as a monophyletic group or not. Thus, in the absence of an assumption of monophyly of the Diphyllidea, the SCCSCS properly formed from Trees 1 and 2 would be completely unresolved above node X. To prevent this behaviour a series of outgroups to the Diphyllidea that bracket this group's position on Tree 1 are needed in the lower-level analysis of the group. In this case, for taxon substitution and SCCSCS construction to produce the same supertree, the outgroups must include two representatives of the sister-taxon to the Diphyllidea, one from each side of the basal split within that

group (i.e., Trypanorhyncha and a leaf descended from node Y, such as the Cyclophyllidea), and a representative of the next group down the tree (the Haplobothridea). If these fall out as outgroups in the analysis of the lower-level taxon then they serve to demonstrate that the members of the taxon are more closely related to each other than to their nearest neighbours on the higher level tree, and thereby greatly constrain the positions of the lower-level taxa relative to the higher level tree. As far as we can determine this is a general requirement for taxon substitution when monophyly is not assumed (with the qualification that only a single representative of the immediate sister-group is required when there is only a single representative). We prefer that monophyly be demonstrated rather than assumed. The importance of this is shown by the analysis of Caira *et al.* (1999), which suggested that several of the higher taxa in Tree 1 are not in fact monophyletic, thereby undermining the use of taxon substitution to merge the trees.

Indirect methods

The second school has its origins in the work of two methodologically minded biologists. Baum (1992) and Ragan (1992) independently suggested combining matrix representations of source trees into a single combined data set, followed by analysis of that data set using parsimony to produce combined trees. This approach has become known as matrix representation with parsimony or MRP, and we refer to the trees that it produces as MRP supertrees. Where there is more than one MRP supertree we refer to a consensus of these as an *MRP consensus supertree*. In matrix representation, as suggested by Baum (1992) and Ragan (1992), each clade of each source tree is represented by a single column or *matrix element* in the combined data with, for example, descendents of the node coded as '1' and non-descendents coded '0'. Leaves that are not present in the tree are coded as missing ('?'). We refer to this coding method as *component coding* because each matrix element in the data corresponds to a component on one of the source trees. The root of the source trees is represented by an MRP outgroup coded as basal for each matrix element. This form of matrix representation of Gordon's (1986) example source trees (Trees 1 and 2, Figure 27.2) is given in Table 27.1. Columns 1–7 and 8–15 represent Trees 1 and 2, respectively.

Ragan (1992) emphasized that MRP applied to a single source tree yields that source tree as a single most-parsimonious tree

Table 27.1 Matrix representation of the two source trees (Trees 1 and 2) in Figure 27.2.

| | Matrix elements | | | | | | | | | | | | | | |
| | Tree 1 | | | | | | | Tree 2 | | | | | | | |
Leaves	1	2	3	4	5	6	7	8	9	10	11	12	13	14	15
MRP outgroup	0	0	0	0	0	0	0	0	0	0	0	0	0	0	0
1	1	1	0	1	0	0	1	1	1	0	0	0	0	0	0
2	1	1	0	1	0	0	1	1	1	0	0	0	0	0	0
3	0	0	0	0	1	1	1	0	0	1	0	1	1	0	1
4	0	0	0	0	1	1	1	0	0	1	0	1	1	0	1
5	0	0	0	0	0	1	1	0	0	0	0	0	0	1	1
6	0	1	0	1	0	0	1	?	?	?	?	?	?	?	?
7	0	0	1	1	0	0	1	?	?	?	?	?	?	?	?
8	0	0	1	1	0	0	1	?	?	?	?	?	?	?	?
9	0	0	0	0	0	0	0	?	?	?	?	?	?	?	?
10	?	?	?	?	?	?	?	0	1	0	0	0	0	0	0
11	?	?	?	?	?	?	?	0	0	0	1	1	1	0	1
12	?	?	?	?	?	?	?	0	0	0	1	1	1	0	1
13	?	?	?	?	?	?	?	0	0	0	0	0	1	0	1
14	?	?	?	?	?	?	?	0	0	0	0	0	0	1	1

(or MRP tree). We conjecture that the MRP supertrees of a set of compatible source trees constitute the span of the source trees. If we are correct, then summarizing the MRP trees with a strict component consensus will give the SCCSCS, the same result as Steel's (1992) direct method, and that Wilkinson and Thorley's (1998) reduced supertrees are given by the reduced cladistic consensus profile of the MRP trees (subject to the limitations of the parsimony implementation). Thus, MRP has the same properties as the known direct methods with respect to sets of compatible trees, but is less efficient because of its reliance on parsimony analysis to enumeration of the set of consistent trees. The potential for exponential increase in the number of consistent supertrees with the number of leaves may render MRP impractical in many cases.

MRP differs from the currently known direct methods of supertree construction in that it can be applied to sets of source trees that conflict. Given that phylogenetic hypotheses often conflict, it is not surprising that MRP has been used in the few studies that have sought to produced supertrees for particular groups (Purvis 1995a; Bininda-Emonds *et al.* 1999; Morand and Müller-Graf 2000). Indeed, the ease of producing matrix representations, the availability of parsimony programs, such as PAUP, for analysing them, and the applicability of MRP to incompatible source trees seem like clear practical advantages of MRP over current direct methods. However, the behaviour of MRP when applied to incompatible source trees is poorly understood. In particular, where MRP resolves conflicts among the original source trees it is not clear whether the resolution is always justified or desirable. Applicability in practice should not be taken as a guide to its acceptability in principle.

Much of the literature on MRP concerns perceived difficulties with, objections to, and/or proposed modifications of the approach. This has led to a number of variants in both matrix representation and in the method of analysis. Purvis (1995b) suggested a modified matrix representation intended to remove redundant information and an associated bias in cases of conflict towards relationships in larger source trees. Another alternative matrix representation makes use of resolved triplets (three-taxon statements) rather than components (Thorley and Wilkinson 2000; see also Nelson and Ladiges 1994). Ronquist (1996) showed that Purvis's coding was problematic, and recommended weighting matrix elements in relation to

some measure of their support (e.g., bootstrap proportions, decay indices of the components that they represent) as opposed to equal weighting of all matrix elements. Such weighting could be used when the matrix elements are resolved triplets also. Bininda-Emonds and Bryant (1998) considered the effects of treating matrix elements as irreversible or not in parsimony analyses, but were unable to reach any firm conclusions. Rodrigo (1996) suggested using compatibility rather than parsimony analysis of the matrix representation but this has never been put into practice with real trees.

The behaviour of MRP: some examples

Which, if any, combination of methodological variants of MRP should be used in any study has not been answered satisfactorily. In the following sections we present results of MRP analyses of four examples drawn from the literature on the phylogeny of the Platyhelminthes in order to illustrate some of the variants of MRP, and to illustrate and draw attention to some of their properties. For simplicity, there are just two source trees in each of our examples.

Example one

The two source trees, Trees 1 and 2 in Figure 27.5 are drawn from the morphology-based phylogeny of Ehlers (1985a; 1986) and from combined morphological and molecular data

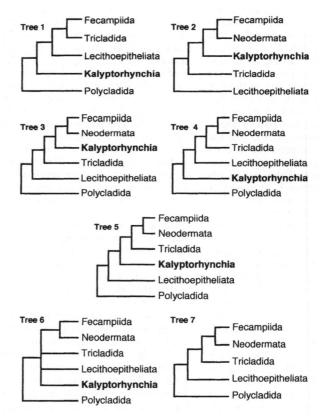

Figure 27.5 Two source trees (Trees 1 and 2), three MRP supertrees (Trees 3–5) and two consensus supertrees (Trees 6 and 7). Trees 1 and 2 are modified from Ehlers (1985a, 1986) and Littlewood *et al.* (1999a), respectively. See text for explanation.

Table 27.2 Standard and resolved triplet matrix representations of the source trees (Trees 1 and 2) in Figure 27.5.

Leaves	Components						Matrix elements — Resolved triplets																							
	Tree 1			Tree 2			Tree 1												Tree 2											
MRP outgroup	0	0	0	0	0	0	0	0	0	0	0	0	0	0	0	0	0	0	0	0	0	0	0	0	0	0	0	0	0	0
Fecampiida	1	1	1	1	1	1	1	1	1	1	1	?	?	1	?	?	1	1	1	1	1	1	1	?	?	1	?	?		
Neodermata	?	?	?	1	1	1	?	?	?	?	?	?	?	?	?	?	1	1	1	?	?	1	1	?	1	?				
Tricladida	1	1	1	0	0	1	1	1	1	?	?	1	1	?	1	?	0	?	?	0	?	0	?	1	1	1				
Lecithoepitheliata	0	1	1	0	0	0	0	?	?	1	1	1	1	?	?	1	?	0	?	?	0	?	0	0	0	0				
Kalyptorhynchia	0	0	1	0	1	1	?	0	?	0	?	0	?	1	1	1	?	?	0	1	1	1	1	?	?	1				
Polycladida	0	0	0	?	?	?	?	?	0	?	0	?	0	0	0	0	?	?	?	?	?	?	?	?	?	?				

analysis of Littlewood *et al.* (1999a). These two trees are in conflict regarding the relationships of the Kalyptorhynchia and otherwise differ in each having a unique leaf. The component and triplet coding of these trees for MRP is shown in Table 27.2. *Standard* MRP (i.e., using component coding and treating the matrix elements as reversible), as proposed by Baum (1992) and Ragan (1992), yields three MRP supertrees (Trees 3–5, Figure 27.5). These differ only in the positions of Kalyptorhynchia. In two of these (Trees 3 and 5), the position of Kalyptorhynchia is the same as in one of the source trees, whereas in the third (Tree 4) it has an intermediate or compromise position that is supported by neither of the source trees. Trees 6 is an MRP consensus supertree formed from the three MRP trees using the SCC method. It contains 3.4594 bits of cladistic information. Application of the RCC method yields an additional, and more informative (CIC = 3.9096 bits), MRP consensus supertree (Tree 7, Figure 27.5). In this case the conflict between the source trees is not resolved by MRP but is removed from the consensus supertrees by either the collapse of branches to polytomies (Tree 6, Figure 27.5) or the pruning of Kalyptorhynchia (Tree 7, Figure 27.5). Also in this case, the MRP consensus supertrees are identical to the consensus supertrees that would be produced using Gordon's consensus-based methods for merging incompatible source trees.

In contrast, MRP using component coding but treating the matrix elements as irreversible leads to a single MRP supertree (Tree 5; Figure 27.4) that resolves the conflict between the source trees in favour of the more basal position of Kalyptorhynchia. Using resolved triplet coding (in which case irreversible or reversible treatments of the matrix elements are analytically equivalent) also leads to a single MRP supertree (Tree 3, Figure 27.5), but one that resolves the conflict between the source trees in favour of the more crownward position of Kalyptorhynchia.

This example serves to demonstrate the (perhaps unsurprising) fact that alternative matrix representations (component, rooted triplet) or analytical treatments (reversible, irreversible) can give different results. It also suggests that some variants of MRP may suffer biases in the way that they resolve conflicts. In our view, biases toward crownward or more basal positions of leaves that are in conflicting positions in the source trees are a discomforting aspect of MRP.

and facilitate our exposition. Our second example is of more complete source trees. The two source trees (Trees 1 and 2; Figure 27.6) are from Ehlers (1984) and Littlewood *et al.* (1999a). Both trees are based on morphology. Standard MRP yields 130 MRP supertrees. The MRP consensus supertree (Tree 3, Figure 27.6) produced using the SCC method presents an interesting combination of the relationships in the source trees. For example, the relationships of Gyrocotylidea, Amphilinidea and the Eucestoda (node r), are unresolved in Tree 2, but consistent with their relationships in Tree 1, and this is nicely resolved in the MRP consensus supertree. In contrast, the relationships of these taxa and the two clades comprising Polyopisthocotylea and Monopisthocotylea and Aspidogastrea and Digenea respectively (node s) are also unresolved in Tree 2, but consistent with their resolution in Tree 1. However, these relationships remain unresolved in the MRP consensus supertree.

The different behaviours here are attributable to differences in the relationships of unique leaves. *Udonella* is present in Tree 2 and not in Tree 1, but its relationships in Tree 2 are incompletely resolved, particularly with respect to the clades subtended by node s. Although each MRP supertree includes the relationships above node s from Tree 1, these are not present in the MRP consensus supertree because of the uncertain relationships of *Udonella* which acts like an underdetermined rogue taxon (Wilkinson 1995). In contrast, Caryophyllidea is not included in Tree 2, but there is no uncertainty regarding its relationships in Tree 1 that obscures the resolution of node r provided by Tree 1.

The source trees conflict over many relationships that are resolved in the MRP consensus supertree. The resolutions are sometimes in favour of relationships in Tree 1, and sometimes in favour of those in Tree 2. For example, compare the positions of Proseriata and Tricladida, and Temnocephalida and Typhloplanida. There also appears to be evidence of a bias toward more crownward positions in the resolution of the relationships of, for example, Macrostomida, Polycladida, Lecithoepitheliata and Prolecithophora.

This example demonstrates the obfuscatory effect that unique leaves can have. They can be responsible for regions of poor resolution in the MRP consensus supertree. This problem affects direct methods also. In contrast, areas where a lack of resolution might be expected, because the source trees conflict, are generally resolved.

Example two

The previous example was somewhat contrived, in that the source trees were subtrees of the published trees, produced by pruning leaves from the published trees in order to simplify

Example three

For our third example we return to somewhat simplified trees, this time from analyses of molecular data. The two source trees

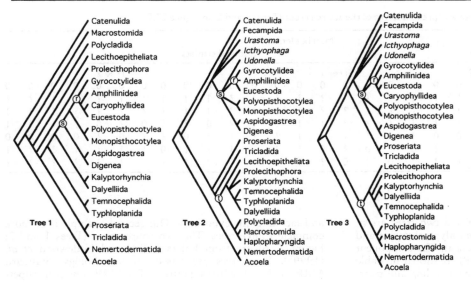

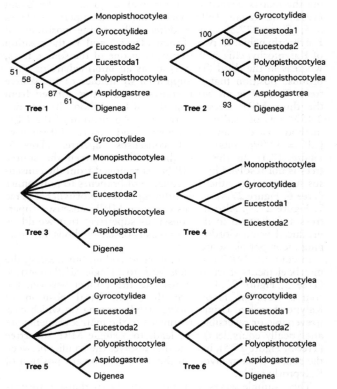

Figure 27.6 Two source trees (Trees 1 and 2) and their standard MRP consensus supertree (Tree 3). Tree 1 is from Ehlers (1984). Tree 2 is from Littlewood *et al.* (1999a). Tree 3 is the SCC of 130 MRP supertrees. See text for explanation.

(Trees 1 and 2, Figure 27.7) are from the analyses of large and small subunit RNA genes by Litvaitis and Rohde (1999) and Littlewood *et al.* (1999a). In this case the source trees have the same leaf sets, and it is possible to apply consensus methods directly to the source trees and to compare these with the results of MRP. In addition, bootstrap proportions from the original analysis allow us to explore the effect of weighting matrix elements to reflect their support. The standard matrix representation of these trees is given in Table 27.3.

The two consensus trees (Trees 3 and 4, Figure 27.7) summarize the agreement among the source trees. The clade comprising Aspidogastrea and Digenea is the only component present in both source trees, but the trees also share a common subtree comprising two four-taxon statements (Tree 4; Figure 27.7). A major disagreement between the source trees is in the relationships of Monopisthocotylea and Polyopisthocotylea. Applied to this example, standard MRP with equal weighting of matrix elements yields three MRP supertrees. Tree 5 (Figure 27.7) is the MRP consensus supertree given by the SCC method. This tree resolves the conflicts in the positions of Monopisthocotylea and Polyopisthocotylea in favour of Tree 1. Indeed, the tree is essentially Tree 1 with a couple of nodes collapsed, and includes no relationships that are in Tree 2 but not in Tree 1. Parallel MRP analysis in which the matrix elements are weighted by their bootstrap proportions yields a single fully resolved MRP supertree (Tree 6, Figure 27.6). Relationships in this tree are a mixture of those in the two source trees. As with equal weighting, the conflict over the positions of Monopisthocotylea and Polyopisthocotylea is resolved in favour of Tree 1 despite the high bootstrap proportion supporting the pairing of Polyopisthocotylea and Monopisthocotylea in Tree 2.

Four of the matrix elements representing Tree 1 (Table 27.3, elements 2–5) conflict with the sister-group pairing of Monopisthocotylea and Polyopisthocotylea in Tree 2. In contrast, only two matrix elements from the matrix representation of Tree 1 (Table 27.3, elements 7 and 8) conflict with the positions of Monopisthocotylea and Polyopisthocotylea in Tree 2. This accounts for why trees in which Monopisthocotylea and Polyopisthocotylea are sister-taxa do not occur amongst the MRP supertrees.

This example suggests to us a possible bias in standard MRP towards relationships in more asymmetric trees like Tree 1. Asymmetric trees maximize the number of separate matrix elements that are treated as independent evidence against

Figure 27.7 Two source trees with identical leaf sets and their bootstrap proportions (Trees 1 and 2), their SCC tree (Tree 3) and an additional RCC tree (Tree 4), the MRP consensus supertree when matrix elements are weighted equally (Tree 5) and the unique MRP supertree when the matrix elements are weighted by their bootstrap proportions (Tree 6). Trees 1 and 2 are from Litvaitis and Rohde (1999) and Littlewood *et al.* (1999a), respectively.

conflicting relationships in more balanced trees, like Tree 2. Although bootstrap support may be high for relationships in more balanced trees, the combined weight of matrix elements representing relatively poorly supported components may still overwhelm relationships in more balanced trees.

Table 27.3 Standard matrix representation of the source trees (Trees 1 and 2) in Figure 27.7.

| | Matrix elements | | | | | | | | | |
| | Tree 1 | | | | | Tree 2 | | | | |
Leaves	1	2	3	4	5	6	7	8	9	10
MRP outgroup	0	0	0	0	0	0	0	0	0	0
Monopisthocotylea	0	0	0	0	0	0	1	1	0	0
Gyrocotylea	0	0	0	0	1	0	1	0	1	0
Eucestoda 1	0	0	0	1	1	0	1	0	1	1
Eucestoda 2	0	0	1	1	1	0	1	0	1	1
Polyopisthocotylea	0	1	1	1	1	0	1	1	0	0
Aspidogastrea	1	1	1	1	1	1	0	0	0	0
Digenea	1	1	1	1	1	1	0	0	0	0

Example four

Earlier, we discussed the effects of taxon selection upon direct methods of supertree construction. As in direct methods MRP requires a minimum of two common leaves to allow any constructive merging of source trees. Our fourth example is used to further emphasize the importance of taxon selection, and is also obtained from the papers by Litvaitis and Rohde (1999) and Littlewood *et al.* (1999a). The two source trees (Trees 1 and 2, Figure 27.8) have 15 and 28 leaves and 47.6009 and

116.9270 bits of cladistic information respectively, and they have a combined leaf set of 37 genera, of which only six are common leaves. The subtrees for these common leaves conflict and have markedly different balance. Standard MRP analysis was aborted after 30 000 MRP supertrees had been found. The large number of trees reflect the instability of many unique leaves, and this is reflected also in the MRP consensus supertree, using the SCC method (Tree 3, Figure 27.8). This is poorly resolved and contains less information (CIC = 101.1205 bits) than the second source tree. The common taxon sampling in this example is simply not dense enough to permit a well-resolved consensus supertree for the combined leaf set. As in the last example, conflict between the two trees is resolved in favour of the more asymmetric relationships among the common leaves in the smaller of the source trees.

Discussion

Many workers have provided syntheses of multiple phylogenies, but formal methods of supertree construction have only recently – and as yet only incompletely – been developed. Both biologists and mathematicians have contributed to the development of supertree methods, but mostly in isolation. The best-understood methods are those developed by mathematicians for producing consensus supertrees from sets of compatible source trees. The consensus supertrees produced show only relationships that are entailed by the source trees. In this sense they are conservative. Unfortunately, none of these methods is implemented in distributed software, and so they are for the most part practically unavailable, and some are limited in other ways (e.g., algorithms for reduced supertrees exist only for pairs of source trees). Of course, the major limitation of these methods is that they cannot be used when source trees conflict, and in the real world this tends to be the case. Although Gordon (1986) outlined approaches that could be applied in cases of conflict, his suggestions have not received the attention they deserve, and this should prove a useful and fruitful area of research. In Gordon's approach the resulting supertree does not resolve the conflict among the source trees, rather it eliminates it by collapsing branches to produce polytomies or by pruning leaves. Just as in the case of compatible source trees, the consensus supertrees only include relationships that are entailed and uncontradicted by the source trees. The consensus supertree provides a conservative summary of relationships in the source trees.

In contrast, matrix representation approaches to supertree construction are readily implemented using existing parsimony software. Just as in parsimony analysis of character data, the operations involved are simple to understand, but the behaviour of the methods is poorly understood. A number of variant matrix representations have been suggested already, as have variations in the weighting and analytical treatment of matrix elements. Indeed, the rate of increase in alternative forms of MRP has far outstripped our growth in understanding of any of the variants. Given our ignorance of the methods, that they are the only ones readily available to researchers seeking to synthesize supertrees, this is an unhappy and potentially dangerous situation.

Some possible biases of of MRP have been discussed previously. Purvis (1995b) presented a simple example of a three-leaf and a four-leaf tree that conflict over relationships of the common leaves. Standard MRP yielded the four-leaf tree, leading Purvis to suggest a bias towards relationships in larger trees and to propose an alternative matrix representation intended to remove the bias. Ronquist (1996) demonstrated that

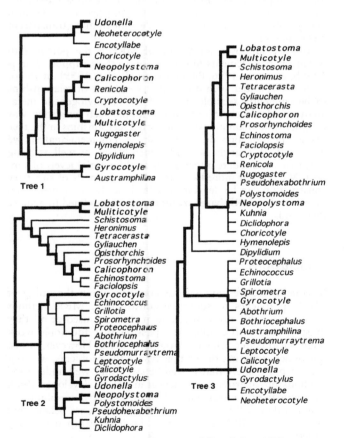

Figure 27.8 Two source trees (Trees 1 and 2) and their MRP consensus supertree (Tree 3). Common leaves of the source trees are in bold, and their relationships are highlighted by thickened branches. The MRP consensus supertree is the SCC of 30 000 MRP supertrees (Tree 3). Trees 1 and 2 are from Litvaitis and Rohde (1999) and Littlewood *et al.* (1999a), respectively.

Purvis's coding does not solve the problem, and proposed weighting the matrix elements representing a source tree (component coding) by the inverse of the number of components in the tree, so that all trees are weighted equally. However, Bininda-Emonds and Bryant (1998) have shown that Ronquist's proposal does not solve the problem and that it is more complex than initially envisaged. Bininda-Emonds and Bryant (1998: 499) concluded that 'a size bias only occurs when the "missing nodes" (missing taxa or polytomies) are located within the region of conflict among the source trees'.

Our examples suggest that the biases of MRP have not yet been adequately characterized. In our first and third examples, the source trees were of the same size, and in the third they included exactly the same leaf sets so there were no 'missing nodes'. These examples are not subject to any kind of size bias, but they yield results that suggest some possible bias. The first example suggests biases towards crownward and basal positions of leaves with resolved triplet coding and irreversible treatments of component coding respectively. The third example suggests a bias toward relationships in more asymmetric source trees in standard MRP. The same bias is seen in our fourth example, where conflicts are resolved in favour of the smaller but more asymmetric source tree, whilst our second example suggests that standard MRP may also be biased towards more crownward positions of leaves. As with previous workers, our investigations of potential biases have depended upon a limited number of examples, and our interpretations do not rest on formal proofs. Thus we hasten to add that whilst our examples demonstrate behaviours of MRP that should be of concern to potential users, our conclusions concerning the precise nature and any generality of biases, in terms of size and balance of source trees, must be considered tentative.

Bininda-Emonds et al. (1999: 146) considered the ability of MRP to resolve conflicts a distinct advantage over more direct methods. They characterized MRP as '...essentially a parsimony analysis of the different phylogenetic signals within each data set stripped of any confounding noise', that it has 'character congruence-like properties', that these are 'analogous to signal enhancement' and that it is 'less sensitive to conflict among source trees in that the composite tree is usually well resolved'. As we have stressed, relationships in consensus supertrees produced by direct methods are those that are entailed and not contradicted by the source trees. Any additional resolution obtained through MRP therefore involves resolution of conflicts among the source trees. Although more resolved supertrees are attractive, the important question must be whether the additional resolution can be justified. In all of our examples we see no good basis for preferring relationships in one of the source trees over any other, and we therefore consider the resolution of conflicts provided by MRP to represent nothing more than a methodological artifact.

If our interpretations of biases are correct, then it seems that MRP may be resolving conflicts on the basis of characteristics of source trees such as their balance. We do not think that this is a good basis for resolving conflict. Fundamentally, we believe that conflicts between trees call for explanation and further work (e.g., further data). In the absence of additional data, resolution of conflict can be justified by explanation and understanding of the conflict. Understanding of the conflict can also help identify where attempts to gather more data should be focused. In an ideal world, conflicts would be understood and resolved by compelling arguments and/or additional data to leave us with source trees that are not in conflict and which can be fitted together using taxon substitution and other direct methods. This general philosophy is in stark contrast to the 'black box' of MRP that will resolve conflicts without promoting any understanding.

These considerations lead us to a fundamental distinction regarding what is considered desirable of a supertree methodology. On one side, supertrees that summarize only agreement among source trees and thereby provide a conservative summary of relationships are the desideratum. Such supertrees identify the best-supported hypotheses and help to identify where more work is needed to understand and resolve conflicts. On the other side, methods capable of resolving conflict are the desideratum. With such methods, which can be thought of as implementing a kind of phylogenetic meta-analysis, the way in which conflicts are resolved needs a rational justification. In the light of the known behaviour of MRP, it remains to be demonstrated that there is any convincing justification for this approach to supertree construction.

In each of our examples, we considered only a pair of source trees. In the absence of additional information (such as the levels of support for relationships in each tree), each tree has equal weight and there is no basis for preferring relationships in one tree over another. However, with larger sets of source trees, relationships that are supported by many trees might reasonably be preferred over conflicting relationships supported by fewer trees. Although MRP might be expected to produce results that are generally consistent with this preference given sufficient source trees, it is far from certain that the biases apparent in our simple examples would not also play a part in producing unjustifiable resolutions of conflict in more complex cases. In contrast, direct methods based, for example on majority-rule consensus approaches (Margush and McMorris 1981) would provide a much more transparent meta-analysis.

Another supposed advantage of MRP is the potential to weight matrix elements in proportion to their support (Purvis 1995b; Ronquist 1996; Bininda-Emonds et al. 1999). In and of itself, we consider such differential weighting of relationships to be eminently sensible in any form of phylogenetic meta-analyses. However, in our third example such weighting was insufficient to maintain well-supported relationships in the face of the apparent bias due to tree shape. For meta-analysis we need methods that do not suffer such biases, as well as being able to accommodate information on differential levels of support. We think it more likely that direct methods of supertree construction can be developed with these desirable properties than it is that MRP-based methods can be modified to achieve these ends.

Conclusions

In conclusion, we have serious concerns about the utility of MRP approaches to supertree construction. MRP methods have undesirable properties, and if they are used at all they must be used with caution. It is possible that additional work might yield better-behaved variants of MRP. However, whilst not underestimating the complexity of the problems, we consider it more likely that if well-behaved and well-understood methods of supertree construction are to be developed (both for summarizing agreement and for meta-analytic resolution of conflict) they will be direct methods. In short, we believe that advocates of MRP methods may be barking up the wrong supertree.

Our real and our hypothetical examples illustrate how taxon selection can impact upon the degree of resolution of supertrees. Indeed, supertrees can be poorly resolved in the absence of conflict among the source trees purely as a result of the distribution of unique leaves. In general, denser sampling of common leaves will enhance prospects for well-resolved supertrees. If platyhelminth workers wish to promote the synthesis of their work through supertree methods, we recommend that they seek to use a common set of reference taxa in their disparate analyses. This may require sharing of resources. Although taxon substitution is an intuitively appealing approach to supertree construction, it is also subject to the rules for combining trees and the impact of unique leaves if we are unwilling simply to assume the monophyly of taxa. Thus, careful selection of outgroups in the manner we have described will help us approach our ultimate goal of a comprehensive supertree for Platyhelminthes.

ACKNOWLEDGEMENTS

J.L.T. was supported by a University of Bristol PhD Scholarship. D.T.J.L. was supported by a Wellcome Trust Senior Research Fellowship (043965/Z/95/Z).

References

Abouheif, E., R. Zardoya, and A. Meyer. 1998. Limitations of metazoan 18S rRNA sequence data: implications for reconstructing a phylogeny of the animal kingdom and inferring the reality of the Cambrian explosion. *Journal of Molecular Evolution*, 47:394–405.

Adlard, R. D., K. B. Sewell, and L. R. G. Cannon. 1996. Systematics within the Temnocephalida – gathering evidence from molecular and morphological characters, *8th International Symposium on the Biology of the Turbellaria*, loose sheet, Brisbane.

Adoutte, A., G. Balavoine, N. Lartillot, and R. de Rosa. 1999. Animal evolution – the end of the intermediate taxa? *Trends in Genetics*, 15:104–108.

Afanas'ev, K. I., and D. G. Tseitlin. 1993. [Isozyme analysis of *Triaenophorus nodulosus* from different sites of its distribution area]. (In Russian). *Trudy gel'mintologicheskoi laboratorii*, 39:3–6.

Afzelius, B. A. 1966. *Anatomy of the cell*. University of Chicago Press, Chicago.

Afzelius, B. A. 1982. The flagellar apparatus of marine spermatozoa. Evolutionary and functional aspects. In W. B. Amos and J. G. Duckett (eds.), *Prokaryotic and eukaryotic flagella*, pp. 495–519. Cambridge University Press, Cambridge.

Aguinaldo, A. M. A., and J. A. Lake. 1998. Evolution of the multicellular animals. *American Zoologist*, 38:878–887.

Aguinaldo, A. M. A., J. M. Turbeville, L. S. Linford, M. C. Rivera, J. R. Garey, R. A. Raff, and J. A. Lake. 1997. Evidence for a clade of nematodes, arthropods and other moulting animals. *Nature*, 387:489–493.

Ali, P. O., A. J. G. Simpson, R. Allen, A. P. Waters, C. J. Humphries, D. A. Johnston, and D. Rollinson. 1991. Sequence of a small subunit rRNA gene of *Schistosoma mansoni* and its use in phylogenetic analysis. *Molecular and Biochemical Parasitology*, 46:201–208.

Allison, R. R. 1980. Sensory receptors of the rosette organ of *Gyrocotyle rugosa*. *International Journal for Parasitology*, 10:341–353.

Amato, J. F. R. 1994. *Pseudobicotylophora atlantica* n. gen., n. sp. (Monogenea: Bicotylophoridae n. fam.), parasite of *Trachinotus* spp. (Osteichthyes: Carangidae) and redescription of *Bicotylophora trachinoti*. *Revista Brasileira de Parasitologia Veterinária*, 3:99–108.

Amato, J. F. R., and J. Pereira. 1995. A new species of *Rugogaster* (Aspidobothrea: Rugogastridae) parasite of the elephant fish, *Callorhinchus callorhynchi* (Callorhinchidae), from the estuary of the La Plata River, coasts of Uruguay and Argentina. *Revista Brasileira Parasitologia Veterinária*, 4:1–7.

Andersen, K., and S. Lysfjord. 1982. The functional morphology of the scolex of two *Tetrabothrius* Rudolphi, 1819 species (Cestoda: Tetrabothriidae) from penguins. *Zeitschrift für Parasitenkunde*, 67:299–307.

Anderson, D. T. 1973. *Embryology and phylogeny in annelids and arthropods*. Pergamon Press, Oxford.

Anderson, G. R. 1999. Identification and maturation of the metacercaria of *Indodidymozoon pearsoni*. *Journal of Helminthology*, 73:21–26.

Anderson, G. R., and S. C. Barker. 1998. Inference of phylogeny and taxonomy within the Didymozoidae (Digenea) from the second internal transcribed spacers (ITS2) of ribosomal DNA. *Systematic Parasitology*, 41:87–94.

Anderson, R. C. 1992. *Nematode parasites of vertebrates: their development and transmission*. CAB International, Wallingford.

Andrews, R. H., and N. B. Chilton. 1999. Multilocus enzyme electrophoresis. a valuable technique for providing answers to problems in parasite systematics. *International Journal for Parasitology*, 29:213–253.

Andrews, R. H., M. Adams, P. R. Baverstock, C. A. Behm, and C. Bryant. 1989. Genetic characterization of three strains of *Hymenolepis diminuta* at 39 enzyme loci. *International Journal for Parasitology*, 19:515–518.

Arme, C., and P. W. Pappas. 1983. *Biology of the Eucestoda*. Academic Press, London.

Arndt, W. 1923. Untersuchungen an Bachtricladen. Ein Beitrag zur Kenntis der Paludicolen Korsikas, Rumäniens und Siberiens. *Zeitschrift für wissenschaftliche Zoologie*, 120:98–146.

Arnolds, W. J. A., J. A. M. van den Biggelaar, and N. H. Verdonk. 1983. Spatial aspects of cell interactions involved in the determination of dorsoventral polarity in equally cleaving gastropods and regulative abilities of their embryos, as studied by micromere deletions in *Lymnaea* and *Patella*. *Roux's Archives of Developmental Biology*, 192:75–85.

Ax, P. 1952. *Ciliopharyngiella intermedia* nov. gen. nov. spec. Repräsentant einer neuen Turbellarien–Famili des marinen Mesopsammon. *Zoologische Jahrbücher Abteilung für Systematik*, 81:286–312.

Ax, P. 1956. Monographie der Otoplanidae (Turbellaria). Morphologie und Systematik. *Abhandlung der Mathematisch–Naturwissenschaftlichen Klasse. Akademie der Wissenschaften und der Literatur*. Mainz, 13:499–796.

Ax, P. 1958. Vervielfachung des Mannlichen Kopulationsapparetes bei Turbellarien. *Verhandlungen der Deutschen Zoologischen Gesellschaft*, 1957:227–249.

Ax, P. 1961. Verwandtschaftsbeziehungen und Phylogenie der Turbellarien. *Ergebnisse der Biologie*, 24:1–68.

Ax, P. 1963. Relationships and phylogeny of the Turbellaria. In E. C. Dougherty (ed.), *The lower Metazoa*, pp. 191–224. University of California Press, Berkeley, CA.

Ax, P. 1971. Neue interstitielle Macrostomida (Turbellaria) der Gattungen *Acanthomacrostomum* und *Haplopharynx*. *Mikrofauna Meeresboden*, 8:298–308.

Ax, P. 1984. *Das phylogenetische System. Systematisierung der lebenden Natur aufgrund ihrer Phylogenese*. Gustav Fischer, Stuttgart & New York.

Ax, P. 1985. The position of the Gnathostomulida and Platyhelminthes in the phylogenetic system of the Bilateria. In S. Conway Morris, J. D. George, R. Gibson and H. M. Platt (eds.), *The origins and relationships of lower invertebrates*, pp. 168–180. Clarendon Press, Oxford.

Ax, P. 1994a. *Japanoplana insolita* n. sp. eine neue Organisation der Lithophora (Seriata, Plathelminthes) aus Japan. *Microfauna Marina*, 9:7–23.

Ax, P. 1994b. *Macrostomum magnicurvituba* n. sp. (Macrostomida, Plathelminthes) replaces *Macrostomum curvituba* in coastal waters of Greenland and Iceland. *Microfauna Marina*, 9:335–338.

Ax, P. 1995. *Das System der Metazoa I. Ein Lehrbuch der phylogenetischen Systematik*. 1st ed. Gustav Fischer, Stuttgart.

Ax, P. 1996. *Multicellular animals. A new approach to the phylogenetic order in nature*. Springer, Berlin.

Ax, P., and W. Armonies. 1987. Amphiatlantic identities in the composition of the boreal brackish water community of Platyhelminthes. *Microfauna Marina*, 3:7–80.

Ax, P., and W. Armonies. 1990. Brackish water Plathelminthes from Alaska as evidence for the existence of a boreal brackish water community with circumpolar distribution. *Microfauna Marina*, 6:7–109.

Ax, P., and E. Schulz. 1959. Ungeschlechliche Fortpflanzung durch Paratomie bei acoelen Turbellarien. *Biologiches Zentralblatt*, 78:613–621.

Bâ, C. T., and B. Marchand. 1995. Spermiogenesis, spermatozoa and phyletic affinities in the Cestoda. In B. G. M. Jamieson, J. Ausio and J.-L. Justine (eds.), *Advances in spermatozoal phylogeny and taxonomy. Memoires du Museum National d'Histoire Naturelle*, Vol. 166, pp. 87–95. Muséum National d'Histoire Naturelle, Paris.

Bâ, C. T., and B. Marchand. 1998. Ultrastructure of spermiogenesis and the spermatozoon of *Vampirolepis microstoma* (Cestoda, Hymenolepididae), intestinal parasite of *Rattus rattus*. *Microscopy Research and Technique*, 42:218–225.

Bâ, C. T., X. Q. Wang, F. Renaud, L. Euzet, B. Marchand, and T. De Meeüs. 1994. Diversity in the genera *Avitellina* and *Thysaniezia* (Cestoda: Cyclophyllidea): genetic evidence. *Journal of the Helminthological Society of Washington*, 61:57–60.

Baccetti, B., and B. A. Afzelius. 1976. *The biology of the sperm cell*. S. Karger, Basel.

Baer, J. G. 1931. Étude monographique du groupe des temnocéphales. *Bulletin Biologique de la France et de la Belgique*, 65:1–59.

Baer, J. G. 1950. Phylogénie et cycles évolutifs des cestodes. *Revue Suisse de Zoologie*, 57:553–558.

Baer, J. G. 1953. Temnocéphales. Zoological results of the Dutch New Guinea Expedition of 1939. Number 4. *Zoologisches Mededelingen (Leiden)*, 32:119–139.

Baer, J. G. 1954. Revision taxonomique et étude biologique des cestodes de la famille des Tetrabothriidae parasites d'oiseaux de haute mer et de mammifères marins. *Mémoires de l'Université de Neuchâtel*, 1:4–122.

Baguñà, J. 1998. Planarians. In P. Ferretti and J. Géraudie (eds.), *Cellular and molecular basis of regeneration: from invertebrates to humans*, pp. 135–165. John Wiley & Sons, Chichester.

Baguñà, J., and R. Ballester. 1978. The nervous system in planarians: peripheral and gastrodermal plexuses, pharynx innervation, and the relationship between central nervous system and the acoelomate organization. *Journal of Morphology*, 155:237–252.

Baguñà, J., and B. C. Boyer. 1990. Descriptive and experimental embryology of the Turbellaria: present knowledge, open questions and future trends. In H.-J. Marthy (ed.), *Experimental embryology in aquatic plants and animals*, pp. 95–128. Plenum Press, New York.

Baguñà, J., E. Saló, R. Romero, J. Garcia–Fernàndez, D. Bueno, A. M. Muñoz–Mármol, J. R. Bayascas–Ramirez, and A. Casali. 1994. Regeneration and pattern–formation in planarians – cells, molecules and genes. *Zoological Science*, 11:781–795.

Baguñà, J., S. Carranza, M. Pala, C. Ribera, G. Giribet, M. A. Arnedo, M. Ribas, and M. Riutort. 1999. From morphology and karyology to molecules. New methods for taxonomical identification of asexual populations of freshwater planarians. A tribute to Professor Mario Benazzi. *Italian Journal of Zoology*, 66:207–214.

Baguñà, J., S. Carranza, J. Paps, I. Ruiz–Trillo, and M. Riutort. 2001. Molecular taxonomy and phylogeny of the Tricladida. In D. T. J. Littlewood and R. A. Bray (eds.), *Interrelationships of the Platyhelminthes*, pp. 49–56. Taylor & Francis, London.

Balavoine, G. 1997. The early emergence of platyhelminths is contradicted by the agreement between 18S rRNA and *Hox* genes data. *Comptes Rendus de l'Académie des sciences Paris, Série III, Sciences de la Vie*, 320:83–94.

Balavoine, G. 1998. Are Platyhelminthes coelomates without a coelom? An argument based on the evolution of *Hox* genes. *American Zoologist*, 38:843–858.

Ball, I. R. 1974a. A contribution to the phylogeny and biogeography of the freshwater triclads (Platyhelminthes: Turbellaria). In N. W. Riser and M. P. Morse (eds.), *The biology of the Turbellaria*, pp. 339–401. McGraw–Hill, New York.

Ball, I. R. 1974b. A new genus and species of freshwater planarian from Australia (Platyhelminthes: Turbellaria). *Journal of Zoology*, 174:149–158.

Ball, I. R. 1974c. A new genus of freshwater triclad from Tasmania, with reviews of the related genera *Cura* and *Neppia* (Turbellaria, Tricladida). *Life Sciences Contributions Royal Ontario Museum*, 99:1–48.

Ball, I. R. 1977a. A monograph of the genus *Spathula* (Platyhelminthes: Turbellaria: Tricladida). *Australian Journal of Zoology*, Supplementary Series no. 47:1–43.

Ball, I. R. 1977b. A new and primitive retrobursal planarian from Australian fresh waters (Platyhelminthes, Turbellaria, Tricladida). *Bijdragen tot de Dierkunde*, 47:149–155.

Ball, I. R., and D. A. Hay. 1977. The taxonomy and ecology of a new monocelid flatworm from Macquarie Island (Platyhelminthes, Turbellaria). *Bijdragen tot de Dierkunde*, 47:205–214.

Ball, I. R., and R. Sluys. 1990. Turbellaria: Tricladida: Terricola. In D. L. Dindal (ed.), *Soil biology guide*, pp. 37–153. John Wiley & Sons, New York.

Bandoni, S. M., and D. R. Brooks. 1987. Revision and phylogenetic analysis of the Amphilinidea Poche, 1922 (Platyhelminthes: Cercomeria: Cercomeromorpha). *Canadian Journal of Zoology*, 65:1110–1128.

Baptista–Farias, M. F. D., A. Kohn, and O. M. Barth. 1995. Ultrastructural aspects of spermatogenesis in *Metamicrocotyla macracantha* (Alexander, 1954) Koratha, 1955 (Monogenea; Microcotylidea). *Memórias do Instituto Oswaldo Cruz*, 90:597–604.

Barker, S. C., and D. Blair. 1996. Molecular phylogeny of *Schistosoma* species supports traditional groupings within the genus. *Journal of Parasitology*, 82:292–298.

Barker, S. C., D. Blair, A. R. Garrett, and T. H. Cribb. 1993a. Utility of the D1 domain of nuclear 28S ribosomal RNA for phylogenetic inference in the Digenea. *Systematic Parasitology*, 26:181–188.

Barker, S. C., D. Blair, T. H. Cribb, and K. Tonion. 1993b. Phylogenetic position of *Heronimus mollis* (Digenea): evidence from 18S ribosomal RNA. *International Journal for Parasitology*, 23:533–536.

Barnes, R. S. K., P. Calow, and P. J. Olive. 1988. *The invertebrates. A new synthesis*. Blackwell Scientific Publications, Oxford.

Baron, P. J. 1971. On the histology, histochemistry and ultrastructure of the cysticercoid of *Raillietina cesticillus* (Molin, 1858) Fuhrmann, 1920 (Cestoda, Cyclophyllidea). *Parasitology*, 62:233–245.

Barriel, V. 1994. Molecular phylogenies and how to code insertion/deletion events. *Comptes Rendus de l'Académie des Sciences, Paris, Série III, Sciences de la Vie*, 317:693–701.

Bartolomaeus, T., and P. Ax. 1992. Protonephridia and metanephridia – their relation within the Bilateria. *Zeitschrift für zoologische Systematik und Evolutionsforschung*, 30:21–45.

Basch, P. F. 1991. *Schistosomes: development, reproduction and host relations*. Oxford University Press, New York.

Batista, C., E. Matos, and C. Azevedo. 1999. Ultrastructural aspects of spermiogenesis of *Temnocephala* sp. (Turbellaria, Rhabdocoela), ectoparasite of *Dilocarcinus septemdentatus* Herbst, 1783 (Crustacea, Decapoda). *Journal of Submicroscopic Cytology and Pathology*, 31:65–72.

Batistoni, R., C. Filippi, A. Salvetti, M. Cardelli, and P. Deri. 1998. Repeated DNA elements in planarians of the *Dugesia gonocephala* group (Platyhelminthes, Tricladida). *Hydrobiologia*, 383:139–146.

Baum, B. R. 1992. Combining trees as a way of combining data sets for phylogenetic inference, and the desirability of combining gene trees. *Taxon*, 41:3–10.

Baverstock, P. R., M. Adams, and I. Beveridge. 1985. Biochemical differentiation in bile duct cestodes and their marsupial hosts. *Molecular Biology and Evolution*, 2:321–337.

Baverstock, P. R., R. Fielke, A. M. Johnson, R. A. Bray, and I. Beveridge. 1991. Conflicting phylogenetic hypotheses for the parasitic platyhelminths tested by partial sequencing of 18S ribosomal RNA. *International Journal for Parasitology*, 21:329–339.

Bayascas, J. R., E. Castillo, A. M. Muñoz-Mármol, and E. Saló. 1997. Planarian Hox genes: novel patterns of expression during regeneration. *Development*, 124:141–148.

Bayascas, J. R., E. Castillo, and E. Saló. 1998. Platyhelminthes have a *Hox* code differentially activated during regeneration, with genes closely related to those of spiralian protostomes. *Development Genes and Evolution*, 208:467–473.

Bayssade–Dufour, C. 1979. L'appareil sensoriel des cercaires et la systématique des Trématodes digénétiques. *Mémoires du Muséum National d'Histoire Naturelle. Série A Zoologie*, 113:1–81.

Bayssade–Dufour, C., and B. Grabda–Kazubska. 1993. Révision systématique d'*Omphalometra flexuosa* (Digenea, Plagiorchiata). Relations avec les genres *Opisthioglyphe*, *Lecithopyge* et *Plagiorchis*. *Annales de Parasitologie Humaine et Comparée*, 68:82–87.

Bayssade–Dufour, C., J.–P. Hugot, and J.–L. Albaret. 1993. Analyse phénétique des Microphalloidea (Trematoda) d'après la chétotaxie des cercaires. *Systematic Parasitology*, 25:1–24.

Beauchamp, P. d. 1932. Biospeologica LVI [LVIII]. Turbellariés, Hirudinées, Branchiobdellidés. Deuxième Série. *Archives de Zoologie Expérimentale et Génerale*, 73:113–380.

Bedini, C., and A. Lanfranchi. 1991. The central and peripheral nervous system of Acoela (Plathelminthes). An electron microscopical study. *Acta Zoologica (Stockholm)*, 72:101–106.

Bedini, C., and A. Lanfranchi. 1998. Ultrastructural study of the brain of a typhloplanid flatworm. *Acta Zoologica (Stockholm)*, 79:243–249.

Bedini, C., and F. Papi. 1970. Peculiar patterns of microtubular organisation in spermatozoa of lower Turbellaria. In B. Baccetti (ed.), *Comparative spermatology*, pp. 363–368. Academic Press, New York.

Bedini, C., and F. Papi. 1974. Fine structure of the turbellarian epidermis. In N. W. Riser and M. P. Morse (eds.), *The biology of the Turbellaria*, pp. 476–492. McGraw–Hill, New York.

Beklemischev, V. N. 1963. On the relationship of the Turbellaria to other groups of the animal kingdom. In E. C. Dougherty (ed.), *The lower Metazoa*, pp. 243–244. University of California Press, Berkeley.

Benham, W. B. 1901. The Platyhelminia, Mesozoa and Nemertini. In R. Lankester (ed.), *Treatise on zoology, part 4*. Adam & Charles Black, London.

Benito, J., and F. Pardos. 1997. Hemichordata. In F. W. Harrison and E. E. Ruppert (eds.), *Microscopic anatomy of invertebrates. Volume 15. Hemichordata, Chaetognatha, and the invertebrate chordates*, pp. 15–101. Wiley–Liss, Inc., New York.

Bennett, J. L., and G. Gianutsos. 1977. Distribution of catecholamines in immature *Fasciola hepatica*: a histochemical and biochemical study. *International Journal for Parasitology*, 7:221–225.

Benz, G. W. 1987. *Dermophthirius penneri* sp. n. (Monogenea: Microbothriidae) an ectoparasite of carcharinid sharks, *Carcharhinus brevipinna* and *Carcharhinus limbatus*. *Proceedings of the Helminthological Society of Washington*, 54:185–190.

Bessho, Y., T. Ohama, and S. Osawa. 1992. Planarian mitochondria I. Heterogeneity of cytochrome C oxidase subunit I gene sequences in the freshwater planarian, *Dugesia japonica*. *Journal of Molecular Evolution*, 34:324–330.

Beveridge, I. 1994. Family Anoplocephalidae Cholodkovsky, 1902. In L. F. Khalil, A. Jones and R. A. Bray (eds.), *Keys to the cestode parasites of vertebrates*, pp. 315–374. CAB International, Wallingford.

Beveridge, I. 1998. Allozyme electrophoresis – difficulties encountered in studies on helminths. *International Journal for Parasitology*, 28:973–979.

Beveridge, I., M. D. Rickard, G. G. Gregory, and B. L. Munday. 1975. Studies on *Anoplotaenia dasyuri* Beddard, 1911 (Cestoda: Taeniidae), a parasite of the Tasmanian devil: Observations on the egg and metacestode. *International Journal for Parasitology*, 5:257–267.

Beveridge, I., R. A. Campbell, and H. W. Palm. 1999. Preliminary cladistic analysis of genera of the cestode order Trypanorhyncha Diesing, 1863. *Systematic Parasitology*, 42:29–49.

Bininda–Emonds, O. R. P., and H. N. Bryant. 1998. Properties of matrix representation with parsimony analyses. *Systematic Biology*, 47:497–508.

Bininda–Emonds, O. R. P., H. N. Bryant, and A. P. Russell. 1998. Supraspecific taxa as terminal in cladistic analysis: implicit assumptions of monophyly and a comparison of methods. *Biological Journal of the Linnean Society*, 94:101–133.

Bininda–Emonds, O. R. P., J. L. Gittleman, and A. Purvis. 1999. Building large trees by combining phylogenetic information: A complete phylogeny of the extant Carnivora (Mammalia). *Biological Reviews*, 74:143–175.

Birstein, V. J. 1991. On the karyotypes of the Neorhabdocoela species and karyological evolution of Turbellaria. *Genetica*, 83:107–120.

Biserkov, V., and T. Genov. 1988. On the life–cycle of *Ophiotaenia europaea* Odening, 1963 (Cestoda: Ophiotaeniidae). *Khelmintologiya*, 25:7–14.

Blair, D. 1993. The phylogenetic position of the Aspidobothrea within the parasitic flatworms inferred from ribosomal RNA sequence data. *International Journal for Parasitology*, 23:169–178.

Blair, D., and S. C. Barker. 1993. Affinities of the Gyliauchenidae: utility of the 18S rDNA gene for phylogenetic inference in the Digenea (Platyhelminthes). *International Journal for Parasitology*, 23:527–532.

Blair, D., A. Campos, M. P. Cummings, and J. P. Laclette. 1996. Evolutionary biology of parasitic platyhelminths: the role of molecular phylogenetics. *Parasitology Today*, 12:66–71.

Blair, D., L. van Herwerden, H. Hirai, T. Taguchi, S. Habe, M. Hirata, K. Lai, S. Upatham, and T. Agatsuma. 1997. Relationships between *Schistosoma malayensis* and other Asian schistosomes deduced from DNA sequences. *Molecular and Biochemical Parasitology*, 85:259–263.

Blair, D., R. A. Bray, and S. C. Barker. 1998. Molecules and morphology in phylogenetic studies of the Hemiuroidea (Digenea: Trematoda: Platyhelminthes). *Molecular Phylogenetics and Evolution*, 9:15–25.

Blair, D., T. H. Le, L. Després, and D. P. McManus. 1999. Mitochondrial genes of *Schistosoma mansoni*. *Parasitology*, 119:303–313.

Blair, K. L., and P. A. V. Anderson, 1996. Physiology and pharmacology of turbellarian neuromuscular systems. *Parasitology*, 113:S73–S82.

Böckerman, I., O. I. Raikova, M. Reuter, and O. Timoshkin. 1994. Ultrastructure of the central nervous system of *Geocentrophora baltica* (Lecithoepitheliata) and the surface sensillae in the *Geocentrophora* group. *Hydrobiologia*, 305:183–185.

Boeger, W. A., and D. C. Kritsky. 1993. Phylogeny and a revised classification of the Monogenoidea Bychowsky, 1937 (Platyhelminthes). *Systematic Parasitology*, 26:1–32.

Boeger, W. A., and D. C. Kritsky. 1997. Coevolution of the Monogenoidea (Platyhelminthes) based on a revised hypothesis of parasite phylogeny. *International Journal for Parasitology*, 27:1495–1511.

Boeger, W. A., and D. C. Kritsky. 2001. Phylogeny of the Monogenoidea: a critical review of hypotheses. In D. T. J. Littlewood and R. A. Bray (eds.), *Interrelationships of the Platyhelminthes*, pp. 92–102. Taylor & Francis, London.

Boeger, W. A., D. C. Kritsky, and E. Belmont-Jégu. 1994. Neotropical Monogenoidea. 20. Two new species of oviparous Gyrodactylidea (Polyonchoinea) from loricariid catfishes (Siluriformes) in Brazil and the phylogenetic status of Ooegyrodactylidae Harris 1983. *Journal of the Helminthological Society of Washington*, 61:34–44.

Boguta, K. K. 1976. [Morphological structure of the nervous system in *Convoluta convoluta* (Turbellaria, Acoela) and its changes under the effect of long–term starvation]. (In Russian). *Zoologicheskii Zhurnal*, 55:815–822.

Böhmig, L. 1887. Zur Kenntnis der Sinnesorgane der Turbellarien. *Zoologischer Anzeiger*, 10:484–488.

Bona, F. V. 1994. Family Dilepididae Railliet and Henry, 1909. In L. F. Khalil, A. Jones and R. A. Bray (eds.), *Keys to the cestode parasites of vertebrates*, pp. 443–554. CAB International, Wallingford.

Boore, J. L. 1999. Animal mitochondrial genes. *Nucleic Acids Research*, 27:1767–1780.

Borradaile, L. A., F. A. Potts, L. E. S. Eastham, and J. T. Saunders. 1948. *The Invertebrata*. Cambridge University Press, Cambridge.

Bowles, J., and D. P. McManus. 1993a. Molecular variation in *Echinococcus*. *Acta Tropica*, 53:291–305.

Bowles, J., and D. P. McManus. 1993b. NADH dehydrogenase 1 gene sequences compared for species and strains of the genus *Echinococcus*. *International Journal for Parasitology*, 23:969–972.

Bowles, J., and D. P. McManus. 1993c. Rapid discrimination of *Echinococcus* species and strains using a polymerase chain reaction-based RFLP method. *Molecular and Biochemical Parasitology*, 57:231–240.

Bowles, J., and D. P. McManus. 1994. Genetic characterization of the Asian *Taenia*, a newly described taeniid cestode of humans. *American Journal of Tropical Medicine and Hygiene*, 50:33–44.

Bowles, J., D. Blair, and D. P. McManus. 1992. Genetic variants within the genus *Echinococcus* identified by mitochondrial DNA sequencing. *Molecular and Biochemical Parasitology*, 54:165–174.

Bowles, J., D. Blair, and D. P. McManus. 1994. Molecular genetic characterization of the cervid strain ('northern form') of *Echinococcus granulosus*. *Parasitology*, 109:215–221.

Bowles, J., D. Blair, and D. P. McManus. 1995a. A molecular phylogeny of the genus *Echinococcus*. *Parasitology*, 110:317–328.

Bowles, J., D. Blair, and D. P. McManus. 1995b. A molecular phylogeny of the human schistosomes. *Molecular Phylogenetics and Evolution*, 4:103–109.

Boyer, B. C. 1971. Regulative development in a spiralian embryo as shown by cell deletion experiments on the acoel, *Childia*. *Journal of Experimental Zoology*, 176:97–106.

Boyer, B. C. 1989. The role of the first quartet micromeres in the development of the polyclad *Hoploplana inquilina*. *Biological Bulletin*, 177:338–343.

Boyer, B. C., and J. Q. Henry. 1998. Evolutionary modifications of the spiralian developmental program. *American Zoologist*, 38:621–633.

Boyer, B. C., and G. W. Smith. 1982. Sperm morphology and development in two acoel Turbellarians from the Philippines. *Pacific Science*, 36:365–380.

Boyer, B. C., J. J. Henry, and M. Q. Martindale. 1996a. Dual origins of mesoderm in a basal spiralian: cell lineage analysis in the polyclad turbellarian *Hoploplana inquilina*. *Developmental Biology*, 179:329–338.

Boyer, B. C., J. J. Henry, and M. Q. Martindale. 1996b. Modified spiral cleavage: the duet cleavage pattern and early blastomere fates in the acoel turbellarian *Neochildia fusca*. *Biological Bulletin*, 191:285–286.

Boyer, B. C., J. J. Henry, and M. Q. Martindale. 1998. The cell lineage of a polyclad turbellarian embryo reveals close similarity to coelomate spiralians. *Developmental Biology*, 204:111–123.

Bozhkov, D. K. 1982. [*Helminth life-cycles and their evolution*]. (In Russian). Durzhavno Izdatelstvo Nauka i Izkustvo, Sofia.

Bråten, T. 1968. The fine structure of the tegument of *Diphyllobothrium latum* (L.) – a comparison of the plerocercoid and adult stages. *Zeitschrift für Parasitenkunde*, 30:104–112.

Bray, R. A. 1987. A revision of the family Zoogonidae Odhner, 1902 (Platyhelminthes, Digenea): subfamily Lepidophyllinae and comments on some aspects of biology. *Systematic Parasitology*, 9:83–123.

Bray, R. A. 1991. Species and speciation in parasitic helminths. In Bulgarian Academy of Sciences Institute of Parasitology (ed.), *Parasite host environment, Second International School*, pp. 174–184. Publishing House of the Bulgarian Academy of Sciences, Sofia, Bulgaria.

Bray, R. A., A. Jones, and K. I. Anderson. 1994a. Order Pseudophyllidea Carus, 1863. In L. F. Khalil, A. Jones and R. A. Bray (eds.), *Keys to the cestode parasites of vertebrates*, pp. 205–247. CAB International, Wallingford, U.K.

Bray, R. A., A. Soto, and D. Rollinson. 1994b. The status and composition of the genus *Steringophorus* Odhner, 1905 (Digenea: Fellodistomidae), based on partial small subunit rRNA sequences. *International Journal for Parasitology*, 24:433–435.

Bray, R. A., T. H. Cribb, and S. C. Barker. 1996. *Cableia pudica* n. sp. (Digenea: Acanthocolpidae) from monacanthid fishes of the southern Great Barrier Reef, Australia. *Parasite*, 3:49–54.

Bray, R. A., A. Jones, and E. P. Hoberg. 1999. Observations on the phylogeny of the cestode order Pseudophyllidea Carus, 1863. *Systematic Parasitology*, 42:13–20.

Bremer, K. 1994. Branch support and tree stability. *Cladistics*, 10:295–304.

Bresciani, J. 1991. Nematomorpha. In F. W. Harrison (ed.), *Microscopic anatomy of invertebrates. Volume 4: Aschelminthes*, pp. 197–218. Wiley–Liss, Inc., New York.

Bresslau, E. 1928–33. Turbellaria. *Handbuch der Zoologie*, 2:52–320.

Briggs, J. C. 1974. *Marine zoogeography*. McGraw–Hill, New York.

Brooks, D. R. 1978. Evolutionary history of the cestode order Proteocephalidea. *Systematic Zoology*, 27:312–323.

Brooks, D. R. 1982. Higher level classification of parasitic Platyhelminthes and fundamentals of cestode classification. In D. F. Mettrick and S. S. Desser (eds.), *Parasites – their world and ours*, pp. 189–193. Elsevier Biomedical Press, Amsterdam.

Brooks, D. R. 1989a. A summary of the database pertaining to the phylogeny of the major groups of parasitic platyhelminths, with a revised classification. *Canadian Journal of Zoology*, 67:714–720.

Brooks, D. R. 1989b. The phylogeny of the Cercomeria (Platyhelminthes, Rhabdocoela) and general evolutionary principles. *Journal of Parasitology*, 75:606–616.

Brooks, D. R. 1992. Origins, diversification, and historical structure of the helminth fauna inhabiting Neotropical freshwater stingrays (Potomotrygonidae). *Journal of Parasitology*, 78:588–595.

Brooks, D. R. 1993. Phylogenetic hypothesis, cladistic diagnoses, and the classification of the Monticellidae (Eucestoda: Proteocephaliformes). *Revista Brasileira de Biologia*, 55:359–367.

Brooks, D. R., and E. P. Hoberg. 2000. Triage for the biosphere: the need and rationale for taxonomic inventories and phylogenetic studies of parasites. *Comparative Parasitology*, 67:1–25.

Brooks, D. R., and D. A. McLennan. 1991. *Phylogeny, ecology and behaviour: a research program in comparative biology*. University of Chicago Press, Chicago.

Brooks, D. R., and D. A. McLennan. 1993a. *Parascript: parasites and the language of evolution*. Smithsonian Institution Press, Washington.

Brooks, D. R., and D. A. McLennan. 1993b. Macroevolutionary patterns of morphological diversification among parasitic flatworms (Platyhelminthes, Cercomeria). *Evolution*, 47:495–509.

Brooks, D. R., T. B. Thorson, and M. A. Mayes. 1981. Freshwater stingrays (Potamotrygonidae) and their helminth parasites: testing hypotheses of evolution & coevolution. In V. A. Funk and D. R. Brooks (eds.), *Advances In Cladistics. Proceedings of the First Willi Hennig Society Meeting*, pp. 147–175. New York Botanical Garden, New York.

Brooks, D. R., R. T. O'Grady, and D. R. Glen. 1985a. Phylogenetic analysis of the Digenea (Platyhelminthes: Cercomeria) with comments on their adaptive radiation. *Canadian Journal of Zoology*, 63:411–443.

Brooks, D. R., R. T. O'Grady, and D. R. Glen. 1985b. The phylogeny of the Cercomeria Brooks, 1982 (Platyhelminthes). *Proceedings of the Helminthological Society of Washington*, 52:1–20.

Brooks, D. R., S. M. Bandoni, C. A. MacDonald, and R. T. O'Grady. 1989. Aspects of the phylogeny of the Trematoda Rudolphi, 1808 (Platyhelminthes, Cercomeria). *Canadian Journal of Zoology*, 67:2609–2624.

Brooks, D. R., E. P. Hoberg, and P. J. Weekes. 1991. Preliminary phylogenetic systematic analysis of the major lineages of the Eucestoda (Platyhelminthes, Cercomeria). *Proceedings of the Biological Society of Washington*, 104:651–668.

Brownlee, D. J. A., and I. Fairweather. 1996. Immunocytochemical localization of glutamate-like immunoreactivity within the nervous system of the cestode *Mesocestoides corti* and the trematode *Fasciola hepatica*. *Parasitology Research*, 82:423–427.

Brüggemann, J. 1984. Ultrastruktur un Differenzierung der prostatoiden Organe von *Polystyliphora filum* (Plathelminthes, Proseriata). *Zoomorphology*, 104:86–95.

Brüggemann, J. 1985. Ultrastruktur und Bildungsweise penialer Hartstrukturen bei freilebenden Plathelminthen. *Zoomorphology*, 105:143–189.

Brüggemann, J. 1986a. Feinstruktur der Protonephridien von *Paromalostomum proceracauda* (Plathelminthes, Macrostomida). *Zoomorphology*, 106:147–154.

Brüggemann, J. 1986b. Ultrastructural investigations on the differentiation of genital hard structures in free-living platyhelminths and their phylogenetic significance. *Hydrobiologia*, 132:151–156.

Brüggemann, J. 1987. Zur Feinstruktur der Protonephridien von *Provortex psammophilus* (Plathelminthes, Rhabdocoela). *Zoomorphology*, 107:96–102.

Brüggemann, J. 1989. Ultrastruktur des Exkretionssystems von *Vejdovskya pellucida* (Plathelminthes, Rhabdocoela). *Microfauna Marina*, 5:227–242.

Brunanska, M. 1999. Ultrastructure of primary embryonic envelopes in *Proteocephalus longicollis* (Cestoda: Proteocephalidea). *Helminthologia*, 36:83–89.

Brusca, R. C., and G. J. Brusca. 1990. *Invertebrates*. Sinauer Associates, Sunderland, MA.

Bryant, D., and M. Steel. 1995. Extension operations of sets of leaf-labelled trees. *Advances in Applied Mathematics*, 16:425–453.

Bryant, H. N. 1994. Comments on the phylogenetic definition of taxon names and conventions regarding the naming of crown clades. *Systematic Biology*, 43:124–130.

Bryant, H. N. 1997. Cladistic information in phylogenetic definitions and designated phylogenetic contexts for the use of taxon names. *Biological Journal of the Linnean Society*, 62:495–503.

Bugge, G. 1902. Zur Kenntnis des Excretionsgefäss-Systems der Cestoden und Trematoden. *Zoologische Jahrbücher Abteilung für Morphologie*, XVI:12–234.

Bullock, T. H., and G. A. Horridge. 1965. *Structure and function in the nervous systems of invertebrates*. W. H. Freeman and Co., San Francisco.

Burt, M. D. B. 1987. Early morphogenesis in the Platyhelminthes with special reference to egg development and development of cestode larvae. *International Journal for Parasitology*, 17:241–253.

Buth, D. G., and R. W. Murphy. 1999. The use of isozyme characters in systematic studies. *Biochemical Systematics and Ecology*, 27:117–129.

Bychowsky, B. E. 1937. [Ontogenesis and phylogenetic interrelationships of parasitic flatworms]. (In Russian). *Izvestiia Akademii Nauk SSSR, Ser Biologiya*, 4:1354–1383 (English translation, Virginia Institute of Marine Science, Translation series 26, 1981).

Cable, J., and R. C. Tinsley. 1993. The ultrastructure of spermatogenesis and spermatozoa in *Pseudodiplorchis americanus* (Monogenea: Polystomatidae). *Canadian Journal of Zoology*, 71:1609–1619.

Cable, R. M. 1954. Studies on marine digenetic trematodes of Puerto Rico. The life cycle in the family Haplosplanchnidae. *Journal of Parasitology*, 40:71–76.

Cable, R. M. 1974. Phylogeny and taxonomy of trematodes with reference to marine species. In W. B. Vernberg (ed.), *Symbiosis in the sea*, pp. 173–193. University of South Carolina Press, Columbia, South Carolina.

Caira, J. N. 1985. *Calliobothrium evani* sp. n. (Tetraphyllidea: Onchobothriidae) from the Gulf of California, with a redescription of the hooks of C. *lintoni* and a proposal for onchobothriid hook terminology. *Proceedings of the Helminthological Society of Washington*, 52:166–174.

Caira, J. N. 1990. Metazoan parasites as indicators of elasmobranch biology. In H. L. Pratt, S. H. Gruber and T. Taniuchi (eds.), *Elasmobranchs as living resources: advances in the biology, ecology, systematics, and the status of the fisheries*, Vol. 90, pp. 71–96. NOAA Technical Report NMFS, Seattle.

Caira, J. N. 1992. Verification of multiple species of *Pedibothrium* in the Atlantic nurse shark with comments on the Australasian members of the genus. *Journal of Parasitology*, 78:289–308.

Caira, J. N., and T. Bardos. 1996. Further information on *Gymnorhynchus isuri* (Trypanorhyncha: Gymnorhynchidae) from the shortfin mako shark. *Journal of the Helminthological Society of Washington*, 63:188–192.

Caira, J. N., and L. S. Runkle. 1993. Two new tapeworms from the goblin shark *Mitsukurina owstoni* off Australia. *Systematic Parasitology*, 26:81–90.

Caira, J. N., C. J. Healy, and J. Swanson. 1996. A new species of *Phoreiobothrium* (Cestoidea: Tetraphyllidea) from the great hammerhead shark *Sphyrna mokarran* and its implications for the evolution of the onchobothriid scolex. *Journal of Parasitology*, 82:458–462.

Caira, J. N., K. Jensen, Y. Yamane, A. Isobe, and K. Nagasawa. 1997. On the tapeworms of *Megachasma pelagios*: Description of a new genus and species of lecanicephalidean and additional information on the trypanorhynch *Mixodigma leptaleum*. In K. Yano, J. F. Morrissey, Y. Yabumoto and K. Nakaya (eds.), *Biology of the megamouth shark*, pp. 181–191. Tokai University Press, Tokyo.

Caira, J. N., K. Jensen, and C. J. Healy. 1999. On the phylogenetic relationships among tetraphyllidean, lecanicephalidean and diphyllidean tapeworm genera. *Systematic Parasitology*, 42:77–151.

Caira, J. N., K. Jensen, and C. J. Healy. 2001. Interrelationships among tetraphyllidean and lecanicephalidean cestodes. In D. T. J. Littlewood and R. A. Bray (eds.), *Interrelationships of the Platyhelminthes*, pp. 135–158. Taylor & Francis, London.

Caley, J. 1974. The functional significance of scolex retraction and subsequent cyst formation in the cysticercoid larva of *Hymenolepis microstoma*. *Parasitology*, 68:207–227.

Caley, J. 1976. Ultrastructural studies of the cysticercoid of *Moniezia expansa* (Anoplocephalidae) with special reference to the development of the cyst. *Zeitschrift für Parasitenkunde*, 48:251–262.

Callaerts, P., A. M. Muñoz–Mármol, S. Glardon, E. Castillo, H. Sun, W.–H. Li, W. J. Gehring, and E. Saló. 1999. Isolation and expression of a *Pax–6* gene in the regenerating and intact planarian *Dugesia (G.) tigrina*. *Proceedings of the National Academy of Sciences USA*, 96:558–563.

Campbell, R. A. 1970. Notes on tetraphyllidean cestodes from the Atlantic Coast of North America, with descriptions of two new species. *Journal of Parasitology*, 56:498–508.

Campbell, R. A. 1975. Tetraphyllidean cestodes from western North Atlantic selachians with descriptions of two new species. *Journal of Parasitology*, 61:265–270.

Campbell, R. A., and I. Beveridge. 1994. Order Trypanorhyncha Diesing, 1863. In L. F. Khalil, A. Jones and R. A. Bray (eds.), *Keys to the cestode parasites of vertebrates*. CAB International, Wallingford.

Campos, A., M. P. Cummings, J. L. Reyes, and J. P. Laclette. 1998. Phylogenetic relationships of Platyhelminthes based on 18S ribosomal gene sequences. *Molecular Phylogenetics and Evolution*, 10:1–10.

Cannon, L. R. G. 1986. *Turbellaria of the world. A guide to families and genera*. Queensland Museum, Brisbane.

Cannon, L. R. G. 1991. Temnocephalan symbionts of the freshwater crayfish *Cherax quadricarinatus* from northern Australia. *Hydrobiologia*, 227:341–347.

Cannon, L. R. G. 1993. New temnocephalans (Platyhelminthes): ectosymbionts of freshwater crabs of shrimps. *Memoirs of the Queensland Museum*, 33:17–40.

Cannon, L. R. G., and B. I. Joffe, 2001. The Temnocephalida. In: D. T. J. Littlewood and R. A. Bray (eds.), *Interrelationships of the Platyhelminthes*, pp. 83–91. Taylor & Francis, London.

Cannon, L. R. G., and K. B. Sewell. 1995. Craspedellinae Baer, 1931 (Platyhelminthes: Temnocephalida) ectosymbionts from the branchial chamber of Australian crayfish (Crustacea: Parastacidae). *Memoirs of the Queensland Museum*, 38:397–418.

Cannon, L. R. G., and N. A. Watson. 1996. Postero-lateral glands of *Temnocephala minor* Haswell, 1888 (Platyhelminthes: Temnocephalida). *Australian Journal of Zoology*, 44:69–73.

Carmichael, A. C. 1984. Phylogeny of and historical biogeography of the Schistosomatidae. PhD thesis, Michigan State University.

Carpenter, K. S., M. Morita, and J. B. Best. 1974. Ultrastructure of the photoreceptor of the planarian *Dugesia dorotocephala*. I. Normal eye. *Cell and Tissue Research*, 148:143–158.

Carranza, S., G. Giribet, C. Ribera, J. Baguñà, and M. Riutort. 1996. Evidence that two types of 18S rDNA coexist in the genome of *Dugesia (Schmidtea) mediterranea* (Platyhelminthes, Turbellaria, Tricladida). *Molecular Biology and Evolution*, 13:824–832.

Carranza, S., J. Baguñà, and M. Riutort. 1997. Are the Platyhelminthes a monophyletic primitive group? An assessment using 18S rDNA sequences. *Molecular Biology and Evolution*, 14:485–497.

Carranza, S., D. T. J. Littlewood, K. A. Clough, I. Ruiz-Trillo, J. Baguñà, and M. Riutort. 1998a. A robust molecular phylogeny of the Tricladida (Platyhelminthes: Seriata) with a discussion on morphological synapomorphies. *Proceedings of the Royal Society of London, Series B, Biological Sciences*, 265:631–640.

Carranza, S., I. Ruiz-Trillo, D. T. J. Littlewood, M. Riutort, and J. Baguñà. 1998b. A reappraisal of the phylogenetic and taxonomic position of land planarians (Platyhelminthes, Turbellaria, Tricladida) inferred from 18S rDNA sequences. *Pedobiologia*, 42:433–440.

Carranza, S., J. Baguñà, and M. Riutort. 1999. Origin and evolution of paralogous rRNA gene clusters within the flatworm family Dugesiidae (Platyhelminthes, Tricladida). *Journal of Molecular Evolution*, 49:250–259.

Carus, J. V. 1863. Vermes. In W. C. H. Peters, J. V. Carus and C. E. A. Gerstaecker (eds.), *Handbuch der Zoologie*, pp. 422–484. Verlag von Wilhelms Engelmann, Leipzig.

Castresana, J., G. Feldmaier–Fuchs, and S. Pääbo. 1998. Codon rearrangement and amino acid composition in hemichordate mitochondria. *Proceedings of the National Academy of Sciences USA*, 95:3703–3707.

Cavalier-Smith, T. 1998. A revised six-kingdom system of life. *Biological Reviews*, 73:203–266.

Chew, M. W. K. 1983. *Taenia crassiceps*: ultrastructural observations on the oncosphere and associated structures. *Journal of Helminthology*, 57:101–113.

Chien, P. K., and G. A. Koopowitz. 1972. The ultrastructure of neuromuscular system in *Notoplana acticola*, a free-living polyclad flatworm. *Zeitschrift für Zellforschung*, 133:277–288.

Chirgwin, J. M., A. E. Przbyla, R. J. MacDonald, and W. Rutter. 1979. Isolation of biologically active ribonucleic acid from sources enriched in ribonuclease. *Biochemistry*, 18:5294–5299.

Chou, T.-C. T., J. L. Bennett, and E. Bueding. 1972. Occurrence and concentrations of biogenic amines in trematodes. *Journal of Parasitology*, 58:1098–1102.

Christensen, A. M., and B. Kanneworff. 1964. *Kronborgia amphipodicola* gen. et sp. nov., a dioecious turbellarian parasitizing ampeliscid amphipods. *Ophelia*, 1:147–166.

Cifrian, B., P. Garcia-Corrales, and S. Martinez–Alós. 1988b. Ultrastructural study of spermatogenesis and mature spermatozoa of *Paravortex cardii* (Platyhelminthes, Dalyellioida). *Acta Zoologica (Stockholm)*, 69:195–204.

Cifrian, B., S. Martinez–Alós, and P. Garcia-Corrales. 1988a. Ultrastructural study of spermatogenesis and mature spermatozoon of *Bothromesostoma personatum* (Rhabdocoela, Typhloplanoida). *Fortschritte der Zoologie*, 36:309–314.

Coil, W. H. 1987. The early egg of *Austramphilina elongata* (Cestodaria). *Parasitology Research*, 73:451–457.

Coil, W. H. 1991. Platyhelminthes: Cestoidea. In F. W. Harrison and B. J. Bogitsh (eds.), *Microscopic anatomy of invertebrates. Volume 3: Platyhelminthes and Nemertinea*, pp. 211–283. Wiley–Liss, Inc., New York.

Collin, W. K. 1970. Electron microscopy of postembryonic stages of the tapeworm, *Hymenolepis citelli*. *Journal of Parasitology*, 56:1159–1170.

Collins, A. G. 1998. Evaluating multiple alternative hypotheses for the origin of Bilateria: an analysis of 18S molecular evidence. *Proceedings of the National Academy of Science USA*, 95:15458–15463.

Combes, C. 1965. *Euzetrema knoepffleri* n. gen., n. sp. (Monogenea) parasite interne d'un amphibien endémique de corse. *Annales de Parasitologie Humaine et Comparée*, 40:451–457.

Combes, C. 1991. The schistosome scandal. *Acta Oecologica*, 12:165–173.

Combes, C., J. Jourdane, and L.-P. Knoepffler. 1974. Le cycle biologique de *Euzetrema knoepffleri* Combes, 1965 (Monogenea, Monopisthocotylea), parasite de l'euprocte de corse. *Bulletin de la Societé Zoologique de France*, 99:219–236.

Conn, D. B. 1985a. Fine structure of the embryonic envelopes of *Oochoristica anolis* (Cestoda: Linstowiidae). *Zeitschrift für Parasitenkunde*, 71:639–648.

Conn, D. B. 1985b. Life cycle and postembryonic development of *Oochoristica anolis* (Cyclophyllidea: Linstowiidae). *Journal of Parasitology*, 71:10–16.

Conn, D. B. 1988. Development of the embryonic envelopes of *Mesocestoides lineatus* (Cestoda: Cyclophyllidea). *International Journal of Invertebrate Reproduction and Development*, 14:119–130.

Conn, D. B., F. J. Etges, and R. A. Sidner. 1984. Fine structure of the gravid paruterine organ and embryonic envelopes of *Mesocestoides lineatus* (Cestoda). *Journal of Parasitology*, 70:68–77.

Constantinescu, M., and D. Sankoff. 1995. An efficient algorithm for supertrees. *Journal of Classification*, 12:101–112.

Costello, D. P., C. Henley, and C. R. Ault. 1969. Microtubules in spermatozoa of *Childia* (Turbellaria, Acoela) revealed by negative staining. *Science*, 163:678–679.

Crandall, K. A., J. W. Fetzner, S. H. Lawler, M. Kinnersley, and C. A. Austin. 1999. Phylogenetic relationships among Australian and New Zealand genera of freshwater crayfishes (Decapoda: Parastacidae). *Australian Journal of Zoology*, 47.

Crandall, R. B. 1960. The life history and affinities of the turtle lung fluke, *Heronimus chelydrae* MacCallum, 1902. *Journal of Parasitology*, 46:289–307.

Crezée, M. 1975. Monograph of the Solenofilomorphidae (Turbellaria, Acoela). *Internationale Revue der Gesamten Hydrobiologie und Hydrographie*, 60:769–845.

Crezée, M., and S. Tyler. 1976. *Hesiolicium* gen. n. (Turbellaria, Acoela) and observations on its ultrastructure. *Zoologica Scripta*, 5:207–216.

Cribb, T. H. 1986. The life cycle and morphology of *Stemmatostoma pearsoni*, gen. et sp. nov., with notes on the morphology of *Telogaster opisthorchis* MacFarlane (Digenea: Cryptogonimidae). *Australian Journal of Zoology*, 34:279–304.

Cribb, T. H. 1987. Studies on gorgoderid digeneans from Australian and Asian freshwater fishes. *Journal of Natural History*, 21:1129–1153.

Cribb, T. H. 1988. Life cycle and biology of *Prototransversotrema steeri* Angel, 1969 (Digenea: Transversotrematidae). *Australian Journal of Zoology*, 36:111–129.

Cribb, T. H., and R. A. Bray. 1999. A review of the Apocreadiidae Skrjabin, 1942 (Trematoda: Digenea) and description of Australian species. *Systematic Parasitology*, 44:1–36.

Cribb, T. H., R. A. Bray, and S. C. Barker. 1992. A review of the family Transversotrematidae (Trematoda: Digenea) with the description of a new genus, *Crusziella*. *Invertebrate Taxonomy*, 6:909–935.

Cribb, T. H., S. P. Pichelin, and R. A. Bray. 1998. *Pholeohedra overstreeti* n. g., n. sp. (Digenea: Haploporidae) from *Girella zebra* (Kyphosidae) in South Australia. *Systematic Parasitology*, 39:95–99.

Cribb, T. H., R. A. Bray, D. T. J. Littlewood, S. P. Pichelin, and E. A. Herniou. 2001. The Digenea. In D. T. J. Littlewood and R. A. Bray (eds.), *Interrelationships of the Platyhelminthes*, pp. 168–185. Taylor & Francis, London.

Crosbie, P. R., S. A. Nadler, E. G. Platzer, C. Kerner, J. Mariaux, and W. M. Boyce. 2000. Molecular systematics of *Mesocestoides* spp. (Cestoda: Mesocestoididae) from domestic dogs and coyotes (*Canis latrans*). *Journal of Parasitology*, 86:350–357.

Crowe, D. G., M. D. B. Burt, and J. S. Scott. 1974. On the ultrastructure of the polycercus larva of *Paricterotaenia paradoxa* (Cestoda: Cyclophyllidea). *Canadian Journal of Zoology*, 52:1397–1405.

Cunningham, C. O., D. M. McGillivray, and K. MacKenzie. 1995. Phylogenetic analysis of *Gyrodactylus salaris* Malmberg, 1957 based on the small subunit (18S) ribosomal RNA gene. *Molecular and Biochemical Parasitology*, 71:139–142.

Cunningham, C. W. 1997. Can three incongruence tests predict when data should be combined? *Molecular Biology and Evolution*, 14:733–740.

Curini-Galletti, M. 1992. Inferences on meiofaunal dispersal from the study of two congeneric platyhelminth species with different adaptive strategies, *VIII International Meiofaunal Conference*. pp. 18, University of Maryland.

Curini-Galletti, M. 1993. *Pseudomonocelis ophiocephala* (Schmidt, 1861) (Platyhelminthes, Proseriata) is a complex of four sibling species, *Seventh International Symposium on the Biology of Turbellaria*. pp. 44, Åbo/Turku, Finland.

Curini-Galletti, M. 1997. Contribution to the knowledge of the Proseriata (Platyhelminthes, Seriata) from eastern Australia: genera *Necia* Marcus, 1950 and *Pseudomonocelis* Meixner, 1938 (partim). *Italian Journal of Zoology*, 64:75–81.

Curini-Galletti, M. 1998. The genus *Polystyliphora* Ax, 1958 (Platyhelminthes: Proseriata) in eastern Australia. *Journal of Natural History*, 32:473–499.

Curini-Galletti, M. 2001. The Proseriata. In D. T. J. Littlewood and R. A. Bray (eds.), *Interrelationships of the Platyhelminthes*, pp. 41–48. Taylor & Francis, London.

Curini-Galletti, M., and L. R. G. Cannon. 1995. Contribution to the knowledge of the Proseriata (Platyhelminthes: Seriata) from eastern Australia: II. genera *Pseudomonocelis* Meixner, 1943 and *Acanthopseudomonocelis* n. g. *Contributions to Zoology*, 65:271–280.

Curini-Galletti, M., and L. R. G. Cannon. 1996a. Five new species of *Monocelis* Ehrenberg, 1831 (Platyhelminthes: Proseriata) from eastern Australia. *Journal of Natural History*, 30:1741–1759.

Curini-Galletti, M., and L. R. G. Cannon. 1996b. The genus *Minona* (Platyhelminthes, Seriata) in eastern Australia. *Zoologica Scripta*, 25:193–202.

Curini-Galletti, M., and L. R. G. Cannon. 1997. *Archimonocelis medusa* n. sp. (Platyhelminthes, Proseriata) from eastern Australia: the first polypharyngeal marine flatworm. *Zoological Journal of the Linnean Society*, 121:485–494.

Curini-Galletti, M., and P. M. Martens. 1990. Karyological and ecological evolution of the Monocelididae (Platyhelminthes, Proseriata). *Marine Ecology: Pubblicazioni della Stazione Zoologica di Napoli*, 11:255–261.

Curini-Galletti, M., and P. M. Martens. 1992. Systematics of the Unguiphora (Platyhelminthes Proseriata) II. Family Nematoplanidae Meixner, 1938. *Journal of Natural History*, 26:285–302.

Curini-Galletti, M., and F. Mura. 1998. Two new species of the genus *Monocelis* Ehrenberg, 1831 (Platyhelminthes: Proseriata) from the Mediterranean, with a redescription of *Monocelis lineata* (O. F. Muller, 1774). *Italian Journal of Zoology*, 65:207–217.

Curry, W. J., C. Shaw, C. F. Johnston, L. Thim, and K. D. Buchanan. 1992. Neuropeptide F: primary structure from the turbellarian, *Artioposthia triangulata*. *Comparative Biochemistry and Physiology*, 101:269–274.

Czubaj, A., and K. Niewiadomska. 1996. Ultrastructure of sensory endings in *Diplostomum pseudospathaceum* Niewiadomska, 1984 cercariae (Digenea, Diplostomidae). *International Journal for Parasitology*, 26:1217–1225.

Daddow, L. Y. M., and B. G. M. Jamieson. 1983. An ultrastructural study of spermiogenesis in *Neochasmus* sp. (Cryptogonimidae: Digenea: Trematoda). *Australian Journal of Zoology*, 31:1–14.

Dailey, M. 1969. *Litobothrium alopias* and *L. coniformis*, two new cestodes representing a new order from elasmobranch fishes. *Proceedings of the Helminthological Society of Washington*, 36:218–224.

Dailey, M. D., and W. Vogelbein. 1982. Mixodigmatidae, a new family of cestode (Trypanorhyncha) from a deep sea, planktivorous shark. *Journal of Parasitology*, 68:145–149.

Damborenea, M. A., and L. R. G. Cannon. In press. On Neotropical *Temnocephala* (Platyhelminthes). *Journal of Natural History*.

Danovaro, R., S. Fraschetti, A. Belgrano, M. Vincx, M. Curini-Galletti, G. Albertelli, and M. Fabiano. 1995. The potential impact of meiofauna on the recruitment of macrobenthos in a subtidal coastal benthic community of the Ligurian Sea (north-western Mediterranean): a field result. In A. Eleftheriou, A. D. Ansell and C. J. Smith (eds.), *Biology and ecology of shallow coastal waters : proceedings of the 28th European marine biology symposium. Institute of Marine Biology of Crete, Iraklio, Crete, 1993*, pp. 115–122. Olsen & Olsen, Denmark.

Davis, G. M. 1980. Snail hosts of Asian *Schistosoma* infecting man: evolution and coevolution. In J. I. Bruce, S. Sornmani, H. L. Asch and K. A. Crawford (eds.), *The Mekong schistosome*, Vol. 2, pp. 195–238. *Malacological Review* Supplement 2, Whitmore Lake, Michigan.

Davis, G. M. 1992. Evolution of prosobranch snails transmitting *Schistosoma*; coevolution with *Schistosoma*; a review. *Progress in Clinical Parasitology*, 3:145–204.

Davis, R. E. 1997. Surprising diversity and distribution of spliced leader RNAs in flatworms. *Molecular and Biochemical Parasitology*, 87:29–48.

Davis, R. E., and L. E. Roberts. 1983. Platyhelminthes – Eucestoda. In K. G. Adiyodi and R. G. Adiyodi (eds.), *Reproductive biology of invertebrates. Vol. II. Spermatogenesis and sperm function*, pp. 131–149. John Wiley and Sons, Chichester.

Davydov, V. G., and B. I. Kuperman. 1993. The ultrastructure of the tegument and the peculiarities of the biology of *Amphilina foliacea* adult (Plathelminthes, Amphilinidae). *Folia Parasitologica*, 40:13–22.

Day, T. A., J. L. Bennett, and R. A. Pax. 1994. Serotonin and its requirement for maintenance of contractility in muscle fibres isolated from *Schistosoma mansoni*. *Parasitology*, 108:425–432.

Day, T. A., A. G. Maule, C. Shaw, and R. A. Pax. 1997. Structure-activity relationships of FMRFamide-related peptides contracting *Schistosoma mansoni* muscle. *Peptides*, 18:917–921.

de Brie, J. 1379. Le bon berger ou le vray régime et gouvernement des bergers et bergéres. Translated passage in Kean, Mott and Russell, 1978, p. 561.

de Queiroz, A., and N. L. Alkire. 1998. The phylogenetic placement of *Taenia* cestodes that parasitize humans. *Journal of Parasitology*, 84:379–383.

de Queiroz, A., and J. Gauthier. 1994. Toward a phylogenetic system of biological nomenclature. *Trends in Ecology and Evolution*, 9:37–31.

de Queiroz, A., M. J. Donoghue, and J. Kim. 1995. Separate versus combined analysis of phylogenetic evidence. *Annual Review of Ecology and Systematics*, 26:657–681.

de Rijk, P., J.-M. Neefs, Y. van de Peer, and R. de Wachter. 1992. Compilation of small ribosomal subunit RNA sequences. *Nucleic Acids Research*, 20:2075–2089.

de Rosa, R., J. K. Grenier, T. Andreeva, C. E. Cook, A. Adoutte, M. Akam, S. B. Carroll, and G. Balavoine. 1999. *Hox* genes in brachiopods and priapulids: implications for protostome evolution. *Nature*, 399:772–776.

de Vos, T., and T. A. Dick. 1989. Differentiation between *Diphyllobothrium dendriticim* and *D. latum* using isozymes, restriction profiles and ribosomal gene probes. *Systematic Parasitology*, 13:161–166.

de Vries, E. J. 1984. On the species of the *Dugesia gonocephala* group (Platyhelminthes, Turbellaria, Tricladida) from Greece. *Bijdragen tot de Dierkunde*, 54:101–126.

de Vries, E. J., and R. Sluys. 1991. Phylogenetic relationships of the genus *Dugesia* (Platyhelminthes, Tricladida, Paludicola). *Journal of Zoology*, 223:103–116.

Dendy, A. 1892. On the presence of ciliated pits in Australian land planarians. *Proceedings of the Royal Society of Victoria*, 4:39–46.

Després, L., and S. Maurice. 1995. The evolution of dimorphism and separate sexes in schistosomes. *Proceedings of the Royal Society of London, Series B, Biological Sciences*, 262:175–180.

Després, L., F. J. Kruger, D. Imbert-Establet, and M. L. Adamson. 1995. ITS2 ribosomal RNA indicates *Schistosoma hippopotami* is a distinct species. *International Journal for Parasitology*, 25:1509–1514.

Dewel, R. A. 2000. Colonial origin for Eumetazoa: major morphological transitions and the origin of bilaterian complexity. *Journal of Morphology*, 243:35–74.

Diensk, H. M. S. 1968. A survey of the metazoan parasites of the rabbit-fish, *Chimaera monstrosa* L. (Holocephali). *Netherlands Journal of Sea Research*, 4:32–58.

Dimitrov, V., J. Busta, and I. Kanev. 1989. Chaetotaxy of cercaria of *Opisthioglyphe ranae* (Frölich, 1791) (Trematoda, Plagiorchiidae). *Folia Parasitologica*, 36:265–274.

Dixon, B. R., and H. P. Arai. 1985. Isoelectric focusing of soluble proteins in the characterization of 3 species of *Hymenolepis* cestoda. *Canadian Journal of Zoology*, 63:1720–1723.

Dixon, B. R., and H. P. Arai. 1989. Differentiation of three species of *Hymenolepis* (Cestoidea) using enzyme isoelectric focusing on thin-layer agarose gels. *Canadian Journal of Zoology*, 67:51–54.

Dixon, M. T., and D. M. Hillis. 1993. Ribosomal RNA secondary structure: compensatory mutations and implications for phylogenetic analysis. *Molecular Biology and Evolution*, 10:256–267.

Doe, D. A. 1981. Comparative ultrastructure of the pharynx simplex in Turbellaria. *Zoomorphology*, 97:133–192.

Doe, D. A. 1982. Ultrastructure of copulatory organs in Turbellaria. I. *Macrostomum* sp. and *Microstomum* sp. (Macrostomidae). *Zoomorphology*, 101:39–59.

Doe, D. A. 1986a. Ultrastructure of the copulatory stylet and accessory spines in *Haplopharynx quadristimulus* (Turbellaria). *Hydrobiologia*, 132:157–163.

Doe, D. A. 1986b. Ultrastructure of the copulatory organ of *Haplopharynx quadristimulus* and its phylogenetic significance (Plathelminthes, Haplopharyngida). *Zoomorphology*, 106:163–173.

Dogiel, V. A. 1954. *[Oligomerisation of homologous organs, one of the principal paths in animal evolution]*. (In Russian), Leningrad.

Dollfus, R. P. 1947. Sur trois espèces de distomes, dont une a 17 ventouses *Enenterum (Jeancadenatia) brumpti* n. sp. parasites du poisson marin *Kyphosus sectatrix* (L.). *Annales de Parasitologie Humaine et Comparée*, 21:119–128.

Dominguez, E., and Q. D. Wheeler. 1997. Taxonomic stability is ignorance. *Cladistics*, 13:367–372.

Dönges, J., and W. Harder. 1966. *Nesolecithus africanus* n. sp. (Cestodaria, Amphilinidea) aus dem Coelom von *Gymnarchus niloticus* Cuvier 1829 (Teleostei). *Zeitschrift für Parasitenkunde*, 28:125–141.

Dörjes, J. 1968. Die Acoela (Turbellaria) der deutschen Nordseeküste und ein neues System der Ordnung. *Zeitschrift für zoologische Systematik und Evolutionsforschung*, 6:56–452.

Doyle, J. J. 1997. Trees within trees: genes and species, molecules and morphology. *Systematic Biology*, 46:537–553.

Dubinina, M. N. 1974. [The development of *Amphilina foliacea* (Rud.) at all stages of its life cycle and the position of Amphilinidea in the system of Platyhelminthes]. (In Russian). *Parazitologicheskii Sbornik*, 26:9–38.

Dubinina, M. N. 1980. [Importance of attachment organs for phylogeny of tapeworms]. (In Russian). *Parazitologicheskii Sbornik*, 29:65–83.

Dubinina, M. N. 1982. [Parasitic worms of the class Amphilinida (Platyhelminthes)]. (In Russian). *Trudy Zoologicheskogo Instituta Akademii Nauk SSSR*, 100:1–142.

Dujardin, F. 1845. *Histoire naturelle des helminthes ou vers intestinaux*. Librarie Encyclopédique de Roret, Paris.

Durand, J. P., and N. Gourbault. 1977. Étude cytologique des organes photorécepteurs de la Planaire australienne *Cura pinguis*. *Canadian Journal of Zoology*, 55:381–390.

Durbin, R., and J. Thierry Mieg. 1991–. A *C. elegans* database. Documentation, code and data available from anonymous FTP servers at lirmm.lirmm.fr, cele.mrc–lmb.cam.ac.uk and ncbi.nlm.nih.gov.

Dutt, S. C., and H. D. Srivastava. 1955. A revision of the genus *Ornithobilharzia* Odhner, 1912 (Trematoda: Schistosomatidae). *Proceedings of the Indian Scientific Congress*, 42:283.

Dutt, S. C., and H. D. Srivastava. 1961. A revision of the genus *Ornithobilharzia* Odhner, 1912 with the creation of two genera: *Orientobilharzia* Dutt and Srivastava 1955 and *Sinobilharzia* Dutt and Srivastava 1955 (Trematoda: Schistosomatidae). *Indian Journal of Helminthology*, 13:61–73.

Dyganova, R. Y., and I. V. Kiseleva. 1988. [New data on the eye morphology of *Dendrocoelum lacteum*]. (In Russian), Conference: *Simple nervous systems and their meaning for theory and practice*, pp. 90–92. Kazan.

Ebrahimzadeh, A., and M. Kraft. 1971. Ultrastrukturelle Untersuchungen zur Anatomie der Cercarien von *Schistosoma mansoni*. II. Das Exkretionssystem. *Zeitschrift für Parasitenkunde*, 36:265–290.

Eernisse, D. J., J. S. Albert, and F. E. Anderson. 1992. Annelida and Arthropoda are not sister taxa: a phylogenetic analysis of spiralian metazoan morphology. *Systematic Zoology*, 41:305–330.

Ehlers, B. 1977. 'Trematoden-artige' Epidermisstrukturen bei einem freilebenden proseriaten Strudelwurm (Turbellaria Proseriata). *Acta Zoologica Fennica*, 154:129–136.

Ehlers, B., and U. Ehlers. 1980. Struktur und Differenzierun penialer Hartebilde von *Carenscolia bidentata*. *Zoomorphologie*, 95:159–167.

Ehlers, U. 1984. Phylogenetisches System der Plathelminthes. *Verhandlungen des Naturwissenschaftlichen Vereins in Hamburg (NF)*, 27:291–294.

Ehlers, U. 1985a. *Das Phylogenetische System der Plathelminthes*. Gustav Fischer, Stuttgart.

Ehlers, U. 1985b. Phylogenetic relationships within the Platyhelminthes. In S. Conway Morris, J. D. George, R. Gibson and H. M. Platt (eds.), *The origins and relationships of lower invertebrates*, pp. 143–158. Oxford University, Oxford.

Ehlers, U. 1986. Comments on a phylogenetic system of the Platyhelminthes. *Hydrobiologia*, 132:1–12.

Ehlers, U. 1988. The Prolecithophora – a monophyletic taxon of the Platyhelminthes. *Fortschritte der Zoologie*, 36:359–365.

Ehlers, U. 1989. The protonephridium of *Archimonotresis limophila* Meixner (Plathelminthes, Prolecithophora). *Microfauna Marina*, 5:261–275.

Ehlers, U. 1992a. On the fine structure of *Paratomella rubra* Rieger & Ott (Acoela) and the position of the taxon *Paratomella* Dörjes in a phylogenetic system of the Acoelomorpha (Plathelminthes). *Microfauna Marina*, 7:265–293.

Ehlers, U. 1992b. Frontal glandular and sensory structures in *Nemertoderma* (Nemertodermatida) and *Paratomella* (Acoela): ultrastructure and phylogenetic implications for the monophyly of the Euplathelminthes (Plathelminthes). *Zoomorphology*, 112:227–236.

Ehlers, U. 1992c. 'Pulsatile bodies' in *Anaperus tvaerminnensis* (Luther, 1912) (Acoela, Plathelminthes) are degenerating epidermal cells. *Microfauna Marina*, 7:295–310.

Ehlers, U. 1992d. Dermonephridia-modified epidermal cells with a probable excretory function in *Paratomella rubra* (Acoela, Plathelminthes). *Microfauna Marina*, 7:253–264.

Ehlers, U. 1995. The basic organization of the Plathelminthes. *Hydrobiologia*, 305:21–26.

Ehlers, U., and B. Ehlers. 1977. Monociliary receptors in interstitial Proseriata and Neorhabdocoela (Turbellaria Neoophora). *Zoomorphologie*, 86:197–222.

Ehlers, U., and B. Sopott-Ehlers. 1986. Vergleichende Ultrastruktur von Protonephridien: ein Beitrag zur Stammesgeschichte der Plathelminthen. *Verhandlungen der Deutschen Zoologischen Gesellschaft*, 79:168–169.

Ehlers, U., and B. Sopott-Ehlers. 1987. Zum Protonephridialsystem von *Invenusta paracnida* (Proseriata, Plathelminthes). *Microfauna Marina*, 3:377–390.

Ehlers, U., and B. Sopott-Ehlers. 1993. The caudal duo-gland adhesive system of *Jensenia angulata* (Plathelminthes, Dalyellidae): ultrastructure and phylogenetic significance (with comments on the phylogenetic position of the Temnocephalida and the polyphyly of the Cercomeria). *Microfauna Marina*, 8:65–76.

Ehlers, U., and B. Sopott-Ehlers. 1997a. Ultrastructure of protonephridial structures within the Prolecithophora (Plathelminthes). *Microfauna Marina*, 11:291–315.

Ehlers, U., and B. Sopott-Ehlers. 1997b. Ultrastructure of the subepidermal musculature of *Xenoturbella bocki*, the adelphotaxon of the Bilateria. *Zoomorphology*, 117:71–79.

Ehlers, U., and B. Sopott-Ehlers. 1997c. *Xenoturbella bocki*: organization and phylogenetic position as sistertaxon of the Bilateria. *Verhandlungen der Deutsche Zoologische Gesellschaft*, 90:68.

Eklu-Natey, D. T., J. Wüest, Z. Swiderski, H. P. Striebel, and H. Huggel. 1985. Comparative scanning electron microscope (SEM) study of miracidia of four human schistosome species. *International Journal for Parasitology*, 15:33–42.

Ellis, C. H., and A. Fausto-Sterling. 1997. Platyhelminthes, the flatworms. In S. F. Gilbert and A. M. Raunio (eds.), *Embryology. Constructing the organism*, pp. 115–130. Sinauer and Associates, Sunderland, MA.

Elvin, M., and H. Koopowitz. 1994. Neuroanatomy of the rhabdocoel flatworm *Mesostoma ehrenbergii* (Focke, 1836). 1. Neuronal diversity in the brain. *Journal of Comparative Neurology*, 343:319–331.

Embley, T. M., and R. P. Hirt. 1998. Early branching eukaryotes? *Current Opinion in Genetics and Development*, 8:624–629.

Eriksson, K. S., and P. A. J. Panula. 1994. Gamma-aminobutyric acid in the nervous system of a planarian. *Journal of Comparative Neurology*, 345:528–536.

Eriksson, K. S., A. G. Maule, D. W. Halton, P. A. J. Panula, and C. Shaw. 1995a. GABA in the nervous system of parasitic flatworms. *Parasitology*, 110:339–346.

Eriksson, K. S., P. A. J. Panula, and M. Reuter. 1995b. GABA in the nervous sytem of the planarian *Polycelis nigra*. *Hydrobiologia*, 305:285–289.

Eriksson, K. S., R. N. Johnston, D. W. Halton, P. A. J. Panula, and C. Shaw. 1996. A widespread distribution of histamine in the nervous system of a trematode flatworm. *Journal of Comparative Neurology*, 373:220–227.

Euzet, L. 1959. Recherches sur les cestodes tétraphyllides des sélaciens des côtes de France. Doctoral thesis, Montpellier, France.

Euzet, L. 1974. Essai sur la phylogénèse des cestodes à la lumière de faits nouveaux. *Third International Congress of Parasitology*, Vol. 1, pp. 378–379. Facta Publication, Munich.

Euzet, L. 1994. Order Tetraphyllidea Carus, 1863. In L. F. Khalil, A. Jones and R. A. Bray (eds.), *Keys to the cestode parasites of vertebrates*, pp. 149–194. CAB International, Wallingford.

Euzet, L., and M. Prost. 1981. Report of meeting on 'Monogenea: problems of systematics, biology and ecology'. In W. Slusarski (ed.), *Review of advances in parasitology*, pp. 1003–1004. Polish Scientific Publishers, Warszawa.

Euzet, L., Z. Swiderski, and F. Mokhtar-Maamouri. 1981. Ultrastructure comparée du spermatozoide des cestodes. Relations avec la phylogénès. *Annales de Parasitologie Humaine et Comparée*, 56:247–259.

Euzet, L., G. Oliver, and M. H. Ktari. 1995. Organisation et développement du système osmorégulateur chez les Monogenea. *Canadian Journal of Fisheries and Aquatic Sciences*, 52:35–51.

Fairweather, I., and L. T. Threadgold. 1981a. *Hymenolepis nana*: the fine structure of the embryonic envelopes. *Parasitology*, 82:429–443.

Fairweather, I., and L. T. Threadgold. 1981b. *Hymenolepis nana*: the fine structure of the 'penetration gland' and nerve cells within the oncosphere. *Parasitology*, 82:445–458.

Falleni, A., and P. Lucchesi. 1992. Ultrastructural and cytochemical aspects of oogenesis in *Castrada viridis* (Platyhelminthes, Rhabdocoela). *Journal of Morphology*, 213:241–250.

Farley, J. 1971. A review of the family Schistosomatidae: excluding the genus *Schistosoma* from mammals. *Journal of Helminthology*, 45:289–320.

Farris, J. S. 1969. A successive approximations approach to character weighting. *Systematic Zoology*, 18:374–385.

Farris, J. S. 1988. Hennig86. Port Jefferson Station, New York.

Farris, J. S., V. A. Albert, M. Källersjö, D. Lipscomb, and A. G. Kluge. 1996. Parsimony jackknifing outperforms neighborjoining. *Cladistics*, 12:99–124.

Faubel, A. 1977. *Bathymacrostomum spirale* n. gen. n. spec., ein Vertreter der Familie Dolichomacrostomidae Rieger, 1971 (Turbellaria) aus dem Sublittoral der Nordsee. *'Meteor' Forschungsergebnisse, Reihe D*, 25:45–48.

Faubel, A., and J. Dörjes. 1978. *Flagellophora apelti* gen. n. sp. n.: a remarkable representative of the order Nemertodermatida (Turbellaria: Archoophora). *Seckenbergiana maritima*, 10:1–13.

Faubel, A., and K. Rohde. 1998. Sandy beach meiofauna of eastern Australia (Southern Queensland and New South Wales). IV. Proseriata, Plathelminthes. *Mitteilungen aus den Hamburgischen Zoologischen Museum und Institut*, 95:7–28.

Faubel, A., D. Blome, and L. R. G. Cannon. 1994. Sandy beach meiofauna of eastern Australia (southern Queensland and New South Wales). 1. Introduction and Macrostomida (Platyhelminthes). *Invertebrate Taxonomy*, 8:989–1007.

Featherston, D. W. 1971. *Taenia hydatigena* III. Light and electron microscope study of spermatogenesis. *Zeitschrift für Parasitenkunde*, 37:148–168.

Featherston, D. W. 1972. *Taenia hydatigena* IV. Ultrastructure study of the tegument. *Zeitschrift für Parasitenkunde*, 38:214–232.

Felsenstein, J. 1978. Cases in which parsimony or compatibility methods will be positively misleading. *Systematic Zoology*, 27:401–410.

Felsenstein, J. 1981. Evolutionary trees from DNA sequences – a maximum-likelihood approach. *Journal of Molecular Evolution*, 17:368–376.

Felsenstein, J. 1985. Confidence limits on phylogenies – an approach using the bootstrap. *Evolution*, 39:783–791.

Felsenstein, J. 1993. PHYLIP (Phylogeny Inference Package). University of Washington, Seattle.

Ferguson, F. F. 1939–1940. A monograph of the genus *Macrostomum* O. Schmidt 1848. *Zoologischer Anzeiger*, 126:7–20; 127:131–144; 128:49–68, 188–205, 274–291; 129:21–48, 120–146, 244–266.

Ferguson, F. F. 1954. Monograph of the macrostomine worms of Turbellaria. *Transactions of the American Microscopical Society*, 73:137–164.

Ferguson, M. A., T. H. Cribb, and L. R. Smales. 1999. Life-cycle and biology of *Sychnocotyle kholo* n. g. n. sp. (Trematoda: Aspidogastrea) in *Emydura macquarii* (Pleurodira: Chelidae) from southern Queensland, Australia. *Systematic Parasitology*, 43:41–48.

Fernández, M., F. J. Aznar, A. Latorre, and J. A. Raga. 1998a. Molecular phylogeny of the families Campulidae and Nasitrematidae (Trematoda) based on mtDNA sequence comparison. *International Journal for Parasitology*, 28:767–775.

Fernández, M., D. T. J. Littlewood, A. Latorre, J. A. Raga, and D. Rollinson. 1998b. Phylogenetic relationships of the family Campulidae (Trematoda) based on 18S rRNA sequences. *Parasitology*, 117:383–391.

Fernando, W. 1934. Studies on the Temnocephaloidea. 1. The females reproductive apparatus in *Caridinicola indica* and *Monodiscus parvus*. *Proceedings of the Zoological Society, London*, 4:251–258.

Ferrer, J. R. 1986. La célula flamígera en trematodos: ultraestructura en el esporocisto y la cercaria de *Proctoeces maculatus* (Digenea, Fellodistomatidae). *Miscellania Zoológica*, 10:45–53.

Ferrero, E. 1973. A fine structure analysis of the statocyst in Turbellaria Acoela. *Zoologica Scripta*, 2:5–16.

Field, K. G., G. J. Olsen, D. G. Lane, S. J. Giovannoni, M. T. Ghiselin, E. C. Raff, N. R. Pace, and R. A. Raff. 1988. Molecular phylogeny of the Animal Kingdom. *Science*, 239:748–753.

Fischer, H., and R. S. Freeman. 1969. Penetration of parenteral plerocercoids of *Proteocephalus ambloplitis* (Leidy) in the gut of small mouth bass. *Journal of Parasitology*, 55:766–774.

Fischer, O. 1926. *Digonopyla (Dolichoplana) harmeri* (Graff), eine Landtriclade aus Celebes mit volkommen getrennten Geschlechtsapparaten. *Zoologischer Anzeiger*, 66:257–261.

Fleming, L. C., and M. D. B. Burt. 1978a. On the genus *Peraclistus* (Turbellaria, Proseriata), with redescription of *P. oofagus* (Friedman). *Zoologica Scripta*, 7:81–84.

Fleming, L. C., and M. D. B. Burt. 1978b. Revision of the turbellarian genus *Ectocotyla* (Seriata, Monocelididae) associated with the crabs *Chionocetes opilio* and *Hyas araneus*. *Journal of the Fisheries Research Board of Canada*, 35:1223–1233.

Fortey, R. A., and R. H. Thomas. 1998. Arthropod relationships. *The Systematics Association special volume series*, Vol. 55, pp. 383. Chapman & Hall, London.

Fournier, A. 1979. Evolution du tégument des *Polystoma* (Monogènes Polystomatidae) au cours du cycle. *Zeitschrift für Parasitenkunde*, 59:169–185.

Fournier, A., and J.-L. Justine. 1994. Ultrastructure of spermiogenesis and spermatozoa in *Euzetrema knoepffleri* (Platyhelminthes, Monogenea). *Parasite*, 1:123–126.

Fransen, M. E. 1980. Ultrastructure of coelomic organization in annelids I. Archiannelids and other small polychaetes. *Zoomorphology*, 95:235–249.

Fransen, M. E. 1988. Coelomic and vascular systems. In W. Westheide and C. O. Hermans (eds.), *The ultrastructure of Polychaeta*, Vol. 4, pp. 199–213. Fischer Verlag, Stuttgart.

Franzén, A., and B. A. Afzelius. 1987. The ciliated epidermis of *Xenoturbella bocki* (Platyhelminthes, Xenoturbellida) with some phylogenetic considerations. *Zoologica Scripta*, 16:9–17.

Fredericksen, D. W. 1978. The fine structure and phylogenetic position of the cotylocidium larva of *Cotylogaster occidentalis* Nickerson 1902 (Trematoda: Aspidogastridae). *Journal of Parasitology*, 64:961–976.

Freeman, R. S. 1957. Life cycle and morphology of *Paruterina rauschi* n. sp. and *P. candelabraria* (Goeze, 1782) (Cestoda) from owls, and significance of plerocercoids in the order Cyclophyllidea. *Canadian Journal of Zoology*, 35:349–370.

Freeman, R. S. 1959. On the taxonomy of the genus *Cladotaenia*, the life histories of *C. globifera* (Batsch, 1786) and *C. circi* Yamaguti, 1935, and a note on distinguishing between the plerocercoides of the genera *Paruterina* and *Cladotaenia*. *Canadian Journal of Zoology*, 37:317–340.

Freeman, R. S. 1973. Ontogeny of cestodes and its bearing on their phylogeny and systematics. *Advances in Parasitology*, 11:481–557.

Friedlander, T. P., J. C. Regier, and C. Mitter. 1992. Nuclear gene sequences for higher level phylogenetic analysis: 14 promising candidates. *Systematic Biology*, 41:483–490.

Froehlich, E. M. 1978. On a collection of Chilean landplanarians. *Boletim de Zoologia*, 3:7–80.

Fuhrmann, O. 1931. Dritte Klasse des Cladus Plathelminthes: Cestoidea. In W. Kükenthal and T. Krumbach (eds.), *Handbuch der Zoologie (1928–1933)*, pp. 141–416. Walter de Gruyter & Co., Berlin.

Fujita, I., I. Kanno, and S. Kobayashi. 1988. *The Paraneuron*. Springer Verlag, London.

Gabrion, C. 1975. Étude expérimentale du développement larvaire d'*Anomotaenia constricta* (Molin, 1858) Cohn, 1900 chez un coléoptère *Pimelia sulcata* Geoffr. *Zeitschrift für Parasitenkunde*, 47:249–262.

Gabrion, C. 1981. Ontogénèse de Cestodes cyclophyllides. Étude morphogenetique du développement post-oncospheral PhD thesis, Université de Montpellier.

Gabrion, C., and J. Gabrion. 1976. Étude ultrastructurale de la larve de *Anomotaenia constricta* (Cestoda, Cyclophyllidea). *Zeitschrift für Parasitenkunde*, 49:161–177.

Gabrion, C., and S. Helluy. 1982. Développement larvaire de *Paricterotaenia porosa* (Cestoda: Cyclophyllidea, Dilepididae) chez des diptères du genre *Chironomus*, hôtes expérimentaux. *Annales de Parasitologie Humaine et Comparée*, 57:33–52.

Galaktionov, K. V., and A. A. Dobrovolskij. 1998. *[The origin and evolution of trematode life cycles]*. (In Russian). Nauka, Saint–Petersburg.

Galkin, A. K. 1987. [On the origin of non-cyclophyllidean cestodes – parasites of gulls]. (In Russian). *Trudy Zoologicheskogo Instituta Akademii Nauk SSSR*, 161:3–23.

Galkin, A. K. 1996. [The postlarval development of the scolex of *Tetrabothrius erostris* (Cestoda: Tetrabothriidea) and phylogenetic essentials of this process]. (In Russian). *Parazitologiya*, 30:315–323.

Gallati, W. W. 1959. Life history, morphology and taxonomy of *Atriotaenia* (*Ershovia*) *procyonis* (Cestoda: Linstowidae), a parasite of the raccoon. *Journal of Parasitology*, 45:363–377.

García Fernandez, J., J. Baguñá, and E. Saló. 1993. Genomic organization and expression of the planarian homeobox genes Dth-1 and Dth-2. *Development*, 118:241–253.

Garcin. 1730. *Hirudinella marina*, or sea-leach. *Philosophical Transactions of the Royal Society*, 36:387–394.

Gasser, R. B., and N. B. Chilton. 1995. Characterisation of taeniid cestode species by PCR-RFLP of ITS2 ribosomal DNA. *Acta Tropica*, 59:31–40.

Gianutsos, G., and J. L. Bennett. 1977. The regional distribution of dopamine and norepinephrine in *Schistosoma mansoni* and *Fasciola hepatica*. *Comparative Biochemistry and Physiology*, 58C:157–159.

Gibson, D. I. 1987. Questions in digenean systematics and evolution. *Parasitology*, 95:429–460.

Gibson, D. I., and R. A. Bray. 1977. The Azygiidae, Hirudinellidae, Ptychogonimidae, Sclerodistomidae and Syncoeliidae of fishes from the north-east Atlantic. *Bulletin of the British Museum (Natural History) (Zoology Series)*, 32:167–245.

Gibson, D. I., and R. A. Bray. 1979. The Hemiuroidea: terminology, systematics and evolution. *Bulletin of the British Museum (Natural History) (Zoology Series)*, 36:35–146.

Gibson, D. I., and R. A. Bray. 1994. The evolutionary expansion and host-parasite relationships of the Digenea. *International Journal for Parasitology*, 24:1213–1226.

Gibson, D. I., and S. Chinabut. 1984. *Rohdella siamensis* gen. et sp. nov. (Aspidogastridae: Rohdellinae subfam. nov.) from freshwater fishes in Thailand, with a reorganization of the classification of the subclass Aspidogastrea. *Parasitology*, 88:383–393.

Gibson, D. I., R. A. Bray, and C. B. Powell. 1987. Aspects of the life history and origins of *Nesolecithus africanus* (Cestoda: Amphilinidea). *Journal of Natural History*, 21:785–794.

Gieysztor, M., and J. Wisniewski. 1947. Sur un Turbellarié vivant sur les branchies de *Gammarus ischnus* G.O. Sars. *Annales Musei Zoologici Polonici*, 14:1–5.

Gilbert, C. 1935. A comparative study of three new American species of the genus *Phaenocora*. *Acta Zoologica*, 16:283–386.

Giribet, G., and W. C. Wheeler. 1999. On gaps. *Molecular Phylogenetics and Evolution*, 13:132–143.

Giribet, G., M. Polz, W. Sterrer, and W. C. Wheeler. 2000. Triploblastic relationships with emphasis on the positions of Gnathostomulida, Cycliophora, Platyhelminthes and Chaetognatha: a combined approach of 18S rDNA sequences and morphology. *Systematic Biology* 49:539–562.

Goloboff, P. A. 1997. [Phylogenetic analysis program]. Published by the author, Túcuman, Argentina.

Golubev, A. I. 1988. Glia and neuroglia relationships in the central nervous system of the Turbellaria (electron microscopic data). *Fortschritte der Zoology*, 36:31–37.

Gordon, A. D. 1986. Consensus supertrees: the synthesis of rooted trees containing overlapping sets of labelled leaves. *Journal of Classification*, 3:335–348.

Gould, S. J. 1989. *Wonderful life: the Burgess Shale and the nature of history*. Norton, New York.

Grabda-Kazubska, B., and A. Lis. 1993. Chaetotaxy of the cercaria of *Telorchis assula* (Dujardin, 1845) (Trematoda, Telorchiidae). *Acta Parasitologica*, 38:96–98.

Grabda-Kazubska, B., and T. Moczon. 1990. The nervous system and chaetotaxy of the cercaria of *Opisthioglyphe ranae* (Frölich, 1791) (Trematoda, Plagiorchiidae). *Bulletin du Museum National d'Histoire Naturelle, Paris. 4er série. Section A. Zoologie, biologie et écologie animales*, 12:375–383.

Grabda-Kazubska, B., K. Niewiadomska, K. Kanev, and C. Bayssade-Dufour. 1991. Système nerveux des trématodes. *Annales de Parasitologies Humaine et Comparée*, 66:24–31.

Grabda-Kazubska, B., P. Borsuk, Z. Laskowski, and H. Moné. 1998. A phylogenetic analysis of trematodes of the genus *Echinoparyphium* and related genera based on sequencing of internal transcribed spacer region of rDNA. *Acta Parasitologica*, 43:116–121.

Graff, L. v. 1882. *Monographie der Turbellarien. I. Rhabdocoela*. Verlag W. Engelmann, Leipzig.

Graff, L. v. 1899. *Monographie der Turbellarien II. Tricladida, Terricola (Landplanarien)*. W. Engelmann, Leipzig.

Graff, L. v. 1904–1908. *Klassen un Ordnung des Thier-Reichs*. C.F. Winter'sche Verlagshandlung.

Graff, L. v. 1912–17. Tricladida. In H. G. Bronn (ed.), *Klassen und Ordnungen des Tier-Reichs. Bd. IV Vermes, Abt. IC: Turbellaria, II Abt.: Tricladida*, pp. 2601–3369+34 plates. C. F. Winter, Leipzig.

Grammeltvedt, A. F. 1973. Differentiation of the tegument and associated structures in *Diphyllobothrium dendriticum* Nitzsch (1824) (Cestoda: Pseudophyllidea). An electron microscopical study. *International Journal for Parasitology*, 3:321–327.

Graybeal, A. 1998. Is it better to add taxa or characters to a difficult phylogenetic problem? *Systematic Biology*, 47:9–17.

Grossman, A. I., R. B. Short, and G. D. Cain. 1981. Karyotype evolution and sex chromosome differentiation in schistosomes (Trematoda, Schistosomatidae). *Chromosoma*, 84:413–430.

Grove, D. I. 1990. *A history of human helminthology*. CAB International, Wallingford.

Gulyaev, V. D. 1997. [Development of the metacestode *Dilepis undula* (Cestoda, Cyclophyllidea, Dilepididae), a primitive non-lacunar cysticercoid]. (In Russian). *Zoologicheskii Zhurnal*, 76:985–991.

Gustafsson, M. K. S. 1984. Synapses in *Diphyllobothrium dendriticum* (Cestoda). An electron microscopical study. *Annales Zoologica Fennica*, 21:167–175.

Gustafsson, M. K. S. 1992. The neuroanatomy of parasitic flatworms. *Advances in Neuroimmunology*, 2:267–286.

Gustafsson, M. K. S., and K. Eriksson. 1991. Localization and identification of catecholamines in the nervous system of *Diphyllobothrium dendriticum* (Cestoda). *Parasitology Research*, 77:498–502.

Gustafsson, M. K. S., and M. Reuter. 1992. The map of neuronal signal substances in flatworms. In R. N. Singh (ed.), *Nervous systems. Priciples of design and function*, pp. 165–188. Wiley Eastern Limited, Delhi.

Gustafsson, M. K. S., and M. C. Wikgren. 1981. Activation of the peptidergic neurosecretory system in *Diphyllobothrium dendriticum* (Cestoda: Pseudophyllidea). *Parasitology*, 83:243–247.

Gustafsson, M. K. S., D. W. Halton, A. G. Maule, M. Reuter, and C. Shaw. 1994. The gull-tapeworm, *Diphyllobothrium dendriticum* and neuropeptide F: an immunocytochemical study. *Parasitology*, 109:599–609.

Gustafsson, M. K. S., H. P. Fagerholm, D. W. Halton, V. Hanzelová, A. G. Maule, M. Reuter, and C. Shaw. 1995. Neuropeptides and serotonin in the cestode, *Proteocephalus exiguus*: an immunocytochemical study. *International Journal for Parasitology*, 25:673–682.

Gustafsson, M. K. S., A. M. Lindholm, K. Mäntylä, M. Reuter, C. A. Lundstrom, and N. Terenina. 1998. NO news on the flatworm front! Nitric oxide synthase in parasitic and free-living flatworms. *Hydrobiologia*, 383:161–166.

Gutell, R. R., B. Weiser, C. R. Woese, and H. F. Noller. 1985. Comparative anatomy of 16S–like ribosomal RNA. *Progress in Nucleic Acid Research and Molecular Biology*, 32:155–216.

Halanych, K. M. 1998. Lagomorphs misplaced by more characters and fewer taxa. *Systematic Biology*, 47:138–146.

Hall, B. K. 1994. Homology: the hierarchical basis of comparative biology, pp. 483. Academic Press, London.

Hall, K. A., T. H. Cribb, and S. C. Barker. 1999. V4 region of small subunit rDNA indicates polyphyly of the Fellodistomidae (Digenea) which is supported by morphology and life–cycle data. *Systematic Parasitology*, 43:81–92.

Hallez, P. 1894. *Catalogue des Rhabdocoelides, Triclades et Polyclades du nord de la France*. L. Daniel, Lille.

Halton, D. W., and M. K. S. Gustaffson. 1996. Functional morphology of the platyhelminth nervous system. *Parasitology*, 113:S47–S72.

Halton, D. W., and A. Hardcastle. 1976. Spermatogenesis in a monogenean, *Diclidophora merlangi*. *International Journal for Parasitology*, 6:43–53.

Halton, D. W., C. Shaw, A. G. Maule, C. F. Johnston, and I. Fairweather. 1992. Peptidergic messengers: a new perspective of the nervous system of parasitic platyhelminths. *Journal of Parasitology*, 78:179–193.

Halton, D. W., C. Shaw, A. G. Maule, and D. Smart. 1994. Regulatory peptides in helminth parasites. *Advances in Parasitology*, 34:163–227.

Halvorsen, O., and H. H. Williams. 1967. *Gyrocotyle* in *Chimaera monstrosa* from Oslo Fjord, with emphasis on its mode of attachment and a regulation in the degree of infection. *Parasitology*, 57 Suppl.:12.

Halvorsen, O., and H. H. Williams. 1968. Studies of the helminth fauna of Norway. IX. *Gyrocotyle* (Platyhelminthes) in *Chimaera monstrosa* from Oslo Fjord, with emphasis on its mode of attachment and a regulation in the degree of infection. *Nytt Magasin for Zoologi*, 15:130–142.

Hamr, P. 1992. A revision of the Tasmanian freshwater crayfish genus *Astacopsis* Huxley (Decapoda: Parasticidae). *The Papers and Proceedings of the Royal Society of Tasmania*, 126:91–94.

Hanström, B. 1926. Uber den feineren Bau des Nervensystems der Tricladed Turbellarien auf Grund von Untersuchungen an *Bdelloura candida*. *Acta Zoologica (Stockholm)*, 7:101–115.

Hanzelová, V., V. Snábel, I. Král'ová, T. Scholz, and S. D'Amelio. 1999. Genetic and morphological variability in cestodes of the genus *Proteocephalus*: geographical variation in *Proteocephalus percae* populations. *Canadian Journal of Zoology*, 77:1450–1458.

Harris, P. D. 1983. The morphology and life cycle of the oviparous *Oögyrodactylus farlowellae* gen. et sp. n. (Monogenea, Gyrodactylidae). *Parasitology*, 87:405–420.

Harrison, L. J. S., R. M. E. Delgado, and R. M. E. Parkhouse. 1990. Differential diagnosis of *Taenia saginata* and *Taenia solium* with DNA probes. *Parasitology*, 100:459–461.

Hassouna, N., B. Michot, and J.-P. Bachellerie. 1984. The complete nucleotide sequence of mouse 28S rRNA gene. Implications for the process of size increase of the large subunit rRNA in higher eukaryotes. *Nucleic Acids Research*, 12:3563–3583.

Haswell, W. A. 1893. A Monograph of the Temnocephaleae. *Linnean Society of New South Wales, MacLeay Memorial Volume*: 93–152.

Haswell, W. A. 1905. Studies on the Turbellaria. Part II. *Anomolocoelus caecus*, a new type of Rhabdocoele. *Quarterly Journal of Microscopical Science*, 49:450–467.

Haszprunar, G. 1996. Plathelminthes and Plathelmintho-morpha – paraphyletic taxa. *Journal of Zoological Systematics and Evolutionary Research*, 34:41–48.

Hathaway, M. A., R. P. Hathaway, and D. C. Kritsky. 1995. Spermatogenesis in *Octomacrum lanceatum* (Monogenoidea, Oligonchoinea, Mazocraeidea). *International Journal for Parasitology*, 25:913–922.

Hathaway, R. P., M. A. Hathaway, and D. C. Kritsky. 1993. Observations on fine structure of sperm development in *Lagarocotyle salamandrae* (Polyonchoinea: Lagarocotylidea). *Second International Symposium on Monogenea*, p. 127, Montpellier, France.

Hauser, M., and H. Koopowitz. 1987. Age–dependent changes in fluorescent neurons in the brain of *Notoplana acticola*, a polyclad flatworm. *Journal of Experimental Zoology*, 241:217–255.

Hedges, S. B., and L. L. Poling. 1999. A molecular phylogeny of reptiles. *Science*, 283:998–1001.

Hein, W. 1904. Beiträge zur Kenntnis von *Amphilina foliacea*. *Zeitschrift für wissenschaftliche Zoologie*, 76:400–438.

Hendelberg, J. 1962. Paired flagella and nucleus migration in the spermiogenesis of *Dicrocoelium* and *Fasciola* (Digenea, Trematoda). *Zoologiska Bidrag från Uppsala*, 35:569–587.

Hendelberg, J. 1969. On the development of different types of spermatozoa from spermatids with two flagella in the Turbellaria with remarks on the ultrastructure of the flagella. *Zoologiska Bidrag fran Uppsala*, 38:1–50.

Hendelberg, J. 1974. Spermiogenesis, sperm morphology, and biology of fertilization in the Turbellaria. In N. W. Riser and M. P. Morse (eds.), *Biology of the Turbellaria*, pp. 148–164. McGraw–Hill, New York.

Hendelberg, J. 1975. Functional aspects of flatworm sperm morphology. In B. A. Afzelius (ed.), *The functional anatomy of the spermatozoon*, pp. 299–309. Pergamon, Oxford.

Hendelberg, J. 1977. Comparative morphology of turbellarian spermatozoa studied by electron microscopy. *Acta Zoologica Fennica*, 154:149–162.

Hendelberg, J. 1983. Platyhelminthes – Turbellaria. In K. G. Adiyodi and R. G. Adiyodi (eds.), *Reproductive biology of invertebrates. Vol. II. Spermatogenesis and sperm function*, pp. 75–104. John Wiley and Sons, Chichester.

Hendelberg, J. 1986. The phylogenetic significance of sperm morphology in the Platyhelminthes. *Hydrobiologia*, 132:53–58.

Hendow, H. T., and B. L. James. 1988. Ultrastructure of the spermatozoon and spermatogenesis in *Maritrema linguilla* (Digenea: Microphallidae). *International Journal for Parasitology*, 18:53–63.

Henley, C. 1968. Refractile bodies in the developing and mature spermatozoa of *Childia groenlandica* (Turbellaria, Acoela) and their possible significance. *Biological Bulletin*, 134:382–397.

Henley, C. 1974. Platyhelminthes ('Turbellaria'). In A. C. Giese and J. S. Pearse (eds.), *Reproduction of marine invertebrates. Volume 1. Acoelomate and pseudocoelomate metazoans*, pp. 267–343. Academic Press, New York.

Henley, C., and D. P. Costello. 1969. Microtubules in spermatozoa of some turbellarian flatworms. *Biological Bulletin*, 137:405.

Henley, C., D. P. Costello, and C. R. Ault. 1968. Microtubules in the axial filament complexes of acoel turbellarian spermatozoa as revealed by negative staining. *Biological Bulletin*, 135:422–423.

Henley, C., D. P. Costello, M. B. Thomas, and W. D. Newton. 1969. The '9 + 1' pattern of microtubules in spermatozoa of Mesostoma (Plathelminthes, Turbellaria). *Proceedings of the National Academy of Sciences USA*, 64:849–856.

Hennig, W. 1950. *Grundzüge einer Theorie der phylogenetischen Systematik*. Deutscher Zentralverlag, Berlin.

Hennig, W. 1966. *Phylogenetic systematics* (English translation). University of Illinois, Urbana.

Henry, J. J., and M. Q. Martindale. 1997. Regulation and the modification of axial properties in partial embryos of the nemertean *Cerebratulus lacteus*. *Development Genes and Evolution*, 207:42–50.

Henry, J. J., and M. Q. Martindale. 1998. Conservation of the spiralian developmental program: cell lineage of the nemertean, *Cerebratulus lacteus*. *Developmental Biology*, 201:253–269.

Henry, J. J., M. Q. Martindale, and B. C. Boyer. 1995. Axial specification of a basal member of the spiralian clade: lineage relationships of the first four cells to the larval body plan in the polyclad turbellarian *Hoploplana inquilina*. *Biological Bulletin*, 189:194–195.

Henry, J. J., M. Q. Martindale, and B. C. Boyer. 2000. The unique developmental program of the acoel flatworm, *Neochildia fusca*. *Developmental Biology*, 220:285–295.

Herde, K. E. 1938. Early development of *Ophiotaenia perspicua* La Rue. *Transactions of the American Microscopical Society*, 57:282–291.

Hertel, L. A. 1993. Excretion and osmoregulation in the flatworms. *Transactions of the American Microscopical Society*, 112:10–17.

Hertzler, P. L., and W. H. Clark. 1992. Cleavage and gastrulation in the shrimp *Sicyonia ingentis*: invagination accompanied by oriented cell division. *Development*, 116:127–140.

Hesse, R. 1897. Untersuchungen über die Organe der Lichtempfindung bei niederen Tieren II. Die Augen der Plathelminthes, insbesondere der tricladen Turbellarien. *Zeitschrift für wissenschaftliche Zoologie*, 65:528–582.

Hett, M. L. 1925. On a new species of *Temnocephala* (*T. chaeropsis*) (Trematoda) from West Australia. *Proceedings of the Zoological Society of London*: 569–575.

Hickman, J. L. 1963. The biology of *Oochoristica vacuolata* Hickman (Cestoda). *Papers and Proceedings of the Royal Society of Tasmania*, 97:81–104.

Hickman, V. V. 1967. Tasmanian Temnocephalidea. *The Papers and Proceedings of the Royal Society of Tasmania*, 101:227–251.

Hidalgo, C., J. Miquel, J. Torres, and B. Marchand. 1999. Ultrastructural study of spermiogenesis and the spermatozoon in *Catenotaenia pusilla*, an intestinal parasite of *Mus musculus*. *Journal of Helminthology*, 74:73–81.

Hillis, D. M. 1987. Molecular versus morphological approaches to systematics. *Annual Review of Ecology and Systematics*, 18:23–42.

Hillis, D. M. 1996. Inferring complex phylogenies. *Nature*, 383:130–131.

Hillis, D. M. 1998. Taxonomic sampling, phylogenetic accuracy, and investigator bias. *Systematic Biology*, 47:3–8.

Hillis, D. M., and M. T. Dixon. 1991. Ribosomal DNA: molecular evolution and phylogenetic inference. *Quarterly Review of Biology*, 66:411–453.

Hillis, D. M., C. Moritz, and B. K. Mable. 1996. *Molecular systematics*. 2nd ed. Sinauer Associates, Sunderland, MA.

Hirai, H., and P. LoVerde. 1995. FISH techniques for constructing physical maps on schistosome chromosomes. *Parasitology Today*, 11:310–314.

Hirt, R. P., J. Logsdon, B. Healy, M. W. Dorey, W. F. Doolittle, and T. M. Embley. 1999. Microsporidia are related to Fungi: evidence from the largest subunit of RNA polymerase II and other proteins. *Proceedings of the National Academy of Sciences, USA*, 96:580–585.

Hoberg, E. P. 1986. Evolution and historical biogeography of a parasite-host assemblage: *Alcataenia* spp. (Cyclophyllidea: Dilepididae) in Alcidae (Charadriiformes). *Canadian Journal of Zoology*, 64:2576–2589.

Hoberg, E. P. 1987. Recognition of larvae of the Tetrabothriidae (Eucestoda): implications for the origin of tapeworms in marine homeotherms. *Canadian Journal of Zoology*, 65:997–1000.

Hoberg, E. P. 1989. Phylogenetic relationships among genera of the Tetrabothriidae (Eucestoda). *Journal of Parasitology*, 75:617–626.

Hoberg, E. P. 1992. Congruent and synchronic patterns in biogeography and speciation among seabirds, pinnipeds, and cestodes. *Journal of Parasitology*, 78:601–615.

Hoberg, E. P. 1994. Order Tetrabothriidea Baer, 1954. In L. F. Khalil, A. Jones and R. A. Bray (eds.), *Keys to the cestode parasites of vertebrates*, pp. 295–304. CAB International, Wallingford.

Hoberg, E. P. 1995. Historical biogeography and modes of speciation across high-latitude seas of the Holarctic: concepts for host-parasite coevolution among the Phocini (Phocidae) and Tetrabothriidae (Eucestoda). *Canadian Journal of Zoology*, 73:45–57.

Hoberg, E. P. 1997. Phylogeny and historical reconstruction: host-parasite systems as keystones in biogeography and ecology. In M. L. Reaka-Kudla, D. E. Wilson and E. O. Wilson (eds.), *Biodiversity II: Understanding and protecting our biological resources*, pp. 243–261. Joseph Henry Press, National Academy of Sciences, Washington, D.C.

Hoberg, E. P., and A. M. Adams. 1992. Phylogeny, historical biogeography, and ecology of *Anophryocephalus* spp. (Eucestoda: Tetrabothriidae) among pinnipeds of the Holarctic during the late Tertiary and Pleistocene. *Canadian Journal of Zoology*, 70:703–719.

Hoberg, E. P., D. R. Brooks, and D. Siegel-Causey. 1997a. Host-parasite co-speciation: history, principles and prospects. In D. H. Clayton and J. Moore (eds.), *Host-parasite evolution, general principles and avian models*, pp. 212–235. Oxford University Press, Oxford.

Hoberg, E. P., S. L. Gardner, and R. A. Campbell. 1997b. Paradigm shifts and tapeworm systematics. *Parasitology Today*, 13:161–162.

Hoberg, E. P., J. Mariaux, J.-L. Justine, D. R. Brooks, and P. J. Weekes. 1997c. Phylogeny of the orders of the Eucestoda (Cercomeromorphae) based on comparative morphology: historical perspectives and a new working hypothesis. *Journal of Parasitology*, 83:1128–1147

Hoberg, E. P., A. Jones, and R. A. Bray. 1999a. Phylogenetic analysis among the families of the Cyclophyllidea (Eucestoda) based on comparative morphology, with new hypotheses for co-evolution in vertebrates. *Systematic Parasitology*, 42:51–73.

Hoberg, E. P., S. L. Gardner, and R. A. Campbell. 1999b. Systematics of the Eucestoda: advances toward a new phylogenetic paradigm, and observations on the early diversification of tapeworms and vertebrates. *Systematic Parasitology*, 42:1–12.

Hoberg, E. P., J. Mariaux, and D. R. Brooks. 2001. Phylogeny among orders of the Eucestoda (Cercomeromorphae): integrating morphology, molecules and total evidence. In D. T. J. Littlewood and R. A. Bray (eds.), *Interrelationships of the Platyhelminthes*, pp. 112–126. Taylor & Francis, London.

Hoeppli, R. 1959. *Parasites and parasitic infections in early medicine and science*. University of Malaya Press, Singapore.

Holmes, S. C., and I. Fairweather. 1984. *Fasciola hepatica*: the effects of neuropharmacological agents upon *in vitro* motility. *Experimental Parasitology*, 58:194–208.

Hori, H., and S. Osawa. 1986. Evolutionary change in 5S rRNA secondary structure and a phylogenic tree of 352 5S rRNA species. *Bio Systems*, 19:163–172.

Horsfall, M. W., and M. F. Jones. 1937. The life history of *Choanotaenia infundibulum*, a cestode parasitic in chickens. *Journal of Parasitology*, 23:435–450.

Howells, R. E. 1969. Observations on the protonephridial system of the cestode, *Moniezia expansa* (Rud., 1805). *Parasitology*, 59:449–459.

Huelsenbeck, J. P., J. J. Bull, and C. W. Cunningham. 1996. Combining data in phylogenetic analysis. *Trends in Ecology and Evolution*, 11:152–158.

Hyman, L. H. 1951. *The Invertebrates: Platyhelminthes and Rhynchocoela. The acoelomate Bilateria. Volume II.* McGraw–Hill Book Company, Inc., New York.

Hyra, G. S. 1993. *Genostoma kozloffi* sp. nov. and *G. inopinatum* sp. nov. (Turbellaria: Neorhabdocoela: Genostomatidae) from leptostracan crustaceans of the genus *Nebalia*. *Cahiers de Biologie Marine*, 34:111–126.

Hyra, G. S. 1994. Spermatozoon ultrastructure and the taxonomy of *Genostoma*, a problematic turbellarian (Prolecithophora ?Rhabdocoela?). *Transactions of the American Microscropical Society*), 113:98.

ICZN. 1999. *International Code of Zoological Nomenclature.* 4th ed. International Trust for Zoological Nomenclature, London.

Iomini, C., and J.-L. Justine. 1997. Spermiogenesis and spermatozoon of *Echinostoma caproni* (Platyhelminthes, Digenea): transmission and scanning electron microscopy, and tubulin immunocytochemistry. *Tissue & Cell*, 29:107–118.

Iomini, C., M. Ferraguti, G. Melone, and J.-L. Justine. 1994. Spermiogenesis in a scutariellid (Platyhelminthes). *Acta Zoologica (Stockholm)*, 75:287–295.

Iomini, C., O. Raikova, N. Noury–Sraïri, and J.-L. Justine. 1995. Immunocytochemistry of tubulin in spermatozoa of Platyhelminthes. In B. G. M. Jamieson, J. Ausio and J.-L. Justine (eds.), *Advances in spermatozoal phylogeny and taxonomy. Memoires du Museum National d'Histoire Naturelle*, Vol. 166, pp. 97–104.

Iomini, C., I. Mollaret, J.-L. Albaret, and J.-L. Justine. 1997. Spermatozoon and spermiogenesis in *Mesocoelium monas* (Platyhelminthes: Digenea): ultrastructure and epifluorescence microscopy of labelling of tubulin and nucleus. *Folia Parasitologica*, 44:26–32.

Iomini, C., M.-H. Bré, N. Levilliers, and J.-L. Justine. 1998. Tubulin polyglycylation in Platyhelminthes: diversity among stable microtubule networks and very late occurrence during spermiogenesis. *Cell Motility and the Cytoskeleton*, 39:318–330.

Ishii, N. 1935. Studies on the family Didymozooidae (Monticelli, 1888). *Japanese Journal of Zoology*, 6:279–335.

Ishii, S. 1980. The ultrastructure of the protonephridial flame cell of the freshwater planarian *Bdellocephala brunnea*. *Cell and Tissue Research*, 206:441–449.

Israelsson, O. 1997. [*Xenoturbella*'s] molluscan embryogenesis. *Nature*, 390:32.

Israelsson, O. 1999. New light on the enigmatic *Xenoturbella* (phylum uncertain): ontogeny and phylogeny. *Proceedings of the Royal Society of London, Series B, Biological Sciences*, 266:835–841.

Ivanov, A. V. 1952a. [*Udonella caligorum* Johnston, 1935 – a representative of a new class of flatworms]. (In Russian). *Zoologicheskii Zhurnal*, 31:173–178.

Ivanov, A. V. 1952b. [The structure of *Udonella caligorum* Johnston, 1835 and the position of Udonellidae in the systematics of flatworms]. (In Russian). *Parazitologicheskii Sbornik*, 14:112–163.

Ivanov, A. V., and Y. V. Mamkaev. 1973. [*Turbellaria, their origin and evolution. Phylogenetic considerations*]. (In Russian). Nauka, Leningrad.

Ivanov, V. A., and R. A. Campbell. 1998. A new species of *Acanthobothrium* van Beneden, 1849 (Cestoda: Tetraphyllidea) from *Rioraja castelnaui* (Chondrichthyes: Rajoidei) in coastal waters of Argentina. *Systematic Parasitology*, 40:203–212.

Ivanov, V. A., and E. P. Hoberg. 1999. Preliminary comments on a phylogenetic study of the order Diphyllidea van Beneden in Carus, 1863. *Systematic Parasitology*, 42:21–27.

Jägersten, G. 1959. Further remarks on the early phylogeny of the Metazoa. *Zoologiska Bidrag från Uppsala*, 33:79–108.

Jägersten, G. 1972. *Evolution of the metazoan life cycle, a comprehensive theory.* Academic Press, New York.

Jamieson, B. G. M., J. Ausio, and J.-L. Justine. 1995. Advances in spermatozoal phylogeny and taxonomy. *Memoires du Museum National d'Histoire Naturelle*, Vol. 166, p. 565.

Jänichen, E. 1896. Beiträge zur Kenntnis des Turbellarienauges. *Zeitschrift für wissenschaftliche Zoologie*, 62:250–288.

Janicki, C. 1908. Über den Bau von *Amphilina liguloidea* Diesing. *Zeitschrift für wissenschaftliche Zoologie*, 89:568–597.

Janicki, C. 1920. 'Grundlinien einer Cercomer Theorie' zur Morphologie der Trematoden und Cestoden. *Festschrift für Zschokke*, 30:1–22.

Janicki, C. 1928. Die Lebensgeschichte von *Amphilina foleacea* G. Wagen., Parasiten des Wolga–Sterlets, nach Beobachtungen und Experimenten. *Arbeiten der Biologischen Wolga-Station, Saratow*, 10:97–134.

Janicki, C. 1930. Über die jüngsten Zustände von *Amphilina foliacea* in der Fischleibeshöhle, sowie Generelles zur Auffassung des Genus *Amphilina* G. Wagner. *Zoologischer Anzeiger*, 90:190–205.

Jarecka, L. 1970. Life cycle of *Valipora campylancristrota* (Wedl, 1855) Baer and Bona 1958–1960 (Cestoda-Dilepididae) and the description of cercoscolex – a new type of cestode larva. *Bulletin de l'Académie Polonaise des Sciences, série des sciences biologiques*, 18:159–163.

Jarecka, L. 1975. Ontogeny and evolution of cestodes. *Acta Parasitologica Polonica*, 23:93–114.

Jarecka, L., W. Michajlow, and M. D. B. Burt. 1981. Comparative ultrastructure of cestode larvae and Janicki's cercomer theory. *Acta Parasitologica Polonica*, 28:65–72.

Jarecka, L., G. N. Bance, and M. D. B. Burt. 1984. On the life cycle of *Anomotaenia micracantha dominicana* (Railliet et Henry, 1912), with ultrastructural evidence supporting the definition cercoscolex for dilepidid larvae (Cestoda, Dilepididae). *Acta Parasitologica Polonica*, 29:27–34.

Jenner, R. A., and F. R. Schram. 1999. The grand game of metazoan phylogeny: rules and strategies. *Biological Reviews of the Cambridge Philosophical Society*, 74:121–142.

Jennings, J. B. 1971. Parasitism and commensalism in the Turbellaria. *Advances in Parasitology*, 9:1–32.

Jennings, J. B. 1974. Symbioses in the Turbellaria and their implications in studies on the evolution of parasitism. In W. B. Vernberg (ed.), *Symbiosis in the sea*, pp. 127–160. University of South Carolina Press, Columbia.

Jennings, J. B. 1997. Nutritional and respiratory pathways to parasitism exemplified in the Turbellaria. *International Journal for Parasitology*, 27:679–691.

Jennings, J. B., L. R. G. Cannon, and A. J. Hick. 1992. The nature and origin of the epidermal scales of *Notodactylus handschini* – an unusual temnocephalid turbellarian ectosymbiotic on crayfish from northern Queensland. *Biological Bulletin*, 182:117–128.

Jeong, K.-H., H.-J. Rim, W.-K. Kim, C.-W. Kim, and H.-Y. Yang. 1980. A study of the fine structure of *Clonorchis sinensis*, a liver fluke. II. The alimentary tract and the excretory system. *Korean Journal of Parasitology*, 18:81–91.

Joffe, B. I. 1981a. Morphology and phylogenetic relations of the Temnocephalida (Turbellaria). *Zoologicheskii Zhurnal*, 60:661–672.

Joffe, B. I. 1981b. Morphology of temnocephalids and morphological alterations during the transition to parasitism in the Platyhelminthes. *Parazitologiya*, 15:209–218.

Joffe, B. I. 1982. The Temnocephalids: their morphology and phylogeny. Ph.D. thesis, Zoological Institute Russian Academy of Science.

Joffe, B. I. 1987. [On the evolution of the pharynx in the flatworms]. (In Russian). *Trudy Zoologicheskogo Instituta Akademii Nauk SSSR*, 221:34–71.

Joffe, B. I. 1988. On the phylogeny of the Temnocephalida. *Fortschritte der Zoologie*, 36:58–62.

Joffe, B. I. 1990. Morphological regularities in the evolution of the nervous system in the Plathelminthes: anatomical variants of the orthogon and their dependence upon the body form. *Trudy Zoologicheskogo Instituta*, 221:87–125.

Joffe, B. I. 1991. On the number and spatial distribution of the catecholamine-containing (Ca-Positive) neurons in some higher and lower turbellarians: a comparison. *Hydrobiologia*, 227:201–208.

Joffe, B. I., and L. R. G. Cannon. 1998a. The organisation and evolution of the mosaic of the epidermal syncytia in the Temnocephalida (Plathelminthes: Neodermata). *Zoologischer Anzeiger*, 237:1–14.

Joffe, B. I., and L. R. G. Cannon. 1998b. The GAIF–positive population of neurons in the evolution of the Temnocephalida. *Acta Zoologica (Stockholm)*, 79:257–265.

Joffe, B. I., and L. R. G. Cannon. 1998c. Anatomy of the sensory nervous system in *Craspedella pedum* (Plathelminthes, Temnocephalida): DiO staining after fixation and *in vivo*. *Zoomorphology*, 118:51–60.

Joffe, B. I., and R. A. Janashvili. 1982. [First finding of temnocephalids (Turbellaria: Scutariellidae) in the territory of the USSR]. (In Russian). *Parazitologiya*, 15:170–174.

Joffe, B. I., and E. E. Kornakova. 1998. *Notentera ivanovi* Joffe et al., 1997: a contribution to the question of phylogenetic relationships between 'turbellarians' and the parasitic Plathelminthes (Neodermata). *Hydrobiologia*, 383:245–250.

Joffe, B. I., and E. E. Kornakova. 2001. Flatworm phylogeneticist: between a molecular hammer and a morphological anvil. In D. T. J. Littlewood and R. A. Bray (eds.), *Interrelationships of the Platyhelminthes*, pp. 279–291. Taylor & Francis, London.

Joffe, B. I., and E. A. Kotikova. 1991. Distribution of catecholamines in turbellarians (with discussion of neuronal homologies in the Plathelminthes). *Studies in Neuroscience*, 13:77–113.

Joffe, B. I., and M. Reuter. 1993. The nervous system of *Bothriomolus balticus* (Proseriata) – a contribution to the knowledge of the orthogon in the Platyhelminthes. *Zoomorphology*, 113:113–127.

Joffe, B. I., I. V. Solovei, and L. R. G. Cannon. 1995a. The structure of the epidermis in *Didymorchis* (Temnocephalida: Platyhelminthes). *Australian Journal of Zoology*, 43:631–641.

Joffe, B. I., I. V. Solovei, K. B. Sewell, and L. R. G. Cannon. 1995b. Organization of the epidermal syncytial mosaic in *Diceratocephala boschmai* (Temnocephalida, Platyhelminthes). *Australian Journal of Zoology*, 43:509–518.

Joffe, B. I., K. M. Valiego Roman, V. Y. Birstein, and A. V. Troitsky. 1995c. 5S rRNA of 12 species of flatworms: implications for the phylogeny of the Platyhelminthes. *Hydrobiologia*, 305:37–43.

Joffe, B. I., I. V. Solovei, and H. C. MacGregor. 1996a. Ends of chromosomes in *Polycelis tenuis* (Platyhelminthes) have telomere repeat TTAGGG. *Chromosome Research*, 4:323–324.

Joffe, B. I., I. Solovei, and L. R. G. Cannon. 1996b. The posttentacular syncytia of temnocephalids: the first indication of a putative osmoregulatory organ in the epidermis of a platyhelminth. *Acta Zoologica (Stockholm)*, 77:241–247.

Joffe, B. I., R. V. Selivanova, and E. E. Kornakova. 1997a. [*Notentera ivanovi* n. gen., n. sp. (Turbellaria, Platyhelminthes), a new parasitic turbellarian]. (In Russian). *Parazitologiya*, 31:126–131.

Joffe, B. I., I. V. Solovei, N. A. Watson, and L. R. G. Cannon. 1997b. Structure and evolution of the pharynx in the Temnocephalida (Platyhelminthes). *Canadian Journal of Zoology*, 75:205–226.

Joffe, B. I., I. V. Solovei, and H. C. MacGregor. 1998a. Ordered arrangement and rearrangement of chromosomes during spermatogenesis in two species of planarians (Plathelminthes). *Chromosoma*, 107:173–183.

Joffe, B. I., L. R. G. Cannon, and E. R. Schockaert. 1998b. On the phylogeny of families and genera within the Temnocephalida. *Hydrobiologia*, 383:263–268.

Joffe, B. I., I. V. Solovei, and L. R. G. Cannon. 1998c. Sensory cells in the sucker of *Craspedella pedum* (Plathelminthes, Temnocephalida): *in vivo* staining with DiO and SEM and TEM observations. *Zoomorphology*, 118:61–68.

Johnson, M. R., and E. P. Hoberg. 1989. Differentiation of *Moniezia expansa* and *Moniezia benedeni* (Eucestoda: Cyclophyllidea) by isoelectric focusing. *Canadian Journal of Zoology*, 67:1471–1475.

Johnston, D. A., R. A. Kane, and D. Rollinson. 1993. Small subunit (18S) ribosomal RNA gene divergence in the genus *Schistosoma*. *Parasitology*, 107:147–156.

Johnston, D. A., M. L. Blaxter, W. M. Degrave, J. Foster, A. C. Ivens, and S. E. Melville. 1999. Genomics and the biology of parasites. *Bioessays*, 21:131–147.

Johnston, G. 1835. Illustrations in British zoology. *Magazine of Natural History, London*, 8:494–498.

Johnston, R. N., C. Shaw, D. W. Halton, P. Verhaert, and J. Baguñà. 1995. GYIRFamide: A novel FMRFamide-related peptide (FaRP) from the triclad turbellarian, *Dugesia tigrina*. *Biochemical and Biophysical Research Communications*, 209:689–697.

Johnston, R. N., C. Shaw, D. W. Halton, P. Verhaert, K. L. Blair, G. P. Brennan, D. Price, and P. A. V. Anderson. 1996. Isolation, localization and bioactivity of the FMRFamide related neuropeptides GYIRFamide and YIRFamide from the marine turbellarian, *Bdelloura candida*. *Journal of Neurochemistry*, 67:814–821.

Jondelius, U. 1992. Sperm morphology in the Pterastericolidae (Platyhelminthes, Rhabdocoela) – phylogenetic implications. *Zoologica Scripta*, 21:223–230.

Jondelius, U. 1997. A new family and three new species of flatworms (Platyhelminthes) from the Darwin Harbour area. In R. H. Hanley, G. Caswell, D. Megirian and H. K. Larson (eds.), *Proceedings of the Sixth International Marine Biological Workshop. The Marine Flora and Fauna of*

Darwin Harbour, Northern Territory, Australia, pp. 187–197.

Jondelius, U. 1998. Flatworm phylogeny from partial 18S rDNA sequences. *Hydrobiologia*, 383:147–154.

Jondelius, U., and M. Thollesson. 1993. Phylogeny of the Rhabdocoela (Platyhelminthes): a working hypothesis. *Canadian Journal of Zoology*, 71:298–308.

Jondelius, U., M. Norén, and J. Hendelberg. 2001. The Prolecithophora. In D. T. J. Littlewood and R. A. Bray (eds.), *Interrelationships of the Platyhelminthes*, pp. 74–80. Taylor & Francis, London.

Jones, H. D., and B. M. Gerard. 1999. A new genus and species of terrestrial planarian (Platyhelminthes; Tricladida; Terricola) from Scotland, and an emendation of the genus *Artioposthia*. *Journal of Natural History*, 33:387–394.

Jones, M. F. 1936. *Metroliasthes lucida*, a cestode of galliform birds, in arthropod and avian hosts. *Proceedings of the Helminthological Society of Washington*, 3:26–30.

Jones, M. F., and M. W. Horsfall. 1936. The life history of a poultry cestode. *Science*, 83:303–304.

Jones, M. K. 1988. Formation of the paruterine capsules and embryonic envelopes in *Cylindrotaenia hickmani* (Jones, 1985) (Cestoda: Nematotaeniidae). *Australian Journal of Zoology*, 36:545–563.

Jourdane, J. 1972. Étude expérimentale du cycle biologique de deux espèces de *Choanotaenia* intestinaux de Soricidae. *Zeitschrift für Parasitenkunde*, 38:333–343.

Jourdane, J. 1977. Ecologie du développement et de la transmission des plathelminthes parasites de Soricidae pyrénéens. *Memoires du Museum national d'Histoire naturelle, Série A, Zoologie*, 103:1–171.

Jousson, O., P. Bartoli, and J. Pawlowski. 1998a. Molecular phylogeny of Mesometridae (Trematoda, Digenea) with its relation to morphological changes in parasites. *Parasite*, 5:365–369.

Jousson, O., P. Bartoli, L. Zaninetti, and J. Pawlowski. 1998b. Use of the ITS rDNA for elucidation of some life–cycles of Mesometridae (Trematoda, Digenea). *International Journal for Parasitology*, 28:1403–1411.

Jousson, O., P. Bartoli, and J. Pawlowski. 1999. Molecular identification of developmental stages in Opecoelidae (Digenea). *International Journal for Parasitology*, 29:1853–1858.

Joyeux, C., and J. G. Baer. 1945. Morphologie, évolution et position systématique de *Catenotaenia pusilla* (Goeze, 1782), cestode parasite de rongeurs. *Revue suisse de Zoologie*, 52:13–51.

Joyeux, C., and J. G. Baer. 1961. Classe des cestodes. In P. P. Grassé (ed.), *Traité de zoologies. Anatomie–systématique biologie*, pp. 347–560. Masson et Cie, Paris.

Justine, J.-L. 1983. A new look at Monogenea and Digenea spermatozoa. In J. Andre (ed.), *The sperm cell, fertilization, surface properties, motility, nucleus and acrosome. Evolutionary aspects*, pp. 454–457. Martinus Nijhoff Publishers, The Hague.

Justine, J.-L. 1986. Ultrastructure of the spermatozoon of the cestode *Duthiersia fimbriata* Diesing, 1854 (Pseudophyllidea, Diphyllobothriidae). *Canadian Journal of Zoology*, 64:1545–1548.

Justine, J.-L. 1991a. Phylogeny of parasitic Platyhelminthes – a critical study of synapomorphies proposed on the basis of the ultrastructure of spermiogenesis and spermatozoa. *Canadian Journal of Zoology*, 69:1421–1440.

Justine, J.-L. 1991b. Cladistic study in the Monogenea (Platyhelminthes), based upon a parsimony analysis of spermiogenetic and spermatozoal ultrastructural characters. *International Journal for Parasitology*, 21:821–838.

Justine, J.-L. 1993. Phylogénie des Monogènes basée sur une analyse de parcimonie des caractères de l'ultrastructure de la spermiogenèse et des spermatozoïdes incluant les résultats récents. *Bulletin Français de la Pêche et de la Pisciculture*, 328:137–155.

Justine, J.-L. 1995. Spermatozoal ultrustructure and phylogeny of the parasitic Platyhelminthes. In B. G. M. Jamieson, J. Ausio and J.-L. Justine (eds.), *Advances in spermatazoal phylogeny and taxonomy. Memoires du Museum National d'Histoire Naturelle*, Vol. 166, pp. 55–86.

Justine, J.-L. 1998a. Non-monophyly of the monogeneans? *International Journal for Parasitology*, 28:1653–1657.

Justine, J.-L. 1998b. Spermatozoa as phylogenetic characters for the Eucestoda. *Journal of Parasitology*, 84:385–408.

Justine, J.-L. 1998c. Systématique des grands groupes de Plathelminthes parasites: quoi de neuf? *Bulletin de la Societé Française de Parasitologie*, 16:34–52.

Justine, J.-L. 1999. Spermatozoa of Platyhelminthes: comparative ultrastructure, tubulin immunocytochemistry and nuclear labelling. In C. Gagnon (ed.), *The male gamete from basic science to clinical applications*, pp. 351–362. Cache River Press, Vienna, Illinois.

Justine, J.-L. 2001. Spermatozoa as phylogenetic characters for the Platyhelminthes. In D. T. J. Littlewood and R. A. Bray (eds.), *Interrelationships of the Platyhelminthes*, pp. 231–238. Taylor & Francis, London.

Justine, J.-L., and X. Mattei. 1981. Étude ultrastructurale du flagelle spermatique des schistosomes (Trematoda: Digenea). *Journal of Ultrastructure Research*, 76:89–95.

Justine, J.-L., and X. Mattei. 1983a. Comparative ultrastructural study of spermiogenesis in monogeneans (Flatworms). 2. *Heterocotyle* (Monopisthocotylea, Monocotylidae). *Journal of Ultrastructure Research*, 84:213–223.

Justine, J.-L., and X. Mattei. 1983b. A spermatozoon with 2 9 +0 axonemes in a parasitic flatworm, *Didymozoon* (Digenea, Didymozoidae). *Journal of Submicroscopic Cytology and Pathology*, 15:1101–1105.

Justine, J.-L., and X. Mattei. 1983c. Étude ultrastructurale comparée de la spermiogenèse des monogenes 1. *Megalocotyle* (Monopisthocotylea Capsalidae). *Journal of Ultrastructure Research*, 82:296–308.

Justine, J.-L., and X. Mattei. 1984a. Atypical spermiogenesis in a parasitic flatworm, *Didymozoon* (Trematoda, Digenea, Didymozoidae). *Journal of Ultrastructure Research*, 87:106–111.

Justine, J.-L., and X. Mattei. 1984b. Comparative ultrastructural study of spermiogenesis in monogeneans (Flatworms). 4. *Diplectanum* (Monopisthocotylea, Diplectanidae). *Journal of Ultrastructure Research*, 88:77–91.

Justine, J.-L., and X. Mattei. 1986. Comparative ultrastructural study of spermiogenesis in monogeneans (Flatworms). 5. *Calceostoma* (Monopisthocotylea, Calceostomatidae). *Journal of Ultrastructure and Molecular Structure Research*, 96:54–63.

Justine, J.-L., and X. Mattei. 1987. Phylogenetic relationships between the families Capsalidae and Dionchidae (Platyhelminthes, Monogenea, Monopisthocotylea) indicated by the comparative ultrastructural study of spermiogenesis. *Zoologica Scripta*, 16:111–116.

Justine, J.-L., A. Lambert, and X. Mattei. 1985a. Spermatozoon ultrastructure and phylogenetic relationships in the monogeneans (Platyhelminthes). *International Journal for Parasitology*, 15:601–608.

Justine, J.-L., N. Lebrun, and X. Mattei. 1985b. The aflagellate spermatozoon of *Diplozoon* (Platyhelminthes, Monogenea, Polyopisthocotylea). A demonstrative case of relationship between sperm ultrastructure and biology of reproduction. *Journal of Ultrastructure Research*, 92:47–54.

Justine, J.-L., R. Ponce de León, and X. Mattei. 1987. Ultrastructural observations on the spermatozoa of two Temnocephalids (Platyhelminthes). *Acta Zoologica (Stockholm)*, 68:1–7.

Justine, J.-L., B. G. M. Jamieson, and V. R. Southgate. 1993. Homogeneity of sperm structure in 6 species of schistosomes (Digenea, Platyhelminthes). *Annales de Parasitologie Humaine et Comparée*, 68:185–187.

Justine, J.-L., C. Iomini, O. I. Raikova, and I. Mollaret. 1998. The homology of cortical microtubules in platyhelminth spermatozoa: a comparative immunocytochemical study of acetylated tubulin. *Acta Zoologica (Stockholm)*, 79:235–241.

Karling, T. G. 1940. Zur Morphologie und Systematik der Alloeocoela Cumulata und Rhabdocoela Lecithophora (Turbellaria). *Acta Zoologica Fennica*, 26:1–260.

Karling, T. G. 1963. *Ulianinia mollissima* Levinsen, 1879, rediscovered (Turbellaria, Prolecithophora). *Videnskabelige Meddelelser Dansk naturhistorisk Forening*, 112:495–510.

Karling, T. G. 1965. *Haplopharynx rostratus* Meixner (Turbellaria) mit den Nemertinen vergleichen. *Zeitschrift für Zoologische Systematik und Evolutionsforschung*, 3:1–18.

Karling, T. G. 1966a. On nematocysts and similar structures in turbellarians. *Acta Zoologica Fennica*, 116:1–28.

Karling, T. G. 1966b. Marine Turbellaria from the Pacific Coast of the North America IV. Coelogynoporidae and Monocelididae. *Arkiv für Zoologie*, 18:493–528.

Karling, T. G. 1966c. On the defecation apparatus in the genus *Archimonocelis* (Turbellaria, Monocelididae). *Sarsia*, 24:37–44.

Karling, T. G. 1974. On the anatomy and affinities of the turbellarian orders. In N. W. Riser and M. P. Morse (eds.), *Biology of the Turbellaria*, pp. 1–16. McGraw-Hill, New York.

Karling, T. G. 1978. Anatomy and systematics of marine Turbellaria from Bermuda. *Zoologica Scripta*, 7:225–248.

Karling, T. G. 1993. Anatomy and evolution in Cylindrostomidae (Plathelminthes, Prolecithophora). *Zoologica Scripta*, 22:325–339.

Karling, T. G., and U. Jondelius. 1995. An East-Pacific species of *Multipeniata* Nasanov and three *Plagiostomum* species (Platyhelminthes, Prolecithophora). *Microfauna Marina*, 10:147–158.

Katayama, T., M. Yamamoto, H. Wada, and N. Satoh. 1993. Phylogenetic position of acoel turbellarians inferred from partial 18S rDNA sequences. *Zoological Science*, 13:529–536.

Katayama, T., H. Wada, H. Furuya, N. Satoh, and M. Yamamoto. 1995. Phylogenetic position of the dicyemid mesozoa inferred from 18S rDNA sequences. *Biological Bulletin* 189:81–90.

Katayama, T., M. Nishioka, and M. Yamamoto. 1996. Phylogenetic relationships among turbellarian orders inferred from 18S rDNA sequences. *Zoological Science*, 13:747–756.

Kaukas, A., E. D. Neto, A. J. G. Simpson, V. R. Southgate, and D. Rollinson. 1994. A phylogenetic analysis of *Schistosoma haematobium* group species based on randomly amplified polymorphic DNA. *International Journal for Parasitology*, 24:285–290.

Kawakatsu, M., J. Hauser, M. G. Friedrich, I. Oki, S. Tamura, and T. Yamayoshi. 1983. Morphological, karyological and taxonomic studies of freshwater planarians from south Brazil. 4. *Dugesia anderlani* sp. nov. (Turbellaria, Tricladida, Paludicola), a new species from São Leopoldo in Estado de Rio Grande do Sul. *Annotationes Zoologicae Japonenses*, 56:196–208.

Kawakatsu, M., O. A. Timoshkin, N. A. Porfireva, and M. Takai. 1994. Taxonomic notes on *Phagocata vivida* (Ijima et Kaburaki, 1916) from South Korea and Primorskiy, Russia (Turbellaria: Tricladida: Paludicola). *Bulletin of the Biogeographical Society of Japan*, 49:1–12.

Kean, B. H., K. E. Mott, and A. J. Russell. 1978. *Tropical medicine and parasitology. Classical investigations. Vols I and II*. Cornell University Press, Ithaca.

Kearn, G. C. 1965. The biology of *Leptocotyle minor*, a skin parasite of the dogfish, *Scyliorhinus canicula*. *Parasitology*, 56.

Kearn, G. C. 1978. Eyes with, and without, pigment shields in the oncomiracidium of the monogenean parasite *Diplozoon paradoxum*. *Zeitschrift für Parasitenkunde*, 157:35–47.

Kearn, G. C. 1994. Evolutionary expansion of the Monogenea. *International Journal for Parasitology*, 24:1227–1271.

Kearn, G. C. 1998. *Parasitism and the platyhelminths*. Chapman & Hall, London.

Kechemir, N. 1978. Démonstration expérimentale d'un cycle biologique à quatre hôtes obligatoires chez les Trématodes Hémiurides. *Annales de Parasitologie Humaine et Comparée*, 53:75–92.

Kedra, A. H., Z. Swiderski, V. V. Tkach, P. Dubinsky, Z. Pawlowski, J. Stefaniak, and J. Pawlowski. 1999. Genetic analysis of *Echinococcus granulosus* from humans and pigs in Poland, Slovakia and Ukraine. A multicenter study. *Acta Parasitologica*, 44:248–254.

Keenan, L., R. Coss, and H. Koopowitz. 1981. Cytoarchitecture of primitive brains: Golgi studies in flatworms. *Journal of Comparative Neurology*, 188:647–679.

Keenan, L., and H. Koopowitz. 1982. Physiology and *in situ* identification of putative aminergic neurotransmitters in the nervous system of *Gyrocotyle fimbriata*, a parasitic flatworm. *Journal of Neurobiology*, 13:9–21.

Keenan, L., H. Koopowitz, and K. Bernardo. 1979. Primitive nervous systems. Action of aminergic drugs and blocking agents on activity in the ventral nerve cord of the flatworm *Notoplana acticola*. *Journal of Neurobiology*, 10:397–408.

Kenk, R. 1930. Beiträge zum System der Probursalier (Tricladida Paludicola). *Zoologischer Anzeiger*, 89:145–162.

Kenk, R. 1975. Fresh-water triclads (Turbellaria) of North America. VII. The genus *Macrocotyla*. *Transactions of the American Microscopical Society*, 94:324–339.

Kenk, R. 1978. The planarians (Turbellaria: Tricladida: Paludicola) of Lake Ohrid in Macedonia. *Smithsonian Contributions to Zoology*, 280:1–56.

Khalil, L. F., A. Jones, and R. A. Bray. 1994. *Keys to the cestode parasites of vertebrates*. CAB International, Wallingford.

Kimura, M. 1980. A simple method for estimating evolutionary rate of base substitution through comparative studies of nucleotide sequences. *Journal of Molecular Evolution*, 16:111–120.

Kiseleva, I. V., and R. Y. Dyganova. 1988. [Morphological details of the ontogenetic development of the eye of *Baikalobia guttata*]. (In Russian), Conference: *Simple nervous systems and their meaning for theory and practice*, pp. 140–144, Kazan.

Kishida, Y. 1967a. Electron microscopic studies on the planarian eye I. Fine structures of the normal eye. *Science Reports of the Kanazawa University*, 12:75–110.

Kishida, Y. 1967b. Electron microscopic studies on the planarian eye II. Fine structures of the regenerating eye. *Science Reports of the Kanazawa University*, 12:111–142.

Klassen, G. J. 1992. Coevolution: A history of the macroevolutionary approach to studying host-parasite associations. *Journal of Parasitology*, 78:573–587.

Klauser, M. D., J. P. S. Smith, and S. Tyler. 1986. Ultrastructure of the frontal organ in *Convoluta* and *Macrostomum* spp.: significance for models of the turbellarian archetype. *Hydrobiologia*, 132:47–52.

Klima, J. 1967. Zur Feinstruktur der acoelen Süsswasser-Turbellars *Oligochoerus limnophilus* (Ax and Dörjes). *Bericht Naturwissenschaftlichen-Med. Vereins fur Innsbrück*, 55:107–124.

Kluge, A. 1989. A concern for evidence and a phylogenetic hypothesis of relationships among *Epicrates* (Boidae, Serpentes). *Systematic Zoology*, 38:7–25.

Kluge, A. G. 1998. Total evidence or taxonomic congruence: cladistics or consensus classification. *Cladistics*, 14:151–158.

Kluge, A., and A. J. Wolf. 1993. Cladistics: what's in a word? *Cladistics*, 9:183–199.

Knauss, E. B. 1979. Indication of an anal pore in Gnathostomulida. *Zoologica Scripta*, 8:181–186.

Kniskern, V. B. 1952. Studies on the trematode family Bucephalidae Poche, 1907, Part II. The life history of *Rhipidocotyle septpapillata* Krull, 1934. *Transactions of the American Microscopical Society*, 71:317–340.

Knoll, A. H., and S. B. Carroll. 1999. Early animal evolution: emerging views from comparative biology and geology. *Science*, 284:2129–2137.

Køie, M. 1985. On the morphology and life-history of *Lepidapedon elongatum* (Lebour, 1908) Nicoll, 1910 (Trematoda, Lepocreadiidae). *Ophelia*, 24:135–153.

Kokaze, A., H. Miyadera, K. Kita, R. Machinami, O. Noya, B. Alarcon De Noya, M. Okamoto, T. Horii, and S. Kojima. 1997. Phylogenetic identification of *Sparganum proliferum* as a pseudophyllidean cestode. *Parasitology International*, 46:271–279.

Koopowitz, H. 1989. Polyclad neurobiology and evolution of central nervous system. In P. A. V. Anderson (ed.), *Evolution of the first nervous systems*, Vol. 188, pp. 315–328. Plenum Press, New York.

Koopowitz, H., and P. Chien. 1974. Ultrastructure of nerve plexus in flatworms. I. Peripheral organizations. *Cell and Tissue Research*, 155:337–351.

Koopowitz, H., and P. Chien. 1975. Ultrastructure of nerve plexus in flatworms. II. Sites of synaptic interactions. *Cell and Tissue Research*, 157:207–216.

Koopowitz, H., M. Elvin, and T. Bae. 1995. Comparison of the nervous system of the rhabdocoel *Mesostoma ehrenbergii* with that of the polyclad *Notoplana acticola*. *Hydrobiologia*, 305:127–133.

Kornakova, E. E. 1988. On morphology and phylogeny of Udonellida. *Fortschritte der Zoologie*, 36:45–49.

Kornakova, E. E., and B. I. Joffe. 1996. Sensory receptors in the head of *Bothrioplana semperi* (Platyhelminthes, Seriata). *Journal of Submicroscopic Cytology and Pathology*, 28:313–317.

Kornakova, E. E., and B. I. Joffe. 1999. A new variant of the neodermatan-type spermiogenesis in a parasitic 'turbellarian', *Notentera ivanovi* (Platyhelminthes) and the origin of the Neodermata. *Acta Zoologica (Stockholm)*, 80:135–151.

Kornet, D. J., and H. Turner. 1999. Coding polymorphism for phylogeny reconstruction. *Systematic Biology*, 48:365–379.

Korneva, J. V., B. I. Kuperman, and V. G. Davydov. 1998. Ultrastructural investigation of the secondary excretory system in different stages of the procercoid of *Triaenophorus nodulosus* (Cestoda, Pseudophyllidea, Triaenophoridae). *Parasitology*, 116:373–381.

Kotikova, E. A. 1986. Comparative characterization of the nervous system of the Turbellaria. *Hydrobiologia*, 132:89–92.

Kotikova, E. A. 1991. [The orthogon of the Plathelminthes [Platyhelminthes] and main trends of its evolution]. (In Russian). In Y. V. Mamkaev (ed.), *Morphological principles of platyhelminth phylogenetics*, Vol. 241, pp. 88–111. Trudy Zoologicheskogo Instituta, Leningrad.

Kotikova, E. A. 1994. Glycoxylic-acid-induced-fluorescence in the nervous system of *Gyratrix hermaphroditus* (Kalyptorhynchia, Polycystidae). *Hydrobiologia*, 305:135–139.

Kotikova, E. A., and B. I. Joffe. 1988. On the nervous system of dalyellioid turbellarians. *Fortschritte der Zoologie*, 36:191–194.

Kotikova, E. A., and B. I. Kuperman. 1977. [The development of the nervous apparatus of *Triaenophorus nodulosus* (Cestoidea, Pseudophyllidea) during ontogenesis]. (In Russian). *Parasitologiya*, 11:252–259.

Kozloff, E. N. 1990. *Invertebrates*. Saunders College Publ., Philadelphia.

Král'ová, I. 1996. A total DNA characterization in *Proteocephalus exiguus* and *P. percae* (Cestoda: Proteocephalidae): random amplified polymorphic DNA and hybridization techniques. *Parasitology Research*, 82:668–671.

Král'ová, I., and M. Spakulová. 1996. Intraspecific variability of *Proteocephalus exiguus* La Rue, 1991 (Cestoda: Proteocephalidae) as studied by the random amplified polymorphic DNA method. *Parasitology Research*, 82:542–545.

Král'ová, I., Y. van de Peer, M. Jirku, M. van Ranst, T. Scholz, and J. Lukes. 1997. Phylogenetic analysis of a fish tapeworm, *Proteocephalus exiguus*, based on the small subunit rRNA gene. *Molecular and Biochemical Parasitology*, 84:263–266.

Krasnoshchekov, G. P., and N. S. Tomilovskaya. 1978. [Morphology and development of the cycticercoid of *Paricterotaenia porosa* (Rud., 1810) (Cestoda: Dilepididae)]. (In Russian). *Parazitologiya*, 12:108–115.

Kritsky, D. C., and W. A. Boeger. 1989. The phylogenetic status of the Ancyrocephalidae Bychowsky, 1937 (Monogenea: Dactylogyroidea). *Journal of Parasitology*, 75:207–211.

Kritsky, D. C., and S. L. H. Lim. 1995. Phylogenetic position of Sundanonchidae (Platyhelminthes: Monogenoidea: Dactylogyridea), with report of two species of *Sundanonchus* from toman, *Channa micropeltes* (Channiformes: Channidae), in Malaysia. *Invertebrate Biology*, 114:285–294.

Kritsky, D. C., E. P. Hoberg, and K. B. Aubry. 1993. *Lagarocotyle salamandrae* n. gen., n. sp. (Monogenoidea, Polyonchoinea, Lagarocotylidea n. ord.) from the cloaca of *Rhyacotriton cascadae* Good and Wake (Caudata, Rhyacotritonidae) in Washington State. *Journal of Parasitology*, 79:322–330.

Krupa, P. L., G. H. Cousineau, and A. K. Bal. 1969. Electron microscopy of the excretory vesicle of a trematode cercaria. *Journal of Parasitology*, 55:985–992.

Kuchiiwa, T., S. Kuchiiwa, and W. Teschirogi. 1991. Comparative morphological studies on the visual systems in a binocular and a multi-ocular species of freshwater planarian. *Hydrobiologia*, 227:241–249.

Kunert, T. 1988. On the protonephridia of *Macrostomum spirale* Ax, 1956 (Macrostomida, Plathelminthes). *Fortschritte der Zoologie*, 36:423–428.

Kuperman, B. I. 1973. *[Tapeworms of the genus Triaenophorus, parasites of fishes]*. (In Russian). Izdatel'stvo Nauka, Leningrad.

Kuperman, B. I., and V. G. Davydov. 1982. The fine structure of the glands in the oncospheres, procercoids and plerocercoids of Pseudophyllidea (Cestoidea). *International Journal for Parasitology*, 12:285–293.

Kuznedelov, K. D., and O. A. Timoshkin. 1993. Phylogenetic relationships of Baikalian species of Prorhynchidae turbellarian worms as inferred by partial 18S rRNA gene sequence comparisons (preliminary report). *Molecular Marine Biology and Biotechnology*, 2:300–307.

Kuznedelov, K. D., and O. A. Timoshkin. 1995. Molecular phylogenetics of Turbellaria deduced from the 18S rRNA sequencing data. *Molecular Biology*, 29:318–325.

L'Hardy, J.-P. 1988. Sperm morphology in Kalyptorhynchia (Platyhelminthes, Rhabdocoela). *Fortschritte der Zoologie*, 36:303–307.

La Rue, G. R. 1914. A revision of the cestode family Proteocephalidae. *Illinois Biological Monographs*, 1:1–350.

La Rue, G. R. 1951. Host-parasite relations among the digenetic trematodes. *Journal of Parasitology*, 37:333–342.

La Rue, G. R. 1957. The classification of digenetic Trematoda: a review and a new system. *Experimental Parasitology*, 6:306–349.

Lacalli, T. C. 1987. The suboral complex in the Müller's larva of *Pseudoceros canadensis* (Platyhelminthes, Polycladida). *Canadian Journal of Zoology*, 66:1893–1895.

Ladurner, P., G. R. Mair, D. Reiter, W. Salvenmoser, and R. M. Rieger. 1997. Serotonergic nervous system of two macrostomid species: recent or ancient divergence? *Invertebrate Biology*, 116:178–191.

Lake, J. A. 1991. Reconstructing evolutionary trees from DNA and protein sequences: paralinear distances. *Proceedings of the National Academy of Sciences USA*, 91:1455–1459.

Lambert, A. 1980. Oncomiracidiums et phylogenèse de Monogenea (Platyhelminthes). 1re Partie: Développement post-larvaire. *Annales de Parasitologie Humaine et Comparée*, 55:165–198.

Lanar, D. E., E. J. Pearce, S. L. James, and A. Sher. 1986. Identification of paramyosin as schistosome antigen recognised by intradermally vaccinated mice. *Science*, 234:593–596.

Lanfranchi, A. 1998. Ultrastructural observations on the development of male gametes in *Allostoma* sp. (Plathelminthes, Prolecithophora). *Hydrobiologia*, 383:227–233.

Lanfranchi, A., and A. Falleni. 1998. Ultrastructural observations on the male gametes in *Austrognathia* sp. (Gnathostomulida, Bursovaginoidea). *Journal of Morphology*, 237:165–176.

Larsen, N., G. J. Olsen, B. L. Maidak, M. J. McCaughey, R. Overbeek, T. J. Mackie, T. L. Marsh, and C. R. Woese. 1993. The ribosomal database project. *Nucleic Acids Research*, 21:3021–3023.

Larson, A. 1994. The comparison of morphological and molecular data in phylogenetic systematics. In B. Schierwater, B. Streit and R. DeSalle (eds.), *Molecular ecology and evolution: approaches and applications*, pp. 371–390. Birkhaüser Verlag, Basel.

Laws, G. F. 1968. The hatching of taeniid eggs. *Experimental parasitology*, 23:1–10.

Le Paslier, M. C., R. J. Pierce, F. Merlin, H. Hirai, W. Wu, D. L. Williams, D. A. Johnston, P. T. LoVerde, and D. Le Paslier. 2000. Construction and characterisation of a *Schistosoma mansoni* bacterial artificial chromosome library. *Genomics*, 65:87–94.

Le Riche, P. D., and M. M. H. Sewell. 1978. Differentiation of taeniid cestodes by enzyme electrophoresis. *International Journal for Parasitology*, 8:479–483.

Le, T. H., D. Blair, T. Agatsuma, P.-F. Humair, N. J. H. Campbell, M. Iwagami, D. T. J. Littlewood, B. Peacock, D. A. Johnston, J. Bartley, D. Rollinson, E. A. Herniou, D. S. Zarlenga, and D. P. McManus. 2000. Phylogenies inferred from mitochondrial gene orders – a cautionary tale from the parasitic flatworms. *Molecular Biology and Evolution*, 17:1123–1125.

Leal-Zanchet, A. M., and J. Hauser. 1999. Penis glands of the dugesiid planarian *Girardia schubarti* (Platyhelminthes, Tricladida, Paludicola). *Invertebrate Biology*, 118:35–41.

Lebedev, B. I. 1987. Is the trematode acetabulum a cercomer? (about one platyhelminth's classification). *Journal of Parasitology*, 73:1250–1251.

Lecointre, G., H. Philippe, H. L. V. Lê, and H. Le Guyader. 1993. Species sampling has a major impact on phylogenetic inference. *Molecular Phylogenetics and Evolution*, 2:205–224.

Lee, M. S. Y. 1998. Phylogenetic uncertainty, molecular sequences, and the definition of taxon names. *Systematic Biology*, 47:719–726.

Lee, S.-U., S. Huh, and C. K. Phares. 1997. Genetic comparison between *Spirometra erinacei* and *S. mansonoides* using PCR-RFLP analysis. *Korean Journal of Parasitology*, 35:277–282.

Lentz, T. L. 1967. Fine structure of nerve cells in a planarian. *Journal of Morphology*, 12:323–338.

LeRoux, P. L. 1958. The validity of *Schistosoma capense* (Harley, 1864) amended as a species. *Transactions of the Royal Society of Tropical Medicine and Hygiene*, 52:12–14.

Leuckart, R. 1881. Zur Entwicklungsgeschichte des Leberegels. *Zoologischer Anzeiger*, 4:641–646 (English translation in Kean, Mott and Russell, 1978, Volume 2, 568–570).

Leuckart, R. 1882. Zur Entwicklungsgeschichte des Leberegels. Zweite Mittheilung. *Zoologischer Anzeiger*, 5:524–528.

Li, M.-M., N. A. Watson, and K. Rohde. 1992. Ultrastructure of spermatogenesis of *Syndisyrinx punicea* (Hickman, 1956) (Platyhelminthes, Rhabdocoela, Umagillidea). *Australian Journal of Zoology*, 40:153–161.

Li, M.-M., Z.-L. Wang, H.-H. Liu, and X.-Y. Wang. 1998. [Ultrastructure of spermatogenesis in *Polystoma* sp. (Monogenea: Polyopisthocotylea: Polystomatidae)]. (In Chinese). *Acta Zoologica Sinica*, 44:126–130.

Lindholm, A. M., M. Reuter, and M. K. S. Gustafsson. 1998. The NADPH-diaphorase staining reaction in relation to the aminergic and peptidergic nervous system and the musculature of adult *Diphyllobothrium dendriticum*. *Parasitology*, 117:283–292.

Lindroos, P. 1983. The excretory ducts of *Diphyllobothrium dendriticum* (Nitzsch 1824) plerocercoids: ultrastructure and marker distribution. *Zeitschrift für Parasitenkunde*, 69:229–237.

Lindroos, P., and T. Gardberg. 1982. The excretory system of *Diphyllobothrium dendriticum* (Nitzsch 1824) plerocercoids as revealed by an injection technique. *Zeitschrift für Parasitenkunde*, 67:289–297.

Linnaeus, C. 1758. *Systema naturae per regna tria naturae, secundrum classes, ordines, genera, species, cum characteribus, differentifs, synonymis, locis*. Editio decima, reformata, Holmiae.

Lipscomb, D., J. S. Farris, M. Källersjö, and A. Tehler. 1998. Support, ribosomal sequences and the phylogeny of the eukaryotes. *Cladistics*, 14:303–338.

Littlewood, D. T. J., and D. A. Johnston. 1995. Molecular phylogenetics of the 4 *Schistosoma* species groups determined with partial 28S ribosomal RNA gene sequences. *Parasitology*, 111:167–175.

Littlewood, D. T. J., and P. D. Olson. 2001. Small subunit rDNA and the Platyhelminthes: signal, noise, conflict and compromise. In D. T. J. Littlewood and R. A. Bray (eds.), *Interrelationships of the Platyhelminthes*, pp. 262–278. Taylor & Francis, London.

Littlewood, D. T. J., and A. B. Smith. 1995. A combined morphological and molecular phylogeny for sea urchins

(Echinoidea, Echinodermata). *Philosophical Transactions of the Royal Society of London, Series B, Biological Sciences*, 347:213–234.

Littlewood, D. T. J., A. B. Smith, K. A. Clough, and R. H. Emson. 1997. The interrelationships of the echinoderm classes: morphological and molecular evidence. *Biological Journal of the Linnean Society*, 61:409–438.

Littlewood, D. T. J., K. Rohde, and K. A. Clough. 1998a. The phylogenetic position of *Udonella* (Platyhelminthes). *International Journal for Parasitology*, 28:1241–1250.

Littlewood, D. T. J., M. J. Telford, K. A. Clough, and K. Rohde. 1998b. Gnathostomulida – an enigmatic metazoan phylum from both morphological and molecular perspectives. *Molecular Phylogenetics and Evolution*, 9:72–79.

Littlewood, D. T. J., K. Rohde, and K. A. Clough. 1999a. The interrelationships of all major groups of Platyhelminthes: phylogenetic evidence from morphology and molecules. *Biological Journal of the Linnean Society*, 66:75–114.

Littlewood, D. T. J., K. Rohde, R. A. Bray, and E. A. Herniou. 1999b. Phylogeny of the Platyhelminthes and the evolution of parasitism. *Biological Journal of the Linnean Society*, 68:257–287.

Littlewood, D. T. J., M. Curini–Galletti, and E. A. Herniou. 2000. The interrelationships of Proseriata (Platyhelminthes: Seriata) tested with molecules and morphology. *Molecular Phylogenetics and Evolution*, 16:449–466.

Litvaitis, M. K., and K. Rohde. 1999. A molecular test of platyhelminth phylogeny: inferences from partial 28S rDNA squences. *Invertebrate Biology*, 118:42–56.

Litvaitis, M. K., M. C. Curini-Galletti, P. M. Martens, and T. D. Kocher. 1996. A reappraisal of the systematics of the Monocelididae (Platyhelminthes, Proseriata) – inferences from rDNA sequences. *Molecular Phylogenetics and Evolution*, 6:150–156.

Liu, D., H. Kato, and K. Sugane. 1997. The nucleotide sequence and predicted secondary structure of small subunit (18S) ribosomal RNA from *Spirometra erinaceieuropaei*. *Gene*, 184:221–227.

Llewellyn, J. 1963. Larvae and larval development of monogeneans. *Advances in Parasitology*, 1:287–326.

Llewellyn, J. 1965. The evolution of parasitic Platyhelminthes. In A. E. R. Taylor (ed.), *Evolution of Parasites (3rd Symposium of the British Society for Parasitology)*, pp. 47–78. Blackwell Publishing, Oxford.

Lockhart, P. J., M. A. Steel, M. D. Hendy, and D. Penny. 1994. Recovering evolutionary trees under a more realistic model of sequence evolution. *Molecular Biology and Evolution*, 11:605–612.

Loennberg, E. 1897. Beiträge zur Phylogenie der Platyhelminthen. *Centralblat Bakteriologie Parasitenkunde und Infektionskrankheiten*, 21:674–684.

Looss, A. 1899. Weitere Beiträge zur Kenntnis der Trematoden–fauna Aegyptens, zugleich Versuch einer natürlichen Gliederung des Genus *Distomum* Retzius. *Zoologischer Jahrbücher*, 12:521–784.

Lotz, J. M., and K. C. Corkum. 1984. Notes on the life cycle of *Dictyangium chelydrae* (Digenea: Microscaphidiidae). *Proceedings of the Helminthological Society of Washington*, 51:353–355.

Lowe, C. Y. 1966. Comparative studies of the lymphatic system of four species of amphistomes. *Zeitschrift für Parasitenkunde*, 27:169–204.

Lumb, S. M., R. A. Bray, and D. Rollinson. 1993. Partial small subunit (18S) rRNA gene sequences from fish parasites of the families Lepocreadiidae and Fellodistomidae (Digenea) and their use in phylogenetic analyses. *Systematic Parasitology*, 26:141–149.

Lumbsch, M., U. Ehlers, and B. Sopott-Ehlers. 1995. Proximal regions of the protonephridial system in *Pseudograffilla arenicola* (Plathelminthes, Rhabdocoela): ultrastructural observations. *Microfauna Marina*, 10:67–78.

Lumsden, R. D., and M. B. Hildreth. 1983. The fine structure of adult tapeworms. In C. Arme and P. W. Pappas (eds.), *Biology of the Eucestoda*, pp. 177–233. Academic Press, London.

Lumsden, R. D., and R. Specian. 1980. The morphology, histology, and fine structure of the adult stage of the cyclophyllidean tapeworm. In H. P. Arai (ed.), *Biology of the tapeworm* Hymenolepis diminuta, pp. 157–280. Academic Press, New York.

Lumsden, R. D., J. A. Oaks, and J. F. Mueller. 1974. Brush border development in the tegument of the tapeworm *Spirometra mansonoides*. *Journal of Parasitology*, 60:209–226.

Lundin, K. 1997. Comparative ultrastructure of the ciliary rootlets and associated structures in species of the Nemertodermatida and Acoela (Plathelminthes). *Zoomorphology*, 117:81–92.

Lundin, K. 1998. The epidermal ciliary rootlets of *Xenoturbella bocki* (Xenoturbellida) revisited: new support for a possible kinship with the Acoelomorpha (Plathelminthes). *Zoologica Scripta*, 27:263–270.

Lundin, K. 2000. Phylogeny of the Nemertodermatida (Acoelomorpha, Plathelminthes). A cladistic analysis. *Zoologica Scripta*, 29:65–74.

Lundin, K., and J. Hendelberg. 1996. Degenerating epidermal bodies ('pulsatile bodies') in *Meara stichopi* (Plathelminthes, Nemertodermatida). *Zoomorphology*, 116:1–5.

Lundin, K., and J. Hendelberg. 1998. Is the sperm type of the Nemertodermatida close to that of the ancestral Platyhelminthes? *Hydrobiologia*, 383:197–205.

Luther, A. 1905. Zur Kenntnis der Gattung *Macrostoma*. *Festschrift für Palmén, Helsingfors*, 5:1–61.

Luther, A. 1947. Untersuchungen an Rhabdocoelen Turbellarien. 6. Macrostomiden aus Finnland. *Acta Zoologica Fennica*, 49:1–40.

Luther, A. 1955. Die Dalyelliiden (Turbellaria: Neorhabdocoela). *Acta Zoologica Fennica*, 87:1–337.

Luther, A. 1960. Die Turbellarien Ostfennoskandiens I. Acoela, Catenulida, Macrostomida, Lecithoepitheliata, Prolecithophora und Proseriata. *Fauna Fennica*, 7:1–155.

Lynch, J. E. 1945. Redescription of the species of *Gyrocotyle* from the ratfish, *Hydrolagus colliei* (Lay and Bennet), with notes on the morphology and taxonomy of the genus. *Journal of Parasitology*, 31:418–446.

Lyons, K. M. 1969a. The fine structure of the body wall of *Gyrocotyle urna*. *Zeitschrifte für Parasitenkunde*, 33:95–109.

Lyons, K. M. 1969b. Sense organs of monogenean skin parasites ending in a typical cilium. *Parasitology*, 59:611–623.

Lyons, K. M. 1969c. Compound sensilla in monogenean skin parasites. *Parasitology*, 59:625–636.

Lyons, K. M. 1972. Sense organs in monogeneans. In E. U. Canning and C. A. Wright (eds.), *Behavioural aspects of parasite transmission*, pp. 181–199. Academic Press, London.

Lyons, K. M. 1973a. The epidermis and sense organs of the Monogenea and some related groups. *Advances in Parasitology*, 11:193–232.

Lyons, K. M. 1973b. Epidermal fine structure and development in the oncomiracidia of *Entobdella soleae* (Monogenea). *Parasitology*, 66:321–333.

Lyons, K. M. 1977. Epidermal adaptations of parasitic platyhelminths. *Symposia of the Zoological Society, London*, 39:97–144.

Mackey, L. Y., B. Winnepenninckx, R. de Wachter, T. Backeljau, P. Emschermann, and J. R. Garey. 1996. 18S rRNA suggests that Entoprocta are protostomes, unrelated to Ectoprocta. *Journal of Molecular Evolution*, 42:553–559.

Mackiewicz, J. S. 1972. Caryophyllidea (Cestoidea): a review. *Experimental Parasitology*, 31:417–512.

MacKinnon, B. M., and M. D. B. Burt. 1984. The comparative ultrastructure of spermatozoa from *Bothrimonus sturionis* Duv. 1842 (Pseudophyllidea), *Pseudanthobothrium hanseni* Baer, 1956 (Tetraphyllidea), and *Monoecocestus americanus* Stiles, 1895 (Cyclophyllidea). *Canadian Journal of Zoology*, 62:1059–1066.

MacKinnon, B. M., and M. D. B. Burt. 1985. Ultrastructure of spermatogenesis and the mature spermatozoon of *Haplobothrium globuliforme* Cooper, 1914 (Cestoda: Haplobothrioidea). *Canadian Journal of Zoology*, 63:1478–1487.

Maddison, D. R., D. L. Swofford, and W. P. Maddison. 1997. Nexus: An extensible file format for systematic information. *Systematic Biology*, 46:590–621.

Maddison, W. P. 1993. Missing data versus missing characters in phylogenetic analysis. *Systematic Biology*, 42:576–581.

Maddison, W. P., and D. R. Maddison. 1992. *MacClade. Version 3.0.1 [Computer software and manual]*. Sinauer Associates, Sunderland, Massachusetts.

Maddison, W. P., M. J. Donoghue, and D. R. Maddison. 1984. Outgroup analysis and parsimony. *Systematic Zoology*, 33:83–103.

Maidak, B. L., J. R. Cole, C. T. Parker, G. M. Garrity, N. Larsen, B. Li, T. G. Lilburn, M. J. McCaughey, G. J. Olsen, R. Overbeek, S. Pramanik, T. M. Schmidt, J. M. Tiedje, and C. R. Woese. 1999. A new version of the RDP (Ribosomal Database Project). *Nucleic Acids Research*, 27:171–173.

Malmberg, G. 1971. On the procercoid protonephridial systems of three *Diphyllobothrium* species (Cestoda, Pseudophyllidea) and Janicki's cercomer theory. *Zoologica Scripta*, 1:43–56.

Malmberg, G. 1974. On the larval protonephridial system of *Gyrocotyle* and the evolution of Cercomeromorphae (Platyhelminthes). *Zoologica Scripta*, 3:65–81.

Malmberg, G. 1979. Ontogeny and fine structure as a basis for a discussion on the relationship Cestoda-Monogenea (Plathelminthes). *Zoologica Scripta*, 8:315.

Malmberg, G. 1986. The major parasitic platyhelminth classes – progressive or regressive evolution? *Hydrobiologia*, 132:23–29.

Malmberg, G., and Y. Lilliemark. 1993. Sperm ultrastructure of *Isancistrum subulatae* (Platyhelminthes: Monogenea: Gyrodactylidae). *Folia Parasitologica*, 40:97–98.

Mamaev, Y. L. 1976. [The system and phylogeny of monogeneans of the family Diclidophoridae]. (In Russian). *Trudy Biologo-Pochvennogo Instituta, Novaya Seriya*, 35:57–80.

Mamkaev, Y. V. 1986. Initial morphological diversity as a criterion in deciphering turbellarian phylogeny. *Hydrobiologia*, 132:31–33.

Mamkaev, Y. V., and V. P. Ivanov. 1970. [Electron microscopy investigation of spermatozoa of *Convoluta convoluta* (Turbellaria, Acoela]. (In Russian). *Proceedings of the Zoological Institute, Leningrad*, 1970:12–13.

Mamkaev, Y. V., and E. A. Kotikova. 1972. [On the morphological characters of nervous system in Acoela]. (In Russian). *Zoologicheskii Zhurnal*, 51:477–489.

Mansour, T. A. G. 1984. Serotonin receptors in parasitic worms. *Advances in Parasitology*, 23:1–36.

Mäntylä, K., D. W. Halton, M. Reuter, A. G. Maule, P. Lindroos, C. Shaw, and M. K. S. Gustafsson. 1998. The nervous system of Tricladida. IV. Neuroanatomy of

Planaria torva (Paludicola, Planaridae): an immunocytochemical study. *Hydrobiologia*, 383:167–173.

Marcus, E. 1946. Sôbre Turbellaria Brasileiros. *Boletin de Faculdade de Filosofia Ciências e Letras, Universidade de São Paulo, seccao Zoologia*, 11:5–254.

Marcus, E. 1949. Turbellaria Brasileiros (7). *Boletins da Faculdade de Filosofia, Ciências e Letras, Zoologia, São Paolo*, 14:7–155.

Marcus, E. 1949. Turbellaria do Brasil. *Boletim da Faculdade de Filosofia, Ciências e Letras. Universidade de São Paulo Tierwelt der Nord- und Ostsee*, 4b:1–146.

Marcus, E. 1950. Turbellaria Brasilieros (8). *Boletins da Faculdade de Filosofia, Ciências e Letras, Zoologia, São Paolo*, 15:5–192.

Marcus, E. 1951. Turbellaria Brasilieros (9). *Boletins da Faculdade de Filosofia, Ciências e Letras, Zoologia, São Paolo*, 16:5–214.

Marcus, E. 1953. Turbellaria Tricladida, *Exploration du Parc National de l'Upemba*, Vol. Fasc. 21, pp. 62, Bruxelles.

Marcus, E. 1955. Turbellaria. In B. Hanstrom, P. Brinck and G. Rudebeck (eds.), *South African animal life, results of the Lund University expedition in 1950–1951*, Vol. I, pp. 101–151. Almqvist & Wicksell, Stockholm.

Margush, T., and F. R. McMorris. 1981. Consensus n-trees. *Bulletin of Mathematical Biology*, 43:239–244.

Mariaux, J. 1996. Cestode systematics: any progress? *International Journal for Parasitology*, 26:231–243.

Mariaux, J. 1998. A molecular phylogeny of the Eucestoda. *Journal of Parasitology*, 84:114–124.

Mariaux, J., and P. D. Olson. 2001. Cestode systematics in the molecular era. In D. T. J. Littlewood and R. A. Bray (eds.), *Interrelationships of the Platyhelminthes*, pp. 127–134. Taylor & Francis, London.

Marks, N. J., S. Johnston, A. G. Maule, D. W. Halton, C. Shaw, T. G. Geary, S. Moore, and D. P. Thompson. 1996. Physiological effects of platyhelminth RFamide peptides on muscle-strip preparations of *Fasciola hepatica* (Trematoda: Digenea). *Parasitology*, 113:393–401.

Marshall, A. G. 1967. The cat flea, *Ctenocephalides felis felis* (Bouché, 1835), as an intermediate host for cestodes. *Parasitology*, 57:419–430.

Martens, E. E. 1984. Ultrastructure of the spines in the copulatory organ of some Monocelididae (Turbellaria, Proseriata). *Zoomorphology*, 104:261–265.

Martens, P. M., and M. C. Curini-Galletti. 1993. Taxonomy and phylogeny of the Archimonocelididae Meixner, 1938 (Platyhelminthes, Proseriata). *Bijdragen tot de Dierkunde*, 63:65–102.

Martens, P. M., and E. R. Schockaert. 1986. The importance of turbellarians in the marine meiobenthos: a review. *Hydrobiologia*, 132:295–303.

Martens, P. M., and E. R. Schockaert. 1988. Phylogeny of the digonoporid Proseriata. *Fortschritte der Zoologie*, 36:399–403.

Martens, P. M., M. Curini-Galletti, and P. van Oostveldt. 1989. Polyploidy in Proseriata (Platyhelminthes) and its phylogenetical implications. *Evolution*, 42:900–907.

Martin, W. E. 1973. Life history of *Saccocoelioides pearsoni* n. sp. and the description of *Lecithobotrys sprenti* n. sp. (Trematoda: Haploporidae). *Transactions of the American Microscopical Society*, 92:80–95.

Matjasic, J. 1990. Monography of the family Scutariellidae (Turbellaria, Temnocephalidea). *Academia Scientiarum et Artium Slovenica Classis IV: Historia Naturalis*, 28:1–167.

Matsuura, T., G. Bylund, and K. Sugane. 1992. Comparison of restriction fragment length polymorphisms of ribosomal

DNA between *Diphyllobothrium nihonkaiense* and *D. latum*. *Journal of Helminthology*, 66:261–266.

Mattis, T. E. 1986. Development of two tetrarhynchidean cestodes from the northern Gulf of Mexico. Ph.D. thesis, University of Southern Mississippi.

Mattison, R. G., R. E. B. Hanna, and W. A. Nizami. 1992. Ultrastructure and histochemistry of the protonephridial system of juvenile *Paramphistomum epiclitum* and *Fischoederius elongatus* (Paramphistomidae: Digenea) during migration in Indian ruminants. *International Journal for Parasitology*, 22:1103–1115.

Maule, A. G., D. W. Halton, J. M. Allen, and I. Fairweather. 1989. Studies on motility *in vitro* of an ectoparasitic monogenean, *Diclidophora merlangi*. *Parasitology*, 98:85–93.

Maule, A. G., D. W. Halton, C. F. Johnston, C. Shaw, and I. Fairweather. 1990. The serotoninergic, cholinergic and peptidergic components of the nervous system of the monogenean parasite, *Diclidophora merlangi*: a cytochemical study. *Parasitology*, 100:255–273.

Maule, A. G., D. W. Halton, C. Shaw, L. Thim, C. F. Johnston, I. Fairweather, and K. D. Buchanan. 1991. Neuropeptide F: a novel parasitic flatworm regulatory peptide from *Moniezia expansa* (Cestoda: Cyclophyllidea). *Parasitology*, 102:309–316.

Maule, A. G., D. W. Halton, C. Shaw, and L. Thim. 1993. GNFFRFamide: a novel FMRFamide–immunoreactive peptide isolated from the sheep tapeworm, *Moniezia expansa*. *Biochemical and Biophysical Research Communications*, 193:1054–1060.

Maule, A. G., D. W. Halton, C. Shaw, W. J. Curry, and L. Thim. 1994. RYIRFamide: a turbellarian FMRFamide-related peptide (FaRP). *Regulatory Peptides*, 50:37–43.

Mayr, E. 1969. *Principles of systematic zoology*. McGraw Hill, New York.

McCullough, J. S., and I. Fairweather. 1987. The structure, composition, formation and possible functions of calcareous corpuscles in *Trilocularia acanthiaevulgaris* (Olsson, 1867) (Cestoda, Tetraphyllidea). *Parasitology Research*, 74:175–182.

McCullough, J. S., and I. Fairweather. 1991. Ultrastructure of the excretory system of *Triocularia acanthiaevulgaris* (Cestoda, Tetraphyllidea). *Parasitology Research*, 77:157–160.

McCully, R. M., J. W. van Niekerk, and S. P. Kruger. 1965. The pathology of bilharziasis in *Hippopotamus amphibius*. *South African Medical Journal*, 39:1026.

McKanna, J. A. 1968. Fine structure of the protonephridial system in planaria. I. Flame cells. *Zeitschrift für Zellforschung*, 92:509–523.

McManus, D. P. 1997. Molecular genetic variation in *Echinococcus* and *Taenia*: an update. *Southeast Asian Journal of Tropical Medicine Public Health*, 28 (Suppl. 1):110–116.

McPhail, M. K. 1997. Late Neogene climates in Australia: fossil pollen– and spore–based estimates in retrospect and prospect. *Australian Journal of Botany*, 45:425–464.

McVicar, A. H. 1972. The ultrastructure of the parasite–host interface of three tetraphyllidean tapeworms of the elasmobranch *Raja naevus*. *Parasitology*, 65:77–88.

Mehlhorn, H. 1988. *Parasitology in focus. Facts and trends*, p. 924. Springer-Verlag, Berlin.

Meixner, J. 1928. Der Genitalapparat der Tricladen und seine Beziehungen zu ihrer allgemeinen Morphologie, Phylogenie, Ökologie un Verbreitung. *Zeitschrift für Morphologie und Ökologie der Tiere*, 11:570–612.

Meixner, J. 1938. Turbellaria (Strudelwürmer) I. (Allgemeiner Teil). *Tierwelt Nord- und Ostsee*, 33, Teil IVb:1–146.

Menizki, Y. L. 1963. [Structure and systematic position of the turbellarian parasitising fish, *Ichthyophaga subcutanea* Syromjatnikova 1949]. (In Russian). *Parasitologicheski Sbornik*, 21:245–258.

Merton, H. 1913. Beiträge zur Anatomie und Histologie von *Temnocephala*. *Abhandlungen der Senkenbergischen Naturforschenden Gesellschaft*, 35:1–58.

Mettrick, D. F., and J. M. Telford. 1963. Histamine in the phylum Platyhelminthes. *Journal of Parasitology*, 49:653–656.

Meuleman, E. A., D. M. Lyaruu, M. A. Khan, P. J. Holzmann, and T. Sminia. 1978. Ultrastructural changes in the body wall of *Schistosoma mansoni* during transformation of the miracidium into the mother sporocyst in the snail host *Biomphalaria pfeifferi*. *Zeitschrift für Parasitenkunde*, 56:227–242.

Michiels, N. K., and L. J. Newman. 1998. Sex and violence in hermaphrodites. *Nature*, 391:647.

Millemann, R. E. 1955. Studies on the life history and biology of *Oochoristica deserti* n. sp. (Cestoda: Linstowidae) from desert rodents. *Journal of Parasitology*, 41.

Miquel, J., and B. Marchard, 1998a. Ultrastructure of spermiogenesis and the spermatozoon of *Anoplocephaloides dentata* (Cestoda, Cyclophyllidea, Anoplocephalidae), intestinal parasite of Arvicolidea rodents. *Journal of Parasitology*, 84:1128–1136

Miquel, J., and B. Marchand. 1998b. Ultrastructure of the spermatozoon of the bank vole tapeworm, *Paranoplocephala omphaldes* (Cestoda, Cyclophyllidea, Anoplocephalidae). *Parasitology Research*, 84:239–245.

Miquel, J., C. T. Bâ, and B. Marchand. 1998a. Striated rootlets in spermatids of *Anoplocephaloides dentata* (Anoplocephalidae) and *Dipylidium caninum* (Dipylidiidae): a new finding in the Cyclophyllidea. *Wiadomosci Parazytologiczne*, 44:597.

Miquel, J., C. T. Bâ and B. Marchand. 1998b. Ultrastructure of spermiogenesis of *Dipylidium caninum* (Cestoda, Cyclophyllidea, Dipylidiidae), an intestinal parasite of *Canis familiaris*. *International Journal for Parasitology*, 28:1453–1458.

Miquel, J., C. Feliu, and B. Marchand. 1999. Ultrastructure of spermiogenesis and the spermatozoon of *Mesocestoides litteratus* (Cestoda, Mescocestoididae). *International Journal for Parasitology*, 29:499–510.

Miyamoto, M. M., and W. M. Fitch. 1995. Testing species phylogenies and phylogenetic methods with congruence. *Systematic Biology*, 44:64–76.

Mokhtar-Maamouri, F. 1982. Étude ultrastructurale de la spermiogenèse de *Acanthobothrium filicolle* var. *filicolle* Zschokke, 1888 (Cestoda, Tetraphyllidea, Onchobothriidae). *Annales de Parasitologie Humaine et Comparée*, 57:429–442.

Mokthar-Maamouri, F., and Z. Swiderski. 1976. Ultrastructure du spermatozoide d'un cestode Tetraphyllidea Phyllobothriidae: *Echeneibothrium beauchampi* Euzet, 1959. *Annales de Parasitologie Humain et Comparée*, 51:673–674.

Mollaret, I., and J.-L. Justine. 1997. Immunocytochemical study of tubulin in the 9+'1' sperm axoneme of a monogenean (Platyhelminthes), *Pseudodactylogyrus* sp. *Tissue & Cell*, 29:699–706.

Mollaret, I., B. G. M. Jamieson, R. D. Adlard, A. Hugall, G. Lecointre, C. Chombard, and J.-L. Justine. 1997. Phylogenetic analysis of the Monogenea and their relationships with Digenea and Eucestoda inferred from 28S rDNA sequences. *Molecular and Biochemical Parasitology*, 90:433–438.

Mollaret, I., L. H. S. Lim, G. Malmberg, B. Afzelius, and J.-L. Justine. 1998. Spermatozoon ultrastructure in two monopisthocotylean monogeneans from Malaysia: *Pseudodactylogyroides marmoratae* and *Sundanonchus micropeltis*. *Folia Parasitologica*, 45:75–76.

Mollaret, I., B. G. M. Jamieson, and J.-L. Justine. 2000. Phylogeny of the Monopisthocotylea and Polyopisthocotylea (Platyhelminthes) inferred from 28S rDNA sequences. *International Journal for Parasitology*, 30:171–185.

Moneypenny, C. 1999. The neuromuscular systems of the monogenean *Diclidophora merlangi* and turbellarian, *Procerodes littoralis*: a comparative analysis. PhD thesis, The Queen's University of Belfast.

Moneypenny, C. G., A. G. Maule, C. Shaw, T. A. Day, R. A. Pax, and D. W. Halton. 1997. Physiological effects of platyhelminth FMRFamide-related peptides (FaRPs) on the motility of the monogenean *Diclidophora merlangi*. *Parasitology*, 115:281–288.

Montgomery, R. E. 1906. Observation on bilharziosis among animals in India. II. *Journal of Tropical Veterinary Science*, 1:138–174.

Monticelli, F. S. 1888. *Saggio di una morfologia dei trematodi*. Stabilimento Tipografico Flli. Ferrante, Napoli.

Mooers, A. O., R. D. M. Page, A. Purvis, and P. H. Harvey. 1995. Phylogenetic noise leads to unbalanced cladistic tree reconstructions. *Systematic Biology*, 44:332–342.

Moore, J., and D. R. Brooks. 1987. Asexual reproduction in cestodes (Cyclophyllidea: Taeniidae): Ecological and phylogenetic influences. *Evolution*, 41:882–891.

Moore, J., and P. Willmer. 1997. Convergent evolution in invertebrates. *Biological Reviews of the Cambridge Philosophical Society*, 72:1–60.

Moquin-Tandon, A. 1846. *Monographie de la famille des Hirudinés*, Paris.

Moraczewski, J. 1977. Asexual reproduction and regeneration of *Catenula* (Turbellaria, Archoophora). *Zoomorphologie*, 88:65–80.

Moraczewski, J. 1981. Fine structure of some Catenulida (Turbellaria, Archophoora). *Zoologica Poloniae*, 28:367–415.

Morand, S., and C. D. M. Müller-Graf. 2000. Muscles or testes? Comparative evidence for sexual competition among dioecious blood parasites (Schistosomatidae) of vertebrates. *Parasitology*, 120:45–56.

Morgan, J. A. T., and D. Blair. 1995. Nuclear rDNA ITS sequence variation in the trematode genus *Echinostoma*: an aid to establishing relationships within the 37-collar-spine group. *Parasitology*, 111:609–615.

Morita, M., and J. B. Best. 1966. Electron microscopic studies of planaria. III. Some observations on the fine structure of planarian nervous tissue. *Journal of Experimental Zoology*, 161:391–413.

Moritz, C., and D. M. Hillis. 1996. Molecular systematics: context and controversies. In D. M. Hillis, C. Moritz and B. K. Mable (eds.), *Molecular systematics*, pp. 1–13. Sinauer Associates, Inc., Sunderland, Massachusetts.

Morseth, D. J. 1966. The fine structure of the tegument of adult *Echinococcus granulosus*, *Taenia hydatigena* and *Taenia pisiformis*. *Journal of Parasitology*, 52:1074–1085.

Mudry, D. R., and M. D. Dailey. 1971. Postembryonic development of certain tetraphyllidean and trypanorhynchan cestodes with a possible alternative life cycle for the order Trypanorhyncha. *Canadian Journal of Zoology*, 49:1249–1253.

Müller, O. F. 1773. *Vermium terrestrium et fluviatilium, seu animalium infusorium, helminthocorum et testaceorum, non marinorum, succincta historia*, Vol. 1., Infusoria, Havniae et Lipsiae.

Müller, O. F. 1788. *Zoologia Danica seu animalium Daniae et Norvegiae rariorum ac minus notorum descriptiones et historia*, Vol. 2. N. Mölleri, Havniae.

Mullis, K. B., and F. A. Faloona. 1987. Specific synthesis of DNA *in vitro* via a polymerase–catalyzed chain reaction. *Methods in Enzymology*, 155:335–350.

Murina, G. V. 1981. Notes on the biology of some psammophile Turbellaria of the Black Sea. *Hydrobiologia*, 84:129–130.

Murrills, R. J., T. A. J. Reader, and V. R. Southgate. 1985. Studies on the invasion of *Notocotylus attenuatus* (Notocotylidae: Digenea) into its snail host *Lymnaea peregra*: the contents of the fully embryonated egg. *Parasitology*, 91:397–405.

Murugesh, M., and R. Madhavi. 1990. Egg and miracidium of *Hirudinella ventricosa* (Trematoda: Hirudinellidae). *Journal of Parasitology*, 76:748–749.

Nadler, S. 1995. Advantages and disadvantages of molecular phylogenetics: a case study of ascaroid nematodes. *Journal of Nematology*, 27:423–432.

Nakajima, K., and S. Egusa. 1969. [Studies on a new trypanorhynchan larva, *Callotetrarhynchus* sp., parasitic on cultured yellowtail – III. On the anchovy worm]. (In Japanese). *Bulletin of the Japanese Society of Scientific Fisheries*, 35:723–729.

Nakajima, K., and S. Egusa. 1973. [Studies on a new trypanorhynchan larva, *Callotetrarhynchus* sp., parasitic on cultured yellowtail – XIII. Morphology of the adult and its taxonomy]. (In Japanese). *Bulletin of the Japanese Society of Scientific Fisheries*, 39:149–158.

Nasin, C., J. N. Caira, and L. Euzet. 1997. A revision of *Calliobothrium* (Tetraphyllidea: Onchobothriidae) with descriptions of three new species and a cladistic analysis of the genus. *Journal of Parasitology*, 83:714–733.

Nasir, P., and Y. Gómez. 1976. *Carassotrema tilapiae* n. sp. (Haploporidae Nicoll, 1914) from the freshwater fish, *Tilapia mossambica* (Peters), in Venezuela. *Rivista di Parassitologia*, 37:207–228.

Naylor, G. J. P., and W. M. Brown. 1997. Structural biology and phylogenetic estimation. *Nature*, 388:527–528.

Neefs, J.-M., Y. van de Peer, L. Hendriks, and R. de Wachter. 1990. Compilation of small ribosomal subunit RNA sequences. *Nucleic Acids Research*, 18:2237–2317.

Neefs, J. M., Y. van de Peer, P. de Rijk, S. Chapelle, and R. de Wachter. 1993. Compilation of small ribosomal subunit RNA structures. *Nucleic Acids Research*, 21:3025–3049.

Neff, N. A. 1986. A rational basis for *a priori* character weighting. *Systematic Zoology*, 35:110–123.

Nei, M., S. Kumar, and T. Kei. 1998. The optimization principle in phylogenetic analysis tends to give incorrect topologies when the number of nucleotides or amino acids is small. *Proceedings of the National Academy of Sciences, USA*, 95:12390–12397.

Nelson, G., and P. Y. Ladiges. 1994. Three-item consensus: empirical tests of fractional weighting. In R. W. Scotland, D. J. Siebert and D. M. Williams (eds.), *Models in phylogeny reconstruction*, pp. 193–209. Clarendon Press, Oxford.

Nielsen, C. 1995. *Animal evolution. Interrelationships of the living phyla*. Oxford University Press, Oxford.

Nielsen, C. 1998. Morphological approaches to phylogeny. *American Zoologist*, 38:942–952.

Niewiadomska, K., and A. Czubaj. 1996. Structure and TEM ultrastructure of the excretory system of the cercaria and daughter sporocyst of *Diplostomum pseudospathaceum* Niew., 1984 (Digenea, Diplostomidae). *Acta Parasitologica*, 41:167–181.

Niewiadomska, K., A. Czubaj, and T. Moczon. 1996. Cholinergic and aminergic nervous systems in developing cercariae and metacercariae of *Diplostomum pseudospathaceum* Niewiadomska, 1984 (Digenea). *International Journal for Parasitology*, 26:161–168.

Nixon, K. C. 1999. The Parsimony Ratchet, a new method for rapid parsimony analysis. *Cladistics*, 15:407–414.

Norén, M., and U. Jondelius. 1997. *Xenoturbella*'s molluscan relatives. *Nature*, 390:31–32.

Norén, M., and U. Jondelius. 1999. Phylogeny of the Prolecithophora (Platyhelminthes) inferred from 18S rDNA sequences. *Cladistics*, 15:103–112.

Noury-Sraïri, N., J.-L. Justine, and L. Euzet. 1989a. Ultrastructure comparée de la spermiogenèse et du spermatozoïde de trois espèces de *Paravortex* (Rhabdocoela, 'Dalyellioida', Graffillidae), Turbellariés parasites intestinaux de mollusques. *Zoologica Scripta*, 18:161–174.

Noury-Sraïri, N., J.-L. Justine, and L. Euzet. 1989b. Implications phylogénétiques de l'ultrastructure de la spermatogenèse, du spermatozoöde et de l'ovogenèse du turbellarié *Urastoma cyprinae* ('Prolecithophora', Urastomidae). *Zoologica Scripta*, 18:175–185.

Novak, M., W. R. Taylor, and E. Pip. 1989. Interspecific variation of isoenzyme patterns in four *Hymenolepis* species. *Canadian Journal of Zoology*, 67:2052–2055.

O'Grady, R. T. 1985. Ontogenetic sequences and the phylogenetics of parasitic flatworm life cycles. *Cladistics*, 1:159–170.

Odening, K. 1959. Das Exkretionssystem von *Omphalometra* und *Brachycoelum* (Trematoda, Digenea) und die Taxonomie der Unterordnung Plagiorchiata. *Zeitschrift für Parasitenkunde*, 19:442–457.

Odening, K. 1960. Der Ansatzmodus des Exkretionsgefäßsystem und die systematische Stellung von *Encylometra* (Trematoda, Digenea). *Monatsberichte der Deutschen Akademie der Wissenschaften zu Berlin*, 2:445–449.

Odening, K. 1964. Exkretionssystem und systematische stellung einiger Fledermaustrematoden aus Berlin und Umgebung nebst Bemerkungen zum lecithodendrioiden Komplex. *Zeitschrift für Parasitenkunde*, 24:453–483.

Odening, K. 1968. Exkretionssystem und systematische Stellung der Trematodengattungen *Anchitrema*, *Cephalogonimus*, *Encyclometra*, *Mesotretes*, *Omphalometra* und *Urotrema*. *Monatsberichte der Deutschen Akademie der Wissenschaften zu Berlin*, 10:492–498.

Odening, K. 1971. Möglichkeiten der Herstllung des bisher unbekannten Zusammenhangs von Cercarien und adulten Trematoden mit Hilfe detailierter Kenntnisse des Exkretionssystems nebst Ausführungen zum weiteren Ausbau des Systems der Plagiorchiata. Perspektiven der Cercarienforschung. *Parasitologische Schriftenreihe*, 21:57–72.

Odening, K. 1974. Verwandtschaft, System und zyklo–ontogenetische Besonderheiten der Trematoden (English translation, 1977, Tunis, Agence Tunisienne de Public-Relations). *Zoologischer Jahrbücher, Systematic*, 101:345–396.

Odhner, T. 1910. *Stichocotyle nephropis* J.T. Cunningham ein aberranter Trematode der Digenenfamilie Aspidogastridae. *Kungliga Svenska Vetenskapsakademiens Handlingar*, 45:1–16.

Ogawa, K., and S. Egusa. 1981. The systematic position of the genus *Anoplodiscus* (Monogenea: Anoplodiscidae). *Systematic Parasitology*, 2:253–260.

Okamoto, M., A. Ito, T. Kurosawa, Y. Oku, M. Kamiya, and T. Agatsuma. 1995. Intraspecific variation of isoenzymes in *Taenia taeniaeformis*. *International Journal for Parasitology*, 25:221–228.

Okamoto, M., T. Agatsuma, T. Kurosawa, and A. Ito. 1997. Phylogenetic relationships of three hymenolepidid species

inferred from nuclear ribosomal and mitochondrial DNA sequences. *Parasitology*, 115:661–666.

Olsen, G. J., and C. R. Woese. 1993. Ribosomal RNA – a key to phylogeny. *FASEB Journal*, 7:113–123.

Olsen, G. J., H. Matsuda, R. Hangstrom, and R. Overbeek. 1994. FastDNAml: a tool for construction of phylogenetic trees of DNA sequences using maximum likelihood. *Computer Applications in the Biosciences*, 10:41–48.

Olson, P. D. 2000. New insights into platyhelminth systematics and evolution. *Parasitology Today*, 16:3–5.

Olson, P. D., and J. N. Caira. 1999. Evolution of the major lineages of tapeworms (Platyhelminthes: Cestoidea) inferred from 18S ribosomal DNA and elongation factor-1 α. *Journal of Parasitology*, 85:1134–1159.

Olson, P. D., T. R. Ruhnke, J. Sanney, and T. Hudson. 1999. Evidence for host–specific clades of tetraphyllidean tapeworms (Platyhelminthes: Eucestoda) revealed by analysis of 18S ssrDNA. *International Journal for Parasitology*, 29:1465–1476.

Olson, P. D., D. T. J. Littlewood, R. A. Bray, and J. Mariaux. Interrelationships and evolution of the tapeworms (Platyhelminthes: Cestoda). *Molecular Phylogenetics and Evolution*, in press.

Orido, Y. 1987. Metamorphosis of the excretory system of *Paragonimus ohirai* (Trematoda), with special reference to its functional significance. *Journal of Morphology*, 194:303–310.

Orii, H., K. Kato, K. Agata, and K. Watanabe. 1998. Molecular cloning of bone morphogenetic protein (BMP) gene from the planarian *Dugesia japonica*. *Zoological Science*, 15:871–877.

Orii, H., K. Kato, Y. Umesono, T. Sakurai, K. Agata, and K. Watanabe. 1999. The planarian HOM/*Hox* homeobox genes (*Plox*) expressed along the anteroposterior axis. *Developmental Biology*, 210:456–468.

Page, R. D. M., and M. A. Charleston. 1997. From gene to organismal tree: reconciled trees and the gene tree/species tree problem. *Molecular Phylogenetics and Evolution*, 7:231–240.

Page, R. D. M., and M. A. Charleston. 1998. Trees within trees: phylogeny and historical associations. *Trends in Ecology and Evolution*, 13:356–359.

Page, R. D. M., and E. C. Holmes. 1998. *Molecular evolution: a phylogenetic approach*. Blackwell Science, Oxford.

Palmberg, E. 1990a. Stem cells in microturbellarians. An autoradiographic and immunocytochemical study. *Protoplasma*, 158:109–120.

Palmberg, E. 1990b. Differentiation in free-living flatworms. Ultrastructural, immunocytochemical and autoradiographic studies of asexually reproducing and regenerating *Microstomum lineare*. Ph.D. thesis, Åbo Academy.

Palmberg, I., M. Reuter, and M. Wikgren. 1980. Ultrastructure of epidermal eyespots of *Microstomum lineare* (Turbellaria, Macrostomida). *Cell & Tissue Research*, 210:21–32.

Palombi, A. 1926. '*Digenobothrium inerme*' n. gen., n. sp. (Crossocoela). *Archivio Zoologico Italiano*, 11:143–175.

Pan, S. C.-T. 1980. The fine structure of the miracidium of *Schistosoma mansoni*. *Journal of Invertebrate Pathology*, 36:307–372.

Panula, P. A. J., K. S. Eriksson, M. K. S. Gustaffson, and M. Reuter. 1995. An immunocytochemical method for histamine: application to the planarians. *Hydrobiologia*, 305:291–295.

Papi, F. 1950. Sulla affinita morphologiche nella fam. Macrostomida (Turbellaria). *Bolletino di Zoologia*, Supplement 17:461–468.

Papi, F. 1953. Beiträge zur Kenntnis der Macrostomiden (Turbellaria). *Annales Zoologica Fennica,* 78:1–32.

Pardos, F. 1988. Fine structure and function of pharynx cilia in *Glossobalanus minutus* Kowalewsky (Enteropneusta). *Acta Zoologica (Stockholm),* 69.

Patterson, C., D. M. Williams, and C. J. Humphries. 1993. Congruence between molecular and morphological phylogenies. *Annual Review of Ecology and Systematics,* 24:153–188.

Pawlak, R. 1969. Zur Systematik un Ökologie (Lebenszyklen, Populationsdynamik) der Turbellarien-Gattung *Paromalostomum. Helgoländer Wissenschaftliche Meeresuntersuchungen,* 19:417–454.

Pawlowski, I. D., K. W. Yap, and R. C. A. Thompson. 1988. Observations on the possible origin, formation and structure of calcareous corpuscles in taeniid cestodes. *Parasitology Research,* 74:293–296.

Pax, R. A., and J. L. Bennett. 1991. Neurobiology of parasitic platyhelminths: possible solutions to the problems of correlating structure with function. *Parasitology,* 102:S31–S39.

Pax, R. A., C. Siefker, and J. L. Bennett. 1984. *Schistosoma mansoni*: differences in acetylcholine, dopamine, and serotonin control of circular and longitudinal parasite muscles. *Experimental Parasitology,* 58:314–324.

Pearson, J. C. 1972. A phylogeny of life–cycle patterns of the Digenea. *Advances in Parasitology,* 10:153–189.

Pearson, J. C. 1992. On the position of the digenean family Heronimidae: an inquiry into a cladistic classification of the Digenea. *Systematic Parasitology,* 21:81–166.

Pedersen, K. J., and L. R. Pedersen. 1986. Fine ultrastructural observations on the extracellular matrix (ECM) of *Xenoturbella bocki* Westblad, 1949. *Acta Zoologica (Stockholm),* 67:103–114.

Pedersen, K. J., and L. R. Pedersen. 1988. Ultrastructural observations on the epidermis of *Xenoturbella bocki* Westblad, 1949; with a discussion of epidermal cytoplasmic filament systems of invertebrates. *Acta Zoologica (Stockholm),* 69:231–246.

Penner, L. R., and A. Wagner. 1956. Concerning the early developmental stages of *Ornithobilharzia canaliculata* (Rudolphi, 1819). *Journal of Parasitology,* 42 (Supplement):37–38.

Petkeviciute, R. 1996. A chromosome study in the progenetic cestode *Cyathocephalus truncatus* (Cestoda: Spathebothriidea). *International Journal for Parasitology,* 26:1211–1216.

Philippe, H., and A. Adoutte. 1998. The molecular phylogeny of Eukaryota: solid facts and uncertainties. In G. H. Coombs, K. Vickerman, M. A. Sleigh and A. Warren (eds.), *Evolutionary relationships among Protozoa,* pp. 25–56. Chapman & Hall, London.

Philippe, H., and E. Douzery. 1994. The pitfalls of molecular phylogeny based on four species as illustrated by the Cetacea/Artiodactyla relationships. *Journal of Mammalian Evolution,* 2:133–152.

Philippe, H., A. Chenuil, and A. Adoutte. 1994. Can the Cambrian explosion be inferred through molecular phylogeny? *Development,* Supplement:S15–S25.

Platt, T. R., and D. R. Brooks. 1997. Evolution of the schistosomes (Digenea: Schistosomatoidea): the origin of dioecy and colonization of the venous system. *Journal of Parasitology,* 83:1035–1044.

Platt, T. R., D. Blair, J. Purdie, and L. Melville. 1991. *Griphobilharzia amoena* n. gen., n. sp. (Digenea: Schistosomatidae), a parasite of the freshwater crocodile *Crocodylus johnstoni* (Reptilia: Crocodylia) from Australia, with the erection of the new subfamily, Griphobilharziinae. *Journal of Parasitology,* 77:65–68.

Pleijel, F. 1999. Phylogenetic taxonomy, a farewell to species, and a revision of Heteropodarke (Hesionidae, Polychaeta, Annelida). *Systematic Biology,* 48:755–789.

Poche, F. 1926. Das System der Platodaria. *Archiv für Naturgeschichte,* 91:1–459.

Poe, S. 1998. Sensitivity of phylogeny estimation to taxonomic sampling. *Systematic Biology,* 47:18–31.

Poppe, G. T., and Y. Goto. 1991. *European seashells. Vol. 1. (Polyplacophora, Caudofoveata, Solenogastra, Gastropoda).* Christa Hemmen, Wiesbaden.

Pospekhova, N. A., G. P. Krasnoshchekov, and V. V. Prospekhov. 1993. [Scolex protonephridia system in Cyclophyllidea]. (In Russian). *Parazitologiya,* 27:48–53.

Powell, E. C. 1972. Optical and electron microscope studies on the excretory bladder of the supposed epitheliocystid cercaria of *Ochetosoma aniarum. Zeitschrift für Parasitenkunde,* 40:19–30.

Powell, E. C. 1973. Studies on the excretory 'bladder' and caudal ducts of the supposed anepitheliocystid cercariae of *Schistosoma mansoni. Zeitschrift für Parasitenkunde,* 43:43–52.

Powell, E. C. 1975. The ultrastructure of the excretory bladder in the supposed anepitheliocystid cercariae of *Posthodiplostomum minimum* (MacCallum, 1921). *Iowa State Journal of Research,* 49:259–262.

Presgraves, D. C., R. H. Baker, and G. S. Wilkinson. 1999. Coevolution of sperm and female reproductive tract morphology in stalk-eyed flies. *Proceedings of the Royal Society of London, Series B, Biological Sciences,* 226:1041–1047.

Price, E. W. 1929. A synopsis of the trematode family Schistosomidae with descriptions of new genera and species. *Proceedings of the United States National Museum,* 75:1–39.

Price, E. W. 1938. North American monogenetic trematodes. II. The families Monocotylidae, Microbothriidae, Acanthocotylidae, and Udonellidae (Capsaloidea). *Journal of Washington Academy of Science,* 28:109–126.

Prudhoe, S. 1985. *A monograph on polyclad Turbellaria.* British Museum (Natural History): Oxford University Press, Oxford.

Prudhoe, S., and R. A. Bray. 1982. *Platyhelminth parasites of the Amphibia.* British Museum (Natural History): Oxford University Press, Oxford.

Purvis, A. 1995a. A composite estimate of primate phylogeny. *Philosophical Transactions of the Royal Society of London, Series B, Biological Sciences,* 348:405–421.

Purvis, A. 1995b. A modification to Baum and Ragan's method for combining phylogenetic trees. *Systematic Biology,* 44:251–255.

Ragan, M. A. 1992. Phylogenetic inference based on matrix representation of trees. *Molecular Phylogenetics and Evolution,* 1:53–58.

Raikova, O. I. 1991. Fine structural organisation in the nervous system and ciliary receptors of acoelan turbellarians. In D. A. Sakharov and W. Winlow (eds.), *Simpler nervous systems,* pp. 37–50. Manchester University Press, Manchester.

Raikova, O. I., and J.-L. Justine. 1994. Ultrastructure of spermiogenesis and spermatozoa in three acoels (Platyhelminthes). *Annales des Sciences Naturelle Zoologie et Biologie Animale,* 15:63–75.

Raikova, O. I., and J.-L. Justine. 1999. Microtubular system during spermiogenesis and in the spermatozoon of *Convoluta saliens* (Platyhelminthes, Acoela): tubulin immunocytochemistry and electron microscopy. *Molecular Reproduction and Development,* 52:74–85.

Raikova, O. I., A. Falleni, and J.-L. Justine. 1997. Spermiogenesis in *Paratomella rubra* (Platyhelminthes,

Acoela): ultrastructural, immunocytochemical, cytochemical studies and phylogenetic implications. *Acta Zoologica (Stockholm)*, 78:295–307.

Raikova, O. I., M. Reuter, E. A. Kotikova, and M. K. S. Gustafsson. 1998a. A commissural brain! The pattern of 5-HT immunoreactivity in Acoela (Plathelminthes). *Zoomorphology*, 118:69–77.

Raikova, O. I., L. P. Flyatchinskaya, and J.-L. Justine. 1998b. Acoel spermatozoa: ultrastructure and immunocytochemistry of tubulin. *Hydrobiologia*, 383:207–214.

Raikova, O. I., M. Reuter, U. Jondelius, and M. K. S. Gustafsson. 2000. The brain of the Nemertodermatida (Platyhelminthes) as revealed by anti-5HT and anti-FMRFamide immunostainings. *Tissue & Cell*, 32:358–365.

Raikova, O. I., M. Reuter, and J.-L. Justine. 2001. Contributions to the phylogeny and systematics of the Acoelomorpha. In D. T. J. Littlewood and R. A. Bray (eds.), *Interrelationships of the Platyhelminthes*, pp. 13–23. Taylor & Francis, London.

Rajpara, S. M., P. D. Garcia, R. Roberts, J. C. Eliassen, D. F. Owens, D. Maltby, R. M. Myers, and E. Mayeri. 1992. Identification and molecular cloning of a neuropeptide Y homolog that produces prolonged inhibition in *Aplysia* neurons. *Neuron*, 9:505–513.

Rannala, B., J. P. Huelsenbeck, Z. Yang, and R. Nielsen. 1998. Taxon sampling and the accuracy of large phylogenies. *Systematic Biology*, 47:702–710.

Rees, G. 1956. The scolex of *Tetrabothrius affinis* (Lönnberg) a cestode from *Balaenopterus musculus* L., the blue whale. *Parasitology*, 46:425–442.

Rees, G. 1969. Cestodes from Bermuda fishes and an account of *Acompsocephalum tortum* (Linton, 1905) gen. nov. from the lizard fish *Synodus intermedius* (Agassiz). *Parasitology*, 59:519–548.

Rees, G. 1973a. The ultrastructure of the cysticercoid of *Tatria octacantha* Rees, 1973 (Cyclophyllidea: Amabiliidae) from the haemocoele of the damsel-fly nymphs *Pyrrhosoma nymphula*, Sulz and *Enallagma cyathigerum*, Charp. *Parasitology*, 66:85–103.

Rees, G. 1973b. Cysticercoids of three species of *Tatria* (Cyclophyllidea: Amabiliidae) including *T. octacantha* sp. nov. from the haemocoele of the damsel-fly nymphs *Pyrrhosoma nymphula*, Sulz and *Enallagma cyathigerum*, Charp. *Parasitology*, 66:423–446.

Rees, G. 1977. The development of the tail and the excretory system in the cercaria of *Cryptocotyle lingua* (Creplin) (Digenea: Heterophyidae from *Littorina littorea* (L.)). *Proceedings of the Royal Society of London, Series B, Biological Sciences*, 195:425–452.

Rees, G. 1979. The ultrastructure of the spermatozoon and spermatogenesis of *Cryptocotyle lingua* (Digenea; Heterophyidae). *International Journal for Parasitology*, 9:405–419.

Rego, A. A. 1994. Order Proteocephalidea Mola, 1928. In L. F. Khalil, A. Jones and R. A. Bray (eds.), *Keys to the cestode parasites of vertebrates*, pp. 257–293. CAB International, Wallingford.

Rego, A. A., A. de Chambrier, V. Hanzelova, E. Hoberg, T. Scholz, P. Weekes, and M. Zehnder. 1998. Preliminary phylogenetic analysis of subfamilies of the Proteocephalidea (Eucestoda). *Systematic Parasitology*, 40:1–19.

Reise, K. 1984. Free-living Plathelminthes (Turbellaria) of a marine sand flat: an ecological study. *Microfauna Marina*, 1:1–62.

Reise, K. 1988. Plathelminth diversity in littoral sediments around the island of Sylt in the North Sea. *Fortschritte der Zoologie*, 36:469–480.

Reisinger, E. 1924. Zur Anatomie von *Hypotrichina* (=*Genostoma*) *tergestina* Cal. nebst einem Beitrag zur Systematik der Alloeocoelen. *Zoologischer Anzeiger*, 60:137–149.

Reisinger, E. 1925. Untersuchungen am Nervensystem von *Bothrioplana semperi* Braun. *Zeitschrift für Morphologie und Ökologie der Tiere*, 5:119–149.

Reisinger, E. 1933. Turbellarien der Deutschen Limnologischen Sunda-Expedition. *Archiv für Hydrobiologie*, Supplement 12, Tropische Binnengewässer 4:239–262.

Reisinger, E. 1960. Was ist *Xenoturbella*? *Zeitschrift für wissenschaftliche Zoologie*, 164:188–198.

Reisinger, E. 1970. Zur Problematik der Evolution der Coelomaten. *Zeitschrift für zoologische Systematik und Evolutionsforschung*, 8:81–109.

Reisinger, E. 1972. Die Evolution des Orthogon der Spiralier und das Archicölomatenproblem. *Zeitschrift für zoologische Systematik und Evolutionsforschung*, 10:1–43.

Reisinger, E. 1976. Zur Evolution des stomatogastrischen Nervensystem bei den Plathelminthen. *Zeitschrift für zoologische Systematik und Evolutionsforschung*, 14:241–253.

Reisinger, E., and S. Kelbetz. 1964. Feinbau und Entladungsmechanismus der Rhabditen. *Zeitschrift für wissenschaftliche Mikroskopie*, 65.

Reisinger, E., I. Cichocki, R. Erlach, and T. Szyskowitz. 1974. Ontogenetische Studien an Turbellarien: ein Beitrag zur Evolution der Dotterverarbeitung im ektolecithalen Ei, II Teil. *Zeitschrift für zoologische Systematik und Evolutionsforschung*, 12:241–278.

Reiter, D., B. Boyer, P. Ladurner, G. Mair, W. Salvenmoser, and R. Rieger. 1996. Differentiation of the body wall musculature in *Macrostomum hystricinum marinum* and *Hoploplana inquilina* (Platyhelminthes), as models for muscle development in lower Spiralia. *Roux's Archives of Developmental Biology*, 205:410–423.

Remane, A. 1933. Verteilung und Organisation der benthonischen Mikrofauna der Kieler Bucht. *Wissenschaftliche Meeresunters der Kommission zur Wissenschaftlichen Untersuchung der Deutschen Meere*, 21:161–221.

Remane, A. 1959. Die Geschichte der Tiere. In G. Heberer (ed.), *Die Evolution der Organismen*, pp. 340–422. G. Fischer Verlag, Stuttgart.

Renaud, F., C. Gabrion, and N. Pasteur. 1983. Le complexe *Bothriocephalus scorpii* (Mueller, 1776): différenciation par électrophorèse enzymatique des espèces parasites du turbot (*Psetta maxima*) et de la barbue (*Scophthalmus rhombus*). *Comptes Rendus de l'Académie des sciences, Paris, Série III, Sciences de la Vie*, 296:127–129.

Renaud, F., C. Gabrion, and N. Pasteur. 1986. Geographical divergence in *Bothriocephalus* (Cestoda) of fishes demonstrated by enzyme electrophoresis. *International Journal for Parasitology*, 16:553–558.

Reuter, M. 1990. From innovation to integration. Trends of the integrative systems in microturbellarians. *Acta Academiae Aboensis Series B*, 50:161–178.

Reuter, M. 1994. Substance P immunoreactivity in sensory structures and the central and pharyngeal nervous system of *Stenostomum leucops* (Catenulida) and *Microstomum lineare* (Macrostomida). *Cell & Tissue Research*, 276:173–180.

Reuter, M., and K. Erikkson. 1991. Catecholamines demonstrated by glyoxylic–acid induced fluorescence and HPLC in some microturbellarians. *Hydrobiologia*, 227:209–219.

Reuter, M., and M. K. S. Gustaffson. 1989. 'Neuroendocrine cells' in flatworms – progenitors to metazoan neurones? *Archives of Histology and Cytology*, 52:253–263.

Reuter, M., and M. K. S. Gustaffson. 1995. The flatworm nervous sytem – pattern and phylogeny. In O. Breidbach and W. Kutsch (eds.), *The nervous systems of invertebrates – an evolutionary and comparative approach*, pp. 25–59. Birkhäuser Verlag, Basel.

Reuter, M., and P. Lindroos. 1979. The ultrastructure of the nervous system of *Gyratrix hermaphroditus* (Turbellaria, Rhabdocoela). *Acta Zoologica (Stockholm)*, 60:153–163.

Reuter, M., and I. Palmberg. 1987. An ultrastructural and immunocytochemical study of gastrodermal cell types in *Microstomum lineare* (Turbellaria, Macrostomida). *Acta Zoologica (Stockholm)*, 68:153–163.

Reuter, M., and I. Palmberg. 1989. Development and differentiation of neuronal subsets in asexually reproducing *Microstomum lineare*. Immunocytochemistry of 5–HT, RF–amide and SCP$_B$. *Histochemistry*, 91:123–131.

Reuter, M., M. Lehtonen, and M. Wikgren. 1986. Immunocytochemical demonstration of 5-HT-like and FMRF-amide-like substances in whole mounts of *Microstomum lineare* (Turbellaria). *Cell & Tissue Research*, 246:7–12.

Reuter, M., M. Lehtonen, and M. Wikgren. 1988. Immunocytochemical evidence of neuroactive substances in flatworms of different taxa – a comparison. *Acta Zoologica (Stockholm)*, 69:29–37.

Reuter, M., B. I. Joffe, and I. Palmberg. 1993. Sensory receptors in the head of *Stenostomum leucops*. II. Localization of catecholaminergic histofluorescence-ultrastructure of surface receptors. *Acta Biologica Hungarica*, 44:25–131.

Reuter, M., A. G. Maule, D. W. Halton, M. K. S. Gustafsson, and C. Shaw. 1995a. The organization of the nervous system in Plathelminthes. The neuropeptide F (NPF)-immunoreactivity pattern in Catenulida, Macrostomida, Proseriata. *Zoomorphology*, 115:83–97.

Reuter, M., M. K. S. Gustafsson, C. Sahlgren, D. W. Halton, A. G. Maule, and C. Shaw. 1995b. The nervous sytem of Tricladida. I. Neuranatomy of *Procerodes littoralis* (Maricola, Procerodidae): an immunocytochemical study. *Invertebrate Neuroscience*, 1:113–122.

Reuter, M., M. K. S. Gustafsson, I. M. Sheiman, N. Terenina, D. W. Halton, A. G. Maule, and C. Shaw. 1995c. The nervous sytem of Tricladida. II. Neuranatomy of *Dugesia tigrina* (Paludicola, Dugesiidae): an immunocytochemical study. *Invertebrate Neuroscience*, 1:133–143.

Reuter, M., M. K. S. Gustafsson, K. Mäntylä, and C. J. P. Grimmelikhuijzen. 1996a. The nervous system of Tricladida. III. Neuroanatomy of *Dendrocoelum lacteum* (Dendrocoelidae) and *Polycelis tenuis* (Planariidae) Plathelminthes, Paludicola : an immunocytochemical study. *Zoomorphology*, 116:111–122.

Reuter, M., I. M. Sheiman, M. K. S. Gustafsson, D. W. Halton, A. G. Maule, and C. Shaw. 1996b. Development of the nervous system in *Dugesia tigrina* during regeneration after fission and decapitation. *Invertebrate Reproduction and Development*, 29:199–211.

Reuter, M., K. Mäntylä, and M. K. S. Gustafsson. 1998a. Organization of the orthogon – main and minor nerve cords. *Hydrobiologia*, 383:175–182.

Reuter, M., O. I. Raikova, and M. K. S. Gustafsson. 1998b. An endocrine brain? The pattern of FMRF-amide immunoreactivity in Acoela (Plathelminthes). *Tissue & Cell*, 30:57–63.

Reuter, M., O. I. Raikova, U. Jondelius, M. K. S. Gustafsson, A. G. Maule, and D. W. Halton. In press. The organisation of the nervous system in the Acoela. An immunocytochemical study. *Tissue & Cell*.

Richards, K. S., and C. Arme. 1984a. Maturation of the scolex syncytium in the metacestode of *Hymenolepis diminuta*, with special reference to microthrix formation. *Parasitology*, 88:341–349.

Richards, K. S., and C. Arme. 1984b. An ultrastructural analysis of cyst wall development in the metacestode of *Hymenolepis diminuta* (Cestoda). *Parasitology*, 89:537–566.

Riedl, R. J. 1978. *Order in living organisms. A systems analysis of evolution*. John Wiley & Sons, New York.

Rieger, R. M. 1969. *Myozonaria bistylifera* nov. gen. nov. spec.: eine Vertreter der Turbellarienordnung Macrostomida aus dem Verwandtschaftskreis von *Dolichomacrostomum* Luther mit einem Muskeldarm. *Zoologischer Anzeiger*, 180:1–22.

Rieger, R. M. 1971a. *Bradynectes sterreri* gen. nov. spec. nov., eine psammobionte Macrostomide (Turbellaria) Muskeldarm. *Zoologische Jahrbücher Abteilung für Systematik*, 98:205–235.

Rieger, R. M. 1971b. Die Turbellarienfamilie Dolichomacrostomidae nov. fam. (Macrostomida) I, Vorbemerkungen und Karlingiinae nov. subfam. 1. *Zoologische Jahrbücher Abteilung für Systematik*, 98:236–314.

Rieger, R. M. 1971c. Die Turbellarienfamilies Dolichomacrostomidae Rieger II, Dolichomacrostominae 1. *Zoologische Jahrbücher Abteilung für Systematik*, 98:569–703.

Rieger, R. M. 1977. The relationship of character variability and morphological complexity in copulatory structure of Turbellaria-Macrostomida and Haplopharyngida. *Mikrofauna Meeresboden*, 61:197–216.

Rieger, R. M. 1980. A new group of interstitial worms, Lobatocerebridae nov. fam. (Annelida) and its significance for metazoan phylogeny. *Zoomorphology*, 95:41–84.

Rieger, R. M. 1981a. Fine structure of the body wall, nervous system and digestive tract in the Lobatocerebridae Rieger and the organization of the gliointerstitial system in Annelida. *Journal of Morphology*, 167:139–165.

Rieger, R. M. 1981b. Morphology of the Turbellaria at the ultrastructural level. *Hydrobiologia*, 84:213–229.

Rieger, R. M. 1985. The phylogenetic status of the acoelomate organization within the Bilateria: a histological perspective. In S. Conway Morris, J. D. George, R. Gibson and H. M. Platt (eds.), *The origins and relationships of lower vertebrates*, pp. 102–122. Clarendon Press, Oxford.

Rieger, R. M. 1986a. Über dem Ursprung der Bilateria: die Bedeutung der Ultrastrukturforschung für ein neues Verstehen der Metazoenevolution. *Verhandlungen der Deutsche Zoologische Gesellschafte*, 79:31–50 [English translation available at http://www.umesci.maine.edu/biology/labs/origin/].

Rieger, R. M. 1986b. Asexual reproduction and the turbellarian archetype. *Hydrobiologia*, 132:35–45.

Rieger, R. M. 1988. Comparative ultrastructure and the Lobatocerebridae: keys to understanding the phylogenetic relationship of Annelida and Acoelomates. In W. Westheide and C. O. Hermans (eds.), *The ultrastructure of Polychaeta*, Vol. 4, pp. 373–382. Fischer Verlag, Stuttgart.

Rieger, R. M. 1994a. Evolution of the lower Metazoa. In S. Bengtson (ed.), *Early life on Earth*, Vol. 84, pp. 475–488. Columbia University Press.

Rieger, R. M. 1994b. The biphasic life cycle – a central theme of metazoan evolution. *American Zoologist*, 34:484–491.

Rieger, R. M. 1998. 100 Years of Research on 'Turbellaria'. *Hydrobiologia*, 383:1–27.

Rieger, R. M. 2001. Phylogenetic systematics of the Macrostomorpha. In D. T. J. Littlewood and R. A. Bray (eds.), *Interrelationships of the Platyhelminthes*, pp. 28–38. Taylor & Francis, London.

Rieger, R., and S. Tyler. 1985. Das Homologietheorem in der Ultrastrukturforschung. In J. A. Ott, G. P. Wagner and F. M. Wuketits (eds.), *Evolution, Ordnung und Erkenntnis*, pp. 21–36. Paul Parey, Berlin.

Rieger, R. M., S. Tyler, J. P. S. Smith, and G. E. Rieger. 1991. Platyhelminthes: Turbellaria. In F. W. Harrison and B. J. Bogitsh (eds.), *Microscopic anatomy of invertebrates. Volume 3: Platyhelminthes and Nemertinea*, pp. 7–140. Wiley–Liss, Inc., New York.

Riek, E. F. 1972. The phylogeny of the Parastacidae (Crustacea: Astacoidea), and a description of a new genus of Australian freshwater crayfishes. *Australian Journal of Zoology*, 20:369–389.

Rinder, H., R. L. Rausch, K. Takahashi, H. Kopp, A. Thomschke, and T. Löscher. 1998. Strain differentiation in *Echinococcus multilocularis*: results of genotypic analyses. *Parasitology International*, 47 (Suppl.):136.

Riser, N. W. 1987. *Nemertinoides elongatus* gen. n., sp. n. (Turbellaria: Nemertodermatida) from coarse sand beaches of the Western North Atlantic. *Proceedings of the Helminthological Society of Washington*, 54:60–67.

Riutort, M., K. G. Field, J. M. Turbeville, R. A. Raff, and J. Baguñà. 1992a. Enzyme electrophoresis, 18S rRNA sequences, and levels of phylogenetic resolution among several species of freshwater planarians (Platyhelminthes, Tricladida, Paludicola). *Canadian Journal of Zoology*, 70:1425–1439.

Riutort, M., K. G. Field, R. A. Raff, and J. Baguñà. 1992b. 18S rRNA sequences and phylogeny of Platyhelminthes. *Biochemical Systematics and Ecology*, 21:71–77.

Roberts, L. S., and J. J. Janovy. 1999. *Gerald D. Schmidt & Larry S. Roberts' foundations of parasitology*. 6th ed. McGraw-Hill, New York.

Robinson, E. S. 1959. Some new cestodes from New Zealand marine fishes. *Transactions of the Royal Society of New Zealand*, 86:381–392.

Rodrigo, A. G. 1996. On combining cladograms. *Taxon*, 45:267–274.

Rogan, M. T., and K. S. Richards. 1987. *Echinococcus granulosus*: changes in the surface ultrastructure during protoscolex formation. *Parasitology*, 94:359–367.

Rohde, K. 1966. A Malayan record of *Temnocephala semperi* Weber (Rhabdocoela, temnocephalidae) an ectocommensal turbellarian on the freshwater crab. *Medical Journal of Malaya*, 20:369–389.

Rohde, K. 1971a. Phylogenetic origin of trematodes. *Parasitologische Schriftenreihe*, 21:17–27.

Rohde, K. 1971b. Untersuchungen an *Multicotyle purvisi* Dawes, 1941 (Trematoda: Aspidogastrea). I. Entwicklung und Morphologie. *Zoologischer Jahrbuch der Anatomie*, 88:320–363.

Rohde, K. 1971c. Untersuchungen an *Multicotyle purvisi* Dawes, 1941 (Trematoda: Aspidogastrea). III. Licht– und electronmikroskopischer Bau des Nervensystems. *Zoologischer Jahrbuch der Anatomie*, 88:320–363.

Rohde, K. 1971d. Untersuchungen an *Multicotyle purvisi* Dawes, 1941 (Trematoda: Aspidogastrea). VIII. Elektronmikroskopischer Bau des Exkretionssytems. *International Journal for Parasitology*, 1:275–286.

Rohde, K. 1972. The Aspidogastrea, especially *Multicotyle purvisi* (Dawes 1941). *Advances in Parasitology*, 10:77–151.

Rohde, K. 1973a. Ultrastructure of the protonephridial system of *Polystomoides malayi* Rohde and *P. renschi* Rohde (Monogenea: Polystomatidae). *International Journal for Parasitology*, 3:329–333.

Rohde, K. 1973b. Structure and development of *Lobatostoma manteri* sp. nov. (Trematoda: Aspidogastrea) from the Great Barrier Reef, Australia. *Parasitology*, 66:63–83.

Rohde, K. 1975a. Fine structure of the Monogenea, especially *Polystomoides* Ward. *Advances in Parasitology*, 13:1–33.

Rohde, K. 1975b. Early development and pathogenesis of *Lobatostoma manteri* Rohde (Trematoda: Aspidogastrea). *International Journal for Parasitology*, 5:597–607.

Rohde, K. 1980. Some aspects of the ultrastructure of *Gotocotyla secunda* and *Hexostoma euthynni*. *Angewandte Parasitologie*, 21:32–48.

Rohde, K. 1982. The flame cells of a monogenean and an aspidogastrean, not composed of two interdigitating cells. *Zoologischer Anzeiger*, 209:311–314.

Rohde, K. 1986a. Ultrastructure of the flame cells and protonephridial capillaries of *Temnocephala*; implications for the phylogeny of parasitic Platyhelminthes. *Zoologischer Anzeiger*, 216:39–47.

Rohde, K. 1986b. Ultrastructural studies of *Austramphilina elongata* (Cestoda, Amphilinidea). *Zoomorphology*, 106: 91–102.

Rohde, K. 1987a. Ultrastucture of flame cells and protonephridial capillaries of *Craspedella* and *Didymorchis* (Plathelminthes, Rhabdocoela). *Zoomorphology*, 106:346–351.

Rohde, K. 1987b. The formation of glandular secretion in larval *Austramphilina elongata* (Amphilinidea). *International Journal for Parasitology*, 17:821–828.

Rohde, K. 1987c. Ultrastructural studies of epidermis, sense receptors and sperm of *Craspedella* sp, and *Didymorchis* sp. (Platyhelminthes, Rhabdocoela). *Zoologica Scripta*, 16:289–295.

Rohde, K. 1988. Phylogenetic relationship of free–living and parasitic Platyhelminthes on the basis of ultrastructural evidence. *Fortschritte der Zoologie*, 36:353–357.

Rohde, K. 1989a. Ultrastructure of the protonephridial system of *Lobatostoma manteri* (Trematoda, Aspidogastrea). *Journal of Submicroscopic Cytology and Pathology*, 21: 599–610.

Rohde, K. 1989b. Ultrastructure of the protonephridial system of *Gyrodactylus* (Monogenea, Gyrodactylidae). *Zoologischer Anzeiger*, 223:311–322.

Rohde, K. 1990. Phylogeny of Platyhelminthes, with special reference to parasitic groups. *International Journal for Parasitology*, 20:979–1007.

Rohde, K. 1991. The evolution of protonephridia of the Platyhelminthes. *Hydrobiologia*, 227:315–321.

Rohde, K. 1993. Ultrastructure of protonephridia in the Monogenea. Implications for the phylogeny of the group. *Bulletin Francais de la Pêche et de la Pisciculture*, 328:115–119.

Rohde, K. 1994a. The minor groups of parasitic Platyhelminthes. *Advances in Parasitology*, 33:145–234.

Rohde, K. 1994b. The origins of parasitism in the Platyhelminthes. *International Journal for Parasitology*, 24:1099–1115.

Rohde, K. 1996. Robust phylogenies and adaptive radiations: a critical examination of methods used to identify key innovations. *American Naturalist*, 148:481–500.

Rohde, K. 1997. Ultrastructure of the protonephridial system of the oncomiracidium of *Zeuxapta seriolae* (Meserve, 1938) (Monogenea, Polyopisthocotylea, Axinidae). *Acta Parasitologica*, 42:127–131.

Rohde, K. 1998. Ultrastructure of the protonephridial system of the oncomiracidium of *Encotyllabe chironemi* (Platyhelminthes, Monopisthocotylea). *Microscopy Research and Technique*, 42:212–217.

Rohde, K. 2001. Protonephridia as phylogenetic characters. In D. T. J. Littlewood and R. A. Bray (eds.), *Interrelationships of the Platyhelminthes*, pp. 203–216. Taylor & Francis, London.

Rohde, K., and A. Faubel. 1997. Spermatogenesis of *Macrostomum pusillum* (Platyhelminthes, Macrostomida). *Invertebrate Reproduction and Development*, 32:209–215.

Rohde, K., and A. Faubel. 1998. Spermatogenesis of *Haplopharynx rostratus* (Platyhelminthes, Haplopharyngida). *Belgian Journal of Zoology*, 128:177–188.

Rohde, K., and P. R. Garlick. 1985a. Two ciliate sense receptors in the larva of *Austramphilina elongata* Johnston, 1931 (Amphilinidea). *Zoomorphology*, 105:30–33.

Rohde, K., and P. R. Garlick. 1985b. Subsurface sense receptors in the larva of *Austramphilina elongata* Johnston, 1931 (Amphilinidea). *Zoomorphology*, 105:34–38.

Rohde, K., and P. R. Garlick. 1985c. Ultrastructure of the posterior sense receptor of larval *Austramphilina elongata* (Amphilinidea). *International Journal for Parasitology*, 15:399–402.

Rohde, K., and M. Georgi. 1983. Structure and development of *Austramphilina elongata* Johnston, 1931 (Cestodaria, Amphilinidea). *International Journal for Parasitology*, 13:273–287.

Rohde, K., and N. Watson. 1986. Ultrastructure of spermatogenesis and sperm of *Austramphilina elongata* (Platyhelminthes, Amphilinidea). *Journal of Submicroscopic Cytology*, 18:361–374.

Rohde, K., and N. Watson. 1987. Ultrastructure of the protonephridial system of larval *Austramphilina elongata* (Platyhelminthes, Amphilinidea). *Journal of Submicroscopic Cytology and Pathology*, 19:113–118.

Rohde, K., and N. Watson. 1988a. Development of the protonephridia of *Austramphilina elongata*. *Parasitology Research*, 74:255–261.

Rohde, K., and N. Watson. 1988b. Ultrastructure of epidermis and sperm of the turbellarian *Syndisyrinx punicea* (Hickman, 1956) (Rhabdocoela: Umagillidae). *Australian Journal of Zoology*, 36:131–139.

Rohde, K., and N. Watson. 1989. Ultrastructural studies of larval and juvenile *Austramphilina elongata* (Platyhelminthes, Amphilinidea); penetration into, and early development in the intermediate host, *Cherax destructor*. *International Journal for Parasitology*, 19:529–538.

Rohde, K., and N. Watson. 1990. Epidermal and subepidermal structures in *Didymorchis* (Platyhelminthes, Rhabdocoela). I. Ultrastructure of epidermis and subepidermal cells. *Zoologischer Anzeiger*, 224:263–275.

Rohde, K., and N. Watson. 1991a. Ultrastructure of the flame bulbs and protonephridial capillaries of *Microstomum* sp. (Platyhelminthes, Macrostomida). *Acta Zoologica (Stockholm)*, 72:137–142.

Rohde, K., and N. Watson. 1991b. Ultrastructure of the flame bulbs and protonephridial capillaries of *Prorhynchus* (Lecithoepitheliata, Prorhynchidae, Turbellaria). *Zoologica Scripta*, 20:99–106.

Rohde, K., and N. A. Watson. 1992a. Ultrastructure of the flame bulbs and protonephridial capillaries of *Artioposthia* sp. (Platyhelminthes, Tricladida, Geoplanidae). *Acta Zoologica (Stockholm)*, 73:231–236.

Rohde, K., and N. A. Watson. 1992b. Ultrastructure of the developing protonephridial system of the cercaria of *Philophthalmus* sp. (Trematoda, Digenea). *Parasitology Research*, 78:368–375.

Rohde, K., and N. A. Watson. 1992c. Ultrastructure of the tegument, ventral sucker and rugae of *Rugogaster hydrolagi* (Trematoda: Aspidogastrea). *International Journal for Parasitology*, 22:967–974.

Rohde, K., and N. A. Watson. 1993a. Ultrastructure of the protonephridial system of regenerating *Stenostomum* sp. (Plathelminthes, Catenulida). *Zoomorphology*, 113:61–67.

Rohde, K., and N. A. Watson. 1993b. Ultrastructure of spermiogenesis and sperm of an undescribed species of Luridae (Platyhelminthes: Rhabdocoela). *Australian Journal of Zoology*, 41:13–19.

Rohde, K., and N. A. Watson. 1993c. Spermatogenesis in *Udonella* (Platyhelminthes, Udonellidea) and the phylogenetic position of the genus. *International Journal for Parasitology*, 23:725–735.

Rohde, K., and N. A. Watson. 1994a. Ultrastructure of the protonephridial system of *Suomina* sp. and *Catenula* sp. (Platyhelminthes, Catenulida). *Journal of Submicroscopic Cytology and Pathology*, 26:263–270.

Rohde, K., and N. A. Watson. 1994b. Ultrastructure of the terminal parts of the protonephridial system of *Baltoplana magna* (Platyhelminths, Kalyptorhynchia, Schizorhynchia, Karkinorhynchidae). *Malaysian Journal of Science*, 15A:13–18.

Rohde, K., and N. A. Watson. 1994c. Spermiogenesis in *Gonoplasius* sp. (Monogenea, Polyopisthocotylea, Microcotylidae). *Acta Parasitologica*, 39:111–116.

Rohde, K., and N. A. Watson. 1994d. Ultrastructure of spermiogenesis and spermatozoa of *Polylabroides australis* (Platyhelminthes, Monogenea, Polyopisthocotylea, Microcotylidae). *Parasite*, 1:115–122.

Rohde, K., and N. A. Watson. 1995. Comparative ultrastructural study of the posterior suckers of four species of symbiotic Platyhelminthes, *Temnocephala* sp. (Temnocephalida), *Udonella caligorum* (Udonellidea), *Anoplodiscus cirrusspiralis* (Monogenea: Monopisthocotylea), and *Philophthalmus* sp. (Trematoda: Digenea). *Folia Parasitologica*, 42:11–28.

Rohde, K., and N. A. Watson. 1996. Ultrastructure of the buccal complex of *Pricea multae* (Monogenea: Polyopisthocotylea, Gastrocotylidae). *Folia Parasitologica*, 43:117–132.

Rohde, K., and N. A. Watson. 1998. The terminal protonephridial complex of *Haplopharynx rostratus* (Platyhelminthes, Haplopharyngida). *Acta Zoologica (Stockholm)*, 79:329–333.

Rohde, K., N. Watson, and P. R. Garlick. 1986. Ultrastructure of three types of sense receptors of larval *Austramphilina elongata* (Amphilinidea). *International Journal for Parasitology*, 16:245–251.

Rohde, K., L. R. G. Cannon, and N. Watson. 1987a. Ultrastructure of epidermis, spermatozoa and flame cells of *Gyratrix* and *Odontorhynchus* (Rhabdocoela, Kalyptorhynchia). *Journal of Submicroscopic Cytology and Pathology*, 19:585–594.

Rohde, K., N. Watson, and L. R. G. Cannon. 1987b. Ultrastructure of the protonephridia of *Mesostoma* sp. (Rhabdocoela, Typhloplanoida). *Journal of Submicroscopic Cytology and Pathology*, 19:107–112.

Rohde, K., L. R. G. Cannon, and N. Watson. 1988a. Ultrastructure of the protonephridia of *Monocelis* (Proseriata, Monocelidae). *Journal of Submicroscopic Cytology and Pathology*, 20:425–435.

Rohde, K., L. R. G. Cannon, and N. Watson. 1988b. Ultrastructure of the flame bulbs and protonephridial capillaries of *Gieysztoria* sp. (Rhabdocoela, Dalyelliida), *Rhinolasius* sp. (Rhabdocoela, Kalyptorhynchia) and *Actinodactylella blanchardi* (Rhabdocoela, Temnocephalida). *Journal of Submicroscopic Cytology and Pathology*, 20:605–612.

Rohde, K., N. Watson, and L. R. G. Cannon. 1988c. Comparative ultrastructural studies of the epidermis, sperm and nerve fibres of *Cleistogamia longicirrus* and *Seritia stichopi* (Rhabdocoela, Umagillidae). *Zoologica Scripta*, 17:337–345.

Rohde, K., N. Watson, and L. R. G. Cannon. 1988d. Ultrastructure of spermiogenesis in *Amphiscolops* (Acoela, Convolutidae) and of sperm in *Pseudactinoposthia* (Acoela, Childiidae). *Journal of Submicroscopic Cytology and Pathology*, 20:595–604.

Rohde, K., N. Watson, and L. R. G. Cannon. 1988e. Ultrastructure of epidermal cilia of *Pseudactinoposthia* sp. (Platyhelminthes, Acoela); implications for the phylogenetic status of the Xenoturbellida and Acoelomorpha. *Journal of Submicroscopic Cytology and Pathology*, 20:759–767.

Rohde, K., J.-L. Justine, and N. Watson. 1989a. Ultrastructure of the flame bulbs of the monopisthocotylean Monogenea *Loimosina wilsoni* (Loimoidae) and *Calceostoma herculanea* (Calceostomatidae). *Annales de Parasitologie Humaine et Comparée*, 64:433–442.

Rohde, K., N. Watson, and F. Roubal. 1989b. Ultrastructure of flame bulbs, sense receptors, tegument and sperm of *Udonella* (Platyhelminthes) and the phylogenetic position of the genus. *Zoologischer Anzeiger*, 222:143–157.

Rohde, K., N. Watson, and F. Roubal. 1989c. Ultrastructure of the protonephridial system of *Dactylogyrus* sp. and an unidentified Ancyrocephaline (Monogenea, Dactylogyridae). *International Journal for Parasitology*, 19:859–864.

Rohde, K., N. Noury-Sraïri, N. Watson, J.-L. Justine, and L. Euzet. 1990a. Ultrastructure of the flame bulbs of *Urastoma cyprinae* (Platyhelminthes, 'Prolecithophora', Urastomidae). *Acta Zoologica (Stockholm)*, 71:211–216.

Rohde, K., N. Watson, and R. Sluys. 1990b. Ultrastructure of the flame bulbs and protonephridial capillaries of *Romankenkius* sp. (Platyhelminthes, Tricladida, Dugesiidae). *Journal of Submicroscopic Cytology and Pathology*, 22:489–496.

Rohde, K., N. A. Watson, and U. Jondelius. 1992a. Ultrastructure of the protonephridia of *Syndisyrinx punicea* (Hickman, 1956) (Rhabdocoela, Umagillidae) and *Pterastericola pellucida* Jondelius, 1989 (Rhabdocoela, Pterastericolidae). *Australian Journal of Zoology*, 40:385–399.

Rohde, K., N. A. Watson, and F. R. Roubal. 1992b. Ultrastructure of the protonephridial system of *Anoplodiscus cirrusspiralis* (Monogenea Monopisthocotylea). *International Journal for Parasitology*, 22:443–457.

Rohde, K., N. A. Watson, and A. Faubel. 1993a. Ultrastructure of the epidermis and protonephridium of an undescribed species of Luridae (Platyhelminthes: Rhabdocoela). *Australian Journal of Zoology*, 41:415–421.

Rohde, K., C. Hefford, J. T. Ellis, P. R. Baverstock, A. M. Johnson, N. A. Watson, and S. Dittmann. 1993b. Contributions to the phylogeny of Platyhelminthes based on partial sequencing of 18S ribosomal DNA. *International Journal for Parasitology*, 23:705–724.

Rohde, K., K. Luton, P. R. Baverstock, and A. M. Johnson. 1994. The phylogenetic relationships of *Kronborgia* (Platyhelminthes, Fecampiida) based on comparison of 18S ribosomal DNA sequences. *International Journal for Parasitology*, 24:657–669.

Rohde, K., A. M. Johnson, P. R. Baverstock, and N. A. Watson. 1995. Aspects of the phylogeny of Platyhelminthes based on 18S ribosomal DNA and protonephridial ultrastructure. *Hydrobiologia*, 305:27–35.

Rohde, K., N. A. Watson, and L. A. Chisholm. 1999. Ultrastructure of the eyes of the larva of *Neoheterocotyle rhinobatidis* (Platyhelminthes, Monopisthocotylea), and

phylogenetic implications. *International Journal for Parasitology*, 29:511–519.

Röhlich, P., and L. J. Török. 1961. Elektronmikroskopische Untersuchungen des Auges von Planarien. *Zeitschrift für Zellforschung und Mikroskopische Anatomie*, 54:362–381.

Rollinson, D., A. Kaukas, D. A. Johnston, A. J. G. Simpson, and M. Tanaka. 1997. Some molecular insights into schistosome evolution. *International Journal for Parasitology*, 27:11–28.

Ronquist, F. 1996. Matrix representation of trees, redundancy, and weighting. *Systematic Biology*, 45:247–253.

Roubal, F. R., and I. D. Whittington. 1990. Observations on the attachment by the monogenean, *Anoplodiscus australis*, to the caudal fin of *Acanthopagrus australis*. *International Journal for Parasitology*, 20:307–314.

Rouse, G. W., and K. Fauchald. 1997. Cladistics and polychaetes. *Zoologica Scripta*, 26:139–204.

Rudolphi, C. A. 1808. *Entozoorum sive vermium intestinalium historia naturalis*. Sumtibus Tabernae Librariae et Artium, Amstelaedami.

Ruhnke, T. R. 1993. A new species of *Clistobothrium* (Cestoda: Tetraphyllidea), with an evaluation of the systematic status of the genus. *Journal of Parasitology*, 79:37–43.

Ruhnke, T. R. 1994a. Resurrection of *Anthocephalum* Linton, 1890 (Cestoda: Tetraphyllidea) and taxonomic information on five proposed members. *Systematic Parasitology*, 29:159–176.

Ruhnke, T. R. 1994b. *Paraorygmatobothrium barberi* n. g., n. sp. (Cestoda: Tetraphyllidea), with amended descriptions of two species transferred to the genus. *Systematic Parasitology*, 28:65–79.

Ruhnke, T. R. 1996. Taxonomic resolution of *Phyllobothrium* van Beneden (Cestoda: Tetraphyllidea) and a description of a new species from the leopard shark *Triakis semifasciata*. *Systematic Parasitology*, 33:1–12.

Ruhnke, T. R., S. S. Curran, and T. Holbert. 2000. Two new species of *Duplicibothrium* Williams & Campbell, 1978 (Tetraphyllidea: Serendipidae) from the Pacific cownose ray *Rhinoptera steindachneri*. *Systematic Parasitology*, 47:135–143.

Ruiz-Trillo, I., M. Riutort, D. T. J. Littlewood, E. A. Herniou, and J. Baguñà. 1999. Acoel flatworms: earliest extant bilaterian metazoans, not members of Platyhelminthes. *Science*, 283:1919–1923.

Ruppert, E. E. 1978. A review of metamorphosis of turbellarian larvae. In F.–S. Chia and M. Rice (eds.), *Settlement and metamorphosis of marine invertebrate larvae*, pp. 65–81. Elsevier, New York.

Ruppert, E. E., and E. J. Balser. 1986. Nephridia in the larvae of hemichordates and echinoderms. *Biological Bulletin*, 171:188–196.

Ruppert, E. E., and P. R. Smith. 1988. The functional organization of filtration nephridia. *Biological Reviews of the Cambridge Philosophical Society*, 63:231–258.

Ruszkowski, J. S. 1928. Etudes sur le cycle évolutif et sur la structure des cestodes de mer. I.-*Echinobothrium benedeni* n. sp., ses larves et son hôte intermédiaire *Hippolyte varians* Leach. *Bulletin de l'Académie Polonaise des Sciences et des Lettres. Classe des Sciences Mathématiques et Naturelles, Série B: Sciences naturelles*, 7:719–738.

Ruszkowski, J. S. 1934. Études sur le cycle évolutif et sur la structure des cestodes de mer. 3ème partie. Le cycle évolutif du tétrarhynque *Grillotia erinaceus* van Beneden. *Mémoires de l'Académie Polonaise des Sciences et des Lettres. Classe des Sciences mathématiques et naturelles. Série B. Sciences naturelles*, 6:1–10.

Rybicka, K. 1966. Embryogenesis in cestodes. *Advances in Parasitology*, 4:107–178.

Saitou, N., and M. Nei. 1987. The neighbor–joining method: a new method for reconstructing phylogenetic trees. *Molecular Biology and Evolution*, 4:406–425.

Sakanari, J. A., and M. Moser. 1989. Complete life cycle of the elasmobranch cestode, *Lacistorhynchus dollfusi* Beveridge and Sakanari, 1987 (Trypanorhyncha). *Journal of Parasitology*, 75:806–808.

Saladin, K. S. 1982. Cercarial responses to irradiance changes. *Journal of Parasitology*, 68:120–124.

Salvini-Plawen, L. v. 1978. On the origin and evolution of the lower Metazoa. *Zeitschrift für zoologische Systematik und Evolutionsforschung*, 16:40–88.

Sandeman, I. M., and M. D. B. Burt. 1972. Biology of *Bothrimonus* (=*Diplocotyle*) (Pseudophyllidea: Cestoda): ecology, life cycle, and evolution; a review and synthesis. *Journal of the Fisheries Research Board of Canada*, 29:1381–1395.

Sanderson, M. J., A. Purvis, and C. Henze. 1998. Phylogenetic supertrees: assembling the trees of life. *Trends in Ecology and Evolution*, 13:105–109.

Santamaría-Fríes, M., L. F. Fajardo L-G, M. Sogin, P. D. Olson, and D. A. Relman. 1996. Lethal infection by a previously unrecognised parasite. *The Lancet*, 347:1797–1801.

Santos, C. P., T. Souto-Padrón. and R. M. Lanfredi. 1997. Ultrastructure of spermatogenesis of *Atriaster heterodus* (Platyhelminthes, Monogenea, Polyopisthocotylea). *Journal of Parasitology*, 83:1007–1014.

Santos, T. M., D. A. Johnston, V. Azevedo, I. L. Ridgers, M. F. Martinez, G. B. Marotta, R. L. Santos, S. J. Fonseca, J. M. Ortega, E. M. Rabelo, M. Saber, H. M. Ahmed, M. H. Romeih, G. R. Franco, D. Rollinson, and S. D. Pena. 1999. Analysis of the gene expression profile of *Schistosoma mansoni* cercariae using the expressed sequence tag approach. *Molecular and Biochemical Parasitology*, 103:79–97.

Schander, C., and M. Tholleson. 1995. Phylogenetic taxonomy – some comments. *Zoologica Scripta*, 24:263–268.

Schell, S. C. 1973. *Rugogaster hydrolagi* gen. et sp. n. (Trematoda: Aspidobothrea: Rugogastridae fam. n.) from the ratfish, *Hydrolagus colliei* (Lay and Bennett, 1839). *Journal of Parasitology*, 59:803–805.

Schell, S. C. 1975. The miracidium of *Lecithaster salmonis* Yamaguti, 1934 (Trematoda: Hemiuroidea). *Journal of Parasitology*, 63:562–563.

Schell, S. C. 1978. *Trematodes of North America*. University Press of Idaho, Moscow, Idaho.

Schell, S. C. 1985. *Handbook of trematodes of North America north of Mexico*. University of Idaho Press, Moscow.

Schierenberg, N. 1997. Nematodes, the roundworms. In S. F. Gilbert and A. M. Raunio (eds.), *Embryology. Constructing the organism*, pp. 131–150. Sinauer and Associates, Sunderland, MA.

Schmidt, G. D. 1986. *Handbook of tapeworm identification*. CRC Press, Boca Raton.

Schmidt, G. D., and L. S. Roberts. 1977. *Foundations of parasitology*. 1st ed. C.V. Mosby Company, Saint Louis.

Schmidt, P., and B. Sopott-Ehlers. 1976. Interstitielle Fauna von Galapagos XV. *Macrostomum* O. Schmidt 1848 und *Siccomacrostomum triviale* nov. gen. nov. spec. (Turbellaria, Macrostomida). *Mikrofauna Meeresboden*, 57:1–45.

Schmidt-Rhaesa, A. 1993. Ultrastructure and development of the spermatozoa of *Multipeniata* (Plathelminthes, Prolecithophora). *Microfauna Marina*, 8:131–138.

Scholz, T. 1999. Life cycles of species of *Proteocephalus* Weinland 1858 (Cestoda: Proteocephalidae), parasites of fishes in the Palearctic Region: a review. *Journal of Helminthology*, 73:1–20.

Schram, F. R. 1991. Cladistic analysis of metazoan phyla and the placement of fossil problematica. In A. M. Simonetta and S. Conway Morris (eds.), *The early evolution of Metazoa and the significance of problematic taxa*, pp. 35–46. Cambridge University Press, Cambridge.

Schuchert, P., and R. M. Rieger. 1990a. Ultrastructural examination of spermatogenesis in *Retronectes atypica* (Catenulida, Platyhelminthes). *Journal of Submicroscopic Cytology and Pathology*, 22:379–387.

Schuchert, P., and R. M. Rieger. 1990b. Ultrastructural observations on the dwarf male of *Bonellia viridis* (Echiura). *Acta Zoologica (Stockholm)*, 71:5–16.

Schulenburg, J. H. G. v. d., and J.-W. Wägele. 1998. Molecular characterization of digenetic trematodes associated with *Cyathura carinata* (Crustacea: Isopoda) with a note on the utility of 18S ribosomal DNA for phylogenetic analysis in the Digenea (Platyhelminthes: Trematoda). *International Journal for Parasitology*, 28:1425–1428.

Schulenburg, J. H. G. v. d., U. Englisch, and J. W. Wägele. 1999. Evolution of ITS1 rDNA in the Digenea (Platyhelminthes: Trematoda): 3′ end sequence conservation and its phylogenetic utility. *Journal of Molecular Evolution*, 48:2–12.

Scotland, R., and T. R. Pennington. 2000. Homology and systematics. Coding characters for phylogenetic analysis. *The systematics association special volume series*, Vol. 58, pp. 217. Taylor & Francis, London.

Scott, J. C., and D. P. McManus. 1994. The random amplification of polymorphic DNA can discriminate species and strains of *Echinococcus*. *Tropical Medicine and Parasitology*, 45:1–4.

Semenov, O. Y. 1991. [Miracidia: their structure, biology and interrelationships with molluscs]. (In Russian). *Trudy Leningradskogo Obshchestva Estestvoispytatelei*, 83:1–204.

Semper, C. 1872. Zoologische Aphorismen II. Über de Gattung *Temnocephala*. *Zeitschrift für Wissenschaftliche Zoologie*, 22:307–310.

Sène, A., C. T. Bâ, and B. Marchand. 1997. Ultrastructure of spermiogenesis and the spermatozoon of *Nomimoscolex* sp. (Cestoda, Proteocephalidea) intestinal parasite of *Clarotes laticeps* (Fish, Teleost) in Senegal. *Journal of Submicroscopic Cytology and Pathology*, 29:1–6.

Sène, A., C. T. Bâ, J. Miquel, and B. Marchand. 1998. Implication of *Phyllobothrium lactuca* (Cestoda, Tetraphyllidea, Phyllobothriidae) spermiogenesis on phylogenesis within the Tetraphyllidea. *Wiadomosci Parazytologiczne*, 44:600.

Sène, A., C. T. Bâ, and B. Marchand. 1999. Ultrastructure of spermiogenesis of *Phyllobothrium lactuca* (Cestoda, Tetraphyllidea, Phyllobothriidae). *Folia Parasitologica*, 46:191–198.

Sereno, P. 1999. Definitions in phylogenetic taxonomy: critique and rationale. *Systematic Biology*, 48:329–351.

Sewell, K. B. 1998. *Craspedella pedum* (Craspedellinae: Temnocephalida): a model for ectosymbiosis. Ph.D. thesis, University of Queensland.

Sewell, K. B., and L. R. G. Cannon. 1995. A scanning electron microscope study of *Craspedella* sp. from the branchial chamber of redclaw crayfish, *Cherax quadricarinatus*, from Queensland, Australia. *Hydrobiologia*, 305:151–158.

Sewell, K. B., and L. R. G. Cannon. 1998a. New temnocephalans from the branchial chamber of Australian *Euastacus* and *Cherax* crayfish hosts. *Proceedings of the Linnean Society of New South Wales*, 119:21–36.

Sewell, K. B., and L. R. G. Cannon. 1998b. The taxonomic status of the ectosymbiotic flatworm *Didymorchis paranephropis* Haswell. *Memoirs of the Queensland Museum*, 42:585–595.

Sewell, K. B., and I. D. Whittington. 1995. A light microscope study of the attachment organs and their role in locomotion in *Craspedella* sp. (Platyhelminthes: Rhabdocoela: Temnocephalidae), an ectosymbiont from the branchial chamber of redclaw crayfish, *Cherax quadricarinatus* (Crustacea: Parastacidae), from Queensland, Australia. *Journal of Natural History*, 29:1121–1141.

Shameem, U., and R. Madhavi. 1991. Observations on the life-cycles of two haploporid trematodes, *Carassotrema bengalense* Rekharani & Madhavi, 1985 and *Saccocoelioides martini* Madhavi, 1979. *Systematic Parasitology*, 20:97–107.

Shapiro, J. E., B. R. Hershenov, and G. S. Tulloch. 1961. The fine structure of *Haematoloechus* spermatozoan tail. *Journal of Biophysical and Biochemical Cytology*, 9:211–217.

Shapkin, V. A., and V. D. Gulyaev. 1973. [On the biology of cestodes of the genus *Lateriporus* Fuhrmann]. (In Russian). *Parazitologiya*, 7:509–512.

Sharpilo, V. P., and N. I. Iskova. 1989. [The fauna of the Ukraine. Trematodes: Plagiorchiata]. (In Russian). *Fauna Ukrainy*, 34:1–278.

Shibata, N., H. Umesono, H. Orii, T. Sakurai, K. Eatanabe, and K. Agata. 1999. Expression of vasa (vas)–related genes in germline cells and totipotent somatic stem cells of planarians. *Developmental Biology*, 206:73–87.

Shinn, A. P., D. I. Gibson, and C. Sommerville. 1998. Chaetotaxy of members of the Gyrodactylidae (Monogenea), with comments upon their systematic relationships with the Monopisthocotylea and Polyopisthocotylea. *Systematic Parasitology*, 39:81–94.

Shishov, B. A. 1991. Aminergic elements in the nervous system of helminths. In D. A. Sakharov and W. Winlow (eds.), *Simpler nervous systems*, Vol. 13, pp. 113–137. Manchester University Press, Manchester.

Short, R. B. 1983. Sex and the single schistosome. *Journal of Parasitology*, 69:3–22.

Short, R. B., and H. T. Gange. 1975. Fine structure of possible photoreceptors in cercariae of *Schistosoma mansoni*. *Journal of Parasitology*, 61:69–74.

Short, R. B., and M. Y. Menzel. 1979. Somatic chromosomes of *Schistosoma mansoni*. *Journal of Parasitology*, 65:471–473.

Short, R. B., J. D. Liberatos, W. H. Teehan, and J. I. Bruce. 1989. Giemsa-stained and C-banded chromosomes of seven strains of *Schistosoma mansoni*. *Journal of Parasitology*, 75:920–926.

Siddall, M. E. 1997. Prior agreement: arbitration or arbitrary? *Systematic Biology*, 46:765–769.

Siddall, M. E. 1998. Success of parsimony in the four–taxon case: long branch repulsion by likelihood in the Farris Zone. *Cladistics*, 14:209–220.

Siewing, R. 1980a. Das Archicoelomatenkonzept. *Zoologische Jahrbücher Abteilung für Systematik*, 103:439–482.

Siewing, R. 1980b. Körpergliederung und phylogenetisches System. *Zoologische Jahrbücher Abteilung für Systematik*, 103:196–210.

Siles-Lucas, M., C. Cuesta-Bandera, and M. César-Benito. 1993. Random amplified polymorphic DNA technique for speciation studies of *Echinococcus granulosus*. *Parasitology Research*, 79:343–345.

Siles-Lucas, M., L. Turcekova, and C. Cuesta-Bandera. 1995. A genetic comparison of *Proteocephalus exiguus* and *P. percae* by RAPD technique. *Helminthologia*, 32:201–203.

Silveira, M. 1967. Formation of structured secretory granules within the Golgi complex in an acoel Turbellarian. *Journal de Microscopie (Paris)*, 6:95–100.

Silveira, M. 1969. Ultrastructural studies on a 'nine plus one' flagellum. *Journal of Ultrastructure Research*, 26:274–288.

Silveira, M. 1975. The fine structure of 9 + 1 flagella in turbellarian flatworm. In B. A. Afzelius (ed.), *The functional anatomy of the spermatozoon*, pp. 289–298. Pergamon Press, Oxford.

Silveira, M. 1993. The protonephridium of catenulids re-evaluated: an ultrastructural study on *Stenostomum grande* Child, 1902 (Platyhelminthes: Catenulida). *Journal of Submicroscopic Research*, 27:525–533.

Simmons, J. E. 1974. *Gyrocotyle*: a century-old enigma. In W. B. Vernberg (ed.), *Symbiosis in the sea*, pp. 195–218. University of South Carolina Press, Columbia.

Simmons, J. E., and J. S. Laurie. 1972. A study of *Gyrocotyle* in the San Juan Archipelago, Puget Sound, U.S.A., with observations on the host, *Hydrolagus colliei* (Lay and Bennet). *International Journal for Parasitology*, 2:59–77.

Simon, C., F. Frati, A. Beckenback, B. Crespi, H. Liu, and P. Flook. 1994. Evolution, weighting, and phylogenetic utility of mitochondrial gene sequences and a compilation of conserved PCR primers. *Annals of the Entomological Society of America*, 87:651–701.

Simpson, A. J. G., A. Sher, and T. F. McCutchan. 1982. The genome of *Schistosoma mansoni*: isolation of DNA, its size, bases, and repetitive sequences. *Molecular and Biochemical Parasitology*, 6:125–137.

Sinitsin, D. F. 1911. [Parthenogenetic generation of trematodes and its progeny in molluscs of the Black Sea]. (In Russian). *Mémoires de l'Académie Impériale des Sciences de St Petersbourg (Sci. Math.-Phys. et Nat.)*, 30:1–127 (English translation by A.M. Bagusin).

Skarbilovich, T. S. 1948. [Family Lecithodendriidae Odhner, 1911]. (In Russian). *Osnovy Trematodologii. Moscow-Leningrad: Izdatel'stvo AN SSSR*, 2:37–590.

Skrjabin, K. I. 1940. [Revision of systematics of cestodes of the order Cyclophyllidea (tapeworms)]. (In Russian). *Zoologicheskii Zhurnal*, 19:3–13.

Skrjabin, K. I., and D. N. Antipin. 1958. [Superfamily Plagiorchioidea Dollfus, 1930]. (In Russian). *Osnovy Trematodologii. Moscow-Leningrad: Izdatel'stvo AN SSSR*, 14:75–634.

Skuce, P. J., C. F. Johnston, I. Fairweather, D. W. Halton, and C. Shaw. 1990. A confocal scanning laser microscopic study of the peptidergic and serotoninergic components of the nervous system in larval *Schistosoma mansoni*. *Parasitology*, 101:227–234.

Šlais, J. 1973. Functional morphology of the cestode larvae. *Advances in Parasitology*, 11:395–480.

Slowinski, J. B., and R. D. M. Page. 1999. How should species phylogenies be inferred from sequence data? *Systematic Biology*, 48:814–825.

Sluys, R. 1989a. Phylogenetic relationships of the triclads (Platyhelminthes, Seriata, Tricladida). *Bijdragen tot de Dierkunde*, 59:3–25.

Sluys, R. 1989b. *A monograph of the marine triclads*. A. A. Balkema, Rotterdam, The Netherlands.

Sluys, R. 1990. A monograph of the Dimarcusidae (Platyhelminthes, Seriata, Tricladida). *Zoologica Scripta*, 19:13–29.

Sluys, R. 1992. A synopsis of Antarctic plagiostomids (Platyhelminthes, Prolecithophora), with the description of a new species and remarks on taxonomy, phylogeny, and biogeography. *Polar Biology*, 12:507–518.

Sluys, R. 1997. An old problem in a new perspective: The enigmatic evolutionary relationships of some Australian freshwater planarians (Platyhelminthes, Tricladida, Paludicola). *Canadian Journal of Zoology*, 75:459–471.

Sluys, R. 2001. Towards a classification and characterization of dugesiid genera (Platyhelminthes, Tricladida, Dugesiidae) – a morphological perspective. In D. T. J. Littlewood and R. A. Bray (eds.), *Interrelationships of the Platyhelminthes*, pp. 57–73. Taylor & Francis, London.

Sluys, R., J. Hauser, and Q. J. Wirth. 1997. Deviation from the groundplan: A unique new species of freshwater planarian from South Brazil (Platyhelminthes, Tricladida, Paludicola). *Journal of Zoology*, 241:593–601.

Sluys, R., M. Kawakatsu, and L. Winsor. 1998a. The genus *Dugesia* in Australia, with its phylogenetic analysis and historical biogeography (Platyhelminthes, Tricladida, Dugesiidae). *Zoologica Scripta*, 27:273–289.

Sluys, R., O. A. Timoshkin, and M. Kawakatsu. 1998b. A new species of giant planarian from Lake Baikal, with some remarks on character states in the Dendrocoelidae (Platyhelminthes, Tricladida, Paludicola). *Hydrobiologia*, 383:69–75.

Smith, A. B. 1994. Rooting molecular trees: problems and strategies. *Biological Journal of the Linnean Society*, 51:279–292.

Smith, J. P. S., and S. Tyler. 1985. The acoel turbellarians: kingpins of metazoan evolution or a specialized offshoot? In S. Conway Morris, J. D. George, R. Gibson and H. M. Platt (eds.), *The origins and relationships of the lower invertebrates*, pp. 123–142. Clarendon Press, Oxford.

Smith, J. P. S., and S. Tyler. 1986. Frontal organs in the Acoelomorpha (Turbellaria): ultrastructure and phylogenetic significance. *Hydrobiologia*, 132:71–78.

Smith, J. P. S., and S. Tyler. 1988. Frontal organs in the Nemertodermatida (Turbellaria). *American Zoologist*, 28:A140.

Smith, J. P. S., S. Tyler, and R. M. Rieger. 1986. Is the Turbellaria polyphyletic? *Hydrobiologia*, 132:13–21.

Smith, J. P. S., M. B. Thomas, R. Chandler, and S. F. Zane. 1988. Granular inclusions in the oocytes of *Convoluta* sp., *Nemertoderma* sp., and *Nemertinoides elongatus* (Turbellaria: Acoelomorpha). *Fortschritte der Zoologie*, 36.

Smith, J. P. S., S. Tyler, D. Boatwright, and K. Lundin. 1994. Rhabdite-like secretions in the Acoelomorpha: evidence for monophyly? *Transactions of the American Microscopical Society*, 28:140A.

Smith, S. W., R. Overbeek, C. R. Woese, W. Gilbert, and P. M. Gillevet. 1994. The Genetic Data Environment: an expandable guide for multiple sequence analysis. *Computer Applications in the Biosciences*, 10:671–675.

Smyth, J. D., and D. W. Halton. 1983. *The physiology of trematodes*. 2nd ed. Cambridge University Press, Cambridge.

Snábel, V., V. Hanzelová, and H.-P. Fagerholm. 1994. Morphological and genetic comparison of two *Proteocephalus* species (Cestoda: Proteocephalidae). *Parasitology Research*, 80:141–146.

Snábel, V., V. Hanzelová, S. Mattiucci, S. D'Amelio, and L. Paggi. 1996. Genetic polymorphism in *Proteocephalus exiguus* shown by enzyme electrophoresis. *Journal of Helminthology*, 70:345–349.

Snyder, S. D., and E. S. Loker. 2000. Evolutionary relationships among the Schistosomatidae (Platyhelminthes: Digenea) and an Asian origin for *Schistosoma*. *Journal of Parasitology*, 86:283–288.

Soboleva, T. N., Z. Zdárská, J. Sterba, and J. Valkounova. 1988. Ultrastructure of the excretory system of *Brachylaimus aequans* (Trematoda: Brachylaimoidea). *Folia Parasitologica*, 35:335–339.

Sokal, R. R., and F. J. Rohlf. 1981. Taxonomic congruence in the Leptopodomorpha re-examined. *Systematic Zoology*, 30:309–325.

Solis-Soto, J. M., and M. de Jong Brink. 1994. Immunocytochemical study on biologically active neurosubstances in daughter sporocysts and cercariae of *Trichobilharzia ocellata* and *Schistosoma mansoni*. *Parasitology*, 108:301–311.

Soltis, P. S., and D. E. Soltis. 1998. Molecular evolution of 18S in rDNA in angiosperms: implications for character weighting in phylogenetic analysis. In D. E. Soltis, P. S. Soltis and J. J. Doyle (eds.), *Molecular systematics of plants II*, pp. 188–210. Kluwer Academic Publishers, Boston.

Sopott, B. 1972. Systematik und Ökologie vor Proseriaten (Turbellaria) der deutschen Nord-seeküste. *Mikrofauna Meeresboden*, 13:1–72.

Sopott, B. 1973. Jahreszeitliche Verteilung und Lebenszyklen der Proseriata (Turbellaria) eines Sandstrandes der Nordseeinseln Sylt. *Mikrofauna Meeresboden*, 15:1–106.

Sopott-Ehlers, B. 1979. Ultrastruktur der Haftapparate von *Nematoplana coelogynoporoides* (Turbellaria, Proseriata). *Helgoländer Wissenschaftliche Meeresuntersuchungen*, 32:365–373.

Sopott-Ehlers, B. 1985. The phylogenetic relationships within the Seriata (Platyhelminthes). In S. Conway Morris, J. D. George, R. Gibson and H. M. Platt (eds.), *The origin and relationships of lower invertebrates*, pp. 159–167. Clarendon Press, Oxford.

Sopott-Ehlers, B. 1986. Fine structural characteristics of female and male germ cells in Proseriata Otoplanidae (Platyhelminthes). *Hydrobiologia*, 132:137–144.

Sopott-Ehlers, B. 1990. Functional aspects of the intercentriolar body in the spermiogenesis of *Nematoplana coelogynoporoides* (Plathelminthes, Proseriata). *Zoomorphology*, 109:245–249.

Sopott-Ehlers, B. 1992. *Coelogynopora faenofurca* nov. spec. (Proseriata, Plathelminthes) aus Wohnröhren des Polychaeten *Arenicola marina*. *Microfauna Marina*, 7:185–190.

Sopott-Ehlers, B. 1993. Ultrastructural observations on the spermiogenesis in the Parotoplaninae (Plathelminthes, Proseriata) with special reference to a striated appendage in the intercentriolar body. *Zoomorphology*, 113:191–197.

Sopott-Ehlers, B. 1994a. Electronmicroscopical observations on the development of male gametes in the Monocelididae *Monocelis lineata* and *Promonotus marci* (Plathelminthes, Proseriata). *Microfauna Marina*, 9:25–43.

Sopott-Ehlers, B. 1994b. Fine structure of spermatozoa in *Anthopharynx sacculipenis* (Plathelminthes, Solenopharyngidae). *Zoomorphology*, 114:33–38.

Sopott-Ehlers, B. 1995. Some data on the ultrastructure of the eyes and spermatozoa in *Provortex psammophilus* (Plathelminthes, Rhabdocoela, 'Dalyellioida'). *Microfauna Marina*, 10:307–312.

Sopott-Ehlers, B. 1996. First evidence of mitochondrial lensing in two species of the 'Typhloplanoida' (Plathelminthes, Rhabdocoela): phylogenetic implications. *Zoomorphology*, 116:95–101.

Sopott-Ehlers, B. 1997a. Submicroscopical anatomy of female gonads in *Ciliopharyngiella intermedia* (Plathelminthes, Rhabdocoela, 'Typhloplanoida'). *Microfauna Marina*, 11:209–221.

Sopott-Ehlers, B. 1997b. Fine-structural features of male and female gonads in *Jensenia angulata* (Plathelminthes,

Rhabdocoela, 'Dalyellioida'). *Microfauna Marina*, 11:251–270.

Sopott-Ehlers, B. 1998. Development and submicroscopic anatomy of male gametes in *Halammovortex nigrofrons* (Plathelminthes, Rhabdocoela, 'Dalyellioida'): phylogenetic implications and functional aspects. *Zoomorphology*, 117:213–222.

Sopott-Ehlers, B., and U. Ehlers. 1986. Differentiation of male and female germ cells in neoophoran Plathelminthes. In M. Porchet, J.-C. Andries and A. Dhainaut (eds.), *Advances in invertebrate reproduction 4*, pp. 187–194. Elsevier Science (Biomedical Division), Amsterdam.

Sopott-Ehlers, B., and U. Ehlers. 1995a. Ultrastructural features of *Bresslauilla relicta* (Plathelminthes, Rhabdocoela). The spermatozoa. *Microfauna Marina*, 10:235–247.

Sopott-Ehlers, B., and U. Ehlers. 1995b. Modified sperm ultrastructure and some data on spermiogenesis in *Provortex tubiferus* (Plathelminthes, Rhabdocoela): phylogenetic implications for the Dalyellioida. *Zoomorphology*, 115:41–49.

Sopott-Ehlers, B., and U. Ehlers. 1997. Electronmicroscopical investigations of male gametes in *Ptychopera westbladi* (Plathelminthes, Rhabdocoela, 'Typhloplanoida'). *Microfauna Marina*, 11:193–208.

Sopott-Ehlers, B., and U. Ehlers. 1998a. Characteristics of spermiogenesis and mature spermatozoa in *Gyratrix hermaphroditus* (Plathelminthes, Rhabdocoela, Kalyptorhynchia). *Zoomorphology*, 118:169–176.

Sopott-Ehlers, B., and U. Ehlers. 1998b. Spermiogenesis and submicroscopic morphology of spermatozoa in *Myozona purpurea* (Platyhelminthes, Macrostomida). *Invertebrate Reproduction and Development*, 34:309–320.

Sopott-Ehlers, B., and U. Ehlers. 1999. Ultrastructure of spermiogenesis and spermatozoa in *Bradynectes sterreri* and remarks on sperm cells in *Haplopharynx rostratus* and *Paramalostomum fusculum*: phylogenetic implication for the Macrostomorpha (Plathelminthes). *Zoomorphology*, 119:105–115.

Sopott-Ehlers, B., and P. Schmidt. 1974. Interstitielle Fauna von Galapagos XII. *Myozona* Marcus (Turbellaria, Macrostomida). *Mikrofauna Meeresboden*, 46:1–19.

Sorensen, R. E., J. Curtis, and D. J. Minchella. 1998. Intraspecific variation in the rDNA ITS loci of 37-collarspined echinostomes from North America: implications for sequence-based diagnoses and phylogenetics. *Journal of Parasitology*, 84:992–997.

Southgate, V. R. 1970. Observations on the epidermis of the miracidium and on the formation of the tegument of the sporocyst of *Fasciola hepatica*. *Parasitology*, 61:177–190.

Spasskii, A. A. 1951. [Anoplocephalate tapeworms of animals and man]. (In Russian). *Osnovy Tsestodologii*, 1:1–735 [English translation, Israel Program for Scientific Translations, 1961].

Spasskii, A. A. 1958. [A short analysis of the system of cestodes]. (In Russian). *Ceskoslovenská Parasitologie*, 5:163–171.

Spasskii, A. A. 1992. [On classification of cestodes]. (In Russian). *Sel'skokhozyaistzennaya Biologiya*, 6:107–114.

Spasskii, A. A., and L. P. Spasskaya. 1973. [Gryporhinchinae n. subf.: Cestoda, Cyclophyllidea]. (In Russian). *Izvestia Akademia Nauk Moldavskoi SSR. Bullitin Akademia Biologii i Khimii, Nauki*, 5:56–58.

Sproston, N. G. 1946. A synopsis of the monogenetic trematodes. *Transactions of the Zoological Society of London*, 45:185–600.

Steel, M. 1992. The complexity of reconstructing trees from qualitative characters and subtrees. *Journal of Classification*, 9:91–116.

Steenstrup, J. J. 1842. *On the alternation of generations, or, the propagation and development of animals through alternate generations: a peculiar form of fostering the young of lower classes of animal.* (Original in Danish, English translation by G. Busk, 1845). Ray Society, London.

Steinböck, O. 1931. Erbebnisse einer von E. Reisinger & O. Steinböck mit Hilfe der Rask-Ørsted Fonds durchgeführten Reise in Grönland 1926. 2. *Nemertoderma bathycola* nov. gen. nov. spec., eine eizigartige Turbellarie aus der Tiefe der Diskobay; nebst einem Beitrag zur Kenntis des Nemertinenepithels. *Videnskablig Meddelende fra Dansk naturhistorisk Forening*, 90:47–84.

Steinböck, O. 1933. Die Turbellarien der Umgebung von Rovigno. *Thalassia*, 1:1–32.

Steinböck, O. 1938. Über die Stellung der Gattung *Nemertoderma* Steinböck im System der Turbellarien. *Acta Societatis pro Fauna et Flora Fennica*, 62:1–26.

Steinböck, O. 1966. Die Hofsteniiden (Turbellaria Acoela). Grundsätzliches zur Evolution der Turbellarien. *Zeitschrift für zoologische Systematik und Evolutionsforschung*, 4:58–195.

Steinböck, O. 1967. Regenerationsversuche mit *Hofstenia giselae* Steinb. (Turbellaria Acoela). *Roux' Archiv für Entwicklungsmechanik*, 158:394–458.

Sterrer, W. 1966. New polylithophorous marine Turbellaria. *Nature*, 210:436.

Sterrer, W. 1970. Turbellaria. In R. Riedl (ed.), *Fauna und Flora der Adria*, pp. 196–201. Parey, Hamburg.

Sterrer, W. 1998. New and known Nemertodermatida (Platyhelminthes-Acoelomorpha) – a revision. *Belgian Journal of Zoology*, 128:55–92.

Sterrer, W., and R. M. Rieger. 1974. Rectronectidae: new cosmopolitan marine family of Catenulida (Turbellaria). In N. W. Riser and M. P. Morse (eds.), *Biology of the Turbellaria*, pp. 63–92. McGraw-Hill, New York.

Stiassny, M. L. J., L. R. Parenti, and G. D. Johnson. 1996. *Interrelationships of fishes*. Academic Press, San Diego.

Stiller, J. W., and B. D. Hall. 1999. Long-branch attraction and the rDNA model of early eukaryotic evolution. *Molecular Biology and Evolution*, 16:1270–1279.

Stimpson, W. 1857. Prodromus descriptionis animalium evertobratorum quae in Expeditione ad Oceanum, Pacificum Septentrionalem a Republica Federata missa, 'Johanne Rodgers Duce, observavit et descripsit. *Proceedings of the Academy of Natural Sciences of Philadelphia*, 9:19–31.

Stoitsova, S. R., B. B. Georgiev, and R. B. Dacheva. 1995. Ultrastructure of spermiogenesis and the mature spermatozoon of *Tetrabothrius erostris* Loennberg, 1896 (Cestoda, Tetrabothriidae). *International Journal for Parasitology*, 25:1427–1436.

Strimmer, K., and A. von Haeseler. 1996. Quartet puzzling: a quartet maximum–likelihood method for reconstructing tree topologies. *Molecular Biology and Evolution*, 13:964–969.

Strong, E. E., and D. Lipscomb. 1999. Character coding and inapplicable data. *Cladistics*, 15:363–371.

Stunkard, H. W. 1961. *Cycloskrjabinia taborensis* (Loewen, 1934), a cestode from the red bat, *Lasiurus borealis* (Müller, 1776), and a review of the family Anoplocephalidae. *Journal of Parasitology*, 41:847–855.

Stunkard, H. W. 1975. Life histories and systematics of parasitic flatworms. *Systematic Zoology*, 24:378–385.

Swiderski, Z. 1972. La structure fine de l'oncosphère du cestode *Catenotaenia pusilla* (Goeze, 1782) (Cyclophyllidea, Catenotaeniidae). *La Cellule*, 69:207–237.

Swiderski, Z. 1981. Reproductive and developmental biology of the cestodes. In W. A. Clark and T. S. Adams (eds.),

Advances in invertebrate reproduction, pp. 365–366. Elsevier/North Holland Biomedical Press, New York.

Swiderski, Z. 1983. *Echinococcus granulosus*: hook-muscle systems and cellular organization of infective oncospheres. *International Journal for Parasitology*, 13:189–299.

Swiderski, Z. 1994. Origin, differentiation and ultrastructure of egg envelopes surrounding the coracidia of *Bothriocephalus clavibothrium* (Cestoda, Pseudophyllidea). *Acta Parasitologica*, 39:73–81.

Swiderki, Z., and D. T. Eklu-Natey. 1978. Fine structure of the spermatozoon of *Proteocephalus longicollis* (Cestoda, Proteocephalidea). *Proceedings of the 9th International Congress of Electron Microscopy*: 572–573.

Swiderski, Z., and F. Mokhtar-Maamouri. 1980. Étude de la spermatogenese de *Bothriocephalus clavibothrium* Ariola, 1899 (Cestoda: Pseudophyllidea). *Archives de l'Institute Pasteur de Tunis*, 57:323–347.

Swiderski, Z., and V. V. Tkach. 1997. Ultrastructural studies on the cellular organisation of the oncosphere of the nematotaeniid cestode, *Nematotaenia dispar* (Goeze, 1782). *Acta Parasitologica*, 42:158–167.

Swiderski, Z., L. Euzet, and N. Schönenberger. 1975. Ultrastructures du système nephridien des cestodes cyclophyllides *Catenotaenia pusilla* (Goeze 1782), *Hymenolepis diminuta* (Rudolphi 1819) et *Inermicapsifer madagascariensis* (Davaine 1870) Baer 1956. *Le Cellule*, 71:7–18.

Swofford, D. L. 1993. PAUP: Phylogenetic Analysis Using Parsimony. Illinois Natural History Survey, Champaign.

Swofford, D. L. 1998. PAUP*. Phylogenetic Analysis Using Parsimony *and other methods. Sinauer Associates Inc., Sunderland, MA.

Swofford, D. L., G. J. Olsen, P. J. Waddell, and D. M. Hillis. 1996. Phylogenetic inference. In D. M. Hillis, C. Moritz and B. K. Mable (eds.), *Molecular systematics*, pp. 407–514. Sinauer Associates Inc., Sunderland, MA.

Syromjatnikova, I. P. 1949. [A new turbellarian parasitizing fish, *Ichthyophaga subcutanea* nov. gen. nov. sp.]. (In Russian). *Doklady Academii Nauk SSSR*, 68:805–808.

Talent, J. A. 1986. Australian biogeography past and present: determinants and implications. In J. J. Veevers (ed.), *Phanerozoic early history of Australia*, pp. 42–47. Clarendon Press, Oxford.

Taliaferro, W. H. 1920. Reactions to light in *Planaria maculata*, with special reference to the function and structure of eyes. *Journal of Experimental Zoology*, 31:59–116.

Tang, Z.-z., and C.-t. Tang. 1980. [Life histories of two species of aspidogastrids and the phylogeny of the group]. (In Chinese). *Acta Hydrobiologica Sinica*, 7:153–174.

Tappenden, T., and G. C. Kearn. 1991. Spermiogenesis and sperm ultrastructure in the monocotylid monogenean parasite *Calicotyle kroyeri*. *International Journal for Parasitology*, 21:57–63.

Taylor, J. D. (ed.) 1996. Origin and evolutionary radiation of the Mollusca. *Centenary Symposium of the Malacological Society of London*, pp. 392. Oxford University Press, Oxford.

Telford, M. J. 1993. Molecular studies of chaetognath evolution. D.Phil. thesis, University of Oxford.

Telford, M. J. 2001. Embryology and developmental genes as clues to flatworm relationships. In D. T. J. Littlewood and R. A. Bray (eds.), *Interrelationships of the Platyhelminthes*, pp. 257–261. Taylor & Francis, London.

Telford, M. J., E. A. Herniou, R. B. Russell, and D. T. J. Littlewood. 2000 Changes in mitochondrial genetic codes as phylogenetic characters: two examples from the flatworms. *Proceedings of the National Academy of Sciences of the United States of America*, 97:11359–11364.

Temirova, S. I., and A. S. Skrjabin. 1978. [The suborder Tetrabothriata (Ariola, 1899) Skrjabin, 1940]. (In Russian). *Osnovy Tsestodologii*, 9:7–117.

Thomas, A. P. 1881. Report of experiments on the development of the liver fluke (*Fasciola hepatica*). *Journal of the Royal Agricultural Society of England*, 17:1–28 (Reprinted in Kean, Mott and Russell, 1978, Volume 2, 570–575).

Thomas, A. P. 1882. Second report of experiments on the development of the liver fluke (*Fasciola hepatica*). *Journal of the Royal Agricultural Society of England*, 18:439–455 (Reprinted in Kean, Mott and Russell, 1978, Volume 2, 575–577).

Thomas, A. P. 1883. The life history of the liver fluke (*Fasciola hepatica*). *Quarterly Journal of Microscopical Science, new series*, 23:99–133.

Thompson, C. S., and D. F. Mettrick. 1989. The effects of 5–hydroxytryptamine and glutamate on muscle contraction in *Hymenolepis diminuta* (Cestoda). *Candian Journal of Zoology*, 67:1257–1262.

Thompson, J. D., T. J. Gibson, F. Plewniak, F. Jeanmougin, and D. G. Higgins. 1997. The CLUSTAL-X windows interface: flexible strategies for multiple sequence alignment aided by quality analysis tools. *Nucleic Acids Research*, 25:4876–4882.

Thompson, J. D., D. G. Higgins, and T. J. Gibson. 1994. CLUSTAL-W – improving the sensitivity of progressive multiple sequence alignment through sequence weighting, position-specific gap penalties and weight matrix choice. *Nucleic Acids Research*, 22:4673–4680.

Thompson, R. C. A. 1998. Species and strains in the genus *Echinococcus* – epidemiological and evolutionary perspectives. *Parasitology International*, 47 (Suppl.):136.

Thompson, R. C. A., and A. J. Lymbery. 1990. Intraspecific variation in parasites – what is a strain? *Parasitology Today*, 6:345–348.

Thompson, R. C. A., and A. J. Lymbery. 1996. Genetic variability in parasites and host-parasite interactions. *Parasitology*, 112 (Suppl.):S7–S22.

Thompson, R. C. A., A. J. Lymbery, and C. C. Constantine. 1995. Variation in *Echinococcus*: towards a taxonomic revision of the genus. *Advances in Parasitology*, 35:145–176.

Thorley, J. L., and R. D. M. Page. 2000. RadCon: Phylogenetic tree comparison and consensus. *Bioinformatics*, 16:486–487.

Thorley, J. L., and M. Wilkinson. 2000. The RadCon Manual. http://taxonomy.zoology.gla.ac.uk/~jthorley/manual/manual.html.

Thorley, J. L., M. Wilkinson, and M. Charleston. 1998. The information content of consensus trees. In A. Rizzi, M. Vichi and H.-H. Bock (eds.), *Advances in data science and classification*, pp. 91–98. Springer, Berlin.

Tian, X., L. Yuan, X. Huo, X. Han, Y. Li, M. Xu, M. Lu, J. Dai, and L. Dong. 1998. Ultrastructural observations on the transformation of the spermatozoon in spermatogenesis of taeniid cestodes. *Chinese Journal of Parasitology and Parasitic Diseases*, 16:269–273.

Timofeeva, T. A., and V. P. Sharpilo. 1979. [*Euzetrema caucasica* sp. n. (Monogenea, Polyopisthocotylidea), a parasite of *Mertensiella caucasica*]. (In Russian). *Parazitologiya*, 8:516–521.

Timoshkin, O. A., Y. V. Mamkaev, and N. A. Osipova. 1989. [Protonephridial system of Prorhynchidae (Lecithoepitheliata)]. (In Russian). *Trudy Zoologicheskogo Instituta, Leningrad*, 195:108–114.

Tkach, V. V., and J. Pawlowski. 1999. A new method of DNA extraction from the ethanol–fixed parasitic worms. *Acta Parasitologica*, 44:147–148.

Tkach, V. V., and Z. Swiderski. 1998. Differentiation and ultrastructure of the oncospheral envelopes in the hymenolepidid cestode *Staphylocystoides stefanskii* (Zarnovski, 1954). *Acta Parasitologica*, 43:222–231.

Tkach, V. V., and Z. Swiderski. 1997. Late stages of egg maturation in the cestode *Pseudhymenolepis redonica* Joyeux et Baer, 1935 (Cyclophyllidea, Hymenolepididae), a parasite of shrews. *Acta Parasitologica*, 42:97–108.

Tkach, V. V., B. Grabda-Kazubska, J. Pawlowski, and Z. Swiderski. 1999. Molecular and morphological evidences for close phylogenetic affinities of the genera *Macrodera*, *Leptophallus*, *Metaleptophallus* and *Paralepoderma* (Digenea, Plagiorchioidea). *Acta Parasitologica*, 44:170–179.

Tkach, V. V., J. Pawlowski, and J. Mariaux. 2000. Phylogenetic analysis of the suborder Plagiorchiata (Platyhelminthes, Digenea) based on partial 28S rDNA sequences. *International Journal for Parasitology*, 30:89–93.

Tkach, V. V., J. Pawlowski, J. Mariaux, and Z. Swiderski. 2001. Molecular phylogeny of the suborder Plagiorchiata and its position in the system of Digenea. In D. T. J. Littlewood and R. A. Bray (eds.), *Interrelationships of the Platyhelminthes*, pp. 186–193. Taylor & Francis, London.

Turceková, L., and I. Králová. 1995. Characterization of DNA restriction profiles and rRNA gene restriction fragment length polymorphisms of *Proteocephalus exiguus* and *P. neglectus* from geographically distinct regions. *Journal of Helminthology*, 69:159–163.

Tyler, G. A., and J. N. Caira. 1999. Two new species of *Echinobothrium* (Cestoidea: Diphyllidea) from myliobatiform elasmobranchs in the Gulf of California, Mexico. *Journal of Parasitology*, 85:327–335.

Tyler, S. 1976. Comparative ultrastructure of adhesive systems in the Turbellaria. *Zoomorphologie*, 84:1–76.

Tyler, S. 1977. Ultrastructure and systematics: an example from turbellarian adhesive organs. *Mikrofauna Meeresboden*, 61:271–286.

Tyler, S. 1988. The role of function in determination of homology and convergence from invertebrate adhesive systems. *Fortschritte der Zoologie*, 36:331–347.

Tyler, S. 2001. The early worm – origins and relationships of the lower flatworms. In D. T. J. Littlewood and R. A. Bray (eds.), *Interrelationships of the Platyhelminthes*, pp. 3–12. Taylor & Francis, London.

Tyler, S., and R. M. Rieger. 1975. Uniflagellate spermatozoa in *Nemertoderma* (Turbellaria) and their phylogenetic significance. *Science*, 188.

Tyler, S., and R. M. Rieger. 1977. Ultrastructural evidence for the systematic position of the Nemertodermatida (Turbellaria). *Acta Zoologica Fennica*, 154:193–207.

Tyler, S., and R. M. Rieger. 1999. Functional morphology of musculature in the acoelomate worm *Convoluta pulchra* (Plathelminthes). *Zoomorphology*, 119:127–141.

Tyler, S., and M. S. Tyler. 1997. Origin of the epidermis in parasitic platyhelminths. *International Journal for Parasitology*, 27:715–738.

Ubelaker, J. E. 1983a. Metacestodes: morphology and development. In C. Arme and P. W. Pappas (eds.), *Biology of the Eucestoda*, pp. 139–176. Academic Press, London.

Ubelaker, J. E. 1983b. The morphology, development and evolution of tapeworm larvae. In C. Arme and P. W. Pappas (eds.), *Biology of the Eucestoda*, pp. 235–296. Academic Press, London.

Uglem, G. L., and J. J. Just. 1983. Trypsin inhibition by tapeworms: antienzyme secretion or pH adjustment? *Science*, 220:79–81.

Umesono, Y., K. Watanabe, and K. Agata. 1999. Distinct structural domains in the planarian brain defined by the expression of evolutionarily conserved homeobox genes. *Development Genes and Evolution*, 209:31–39.

Valentine, J. W. 1997. Cleavage patterns and the topology of the metazoan tree of life. *Proceedings of the National Academy of Sciences USA*, 94:8001–8005.

van Beneden, P. J. 1858. *Mémoire sur les vers intestinaux*. J.B. Bailliére, Paris.

van Beneden, P. J. 1876. *Animal parasites and messmates*. International ed. H. S. King, London.

van de Peer, Y., J. Jansen, P. de Rijk, and R. de Wachter. 1997. Database on the structure of small ribosomal subunit RNA. *Nucleic Acids Research*, 25:111–116.

van de Peer, Y., A. Caers, P. de Rijk, and R. de Wachter. 1998. Database on the structure of small ribosomal subunit RNA. *Nucleic Acids Research*, 26:179–182.

van de Peer, Y., P. de Rijk, J. Wuyts, T. Winkelmans, and R. de Wachter. 2000. The European small subunit ribosomal RNA database. *Nucleic Acids Research*, 28:175–176.

van den Biggelaar, J. A. M., W. J. A. G. Dictus, and A. E. van Loon. 1997. Cleavage patterns, cell-lineages and cell specification are clues to phyletic lineages in Spiralia. *Seminars in Cell & Developmental Biology*, 8:367–378.

van der Land, J., and H. Dienske. 1968. Two new species of *Gyrocotyle* (Monogenea) from chimaerids (Holocephali). *Zoologische Medelingen, Leiden*, 43:97–105.

van der Land, J., and W. Templeman. 1968. Two new species of *Gyrocotyle* (Monogenea) from *Hydrolagus affinis* (Brito Capello) (Holocephali). *Journal of the Fisheries Research Board of Canada*, 11:2365–2385.

van Herwerden, L., D. Blair, and T. Agatsuma. 1999. Intra- and interindividual variation in ITS1 of *Paragonimus westermani* (Trematoda: Digenea) and related species: implications for phylogenetic studies. *Molecular Phylogenetics and Evolution*, 12:67–73.

Vawter, L., and W. M. Brown. 1993. Rates and patterns of base change in small subunit ribosomal RNA gene. *Genetics*, 134:597–608.

Venard, J. E. 1938. Morphology, bionomics, and taxonomy of the cestode *Dipylidium caninum*. *Annals of the New York Academy of Science*, 37:273–328.

Verneau, O., F. M. Catzeflis, and F. Renaud. 1997a. Molecular relationships between closely related species of *Bothriocephalus* (Cestoda: Platyhelminthes). *Molecular Phylogenetics and Evolution*, 7:201–207.

Verneau, O., F. Renaud, and F. Catzeflis. 1997b. Evolutionary relationships of sibling tapeworm species (Cestoda) parasitizing teleost fishes. *Molecular Biology and Evolution*, 14:630–636.

Verneau, O., F. Thomas, A. de Mees, F. Catzeflis, and F. Renaud. 1995. Evidence of two genetic entities in *Bothriocephalus funiculus* (Cestoda) detected by arbitrary-primer polymerase chain reaction random amplified polymorphic DNA fingerprinting. *Parasitology Research*, 1995:591–594.

Voge, M. 1960. Studiees (sic) in cysticercoid histology. III. Observations on the fully developed cysticercoid of *Raillietina cesticillus* (Cestoda: Cyclophyllidea). *Proceedings of the Helminthological Society of Washington*, 27:271–274.

Voge, M. 1967. The post-embryonic developmental stages of cestodes. *Advances in Parasitology*, 11:707–730.

Vogel, H., and W. Minning. 1940. Bilharziose bei Elefanten. *Archives Schiffs- und Tropenhygiene*, 44:562–574.

von Nickisch-Rosenegk, M., R. Lucius, and B. Loos-Frank. 1999. Contributions to the phylogeny of the

Cyclophyllidea (Cestoda) inferred from mitochondrial 12S rDNA. *Journal of Molecular Evolution*, 48:586–596.

Vreys, C., and N. K. Michiels. 1997. Flatworms flatten to size up each other. *Proceedings of the Royal Society of London, Series B, Biological Sciences*, 264:1559–1564.

Vreys, C., E. R. Schockaert, and N. K. Michiels. 1997a. Unusual pre-copulatory behaviour in the hermaphroditic planarian flatworm, *Dugesia gonocephala* (Tricladida, Paludicola). *Ethology*, 103:208–221.

Vreys, C., E. R. Schockaert, and N. K. Michiels. 1997b. Formation, transfer, and assimilation of the spermatophore of the hermaphroditic flatworm *Dugesia gonocephala* (Tricladida, Paludicola). *Canadian Journal of Zoology*, 75:1479–1486.

Vreys, C., N. Steffanie, and H. Gevaerts. 1997c. Digestion of the spermatophore contents in the female genital system of the hermaphroditic flatworm *Dugesia gonocephala* (Tricladida, Paludicola). *Invertebrate Biology*, 116:286–293.

Waddell, P. J., and M. A. Steel. 1997. General time–reversible distances with unequal rates across sites: mixing gamma and inverse Gaussian distributions with invariant sites. *Molecular Phylogenetics and Evolution*, 8:398–414.

Waggoner, B. M. 1996. Phylogenetic hypotheses of the relationships of arthropods to Precambrian and Cambrian problematic fossil taxa. *Systematic Biology*, 45:190–222.

Wall, L. D. 1941. *Spirorchis parvus* (Stunkard) its life history and the development of its excretory system (Trematoda: Spirorchiidae). *Transactions of the American Microscopical Society*, 60:221–260.

Wall, L. D. 1951. The life history of *Vasotrema robustum* (Stunkard 1928), Trematoda: Spirorchiidae. *Transactions of the American Microscopical Society*, 70:173–184.

Wang, D. Y.-C., S. Kumar, and S. B. Hedges. 1999. Divergence time estimates for the early history of animal phyla and the origin of plants, animals and fungi. *Proceedings of the Royal Society, London*, 266B:163–171.

Wardle, R. A., and A. J. McLeod. 1952. *The zoology of tapeworms*. North Central Publishing Company, St Paul.

Watrous, L. E., and Q. D. Wheeler. 1981. The outgroup comparison method of character analysis. *Systematic Zoology*, 30:1–11.

Watson, N. A. 1997a. Proximo-distal fusion of flagella during spermiogenesis in the 'turbellarian' platyhelminth *Urastoma cyprinae*, and phylogenetic implications. *Invertebrate Reproduction and Development*, 32:107–117.

Watson, N. A. 1997b. Spermiogenesis and sperm ultrastructure in *Troglocephalus rhinobatidis*, *Neoheterocotyle rhinobatidis* and *Merizocotyle australensis* (Platyhelminthes, Monogenea, Monopisthocotylea, Monocotylidae). *International Journal for Parasitology*, 27:389–401.

Watson, N. A. 1998a. Characteristics of protonephridial terminal organs and the eyes of four species of Trigonostominae (Platyhelminthes: Rhabdocoela). *Australian Journal of Zoology*, 46:251–265.

Watson, N. A. 1998b. Spermiogenesis and eyes with lenses in two kalyptorhynch flatworm species, *Toia calceformis* and *Nannorhynchides herdlaensis* (Eukalyptorhynchia, Platyhelminthes). *Invertebrate Biology*, 117:9–19.

Watson, N. A. 1999a. Clue to the origin of anucleate flame bulbs in some flatworms. *Invertebrate Biology*, 118:18–23.

Watson, N. A. 1999b. Platyhelminthes. In B. G. M. Jamieson (ed.), *Progress in male gamete ultrastructure and phylogeny*, Vol. IX, Part A, pp. 97–142. John Wiley, London.

Watson, N. A. 1999c. Intrageneric variation in spermiogenesis, spermatozoa, and ocelli in *Pogaina* (Platyhelminthes, Rhabdocoela, Provorticidae). *Invertebrate Biology*, 118:243–257.

Watson, N. A. 2001. Insights from comparative spermatology in the 'turbellarian' Rhabdocoela. In D. T. J. Littlewood and R. A. Bray (eds.), *Interrelationships of the Platyhelminthes*, pp. 217–230. Taylor & Francis, London.

Watson, N. A., and L. A. Chisholm. 1998. Spermatozoa and spermiogenesis in the monocotylid *Heterocotyle capricornensis* (Platyhelminthes, Monogenea, Monopisthocotylea), including observations of aberrant folding and fusing of spermatozoa in one individual. *Folia Parasitologica*, 45:211–220.

Watson, N. A., and U. Jondelius. 1995. Comparative ultrastructure of spermiogenesis and sperm in *Maehrenthalia* sp. and *Bresslauilla relicta* (Platyhelminthes, Rhabdocoela). *Invertebrate Reproduction and Development*, 28:103–112.

Watson, N. A., and U. Jondelius. 1997. Spermiogenesis and sperm ultrastructure in *Cylindrostoma fingalianum* (Platyhelminthes, Prolecithophora) with notes on a protozoan (kinetoplastid) symbiont closely associated with allosperm. *Invertebrate Reproduction and Development*, 32:273–282.

Watson, N. A., and J. P. L'Hardy. 1995. Origin of the uniflagellate spermatozoon of *Baltoplana magna* (Platyhelminthes, Kalpytorhynchia). *Invertebrate Reproduction and Development*, 28:185–192.

Watson, N. A., and K. Rohde. 1992. Ultrastructure of the flame bulbs and protonephridial capillaries of *Rugogaster hydrolagi* (Platyhelminthes, Trematoda, Aspidogastrea). *Annales de Parasitologie Humaine et Comparée*, 67:67–74.

Watson, N. A., and K. Rohde. 1993a. Ultrastructural evidence for an adelphotaxon (sister group) to the Neodermata (Platyhelminthes). *International Journal for Parasitology*, 23:285–289.

Watson, N. A., and K. Rohde. 1993b. Ultrastructure of sperm and spermiogenesis of *Kronborgia isopodicola* (Platyhelminthes, Fecampiidae). *International Journal for Parasitology*, 23:737–744.

Watson, N. A., and K. Rohde. 1994a. Ultrastructure of spermiogenesis and spermatozoa in *Phaenocora anomalocoela* (Platyhelminthes, Typhloplanida, Phaenocorinae). *Invertebrate Reproduction and Development*, 25:237–246.

Watson, N. A., and K. Rohde. 1994b. Ultrastructure of sperm and spermiogenesis in the monocotylid monogeneans *Monocotyle helicophallus* and *Calicotyle australiensis* (Platyhelminthes). *International Journal for Parasitology*, 24:1019–1030.

Watson, N. A., and K. Rohde. 1995a. Ultrastructure of spermiogenesis and spermatozoa in the platyhelminths *Actinodactylella blanchardi* (Temnocephalida, Actinodactylellidae), *Didymorchis* sp. (Temnocephalida, Didymorchidae) and *Gieysztoria* sp. (Dalyelliida, Dalyelliidae), with implications for the phylogeny of the Rhabdocoela. *Invertebrate Reproduction and Development*, 27:145–158.

Watson, N. A., and K. Rohde. 1995b. Sperm and spermiogenesis of the 'Turbellaria' and implications for the phylogeny of the phylum Platyhelminthes. In B. G. M. Jamieson, J. Ausio and J.-L. Justine (eds.), *Advances in spermatozoal phylogeny and taxonomy. Memoires du Museum National d'Histoire Naturelle*, Vol. 166, pp. 37–54.

Watson, N. A., and K. Rohde. 1995c. Ultrastructure of spermiogenesis and spermatozoa of *Neopolystoma spratti* (Platyhelminthes, Monogenea, Polystomatidae). *Parasitology Research*, 81:343–348.

Watson, N. A., and K. Rohde. 1997. Novel protonephridial filtration apparatus in *Cylindrostoma fingalianum*, *Allostoma* sp. and *Pseudostomum quadrioculatum* (Platyhelminthes: Prolecithophora). *Australian Journal of Zoology*, 45:621–630.

Watson, N. A., and E. R. Schockaert. 1996. Spermiogenesis and sperm ultrastructure in *Thylacorhynchus ambronensis* (Schizorhynchia, Kalyptorhynchia, Platyhelminthes). *Invertebrate Biology*, 115:263–272.

Watson, N. A., and E. R. Schockaert. 1997. Divergent protonephridial architecture within the Kalyptorhynchia (Platyhelminthes) and implications for the phylogeny of the Rhabdocoela. *Belgian Journal of Zoology*, 127:139–158.

Watson, N. A., K. Steiner, and K. Rohde. 1991. Ultrastructure of the flame bulbs and protonephridial capillaries of *Macrostomum tuba* Graff 1882 (Platyhelminthes, Macrostomida). *Journal of Submicroscopic Cytology and Pathology*, 23:255–260.

Watson, N. A., K. Rohde, and A. Lanfranchi. 1992a. The ultrastructure of the protonephridial system of Götte's larva of *Stylochus mediterraneus* (Polycladida, Platyhelminthes). *Zoologica Scripta*, 21:217–221.

Watson, N., K. Rohde, and J. B. Williams. 1992b. Ultrastructure of the protonephridial system of larval *Kronborgia isopodicola* (Platyhelminthes). *Journal of Submicroscopic Cytology and Pathology*, 24:43–49.

Watson, N. A., K. Rohde, and U. Jondelius. 1993. Ultrastructure of sperm and spermiogenesis of *Pterastericola astropectinis* (Platyhelminthes, Rhabdocoela, Pterastericolidae). *Parasitology Research*, 79:322–328.

Watson, N. A., K. Rohde, and K. B. Sewell. 1995a. Ultrastructure of spermiogenesis and spermatozoa of *Decadidymus gulosus*, *Temnocephala dendyi*, *T. minor*, *Craspedella* sp., *C. spenceri* and *Diceratocephala boschmai* (Platyhelminthes, Temnocephalida, Temnocephalidae), with emphasis on the intercentriolar body and zone of differentiation. *Invertebrate Reproduction and Development*, 27:131–143.

Watson, N. A., L. D. Whittington, and K. Rohde. 1995b. Ultrastructure of spermiogenesis and spermatozoa in the monogeneans *Concinnocotyla australensis* (Polystomatidae) and *Pricea multae* (Gastrocotylidae). *Parasite*, 2:357–366.

Watzin, M. C. 1983. The effects of meiofauna on settling macrofauna: meiofauna may structure macrofaunal communities. *Oecologia*, 59:163–166.

Webb, R. A., and H. Eklove. 1989. Demonstration of intense glutamate-like immunoreactivity in the longitudinal nerve cords of the cestode *Hymenolepis diminuta*. *Parasitology Research*, 75:545–548.

Wenzel, C., U. Ehlers, and A. Lanfranchi. 1992. The larval protonephridium of *Stylochus mediterraneus* Galleni (Polycladida, Plathelminthes): an ultrastructural analysis. *Microfauna Marina*, 7:323–340.

Westblad, E. 1923. Zur Physiologie der Turbellarien. I. Die Verdauung. II. Die Exkretion. Doctoral thesis, Lund University.

Westblad, E. 1926. Parasitische Turbellarien von der Westküste Skandinaviens. *Zoologischer Anzeiger*, 68:212–216.

Westblad, E. 1937. Die Turbellarien-Gattung *Nemertoderma* Steinböck. *Acta Societatis pro Fauna et Flora Fennica*, 60:45–89.

Westblad, E. 1949. *Xenoturbella bocki* n. g., n. sp. a peculiar, primitive turbellarian type. *Arkiv för Zoologi*, 1:11–29.

Westblad, E. 1950. On *Meara stichopi* (Bock) Westblad, a new representative of Turbellaria Archoophora. *Arkiv för Zoologi*, 1:43–57.

Westblad, E. 1952. Some new 'Alloecoels' (Turbellaria) from the Scandinavian West Coast. *Bergens Museums årbok Naturvitenskapelig rekke*, 5:1–27.

Westblad, E. 1953. Marine Macrostomida (Turbellaria) from Scandinavia and England. *Arkiv för Zoologi*, 4:391–408.

Westblad, E. 1956. Marine 'Alleocoels' (Turbellaria) from North Atlantic and Mediterranean coasts. II. *Arkiv för Zoologi*, 9:131–174.

Westheide, W. 1987. Progenesis as a principle in meiofauna evolution. *Journal of Natural History*, 21:843–854.

Westheide, W., and R. Rieger. 1996. *Spezielle Zoologie. Erster Teil: Einzeller und Wirbellose Tiere*. Gustav Fischer, Stuttgart.

Wetzel, R. 1932. Zur Kenntnis der weniggliedrigen Hühnerbandwurmes *Davainea proglottina*. *Archiv für wissenschaftliche und praktische Teilheilkunde*, 65:595–625.

Wetzel, R. 1934. Untersuchungen über den Entwicklungskreis des Hühnerbandwurmes *Raillietina cesticillus* (Molin, 1858). *Archiv für wissenschaftliche und praktische Tierheilkunde*, 68:221–232.

Wheeler, T. A., and L. A. Chisholm. 1995. Monogenea versus Monogenoidea: the case for stability in nomenclature. *Systematic Parasitology*, 30:159–164.

Whitfield, P. J., R. M. Anderson, and N. A. Moloney. 1975. The attachment of cercariae of an ectoparasitic digenean, *Transversotrema patialensis*, to the fish host: behavioural and ultrastructural aspects. *Parasitology*, 70:311–329.

Whittington, I. D., L. A. Chisholm, and K. Rohde. 2000. The larvae of Monogenea (Platyhelminthes). *Advances in Parasitology*, 44:139–232.

Widmer, D. E., and O. W. Olsen. 1967. The life history of *Oochoristica osheroffi* Meggitt, 1934 (Cyclophyllidea: Anoplocephalidae). *Journal of Parasitology*, 53:343–349.

Wiens, J. J. 1998. The accuracy of methods for coding and sampling higher–level taxa for phylogenetic analysis: a simulation study. *Systematic Biology*, 47:397–413.

Wikgren, M. C. 1986. The nervous system of early larval stages of the cestode *Diphyllobothrium dendriticum*. *Acta Zoologia (Stockholm)*, 67:155–163.

Wikgren, M. C. 1987. Neuroendreocrine systems in flatworms. Doctoral thesis. Åbo Akademi University.

Wikgren, M., and M. Reuter. 1985. Neuropeptides in a microturbellarian – whole mount immunocytochemistry. *Peptides*, 6:471–475.

Wikgren, M., M. Reuter, M. K. S. Gustafsson, and P. Lindroos. 1990. Immunocytochemical localization of histamine in flatworms. *Cell & Tissue Research*, 260:479–484.

Wiley, E. O. 1981. *Phylogenetics, the principles and practice of phylogenetic systematics*. John Wiley & Sons, New York.

Wiley, E. O., D. Siegel-Causey, D. R. Brooks, and V. A. Funk. 1991. *The compleat cladist: a primer of phylogenetic procedures*. University of Kansas Museum of Natural History, Lawrence.

Wilkinson, M. 1994. Common cladistic information and its consensus representation: Reduced Adams and Reduced Cladistic Consensus Trees and Profiles. *Systematic Biology*, 43:343–368.

Wilkinson, M. 1995. Coping with abundant missing entries in phylogenetic inference using parsimony. *Systematic Biology*, 44:501–514.

Wilkinson, M., and J. L. Thorley. 1998. Reduced supertrees. *Trends in Evolution and Ecology*, 13:283.

Williams, C. O. 1942. Observations on the life history and taxonomic relationships of the trematode *Aspidogaster conchicola*. *Journal of Parasitology*, 28:467–475.

Williams, H. H., and A. Jones. 1994. *Parasitic worms of fish*. Taylor & Francis, London.

Williams, J. B. 1975. Studies on the epidermis of *Temnocephala* I. Ultrastructure of the epidermis of *Temnocephala novaezealandiae*. *Australian Journal of Zoology*, 23:321–331.

Williams, J. B. 1981a. Classification of the Temnocephaloidea (Platyhelminthes). *Journal of Natural History*, 15:277–299.

Williams, J. B. 1981b. The protonephridial system of *Temnocephala novaezealandiae*: structure of the flame cells and main vessels. *Australian Journal of Zoology*, 29:131–146.

Williams, J. B. 1983. The genital system of *Temnocephala* I. Ultrastructural features of the differentiating spermatid of *Temnocephala novaezealandiae*, including notes on a possible correlation between cellular autophagy and mitochondrial function. *Australian Journal of Zoology*, 31:317–331.

Williams, J. B. 1984. The genital system of *Temnocephala* II. Further observations on the spermatogenesis of *Temnocephala novaezealandiae*, with particular reference to the mitochondria. *Australian Journal of Zoology*, 32:447–461.

Williams, J. B. 1994. Comparative study of sperm morphology and reproductive biology of *Kronborgia* and *Temnocephala* (Platyhelminthes, Neoophora): implications for platyhelminth phylogeny. *New Zealand Journal of Zoology*, 21:179–194.

Williams, S. A., and D. A. Johnston. 1999. Helminth genome analysis: the current state of the filarial and schistosome genome projects. *Parasitology*, 113 (Suppl.):19–38.

Willmer, P. 1990. *Invertebrate relationships: patterns in animal evolution*. Cambridge University Press, Cambridge.

Wilrich, R. T., and H. J. Hennig. 1987. Formeln und Tabellen der angewandten matematischen Statistik. Springer, Berlin.

Wilson, A. C., S. S. Carlson, and T. J. White. 1977. Biochemical evolution. *Annual Review of Biochemistry*, 46:573–639.

Wilson, R. A. 1969a. Fine structure of the tegument of the miracidium of *Fasciola hepatica*. *Parasitology*, 59:461–467.

Wilson, R. A. 1969b. The fine structure of the protonephridial system in the miracidium of *Fasciola hepatica*. *Journal of Parasitology*, 55:124–133.

Wilson, R. A. 1970. Fine structure of the nervous system and specialised nerve endings in the miracidium of *Fasciola hepatica*. *Parasitology*, 60:399–410.

Wilson, R. A. 1971. Gland cells and secretions in the miracidium of *Fasciola hepatica*. *Parasitology*, 63:225–231.

Wilson, R. A., and L. A. Webster. 1974. Protonephridia. *Biological Review*, 49:127–160.

Wilson, S. J. 1999. Building phylogenetic trees from quartets by using local inconsistency measures. *Molecular Biology and Evolution*, 16:685–693.

Winnepenninckx, B., and T. Backeljau. 1996. 18S rRNA alignments derived from different secondary structure models can produce alternative phylogenies. *Journal of Zoological Systematics and Evolutionary Research*, 34:135–143.

Winnepenninckx, B., T. Backeljau, and R. de Wachter. 1995. Phylogeny of protostome worms derived From 18S ribosomal RNA sequences. *Molecular Biology and Evolution*, 12:641–649.

Winnepenninckx, B., T. Backeljau, and R. de Wachter. 1996. Investigation of molluscan phylogeny on the basis of 18S rRNA sequences. *Molecular Biology and Evolution*, 13:1306–1317.

Winsor, L. 1998. Aspects of taxonomy and functional histology in terrestrial flatworms (Tricladida: Terricola). *Pedobiologia*, 42:412–432.

Winsor, L., P. M. Johns, and G. W. Yeates. 1998. Introduction, and ecological and systematic background, to the Terricola (Tricladida). *Pedobiologia*, 42:389–404.

Wood, D. E. 1965. Nature of the end organ in *Ophiotaenia filaroides* (La Rue). *Journal of Parasitology*, 51:541–544.

Xylander, W. E. R. 1984. A presumptive ciliary photoreceptor in larval *Gyrocotyle urna* Grube and Wagener (Cestoda). *Zoomorphology*, 104:21–25.

Xylander, W. E. R. 1986a. Zur Biologie und Ultrastruktur der Gyrocotylida un Amphilinida sowie ihre Stellung im phylogenetischen System der Plathelminthes. Ph.D. thesis, Göttingen.

Xylander, W. E. R. 1986b. Ultrastrukturelle Befunde zur Stellung von *Gyrocotyle* im System der parasitischen Plathelminthen. *Verhandlungen der Deutschen Zoologischen Gesellschaft*, 79:193.

Xylander, W. E. R. 1987a. Ultrastructure of the lycophora larva of *Gyrocotyle urna* (Cestoda, Gyrocotylidea). III. The protonephridial system. *Zoomorphology*, 107:88–95.

Xylander, W. E. R. 1987b. Das Protonephridialsystem der Cestoda: evolutive Veräderungen und ihre mögliche funktionelle Bedeutung. *Verhandlungen der Deutschen Zoologischen Gesellschaft*, 80:257–258.

Xylander, W. E. R. 1987c. Ultrastructure of the lycophora larva of *Gyrocotyle urna* (Cestoda, Gyrocotylidea). 1. Epidermis, neodermis anlage and body musculature. *Zoomorphology*, 106:352–360.

Xylander, W. E. R. 1987d. Ultrastructure of the lycophora larva of *Gyrocotyle urna* (Cestoda, Gyrocotylidea). 2. Receptors and nervous system. *Zoologischer Anzeiger*, 219:239–255.

Xylander, W. E. R. 1988a. Ultrastructural studies on the reproductive system of Gyrocotylidea and Amphilinidea (Cestoda). 1. Vitellarium, vitellocyte development and vitelloduct in *Amphilina foliacea* (Rudolphi, 1819). *Parasitology Research*, 74:363–370.

Xylander, W. E. R. 1988b. Ultrastructural studies on Udonellidae: evidence for a position within the Neodermata. *Fortschritte der Zoologie*, 36:51–57.

Xylander, W. E. R. 1989a. Investigations on the biology of *Gyrocotyle urna* (Cestoda) and considerations on its life cycle. *Verhandlungen der Deutschen Zoologischen Gesellschaft*, 82:251.

Xylander, W. E. R. 1989b. Ultrastructural studies on the reproductive system of Gyrocotylidea and Amphilinidea (Cestoda): spermatogenesis, spermatozoa, testes and vas deferens of *Gyrocotyle*. *International Journal for Parasitology*, 19:897–905.

Xylander, W. E. R. 1990. Ultrastructure of the lycophora larva of *Gyrocotyle urna* (Cestoda, Gyrocotylidea). 4. The glandular system. *Zoomorphology*, 109:319–328.

Xylander, W. E. R. 1991. Ultrastructure of the lycophora larva of *Gyrocotyle urna* (Cestoda, Gyrocotylidea). 5. Larval hooks and associated tissues. *Zoomorphology*, 111:59–66.

Xylander, W. E. R. 1992a. Sinneszellen von *Gyrocotyle urna*: Rezeptorenvielfalt bei einem ursprünglichen Cestoden. *Verhandlungen der Deutschen Zoologischen Gesellschaft*, 85:230.

Xylander, W. E. R. 1992b. Investigations on the protonephridial system of postlarval *Gyrocotyle urna* and *Amphilina foliacea* (Cestoda). *International Journal for Parasitology*, 22:287–300.

Xylander, W. E. R. 1993. Ultrastructural investigations of spermatogenesis and morphology of spermatozoa, vas efferens and receptaculum seminis of *Amphilina foliacea*. *Verhandlungen der Deutschen Zoologischen Gesellschaft*, 86:184.

Xylander, W. E. R. 1996. Neodermata. In W. Westheide and R. M. Rieger (eds.), *Spezielle Zoologie, Teil 1: Einzeller und Wirbellose*, pp. 230–258. Gustav Fischer, Stuttgart.

Xylander, W. E. R. 1998. The systematic position of Gyrocotyloidea and Amphilinidea within the Neodermata. *Wiadomosci Parazytologiczne*, 44:493.

Xylander, W. E. R., and T. Bartolomaeus. 1995. Protonephridien: neue Erkenntnisse über Funktion und Evolution. *Biologie in unserer Zeit*, 25:107–114.

Yamaguti, S. 1958. *Systema helminthum. Volume I. The digenetic trematodes of vertebrates*. Interscience Publishers, Inc., New York.

Yamaguti, S. 1959. *Systema helminthum. Volume II. The cestodes of vertebrates*. Interscience Publishers, Inc., New York.

Yamaguti, S. 1963. *Systema helminthum. Volume IV. Monogenea and Aspidocotylea*. Intersciences Publishers, London & New York.

Yamaguti, S. 1971. *Synopsis of digenetic trematodes of vertebrates*. Keigaku Publishing Co., Tokyo.

Yamaguti, S. 1975. *A synoptical review of life histories of digenetic trematodes of vertebrates*. Keigaku Publishers, Tokyo.

Yeates, D. K. 1995. Groundplans and exemplars: paths to the tree of life. *Cladistics*, 11:343–357.

Zabusov, H. 1911. [Untersuchungen über die Morphologie und Systematik der Planarien aus dem Baikalsee. I. Die Gattung *Sorocelis* Grube]. (In Russian). *Trudy Obscestva Estestvoispytatelej pri Kazanskom Universitete*, 43:1–422.

Zabusov, H. 1929. Die Turbellarien der Kamschatka-Halbinsel nach den Sammlungen der Rajbuschinsky-Expedition 1908–1909. *Zoologische Jahrbücher Abteilung für Systematik, Oekologie und Geographie der Tiere*, 57:497–536.

Zardoya, R., and A. Meyer. 1998. Complete mitochondrial genome suggests diapsid affinities of turtles. *Proceedings of the National Academy of Sciences USA*, 95:14226–14231.

Zarlenga, D. S., and M. George. 1995. *Taenia crassiceps*: cloning and mapping of mitochondrial DNA and its application to the phenetic analysis of a new species of *Taenia* from Southeast Asia. *Experimental Parasitology*, 81:604–607.

Zarlenga, D. S., D. P. McManus, P. C. Fan, and J. H. Cross. 1991. Characterization and detection of a newly described Asian taeniid using cloned ribosomal DNA fragments and sequence amplification by the polymerase chain reaction. *Experimental Parasitology*, 72:174–183.

Zdárská, Z., V. Nasincová, and J. Valkounová. 1988. Multiciliate sensory endings in the redia of *Echinostoma revolutum* (Trematoda, Echinostomatidae). *Folia Parasitologica*, 35:17–20.

Zehnder, M. P., and A. de Chambrier. 2000. Morphological and molecular analyses of the genera *Peltidocotyle* Diesing 1850 and *Othinoscolex* Woodland 1933, and morphological study of *Woodlandiella* Freze, 1965 (Eucestoda, Proteocephalidea), parasites of South American siluriform fishes (Pimelodidae). *Systematic Parasitology*, 46:33–43.

Zehnder, M. P., A. de Chambrier, C. Vaucher, and J. Mariaux. 2000. *Nomimoscolex suspectus* n. sp. (Eucestoda: Proteocephalidea: Zygobothriinae) with morphological and molecular phylogenetic analysis of the genus. *Systematic Parasitology*, 47:157–172.

Zrzavý, J., S. Mihulka, P. Kepka, A. Bezdek, and D. Tietz. 1998. Phylogeny of the Metazoa based on morphological and 18S ribosomal DNA evidence. *Cladistics*, 14:249–285.

Index

Taxa listed are at the generic level and higher. Page references in **bold** indicate illustrations. Page numbers within [square brackets] indicate reference to complete chapter dealing with topic.

D

Systematics Association Publications

1. Bibliography of key works for the identification of the British fauna and flora, 3rd edition (1967)†
Edited by G.J. Kerrich, R.D. Meikie and N. Tebble
2. Function and taxonomic importance (1959)†
Edited by A.J. Cain
3. The species concept in palaeontology (1956)†
Edited by P.C. Sylvester-Bradley
4. Taxonomy and geography (1962)†
Edited by D. Nichols
5. Speciation in the sea (1963)†
Edited by J.P. Harding and N. Tebble
6. Phenetic and phylogenetic classification (1964)†
Edited by V.H. Heywood and J. McNeill
7. Aspects of Tethyan biogeography (1967)†
Edited by C.G. Adams and D.V. Ager
8. The soil ecosystem (1969)†
Edited by H. Sheals
9. Organisms and continents through time (1973)†
Edited by N.F. Hughes
10. Cladistics: a practical course in systematics (1992)*
P.L. Forey, C.J. Humphries, I.J. Kitching, R.W. Scotland, D.J. Siebert and D.M. Williams
11. Cladistics: the theory and practice of parsimony analysis (2nd edition) (1998)*
I.J. Kitching, P.L. Forey, C.J. Humphries and D.M. Williams

* Published by Oxford University Press for the Systematics Association
† Published by the Association (out of print)

Systematics Association Special Volumes

1. The new systematics (1940)
Edited by J.S. Huxley (reprinted 1971)
2. Chemotaxonomy and serotaxonomy (1968)*
Edited by J.C. Hawkes
3. Data processing in biology and geology (1971)*
Edited by J.L. Cutbill
4. Scanning electron microscopy (1971)*
Edited by V.H. Heywood
5. Taxonomy and ecology (1973)*
Edited by V.H. Heywood
6. The changing flora and fauna of Britain (1974)*
Edited by D.L. Hawksworth
7. Biological identification with computers (1975)*
Edited by R. J. Pankhurst
8. Lichenology: progress and problems (1976)*
Edited by D.H. Brown, D.L. Hawksworth and R.H. Bailey
9. Key works to the fauna and flora of the British Isles and northwestern Europe, 4th edition (1978)*
Edited by G.J. Kerrich, D.L. Hawksworth and R.W. Sims
10. Modern approaches to the taxonomy of red and brown algae (1978)
Edited by D.E.G. Irvine and J.H. Price
11. Biology and systematics of colonial organisms (1979)*
Edited by C. Larwood and B.R. Rosen
12. The origin of major invertebrate groups (1979)*
Edited by M.R. House
13. Advances in bryozoology (1979)*
Edited by G.P. Larwood and M.B. Abbott
14. Bryophyte systematics (1979)*
Edited by G.C.S. Clarke and J.G. Duckett
15. The terrestrial environment and the origin of land vertebrates (1980)
Edited by A.L. Pachen
16. Chemosystematics: principles and practice (1980)*
Edited by F.A. Bisby, J.G. Vaughan and C.A. Wright

17. The shore environment: methods and ecosystems (2 volumes) (1980)*
Edited by J.H. Price, D.E.C. Irvine and W.F. Farnham
18. The Ammonoidea (1981)*
Edited by M.R. House and J.R. Senior
19. Biosystematics of social insects (1981)*
Edited by P.E. House and J-L. Clement
20. Genome evolution (1982)*
Edited by G.A. Dover and R.B. Flavell
21. Problems of phylogenetic reconstruction (1982)
Edited by K.A. Joysey and A.E. Friday
22. Concepts in nematode systematics (1983)*
Edited by A.R. Stone, H.M. Platt and L.F. Khalil
23. Evolution, time and space: the emergence of the biosphere (1983)*
Edited by R.W. Sims, J.H. Price and P.E.S. Whalley
24. Protein polymorphism: adaptive and taxonomic significance (1983)*
Edited by G.S. Oxford and D. Rollinson
25. Current concepts in plant taxonomy (1983)*
Edited by V.H. Heywood and D.M. Moore
26. Databases in systematics (1984)*
Edited by R. Allkin and F.A. Bisby
27. Systematics of the green algae (1984)*
Edited by D.E.G. Irvine and D.M. John
28. The origins and relationships of lower invertebrates (1985)‡
Edited by S. Conway Morris, J.D. George, R. Gibson and H.M. Platt
29. Infraspecific classification of wild and cultivated plants (1986)‡
Edited by B.T. Styles
30. Biomineralization in lower plants and animals (1986)‡
Edited by B.S.C. Leadbeater and R. Riding
31. Systematic and taxonomic approaches in palaeobotany (1986)‡
Edited by R.A. Spicer and B.A. Thomas
32. Coevolution and systematics (1986)‡
Edited by A.R. Stone and D.L. Hawksworth
33. Key works to the fauna and flora of the British Isles and northwestern Europe, 5th edition (1988)‡
Edited by R.W. Sims, P. Freeman and D.L. Hawksworth
34. Extinction and survival in the fossil record (1988)‡
Edited by G.P. Larwood
35. The phylogeny and classification of the tetrapods (2 volumes) (1988)‡
Edited by M.J. Benton
36. Prospects in systematics (1988)‡
Edited by J.L. Hawksworth
37. Biosystematics of haematophagous insects (1988)‡
Edited by M.W. Service
38. The chromophyte algae: problems and perspective (1989)‡
Edited by J.C. Green, B.S.C. Leadbeater and W.L. Diver
39. Electrophoretic studies on agricultural pests (1 989)‡
Edited by H.D. Loxdale and J. den Hollander
40. Evolution, systematics, and fossil history of the Hamamelidae (2 volumes) (1989)‡
Edited by P.R. Crane and S. Blackmore
41. Scanning electron microscopy in taxonomy and functional morphology (1990)‡
Edited by D. Claugher
42. Major evolutionary radiations (1990)‡
Edited by P.D. Taylor and G.P. Larwood
43. Tropical lichens: their systematics, conservation and ecology (1991)‡
Edited by G.J. Galloway
44. Pollen and spores: patterns of diversification (1991)‡
Edited by S. Blackmore and S.H. Barnes
45. The biology of free-living heterotrophic flagellates (1991)‡
Edited by D.J. Patterson and J. Larsen
46. Plant-animal interactions in the marine benthos (1992)‡
Edited by D.M. John, S.J. Hawkins and J.H. Price
47. The Ammonoidea: environment, ecology and evolutionary change (1993)‡
Edited by M.R. House
48. Designs for a global plant species information system (1993)‡
Edited by F.A. Bisby, G.F. Russell and R.J. Pankhurst

* Published by Academic Press for the Systematics Association
† Published by the Palaeontological Association in conjunction with Systematics Association
‡ Published by the Oxford University Press for the Systematics Association
** Published by Chapman & Hall for the Systematics Association